Lecture Notes of the Institute for Computer Sciences, Social Informatics and Telecommunications Engineering 74

Xi Zhang Daji Qiao (Eds.)

Quality, Reliability, Security and Robustness in Heterogeneous Networks

7th International Conference on Heterogeneous Networking for Quality, Reliability, Security and Robustness, QShine 2010, and Dedicated Short Range Communications Workshop, DSRC 2010, Houston, TX, USA, November 17–19, 2010
Revised Selected Papers

Springer

Volume Editors

Xi Zhang
Texas A&M University
Department of Electrical and Computer Engineering
College Station,TX 77843-3259, USA
E-mail: xizhang@ece.tamu.edu

Daji Qiao
Iowa State University
Department of Electrical and Computer Engineering
Ames, IA 50011, USA
E-mail: daji@iastate.edu

ISSN 1867-8211
e-ISSN 1867-822X
ISBN 978-3-642-29221-7
ISBN 978-3-642-29222-4 (eBook)
DOI 10.1007/978-3-642-29222-4

Springer Heidelberg Dordrecht London New York

Library of Congress Control Number: 2012934532

CR Subject Classification (1998): C.2, H.4, D.2, H.3, F.2, C.2.4-5

Typesetting: Camera-ready by author, data conversion by Scientific Publishing Services, Chennai, India

Printed on acid-free paper

Springer is part of Springer Science+Business Media (www.springer.com)

Preface

The 7th International ICST Conference on Heterogeneous Networking for Quality, Reliability, Security and Robustness (QShine 2010) was held during November 17–19, 2010 at the Doubletree Hotel Houston Downtown in Houston, USA. Following the tradition of excellence in the previous QShine editions, QShine 2010 continued to focus on the forefront research challenges associated with the designs and implementations of large-scale wired and wireless networks and distributed systems.

QShine 2010 was a successful three-day event. We were delighted to have Behnaam Aazhang, IEEE Fellow, from Rice University, USA, and Parameswaran Ramanathan, IEEE Fellow, from the University of Wisconsin, Madison, USA, to give keynote speeches on "Context-Aware Wireless Networks: A Physical Layer Perspective" and "Seamless Support for End-to-End Gigabit Throughput in Cognitive Radio Mesh Networks," respectively. Both keynote speeches provided great insights on wireless networks and distributed systems, and were very well received by the conference participants.

QShine 2010 had a very exciting and diverse technical program, which included 37 papers for presentation at the main conference and seven papers at the Dedicated Short Range Communications (DSRC) Workshop. Conference papers were organized into nine technical sessions, covering the topics of cognitive radio networks, security, resource allocation, wireless protocols and algorithms, advanced networking systems, sensor networks, scheduling and optimization, routing protocols, and multimedia and stream processing. Workshop papers were organized into two sessions: DSRC Networks and DSRC Security.

QShine 2010 had participants from both industrial and academic sectors from various parts of the world such as Germany, France, Italy, Australia, and Canada. The conference provided tea and coffee breaks as well as lunch at the hotel for all participants throughout the conference. A Conference Banquet event was also organized on the first day of the conference. The quality of the venue, services provided by the hotel staff, and the quality of food and drinks were all highly appreciated by the participants.

The success of QShine 2010 would not have been possible without the support and dedication of many people. We would like to thank the QShine Steering Committee Chairs: Imrich Chlamtac and Sherman (Xuemin) Shen, Workshop Co-chairs: Yunxin (Jeff) Li and Vojislav B. Misic, Invited-Paper Co-chairs: Hsiao-Chun Wu, Richard Yu, and Lijun Qian, Local Arrangements Co-chairs: Zhu Han and Rong Zheng, Publication Chair: Richard Yu, Publicity Co-chairs: Qinghe Du and Jose Saldana, Web Chair: Hang Su, Conference Coordinator: Mona Hezso, and the ICST Conference Director: Edit Marosi. This conference was organized under the sponsorship of ICST and CREATE-NET. Our special thanks to the members of the Technical Program Committee and many additional reviewers for their professional and timely reviews. We would also like to thank all the authors for their submissions.

We very much look forward to the continuous success of future QShine conferences in the forthcoming years.

November 2010

Xi Zhang
Daji Qiao

Organization

General Chair

Xi Zhang	Texas A&M University, USA

Steering Committee Co-chairs

Imrich Chlamtac	Create-Net, Italy
Xi Zhang	Texas A&M University, USA
Sherman Xuemin Shen	University of Waterloo, Canada

Technical Program Committee Chair

Daji Qiao	Iowa State University, USA

Workshops Co-chairs

Yunxin (Jeff) Li	NICTA, Australia
Vojislav B. Misic	Ryerson University, Canada

Invited-Paper Co-chairs

Hsiao-Chun Wu	Louisiana State University, USA
Richard Yu	Carleton University, Canada
Lijun Qian	Prairie View A&M University, USA

Local Arrangements Co-chairs

Zhu Han	University of Houston, USA
Rong Zheng	University of Houston, USA

Publication Chair

Richard Yu	Carleton University, Canada

Publicity Co-chairs

Qinghe Du	Texas A&M University, USA
Jose Saldana	University of Zaragoza, Spain

Web Chair

Hang Su	Texas A&M Univeristy, USA

Conference Coordinator

Mona Hezso	ICST, Hungary

Technical Program Committee

Rebecca Braynard	Palo Alto Research Center, USA
Ioannis Broustis	Alcatel-Lucent, USA
Lin Cai	University of Victoria, Canada
Shigang Chen	University of Florida, USA
Xiuzhen (Susan) Cheng	George Washington University, USA
Chun-Ting Chou	National Taiwan University
Yi Cui	Vanderbilt University, USA
Zhu Han	University of Houston, USA
Xiaoxia Huang	Shenzhen Institute of Advanced Technology, China
Juan Jose Jaramillo	Iowa State University, USA
Sunggeun Jin	ETRI, Korea
Christine Julien	University of Texas at Austin, USA
Pan Li	Mississippi State University, USA
Qilian Liang	University of Texas at Arlington, USA
Ben Liang	University of Toronto, Canada
Xiaodong Lin	University of Ontario Institute of Technology, Canada
Fang Liu	University of Texas - Pan American, USA
Brian Mark	George Mason University, USA
Vojislav Misic	Ryerson University, Canada
Jianping Pan	University of Victoria, Canada
Kui Ren	Illinois Institute of Technology, USA
Jose Saldana	University of Zaragoza, Spain
Srinivas Shakkottai	Texas A&M University, USA
Wee-Seng Soh	National University of Singapore
Hui Song	Frostburg State University, USA

Alex Sprintson	Texas A&M University, USA
Aaron Striegel	University of Notre Dame, USA
Ilenia Tinnirello	University of Palermo, Italy
Damla Turgut	University of Central Florida, USA
Jianfeng Wang	Philips Research North America, USA
Ping Wang	Nanyang Technological University, Singapore
Hongyi Wu	University of Louisiana at Lafayette, USA
Lei Wu	University of Houston - Clear Lake, USA
Jiang (Linda) Xie	University of North Carolina at Charlotte, USA
Zhenyu Yang	Florida International University
Lei Ying	Iowa State University, USA
Kai Zeng	UC Davis, USA
Hongqiang Zhai	Philips Research North America, USA
Hongwei Zhang	Wayne State University, USA

Table of Contents

Session 1: Cognitive Radio Networks

Session 2: Networks Security

Session 3: Resource Allocation

Session 4: Wireless Protocols and Algorithms

Session 5: Advanced Networking Systems

Session 6: Sensor Networks

Session 7: Scheduling and Optimization

Session 8 : Routing Protocols

Session 9 : Multimedia and Stream Processing

DSRC Workshop Session 1: DSRC Networks

DSRC Workshop Session 2: DSRC Security

Transport Control Based on Spectrum Switching Detection in Cognitive Radio Ad Hoc Networks

Ren Han[1,2,3], Xiaoxia Huang[3], and Pan Li[4]

[1] Institute of Computing Technology, Chinese Academy of Sciences, Beijing, China
[2] Graduate University of Chinese Academy of Sciences, Beijing, China
[3] Shenzhen Institutes of Advanced Technology, Chinese Academy of Sciences, Shenzhen, China
[4] Department of Electrical and Computer Engineering, Mississippi State University, MS 39762
{ren.han,xx.huang}@sub.siat.ac.cn, li@ece.msstate.edu

Abstract. Cognitive Radio is an intelligent wireless communication technology which can improve the radio spectrum utilization through opportunistically accessing the vacant portions of the spectrum. Due to the time-varying spectrum availability, a cognitive node has to perform spectrum switching frequently, which causes severe delay and high packet loss. Therefore, network protocol design for cognitive radio ad hoc networks (CRAHNs) should address this issue to ensure reliable end-to-end packet delivery. In this paper, we first propose a novel spectrum switching detection scheme for CRAHNs in which intermediate nodes monitor the variation of the Round-Trip Time (RTT) and the arrival interval time of packets to detect the spectrum switching over succeeding links. Based on the scheme, a new transport control mechanism is then designed to deal with spectrum switching in CRAHNs. We implement the proposed mechanisms on a USRP2-based testbed. The experiment results demonstrate that the mechanisms can detect the spectrum switching and improve end-to-end throughput by up to 45%.

Keywords: Cognitive radio, ad hoc networks, spectrum switching, transport control.

1 Introduction

A Wireless ad hoc network (WANET) is characterized by its self-organized nature that does not rely on any infrastructure and centralized administration. In WANETs, each node can dynamically select proper routing with multiple hops to forward data to its destination. Currently, WANETs are often limited to operate on industrial, scientific and medical (ISM) frequency bands. With the rapid development and deployment of wireless networks, ISM bands have become increasingly congested [1], [3]. Cognitive Radio (CR) technology is believed to be an intelligent wireless communication technology for improving the radio spectrum utilization by opportunistically exploiting the existence of spectrum holes. In CR networks, the CR users are capable of utilizing these licensed bands unoccupied by the primary users (PUs) while not causing

X. Zhang and D. Qiao (Eds.): QShine 2010, LNICST 74, pp. 1–15, 2011.

interference to the primary service. It is envisioned that the CR technology can enhance the capacity of wireless networks by expanding the range of current wireless networks and making better use of spectrum resources. In traditional WANETs, the mobility of intermediate nodes and inherent wireless channel instability are the key factors that jeopardize the reliable end-to-end delivery of data. While in cognitive radio ad hoc networks (CRAHNs), since the CR users have to evacuate the licensed band once PUs return, some other issues, such as spectrum sensing, spectrum switching and awareness of the PUs' activities, are the major problems that account for unreliable data delivery[2], [3]. In particular, in CRAHNs, when PUs arrive, CR users have to give up the current channel and trigger the spectrum switching. In this process, CR users search for a set of available channels on various spectrum bands through spectrum sensing and negotiate with each other to get a mutually acceptable channel to recover the link between them. From the standpoint of end-to-end transmissions, there is a temporary link breakage on the path during the spectrum switching that could cause high packet loss and large delay. Thus, in CRAHNs, how to detect the spectrum switching in time and react to it efficaciously is a challenge to ensure reliable end-to-end data delivery.

In this paper, we first attempt to find an efficient spectrum switching detection scheme. Since spectrum switching causes a temporary link breakage, during this time, all packets have to be buffered in intermediate nodes and cannot be delivered to source or destination nodes. Therefore, Round-Trip Time (RTT) and arrival interval time of these delayed packets should manifest greater value than their counterparts in normal transmissions. Along an end-to-end path, the intermediate nodes near the links performing the spectrum switching can obtain more accurate information in time than the source node. We resort to these intermediate nodes to monitor RTT and arrival interval time of packets and design a spectrum switching detection scheme based on these two parameters. Based on this scheme, we then propose a new transport control protocol which prevents source nodes from injecting more packets into the network during the spectrum switching process in order to alleviate MAC layer contentions and packet bursts.

The rest of the paper is organized as follows. Section 2 reviews related work in this area. In Section 3, we describe our system model. A new MAC protocol is proposed in Section 4, which has been implemented on our testbed. Section 5 elaborates the spectrum switching detection scheme. A new transport control mechanism is introduced in Section 6. In Section 7, we describe our testbed and analyze the experiment results. Section 8 concludes our work.

2 Related Works

Considerable work has been done on transport control protocols (TCPs) to explore the reason for message loss. Ramani and Karandikar [4] employ Explicit Congestion Notification (ECN) to differentiate whether the loss is caused by congestion or not. Casetti et al. [5] introduce the TCP Westwood scheme relying on end-to-end bandwidth to discriminate the cause of packet loss (congestion or degraded wireless channel). Brakmo, Malley and Peterson [6] monitor the changes in the estimated amount of extra data in the network based on an explicit congestion model between throughput and RTT. Barman and Matta [7] assume that the variations in RTT and packet loss are correlated. By estimating the average RTT, TCP is able to distinguish the loss owing to congestion from that due to unreliable wireless transmissions. Biaz and Vaidya [8], Tobe et al. [9] investigate a set of "loss predictors" involving interval-arrival time and Relative One-way Trip Time (ROTT) to predict the underlying packet loss reasons. However, Cen, Cosman, Voelker [10] have pointed out that the classification indicators, as described above, are constrained by the network topology and each indicator can only be effective in certain topologies.

In wireless networks, data deliveries are plagued by problems such as high bit error rate (BER), route failure, multipath routing and network partition. Apart from these problems, dynamic spectrum access is another serious problem in CRAHNs which leads to route disruption. For example, route disconnection and longer packet RTT may be incurred if an intermediate node on the route is engaged in spectrum sensing or switching. A window-based transport protocol for CRAHNs called TP-CRAHN is first proposed by Chowdhury, Felice and Akyildiz [2], using the ECN mechanism to determine the loss causes. If the time lag contained in ECN is within the threshold Lmax (Lmax = 1.5× RTT in [2]), and no prior action has been taken on an earlier ECN from the same node, TP-CRAHN assumes that congestion occurs; while any further delay (i.e., time lag > Lmax) indicates that the path was temporarily disconnected due to a spectrum sensing or channel switching event. However, TP-CRAHN is just one modified TCP scheme which enables intermediate nodes to send or piggyback ECN to notify the source node. Felice, Chowdhury and Bononi [11] exploit a CRAHN model integrating three impact factors: spectrum sensing cycle, interference from PUs and channel heterogeneity. The performances of different TCP variants in CRAHNs are evaluated under this model using NS-2. Simulation results show that existing TCP schemes proposed for wireless ad hoc networks might not work well over CRAHNs. Sarkar and Narayan [12] propose another modular architecture chartered two modules: knowledge module and cognitive module. The approach is separated from standard transport protocols and adapts better to the unique characteristics of cognitive radio networks.

3 System Model

In our model, each CR node is equipped with a single transceiver that is capable of tuning to any channel among a licensed spectrum set. We assume that activities over the licensed channels can be described as two states, ON and OFF. The ON (Busy) state represents the channel is being used by PUs. The OFF (Idle) state indicates that the channel is free of PU activities.

In CRAHNs, a CR user needs to detect whether any PU appears on current channel before it gets ready to transmit or receive data packets. As shown in Fig. 1, if PU_{C3} occupies channel C3, CR user 3 has to suspend its transmission and starts a spectrum sensing operation. Spectrum sensing procedure searches for available channels over various spectrum bands that are not occupied by PUs. When CR user 3 finishes spectrum sensing and obtains a new available channel set, it triggers spectrum switching operation to negotiate with its neighbors to guarantee a mutually acceptable channel. If the spectrum switching succeeds, the link breakage is recovered and data transmission resumes. In this paper, we address three issues: an implementable medium

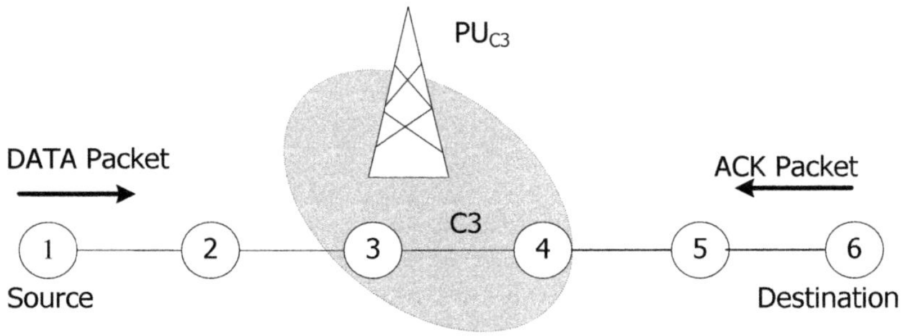

Fig. 1. PU occupies its licensed channel

access control (MAC) mechanism, an efficient spectrum switching detection scheme, and a reliable transport control scheme, in CRAHNs. In the following sections, we will elaborate these three design issues, respectively.

4 An Implementable MAC Protocol

In CRAHNs, CR users are capable of transmitting data on different channels via dynamic spectrum access. The CR MAC layer should provide an efficient mechanism considering two issues: (i) sensing the channel to detect the presence of PUs and switching to another available channel when PUs appear on the current channel; (ii) sharing the channel with neighboring CR users with a tolerable interference level to PUs [13]. MAC protocol design is a challenging issue in CRAHNs. IEEE 802.22 makes a standardization effort, while other MAC protocols are specially designed for various environments and applications. In our study, we attempt to design an implementable MAC protocol to balance between the limitation of hardware support and the performance of the protocol. To this end, we adopt a random access methodology based on carrier sense multiple access with collision avoidance (CSMA/CA).

Our implementable MAC is a special application of CSMA/CA in CRAHNs, as shown in Fig. 2. Due to the dynamic spectrum access, the MAC mechanism employs two types of channels, a common control channel and data channels. Two neighboring CR users exchange RTS/CTS to carry out handshake on the common control channel. If the process is successful, they turn to a data channel negotiated in advance and perform spectrum sensing to determine if any PU occupies the channel. Guaranteeing the absence of PUs at that time, two CR users can complete the data frame transmission. If any PU is active on the data channel, CR users have to give up the current channel and trigger spectrum switching to find another idle channel. Then, CR users negotiate a common idle channel to rebuild a new data channel.

Consider a network in which all nodes are deployed at the same distance to its neighboring nodes. Every node can only communicate with its one-hop neighbors, sense the transmissions of its two hop neighbors, and cannot detect other nodes. When a node is transmitting a packet on a channel, all nodes within its sensing range cannot transmit over the same channel. A chain topology is shown in Fig. 3. Assume there is

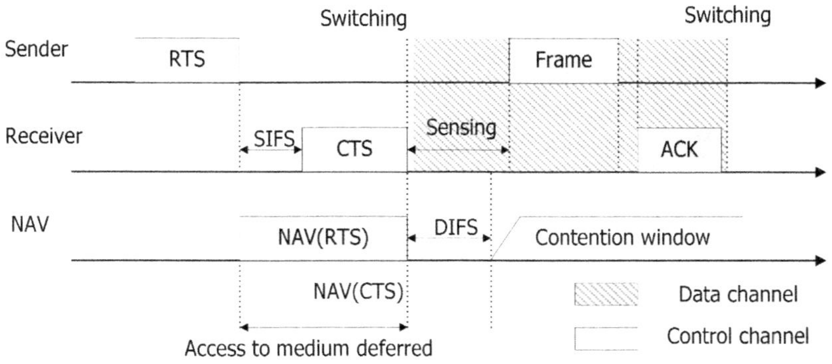

Fig. 2. An implementable MAC protocol

only one channel available. If node 1 and node 2 are exchanging RTS/CTS, node 3 may detect it and hence cannot transmit or receive any packet. It will wait for a time interval of NAV (Network Allocation Vector). However, node 4 is a hidden terminal to the RTS transmission. It is possible that node 4 considers the channel to be idle and attempts to transmit to node 5. The transmission between node 4 and node 5 may collide with the reception at node 2 and cause packet loss.

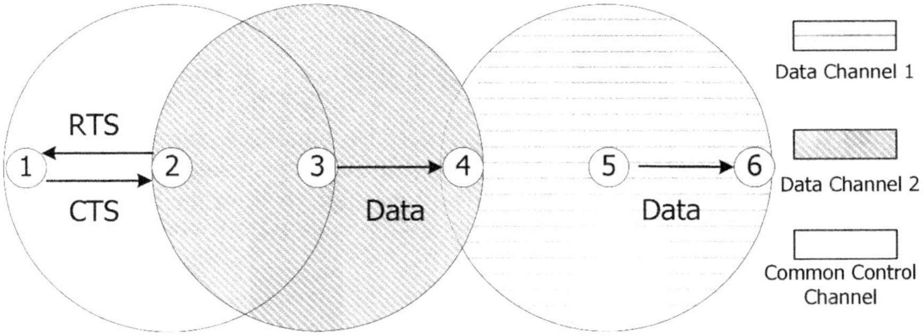

Fig. 3. A chain topology

Chen et al. [14] show that the bandwidth-delay product (BDP-UB) of a path can be upper bounded by kN with $k = 1/5$, where N is the number of round-trip hops. Our proposed MAC protocol utilizes one control channel and several data channels. We assume that there are enough available channels in the network and each link on an end-to-end path is built on a different data channel. Therefore, CR users only contend to access the common control channel. Although any pair of nodes can communicate on a different data channel simultaneously, as shown in Fig. 3, the BDP-UB of the path in CRAHNs cannot exceed $L/2$ (ideally, every two neighboring nodes compose a pair and transmit data simultaneously), where L is the length of the path.

5 A Spectrum Switching Detection Scheme

Since the link congestion leads to packet loss and timeout, TCP variants propose to control congestion window size to alleviate the congestion on the path. The conventional TCP approaches decrease the window size multiplicatively if the network is congested, e.g. the AIMD algorithm. This method is both costly and inefficient in multihop wireless networks. Especially in CRAHNs, since the CR users have to evacuate the licensed band once PUs return, some other issues, such as spectrum sensing, spectrum switching and awareness of the PUs' activities, are the major factors that account for unreliable data delivery. Thus, in CRAHNs, how to detect the spectrum switching in time and react to it efficaciously is a big challenge to ensure reliable end-to-end data delivery.

In particular, the appearance of PUs may cause spectrum switching which results in packet loss and path breakage in CRAHNs. The path breakage due to the spectrum switching is a new cause of packet loss and congestion for CRAHNs. For example, as shown in Fig. 1, the data channel between node 3 and node 4 is occupied by a PU. If the PU_{C3} appears on channel C3 and interrupts the data transmission between node 3 and node 4, the two nodes have to trigger spectrum switching to negotiate a new data channel. During this period, node 3 and 4 cannot communicate with other nodes. Source node 1 is unaware of the situation and will keep injecting packets into the network. As a result, node 2 will queue many packets and cannot transmit them to the next hop, node 3. When the spectrum switching between node 3 and 4 succeeds and the path recovers, the heavy traffic load at node 2 could lead to severe contentions, thus causing high packet loss.

The spectrum switching occurs frequently due to the dynamic spectrum access which greatly affects the performance of end-to-end throughput. In CRAHNs, predicting the spectrum switching and reacting to it as quickly as possible are critical to improving the end-to-end transmission throughput. In conventional transport layer control mechanisms, a source node needs a long time to learn the change of a path. Moreover, the information obtained by the source node becomes less accurate as the path length increases.

Compared with the source node, the intermediate nodes near the links performing spectrum switching, called detection nodes in this paper, can get more accurate information and react to the occurrence of spectrum switching effectively. It is important that the detection node could choose some appropriate features to detect the spectrum switching. In this paper, we investigate whether RTT and packet arrival interval time can be used as the indication of the spectrum switching. RTT measured at an intermediate node is defined as the time interval between the data packet arrival time and its corresponding ACK packet arrival time. Packet arrival interval time is defined as the interval between successive packets arrived at the intermediate node. During the process of end-to-end transmission, there are two types of packets, forward data packets and reverse ACK packets which are necessary to the reliable transmission. In general conditions, the throughput of data flow and ACK flow could be stable and have less sudden changes. In these situations, RTT and packet arrival interval time of these two types of packets are small and within a certain range. In CRAHNs, the occurrence of the spectrum switching could cause sudden increase of RTT and packet arrival interval time. As Fig. 3 shows, the link between node 3 and node 4 is unavailable because PU_{C3} occupies channel C3. During the temporary path breakage, forward data packets and reverse ACK packets cannot be exchanged between node 3 and 4. Instead, they are buffered in the queues of intermediate nodes near node 3 and 4 until the spectrum switching finishes and the link is recovered. In this process, significant delay occurs on these buffered packets. Consequently, RTT and arrival interval time would demonstrate greater values beyond the normal range. This shows the distinct features of the spectrum switching. Therefore, we observe the RTT and packet arrival interval time at the intermediate nodes on the path to unveil the effect of the spectrum switching on these two factors.

6 A New Transport Control Mechanism

CRAHN differs from existing networks in that CR users access the spectrum opportunistically. If two nodes in different types of networks attempt to realize data transmission, a transport control mechanism across heterogeneous networks is necessary. In general, the end-to-end connection mechanism could control the whole path, but it needs to modify existing protocols or create a new protocol to make it compatible with the two types of networks. This method is costly and inefficient. Every transport control mechanism can only work efficiently in certain situations. The cross-network mechanism considering all factors of different networks is too complicated and difficult to deploy. A feasible approach is to divide the end-to-end connection into separated connections at the border of any two different networks. As Fig. 4 shows, an end-to-end connection from node S to node 4 is established. Node 1 at the border of the CRAHN can be selected as the interface between two networks. The approach also makes the transport control mechanism of CRAHN transparent to the other network. Therefore, the nodes in the left side network can apply its original transport control mechanism without modifications. All the packets destined to node 4 are received, buffered and acknowledged by node 1. In the view of node 2, 3 and 4, node 1 is like a source node of the path.

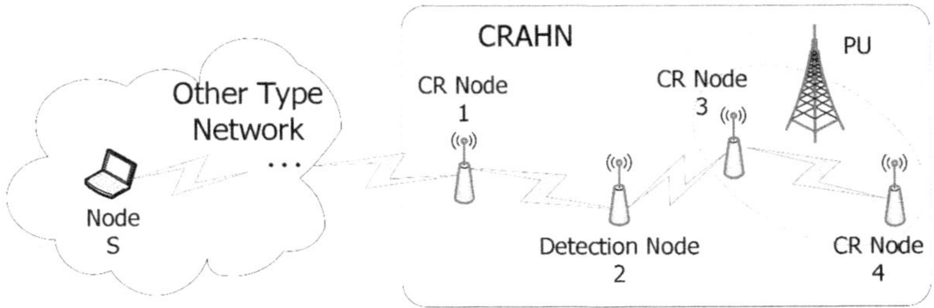

Fig. 4. End-to-end transmission

In CRAHNs, detection nodes are used to assist detecting the spectrum switching. As Fig. 4 shows, detection node 2 is near the link needing the spectrum switching. It can obtain accurate information of changes on the link through monitoring RTT and packet arrival interval time. Based on the discussion in Section 5, node 2 can monitor the arrival interval time of reverse ACK packets to detect the spectrum switching happened on its succeeding nodes, e.g. node 3 and 4. For the path without spectrum switching, the packet arrival interval time is within a certain range. If it suddenly grows beyond the range, probably spectrum switching has happened over the path. When receiving an ACK packet, the detection node starts a timer and waits for the next packet. If the timer expires and the next packet does not arrive, the detection node concludes that the spectrum switching has occurred and sends a message to inform node 1. We will find proper thresholds of RTT and packet arrival interval that indicate the spectrum switching has occurred. Node 2 can piggyback the message in ACK packet to node 1. Being aware of spectrum switching, node 1 runs a simple flow control mechanism to deal with it. The mechanism employs a fixed window transmission which is constrained by BDP-UB. Node 1 can either monitor RTT and packet arrival interval time to detect the spectrum switching or get more accurate information from node 2. When node 1 believes that the spectrum switching occurs, it checks data packets that are not acknowledged. If the number of these packets is greater than 2, node 1 will not send packets until it recognizes that the spectrum switching finished. In order to deal with ACK packet loss, the mechanism also employs retransmission timeout. The retransmission timeout does not increase exponentially as that in TCP when retransmission happens, because the spectrum switching and MAC contention are the main reasons for packet loss rather than the congestion in CRAHNs.

7 Experiment Environment and Results

In this section we demonstrate the variation of RTT and packet arrival interval time at intermediate nodes during spectrum switching in CRHANs based on the experiments. Our testbed is built with software defined radio devices. We analyze the experiment results and prove the efficiency of the spectrum switching detection scheme. The transport control mechanism based on the detection scheme also runs on our testbed and manifests a higher throughput than conventional TCP in CRAHNs.

7.1 Testbed Description

The software defined radio is used to create a flexible radio platform with software instead of traditional hardware to perform signal processing. The ideal software defined radio platform would use as little hardware as possible and let software deal with all of the processing. The receiver might have just an antenna connected to an analog-to-digital converter (ADC). Samples would then be read from the ADC and software would handle all signal processing, as shown in Fig. 5(a) [15]. Our testbed is built based on the GNU Radio and the 2nd version of Universal Software Radio Peripheral (USRP2), as shown in Fig. 5(b).

GNU Radio is a signal processing package with a free software development toolkit and provides an open-source library of common signal processing blocks [16]. The 2nd version USRP offers higher performance than the original USRP which leads us to choose USRP2 as a component of our testbed.

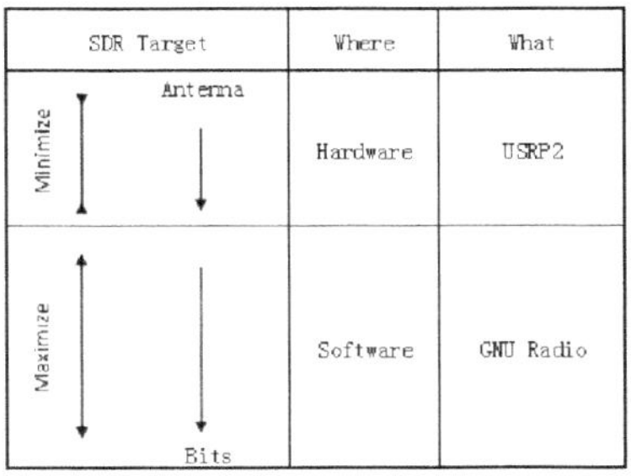

(a)

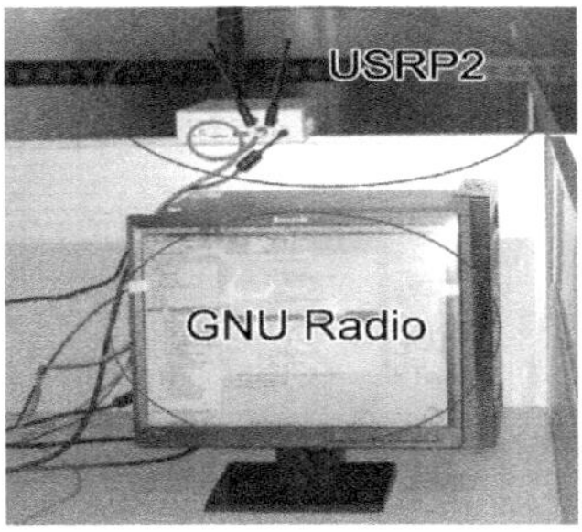

(b)

Fig. 5. (a) Software defined radio design principle. (b) USRP2 and GNU Radio

7.2 Experiment Results

Each USRP2 is equipped with one transceiver and communicates with bandwidth of 0.5Mbps. We implement our MAC protocol based on GNU Radio that runs on USRP2. In order to study the influence of spectrum switching on the path, we set fixed routing table at every node from the source node to the destination node. In this experiment, the CR node communicates over frequency range from 2.4 to 2.5 GHz divided into 9 channels. The channel set C is {2.40GHz, 2.41GHz,...,2.48GHz} in which 2.48GHz is used for the common control channel and the others are data channels. The channel division guarantees that every link is allocated a different data channel with others over a connected path. A flow runs over a 3-hop chain topology as shown in Fig. 4. The link between node 3 and node 4 is subject to the appearance of PUs. In the experiment, PU appears periodically which leads to the spectrum switching of node 3 and node 4. As a detection node, node 2 records the RTT and packet arrival interval of the packets.

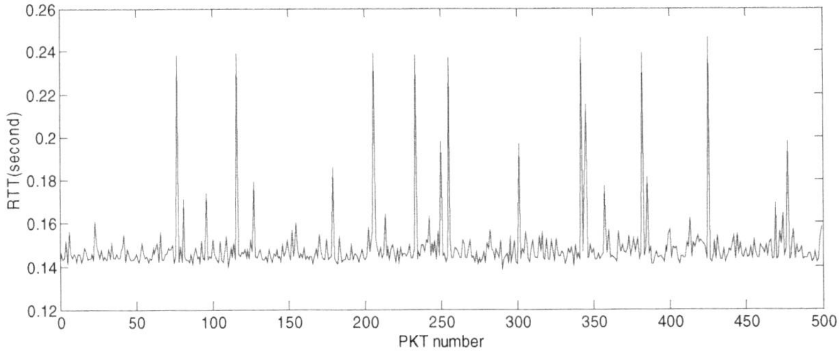

Fig. 6. RTT at detection node (no PU)

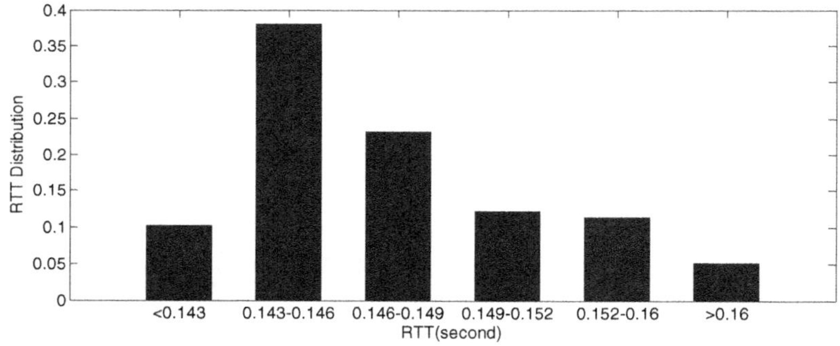

Fig. 7. RTT distribution in various time ranges at detection node (no PU)

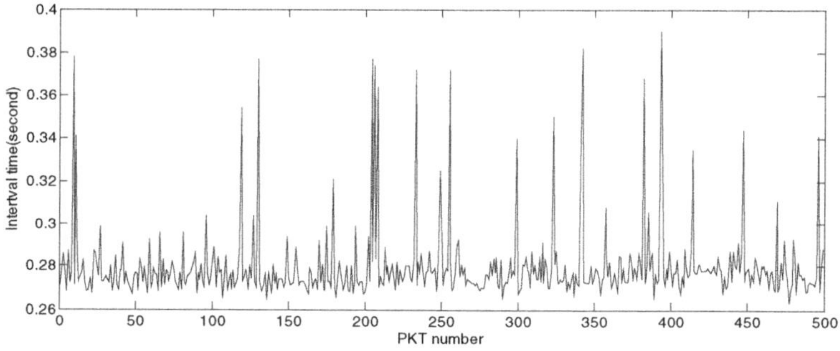

Fig. 8. Packet arrival interval time at detection node (no PU)

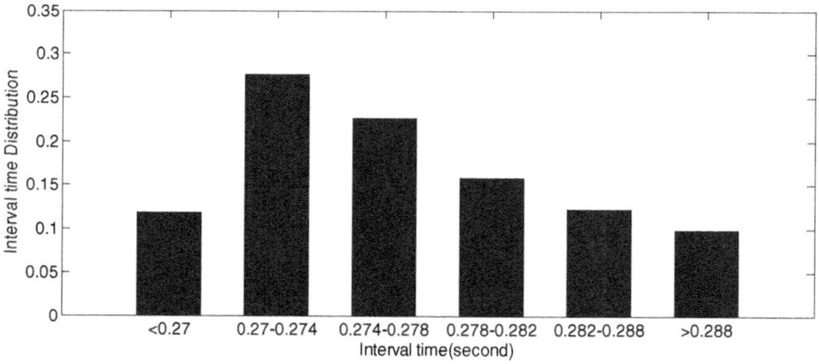

Fig. 9. Arrival interval time distribution in various time ranges at detection node (no PU)

We need to explore the distributions of the RTT and packet arrival interval time without spectrum switching as the benchmark for the comparison in various conditions. Fig. 6 manifests the RTT values of 500 packets at detection node with no PU. Most of these values are limited in a narrow range which is shown more evidently in Fig. 7. Almost 85% values of RTT are distributed in the range from 0.143 seconds to 0.16

seconds. Only 5% values are greater than 0.16 seconds which are demonstrated as spikes in Fig. 6. We investigate the packet arrival interval time under the same condition. As can be seen in Fig. 8, the values of the packet arrival interval remain relatively stable. Fig. 9 shows that about 90% of packet arrival intervals are below 0.288 seconds and the max value is about 0.39 seconds. Notice that a few spikes in Fig. 6 and 8 appear irregularly. They are caused by unreliable wireless channel condition. According to the experimental results of Fig. 6-9, we can conclude that the distributions of RTT and packet arrival interval time are relatively stable without spectrum switching.

We explore the impact of the spectrum switching on RTT and packet arrival interval time on various conditions. In the scenario, we set the sensing time to 0.2 seconds. PU arrives every 6 seconds in the experiment. As can be seen in Fig. 10 and Fig. 11, the values of RTT and packet arriving interval time experience sharp changes. These peaks in the two figures manifest that some packets are seriously impacted by the spectrum switching and have greater values of RTT and packet arrival interval time which are easily distinguished from normal values. Differed from these greater RTT values incurred by performing spectrum switching, many values of packet arrival interval time are less than normal values. When the spectrum switching finishes and the link recovers, certain node continuously sends packets buffered in its queue which leads to lower packet arrival interval time.

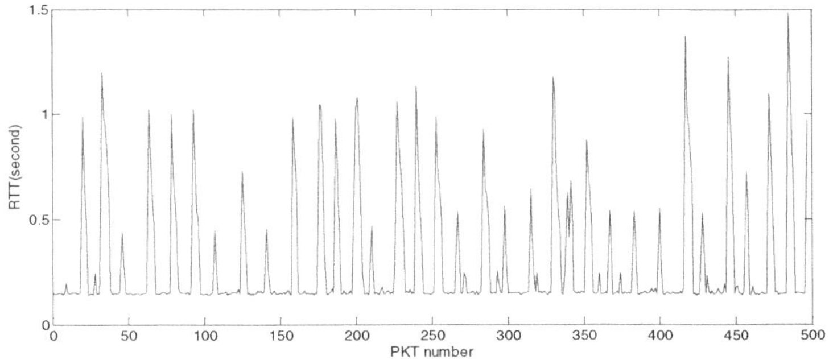

Fig. 10. RTT at detection node (sensing time is 0.2 seconds, PU arrives every 6 seconds)

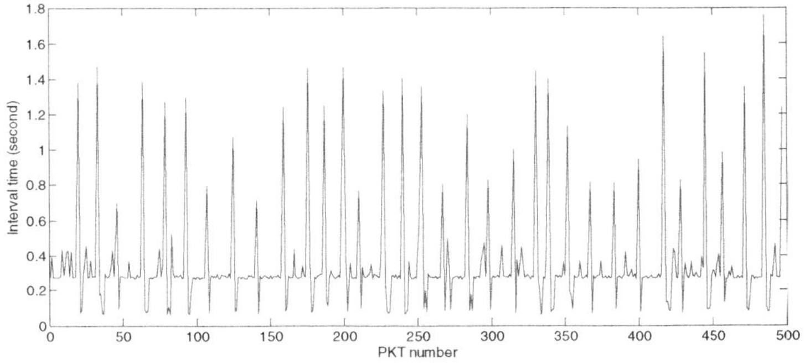

Fig. 11. Packet arrival interval time at detection node (sensing time is 0.2 seconds, PU arrives every 6 seconds)

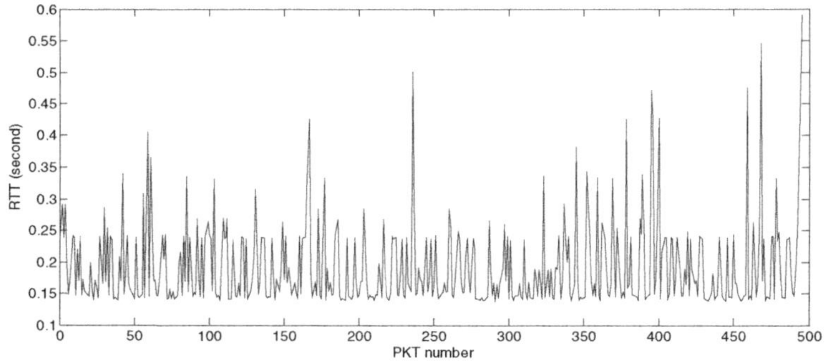

Fig. 12. RTT at detection node (sensing time is 0.2 seconds, PU arrives every 3 seconds)

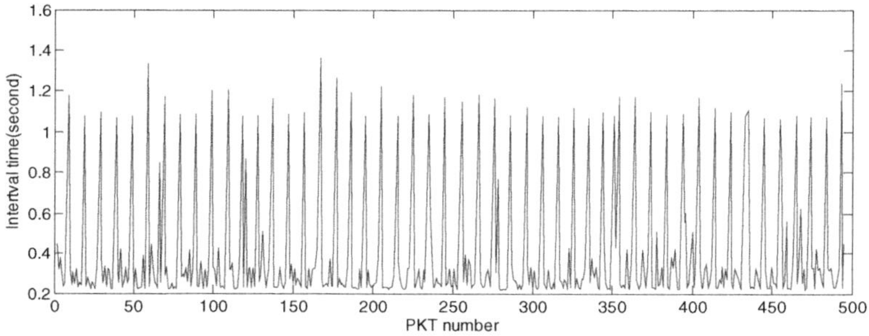

Fig. 13. Packet arrival interval time at detection node (sensing time is 0.2 seconds, PU arrives every 3 seconds)

In Fig. 12 and 13, we set the sensing time to 0.2 seconds and let a PU arrive every 3 seconds. Frequent spectrum switching causes the RTT of some packets increasing abruptly and also leads to longer packet delay and more serious MAC contention which are explicitly demonstrated by spikes in Fig. 12. These regular peaks shown in Fig. 13 manifest that packet arrival interval time can indicate frequent spectrum switching.

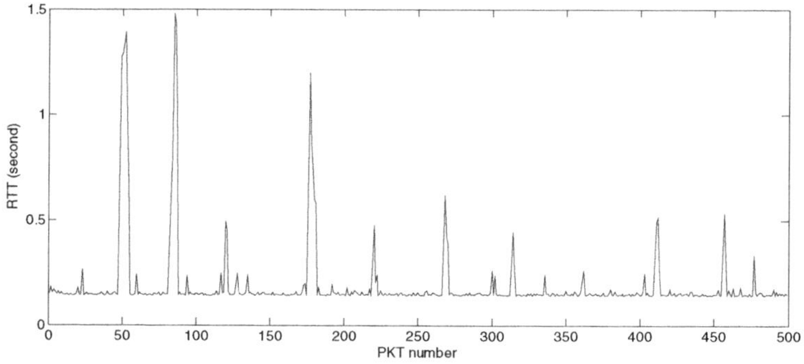

Fig. 14. RTT at detection node (sensing time is 0.1 seconds, PU arrives every 15 seconds)

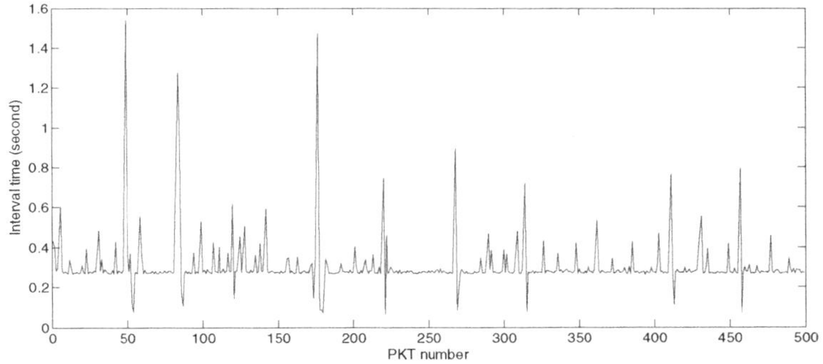

Fig. 15. Packet arrival interval time at detection node (sensing time is 0.1 seconds, PU arrives every 15 seconds)

We further adjust the sensing time from 0.2 seconds to 0.1 seconds and let a PU arrive every 15 seconds. As shown in Fig. 14 and 15, while the sensing time becomes shorter, these peaks still can indicate the occurrence of the spectrum switching. There are some spikes in the two figures which are caused by the variation of wireless channel condition even when no spectrum switching happens. As can be seen from Fig. 6-15, the packets affected by the spectrum switching experience longer RTT and packet arrival interval time compared with packets in normal transmissions. It is evident that the detection of spectrum switching via these two factors at intermediate nodes is reasonable and feasible in CRAHNs.

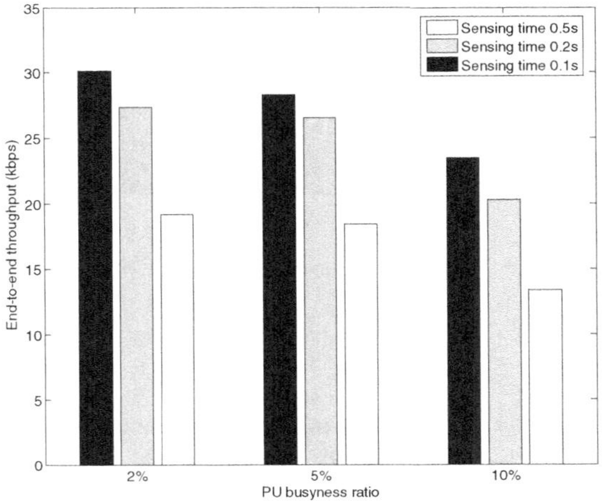

Fig. 16. End-to-end throughput in various conditions

According to the above analysis of the experiment results, the detection scheme depending on monitoring the variations of RTT and packet arrival interval time can indicate the spectrum switching accurately. Based on the scheme, we run the transport control mechanism discussed in Section 6 on our testbed. Fig. 16 shows the end-to-end throughput with respect to various sensing time and PU busyness ratio. With short sensing time (e.g. 0.1 seconds or 0.2 seconds), the differences of throughput are relatively small. The total time of the spectrum switching includes sensing time and negotiation time which is uncertain. When the sensing time is short, the negotiation time dominates the total time of spectrum switching. However, when the sensing time increases (e.g. 0.5 seconds), the ratio of the sensing time to the total time rises. Therefore, there is a noticeable decline of the end-to-end throughput. PU busyness ratio can also affect the throughput along the path. Frequent PU arrivals increase the number of spectrum switching. So the source node has to restrain from sending too many packets. Compared to the throughput under the other two PU busyness ratios, the throughput drops visibly when the PU busyness ratio is 10%. We also compare the throughput between our transport control mechanism and TCP. In Fig. 17, our mechanism achieves higher throughput than TCP by up to 45%. The reason is that our mechanism can detect the occurrence of the spectrum switching and slow down injecting packets into the path to avoid MAC layer contentions and packet bursts.

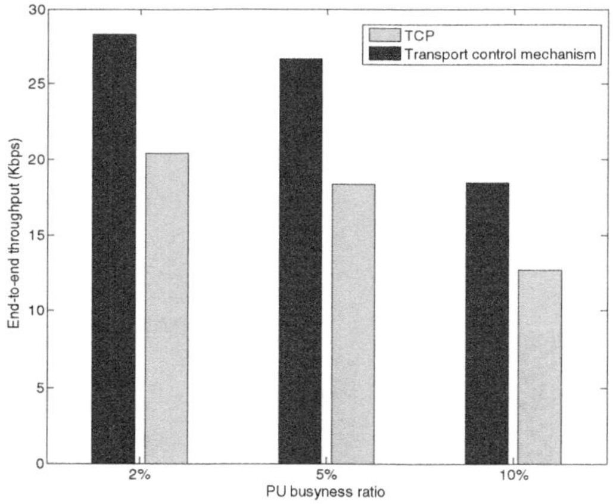

Fig. 17. End-to-end throughput (sensing time is 0.2 seconds)

8 Conclusions

In this paper, we first study the detection of the spectrum switching via RTT and packet arrival interval time in CRAHNs. The proposed novel detection scheme resorts to the intermediate node to monitor the changes of these two factors which can indicate the occurrence of the spectrum switching. Based on the scheme, a new transport control mechanism is designed for CRAHNs. We implement the system on our testbed and collect experiment data for analysis. The experiment results demonstrate that the proposed mechanisms can detect the spectrum switching and improve end-to-end throughput efficiently.

Acknowledgments. This work was supported by the National Natural Science Foundation of China under grant No. 60903192.

References

1. Akyildiz, I.F., Lee, W.F., Vuran, M.C., Mohanty, S.: NeXt Generation/Dynamic Spectrum Access/Cognitive Radio Wireless Networks: A Survey. Computer Networks 50, 2127–2159 (2006)
2. Chowdhury, K.R., Felice, M.D., Akyildiz, I.F.: TP-CRAHN: A Transport Protocol for Cognitive Radio Ad-Hoc Networks. In: 28th Conference on Computer Communications, Rio de Janeiro, Brazil, pp. 2482–2490 (2009)
3. Akyildiz, I.F., Lee, W.Y., Chowdhury, K.R.: CRAHNs: Cognitive Radio Ad-hoc Networks. Computer Networks 7, 810–836 (2009)
4. Ramani, R., Karandikar, A.: Explicit Congestion Notification (ECN) in TCP over Wireless Network. In: 2000 IEEE International Conference on Personal Wireless Communications, Hyderabad, India, pp. 495–499 (2000)
5. Casetti, C., Gerla, M., Mascolo, S., Sanadidi, M.Y., Wang, R.: TCP Westwood: Bandwidth Estimation for Enhanced Transport over Wireless Links. In: 7th Annual International Conference on Mobile Computing and Networking, Rome, Italy, pp. 287–297 (2001)
6. Brakmo, L.S., O'Malley, S.W., Peterson, L.L.: TCP Vegas: New Techniques for Congestion Detection and Voidance. In: ACM SIGCOMM Computer Communication Review, London, U.K, pp. 24–35 (1994)
7. Barman, D., Matta, I.: Effectiveness of Loss Labeling in Improving TCP Performance in Wire/Wireless Networks. In: 10th IEEE International Conference on Network Protocols, pp. 24–35. IEEE Computer Society, Washington (1994)
8. Biaz, S., Vaidya, N.H.: Discriminating Congestion Losses from Wireless Losses Using Inter-Arrival Times at the Receiver. In: Proceedings of the 1999 IEEE Symposium on Application - Specific Systems and Software Engineering and Technology, pp. 10–17. IEEE Computer Society, Washington (2002)
9. Tobe, Y., Tamura, Y., Molano, A., Ghosh, S., Tokuda, H.: Achieving Moderate Fairness for UDP Flows by Path-Status classification. In: 25th Annual IEEE Conference on Local Computer Networks, pp. 252–261. IEEE Computer Society, Washington (2000)
10. Cen, S., Cosman, P.C., Voelker, G.M.: End-to-End Differentiation of Congestion and Wireless Losses. IEEE/ACM Transaction on Networking 11, 703–717 (2003)
11. Felice, D., Chowdhury, K.R., Bononi, L.: Modeling and Performance Evaluation of Transmission Control Protocol over Cognitive Radio Ad Hoc Networks. In: 12th ACM International Conference on Modeling, Analysis and Simulation of Wireless and Mobile Systems, Tenerife, Canary Islands, Spain, pp. 4–12 (2009)
12. Sarkar, D., Narayan, H.: Transport Layer Protocols for Cognitive Networks. In: IEEE Conference on Computer Communications Workshops, pp. 1–6 (2010)
13. Cormio, C., Chowdhury, K.R.: A Survey on MAC Protocols for Cognitive Radio Networks. Ad Hoc Networks 7, 1315–1329 (2009)
14. Chen, K., Xue, Shah, S.H.: Understanding Bandwidth-Delay Product in Mobile Ad Hoc Networks. Computer Communications 27, 923–934 (2004)
15. Verduin, A.: Gnu Radio Wireless Protocol Analysis Approach. Master Program System and Network Engineering, Universiteit Van Amsterdam (2008)
16. Welcome to GNU Radio, `http://gnuradio.org/redmine/wiki/gnuradio`

Spectrum Prediction via Temporal Conditional Gaussian Random Field Model in Wideband Cognitive Radio Networks

Zhenghao Zhang[1,2], Husheng Li[1], Hannan Ma[1], Kun Zheng[1], Depeng Yang[1], and Changxing Pei[2]

[1] Department of EECS, The University of Tennessee, Knoxville TN 37996, USA
chancejack@gmail.com, husheng@eecs.utk.edu
http://www.ece.utk.edu/ husheng/
[2] Xidian University,
No.2 Taibai Road, Xi'an, China, 710071

Abstract. Wideband spectrum sensing remains an open challenge for cognitive radio networks due to the insufficient wideband sensing capability. This paper introduces the theory of Gaussian Markov Random Field to estimate the un-sensed sub-channel status. We set up a measurement system to capture the WiFi spectrum data. With the measurement data, we verify that the proposed model of Temporal Conditional Gaussian Random Field can efficient estimate the sub-channel status.

Keywords: cognitive radio networks, net components, dynamic software architecture, modeling, agents, software development approach.

1 Introduction

Cognitive radio (CR) [1][3] has been considered to be a promising technology to tackle the problem of spectrum resource scarcity. As one of the most important requirements for CR, perceiving the activities of primary users (PU) has been widely studied [9]. However, in wideband CR networks, wideband spectrum sensing still remains an open challenge [2][6]. Due to the radio frequency front-end hardware limitation, it is difficult to capture the change of spectrum with a limited sensing time in a wide spectrum range, subject to the contamination of noise. With limited wideband sampling capacity, [10] has proposed an opportunistic spectrum sensing scheme, in which the authors offered an optimal scheme to select sub-channels to sense.

Since the shared frequency band may have different division scheme for the PU network and the CR network, two adjacent sub-channels in the CR network may be affected by the same channel in the PU network. Fig. 1 illustrates the different spectrum division schemes of PU network and CR network. It is observed that, when the sub-channel 1 of the PU network is active, both sub-channels 1 and 2 of the CR network cannot be utilized. Thus, the channel usage statuses of sub-channel 1 and sub-channel 2 of the CR network are highly correlated since they are affected by the same PU sub-channel. However, the CR

X. Zhang and D. Qiao (Eds.): QShine 2010, LNICST 74, pp. 16–27, 2011.

network typically has no priori knowledge of the sub-channel division scheme of the PU network. In addition, from the temporal viewpoint, two adjacent sensing periods may also have correlation. Based on the binary time series analysis, [8] proposed a spectrum occupancy prediction algorithm where the subchannel occupancy status is modeled as time series and the model parameters are learned from previous known subchannel occupancy status using the logistic regression algorithm. However, it is assumed that the time series representing subchannel occupancy status is stationary, which is a very strong assumption. Our measurement data is verified not to satisfy this requirement. According to the correlation existing among sub-channels and sensing periods, in this paper, we propose to use the Temporal Conditional Gaussian Random Field (TCGRF) [4][7] to model the wideband spectrum activity.

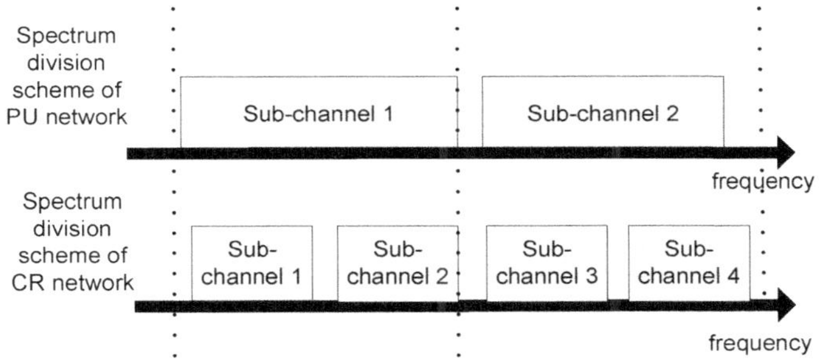

Fig. 1. An illustration of different sub-channel division scheme between PU network and CR network

As Fig. 2 illustrates, each pixel[1] represents a channel in a certain sensing period. Pixels within a certain neighborhood have correlation with each other. These pixels can be considered as the realization of a Gaussian Markov random field (GMRF). In each sensing period, not all the sub-channels are sensed. In this paper, we propose a TCGRF model to depict the sub-channel usage status within certain sensing periods. Based on the TCGRF model, a fraction of sub-channel usage status in current sensing period can be predicted without actual spectrum sensing.

The contribution of this paper is as follows. *We propose a TCGRF model to describe the sub-channel usage status, and provide the prediction of current sub-channel usage which has not been assigned for spectrum sensing. Furthermore, we set up a spectrum measurement to measure the spectrum activities in a WiFi network. The TCGRF model based prediction is verified using the measurement data.*

The remainder of the paper is organized as follows. The system model is defined in Section 2. The concept of GMRF is briefly introduced and the sub-channel usage

[1] Here we use the terminology of image processing since the 2-dimensional spectrum occupancy is similar to an image.

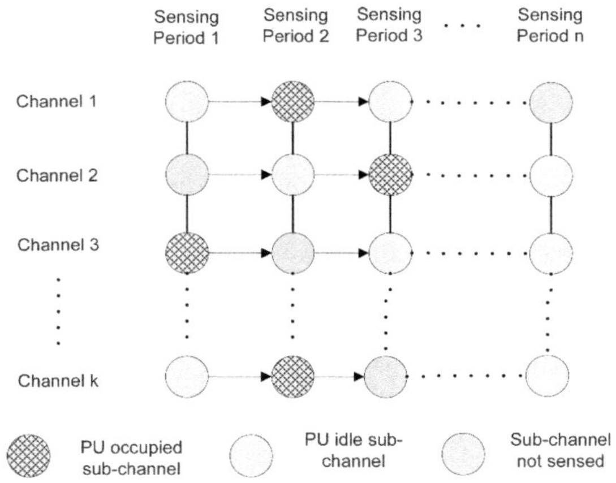

Fig. 2. An illustration of different sub-channel division schemes of PU network and CR network

status prediction based on TCGRF model is proposed in Section 3. We present the WiFi data measurement setup in Section 4 and provide the simulation results in Section 5. Conclusions and future works are provided in Section 6.

2 System Model

In this section, we introduce the system model for the wideband spectrum sensing in CR network. We assume that the CR network has different sub-channel division schemes from the PU network and the CR network does not have the priori information about how the PU network divides the entire shared frequency band. We assume that the CR network divides the shared spectrum into K sub-channels. In sensing period j, we define a sub-channel sensing set χ^j, which includes all the selected sub-channels to be sensed at sensing period j. For the ith ($i \in \chi^j$) sub-channel, the received signal is represented by $r_{i,j}(t)$. Applying the energy detection to each sub-channel, we obtain the sub-channel energy $y_{i,j}$ on the ith sub-channel at sensing period j, which is given by

$$y_{i,j} = \int_{\mathcal{T}_j} (r_{i,j}(t))^2 dt, \tag{1}$$

where $\mathcal{T}_j$ denotes the sensing interval of the jth sensing period. Thus, the secondary user carries out the binary hypotheses testing based on the sub-channel energy $y_{i,j}$. And we define an $K \times N$ spectrum usage indicator matrix, denoted by $\mathbf{X}$ where its element $x_{i,j}$, $(i = 1, 2, ..., K\ j = 1, 2, ..., N)$ represents the primary users' spectrum usage status on the ith sub-channel in the jth sensing period. The binary hypotheses testing is denoted by

$$x_{i,j} = \begin{cases} 1, & \text{if } y_{i,j} \geq \tau; \\ -1, & \text{if } y_{i,j} < \tau. \end{cases}, \tag{2}$$

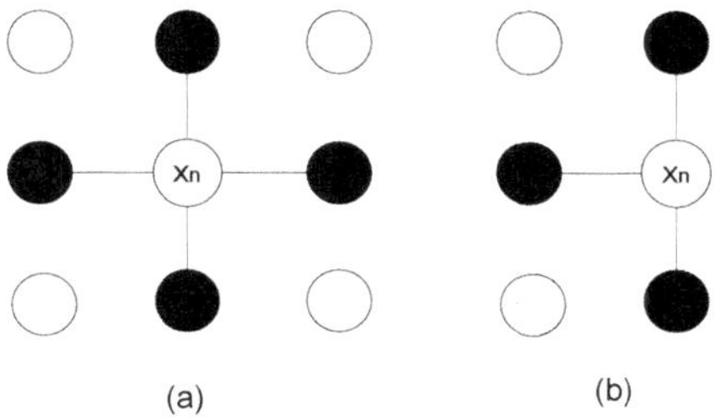

Fig. 3. Two kinds of neighborhood of pixel x_n

where τ is the energy detection threshold.

Therefore, the spectrum usage indicator matrix $\mathbf{X}$ can be considered as a binary-value image and its element $x_{i,j}$ represents the pixel in this image. The double index for each pixel $x_{i,j}$ can be transformed into a univariate index, which is given by

$$(i, j) \mapsto n = i + (j - 1)K. \tag{3}$$

Since we try to apply the GMRF, we define a neighborhood N_n for element x_n. We define two types of neighborhoods in this paper, which is illustrated in Fig. 3.

3 Gaussian Markov Random Field and Sub-channel Status Prediction by TCGRF

In this section, we briefly introduce the GMRF [5], which has been widely used in image processing, biologic information processing, etc. Then we will apply TCGRF to predict the sub-channel status.

3.1 Gaussian Markov Random Field

A GMRF is defined as a random vector $\mathbf{z} = (z_1, z_2, ..., z_n)^T \in \mathbb{R}^n$, whose conditional probability of each element z_n satisfies Markov property, which is given by

$$\pi(z_n|\mathbf{z}_{-n}) = \pi(z_n|z_i \in N_n), \tag{4}$$

where $\mathbf{z}_{-n}$ represents the set of all the elements in $\mathbf{z}$ excluding z_n. Furthermore, the probability density function (pdf) of $\mathbf{z}$ is given by

$$\begin{aligned}\pi(\mathbf{z}) &= (2\pi)^{-\frac{n}{2}}|\Sigma|^{-1/2} \exp -\frac{1}{2}(\mathbf{z}-\mu)^T \Sigma^{-1}(\mathbf{z}-\mu) \\ &= (2\pi)^{-\frac{n}{2}}|Q|^{1/2} \exp -\frac{1}{2}(\mathbf{z}-\mu)^T Q(\mathbf{z}-\mu),\end{aligned} \tag{5}$$

where μ and Σ are the expected trend and the covariance matrix of the GMRF, respectively. Due to the fact that z_i and z_j are conditionally independent for $z_i \notin$

$\mathcal{N}_j$, Σ^{-1} has a sparse representation [5]. Thus Σ^{-1} is defined as precision matrix Q for mathematical convenience. Furthermore, the elements of $\mathbf{z}$ is regarded as a graph $\mathcal{G} = (\mathcal{V}, \mathcal{E})$. Graph $\mathcal{G}$ depicts the neighborhood for each element z_n, each edge in the graph represents a pair elements in a neighborhood. And the precision matrix Q has the following property:

$$Q_{i,j} \begin{cases} \neq 0, & \text{if } i,j \text{ and } z_j \text{ is neighbor} \\ = 0, & \text{if } z_i \text{ and } z_j \text{ isn't neighbor} \end{cases}. \tag{6}$$

In this paper, we adopt the neighborhood defined in Fig. 3; hence the precision matrix Q is a sparse matrix.

3.2 Intrinsic GMRF

Given the precision matrix Q, we can attain the conditional distribution for some elements in the GMRF $\mathbf{z}$ given the remaining elements, which is given by

$$\mathbf{X}|\mathbf{Y} \sim \mathcal{N}(\mu_X - Q_{X,X}^{-1} Q_{X,Y}(Y - \mu_Y), Q_{X,X}^{-1}), \tag{7}$$

where $\mathbf{X}$ and $\mathbf{Y}$ represent the unknown elements and the known elements in the GMRF, respectively. Meanwhile, $Q_{X,X}$ and $Q_{X,Y}$ represent the inverse of the covariance matrix of $\Sigma_{X,X}$ and $\Sigma_{X,Y}$, respectively. We apply the Cholesky factorization of $Q_{X,X}$ and attain the conditional expectation of $\mathbf{X}$ given $\mathbf{Y}$, which is given by

$$E(X|Y) = \mu_X - R_{X,X}^{-1}((R_{X,X}^{-1})^T Q_{X,Y}(Y - \mu_Y)), \tag{8}$$

where

$$Q_{X,X} = R_{X,X}^T R_{X,X}. \tag{9}$$

For attaining the construction of Q according to the known data, we introduce the normal increment which is given by

$$z_i - z_j \sim \mathcal{N}(0, \kappa^{-1}). \tag{10}$$

According to the conditional independence provided by the graph $\mathcal{G} = (\mathcal{V}, \mathcal{E})$, we attain the intrinsic GMRF model, which is given by

$$\pi(\mathbf{z}) \propto \kappa^{(n-1)/2} \exp(-\frac{\kappa}{2} \sum_{i \sim j} (z_i - z_j)^2), \tag{11}$$

where the element pair $i \sim j$ represents an edge in the graph. We define the number of neighbors of z_i as n_i. The precision matrix Q has elements given by

$$Q_{i,j} = \begin{cases} \kappa n_i, & \text{if } i = j \\ -\kappa, & \text{if } i \sim j \\ 0, & \text{otherwise.} \end{cases} \tag{12}$$

Given the precision matrix Q, we can apply (8) to estimate the mean value of the unknown data in the intrinsic GMRF.

3.3 Temporal Conditional Gaussian Random Field

In the learning stage, we apply the neighborhood structure of (a) in Fig. 3. Suppose that we assign N_0 spectrum sensing periods for learning, which means that we try to reconstruct an $K \times N_0$ spectrum usage indicator matrix. Due to the limitation of the wideband spectrum sensing, we can obtain only a fraction of the spectrum usage information within this N_0 sensing period. We denote the k_0 known spectrum status vector by $\mathbf{z}_{k0}$ and the unknown spectrum status vector by $\mathbf{z}_{\tilde{k0}}$. We define the sensing ratio as

$$\rho = \frac{k_0}{K \times N_0}. \tag{13}$$

According to (8), we attain the learning spectrum status matrix $\mathbf{z}_0$, which is given by

$$\mathbf{z}_0 = \mathbf{z}_{k0} + \mathbf{z}_{\tilde{k0}}. \tag{14}$$

After we attain the learning spectrum status matrix $\mathbf{z}_0$, we further apply the neighborhood structure (b) in Fig. 3 to estimate the sub-channel usage status in $(N_0 + 1)$ sensing period. Implementing spectrum sensing on m sub-channels, we attain $K \times N_0 + m$ known spectrum usage status. The unknown sub-channel usage status $z_i, (i = 1, 2, ..., K - m)$ can be estimated by

$$\begin{aligned} z_i &= E(z_i | \mathbf{z}_{-i}) \\ &= -\frac{1}{Q_{ii}} \sum_{j:j\sim i} Q_{ij} x_j. \end{aligned} \tag{15}$$

Then, z_i can be added to the known set of the sub-channel usage status to estimate the unknown sub-channel status in next sensing period.

4 Measuring WiFi Spectrum

In this section, we will introduce our spectrum measurement for WiFi signal which is considered as a PU. Fig. 4 shows the experiment setup. A photo of the equipment is shown in Fig. 5. The measurement is carried out inside or outside the Ferris Hall of the University of Tennessee.

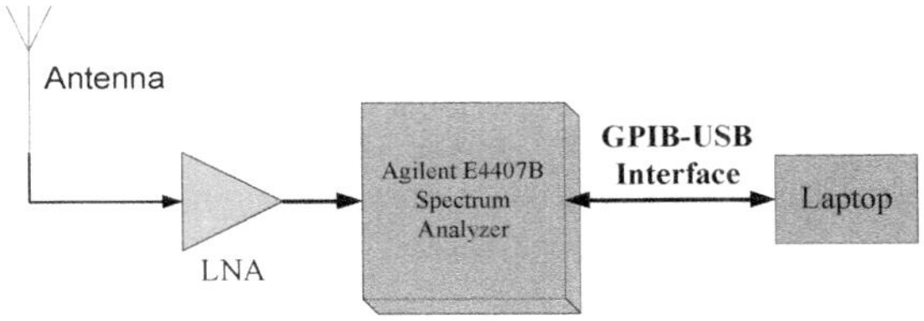

Fig. 4. An illustration of the setup for measuring WiFi spectrum signal

Fig. 5. A photo of the equipment for measuring the WiFi spectrum signal

The key instrument is the Agilent E4407B (9kHz to 26.5GHz). Through the GPIB-USB cable, the laptop can control the spectrum measurement by setting the configurations. The GPIB-USB communication bus is fast and instant. Hence the spectrum activity data can be quickly updated from the spectrum analyzer to the laptop. Also through the GPIB-USB, the spectrum measurement is handled by Matlab, which provides convenience and flexibility for realtime operation and online data storage.

The antenna in the measurement system is the Vivaldi wideband antenna. The frequency range of the antenna is from 1GHz-4GHz with the return loss lower than -10dB. The wideband antenna provides more flexibility for sensing different bands than a narrow band one. Although the antenna is not omni-directional, the testing environment is on campus where dense buildings, trees and other objects make the far-field electrical-magnetic wave reflect and refract. Hence the antenna has a capability to sense the spectrum activities from all directions.

Note that, in order to improve the sensitivity of the experiment, we add an external LNA. Generally, modern complicated spectrum analyzer is able to sense a very wide band spectrum. However, the wideband spectrum sensing incurs a lot of thermal noise, which buries the weak signal. Therefore, we utilize an external LNA close to the antenna to further lower the noise floor for sensing the weak signal activity. The LNA has a 20dB gain with 1.7dB Noise Figure (NF). The LNA can amplify the signal from 1GHz-3GHz.

Since we focus on the WiFi band, we set frequency measurement range from $2.4GHz$ to $2.5GHz$, applying 1024 points sampling for each spectrum measurement. Each sensing period is $20ms$ including spectrum measurement and data recording.

After we attained the spectrum data, an energy detection threshold is set to $-65dBm$ and $-70dBm$ for outdoor and indoor measurements, respectively. We divide the whole spectrum measurement band into 16 equal bandwidth sub-channels. Thus, if the energy of a sub-channel exceeds the threshold, we decide that this sub-channel is currently occupied by primary users, which is represented by a black strip in a time-frequency chart. If the energy of a sub-channel is

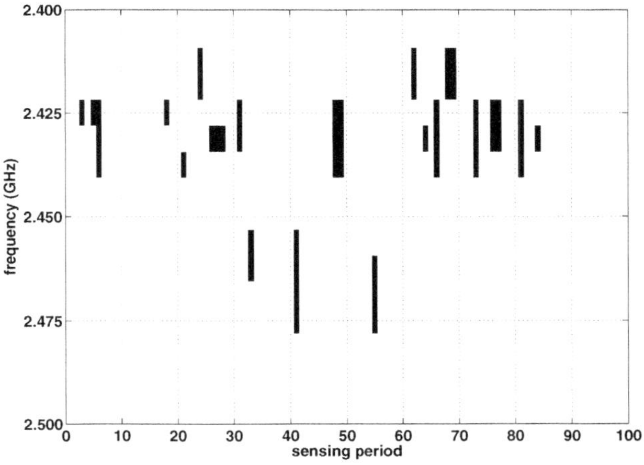

Fig. 6. Time-frequency chart of WiFi spectrum measurement in an indoor environment

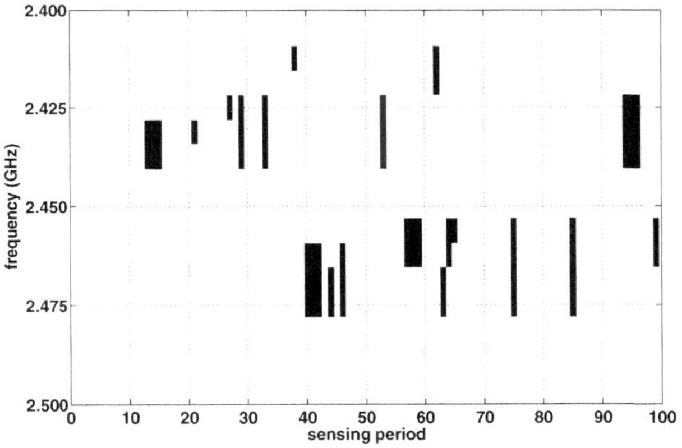

Fig. 7. Time-frequency chart of WiFi spectrum measurement in an outdoor environment

smaller than the threshold, we decide that this sub-channel is idle, represented by a white strip. Fig. 6 illustrates one of the spectrum measurement in an indoor environment and Fig. 7 illustrates the case in an outdoor environment.

5 Simulation Results

In this section, we demonstrate the performance of sub-channel usage status prediction by comparing the true spectrum activity measured in the WiFi frequency band. According to the neighborhood structure depicted in Fig. 3, we

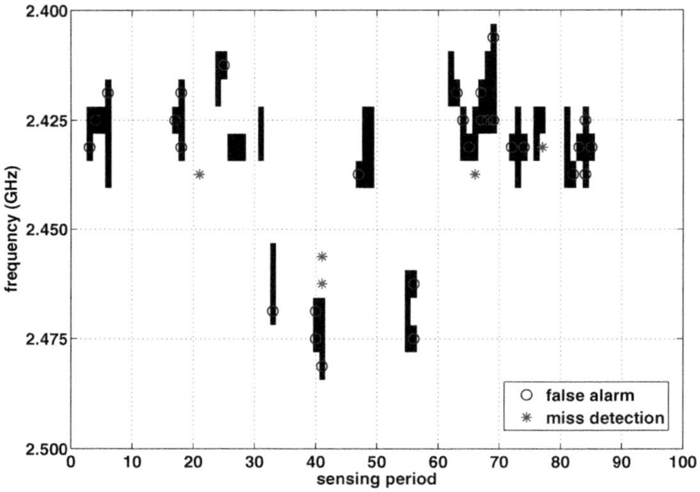

Fig. 8. Predicted time-frequency chart of WiFi spectrum usage status in an indoor environment

set the sensing ratio $\rho = 0.6$ and apply the TCGRF algorithm to estimate the status of the unsensed sub-channels. Fig. 8 and Fig. 9 demonstrate the estimated time-frequency chart of the WiFi spectrum usage status in indoor and outdoor environments, respectively.

It can be observed that the false alarm[2] probability of an outdoor measurement is higher than that of indoor environment. This may be due to the higher background noise level of an outdoor environment.

Fig. 10 illustrates the prediction error rate along with the sensing period. We apply an adaptive learning scheme, when the prediction error rate exceeds 0.02, we will implement the learning procedure according to the neighborhood structure of Fig. 3 (*a*). The spikes in the graph denote the sensing period when the prediction error exceeds 0.02; hence a learning procedure will be carried out, and the prediction error will be controlled within the threshold again.

We introduce the receiver operating characteristic (ROC) curve to demonstrate the detection performance of various spectrum occupancy prediction algorithms. We compare the proposed TCGRF algorithm with the logistic regression algorithm [8] and a traditional image filter algorithm. The logistic regression algorithm assumes that the subchannel occupancy is a stationary time series and the logistic regression model parameter can be learned from the previous subchannel occupancy status. However, in our practical spectrum measurements for

[2] Here we define the false alarm as the event that the spectrum sensor claims that an idle sub-channel is detected while the sub-channel is actually occupied by primary users.

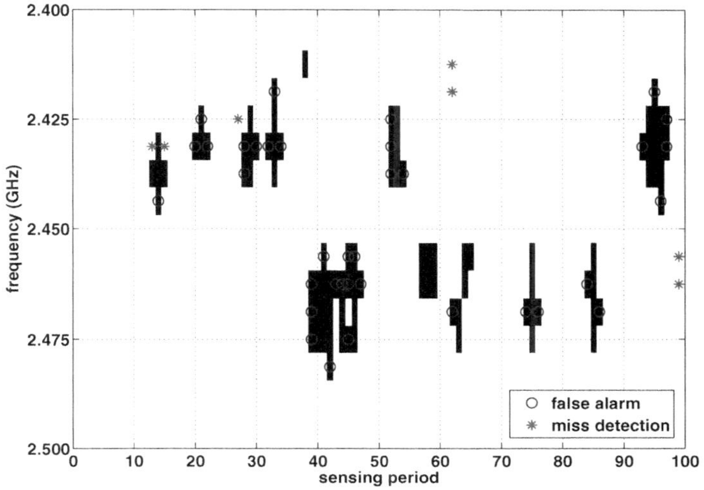

Fig. 9. Predicted time-frequency chart of WiFi spectrum usage status in an outdoor environment

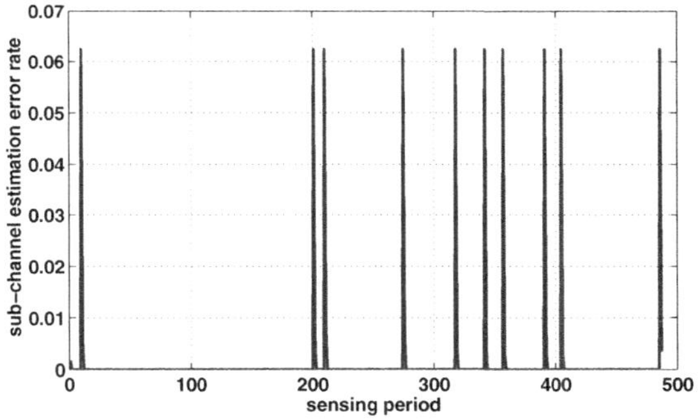

Fig. 10. Sub-channel status prediction error rate along with different sensing period

WiFi band, we found that it is difficult to meet this assumption. Consequently, the subchannel occupancy prediction performance of logistic regression degrades severely, which is illustrated in the following simulation results. Fig. 11 and Fig. 12 show the ROC curves of the indoor environment measurement and the outdoor environment, respectively.

The probability of false alarm is defined as the probability of the subchannel is idle but report busy. And the probability of detection is defined as the probability of the busy subchannels are successfully detected. It can observed that, in the low false alarm area, TCGRF based spectrum detection performance is significantly outperform the logistic algorithm and the image filter method. Under the same

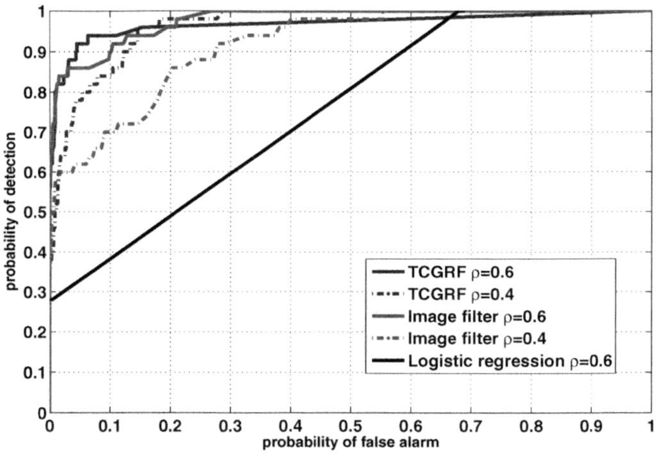

Fig. 11. ROC curve with various sensing ratios in an indoor environment

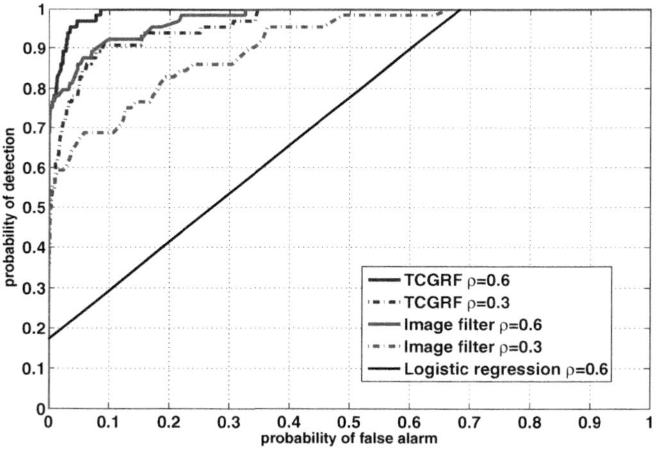

Fig. 12. ROC curve with various sensing ratios in an outdoor environment

probability of false alarm and sensing ratio ρ, the TCGRF based method can achieve substantially higher probability of detection. It is noticed that, when the sensing rate $\rho = 0.6$, the probability of detection can is higher than 90% while the probability of false alarm is only 0.1. This implies that the TCGRF algorithm can offer an efficient sub-channel status prediction. The insight of the TCGRF algorithm is that it utilizes the temporal correlation of the the sub-channel status between two adjacent sensing periods.

6 Conclusion

In this paper, we have used the theory of GMRF to predict the sub-channel usage status when not all sub-channel statuses can be sensed due to the limited wideband spectrum sensing capability. We have proposed a TCGRF model to predict the statuses of the unsensed sub-channels. In addition, we have carried out a measurement experiment over the WiFi band. The practical spectrum activity measured in the WiFi band is used to verify the TCGRF model. By applying the correlation between two adjacent sensing periods, the TCGRF algorithm provides an efficient sub-channel status prediction method, which can be used for channel selection in wideband spectrum sensing.

Acknowledgements. This work is supported by National Science Foundation under grants CCF-0830451 and ECCS-0901425, the "973" program of China under Grant 2007CB311201, China scholarship Council (CSC), National Natural Science Foundation of China (61072067, 60970119), and State Key Laboratory of Integrated Service Networks Project Grant (ISN02080002).

References

1. Akyildiz, I., Lee, W., Vuran, M., Mohanty, S.: Next generation dynamic spectrum access/cognitive radio wireless networks: A survey. Elsevier Computer Networks 50, 2127–2159 (2006)
2. Cabric, D., Mishra, S.M., Brodersen, R.W.: Implementation issues in spectrum sensing for Cognitive Radio. In: Proc. of Asilomar Conference on Signals, Systems, and Computers (2004)
3. Haykin, S.: Cognitive radio: Brain-empowered wireless communications. IEEE J. Sel. Areas Commun. 23(2), 201–220 (2005)
4. Lafferty, J., McCallum, A., Pereira, F.: Conditional random Fields: Probabilistic models for segmenting and labeling sequence data. In: In Proc. ICML, pp. 282–289 (2001)
5. Rue, H., Held, L.: Gaussian Markov random fields: theory and applications. Chapman & Hall, Boca Raton (2005)
6. Tang, H.: Some physical layer issues of wide-band cognitive radio systems. In: Proc. IEEE Dynamic Spectrum Access Networks (DySPAN), pp. 151–159 (November 2005)
7. Winkler, G.: Image Analysis, Random Fields and Markov Chain Mone Carlo Methods. Springer, Heidelberg (2003)
8. Yarkan, S., Arslan, H.: Binary time series approach to spectrum prediction for cognitive radio. In: IEEE 66th Vehicular Technology Conference, VTC 2007 Fall, pp. 1563–1567 (2007)
9. Zhao, Q., Sadler, B.: A survey of dynamic spectrum access. IEEE Signal Processing Mag. 24, 79–89 (2007)
10. Zhao, Q., Krishnamachari, B., Liu, K.: On myopic sensing for multi-channel opportunistic access: Structure, optimality, and performance. IEEE Trans. Wireless Communications 7, 5431–5440 (2008)

Cross-Layer Control for Utility Maximization in Multihop Cognitive Radio Networks

Long Bao Le

INRS-EMT, University of Quebec
Montreal, Quebec, Canada
long.le@emt.inrs.ca

Abstract. We investigate the cross-layer control problem for utility maximization in a multihop cognitive radio network. Specifically, we consider a scenario where wireless links of a secondary multihop wireless network opportunistically exploit a frequency band for data delivery when their neighboring primary users do not access it. We assume that the busy/idle status of the underlying channel in view of any particular secondary wireless link follows a two-state Markov chain and this information is only available at each secondary wireless link within one time slot delay. We develop a joint flow control, routing and scheduling algorithm that can achieve the total network utility arbitrarily close to optimality. In addition, we show that the proposed algorithm can maintain stability of all network queues while keeping collision probabilities with primary users below predetermined desirable values. To the best of our knowledge, this paper is the first attempt to design optimal control algorithms for multihop cognitive radio networks.

Keywords: Network control, routing, scheduling, utility and backlog tradeoff, stability/throughput region, cognitive radio, multihop wireless networks.

1 Introduction

Recent measurements have shown that spectrum utilization on many frequency bands is very low [1]. This has motivated a great deal of research interests from FCC, wireless industry as well as academia [2], [3], [4], [5], [6]. These research interests are mostly driven by growing bandwidth demands of emerging broadband wireless applications. In general, wireless technologies that aim at improving spectrum utilization through efficient spectrum sharing/allocation techniques are commonly referred to as *cognitive radio technologies*. In fact, the first standard that specifies physical, MAC and air interface for spectrum sharing in the TV broadcast band has been under active development [4]. There have been growing research activities on information theoretic, protocol and system engineering issues of cognitive radio systems.

Although the cognitive radio can be very broad and abstract in concept [3], research activities in this area mainly focus on developing efficient hierarchical spectrum sharing techniques between primary and secondary users/networks

X. Zhang and D. Qiao (Eds.): QShine 2010, LNICST 74, pp. 28–40, 2011.

[5], [16]. In particular, primary users usually have strictly higher priority than secondary users in accessing one or several frequency bands. Here, there are two important spectrum sharing paradigms between primary and secondary users, namely spectrum underlay and spectrum overlay [5]. In the spectrum underlay paradigm, secondary users are allowed to transmit simultaneously with primary users on the same frequency band. However, transmission powers of secondary users should be carefully controlled so that the total interference they create to primary users must be smaller some allowable limit. The spectrum underlay can be realized by CDMA or Ultra-wide bandwidth (UWB) radio access technologies [7], [8]. In the spectrum overlay paradigm, secondary users can only access the channels that are not being used by primary users [9]. Here, secondary users have to detect or sense the presence of primary users by employing some form of spectrum sensing [10], [11].

Several spectrum sharing protocols based on the spectrum overlay paradigm have been proposed for both single-hop and multihop cognitive radio networks. In particular, significant efforts have been made to develop efficient contention-based medium access control (MAC) protocols for opportunistic spectrum access [18]-[23]. These MAC protocols aim at incorporating spectrum sensing with spectrum sharing functionalities in an intelligent manner. There have been also some recent works that proposed scheduling-based spectrum sharing solutions for multihop cognitive radio networks [17]. However, these spectrum sharing solutions consider static network settings, which, therefore, do not consider the network stability issue.

In this paper, we investigate the network control problem for a secondary multihop cognitive radio network using spectrum overlay paradigm within the stability framework proposed by Tassiulas and Ephremides [12]. There are several research challenges in designing network control protocols for this research problem. First, a secondary user can only access a channel if no primary users in its neighborhood is using the channel. Therefore, new cognitive interference constraints need to be defined, which must capture the conflict relationship among secondary users and between primary and secondary users. Second, secondary users have to periodically sense the channel to detect the presence of primary users, which will introduce several forms of imperfection, namely delayed and/or erroneous spectrum sensing.

We will consider the scenario where spectrum sensing outcomes are available to secondary users within one time slot delay. Due to this delayed spectrum sensing, spectrum access of secondary users should be controlled to keep collision rates with primary users to be within a tolerable limit. We investigate the utility optimization problem for the secondary network and propose a cross-layer control algorithm, which is proved to achieve optimal network utility. The considered model captures conflict relationships between primary and secondary users as well as imperfect aspects of spectrum sensing. The most similar work to ours was published in [14]. However, this paper considered a special network setting where primary and secondary users communicate with their access points. In addition, delayed spectrum sensing was not explicitly considered in their paper.

The remaining of this paper is organized as follows. System model is described in section II. We propose optimal control algorithm in section III and analyze its performance in section IV. Then, we conclude the paper in section V.

2 System Model

Consider a scenario where secondary users form a multihop wireless network, in which they share a single frequency band with primary users for multihop communications. We assume a hierarchical spectrum sharing model where a secondary user can only transmit to its intended receiver when the channel is not used by any primary users in its local neighborhood. In essence, this local neighborhood determines the conflict relationship among primary and secondary users. We model the secondary multihop wireless network as a network graph $G = (V, E)$ where V is the set of secondary nodes and E is the set of secondary links. Assume that the cardinality of E is N. Also, suppose there are M primary links in the network, which will be referred to as primary users in the following.

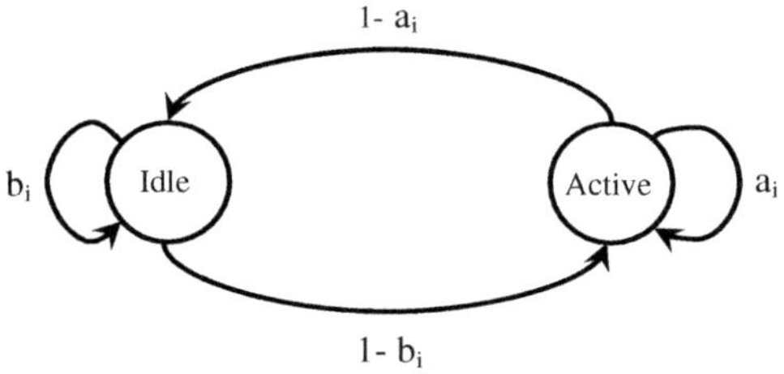

Fig. 1. Two-state Markov chain to capture idle/active state of primary user

Consider a time slotted system where a secondary user can transmit one packet/slot to its intended neighbor if there is no conflict transmissions from other secondary links and primary users. We model the status of a particular primary user i as two-state Markov chain, which is illustrated in Fig. 1. This model has been justified by several recent practical measurements [24]. We assume that the statuses of different primary users are independent from one another. For the secondary network, we assume that there are a set of traffic flows, each of which corresponds to a source and destination node. We will allocate different buffers at each secondary node to queue secondary traffic flows with the corresponding destination (i.e., per destination queueing). Traffic with the same destination is said to belong to the same commodity. It is assumed that arrival traffic waits outside the network in overflow buffers for being admitted into the network by flow controllers.

Now, we describe the conflict relationship of different transmissions in the network. Let Ψ_i be the set of secondary links that can cause collision with primary user i if any of these secondary links and primary user i transmit at the same time. Let Π_{mn} be the set of primary users, which are impacted by the transmission of link (m,n), i.e., if $i \in \Pi_{mn}$ then $(m,n) \in \Psi_i$. In addition, there is a finite number of feasible activation sets of secondary links, where secondary links in each feasible activation set can be activated simultaneously. We denote the set of feasible activation sets as Δ. In general, the set Δ is determined by some underlying interference model. Examples of interference models are k-hop interference model [26] and SINR interference model [25]. We denote a schedule as an N-dimensional vector $\boldsymbol{I}$ whose element is equal to one if the corresponding link is activated and equal to zero, otherwise.

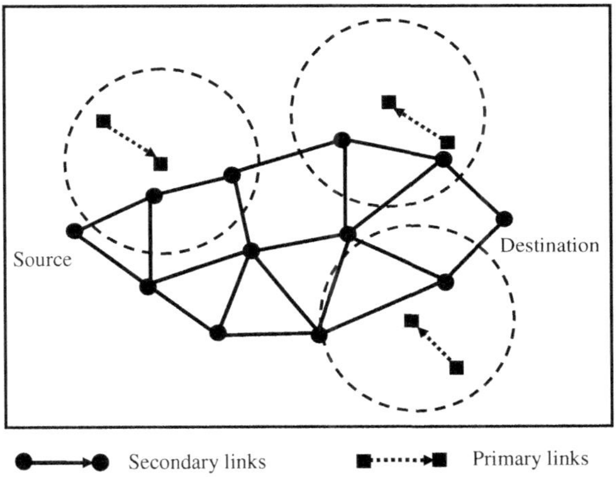

Fig. 2. Multihop cognitive radio network

2.1 Sensing Model and System Constraints

We assume that each secondary link (m,n) will sense the active/idle status of primary users in set Π_{mn} in each time slot. Ideally, link (m,n) is only allowed to transmit if all primary users in Π_{mn} are idle in that time slot. We assume that each secondary user performs sensing continuously without errors. However, secondary users can only report the sensing results to the network controller within one time slot delay, based on which control actions are determined. Due to delayed sensing, collisions between primary and secondary users can occur. Specifically, this collision occurs when secondary links mis-detect the presence of a conflict primary user, i.e., secondary links transmit when conflict primary users are in active state. Let $X_i(t)$ represent a collision variable, which captures the collision event between primary user i and some secondary users in Ψ_i. Specifically, we have

$$X_i(t) = \begin{cases} 1 \text{ if there is a collision with primary} \\ \quad \text{user } i \text{ in slot } t \\ 0 \text{ otherwise.} \end{cases} \tag{1}$$

Let $\overline{X}_i$ be the time average collision rate experienced by primary user i. We have

$$\overline{X}_i = \lim_{t \to \infty} \frac{1}{t} \sum_{\tau=0}^{t-1} E\left\{X_i(\tau)\right\}. \tag{2}$$

To protect primary users, we require that the time average collision rate of primary user i be smaller than some desired value δ_i, i.e.,

$$\overline{X}_i \leq \delta_i, \; \forall i = 1, 2, \cdots, M \tag{3}$$

where δ_i is small number chosen in advance.

2.2 Queueing Dynamics and Problem Formulation

Let $S_i(t)$ denote the status of primary user i in time slot t where $S_i(t) = 1$ if primary user i is active in time slot t, and $S_i(t) = 0$, otherwise. Let us define the following quantity

$$\omega_{mn}(t) = \prod_{i \in \Pi_{mn}} (1 - S_i(t)). \tag{4}$$

Then, to avoid collisions with primary users, a particular secondary link (m, n) should be silent in time slot t if $\omega_{mn}(t) = 0$. Let $Q_n^{(c)}(t)$ be the backlog at secondary node n for commodity c in time slot t. Also, let $R_n^{(c)}(t)$ be the number packets admitted into the network at secondary node n for commodity c in time slot t. We assume the constantly backlogged scenario where there are always enough packets to admit into the secondary network at all times. Let $\mu_l^{(c)}(t)$ be the number of commodity c packets transmitted over secondary link l in time slot t. For brevity, we sometimes use a single letter l to denote a wireless link. The queue evolution can be written as

$$Q_n^{(c)}(t+1) = Q_n^{(c)}(t) - \sum_{l \in \Omega_n^{\mathsf{out}}} \mu_l^{(c)}(t)\omega_l(t) + \sum_{l \in \Omega_n^{\mathsf{in}}} \mu_l^{(c)}(t)\omega_l(t) + R_n^{(c)}(t) \tag{5}$$

where Ω_n^{out} and Ω_n^{in} denote the set of outgoing and incoming links at node n, respectively. The equation (5) can be interpreted as follows. The backlog for commodity c at node n decreases by a value, which is equal to the number of packets successfully transmitted over all outgoing links and increases by a value, which is equal to the total number of admitted packets and successfully transmitted packets over incoming links. In particular, transmission over link l is only successful if $\omega_l(t) = 1$ (i.e., no collision with active primary users occurs).

Let $\overline{R}_n^{(c)}(t)$ be the time average rate of admitted traffic for commodity c at node n up to time t, that is

$$\overline{R}_n^{(c)}(t) \triangleq \frac{1}{t} \sum_{\tau=0}^{t-1} E\left\{R_n^{(c)}(\tau)\right\}. \tag{6}$$

The long-term time-average admitted rate for commodity c at node n is defined as

$$\overline{R}_n^{(c)} \triangleq \lim_{t\to\infty} \overline{R}_n^{(c)}(t). \tag{7}$$

Now, we recall the definitions of network stability and the maximum throughput region (or throughput region for brevity) [12], which will be used in our analysis. A queue for a particular commodity c at node n is called strongly stable if

$$\limsup_{t\to\infty} \frac{1}{t} \sum_{\tau=0}^{t-1} E\left\{Q_n^{(c)}(\tau)\right\} < \infty. \tag{8}$$

In addition, the network is called strongly stable (or stable for simplicity) if all individual queues in the network are stable. The maximum throughput region Λ contains the union of all traffic arrival/admitted rate vectors such that there exists a network control algorithm to stabilize all individual queues in the network.

Let $U_n^{(c)}\left(\overline{R}_n^{(c)}\right)$ be the utility achieved by admitting an average rate $\overline{R}_n^{(c)}$ for commodity c at node n. We assume that utilities functions $U_n^{(c)}(.)$ are concave, increasing, and differentiable. We seek to optimize the total network utility subject to constraints on network throughput and collisions with primary users due to delayed sensing performed by secondary links. Specifically, we are interested in solving the following optimization problem

$$\text{maximize} \quad \sum_{n,c} U_n^{(c)}\left(\overline{R}_n^{(c)}\right) \tag{9}$$

$$\text{subject to} \quad \left(\overline{R}_{n_c}^{(c)}\right) \in \Lambda \tag{10}$$

$$\overline{X}_i \leq \delta_i, \; \forall i = 1, 2, \cdots, M \tag{11}$$

where $\overline{R}_n^{(c)}$ is the time average admitted rate for commodity c at node n, and δ_i are the desired collision rates. Constraints (11) ensure that long-term collision rates with primary users are below desired levels.

2.3 Discussion of Formulated Problem

The optimization problem (9)-(11) is the network utility maximization (NUM) problem, which seeks to achieve a fair resource sharing for different traffic flows. Here, the desired fairness for radio resource sharing can be achieved by choosing appropriate utility functions. One popular class of utility functions is the α-fair utilities for which different fairness criteria can be achieved by changing a parameter α [27]. Consideration of NUM under the stability framework of Tassiulas and Ephremides [12] has been done in [13], [15]. Investigation of this problem in the cognitive radio setting has been recently performed in [14]. However, this paper considers a simple setting with single-hop traffic flows. In the current paper, we extend this problem to the multihop setting, where we need to design a joint flow control, routing, and scheduling algorithm. In addition, we consider the realistic scenario where only delayed spectrum sensing information is available at the network controller.

3 Optimal Control Algorithm

To solve the optimization problem (9)-(11), we employ the Lyapunov optimization technique developed in [15]. In particular, to capture the collisions experienced by primary user i, we define a virtual queue for each primary user i with the following evolution

$$Z_i(t+1) = \max\left[Z_i(t) - \delta_i, 0\right] + X_i(t). \tag{12}$$

We call these queues as virtual queues because their values can be maintained in software counters (i.e., no physical buffers are needed for implementation). In addition, $Z_i(t)$ captures the "backlog" in the virtual queue with "arrival processs" $X_i(t)$ and constant service rate δ_i. Therefore, it can be shown that if all virtual queues $Z_i(t)$ are stable then all collision requirements in (3) are satisfied. This is because the average arrival rate should be smaller than the service rate for a stable queue.

Specifically, an optimal control algorithm can be developed by minimizing the Lyapuniv drift minus total utility for an appropriate Lyapunov function. Now, let $\boldsymbol{\Theta}(t) = (\boldsymbol{Q}(t), \boldsymbol{Z}(t), \boldsymbol{S}(t-1))$ denote the system states. We describe the optimal cognitive control algorithm in the following, whose performance is analyzed in the next section.

Cross-Layer Cognitive Control Algorithm

- *Flow Control:* Each node n injects an amount of traffic of commodity c into the network equal to $r_n^{(c)}$, which is the optimal solution of the following optimization problem

$$\begin{array}{l} \text{maximize } V \sum_{n,c} U_n^{(c)}(r_n^{(c)}) - 2\sum_{n,c} Q_n^{(c)}(t) r_n^{(c)} \\ \text{subject to } r_n^{(c)} \leq R_n^{\max} \end{array} \tag{13}$$

  where $R_n^{\max} > 0$ is a positive number, which controls the burstiness of the admitted traffic and V is a control parameter.
- *Routing and Scheduling:* For each link (m,n), find the commodity satisfying the following

$$c^* = \underset{c}{\operatorname{argmax}}\, E\left\{\omega_{mn}(t)|\boldsymbol{\Theta}(t)\right\}\left[Q_m^{(c)}(t) - Q_n^{(c)}(t)\right]. \tag{14}$$

  The weight for link (m,n) is defined as

$$W_{mn}(t) \triangleq E\left\{\omega_{mn}(t)|\boldsymbol{\Theta}(t)\right\}\left[Q_m^{(c^*)}(t) - Q_n^{(c^*)}(t)\right] - \sum_{i\in\Pi_{mn}} Z_i(t) E\left\{S_i(t)|\boldsymbol{\Theta}(t)\right\}. \tag{15}$$

Using these weights, we find a feasible schedule $\boldsymbol{I}^*(t)$ as follows:

$$\boldsymbol{I}^*(t) = \operatorname*{argmax}_{\boldsymbol{I} \in \Delta} \sum_{(m,n)} I_{mn} W_{mn}(t) \tag{16}$$

where recall that Δ denotes the set of all feasible schedules. For each scheduled link (m, n) in $\boldsymbol{I}^*$, we transmit one packet of flow c^* that satisfies (14). After the scheduled transmissions occur, based on the feedbacks of the "collisions outcomes" $X_i(t)$ from the primary users, the control queues $Z_i(t)$ are updated according to (12).

3.1 Discussion of Proposed Control Algorithm

The scheduling policy described in (16) has the max-weight structure similar to that proposed by Tassiulas and Ephremides in [12]. However, the weight in (15) is different from that in [12] to capture two important aspects of the current model, i.e., the delayed channel sensing information and the potential collisions with primary users. The modified differential backlog measure in (14) is the scaled version of the traditional measure of [12]. Here, the scaling factors are $E\{\omega_{mn}(t)|\boldsymbol{\Theta}(t)\}$, which capture the expected number of packets that can be transmitted over link (m, n) given queue length and delay sensing information.

In addition, the link weight in (15) has the term $\sum_{i \in \Pi_{mn}} Z_i(t) E\{S_i(t)|\boldsymbol{\Theta}(t)\}$, which captures the collision measure with primary users due to secondary link (m, n). In particular, secondary links with large $\sum_{i \in \Pi_{mn}} Z_i(t) E\{S_i(t)|\boldsymbol{\Theta}(t)\}$ will achieve small weights, which are therefore less likely to be scheduled. This weight structure helps avoid excessive collisions with active primary users.

Note that given $\boldsymbol{S}(t-1)$, we can easily calculate $E\{\omega_{mn}(t)|\boldsymbol{\Theta}(t)\}$ and $E\{S_i(t)|\boldsymbol{\Theta}(t)\}$ using the transition probabilities of the corresponding Markov chains. Specifically, we have

$$\begin{aligned} E\{\omega_{mn}(t)|\boldsymbol{\Theta}(t)\} &= \prod_{i \in \Pi_{mn}} \left(1 - \Pr\{S_i(t) = 1 | S_i(t-1)\}\right) \\ &= \prod_{i \in \Pi_{mn}} \Pr\{S_i(t) = 0 | S_i(t-1)\} \\ E\{S_i(t)|\boldsymbol{\Theta}(t)\} &= \Pr\{S_i(t) = 1 | S_i(t-1)\}. \end{aligned}$$

This is because $S_i(t)$ is independent of queue length $\boldsymbol{Q}(t)$ and $\boldsymbol{X}(t)$. The performance of the proposed cross-layer control algorithm is stated in the following theorem.

Theorem: Let $\boldsymbol{R}^*$ be the optimal solution of the considered optimization problem. The proposed control algorithm achieves the following performance bounds:

$$\liminf_{M \to \infty} \sum_{n,c} U_n^{(c)} \left(\overline{R}_n^{(c)}(M)\right) \geq \sum_{n,c} U_n^{(c)} \left(R_n^{(c)*}\right) - \frac{B}{V} \tag{17}$$

$$\limsup_{M\to\infty} \frac{1}{M}\sum_{\tau=0}^{M-1}\sum_{n,c} E\left\{Q_n^{(c)}(\tau)\right\} \leq \frac{B+VG_{\mathsf{max}}}{2\lambda_{\mathsf{max}}} \tag{18}$$

where B is a finite number, V is a control parameter of the control algorithm, G_{max} is maximum achievable utility, and λ_{max} is the largest value such that $\boldsymbol{\lambda}_{\mathsf{max}} \in \Lambda$.

It can be observed that the control parameter V can be used to control the utility and backlog tradeoff. Specifically, larger V leads to better achievable utility with the penalty on the average backlog bound in the network.

3.2 Further Discussion and Possible Extension

The flow controller of the proposed control algorithm can be implemented in a distributed manner by each source node. This is because to make the flow control decision in each time slot, flow controllers only need to know the backlog information of its own buffer for the corresponding commodity. In addition, the link weight in (15) can also be calculated by each link using the updated "virtual queue value" $Z_i(t)$ upon collecting collision outcomes $X_i(t)$ from the conflict primary users. However, the scheduling scheme in (16) requires centralized implementation in general. This is because we need to find one feasible schedule that achieves the maximum total weight among all possible feasible schedules in each time slot.

Fortunately, there are several techniques available in the literature, which enable us to decentralize the scheduling operation [28], [29], [30]. In particular, there is a tradeoff between performance and complexity in implementing these decentralization techniques. Specifically, the PICK-COMPARE based scheduling schemes [29], [30] can achieve very close to optimum throughput but requires high communication complexity. Other greedy scheduling schemes such as the one proposed in [28] have low complexity but only achieve a fraction of the throughput region. Overall, it is feasible to apply any of these available scheduling techniques instead of the max-weight scheme to our proposed cross-layer control algorithm.

4 Performance Analysis

We analyze the performance of the proposed control algorithm and prove the main theorem of the paper in this section. As mentioned above, the analysis is based on the Lyapunov optimization technique proposed in [13], [15]. Now, consider the following Lyapunov function

$$L(\boldsymbol{Q}) \triangleq \sum_{n,c}\left(Q_n^{(c)}(t)\right)^2 + \sum_i \left(Z_i(t)\right)^2. \tag{19}$$

Consider the Lyapunov drift defined as follows:

$$\Delta(t) \triangleq E\left\{L(\boldsymbol{Q})(t+1) - L(\boldsymbol{Q})(t)|\boldsymbol{\Theta}(t)\right\} \tag{20}$$

where recall that $\boldsymbol{\Theta}(t) = (\boldsymbol{Q}(t), \boldsymbol{Z}(t), \boldsymbol{S}(t-1))$ denotes the system states. Using the queue evolution equations in (5) and (12), we have

$$\Delta(t) \leq B + 2E\left\{\sum_i Z_i(t)\left(X_i(t) - \delta_i\right)|\boldsymbol{\Theta}(t)\right\} + 2E\left\{\sum_{n,c} Q_n^{(c)}(t)R_n^{(c)}(t)|\boldsymbol{\Theta}(t)\right\}$$
$$+2\sum_{n,c} Q_n^{(c)}(t)E\left\{-\sum_{l\in\Omega_n^{\text{out}}} \mu_l^{(c)}(t)\omega_l(t) + \sum_{l\in\Omega_n^{\text{in}}} \mu_l^{(c)}(t)\omega_l(t)|\boldsymbol{\Theta}(t)\right\} \quad (21)$$

where B is a finite number. Now, we can bound the collision variable $X_i(t)$ as follows:

$$X_i(t) \leq \sum_{c,(m,n)\in\Psi_i} \mu_{m,n}^{(c)}(t)S_i(t). \quad (22)$$

This is because collision with primary user i occurs if some conflict secondary links in Ψ_i transmit while primary user i is in active state. Substitute this relationship into (21), we have

$$\Delta(t) - V\sum_{n,c} E\left\{U_n^{(c)}(R_n^{(c)}(t))|\boldsymbol{\Theta}(t)\right\} \leq B - 2\sum_i \delta_i Z_i(t)$$
$$+E\left\{2\sum_{n,c} Q_n^{(c)}(t)R_n^{(c)}(t) - V\sum_{n,c} U_n^{(c)}(R_n^{(c)}(t))|\boldsymbol{\Theta}(t)\right\}$$
$$-2\sum_{(m,n),c} E\left\{\mu_{mn}^{(c)}(t)\omega_{mn}(t)|\boldsymbol{\Theta}(t)\right\}\left[Q_m^{(c)}(t) - Q_n^{(c)}(t)\right]$$
$$+2\sum_i Z_i(t)E\left\{S_i(t)\sum_{(mn)\in\Psi_i,c} \mu_{mn}^{(c)}(t)|\boldsymbol{\Theta}(t)\right\}$$
$$= B - 2\sum_i \delta_i Z_i(t) - E\left\{V\sum_{n,c} U_n^{(c)}(R_n^{(c)}(t)) - 2\sum_{n,c} Q_n^{(c)}(t)R_n^{(c)}(t)|\boldsymbol{\Theta}(t)\right\} \quad (23)$$
$$-2\left[\sum_{(m,n),c} E\left\{\mu_{m,n}^{(c)}(t)\omega_{mn}(t)|\boldsymbol{\Theta}(t)\right\}\left[Q_m^{(c)}(t) - Q_n^{(c)}(t)\right] \quad (24)\right.$$
$$\left. - \sum_{(m,n),c} E\left\{\mu_{m,n}^{(c)}(t)\sum_{i\in\Pi_{mn}} Z_i(t)S_i(t)|\boldsymbol{\Theta}(t)\right\}\right] \quad (25)$$

It can be observed that the proposed control algorithm minimizes the RHS of the above inequality. Specifically, the flow controller minimizes the third term of (23) and the routing/scheduling algorithm minimizes (24) and (25) in the RHS of the above inequality. Now, to quantify the performance of the proposed control algorithm, we need some more definitions. First, let us define the ϵ-stripped throughput region as follows:

$$\Lambda_\epsilon \triangleq \left\{ \left(r_n^{(c)} \right) \mid \left(r_n^{(c)} + \epsilon \right) \in \Lambda \right\} \tag{26}$$

where $\left(r_n^{(c)} \right)$ denotes the vector of admitted rates for all commodities. Also, let $(R_n^{*(c)}(\epsilon))$ be the optimal solution of the following optimization problem

$$\text{maximize} \quad \sum_c U_n^{(c)} \left(R_n^{(c)} \right) \tag{27}$$

$$\text{subject to} \quad \left(R_n^{(c)} \right) \in \Lambda_\epsilon \tag{28}$$

$$\overline{X}_i \leq \delta_i, \; \forall i = 1, 2, \cdots, M. \tag{29}$$

We will quantify the performance of the considered control algorithms in terms of $(R_n^{*(c)}(\epsilon))$. Note that $(R_n^{*(c)}(\epsilon))$ tends to the optimal solution $(R_n^{*(c)})$ as $\epsilon \to 0$ where $(R_n^{*(c)})$ is the optimal solution of the optimization problem (27)-(29) where Λ_ϵ is replaced by Λ (i.e., the original throughput region). Because $R_n^{(c)*}(\epsilon)$ is inside the ϵ-stripped throughput region, there exists randomized stationary scheduling and routing scheme to support this rate, i.e., we have

$$V \sum_{n,c} U_n^{(c)}(R_n^{(c)}(t)) - 2 \sum_{n,c} Q_n^{(c)}(t) R_n^{(c)}(t)$$
$$\geq V \sum_{n,c} U_n^{(c)}(R_n^{(c)*}) - 2 \sum_{n,c} Q_n^{(c)}(t) R_n^{(c)*} \tag{30}$$

$$E\left\{ X_i(t) | \boldsymbol{\Theta}(t) \right\} \leq \delta_i \tag{31}$$

$$\sum_{n,c} Q_n^{(c)}(t) \left[\sum_{l \in \Omega_n^{\mathsf{out}}} \mu_l^{(c)}(t) \omega_l(t) - \sum_{l \in \Omega_n^{\mathsf{in}}} \mu_l^{(c)}(t) \omega_l(t) \right] \geq \sum_{n,c} Q_n^{(c)}(t) \left[R_n^{(c)*} + \epsilon \right]. \tag{32}$$

Using the results of (30), (31), and (32) in (23)-(25), we can obtain the following

$$\Delta(t) - V \sum_{n,c} E\left\{ U_n^{(c)}(R_n^{(c)}(t)) | \boldsymbol{\Theta}(t) \right\} \leq B - 2\epsilon \sum_{n,c} Q_n^{(c)}(t) - V \sum_{n,c} U_n^{(c)}(R_n^{(c)*}). \tag{33}$$

Taking the expectations over the distribution of $\boldsymbol{\Theta}(t)$ and summing over $t \in \{1, 2, \cdots, M\}$, we have

$$E\left\{ L(\boldsymbol{Q}(M)) \right\} - E\left\{ L(\boldsymbol{Q}(0)) \right\} - V \sum_{\tau=0}^{M-1} \sum_{n,c} E\left\{ U_n^{(c)} \left(R_n^{(c)}(\tau) \right) \right\}$$
$$\leq MB - VM \sum_{n,c} U_n^{(c)} \left(R_n^{(c)*}(\epsilon) \right) - 2\epsilon \sum_{\tau=0}^{M-1} \sum_{c} E\left\{ Q_n^{(c)}(\tau) \right\}. \tag{34}$$

To prove the backlog bound, we arrange the inequality (34) appropriately and divide both sides by M, we have

$$\frac{1}{M} \sum_{\tau=0}^{M-1} \sum_{n,c} E\left\{ Q_n^{(c)}(\tau) \right\} - \frac{E\left\{ L(\boldsymbol{Q}(0)) \right\}}{M} \leq \frac{B + V G_{\mathsf{max}}}{2\epsilon}. \tag{35}$$

Note that the above inequality holds for any $0 < \epsilon \leq \lambda_{\max}$. Hence, by choosing $\epsilon = \lambda_{\max}$ and taking the limit for $M \to \infty$ in (35), we can obtain the backlog bound. By arranging the terms of (34) appropriately and dividing both sides by VM, we have

$$\frac{1}{M}\sum_{\tau=0}^{M-1}\sum_{n,c} E\left\{U_n^{(c)}\left(R_n^{(c)}(\tau)\right)\right\} \geq \sum_{n,c} U_n^{(c)}\left(R_n^{(c)*}(\epsilon)\right)$$
$$-\frac{B + E\left\{L(\boldsymbol{Q}(0))\right\}/M}{V} + \frac{2\epsilon}{VM}\sum_{\tau=0}^{M-1}\sum_{n,c} E\left\{Q_n^{(c)}(\tau)\right\} \tag{36}$$

where we have used the fact that $L(\boldsymbol{Q}(M)) \geq 0$ to obtain (36). Using the Jensen's inequality and taking the limit $M \to \infty$ in (36), we have

$$\liminf_{M\to\infty} \sum_{c} U_n^{(c)}\left(\overline{R}_n^{(c)}(M)\right) \geq \sum_{n,c} U_n^{(c)}\left(R_n^{(c)*}(\epsilon)\right) - \frac{B}{V}. \tag{37}$$

Hence, we can obtain the utility bound by letting $\epsilon \to 0$. Therefore, we have completed the proof for the main theorem of the paper.

5 Conclusion

We investigate the optimal control problem for utility maximization in multihop cognitive radio networks in this paper. Specifically, we seek to maximize the total utility achieved by different traffic flows of the secondary network subject to network stability and collision constraints with primary users. We propose a cross-layer control algorithm that is proved to achieve utility arbitrarily close to optimality and derive the corresponding utility-backlog tradeoff.

References

1. FCC. Spectrum policy task force report, FCC 02-155 (November 2002)
2. FCC. Facilitating opportunities for flexible, efficient, and reliable spectrum use employing cognitive radio technologies, notice of proposed rule making and order, FCC 03-322 (December 2003)
3. Haykin, S.: Cognitive radio: Brain-empowered wireless communications. IEEE J. Sel. Areas Commun. 23(2), 201–220 (2005)
4. Cordeiro, C., Challapali, K., Birru, D., Shankar, N.S.: IEEE 802.22: The first world-wide wireless standard based on cognitive radios. In: DySpan 2005 (November 2005)
5. Zhao, Q., Sadler, B.M.: A survey of dynamic spectrum access: Signal processing, networking, and regulatory policy. IEEE Signal Processing Mag., 79–89 (May 2007)
6. Etkin, R., Parekh, A., Tse, D.: Spectrum sharing for unlicensed bands. IEEE J. Sel. Areas Commun. 25(3), 517–528 (2007)
7. Le, L.B., Hossain, E.: Resource allocation for spectrum underlay in cognitive radio networks. IEEE Trans. Wireless Commun. 7(12), 5306–5315 (2008)

8. Kim, D.I., Le, L.B., Hossain, E.: Joint rate and power allocation for cognitive radios in dynamic spectrum access environment. IEEE Trans. Wireless Commun. 7(12), 5517–5527 (2008)
9. Chou, C.T., Shankar, N.S., Kim, H., Shin, K.G.: What and how much to gain by spectrum agile? IEEE J. Sel. Areas Commun. 25(3), 576–588 (2007)
10. Gandetto, M., Regazzoni, C.: Spectrum sensing: A distributed approach for cognitive terminals. IEEE J. Sel. Areas Commun. 25(3), 546–557 (2007)
11. Unnikrishnan, J., Veeravalli, V.V.: Cooperative sensing for primary detection in cognitive radio. IEEE J. Sel. Signal Processing 2(1), 18–27 (2008)
12. Tassiulas, L., Ephremides, A.: Stability properties of constrained queueing systems and scheduling policies for maximum throughput in multihop radio networks. IEEE Trans. Automatic Control 37(12), 1936–1948 (1992)
13. Georgiadis, L., Neely, M.J., Tassiulas, L.: Resource allocation and cross-layer control in wireless networks. Foundations and Trends in Networking 1(1), 1–144 (2006)
14. Urgaonkar, R., Neely, M.J.: Opportunistic scheduling with reliability guarantees in cognitive radio networks. In: IEEE INFOCOM (2008)
15. Neely, M.J.: Energy optimal control for time varying wireless networks. IEEE Trans. Information Theory 52(7), 2915–2934 (2006)
16. Devroye, N., Mitran, P., Tarokh, V.: Achievable rates in cognitive radio channels. IEEE Trans. Inf. Theory 52(5), 1813–1827 (2006)
17. Hou, Y.T., Shi, Y., Sherali, H.D.: Optimal spectrum sharing for multi-hop software defined radio networks. In: IEEE INFOCOM 2007 (May 2007)
18. Cao, L., Zheng, H.: Stable and efficient spectrum access in next generation dynamic spectrum networks. In: IEEE INFOCOM 2008 (2008)
19. Zhao, Q., Tong, L., Swami, A., Chen, Y.: Decentralized cognitive MAC for opportunistic spectrum access in ad hoc networks: A POMDP framework. IEEE J. Sel. Areas Commun. 25(3), 589–600 (2007)
20. Chang, N.B., Liu, M.: Optimal channel probing and transmission scheduling for opportunistic spectrum access. In: ACM MOBICOM 2007 (2007)
21. Huang, S., Liu, X., Ding, Z.: Opportunistic spectrum access in cognitive radio networks. In: IEEE INFOCOM 2008 (2008)
22. Le, L., Hossain, E.: OSA-MAC: A MAC protocol for opportunistic spectrum access in cognitive radio networks. In: IEEE WCNC 2008 (March 2008)
23. Su, H., Zhang, X.: Cross-layer based opportunistic MAC protocols for QoS provisionings over cognitive radio mobile wireless networks. IEEE J. Sel. Areas Commun. 26(1), 118–129 (2008)
24. Geirhofer, S., Tong, L., Sadler, B.M.: Dynamic spectrum access in the time domain: Modeling and exploiting white space. IEEE Commun. Mag. 45(5), 66–72 (2007)
25. Le, L.B., Modiano, E., Joo, C., Shroff, N.B.: Longest-queue-first scheduling under SINR interference model. In: ACM MobiHoc 2010 (2010)
26. Sharma, G., Mazumdar, R., Shroff, N.B.: On the complexity of scheduling in multihop wireless systems. In: ACM MOBICOM 2006 (2006)
27. Chiang, M., Low, S.H., Calderbank, A.R., Doyle, J.C.: Layering as optimization decomposition: A mathematical theory of network architectures. Proc. of the IEEE 95(1), 255–312 (2007)
28. Joo, C., Shroff, N.B.: Performance of random access scheduling schemes in multihop wireless networks. In: IEEE INFOCOM 2007 (2007)
29. Modiano, E., Shah, D., Zussman, G.: Maximizing throughput in wireless networks via gossiping. In: ACM SIGMETRICS (2006)
30. Sanghavi, S., Bui, L., Srikant, R.: Distributed link scheduling with constant overhead. In: ACM SIGMETRICS 2007 (June 2007)

Reliable Channel Selection and Routing for Real-Time Services over Cognitive Radio Mesh Networks

Lang Xie[1], P.E. Heegaard[1], Yan Zhang[2,3], and Jie Xiang[2]

[1] Department of Telematics, NTNU, Trondheim, Norway
[2] Simula Research Laboratory, Norway
[3] Department of Informatics, University of Oslo, Norway
{langxie,poul.heegaard}@item.ntnu.no, {yanzhang,jxiang}@simula.no

Abstract. In Cognitive Radio Mesh Networks (CogMesh), Secondary Mesh Routers (SMRs) can opportunistically utilize the licensed spectrums for the traffic of the Secondary Mesh Users (SMUs). How to guarantee Quality of Service (QoS) for real-time services over CogMesh is still an opening issue. In this paper, we present a discrete-time vacation queueing system to abstract the Primary User (PU) interruption to SMR data transmission. Moreover, we formulate the optimization problem of joint channel selection and routing, to achieve minimum end-to-end delay for SMR while guaranteeing the channel unavailability. A heuristic method is proposed to solve this problem and results show our proposed method performs the closest to the scheme using optimization tool, and outperforms the minimal unavailability scheme and minimal delay scheme in terms of end-to-end delay and solution rate.

Keywords: cognitive radio mesh networks, real-time services, vacation queueing, channel selection.

1 Introduction

Wireless mesh networks have emerged as a highly promising technology to extend the network access area in an economical and convenient way [1] [2]. To further improve the network flexibility and increase spectrum utilization, there is a strong motivation to utilize the unused spectrum to deliver the mesh network traffic flows [3]. Cognitive Radio (CR) [4], an agile technology enables Secondary Users (SUs) to intelligently access the spectrum bands licensed to Primary Users (PUs), is come forth for this critical requirement. Furthermore, the Cognitive Radio Mesh Networks (CogMesh), which combines CR and mesh technologies, is proposed to improve the spectrum utilization and expand the network access area simultaneously [3] [5] [6].

A significant challenge in CogMesh is the real-time service communication, which has strict Quality of Service (QoS) constraints on end-to-end delay, jitter, packet loss, etc. Channel selection and routing are two ever important mechanisms in the provision of QoS in CogMesh. The integration of channel selection

X. Zhang and D. Qiao (Eds.): QShine 2010, LNICST 74, pp. 41–57, 2011.

in the establishment of end-to-end routes is of critical importance. However, it is still a tough problem.

There are some related works which focus on channel assignment and end-to-end delay guarantee for wireless mesh networks [7] [8]. However, the schemes for traditional wireless mesh networks cannot be applied directly to CogMesh scenario due to the channel unavailability caused by primary system. Therefore, some research devotes to channel selection and dynamic spectrum access in CR networks. Y. Hou *et al.* [9] studied the channel selection and routing in multi-hop CR networks, with the objective of minimizing the total bandwidth used in the network. Y. Song *et al.* [10] proposed a stochastic channel selection algorithm based on learning techniques. Each secondary node selects one channel with a probability which is recorded in a list and updated according to the result of each selection. And then a packet will be sent once the channel selected is available to use. M. Rehmani *et al.* [11] proposed a channel selection scheme to select the channel with the highest channel weight which is defined as $e^{-p}(1-p)$, where p is the occupancy rate of PUs.

However, these previous works do not consider the end-to-end service requirement. In this paper, we will address the problem of joint channel selection and routing, which is crucial in QoS guarantee for real-time services over CogMesh. The major contributions of this paper are threefold. First, we adopt a vacation queueing system to abstract the channel vacation due to PU interruption in CogMesh. The analytical results based on the vacation queueing system are used to derive the channel asymptotic unavailability and the expected end-to-end delay for SMR. Second, we formulate the optimization problem to minimize the end-to-end delay for SMR while guaranteeing the channel unavailability and propose the joint channel selection and routing scheme. Third, numerical results show our proposed scheme performs close to the scheme using optimization tool and outperforms two other greedy heuristic schemes, i.e., the minimal unavailability scheme and minimal delay scheme.

The rest of the paper is organized as follows. We introduce the system model in Section 2. In Section 3, we formulate and analyze the optimization problem. In Section 4, we describe our proposed scheme. In Section 5, we evaluate the performance of the proposed scheme. Finally, we draw the conclusions in Section 6.

2 System Model and Assumptions

Each SMR and SMG are equipped with one CR transceiver and one normal radio transceiver with a dedicated control channel. Several secondary mesh users (SMUs) access their nearby SMRs to communicate with the users in not only the CogMesh but also the Internet through the SMG. In this paper, we use pre-computed paths for each source-destination pair to guarantee the QoS for each service session. $\mathcal{R}$ is the route set from source SMR to the gateway SMG. For each link l on one route r, the available channels consist a subset $\mathcal{M}_{r,l}$ of $\mathcal{M}$.

Different channels at the same link may have different quality metrics, such as channel fading parameters, interference, channel bandwidth, and so on. We

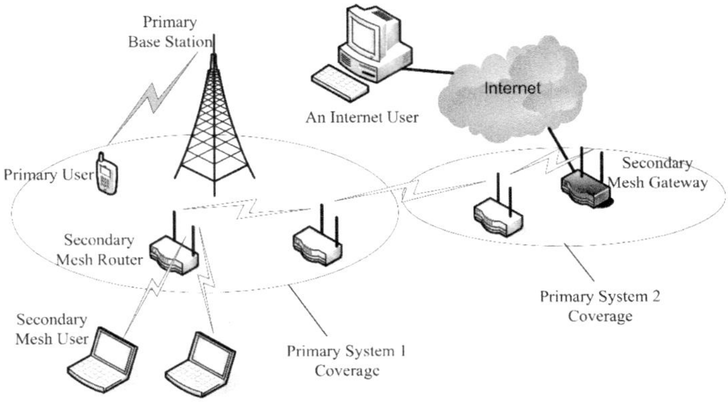

Fig. 1. An example of Cognitive Radio Mesh Network

use the Finite-State Markov Channel (FSMC) model [12] to represent the time- and frequency-selective slow fading Channels. Assume that all channels have S states. In each state, the received Signal to Interference plus Noise Ratio (SINR) is different. We define Γ_s $(s = 1, ..., S)$ as the lower bound threshold of the state s, where $0 < \Gamma_1 < ... < \Gamma_S < \infty$. We say link e_i is in state s, if the SINR is between Γ_s and Γ_{s+1}. Adaptive Modulation Coding (AMC) technique is used in our system model. Where, channel's quality can be estimated by the SINR measured on the receiving node. Different modulation schemes can bring out different data transmission rates.

The mode of access mechanism in this paper is 802.11 distributed coordination function (DCF) medium access control (MAC) protocol. Each SMR maintains a separate queue for real-time data packets on the network layer. We consider a time-slotted packet transmission scenario, where the slot length is fixed as T_{slot}.

Table 1. List of Notations

Symbol	**Meaning**
$\mathcal{R}$	the set of routes
$\mathcal{N}$	the set of SMRs
$\mathcal{L}$	the set of links
$\mathcal{M}$	the set of channels
$\mathcal{N}_r$	the set of SMRs in route r
$\mathcal{L}_r$	the set of links in route r
$\mathcal{M}_l$	the set of channels of link l on route r
r	a route
l	a link
n	a SMR
m	a channel
C	the average packet length
x_{rlm}	the binary indicator of channel m at link i in route r

The basic slot structure consists of sensing, data transmission, and acknowledge periods, as illustrated in Fig. 3. At the beginning of a slot, the SMR senses the channel. If the channel is identified idle, the SMR transmits data. At the end of the slot, the receiver acknowledges a successful transmission. For the sake of simplicity, the transmission of acknowledgement is assumed to be error-free. It is natural for us to model the PUs interruption as a discrete-time queueing system. For this reason we devote the next section to the PU interruption modeling. Table 1 describes the main notations used in this paper.

3 Problem Formulation and Analysis

In this section, the optimal joint channel selection and routing for real-time services in CogMesh is formulated as an optimization problem. We will first introduce the PU interruption model and discuss end-to-end delay. Afterwards, we formulate the optimization problem and solve it.

Note our further discussion will take the example case of channel m at link l in route r. For description simplicity, we employ a binary variable x_{rlm} ($x_{rlm} \in \{0,1\}, \forall r \in \mathcal{R}; \forall l \in \mathcal{L}; \forall m \in \mathcal{M}$) to indicate whether channel m is selected for the link l or not. If x_{rlm} is equal to 1, channel m is selected for the link l, 0 otherwise.

3.1 PU Interruption Model

In CogMesh, SMR's data transmission process is error-prone since data packets are transmitted over unreliable wireless channels and PUs may interrupt SMRs data transmission randomly. How to abstract the PU interruption process? Our choice is to use a discrete-time queueing model subjected to vacations. Before we proceed with the model, it is clear that we should understand why we do in this fashion. First, the discrete-time scale reflects the nature of the underlying wireless communication, i.e., time-slotted data transmission. Moreover, "vacation" refers to the process that PU interruption such that channel becomes unavailable for SMR, and "random" refers to the fact that vacation occurs independently of the system state. Without loss of generality, we consider "continue after vacation" operation mode, which means a packet continues to transmit after vacation times. It can be easily extended to other operation modes in vacation queueing system, such as "repeat after interruption", "repeat after interruption with resampling" [13]. The concepts of vacation queueing system will be found useful throughout the remainder analysis.

As noted above, one of the most important factors causing the channel unavailable is PU behavior, which is modelled as a Markovian ON-OFF process as illustrated in Fig. 2. The ON-OFF process is used most frequently to model the PU behaviors, slots during which the channel is available for SMRs are called ON slots, and analogously, slots during which the channel takes a vacation for SMRs due to PU interruption are called OFF slots. Let α_{rlm} and β_{rlm} denote the probabilities that the channel m remains in ON and OFF state, respectively,

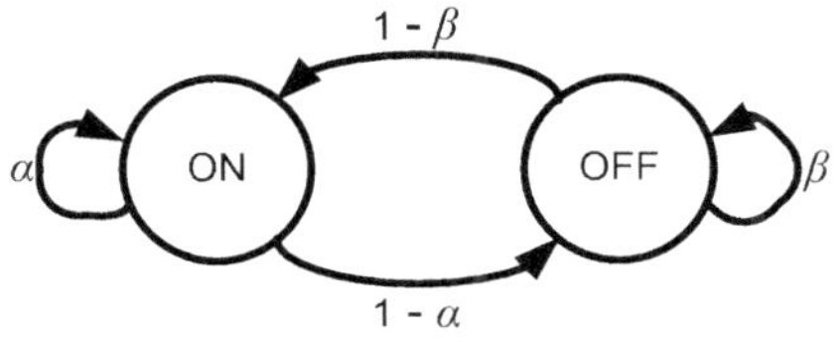

Fig. 2. Transition diagram of the channel vacation

in addition, μ_{rlm}^{ON} and μ_{rlm}^{OFF} represent the mean time in the ON and OFF state, respectively. Accordingly, the fraction of ON slots A_{rlm} is calculated by Eq. (1)

$$A_{rlm} = \frac{\mu_{rlm}^{ON}}{\mu_{rlm}^{ON} + \mu_{rlm}^{OFF}} = \frac{1 - \beta_{rlm}}{2 - \alpha_{rlm} - \beta_{rlm}} \tag{1}$$

We have considerable freedom in constructing a large number of vacation models through the choice of α_{rlm} and β_{rlm} corresponding to different PU behaviors. In a similar fashion, we obatin the fraction of OFF slots

$$U_{rlm} = \frac{\mu_{rlm}^{OFF}}{\mu_{rlm}^{ON} + \mu_{rlm}^{OFF}} = \frac{1 - \alpha_{rlm}}{2 - \alpha_{rlm} - \beta_{rlm}} \tag{2}$$

In the view of dependability engineering, Eq. (2) matches the definition of asymptotic unavailability. For this reason we immediately obatin the channel asymptotic unavailability due to PU interruption for the link l

$$U_{rl} = \sum_{m \in \mathcal{M}_l} U_{rlm} x_{rlm}, \quad \forall l \in \mathcal{L}_r. \tag{3}$$

Furthermore, let U_r denote the channel asymptotic unavailability for PU interruption over route r. Since all the links over the route consist a series system and channels on each link fail independently, we can obtain

$$U_r = 1 - \prod_{l \in \mathcal{L}_r} (1 - \sum_{m \in \mathcal{M}_l} U_{rlm} x_{rlm}), \quad \forall r \in \mathcal{R}. \tag{4}$$

3.2 End-to-End Delay

End-to-end delay is one important element of the network performance experienced by a user. In particular, it is of concern to the real-time service. The end-to-end delay over one route is the summation of delays of all links along the route. In our study we shall only consider access delay and transmission delay, for the sake of simplicity.

Since the 802.11 DCF MAC protocol is adopted as our distributed access scheme, the access delay is mainly caused by SMR packets transmission backoff. Let K denote the maximum number of transmission retries, and W_j denote the contention window at the jth $(1 \leq j \leq K+1)$ backoff stage. According to the 802.11 standard, the set of contention window W_j shall be sequentially ascending integer powers of 2, minus 1, ranging from the specified minimum value W_{min} to the maximal value W_{max}, that is, $W_j = 2^{j-1}(W_{min}+1) - 1$. Based on the contention window parameter, SMR backoff time T_j^Q in jth backoff stage can be calculated as below

$$T_j^Q = W_j^{'} \times T_{slot} \tag{5}$$

where $W_j^{'}$ is a random integer drawn from a uniform distribution over the interval $[0,\, W_j]$, and T_{slot} is the slot length. For the sake of simplicity, we set $T_{slot} = 1$. Then, the mean value of jth backoff time can be expressed $E[T_j^Q] = \frac{W_j}{2}$.

Now it satisfies the condition to calculate the mean backoff access delay. We still take the channel m at link l over route r for example. Let p_{rlm} represent the transmission failure probability which remains the same at all backoff stages. To guarantee the selected channel quality, we consider a predefined threshold value I for p_{rlm}, that is p_{rlm} of the selected channel should not exceed the upper bound I. Note, each SMR needs to measure the transmission failure probability periodically. Then, the mean backoff access delay

$$\begin{aligned} D_{rlm}^Q &= \sum_{i=1}^{K} p_{rlm}^{i-1}(1-p_{rlm}) \sum_{j=1}^{i} E[T_j^Q] + p_{rlm}^K \sum_{j=1}^{K+1} E[T_j^Q] \\ &= \frac{1-(2p_{rlm})^{K+1}}{2(1-2p_{rlm})}(W_{min}+1) \end{aligned} \tag{6}$$

Similar results on access delay can be found in [8], we extend the related work on transmission delay to take the PU interruption into account. In fact, the transmission delay can also be interpreted as the packet effective service time, which is defined as the number of slots between the beginning of the slot where the packet enters the channel and the end of the slot where the packet leaves the channel as illustrated in Fig. 3. Based on this definition, the transmission delay for one packet should include the actual transmission time and the channel vacation time.

In the "continue after vacation" operation mode, a packet's transmission continues after PU interruption. The unfinished part can be seen as a new packet with length equals to the remaining transmission time. Therefore, we can derive the mean transmission delay D_{rlm}^T per packet for SMR as follows

$$D_{rlm}^T = \frac{C}{B_{rlm}A_{rlm}} \tag{7}$$

here C represents the mean packet length including MAC control overhead, in bits, B_{rlm} is the link transmission capacity, in bits per sec, and A_{rlm} is the fraction of ON slots calculated from Eq. (1).

As noted above, the delay over one link D_{rlm} consists of access delay D_{rlm}^Q and transmission delay D_{rlm}^T. From Eq. (6) and (7) it is apparent that the end-to-end delay D_r in one route r can be expressed as the sum of mean delay of all links in route r

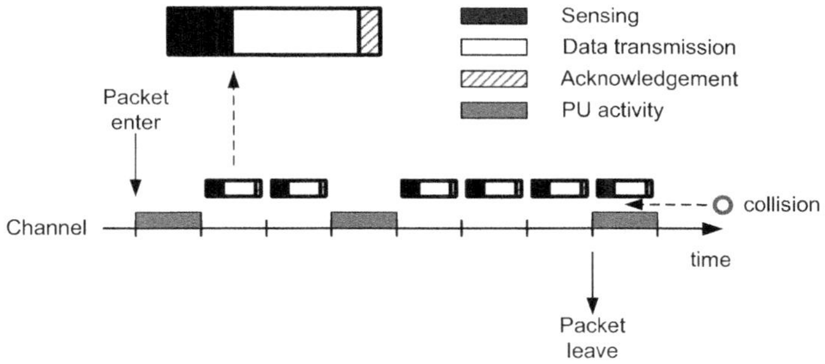

Fig. 3. Transmission scenario

$$D_{rlm} = D^{Q}_{rlm} + D^{T}_{rlm} \tag{8}$$

$$D_r = \sum_{l \in \mathcal{L}_r} \sum_{m \in \mathcal{M}} (Q_{rl} + 1) D_{rlm} x_{rlm} \tag{9}$$

where Q_{rl} is the number of packets buffered on link l. It should be monitored by SMR in each transmission.

3.3 Formulation of Channel Selection for Each Route

The problem we study is how to enable every link choose an optimal channel for data transmission. The route selection is based on the result of the optimization formulation. Here, the objective function is to minimize the expected end-to-end delay of real-time service. We assume that each link has at least one channel that can be used. For each route, we select channels to minimize the objective function while guaranteeing the channel asymptotic unavailability due to PU interruption not exceed the threshold, i.e.,

P1

$$\text{Minimize } D_r = \sum_{l \in \mathcal{L}_r} \sum_{m \in \mathcal{M}_l} (Q_{rl} + 1) D_{rlm} x_{rlm} \tag{10}$$

Subject to:

$$U_r \leq U \tag{11}$$

$$p_{rlm} x_{rlm} \leq I \tag{12}$$

$$\sum_{m \in \mathcal{M}_l} x_{rlm} = 1, \quad \forall l \in \mathcal{L}_r \tag{13}$$

$$x_{rim} + x_{rjm} \leq 1, \quad \forall m \in \mathcal{M}_l; i, j \in \mathcal{L}_r; j \in \mathcal{L}_{I,i} \tag{14}$$

$$x_{rlm} \in \{0, 1\}, \quad \forall l \in \mathcal{L}_r; m \in \mathcal{M}_l. \tag{15}$$

where U is the route channel unavailability threshold, I is the required transmission failure probability threshold. Constraint (11) represents the route channel unavailability in Eq. (2) can not exceed the threshold U. Constraint (12) represents the transmission failure probability in Eq. (6) can not exceed the threshold I. Constraint (13) indicates that each link should work on one and only one data channel. Constraint (14) means the link in interference range area can not work on the same channel, thus there is no additional delay in the data channel. The solution is to find out every x_{rim}, so that all the constraints are satisfied and the objective function is minimized. Obviously, this is a nonlinear integer problem.

3.4 Transformation

According to (11) and (4), we have

$$
\begin{aligned}
& 1 - \prod_{l \in \mathcal{L}_r} (1 - \sum_{m \in \mathcal{M}_l} U_{rlm} x_{rlm}) \leq U \\
\Leftrightarrow & \prod_{l \in \mathcal{L}_r} (1 - \sum_{m \in \mathcal{M}_l} U_{rlm} x_{rlm}) \geq 1 - U \\
\Leftrightarrow & \ln \left(\prod_{l \in \mathcal{L}_r} (1 - \sum_{m \in \mathcal{M}_l} U_{rlm} x_{rlm}) \right) \geq \ln(1 - U) \\
\Leftrightarrow & \sum_{l \in \mathcal{L}_r} \ln(1 - \sum_{m \in \mathcal{M}_l} U_{rlm} x_{rlm}) \geq \ln(1 - U) \\
\Leftrightarrow & \sum_{l \in \mathcal{L}_r} \sum_{m \in \mathcal{M}_l} \ln(1 - U_{rlm}) x_{rlm} \geq \ln(1 - U)
\end{aligned}
$$

The last transformation is ture since x_{rlm} is either 0 or 1, and only one channel for each link is selected. Moreover, since U is less than 1, $\ln(1 - U)$ is negative, the above inequality can be transformed to the following inequality by deviding $\ln(1 - U)$ on both sides.

$$
\begin{aligned}
& \sum_{l \in \mathcal{L}_r} \sum_{m \in \mathcal{M}_l} \ln(1 - U_{rlm}) x_{rlm} \geq \ln(1 - U) \\
\Leftrightarrow & \sum_{l \in \mathcal{L}_r} \sum_{m \in \mathcal{M}_l} \frac{\ln(1 - U_{rlm})}{\ln(1 - U)} x_{rlm} \leq 1 \\
\Leftrightarrow & \sum_{l \in \mathcal{L}_r} \sum_{m \in \mathcal{M}_l} \log_{(1-U)}(1 - U_{rlm}) x_{rlm} \leq 1
\end{aligned}
$$

From (11) and (12), the number of binary varialbles can be reduced for each link by removing any channel m where $U_{rlm} \geq U$ or $p_{rlm} > I$. Assume the new channel set for each link l is $\mathcal{M}'_L$.

Let $\hat{D}$ denote the maximum link delay. The objective function in (10) can be transformed as follows.

$$Minimize \sum_{l\in\mathcal{L}_r}\sum_{m\in\mathcal{M}'_l}(Q_{rl}+1)D_{rlm}x_{rlm}$$

$$\Leftrightarrow Maximize \sum_{l\in\mathcal{L}_r}\sum_{m\in\mathcal{M}'_l}\left(\hat{D}-(Q_{rl}+1)D_{rlm}\right)x_{rlm}$$

We introduce positive variable v_{rlm} and w_{rlm} to denote the coefficience in the modified objective function and constraint function as follows

$$v_{rlm}=\hat{D}-(Q_{rl}+1)D_{rlm}, \forall l\in\mathcal{L}_r; m\in\mathcal{M}'_l \tag{16}$$

$$w_{rlm}=\log_{(1-U)}(1-U_{rlm}), \forall l\in\mathcal{L}_r; m\in\mathcal{M}'_l \tag{17}$$

For any channel m in link l on route r, the analog meaning of v_{rlm} is the value (profit), while the meaning of w_{rlm} is the weight (cost). The retransformed problem can be defined as follows *P2*

$$\text{Maximize} \sum_{l\in\mathcal{L}_r}\sum_{m\in\mathcal{M}'_l} v_{rlm}x_{rlm} \tag{18}$$

Subject to:

$$\sum_{l\in\mathcal{L}_r}\sum_{m\in\mathcal{M}'_l} w_{rlm}x_{rlm}\leq 1 \tag{19}$$

$$\sum_{m\in\mathcal{M}'_l} x_{rlm}=1, \quad \forall l\in\mathcal{L}_r \tag{20}$$

$$x_{rim}+x_{rlm}\leq 1, \quad \forall l\in\mathcal{L}_r; m\in\mathcal{M}'_l; i\in\mathcal{L}_{I,l} \tag{21}$$

$$x_{rlm}\in\{0,1\}, \quad \forall l\in\mathcal{L}_r; m\in\mathcal{M}'_l. \tag{22}$$

This is a $0-1$ integer linear problem, which is in general NP-complete [14]. Moreover, without constraint (21), this problem can be viewed as an instance of Multiple-Choice Knapsack Problem, where we have $|\mathcal{L}_r|$ mutually disjoint classes (links) of items (channels) to be packed into a knapsack of capacity 1. Each item m ($m\in\mathcal{M}_l$) has a profit v_{rlm} and a cost w_{rlm}. The problem is to choose exactly one item from each class such that the total profit is maximized without exceeding the capacity. In addition, the item in interfered classes should be varied from each other.

Using optimization tools such as MOSEK [15] and CPLEX [16], we can get the optimal solution for the above problem. In this paper, we are interested in finding a heuristic method to get an accepted result within affordable time complexity.

4 Proposed Channel Selection and Routing Schemes

In this section, we propose a heuristic channel selection scheme for each route, and then choose the route with minimum end-to-end delay D_r.

For any link l, we sort the channels according to increasing weights w_{rlm}, and derive R_l. Therefore, the index of channels in $\mathcal{R}_l$ is different from that in $\mathcal{M}_l$. We then construct an instance of knapsack by setting

$$\tilde{v}_{rlm} = v_{rlm} - v_{rl,m-1}, \forall l \in \mathcal{L}_r, m = 2, 3, ..., |\mathcal{M}_l^{'}|.$$

and

$$\tilde{w}_{rlm} = w_{rlm} - w_{rl,m-1}, \forall l \in \mathcal{L}_r, m = 2, 3, ..., |\mathcal{M}_l^{'}|.$$

and the residual capacity is

$$\bar{c} = 1 - \sum_{l \in \mathcal{L}_r} w_{rl1}$$

Algorithm 1. Channel selection algorithm

Input: $\{\mathcal{N}_i\}$, $\{\mathcal{M}_i\}$, $\{Q_{rlm}\}$, α, β.
Output: $\{x_{ijc}\}$.

Initialization: $\bar{c} \leftarrow 1$
Calculate w_{rlm}, v_{rlm}, μ_{rlm} for all route link and channels.
for $l \in \mathcal{L}_r$ **do**
  Remove the channels where $U_{rlm} \geq U$ or $p_{rlm} > I$.
  Sort the channels according to increasing w_{rlm} and derive $\mathcal{M}_l^*$.
  for $i = 2; i \leq |\mathcal{M}_l^*|; i++$ **do**
    $\tilde{v}_{rli} \leftarrow v_{rli} - v_{rl,i-1}$
    $\tilde{w}_{rli} \leftarrow w_{rli} - w_{rl,i-1}$
  end for
  $\bar{c} \leftarrow \bar{c} - w_{rl1}$
end for
Sort the link-channel pairs according to decreasing incremental efficiencies $\tilde{\eta}$
while 1 **do**
  Get the index of link-channel pair $\{i, j\}$ with the maximal $\tilde{\eta}$: $\{i, j\} \leftarrow \arg \max_{l \in \mathcal{L}_r, m \in \mathcal{M}_l^*} \tilde{\eta}_{rlm}$
  $\bar{c} \leftarrow \bar{c} - \tilde{w}_{ri1}$
  if $\bar{c} < 0$ **then**
    Break;
  else
    Record the channel index as m' for link i in the knapsack.
    Mark channel m' for all the interfering links as active.
    $x_{rim'} \leftarrow 0$
    Mark channel j for all the interfering links as inactive.
    $x_{rij} \leftarrow 1$
  end if
end while

Then, we sort all the link-channel pair according to decreasing incremental efficiencies defined as follows

$$\tilde{\eta}_{rlm} = \frac{\tilde{v}_{rlm}}{\tilde{w}_{rlm}}$$

We then fill the knapsack up to capacity $\bar{c}$ according to the order of the link-channel pair sorting in terms of incremental efficiencies $\tilde{\eta}_{rlm}$. Capacity constraint is checked before adding a link-channel pair. After adding a link-channel pair, the channel m used in this link l is marked as inactive from any interfering links $\mathcal{I}_l$, and the previous channel m' from the same link l in the knapsack is taken out, which means $x_{rlm'} = 0$, and channel m' in interfering links $\mathcal{I}_l$ is marked to be active. The residual capacity $\bar{c}$ will updated by

$$\bar{c} = \bar{c} - \tilde{w}_{rlm}$$

Following this approach until the capacity constraints break, the channel selection for all links in route r is finished. The details of this scheme is shown in Algorithm 1.

We shall select the route with the minimum end-to-end delay, while the other constraints are guaranteed.

$$r^* = \arg\min_{r \in \mathcal{R}} D_r \tag{23}$$

5 Numerical Results and Analysis

We have implemented the algorithms in MATLAB and evaluated the performance. We consider a grid topology similar to the topology used in [8], where SMRs are uniformly placed. The interference range of any SMR is one neighbouring hop. The number of available routes is 10. For each route, the number of hops changes from 1 to 10. For each link, the number of available channels changes from 2 to 10. Each channel on every link selects a data rate from {11, 5.5, 2, 1} Mbps according to the quality of that channel. The ON-OFF changing rate for each channel is randomly generated between 0 and 1. We assume the traffic follows the constant bit rate with the packet size of 128 bytes. The minimum contention window size W_{min} is 0.02ms, and the maximum number of retransmissions K is 4. The transmission failure probability p_{rlm} is randomly generated in [0,0.1]. The threashold of transmission failure probability for each link is 0.1. The threashold of channel asymptotic unavailability for PU interruption over any route is set as 0.9. For each study we randomly generated for 100 times (seeds).

We present the results of solution rate and end-to-end delay. Solution rate is defined as the rate of seeds with valid solutions from all the 100 seeds. End-to-end delay only with valid solution is presented. To make the discussion more concrete, a series of comparative performance evaluation between the proposed scheme and other schemes, including optimization tool based channel selection, minimal unavailability channel selection, and minimal delay channel selection are carried out.

5.1 Solution Rate

We study the solution rates for all the schemes. Figure 4, 5, 6, and 7 show the solution rate for channel selection schemes, MOSEK, proposed, minimal unavailability and minimal delay, respectively. Then, we can see for all schemes, when the number of available channels is less than 6, the MOSEK solution has the highest solution rate among others.

As the number of links increases, some cases will reduce the solution rate. For example in the MOSEK results, the solution rate for 2-channel case starts reduction from 4 hops, 3-channel case starts reduction from 5, 4-channel case starts reduction from 7, and 5-channel case starts reduction from 9. Other schemes have similar results. The reason behind it is that when the number of hops grows, more channels are required to avoid the interference.

Figure 8 compares the solution rate with different schemes in the case of 3 channels per link. The minimal delay scheme always achieves the lowest solution rate since it doesn't consider the unavailability constraints for the route which may cause invalid solutions.

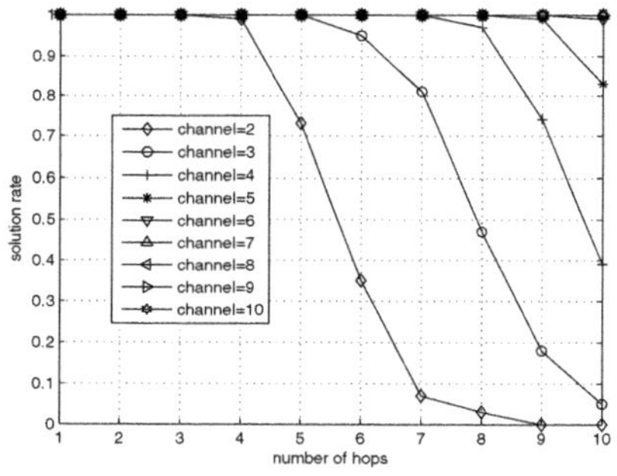

Fig. 4. Solution rate for MOSEK solution

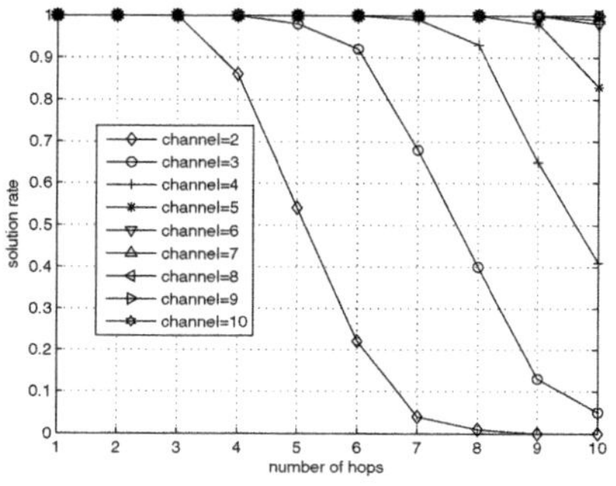

Fig. 5. Solution rate for proposed channel selection

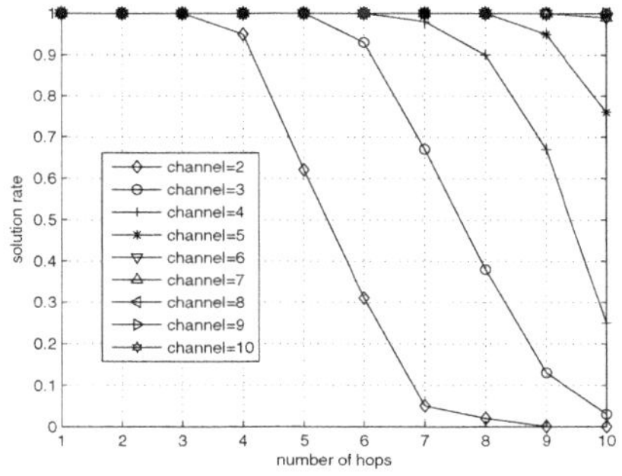

Fig. 6. Solution rate for channel selection based on minimal unavailability

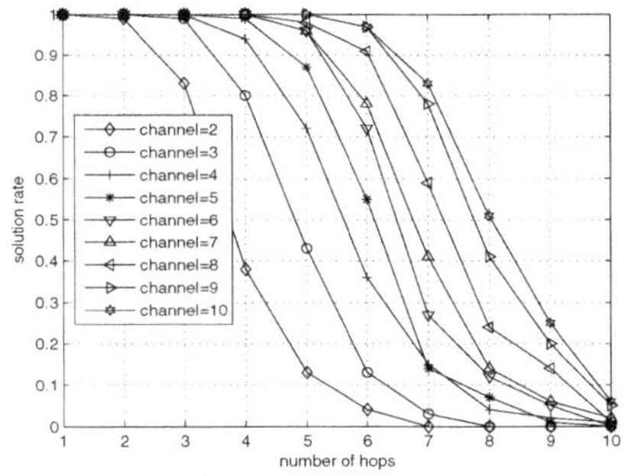

Fig. 7. Solution rate for channel selection based on minimal delay

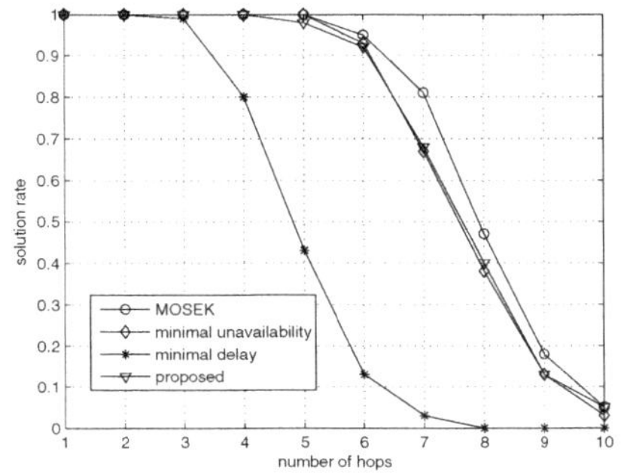

Fig. 8. Solution rate comparison for different channel selection schemes (when the number of available channel is 3)

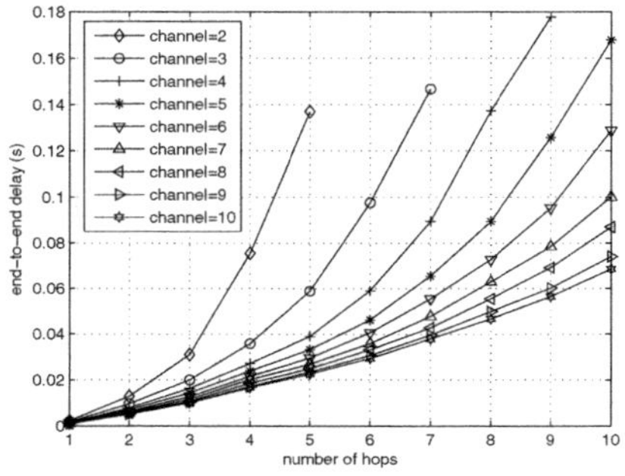

Fig. 9. End-to-end delay for MOSEK solution

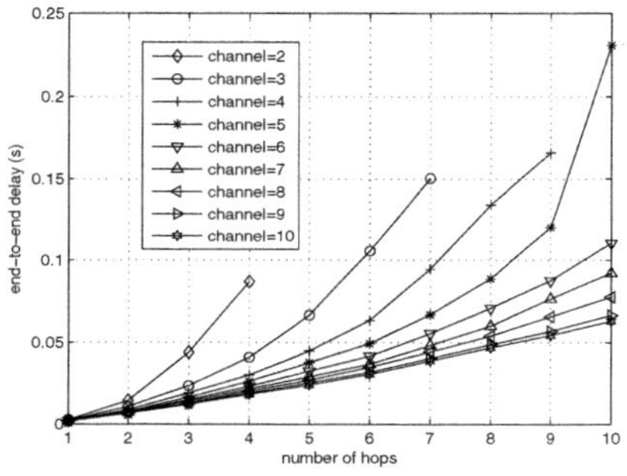

Fig. 10. End-to-end delay for proposed channel selection

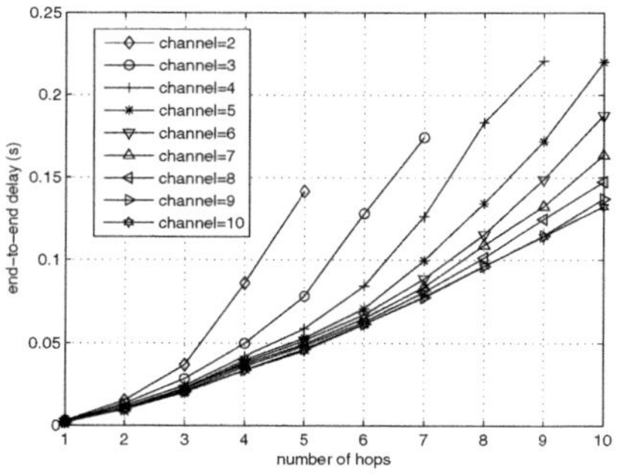

Fig. 11. End-to-end delay for channel selection based on minimal unavailability

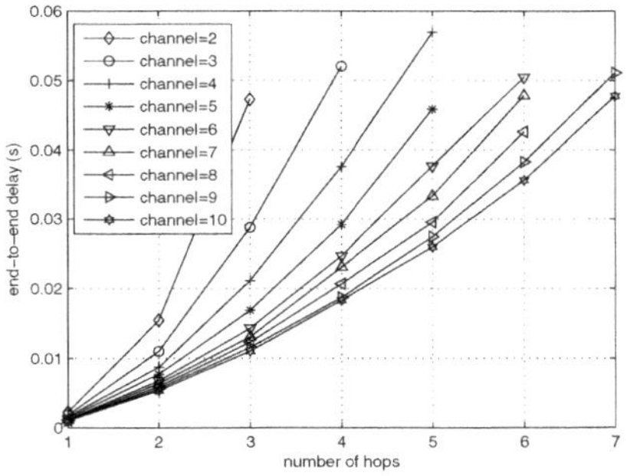

Fig. 12. End-to-end delay for channel selection based on minimal delay

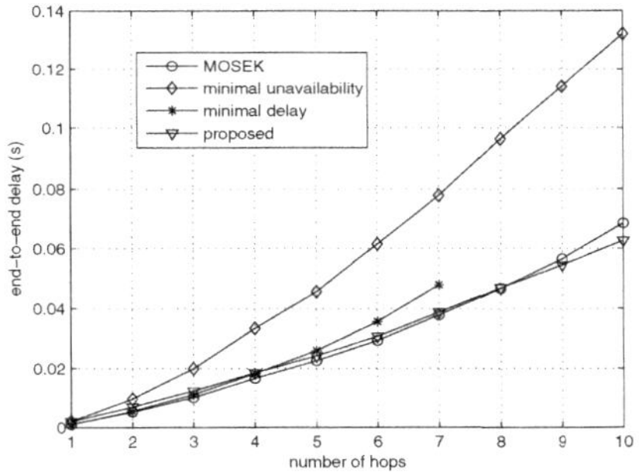

Fig. 13. End-to-end delay comparison for different channel selection schemes (when the number of available channel is 10)

5.2 End-to-End Delay

We show the performance in terms of average end-to-end delay in the following. For those cases of no valid solution by a certain scheme, their end-to-end delay are not taken into account the caculation of the average end-to-end delay. Thus, we introduce a solution rate threshold, above which the result is shown for every scheme. In Figure 9, 10, 11, 12, and 13, the solution rate threshold is set as 0.6.

As shown in Figure 9, 10, 11, and 12, the end-to-end delay increases with the number of hops increases. In the same case of number of hops, for more channels, the end-to-end delay decreases.

Figure 13 compares the end-to-end delay with different schemes in the case of 10 available channels for each link. The minimal unavailability scheme always achieves the highest delay since it doesn't consider delay while choosing channels for each link.

6 Conclusion

In this paper, we have investigated the joint channel selection and routing problem in cognitive radio mesh networks. We formulated this problem and transformed it to a variant of multiple-choice knapsack problem, then we proposed a heuristic method to solve this problem. Simulation results showed that our proposed heuristic method and MOSEK solution can achieve the closest in performance. It outperforms the minimal unavailability scheme and minimal delay scheme in terms of end-to-end delay and solution rate.

References

1. Zhang, Y., Luo, J., Hu, H.: Wireless Mesh Networking: Architectures, Protocols and Standards (Wireless Networks and Mobile Computing). Auerbach Publications (December 2006)
2. Walke, B.H., Mangold, S., Berlemann, L.: IEEE 802 Wireless Systems: Protocols, Multi-Hop Mesh/Relaying, Performance and Spectrum Coexistence. Wiley, Chichester (2007)
3. Chowdhury, K.R., Akyildiz, I.F.: Cognitive wireless mesh networks with dynamic spectrum access. IEEE Journal on Selected Areas in Communications 26(1), 168–181 (2008), http://dx.doi.org/10.1109/JSAC.2008.080115
4. Mitola, J., Maguire, G.: Cognitive radio: making software radios more personal. IEEE Personal Communications 6(4), 13–18 (1999)
5. Chen, T., Zhang, H., Maggio, G., Chlamtac, I.: Topology management in cogmesh: A cluster-based cognitive radio mesh network. In: IEEE International Conference on Communications (ICC 2007), pp. 6516–6521 (June 2007)
6. Pereira, R.C., Souza, R.D., Pellenz, M.E.: Overlay cognitive radio in wireless mesh networks (2008), http://www.citebase.org/abstract?id=oai:arXiv.org:0805.3643
7. Dutta, P., Jaiswal, S., Panigrahi, D., Rastogi, R.: A new channel assignment mechanism for rural wireless mesh networks. In: IEEE INFOCOM 2008, pp. 2261–2269 (April 2008)
8. Li, H., Cheng, Y., Zhou, C., Zhuang, W.: Minimizing end-to-end delay: A novel routing metric for multi-radio wireless mesh networks. In: IEEE INFOCOM 2009, pp. 46–54 (April 2009)
9. Hou, Y., Shi, Y., Sherali, H.: Optimal spectrum sharing for multi-hop software defined radio networks. In: IEEE INFOCOM 2007, pp. 1–9 (May 2007)
10. Song, Y., Fang, Y., Zhang, Y.: Stochastic channel selection in cognitive radio networks. In: Global Telecommunications Conference, GLOBECOM 2007, pp. 4878–4882. IEEE, Los Alamitos (2007)
11. Rehmani, M.H., Viana, A.C., Khalife, H., Fdida, S.: Adaptive and occupancy-based channel selection for unreliable cognitive radio networks. In: Rencontres Francophones sur les Aspects Algorithmiques des Telecommunications, ALGOTEL 2009 (juin 2009)
12. Wang, H.S., Moayeri, N.: Finite-state markov channel-a useful model for radio communication channels. IEEE Transactions on Vehicular Technology 44(1), 163–171 (1995)

13. Fiems, D., Bruneel, H.: Analysis of a discrete-time queueing system with timed vacations. Queueing Syst. Theory Appl. 42(3), 243–254 (2002)
14. Bertsekas, D.P.: Network Optimization: Continuous and Discrete Models. Athena Scientific, Belmont (May 2008)
15. The mosek optimization software, http://www.mosek.com/
16. ILOG CPLEX, http://www-01.ibm.com/software/integration/optimization/cplex-optimizer/(2010)

Studying Non-intrusive Tracing in the Internet

Alina Olteanu[1], Yang Xiao[1,*], Jing Liu[1], and Thomas M. Chen[2]

[1] Dept. of Computer Science, University of Alabama, Tuscaloosa, AL 35487 USA
aolteanu@cs.ua.edu,yangxiao@ieee.org,jliu39@crimson.ua.edu
[2] School of Engineering, Swansea University, Swansea, Wales, UK SA2 8PP
t.m.chen@swansea.ac.uk

Abstract. Intruders which log-in through a series of machines when conducting an attack are hard to trace because of the complex architecture of the Internet. The thumbprinting method provides an efficient way to tracing such intruders by determining whether two connections are part of the same connection chain. Since many connections are transient, and therefore short in length, choosing the best time interval to thumbprint over can be an issue. In this paper, we provide a way to shorten the time interval used for thumbprinting. We then study some special properties of the thumbprinting function. We also study another mechanism for tracing intruders in the Internet, based on a timestamping approach of passively monitoring flows between source and destination pairs. Given a potentially suspicious source, we identify the true destination of this source. We compute the error probability of our algorithm and show that its value decreases exponentially as the observation time increases. Our simulation results show that our approach performs well.

Keywords: Security, Tracing, Thumbprinting.

1 Introduction

Constant change is perhaps one major principle that characterizes the Internet. Recent advances in technology lead to a significant growth of the Internet by factors of 10^3 and 10^6 in the backbone speed and in the number of hosts, respectively [7]. Since the public expansion of the Internet in 1990, many new challenges have surfaced; among them, operations between un-trusted end-points, more demanding applications and less sophisticated users have posed severe stress on the Internet requirements. Furthermore, the number of attacks on networked computer systems has been growing exponentially from year to year. When considering the task of tracing intruders in the Internet, we have to take into account three main challenges. First, attackers hide their origin by making use of the Internet's architecture. By using different hosts, belonging to different countries and administrative domains, to route malicious acts, intruders' actions become extremely difficult to trace back. Second, the data collected from an Internet trace is usually incomplete or has missing values. For example, different domains of Internet service providers (ISPs) may not share

[*] Corresponding author.

X. Zhang and D. Qiao (Eds.): QShine 2010, LNICST 74, pp. 58–74, 2011.

data due to access issues, such as in different counties. Finally, routes change frequently, packets are lost and the network latency and time to convergence can be significantly increased due to routing instability.

To deal with such problems, two tracing mechanisms exist [4]: 1) methods of keeping track of all individuals and accounting all activities, and 2) reactive tracing, in which no global accounting is attempted until a problem arises and then tracing back its source is attempted. In own view, the first mechanism is much related to network accountability and the second mechanism is much related to network forensics. Network-based tracing and host-based tracing are two main approaches for reactive tracing in connection chains. Host-based tracing involves one tracing system per network host [4], and 1) a chain of connection hosts can be known via host communications [3] or 2) revering the attack chain by breaking the hosts in the reversed order [5]. Host-based tracing schemes suffer when an extended connection crosses a host not running the system [4]. Network-based tracing has the advantage that it does not rely on hosts which can be untrusty and it does not require host participation. Instead, network-based tracing uses the invariance of connections at higher protocol layers (such as the transport layer) and based on this observation can establish whether two connections are part of the same connection chain. We study two approaches for network-based tracing. The first approach is based on the idea of a thumbprint of a connection. A thumbprint is very similar to a checksum or the computed summary of the content of a connection [2], [4]. The thumbprint summarizes the content of a connection using only a small amount of data while at the same time preserving the uniqueness of the connection. Therefore, thumbprints of related connections are similar and can lead to constructing a connection chain. However, the thumbprinting method is dependent on the duration of the connection so the time period used for thumbprinting is very important especially in the case of transient flows. In this paper, we provide a way to shorten the time interval used for thumbprinting. We study the tradeoff introduced by using such a smaller time interval and how it affects basic thumbprint properties such as sensitivity and robustness.

Since thumbprinting relies on the content of the packages and therefore can be counterattacked by encryption, we study a second mechanism for tracing intruders in the Internet, involving passively monitoring flows between source and destination pairs in the Internet, similar to [1]. This approach is based on monitoring transmission activities of nodes and does not interfere with network operations. A one-hop communication graph can be constructed by matching transmission timestamps and acknowledgements. Based on this graph, the nodes that are part of a connection chain are those who communicate at a *sufficiently high rate*.

Our approach is different from the approach in [1] in that we make fewer statistical assumptions and we use clearer, simpler derivations. In addition, by working in the Internet, we face other challenges, like the availability of only partial information and route instability mentioned above. To account for this, we assume that the information on every link is available with a certain probability. There are many related research papers in security [12-111].

The remainder of the paper is organized as follows. Section 2 presents significant work related to our problem. In Section 3 we present our results concerning minimizing the thumbprinting interval. Section 4 contains a detailed introduction to the tracing algorithm followed by an evaluation of its consistency in terms of the error probability. We conclude the paper in Section 5.

2 Related Work

The paper [4] presents an IP traceback method based on the idea of a thumbprint. Often attacks are conducted by logging through a chain of host machines. In this way, intruders make use of the Internet architecture to hide their true origin. The thumbprint technology, which is a summary of a connection, is used to compare connections and therefore trace the source of such an attack. A thumbprint is computed for each time interval, typically a minute, of each connection [4]. Thumbprinting has the advantage that it preserves the characteristics of a connection while at the same time being cheap to compute and requiring little storage. Similar methods like checksum and message digest have two main disadvantages. Firstly, an error in the content leads to a different value being computed, and thus, these methods are not robust. Secondly, they are not additive, i.e., two successive values cannot be combined to provide a new value for a longer interval. Other compression schemes are by far costier than thumbprinting in terms of storage space.

As the thumbprinting method is dependent on the duration of the connection, it is of importance to shorten the time interval used for thumbprinting, especially in the case of transient flows. This is the purpose of our derivation from section 3.

However, fingerprinting techniques are vulnerable and may not be of interest when the attacker uses encryption at every hop in the chain. To deal with such kind of problems, techniques based on packet timing information are more suitable.

The paper [1] presents a mechanism for tracing intruders in anonymous MANETs, based on passively monitoring flows between source and destination pairs. The algorithm is concerned with tracing the destination of a certain source, considered suspicious according to a higher layer intrusion detection protocol. Because anonymity of nodes in the MANET is assumed, and the monitoring is done in a non-intrusive way, this research only considers the timestamps of packets on adjacent links, and based on this, establishes causal relationships between packets. A graph is used to model the communications among nodes and based on traffic analysis and information carried in the packets, the graph is partitioned into two parts, one containing the set of possible destinations, and one with which the source does not communicate. The assumptions that are made are related to the distribution of traffic: the transmission activities on each link are assumed to follow a Poisson process. In addition, the rate of the flow to be traced must be sufficiently high, and the duration of the observation must be long enough. Therefore, the method will not work for low-rate flows, or for transient flows.

Much like [1], we focus on tracing flows between pairs (source, destination), in the Internet this time. Unlike [1], we assume differences in data due to propagation delays, skew of clocks during observation at multiple nodes at the same time and data missing from observations, or observed erroneously. We account for this by using a certain probability with which information is available on every link. In addition, we make fewer assumptions about the rate of the Poisson process and the calculations are significantly simplified.

3 Thumbprinting

The thumbprinting approach is based on the fact that, at higher protocol layers, the content of a connection is invariant at all points of the chain. The thumbprint summarizes the content of the connection using only a small amount of data while at the same time preserving the uniqueness of the connection. We first define the terminology and then present a method for shortening the time interval used for thumbprinting and some interesting properties of the thumbprint function.

3.1 Algorithm

A thumbprint is a function of a connection which preserves the unique characteristics of the connection. We use the notations and definition of a thumbprint from [4]. Consider a sequence of transmitted characters $a_1, a_2, \ldots, a_L$ to thumbprint. Consider also the function $\alpha : \mathcal{A} \rightarrow \mathbf{R}^K$, which takes a character and returns a vector of K real numbers. Let $\alpha_k(\bullet)$ denote the k-th component of the function. Here K is a short fixed number representing the number of thumbprint components. If we consider L to be the period of a connection, we can associate frequencies $\overline{f} = (f_1, f_2, \ldots, f_L)$ to our character sequence. Then the thumbprint is defined as [4]:

$$T_k = \sum_{i=1}^{L} \alpha_k(a_i) f_i, \qquad k \in \{1,2,\ldots,K\}. \tag{1}$$

The thumbprint is thus a linear combination of the frequencies of characters and their corresponding weights. T_k represents the k th component of the K-dimensional thumbprint vector.

Each thumbprint component $T_k(C,t)$, is a function of a specific connection C and the time interval t, in which the thumbprints have been computed. From [4], we also know that the comparison of two sets of thumbprints, $T_k(C,t)$ and $T_k(C',t)$, $k \in \{1,2,\ldots,K\}$, for two different connections C and C', is given by the following formula:

$$\delta_t(C,C') = \log\left(\prod_{k=1}^{K} \left|T_k(C',t) - T_k(C,t)\right|\right). \tag{2}$$

δ_t represents the logarithm of the product of component differences for two thumbprints in a specific time interval t. A large absolute value of δ_t implies that the two connections are related, while a small absolute value suggests independent connections.

3.2 Minimizing the Thumbprinting Interval

In this section, based on the results in [4], we try to find the best period of time, T, to thumbprint over. The length of time divisions for the experiments in [4] is 1 time unit. However, many connections in the Internet are transient, and therefore short in length; in these cases a shorted thumbprinting interval is needed.

We start with the thumbprint function given by (1) and in addition, we account for the time in which the thumbprint has been computed. This gives us the following function:

$$\sum_{a,t} \alpha_k(a,t) f(a,t) \,. \tag{3}$$

We will use the function given by (3) above as our thumbprinting function.

Following, we do the changes of variable $a = Ly$ and $t = T\tau$. This transforms our time interval from $[0,T]$ to $[0,1]$ and the samples from every time unit to every $1/T$ time units. It can be seen that the thumbprinting function in (3) can be approximated be the following integral:

$$\iint_{[0,L]\times[0,T]} f(a,t)\alpha_k(a,t)\,da\,dt \,. \tag{4}$$

We have:

$$\begin{aligned}
\iint_{[0,L]\times[0,T]} f(a,t)\alpha_k(a,t)\,da\,dt &\approx LT \sum_{a,t} \alpha_k(a,t) f(a,t) \Big/ (LT) \\
&\overset{\substack{a=Ly \\ t=T\tau}}{=} LT \sum_{y,\tau} \alpha_k(Ly,T\tau) f(Ly,T\tau) \Big/ (LT) \\
&= LT \iint_{[0,1]\times[0,1]} f(Ly,T\tau)\alpha_k(Ly,T\tau)\,dy\,d\tau
\end{aligned}$$

The time period has become interval $[0,1]$.

Since thumbprinting over a one unit time interval may not provide the best results, we further provide a way to transform interval $[0,1]$ into an arbitrary time interval $[p,q]$. In the following calculus, we use $y = (u-m)/(n-m)$ and $\tau = (v-p)/(q-p)$. We have:

$$\begin{aligned}
&LT \iint_{[0,1]\times[0,1]} f(Ly,T\tau)\alpha_k(Ly,T\tau)\,dy\,d\tau \\
&= LT \iint_{[m,n]\times[p,q]} f(L(u-m)/(n-m), T(v-p)/(q-p)) \cdot \\
&\quad \alpha_k(L(u-m)/(n-m), T(v-p)/(q-p)) \left| \frac{\partial(y,\tau)}{\partial(u,v)} \right| du\,dv \\
&= LT \iint_{[m,n]\times[p,q]} \frac{f(L(u-m)/(n-m), T(v-p)/(q-p))}{(n-m)(q-p)} \cdot \\
&\quad \frac{\alpha_k(L(u-m)/(n-m), T(v-p)/(q-p))}{(n-m)(q-p)} du\,dv \,.
\end{aligned} \tag{5}$$

The time period has become interval $[p,q]$. This interval can be suitably chosen to accommodate connections with low/high data rates and transient connections in order to give the best performance possible under different scenarios.

3.3 Properties of the Thumbprinting Function

In the following, we describe a way to easily compute the value of the integral given by (4) by the use of the mean value theorem [10]. We further study some special properties of the thumbprinting function.

Consider the right-hand side of equation (5).

Let us choose u_0 and v_0 to be the points toward which we concentrate intervals $[m,n]$ and $[p,q]$, respectively. By making $m,n \to u_0$ and $p,q \to v_0$ (i.e. $n-m \to 0$ and $q-p \to 0$), we have:

$$\iint_{[0,L]\times[0,T]} f(a,t)\alpha_k(a,t)dadt = LT \iint_{[0,1]\times[0,1]} f(Ly,T\tau)\alpha_k(Ly,T\tau)dyd\tau$$

$$= \frac{LT \iint_{[m,n]\times[p,q]} f\left(L\frac{u-m}{n-m},T\frac{v-p}{q-p}\right)\alpha_k\left(L\frac{u-m}{n-m},T\frac{v-p}{q-p}\right)dudv}{(n-m)(q-p)}$$

$$= LTf\left(L\frac{u_0-m}{n-m},T\frac{v_0-p}{q-p}\right)\alpha_k\left(L\frac{u_0-m}{n-m},T\frac{v_0-p}{q-p}\right)$$

$$= LT\,h\left(L\frac{u_0-m}{n-m},T\frac{v_0-p}{q-p}\right). \tag{6}$$

We have considered $h = f\alpha_k$ where h is a continuous function. Let $I = LT \iint_{[0,1]\times[0,1]} f(Ly,T\tau)\alpha_k(Ly,T\tau)dyd\tau$. If h is a C^n function, $n \geq 1$, then the set of points $\gamma = \{(u_0,v_0); I = LT(f\alpha_k)(u_0,v_0)\}$ represents a C^n curve which, according to the Implicit Function Theorem [6], ensures the existence of a function ϕ such that $v_0 = \phi(u_0)$, $u_0 \in J$, where J is an interval included in $[m,n]$. This way we can express one of the variables u_0, v_0 as a function of the other.

The last equality in (6) is true for any $(u_0,v_0) \in [m,n]\times[p,q]$ on the curve γ.

We have thus shown that the value of the integral in (4) can be found by computing the value of the function $f\alpha_k$ in a certain point, in fact, in any point belonging to curve γ.

Furthermore, if we choose:

$$n = m+\varepsilon, u_0 = m+\varepsilon t, 0<t<1, q = p+\delta,$$
$$\phi(u_0) = p+\lambda(t)\delta,$$

we have:

$$I = LT\left(f\alpha_k\right)\left(L\frac{\varepsilon t}{\varepsilon}, T\frac{\lambda(t)\delta}{\delta}\right)$$
$$= LTf\left(Lt, T\lambda(t)\right)\alpha_k\left(Lt, T\lambda(t)\right), t \in J.$$

We have shown that on the curve $\gamma = \{(u_0, \phi(u_0)); u_0 \in J\}$ we have:

$$\alpha_k\left(u_0, \phi(u_0)\right) = \frac{I}{LTf\left(u_0, \phi(u_0)\right)} = const..$$

In order to determine the constant, it is sufficient to compute I. Therefore α_k and f are inversely proportional on curve γ.

Following we introduce an abstraction of the derivations used above by the use of a mean value operator. The use of this operator generalizes the procedure of associating the function h with its mean value. Such an operator can be defined as follows:

$$T(h)(0) = h(0) = \lim_{x\downarrow 0}\frac{\int_0^x h(t)dt}{x} = \lim_{x\downarrow 0} T(h)(x), h \in C([0,b])$$

, for one variable and

$$T(h)(0,y) = \lim_{x\downarrow 0}\frac{\iint_{[0,x]\times[0,y]} h(t_1,t_2)dt_1dt_2}{xy} = \frac{\int_0^y h(0,t_2)dt_2}{y}, y \neq 0$$

,

$$T(h)(x,0) = \frac{\int_0^x h(t_1,0)dt_1}{x}$$

,

$T(h)(0,0) = h(0,0)$, for two variables.

Operator T associates function h with another function that is equal in every point x to the mean value of function h, on interval $[0,x]$.

4 Tracing the True Destination of a Source

Since many attackers use encryption at every hop in the chain, the same packet will have different content on every link and fingerprinting techniques are not appropriate in this case. Therefore, we present a mechanism from paper [1], for tracing intruders in anonymous MANETs, based on using timestamps of transmission to passively monitor flows between source and destination pairs. Such an approach can be implemented by using sensors equipped with energy detectors to measure transmission timestamps and distributing them on the field of interest. All measurements are fused and processed by a centralized monitor at a fusion center.

We first introduce the algorithm and its main features and then analyze the algorithm in terms of the error probability of detecting the true destination of a source. We show that the error probability decays exponentially with the observation time and also prove a similar result when the number of observed graphs goes to infinity.

4.1 Algorithm

Our derivation is based on a variation of the tracing algorithm from [1]. In [1], the approach for tracing the destination of a source is twofold. First, based on the transmission activities on adjacent links, it can be established whether these links are part of the same flow. Then, a set of possible destinations is computed. Second, the intersection method is used: using the changes in topology, some nodes are eliminated, leading to a smaller set of possible destinations.

It is assumed that the transmission activities on each link follow a Poisson process $\boldsymbol{S}$ and its realization is denoted by $\boldsymbol{s}$. Therefore a realization of the transmission timestamps of data packets on link 1 will be $\boldsymbol{s}_1 = (s_1(1), s_1(2), s_1(2), \ldots)$. Here uppercase letters denote random variables, lowercase letters realizations, boldface letters vectors and plain letters scalars.

Following, we briefly introduce the idea behind the traffic analysis method and the intersection of different topologies method from [1].

For traffic analysis, consider two realizations $(\boldsymbol{s}_1, \boldsymbol{s}_2)$ of the transmission activities (timestamps) on two adjacent links. Let m and n be the indices of the current timestamps on links 1 and 2, respectively ($0 \leq m \leq \delta_1, 0 \leq n \leq \delta_2$). These sets of timestamps are matched against each other sequentially by assessing the difference $s_2(n) - s_1(m)$. If $s_2(n) - s_1(m) \leq \Delta$, where Δ is a predefined maximum delay, and if at the same time this difference is nonnegative, then the two timestamps match. The matching timestamps from the two connections form a pair of sequences $(\boldsymbol{f}_1, \boldsymbol{f}_2)$. Given the number of matching timestamps on a link, the empirical rate on that link can be now estimated. Specifically, if $\boldsymbol{f}_i$ contains $|F_i|$ timestamps over time T, then the empirical rate is $|F_i|/T$.

Let $\tau \geq 0$ be a given rate. Given a graph sourced at j, by repeatedly applying the matching timestamps algorithm for pairs of adjacent links, and selecting only the flows which support rate at least τ, a sub-graph of the initial graph can be obtained. This new graph will contain only nodes to which node j can talk at a rate higher or equal to τ. The method, called Trace Destination (TD), is applied for the graph sourced at O and a set of possible destinations is obtained. However, node O can be only a relay node for some of these flows. In order to trace down the flows that are going through O, and did not originate in O, TD is applied for all immediate predecessors of O and the corresponding graphs sourced at these nodes. The set of obtained destinations is then subtracted from the previously obtained set of destinations of O.

To speed up the algorithm's convergence, a second feature is used, that of the changing topology. Basically, the algorithm TD is applied for every communication graph that is observed. For each observed graph, a set of destinations if obtained. The final set of possible destinations is the intersection over all sets from all observed graphs. Out of this final set, the node which appears most frequently in all topologies is selected. If there are more possible destinations with the same number of appearances across all different topologies, then one node is selected at random.

4.2 Analysis

Our tracing algorithm is a variation of the one from section 4.1. in which we account for the lack of information in the observations. We model the lack of information by considering that the information on each link is available with a certain probability. We denote this probability by $p_k, k \in \{1,2,...,K\}$, where K is the number of links in the considered path. In this modified algorithm, only flows which support rate at least $r_k / p_k \geq \tau$ are selected from the initial graph.

It is interesting now to study how well our variation of the algorithm from section 4.1 converges. We thus study the probability P_e with which the algorithm finds the destination erroneously (called error probability) and analyze its asymptotic behavior over time T. We would like to mention that even though we assume Poisson distribution of packets on a link for our specific derivation, the algorithm and analysis here can be applied to different types of traffic.

Our error probability derivation is also inspired from [1]. Assume that the transmission activities on each link follow a Poisson process of rate $R < 1$. It is well-known that the Poisson distribution of the number of events in a time interval $(t, t+t']$ is given by the following relation:

$$P[(m_{t+t'} - m_t) = n] = \frac{e^{-Rt'}(Rt')^n}{n!},\ n = 0,1,...,$$

where $m_{t+t'} - m_t$ represents the number of events in time interval $(t, t+t']$ (see for example [11]).

Let O be the source whose destination we are trying to find and let θ be the true destination of O. We denote the destination rendered by our algorithm with $\hat{\theta}$. Following, we derive the probability that $\hat{\theta}$ is not the true destination of O. Let n_k be the number of timestamps taken in time interval (m_k, m_{k+1}), corresponding to link $(k, k+1)$. We denote the total number of timestamps taken for one flow by N_1.

By definition,

$$\begin{aligned} P_e &:= P_e(TD) = P(\theta \neq \hat{\theta}) \\ &= P(\theta \text{ is not correctly detected by algorithm TD}). \end{aligned}$$

According to [1], Theorem 4.4, the fact that the algorithm does not find the true destination is due to one of three possibilities:

A) θ is not detected because the empirical rate along the path from O to θ is less than τ; we will denote the probability of this event by $P(A)$;

B) θ is mistaken for a relay node because there is some node j which is a successor of θ, for which the empirical rate from O to j is greater than τ. We denote the probability of this event by $P(B)$;

C) θ is incorrectly detected as being the destination of a flow originating in some predecessor of O. We denote the probability of this event by $P(C)$;

In both B) and C) cases, all the empirical probabilities are greater or equal to τ . So

$$P_e \leq P(A) + P(B) + P(C) \tag{7}$$

Following we establish an upper bound for $P(B)$. The empirical rate on the path from O to j is given by $N_1(T)/T$.

$$P(B) \leq Ke^{-\frac{N_1(T)}{T}T},$$

where K is a constant related to $\mathrm{card}(G)$.

$$\frac{N_1(T)}{T} \geq r_k \geq p_k \tau \text{ , for all } k$$

$$\Rightarrow \frac{N_1(T)}{T} \geq \tau \frac{\sum p_k}{\#\text{of links}} = \tau \tilde{p}.$$

Using the fact that $-\frac{N_1(T)}{T} \leq -\tau \tilde{p}$, we obtain:

$$P(B) \leq Ke^{-N_1(T)} \leq Ke^{-\tau T \tilde{p}} .$$

Similarly, for $P(C)$ we have: $P(C) \leq Ke^{-\tau T \hat{p}}$.

For $P(A)$, the assumption is that there is at least one link along the path from O to θ with empirical rate less than τ . Suppose j_0 is the first node on the path connected to a link with rate smaller than τ . Therefore all links along the path up to j_0 have rate greater or equal to τ . Let N_1 be the total number of timestamps taken during the observation period T . We have:

We denote by λ the rate of the Poisson process defined by the recorded timestamps. Then the rate on an arbitrary interval $[m_k, m_{k+1}]$ is $\lambda(m_{k+1} - m_k)$.

$$P(A) \leq \prod_{k=0}^{j_0-1} \frac{[r_k(m_{k+1} - m_k)]^{n_k}}{n_k!} e^{-r_k(m_{k+1}-m_k)} \prod_{k' \in N^+(j_0)} e^{-r_{k'}(m_{k'+1}-m_{k'})}.$$

$N^+(j_0)$ denotes all successors of node j_0 on links with rate at least R . Notice that in the first product $r_k \geq p_k \tau$ for all k , while all $r_{k'}$ from the second product are less than $p_k \tau$. On the other hand, $r_{k'} \geq (N_1(T) - N(j_0))/(T - m(j_0))$. We thus have:

$$P(A) \leq C_{j_0} \exp[-\tau(\sum_{k=0}^{j_0-1} p_k(m_{k+1} - m_k)]\exp[-\frac{N_1(T) - N(j_0)}{T - m(j_0)}$$

$$(T - m(j_0)] = C_{j_0} \exp(N(j_0)\exp[-\tau(\sum_{k=0}^{j_0-1} p_k(m_{k+1} - m_k)]$$

$$\exp(-N_1(T)) = K_1(j_0, \tau)\exp(-N_1(T))$$

$$= K_1(j_0, \tau)\exp(-T\rho(T)),$$

where

$$C(j_0)=\prod_{k=0}^{j_0-1}\frac{r_k(m_{k+1}-m_k)^{n_k}}{n_k!} \text{ and } K'(j_0,\tau)=C(j_0)e^{N(j_0)-\tau m_{j_0}}.$$

Here we let $\rho=\lim_{T\to\infty} N_1(T)/T>0$.

Therefore $P(A)$ decreases exponentially with the observation time.

Using equation (7), we now have:

$$\begin{aligned} P_e &\le K'(j_0,\tau)e^{-\rho T}+\text{card}(G)e^{-\tau T\tilde{p}}+\text{card}(G)e^{-\tau T\hat{p}} \\ &= e^{-T\tau\min\{\tilde{p},\hat{p}\}}(K'(j_0,\tau)+2K). \end{aligned}$$

We can now investigate the convergence rate of our algorithm by showing the asymptotic decay rate of the error probability as time increases.

By logarithming the above inequality we obtain:

$$\lim_{T\to\infty}(-\ln P_e/T)\ge\tau\min(\tilde{p},\hat{p})>0.$$

Compared to the upper bound for the error of algorithm TD, which is τ (corresponding to the case $p_k=1$ for all k), the modified algorithm has upper bound: $\tau\min(\tilde{p},\hat{p})$.

The minimum is strictly positive as the minimum of a finite number of strictly positive values.

We can see that the algorithm has an exponential convergence rate.

5 Conclusion

We provided a way to shorten the time interval used for thumbprinting and tune it suitably depending on network conditions. We have found interesting properties of the thumbprinting function using the mean value and provided a general method to compute the value of the function. The method works for any function that satisfies certain properties. We have also studied another mechanism for tracing intruders in the Internet, based on an approach of passively monitoring flows between source and destination pairs. We computed the error probability of our algorithm and show that its value decreases exponentially as the observation time increases.

Acknowledgments. This work is supported in part by the US National Science Foundation (NSF) under the grant numbers CCF-0829827, CNS-0716211, and CNS-0737325.

References

1. He, T., Wong, H.Y., Lee, K.-W.: Traffic Analysis in Anonymous MANETs. In: Proc. IEEE MILCOM, San Diego, pp. 1–7 (2008)
2. Heberlein, L.T., Levitt, K., Mukherjee, B.: Internetwork Security Monitor: An Intrusion-Detection System for Large Scale Networks. In: Proc. 15th National Computer Security Conference, pp. 262–271 (1992)

3. Tae, H., Kim, H.L., Seo, Y.M., Choe, G., Min, S.L., Kim, C.S.: Caller Identification System in the Internet Environment. In: Proc. of 4th USENIX Security Symposium, pp. 69–78 (1993)
4. Staniford-Chen, S., Heberlein, L.T.: Holding Intruders Accountable on the Internet. In: Proc. the 1995 IEEE Symposium on Security and Privacy, Oakland, CA, pp. 39–49 (1995)
5. Wadell, S.: Private Communication (1994)
6. Implicit Function Theorem, `http://en.wikipedia.org/wiki/Implicit_function_theorem`
7. Internet, `http://en.wikipedia.org/wiki/Internet`
8. Jensen's Inequality, `http://en.wikipedia.org/wiki/Jensen'sinequality`
9. Least Squares, `http://en.wikipedia.org/wiki/Least_squares`
10. Mean Value Theorem, `http://en.wikipedia.org/wiki/Mean_value_theorem`
11. Poisson Process, `http://en.wikipedia.org/wiki/Poisson_proce`
12. Xiao, Y.: Editorial. International Journal of Security and Networks 1(1/2), 1 (2006)
13. Shehab, M., Bertino, E., Ghafoor, A.: Workflow Authorization in Mediator-free Environments. International Journal of Security and Networks 1(1/2), 2–12 (2006)
14. Jung, E., Gouda, M.G.: Vulnerability Analysis of Certificate Graphs. International Journal of Security and Networks 1(1/2), 13–23 (2006)
15. Kiayias, A., Yung, M.: Secure Scalable Group Signature with Dynamic Joins and Separable Authorities. International Journal of Security and Networks 1(1/2), 24–45 (2006)
16. Franklin, M.: A Survey of Key Evolving Cryptosystems. International Journal of Security and Networks 1(1/2), 46–53 (2006)
17. Hamadeh, I., Kesidis, G.: A Taxonomy of Internet Traceback. International Journal of Security and Networks 1(1/2), 54–61 (2006)
18. Jhumka, A., Freiling, F., Fetzer, C., Suri, N.: An Approach to Synthesize Safe Systems. International Journal of Security and Networks 1(1/2), 62–74 (2006)
19. Evans, J.B., Wang, W., Ewy, B.J.: Wireless Networking Security: Open Issues in Trust, Management, Interoperation and Measurement. International Journal of Security and Networks 1(1/2), 84–94 (2006)
20. Englund, H., Johansson, T.: Three Ways to Mount Distinguishing Attacks on Irregularly Clocked Stream Ciphers. International Journal of Security and Networks 1(1/2), 95–102 (2006)
21. Zhu, B., Jajodia, S., Kankanhalli, M.S.: Building Trust in Peer-to-Peer Systems: A Review. International Journal of Security and Networks 1(1/2), 103–112 (2006)
22. Ramkumar, M., Memon, N.: Secure Collaborations Over Message Boards. International Journal of Security and Networks 1(1/2), 113–124 (2006)
23. Xiao, Y., Jia, X., Sun, B., Du, X.: Editorial: Security Issues on Sensor Networks. International Journal of Security and Networks 1(3/4), 125–126 (2006)
24. Wang, H., Sheng, B., Li, Q.: Elliptic Curve Cryptography-based Access Control. International Journal of Security and Networks 1(3/4), 127–137 (2006)
25. Zheng, J., Li, J., Lee, M.J., Anshel, M.: A Lightweight Encryption and Authentication Scheme for Wireless Sensor Networks. International Journal of Security and Networks 2006 1(3/4), 138–146 (2006)
26. Al-Karaki, J.N.: Analysis of Routing Security-Energy Trade-offs in Wireless Sensor Networks. International Journal of Security and Networks 1(3/4), 147–157 (2006)

27. Araz, O., Qi, H.: Load-balanced Key Establishment Methodologies in Wireless Sensor Networks. International Journal of Security and Networks 1(3/4), 158–166 (2006)
28. Deng, J., Han, R., Mishra, S.: Limiting DoS Attacks During Multihop Data Delivery in Wireless Sensor Networks. International Journal of Security and Networks 1(3/4), 167–178 (2006)
29. Hwu, J., Hsu, S., Lin, Y.-B., Chen, R.: End-to-End Security Mechanisms for SMS. International Journal of Security and Networks 1(3/4), 177–183 (2006)
30. Wang, X.: The Loop Fallacy and Deterministic Serialisation in Tracing Intrusion Connections through Stepping Stones. International Journal of Security and Networks 1(3/4), 184–197 (2006)
31. Jiang, Y., Lin, C., Shi, M., Shen, X.: A Self-Encryption Authentication Protocol for Teleconference Services. International Journal of Security and Networks 1(3/4), 198–205 (2006)
32. Owens, S.F., Levary, R.R.: An Adaptive Expert System Approach for Intrusion Detection. International Journal of Security and Networks 1(3/4), 206–217 (2006)
33. Chen, Y., Susilo, W., Mu, Y.: Convertible Identity-based Anonymous Designated Ring Signatures. International Journal of Security and Networks 1(3/4), 218–225 (2006)
34. Teo, J., Tan, C., Ng, J.: Low-power Authenticated Group Key Agreement for Heterogeneous Wireless Networks. International Journal of Security and Networks 1(3/4), 226–236 (2006)
35. Tan, C.: A New Signature Scheme without Random Oracles. International Journal of Security and Networks 1(3/4), 237–242 (2006)
36. Liu, Y., Comaniciu, C., Man, H.: Modelling Misbehaviour in Ad Hoc Networks: A Game Theoretic Approach for Intrusion Detection. International Journal of Security and Networks 1(3/4), 243–254 (2006)
37. Karyotis, V., Papavassiliou, S., Grammatikou, M., Maglaris, V.: A Novel Framework for Mobile Attack Strategy Modelling and Vulnerability Analysis in Wireless Ad Hoc Networks. International Journal of Security and Networks 1(3/4), 255–265 (2006)
38. Chen, H., Guizani, M.: Editorial. International Journal of Security and Networks 2(1/2), 1–2 (2007)
39. Li, R., Li, J., Chen, H.: DKMS: Distributed Hierarchical Access Control for Multimedia Networks. International Journal of Security and Networks 2(1/2), 3–10 (2007)
40. Sakarindr, P., Ansari, N.: Adaptive trust-based anonymous network. International Journal of Security and Networks 2(1/2), 11–26 (2007)
41. Malaney, R.A.: Securing Wi-Fi Networks with Position Verification: Extended Version. International Journal of Security and Networks 2(1/2), 27–36 (2007)
42. Sun, F., Shayman, M.A.: On Pairwise Connectivity of Wireless Multihop Networks. International Journal of Security and Networks 2(1/2), 37–49 (2007)
43. Erdogan, O., Cao, P.: Hash-AV: Fast Virus Signature Scanning by Cache-Resident Filters. International Journal of Security and Networks 2(1/2), 50–59 (2007)
44. Rabinovich, P., Simon, R.: Secure Message Delivery in Publish/Subscribe Networks using Overlay Multicast. International Journal of Security and Networks 2(1/2), 60–70 (2007)
45. Chen, Z., Ji, C.: Optimal worm-scanning method using vulnerable-host distributions. International Journal of Security and Networks 2(1/2), 71–80 (2007)
46. Pan, J., Cai, L., Shen, X.: Vulnerabilities in Distance-indexed IP Traceback Schemes. International Journal of Security and Networks 2(1/2), 81–94 (2007)

47. Korkmaz, T., Gong, C., Sarac, K., Dykes, S.G.: 8 Single Packet IP Traceback in AS-Level Partial Deployment Scenario. International Journal of Security and Networks 2(1/2), 95–108 (2007)
48. Ling, H., Znati, T.: End-to-end Pairwise Key Establishment using Node Disjoint Secure Paths in Wireless Sensor Networks. International Journal of Security and Networks 2(1/2), 109–121 (2007)
49. Artan, N.S., Chao, H.J.: Design and Analysis of A Multipacket Signature Detection System. International Journal of Security and Networks 2(1/2), 122–136 (2007)
50. Zhu, Y., Fu, X., Bettati, R., Zhao, W.: Analysis of Flow-correlation Attacks in Anonymity Network. International Journal of Security and Networks 2(1/2), 137–153 (2007)
51. Gu, Q., Liu, P., Chu, C., Zhu, S.: Defence Against Packet Injection in Ad Hoc Networks. International Journal of Security and Networks 2(1/2), 154–169 (2007)
52. Mu, Y., Chen, L., Chen, X., Gong, G., Lee, P., Miyaji, A., Pieprzyk, J., Pointcheval, D., Takagi, T., Traore, J., Seberry, J., Susilo, W., Wang, H., Zhang, F.: Editorial. International Journal of Security and Networks 2(3/4), 171–174 (2007)
53. Tartary, C., Wang, H.: Efficient Multicast Stream Authentication for the Fully Adversarial Network Model. International Journal of Security and Networks 2(3/4), 175–191 (2007)
54. Bhaskar, R., Herranz, J., Laguillaumie, F.: Aggregate Designated Verifier Signatures and Application to Secure Routing. International Journal of Security and Networks 2(3/4), 192–201 (2007)
55. Hsu, H., Zhu, S., Hurson, A.R.: LIP: A Lightweight Interlayer protocol for Preventing Packet Injection Attacks in Mobile Ad Hoc Network. International Journal of Security and Networks 2(3/4), 202–215 (2007)
56. Oliveira, L.B., Wong, H., Loureiro, A.A.F., Dahab, R.: On the Design of Secure Protocols for Hierarchical Sensor Networks. International Journal of Security and Networks 2(3/4), 216–227 (2007)
57. Michail, H.E., Panagiotakopoulos, G.A., Thanasoulis, V.N., Kakarountas, A.P., Goutis, C.E.: Server Side Hashing Core Exceeding 3 Gbps of Throughput. International Journal of Security and Networks 2(3/4), 228–238 (2007)
58. Hoeper, K., Gong, G.: Preventing or Utilizing Key Escrow in Identity-based Schemes Employed in Mobile Ad Hoc Networks. International Journal of Security and Networks 2(3/4), 239–250 (2007)
59. Cheng, Z., Chen, L.: On Security Proof of McCullagh–Barreto's Key Agreement Protocol and Its Variants. International Journal of Security and Networks 2(3/4), 251–259 (2007)
60. Finnigin, K.M., Mullins, B.E., Raines, R.A., Potoczny, H.B.: Cryptanalysis of An Elliptic Curve Cryptosystem for Wireless Sensor Networks. International Journal of Security and Networks 2(3/4), 260–271 (2007)
61. Huang, D.: Pseudonym-based Cryptography for Anonymous Communications in Mobile Ad Hoc Networks. International Journal of Security and Networks 2(3/4), 272–283 (2007)
62. Abdalla, M., Bresson, E., Chevassut, O., Moller, B., Pointcheval, D.: Strong Password-based Authentication in TLS Using the Three-party Group Diffie–Hellman Protocol. International Journal of Security and Networks 2(3/4), 284–296 (2007)
63. Kotzanikolaou, P., Vergados, D.D., Stergiou, G., Magkos, E.: Multilayer Key Establishment for Large-scale Sensor Networks. International Journal of Security and Networks 3(1), 1–9 (2008)

64. Wang, W., Kong, J., Bhargava, B., Gerla, M.: Visualization of Wormholes in Underwater Sensor Networks: A Distributed Approach. International Journal of Security and Networks 3(1), 10–23 (2008)
65. Scheirer, W., Chuah, M.: Syntax vs. Semantics: Competing Approaches to Dynamic Network Intrusion Detection. International Journal of Security and Networks 3(1), 24–35 (2008)
66. Burt, A.L., Darschewski, M., Ray, I., Thurimella, R., Wu, H.: Origins: An Approach to Trace Fast Spreading Worms to Their Roots. International Journal of Security and Networks 3(1), 36–46 (2008)
67. Zou, X., Karandikar, Y.: A Novel Conference Key Management Solution for Secure Dynamic Conferencing. International Journal of Security and Networks 3(1), 47–53 (2008)
68. Asadpour, M., Sattarzadeh, B., Movaghar, A.: Anonymous Authentication Protocol for GSM Networks. International Journal of Security and Networks 3(1), 54–62 (2008)
69. Hu, F., Rughoonundon, A., Celentano, L.: Towards a Realistic Testbed for Wireless Network Reliability and Security Performance Studies. International Journal of Security and Networks 3(1), 63–77 (2008)
70. Memon, N., Goel, R.: Editorial. International Journal of Security and Networks 3(2), 79 (2008)
71. Ray, I., Poolsappasit, N.: Using Mobile Ad Hoc Networks to Acquire Digital Evidence from Remote Autonomous Agents. International Journal of Security and Networks 3(2), 80–94 (2008)
72. Kilpatrick, T., Gonzalez, J., Chandia, R., Papa, M., Shenoi, S.: Forensic Analysis of SCADA Systems and Networks. International Journal of Security and Networks 3(2), 95–102 (2008)
73. Cronin, E., Sherr, M., Blaze, M.: On the (Un)reliability of Eavesdropping. International Journal of Security and Networks 3(2), 103–113 (2008)
74. Okolica, J.S., Peterson, G.L., Mills, R.F.: Using PLSI-U to Detect Insider Threats by Datamining E-mail. International Journal of Security and Networks 3(2), 114–121 (2008)
75. Lin, X., Ling, X., Zhu, H., Ho, P., Shen, X.: A Novel Localised Authentication Scheme in IEEE 802.11 based Wireless Mesh Networks. International Journal of Security and Networks 3(2), 122–132 (2008)
76. Challal, Y., Gharout, S., Bouabdallah, A., Bettahar, H.: Adaptive Clustering for Scalable Key Management in Dynamic Group Communications. International Journal of Security and Networks 3(2), 133–146 (2008)
77. Xu, H., Ayachit, M., Reddyreddy, A.: Formal Modelling and Analysis of XML Firewall for Service-oriented Systems. International Journal of Security and Networks 3(3), 147–160 (2008)
78. Bouhoula, A., Trabelsi, Z., Barka, E., Benelbahri, M.: Firewall Filtering Rules Analysis for Anomalies Detection. International Journal of Security and Networks 3(3), 161–172 (2008)
79. Li, F., Srinivasan, A., Wu, J.: PVFS: A Probabilistic Voting-based Filtering Scheme in Wireless Sensor Networks. International Journal of Security and Networks 3(3), 173–182 (2008)
80. Ma, X., Cheng, X.: Verifying Security Protocols by Knowledge Analysis. International Journal of Security and Networks 3(3), 183–192 (2008)

81. Uphoff, B., Wong, J.S.: An Agent-based Framework for Intrusion Detection Alert Verification and Event Correlation. International Journal of Security and Networks 3(3), 193–200 (2008)
82. Tripathy, S., Nandi, S.: Secure User-identification and Key Distribution Scheme Preserving Anonymity. International Journal of Security and Networks 3(3), 201–205 (2008)
83. Li, F., Xin, X., Hu, Y.: ID-based Threshold Proxy Signcryption Scheme from Bilinear Pairings. International Journal of Security and Networks 3(3), 206–215 (2008)
84. Ma, L., Teymorian, A.Y., Xing, K., Du, D.: An One-way Function Based Framework for Pairwise Key Establishment in Sensor Networks. International Journal of Security and Networks 3(4), 217–225 (2008)
85. Srinivasan, A., Li, F., Wu, J., Li, M.: Clique-based Group Key Assignment in Wireless Sensor Networks. International Journal of Security and Networks 3(4), 226–239 (2008)
86. Hsieh, C., Chen, J., Lin, Y.-B., Chen, K., Liao, H., Liang, C.: NTP-DownloadT: A Conformance Test Tool for Secured Mobile Download Services. International Journal of Security and Networks 3(4), 240–249 (2008)
87. Sadowitz, M., Latifi, S., Walker, D.: An Iris and Retina Multimodal Biometric System. International Journal of Security and Networks 3(4), 250–257 (2008)
88. Kandikattu, R., Jacob, L.: Secure Hybrid Routing with Micro/Macro-mobility Handoff Mechanisms for Urban Wireless Mesh Networks. International Journal of Security and Networks 3(4), 258–274 (2008)
89. Mayrhofer, R., Nyberg, K., Kindberg, T.: Foreword. International Journal of Security and Networks 4(1/2), 1–3 (2009)
90. Scannell, A., Varshavsky, A., LaMarca, A., De Lara, E.: Proximity-based Authentication of Mobile Devices. International Journal of Security and Networks 4(1/2), 4–16 (2009)
91. Soriente, C., Tsudik, G., Uzun, E.: Secure Pairing of Interface Constrained Devices. International Journal of Security and Networks 4(1/2), 17–26 (2009)
92. Buhan, I., Boom, B., Doumen, J., Hartel, P.H., Veldhuis, R.N.J.: Secure Pairing with Biometrics. International Journal of Security and Networks 4(1/2), 27–42 (2009)
93. McCune, J.M., Perrig, A., Reiter, M.K.: Seeing-Is-Believing: Using Camera Phones for Human-verifiable Authentication. International Journal of Security and Networks 4(1/2), 43–56 (2009)
94. Goodrich, M.T., Sirivianos, M., Solis, J., Soriente, C., Tsudik, G., Uzun, E.: Using Audio in Secure Device Pairing. International Journal of Security and Networks 4(1/2), 57–68 (2009)
95. Laur, S., Pasini, S.: User-aided Data Authentication. International Journal of Security and Networks 4(1/2), 69–86 (2009)
96. Suomalainen, J., Valkonen, J., Asokan, N.: Standards for Security Associations in Personal Networks: A Comparative Analysis. International Journal of Security and Networks 4(1/2), 87–100 (2009)
97. Kuo, C., Perrig, A., Walker, J.: Designing User Studies for Security Applications: A Case Study with Wireless Network Configuration. International Journal of Security and Networks 409(1/2), 101–109 (2009)
98. Berthier, R., Cukier, M.: An Evaluation of Connection Characteristics for Separating Network Attacks. International Journal of Security and Networks 4(1/2), 110–124 (2009)
99. Wu, B., Wu, J., Dong, Y.: An Efficient Group Key Management Scheme for Mobile Ad Hoc Networks. International Journal of Security and Networks 4(1/2), 125–134 (2009)
100. Chen, Z., Chen, C., Li, Y.: Deriving a Closed-form Expression for Worm-scanning Strategies. International Journal of Security and Networks 4(3), 135–144 (2009)

101. Lee, S., Sivalingam, K.M.: An Efficient One-Time Password Authentication Scheme Using a Smart Card. International Journal of Security and Networks 4(3), 145–152 (2009)
102. Watkins, L., Beyah, R., Corbett, C.: Using Link RTT to Passively Detect Unapproved Wireless Nodesd. International Journal of Security and Networks 4(3), 153–163 (2009)
103. Drakakis, K.E., Panagopoulos, A.D., Cottis, P.G.: Overview of Satellite Communication Networks Security: Introduction of EAP. International Journal of Security and Networks 4(3), 164–170 (2009)
104. Chakrabarti, S., Chandrasekhar, S., Singhal, M.: An Escrow-less Identity-based Group-key Agreement Protocol for Dynamic Peer Groups. International Journal of Security and Networks 4(3), 171–188 (2009)
105. Ehlert, S., Rebahi, Y., Magedanz, T.: Intrusion Detection System for Denial-of-Service Flooding Attacks in SIP Communication Networks. International Journal of Security and Networks 4(3), 189–200 (2009)
106. Bai, L., Zou, X.: A Proactive Secret Sharing Scheme in Matrix Projection Method. International Journal of Security and Networks 4(4), 201–209 (2009)
107. Bettahar, H., Alkubeily, M., Bouabdallah, A.: TKS: A Transition Key Management Scheme for Secure Application Level Multicast. International Journal of Security and Networks 4(4), 210–222 (2009)
108. Huang, H., Kirchner, H., Liu, S., Wu, W.: Handling Inheritance Violation for Secure Interoperation of Heterogeneous Systems. International Journal of Security and Networks 4(4), 223–233 (2009)
109. Rekhis, S., Boudriga, N.A.: Visibility: A Novel Concept for Characterizing Provable Network Digital Evidences. International Journal of Security and Networks 4(4), 234–245 (2009)
110. Djenouri, D., Bouamama, M., Mahmoudi, O.: Black-hole-resistant ENADAIR-based Routing Protocol for Mobile Ad Hoc Networks. International Journal of Security and Networks 4(4), 246–262 (2009)
111. Hu, F., Dong, D., Xiao, Y.: Attacks and Countermeasures in Multi-hop Cognitive Radio Networks. International Journal of Security and Networks 4(4), 263–271 (2009)

A Sparse Bayesian Framework for Anomaly Detection in Heterogeneous Networks

Jian Zhang and Rajgopal Kannan

Dept. Computer Science
Louisiana State University
Baton Rouge, LA 70803
{zhang,rkannan}@csc.lsu.edu

Abstract. The capability to detect anomalous states in a network is important for both the smooth operation of the network and the security of the network. Modern networks are often heterogeneous. This raises a new challenge for anomaly detection, as there may be a wide variety of anomalous activities across the heterogeneous components of a network. We often seek a detection system that not only performs accurate anomaly detection but also provides mechanisms for human expert to understand the decision making process inside the system. In this paper, we investigate the application of sparse Bayesian methods for anomaly detection in such scenario. By taking advantage of the sparse Bayesian framework's capability to conduct automatic relevance discovery, we construct a detection system whose decision making is mostly based on a few representative examples from the training set. This provides human interpretability as expert can analyze the representative examples to understand the detection mechanism. Our experiment results show the potential of this approach.

Keywords: Anomaly Detection, Sparse Bayesian Classification.

1 Introduction

In a complex system, it is crucial to monitor the system activities for the normal operation of the system. Techniques that can identify anomalous state in a system have been the subject of many researches [7,1,11,3]. Machine learning methods are a particularly promising approach to the problem. However, in this case, it is often not enough to just classify a system event or activity to be normal or abnormal. We also want to know why the event is abnormal. And when there are many different types of abnormal events, we want to know what the types are. This is important because detecting the anomaly is just a start. To deal with the problem, e.g., to recover from the abnormal state or to fight against security breach, one need to know more.

Many off-the-shelf machine learning methods lack the capability to explain the decision made by the methods. For example, given a particular trained neural network for anomaly detection, it is not easy to derive a human-understandable

X. Zhang and D. Qiao (Eds.): QShine 2010, LNICST 74, pp. 75–87, 2011.

model on how the neural network judges some event to be normal and the others to be abnormal. On the other hand, in many cases, the detection problem can be complex. There may be many different types of anomalies. And we may need to combine many features of an event to specify the anomaly. In these scenarios, human analysis is often needed to explain the anomalous event and identify the type of the anomaly. However, automated data analysis program can still help. In particular, we can use automated programs to select a few "representative" cases and present them to the human experts. The expert then analyze the "representative" cases and build models for the (different types of) anomalies. In this way, automated analysis facilitates human analysis by reducing the amount of cases that the experts need to look through.

In this paper, we propose a sparse Bayes framework for anomaly detection. Given a set of normal and abnormal examples, we build a classifier that detects future anomalies. The classification is based on a very small set of examples and these examples can be viewed as the "representative" cases of the abnormal activities. By analyzing this small set of examples, a human expert may identify the mechanism that causes the anomaly.

One way to build a small set of "representative" cases from a collection is to perform clustering. Clustering groups the data into clusters and the cluster centers can serve as the "representative" cases. However, several factors make clustering less applicable in our scenario: 1) We may not know how many number of clusters exist in our collection of events. 2) In many cases, the "center" of a cluster is not well-defined when the attributes describing the event involves nominal attributes. For example, if a cluster contains some events that involves TCP as the connection protocol and some other involves UDP. There is no center that "averages" over the two protocols. 3) We seek "representative" cases that not only represent the data in the collection, but also can help in making the distinction between the abnormal and the normal. i.e., The "representative" cases should play an important role in the classification of the events.

In the sparse Bayes framework, the search for the representative cases (events) is performed while the classifier for anomaly detection is being constructed. The representative cases (events) are also the ones that are important for the classification. In fact, to classify an unknown event, one only needs to compare the event to the representative events and the result of such comparison decides whether the unknown event is normal or abnormal. When the representative events can lead to accurate classification, they represent well the data in the collection of examples.

We test our sparse Bayes framework for anomaly detection using the KDD intrusion detection dataset. Experiment results show that a classifier constructed following the sparse Bayes framework gives accurate classification between the normal and the abnormal events. The performance of the classifier is at the same level as that of a classifier based on support vector machine. Our classification framework also generates a very small set of representative cases, which human expert can analyze to understand the anomaly or determine the types of the anomaly.

The rest of the paper is organized as follows: In Section 2, we discuss related work. In Section 3, we give a detailed description of our sparse Bayesian framework for anomaly detection. In Section 4, we present and discuss our experiment results. We conclude the paper in Section 5.

2 Related Work

Anomaly detection has been the subject of many researches. There are excellent surveys on the problem, the approaches considered and the applications. We refer to [7,1,11,3] for comprehensive elaboration on the topic. Intrusion detection is a major application area of anomaly detection and many systems have been proposed, for example, [13,12,2]. Statistical models and machine learning methods are heavily employed in anomaly and intrusion detection systems.

For example, one popular approach in intrusion detection is to use rule-based systems. A set of rules specifying abnormal events (intrusions) is extracted and new intrusion events are identified by matching them to these rules. The rules can be extracted either manually, or through data mining or machine learning techniques. Decision trees [10] and fuzzy logic [5] have been used for intrusions detection. Rule learning algorithms such as RIPPER have also been tested for such purpose [14]. The advantage of these techniques is that the rules they use to make decisions can be easily interpreted by humans. Pure machine learning techniques such as neural networks [8] and support vector machines [4] are also employed in intrusion and anomaly detection.

Sparse Bayesian learning is a learning technique developed by Tipping [15,6]. The goal of this framework is to construct a parsimonious model that can give good prediction. The framework achieves this by select features using an automatic relevance detection technique. A classifier constructed in this framework is called a relevance vector machine. The relevance vector machine has been applied to help dynamically tracking faces in video sequences [16].

3 Sparse Bayesian Framework for Anomaly Detection

We specify a network event as a vector of attributes. Some are numerical and the others are nominal. We have a collection of N examples of normal and abnormal events, which we denote as $\{x_i, t_i\}^N$, where $x_i = \{x_i^1, x_i^2, \ldots, x_i^k\}$ is the vector of k attributes describing the i-th example event. And t_i is the class label (normal or abnormal) of the event.

Our goal is to construct a classifier to classify future unknown events. Normally, the classification would be based on a function of attributes. For example, one may learn a function in the form of $\sum_j w_j x^j + b$ and classifies the unknown event x by the sign of the function values. Our framework takes a different approach. The classification function is based on the similarity measure between the unknown event and the events in the training collection. Our approach bears some similarity to the nearest-neighbor classification. However, instead of compare the unknown event to the nearest neighbors, we compare it to a small set of examples that are selected according to a statistical process that automatically discovers the relevance between the events.

3.1 Similarity Based on Normalized Features

The first step in our sparse Bayesian framework is to calculate similarity between network events. Given two network event x_i and x_j, we measure the similarity between the two event by the "extended" dot product of the two corresponding feature vectors. i.e., $S(x_i, x_j) = \sum_k x_i^k \circ x_j^k$. We call it the "extended" dot product because some features are nominal and the others are numerical. For a numerical feature (x_i^k and x_j^k), $x_i^k \circ x_j^k$ is the product of x_i^k and x_j^k. When the feature is nominal, we define

$$x_i^k \circ x_j^k = \begin{cases} 1 & \text{if } x_i^k = x_j^k; \\ 0 & \text{otherwise;} \end{cases} \tag{1}$$

One problem in calculating similarity in such fashion is that the result may be dominated by one or a few features that has very large values compare to the other features. (In fact, this is a common problem for similarity calculation, not just for our extended dot product.) To deal with this situation, we normalize the features before performing the extended dot product. Clearly, only the numerical features require normalization. Consider the vector of values for the k-th feature $k = \{x_1^k, x_2^k, \ldots, x_N^k\}$. We view that the values follow a normal distribution. We then transform each value x_i^k to be

$$\phi(x_i^k) = \frac{x_i^k - \mu_k}{\sigma_k}. \tag{2}$$

where $\mu_k = \frac{1}{N}\sum_i x_i^k$ is the mean of the values for the k-th feature and σ_k is the standard deviation of the values. Once the feature values are normalized, the similarity calculation is done with the transformed values. Let O be the set of nominal features and U be the set of numerical features, we have

$$S(i,j) = \sum_{k \in U} \phi(x_i^k)\phi(x_j^k) + \sum_{k \in O} x_i^k \circ x_j^k. \tag{3}$$

3.2 Sparse Bayesian Classification

We follow a logistic regression model to define the probability of an event being normal or abnormal. Given the similarity measure S and a network event x, let t be the type ("normal" and "abnormal") of the event, we define the probability of the event being normal or abnormal to be

$$p(t = \text{“abnormal”}|x, w) = \frac{1}{1 + \exp(\sum_i w_i S(x, x_i) + w_0)}$$

and

$$p(t = \text{“normal”}|x, w) = 1 - p(t = \text{“normal”}|x, w) = \frac{\exp(\sum_i w_i S(x, x_i) + w_0)}{1 + \exp(\sum_i w_i S(x, x_i) + w_0)}.$$

$w = \{w_0, w_1, \ldots, w_N\}$ is a vector of parameters for the probability model. In a general logistic regression model, the probability is defined as a function of features, i.e., it involves a quantity in the form of $\sum_k \hat{w}_k x^k$. Our model differs from the standard logistic regression in that the probability is defined as a function of training instances. Therefore, each w_i, $i \neq 0$ corresponds to an instance (event) in the training set (rather than a feature as is in a standard logistic regression model). If we encode the "abnormal" case as $t = 1$ and the "normal" case as $t = 0$, we have the following likelihood of the event given the parameter vector:

$$p(t|x,w) = \left(\frac{1}{1+\exp(\sum_i w_i S(x,x_i) + w_0)}\right)^t \left(1 - \frac{1}{1+\exp(\sum_i w_i S(x,x_i) + w_0)}\right)^{(1-t)} \quad (4)$$

Let $y(x) = \frac{1}{1+\exp(\sum_i w_i S(x,x_i)+w_0)}$, the likelihood can be simplified as $y(x)^t(1 - y(x))^{(1-t)}$.

One may estimate the parameters using a maximum likelihood approach. In Bayesian framework, rather than selecting a particular vector of values for w, we often model the parameter vector as a random variable draw from a prior distribution $p(w)$. Note that in many cases, as is in this paper, it is not necessarily to use a distribution that matches exactly the actual prior distribution of the problem. We often treat the model prior as a way to encode our bias in choosing the model parameters.

Following the sparse Bayesian framework of [15], we consider a particular prior distribution for w. We model that the i-th parameter w_i is draw from a normal distribution with zero mean and variance a_i, i.e., $p(w_i) \sim N(0, a_i)$. The distribution for the whole parameter vector w is a multivariate normal distribution with zero mean and a covariance matrix A, i.e.,

$$p(w|A) = N(w|\mathbf{0}, A). \quad (5)$$

where $\mathbf{0}$ is a $(N+1)$-dimensional all zero vector and A is a diagonal matrix with the diagonal $A_{ii} = a_i$.

The normal and abnormal probability of a new network event x now can be calculated as

$$p(t|x, t_{\text{train}}, X_{\text{train}}, A) = \int p(t|x,w)p(w|t_{\text{train}}, X_{\text{train}}, A)dw \quad (6)$$

with $p(t|x,w)$ defined by Eq. 4 and $p(w|t_{\text{train}}, X_{\text{train}}, A)$ being the posterior distribution of w given the prior distribution $p(w|A)$ and the set of training example $\{t_i, x_i\}^N$. ($t_{\text{train}} = \{t_1, t_2, \ldots, t_N\}^t$ is the N-dimensional vector of class labels for the training examples. $X_{\text{train}} = \{S_{\text{train}}, \mathbf{1}\}$ is constructed by appending an N-dimensional all ones column vector to the right of the similarity matrix S_{train} of the training examples.) The probability of the new network event being normal or abnormal is determined by both the prior distribution (specified by A) and the training data.

3.3 Parameter Estimation

We follow the approach in [15] to estimate the parameters a_i. We give here a brief introduction to the parameter estimation process. Detailed description can be found in [15]. We first consider the posterior distribution of w, supposing that the values in A is given. Following the Bayes rule, we can write the posterior distribution as

$$p(w|t_{\text{train}}, X_{\text{train}}, A) = \frac{p(t_{\text{train}}|X_{\text{train}}, w)p(w|A)}{p(t_{\text{train}}|X_{\text{train}}, A)}.$$

Instead of considering the exact posterior distribution, we apply Laplace approximation to the distribution, i.e., we construct a normal distribution centered at the mode of the posterior distribution and use the normal distribution to approximate the posterior distribution.

The mode of the posterior distribution can be found by maximizing the log probability of the distribution:

$$\begin{aligned} &\ln p(w|t_{\text{train}}, X_{\text{train}}, A) \\ &= \ln p(t_{\text{train}}|X_{\text{train}}, w)p(w|A) - \ln p(t_{\text{train}}|X_{\text{train}}, A) \\ &= \sum_{i=1}^{N}\{t_i \ln y_i + (1-t_i)\ln(1-y_i)\} - \frac{1}{2}w^t Aw + C \end{aligned}$$

where $y_i = y(x_i) = \frac{1}{1+\exp(\sum_j w_j S(x_i,x_j)+w_0)}$ and C is some quantity not dependent on w.

Suppose A is known. The log probability is a function of w. To find the w that maximize the log probability, we set the gradient of the log probability to zero, i.e.,

$$\bigtriangledown \ln p(w|t_{\text{train}}, X_{\text{train}}, A) = X_{\text{train}}^t(t_{\text{train}} - y) - Aw = 0$$

where $y = \{y_1, y_2, \ldots, y_N\}^t$. This is a system of N nonlinear equations. One can obtain the solution (roots) of the equations numerically using Newton-Raphson iterative method, which leads to a type of iterative reweighted least square (IRLS) problem. In particular, the Hessan of the log likelihood is $\bigtriangledown\bigtriangledown \ln p(w|t_{\text{train}}, X_{\text{train}}, A) = -(X_{\text{train}}^t BX_{\text{train}} + A)$ where B is a diagonal matrix with $B_{ii} = y_i(1-y_i)$. The update rule for the Newton-Raphson method would then be:

$$w_{\text{new}} = w_{\text{old}} - (X_{\text{train}}^t BX_{\text{train}} + A)^{-1}(X_{\text{train}}^t(y - t_{\text{train}}) + Aw_{mathrmold})$$

with y evaluated using w_{old} in each step.

Once we obtain $w^* \approx \arg\max_w \ln p(w|t_{\text{train}}, X_{\text{train}}, A)$, we approximate $p(w|t_{\text{train}}, X_{\text{train}}, A)$ by a normal distribution $N(w^*, (X_{\text{train}}^t B^* X_{\text{train}} + A)^{-1})$, where B^* is B with y evaluated using w^*.

The above steps assume that we know the values of a_i in A. The parameters a_i can be estimated following an empirical Bayes approach. That is, we set A to be $A = \arg\max_A \int p(t_{\text{train}}|X_{\text{train}}, w)p(w|A)dw$ where $p(t_{\text{train}}|X_{\text{train}}, w)$ follows Eq. 4 and $p(w|A)$ follows Eq. 5. Unfortunately, there is no close-form solution to this integration. We again apply Laplace approximation. Note that the w^* that maximize the function $f(w) = p(t_{\text{train}}|X_{\text{train}}, w)p(w|A)$ is exactly the w^* we obtain in the approximation of the posterior distribution. With this, the probability of the training examples can be written as:

$$\begin{aligned} & p(t_{\text{train}}|X_{\text{train}}, A) \\ = & \int p(t_{\text{train}}|X_{\text{train}}, w)p(w|A)dw \\ \approx & \; p(t_{\text{train}}|X_{\text{train}}, w^*)p(w^*|A)(2\pi)^{(N+1)/2}|\Sigma^*|^{1/2}. \end{aligned}$$

When we know w^* (and Σ^*), we can obtain A by maximizing the above probability. Set the derivative of this probability to zero, one can obtain the update rule for the parameters a_i as:

$$a_i^{(\text{new})} = \frac{1 - a_i^{(\text{old})}\Sigma^*_{ii}}{(w_i^*)^2} \tag{7}$$

Combining all the above, we have that, given an initial set of values of a_i, we can estimate w^*. With the values of w^*, we can then obtain an updated estimate of a_i. One can repeat this estimation process until a good estimate of the a_i is obtained. The algorithm for parameter estimation is summarized in Algorithm 1.

Input : A set of N examples $\{x_i, t_i\}$; Convergence criteria δ_a, δ_w and M
Output: The parameters $\{a_i\}_{i=1}^N$

Process the training examples to obtain the similarity matrix.
$S(i,j) \leftarrow \sum_{k\in U}\phi(x_{ik})\phi(x_{jk}) + \sum_{k\in O} x_{ik} \circ x_{jk}$, where ϕ and $\circ$ are defined in Eq. 1 and Eq. 2 respectively. O is the set of nominal attributes and U is the set of numerical attributes.

$X \leftarrow \{S, \mathbf{1}\}$;

while *max(*a_i*)* $< M$ *and change of* a_i *larger than* δ_a **do**
    $w \leftarrow \mathbf{0}$;
    while *change of* w_i *large than* δ_w **do**
        $y_i \leftarrow \frac{1}{1+\exp(\sum_{j\neq i} w_j S(x_i,x_j) + w_0)}$;
        $B \leftarrow \text{diag}(y)$, diag($y$) constructs a diagonal matrix whose diagonal is y;
        $w \leftarrow w - (X^tBX + A)^{-1}(X^t(y-t) + Aw)$;
    end
    $\Sigma \leftarrow (X^tBX + A)^{-1}$;
    $a_i \leftarrow \frac{1-a_i\Sigma_{ii}}{(w_i)^2}$, where Σ_{ii} is the i-th element on the diagonal of the matrix Σ;
end

Algorithm 1. Parameter Estimate

3.4 Simplified Classification

In a full Bayesian framework, the probability of a new network event being normal or abnormal is calculated using Eq. 6. Since there is no close-form solution to the integration, one either employs numerical methods or applies an approximation (e.g., the one we used for parameter estimation). This will either increase computational cost or produce less accurate result.

To classify new network events (after we have obtained the values for a_i), instead of following Eq. 6, we can calculate the probability following Eq. 4 using a $w^* = \arg\max_w p(w|t_{\text{train}}, X_{\text{train}}, A)$. This is similar to making prediction by a MAP (maximum a posterior) approach. We now show that if the posterior distribution of w is approximated using a Normal distribution, classifying new network events following the Bayesian approach and the MAP approach generates the same result.

Theorem 1. *Assume that the posterior distribution $p(w|t_{\text{train}}, X_{\text{train}}, A)$ is approximated by a normal distribution $N(w^*, \Sigma^*)$. Classification using $p^{\text{MAP}}(t) = p(t|x, w^*) = y(x)^t(1 - y(x))^{(1-t)}$ and $p^{\text{Bayes}}(t) = p(t|x, t_{\text{train}}, X_{\text{train}}, A) = \int p(t|x, w)p(w|t_{\text{train}}, X_{\text{train}}, A)dw \approx \int p(t|x, w)N(w|w^*, \Sigma^*)dw$ yields the same result.*

Proof. Consider classification using $p^{\text{Bayes}}(t)$. We say a network event x is abnormal if $p^{\text{Bayes}}(1) > p^{\text{Bayes}}(0)$ and normal otherwise. We calculate

$$\begin{aligned}
&\int p(1|x, w)N(w|w^*, \Sigma^*)dw - \int p(0|x, w)N(w|w^*, \Sigma^*)dw \\
&= \int [p(1|x, w) - p(0|x, w)]N(w|w^*, \Sigma^*)dw \\
&= \int [\frac{2}{1 + exp(-u)} - 1]N(u|\mu, \sigma)du
\end{aligned}$$

In the last step, we set $u = w^t x$. For a fixed x, because w follows a normal distribution $N(w^*, \Sigma^*)$, $u = w^t x$ also follows a normal distribution with the mean $\mu = w^{*t}x$.

Let $f(u) = \frac{2}{1+exp(-u)} - 1$. Note that $f(u)$ is symmetric with respect to the origin, i.e., $f(-u) = -f(u)$ and $f(0) = 0$. Also, $f(u)$ is a monotonically increasing function. The classification depends on $\int f(u)N(u|\mu, \sigma)du$. Let x_0 be an event such that $\mu(x_0) = w^{*t}x_0 > 0$. For every $u_0 = \mu(x_0) - \epsilon$, there is $u_0' = \mu(x_0) + \epsilon$, such that $p(u_0) = p(u_0')$ and $|f(u_0)| < |f(u_0')|$ where $|\cdot|$ denotes the absolute value. Therefore, $\int f(u)N(u|\mu(x_0), \sigma)du >> 0$ and the event x_0 will be classified as abnormal using p^{Bayes}. When using p^{MAP}, classification depends on $f(u)$ with $u = w^{*t}x$. Because $w^{*t}x_0 > 0$, $f(w^{*t}x_0) > 0$. Therefore, x_0 will be classified as abnormal too. The same argument holds for an event x_0 such that $w^{*t}x_0 \leq 0$. In this case, both methods will classify the event as normal. □

4 Experiments

We test our framework using the KDD Cup anomaly detection dataset [9]. The dataset is the most widely used data set for evaluation in anomaly detection research. It is based on the data captured in DARPA98 IDS evaluation program. The dataset is a collection of connection vectors. Each vector has 41 features and is labeled either as normal or an attack. The dataset we use is a subsample of the original KDD Cup dataset called the 10-percent KDD Cup dataset. There are 22 attack types in the 10-percent subset.

We first conduct experiments to test the detection performance of our framework and compare it to that of an SVM-based detection system. We then analyze the representative instances selected by our framework.

4.1 Detection Performance

From the 10-percent KDD Cup dataset, we randomly sample a subset of normal and attack instances as training examples. From the remaining of the instances, we then randomly sample another subset to use as test data. We train a SVM-based classifier and a sparse Bayes classifier using the training examples. The performance of the classifiers are then tested on the test data. For each experiment condition, we repeat the experiment 20 times. Each data point reported in the figures is the average of the results from the 20 repetitions.

We measure the performance of the classifiers using the standard F-score. There are only two classes in our data: normal and attack. If a classifier label a true attack as attack, we count it as a true positive. Otherwise if it labels a true attack as normal, we count the case as a false negative. Similarly, we have true negative (label a normal instance as normal) and false positive (label a normal instance as attack). We denote by TP the number of true positives, FN the number of false negatives, TN true negatives and FP false positive. The precision of a classification is define as $TP/(TP+FP)$ and the recall as $TP/(TP+FN)$. The harmonic mean of the precision and the recall is the F-score of the classification performance, i.e., $2\frac{\cdot \text{precision} \times \text{recall}}{\text{precision}+\text{recall}}$. If a classifier gives perfect classification, the F-score should take the value 1 and the worst classification will give an F-score zero.

Fig. 1 shows the performance of the SVM-based classifier and the sparse Bayes classifier when given different numbers of training examples. 4 experiments are conducted using training set of different sizes. In experiment 1, the collection of training examples contain 9 normal connection vectors and 11 attack vectors. In experiment 2, 19 normal and 23 attack vectors. Experiment 3, 48 normal and 59 attacks and experiment 4, 97 normal and 118 attacks. The result shows that for both classifiers, the performance becomes better if more training examples are used. Overall, the SVM-based classifier and the sparse Bayes classifier have similar level of performance.

In the above experiment, attack instances of all types are grouped into one collection. (We omitted some attack types in the original data set that only have few instances.) The training and the testing instances for the attack class

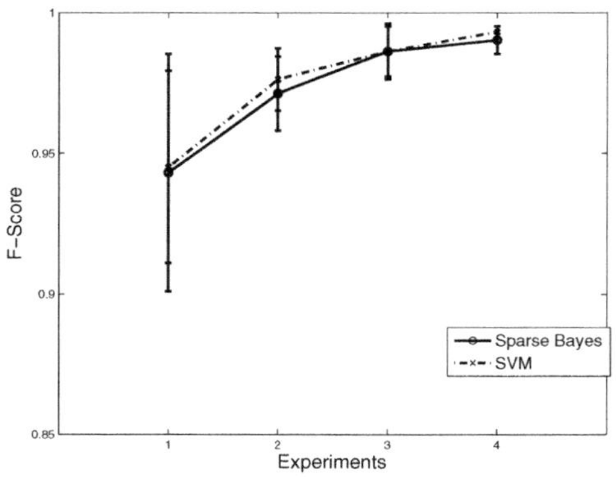

Fig. 1. Detection Performance Given Training Set of Different Size

are sampled from this collection. Intuitively, one may think that the detection task would be more difficult when there are many different types of attacks. The classifier need to recognize many types rather than dealing with one type of attack (anomaly). The number of attack types in the dataset then reflects, to some extent, the complexity of the detection task. To test the performance of the two classifiers on tasks of different complexity, we constructed datasets in which the attack class contains limited types of attacks. We then conducted experiments using these datasets.

Fig. 2 plots the results of the experiments. In experiment 1, the attack data only contain one type of attacks (portsweep). In experiment 2, the data contain two types of attacks (portsweep, satan). In the experiment 3, 4 types of attacks are used (portsweep, satan, ipsweep and warezclient). The results show that both classifiers perform slightly but not significantly better when the task is less complex (involving less types of attacks). The results again show that the performance of the two classifiers are comparable across detection tasks of different complexity.

4.2 The Representative Instances

As discussed before, our goal of applying sparse Bayes framework to anomaly detection is not simply to construct a detection system. We also want the system to help human analysis of the anomaly. In particular, we use the sparse Bayes framework to identify a small set of representative anomaly cases such that by examining this small set, human expert can gain better understanding of the anomalous activities and better understanding of the subtypes inside the abnormal class, if there is any.

We constructed a collection of normal and attack vectors in which the attack class contains 4 types of attacks (portsweep, satan, warezclient and ipsweep). We then train a sparse Bayes classifier from the collection. Recall that the probability of an unknown vector x belonging to normal or abnormal (attack) class

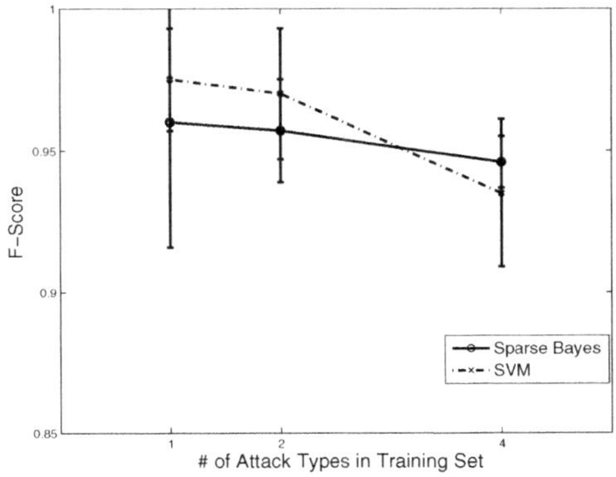

Fig. 2. Detection Performance with Different Number of Attack Types in Training Set

is essentially determined by $\sum_{i\in \text{TraingSet}} w_i S(x, x_i)$. Therefore, the training examples x_i whose corresponding weight w_i has a large magnitude decides the outcome of the classification. We thus view these examples as representative instances. Fig. 3 plot the w vector obtained after training. The dash-doted vertical lines on the figure separates different classes. For example, the first section (from instance 1 to instance 97) contains the weights for the training instances in the normal class. The second section contains the weights for the attacks of the type portsweep. (third section, satan; 4th section, warezclient and the last section ipsweep.) The plot shows that, except for the normal class, most w_i has a value zero or close to zero. There are only few w_i that has large magnitude. The corresponding instances are the representative instances.

We select a few representative instances and plot the similarity measure S between the representative instances and the other instances in the training data. Fig. 4 shows the similarity matrix. Each row of the similarity matrix corresponds to an instance in the training data and each column corresponds to a representative instance. The (i, j)-entry of the matrix is the similarity measure between the i-th training example and the j-th representative instance. We plot the matrix

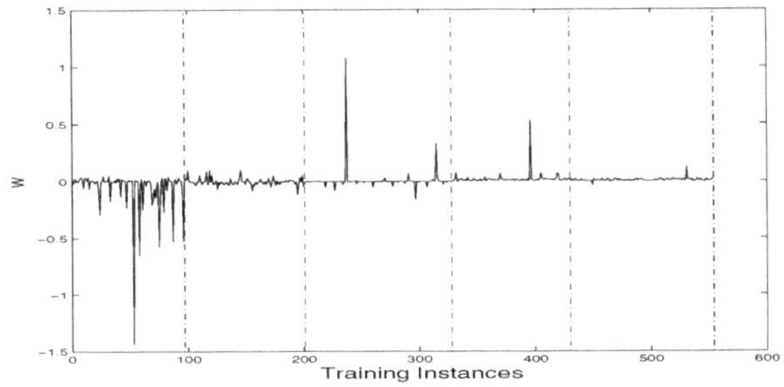

Fig. 3. Weight of the Training Instances

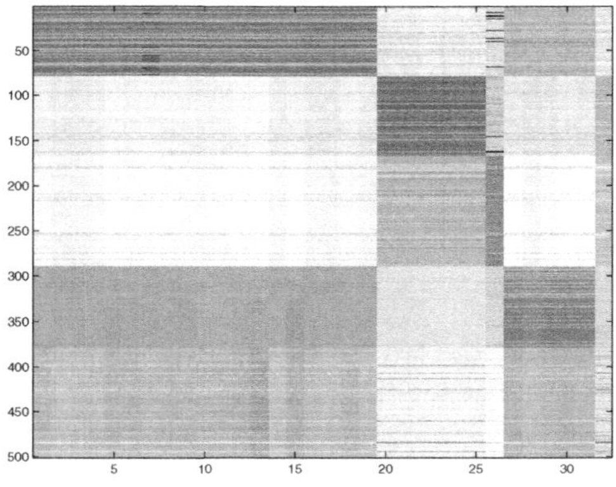

Fig. 4. Similarity between the Representative Instances and the Other Training Examples

using grayscale, with dark indicating high similarity and white low similarity. We observe that the similarity matrix divides into several blocks. Columns 1-19 correspond to the representative instances from the normal class and they have the strongest similarity to the normal examples (row 1-78). The representative instances for the second class (portsweep) lie on column 20-25. Again they have the strongest similarity to the training examples in the portsweep class (row 79-166). In general (except the last group) the representative instances have the highest similarity to the instances in the group to which they belong.

5 Conclusion

We investigate the application of sparse Bayes classification in anomaly detection where there are multiple types of anomalies. Our goal is to construct a detection system that not only identifies anomaly but also facilitates human analysis of the anomalous activities. It does so by selecting a small set of instances that are important for making the detection. These instances can be viewed as representative instances for the anomalies. By analyzing the representative instances, an expert may determine the type and the mechanism of the anomaly. When an unknown event is detected as anomaly due to its connection to a representative instance, corresponding response can be applied to deal with the new anomaly. We conduct experiments to test our framework. The results show that the system based on our framework has similar detection performance comparing to an SVM-based detection system. Furthermore, a few representative instances can be identified by our framework that can be potentially helpful for expert analysis.

Acknowledgment. This work was supported in part by NSF grant IIS-0905478 and Louisiana Board of Regents grant LEQSF(2009-11)-RD-A-09.

References

1. Agyemang, M., Barker, K., Alhajj, R.: A comprehensive survey of numeric and symbolic outlier mining techniques. Intell. Data Anal. 10(6), 521–538 (2006)
2. Beale, J., Caswell, B., Poor, M.: Snort 2.1 intrusion detection. Syngress Publishing (2004)
3. Chandola, V., Banerjee, A., Kumar, V.: Anomaly detection: A survey. ACM Comput. Surv. 41(3) (2009)
4. Chen, W.-H., Hsu, S.-H., Shen, H.-P.: Application of SVM and ANN for intrusion detection. Computers & Oper. Res. 32, 2617–2634 (2005)
5. Dickerson, J.E., Dickerson, J.A.: Fuzzy network profiling for intrusion detection. In: 19th International Conference of the North American Fuzzy Information Processing Society (2000)
6. Faul, A.C., Tipping, M.E.: Analysis of sparse bayesian learning. In: Advances in Neural Information Processing Systems, pp. 383–389 (2001)
7. Hodge, V., Austin, J.: A survey of outlier detection methodologies. Artificial Intelligence Review 22(2), 85–126 (2004)
8. Kayacik, H.G., Zincir-Heywood, A.N., Heywood, M.I.: A hierarchical SOM-based intrusion detection system. Eng. Appl. of AI 20(4), 439–451 (2007)
9. KDD. Kdd cup intrusion detection dataset (1999), `http://kdd.ics.uci.edu/databases/kddcup99/kddcup99.html`
10. Levin, I.: KDD-99 classifier learning contest: LLSoft's results overview. SIGKDD Explorations 1(2), 67–75 (2000)
11. Patcha, A., Park, J.-M.: An overview of anomaly detection techniques: Existing solutions and latest technological trends. Computer Networks 51(12), 3448–3470 (2007)
12. Paxson, V.: Bro: a system for detecting network intruders in real-time. In: Proceedings of the 7th USENIX Security Symposium (1998)
13. Porras, P.A., Neumann, P.G.: Emerlad. In: Proceedings of 20th National Information Systems Security Conference, pp. 353–365 (1997)
14. Stolfo, S., Prodromidis, A., Tselepsis, S., Lee, W., Fan, D., Chan, P.: JAM: Java agents for meta-learning over distributed databases. In: Workshop on Fraud Detection and Risk Management AAAI 1997 (1997)
15. Tipping, M.E.: Sparse bayesian learning and the relevance vector machine. Journal of Machine Learning Research 1, 211–244 (2001)
16. Williams, O., Blake, A., Cipolla, R.: Sparse bayesian learning for efficient visual tracking. IEEE Trans. Pattern Analysis and Machine Intelligence 27(8), 1292–1304 (2005)

Mobile Sensing Enabled Robust Detection of Security Threats in Urban Environments

Jie Yang[1], Jerry Cheng[2], and Yingying Chen[1]

[1] Department of Electrical and Computer Engineering
Stevens Institute of Technology
Castle Point On Hudson, Hoboken, NJ, 07030, USA
{jyang,yingying.chen}@stevens.edu
[2] Department of Medicine
Robert Wood Johnson Medical School, UMDNJ
New Brunswick, NJ, 08901
jcheng@stat.columbia.edu

Abstract. Mobile sensing enables data collection from large numbers of participants in ways that previously were not possible. In particular, by affixing a sensory device to a mobile device, such as smartphone or vehicle, mobile sensing provides the opportunity to not only collect dynamic information from environments but also detect the environmental hazards. In this paper, we propose a mobile sensing wireless network for surveillance of security threats in urban environments, e.g., environmental pollution sources or nuclear radiation materials. We formulate the security threats detection as a significant cluster detection problem. To make our approach robust to unreliable sensing data, we propose an algorithm based on the Mean Shift method to identify the significant clusters and determine the locations of threats. Extensive simulation studies are conducted to evaluate the effectiveness of the proposed detection algorithm.

Keywords: Mobile sensing, security threats, Mean Shift Clustering.

1 Introduction

Mobile sensing has become increasingly popular in recent years as it enables data collection from large number of participants in ways that previously were not possible [1]. In particular, by affixing sensory devices to mobile devices, such as smartphones or vehicles, mobile sensing provides an opportunity to not only collect and share dynamic information at an urban-scale but also perform data analysis to detect security threats presenting in urban environments. In this work, we consider security threats as environmental sources hazardous to civilian's daily lives, for example, air pollution sources and nuclear radiation sources.

We show the importance of detecting the hazardous environmental sources by using the following two examples. The environmental air pollution directly affects the public health in a metropolitan area. According to reports of the

X. Zhang and D. Qiao (Eds.): QShine 2010, LNICST 74, pp. 88–104, 2011.

World Health Organization, over 4.6 million people are estimated to die annually from the direct impact of air pollution - more than those from car accidents every year [2]. Meanwhile radiological sources have become commonplace at research, industrial and medical facilities. As a result, we are facing a growing danger that nuclear materials might be acquired by terrorist organizations. For example, there are numerous cases of lost or stolen radioactive materials have been reported [3, 4]. Moveover, major metropolitan areas are attractive targets for placement of radiation materials by attackers due to their dense population and economic importance. Previous detection of such security threats are carried out mostly by specialized teams or fixed sensors, and may not perform frequently in a large-scale due to the limited resources and high cost involved. However, because of their growing danger and severe harm, detecting these security threats through daily activities is not only desirable but also feasible by the increasing popularity of mobile sensors. In this paper, we develop a mechanism that exploits sensing data obtained from mobile sensing to detect such security threats.

In our framework, real time sensor readings together with participants' locations will be reported back through existing wireless networks to a central surveillance center, where the data is analyzed to detect and localize the security threats sources. The ubiquitous nature of smartphones, vehicles and other portable electronic devices makes it possible to constantly sense threats sources. However, there are a number of underlying challenges to be overcome before the accurate detection can occur. For example, sensitivity of a sensor is diminishing with distance from the source; a mobile sensor may fail to correctly sense an existing threat (causing false negative), or report a reading when there is none threat source (causing false positive). In addition, the background noise may vary significantly from one region to another. For instance, a person who just came out from a radioactive therapy or a bag of cat litter may even set off false alert of a mobile sensor.

To enable robust detection of security threat sources under the presence of unreliable sensor readings, we propose to detect these sources using sensor readings collaboratively. In our work, the security threats source detection is formulated as significant cluster detection problem, in which we aim to identify whether one or more spatial clusters exist in the area significantly from background noise. When a source emits energy at a certain location, the mobile sensors within the detection range of the source would have higher probability to be activated, hence create a cluster of points in the physical space, while the background false positive readings would be likely randomly distributed throughout the area. We develop a detection mechanism grounded on Mean Shift Clustering procedure to both detect the significant clusters and localize the threat sources.

To validate the effectiveness of the proposed security threat detection mechanism, we simulated a mobile sensing wireless network in an area of similar size as metropolitan Manhattan in New York City. Our simulation results show that our proposed framework can achieve over 90% detection rate for both single and multiple sources with small localization errors. Thereby this shows that our approach is effective in detecting treat sources.

The rest of the paper is organized as follows. We place our work in the context of related research in Section 2. The system overview of the mobile sensing enabled security threat detection is presented in Section 3. We detail our detection mechanism by focusing on the Mean Shift procedure in Section 4 and describing robust threat source localization in Section 5. We then present our simulation results in Section 6 and conclude our work in Section 7.

2 Related Work

Although mobile sensing has been gaining popularity in various applications [5, 6, 7], there is relatively less work focusing on detecting security threat sources. The PEIR project [7] proposed by UCLA is prototyped to measure how often people are exposed to high levels of air pollution by using sensing data collected by mobile phones. As a part of the project of Use of Mobile Technology for Social Change [1], taxis equipped with a carbon monoxide sensor and a GPS device are used to collect environmental data such as readings of carbon monoxide levels, which is then displayed on a city map for visulization. However, none of these work performed systematic data analysis to detect the presence of security threat sources.

In the area of radiation detection, using individual and fixed sensors to detect radiation sources have been well studied [8,9]. These approaches have limited application to identify portable nuclear radiation sources placed by attackers. The recent progress has shown some success of using a network of sensors for detecting and tracking radiation sources [10,11,12,13]. In [11,12], a linear arrangement of fixed detectors has been considered to detect the radioactive sources. In [13], a latent model is proposed to detect multiple nuclear materials by using a mobile sensor network approach. In addition, the Radiation Laboratory at Purdue University used a network of cell phones with GPS capabilities to detect and track radiation [14]. The difference between our work to previous studies is that we propose to use mobile sensing that enables an extensive coverage in a metropolitan area to detect and localize the security threats sources with the presence of unreliable sensor readings.

Our approach is based on Mean Shift Clustering method. Traditional statistical methods for detecting a cluster of events in spatial data is using a class of Scan Statistics [15, 16, 17]. The most commonly used scan statistics is the maximum number of cases within a fixed size window that scans through the study area. Based on this scan statistics, a generalized likelihood ratio test has been developed to test the null hypothesis whether all the information are uniformly distributed in the area (no cluster). Scan statistics procedures are mainly used in detecting a single significant cluster, and they also have had some success in detecting multiple clusters of fixed sizes.

However, problems arise for detecting multiple clusters of varying sizes. In recent years, procedures have been developed to overcome the difficulty. One of the well known approaches is a stepwise regression model combined with model selection procedures to locate and determine the number of clusters [18]. These approaches rely on a weighted least square formulation, although the response variable (gaps between incidents) is typically non-Gaussian. In our approach, we utilized Mean Shift Clustering algorithm, a nonparametric clustering technique, does not assume any prior knowledge of the number of clusters, and can handle arbitrarily shaped clusters. Thus, it is suitable for handling clusters of arbitrary shape and number for detecting security threats using mobile sensing data.

3 System Overview

In this section, we first present the threat model that describes the transmission behavior of the threat source. We then provide our network model and sensing model enabled by mobile sensors. Finally, we overview the key components of our threat source detection mechanism.

3.1 Threat Model

We consider security threats in urban environments that can cause biological hazards to civilians, for example, environmental pollution sources or nuclear radiation materials. A threat source can be either static or mobile. For simplicity, we start with assuming the impact of the threat source travels in spherical waves. Thus, the impact intensity T decreases by the inverse square of the distance r: $T(r) = c/r^2$, where the constant c is a factor related to the energy of the source. This simple model is also used to describe nuclear radiation emitting from a threat source [19].

In addition, there might be multiple threat sources, whose impact region may overlap. In this work, we assume we know the number of threat sources and consider the sources of the same type, e.g., environmental pollution sources that can trigger the reading of mobile sensors measuring pollution level; or nuclear radiation materials that can trigger the reading of mobile sensors measuring nuclear radiation. When these sources assume the same energy spectrum, the overall impact intensity at a particular location is an aggregation of individual ones: $T_{total} = \sum_{i=1}^{N} c_i/r_i^2$ where N is the total number of threat sources of the same type. The sources of different types will be considered in our future work.

3.2 Network Model

We build our solution of detecting threat sources in urban environments upon the wireless mobile sensor networks with the following characteristics.

Mobility. Mobile sensors can be either mounted or built-in on smart phones or vehicles. This enables the mobile sensors to move randomly or in some routine patterns in urban cities. For example, the participating vehicles can be taxicabs, police patrol vehicles, or city buses.

Location-Aware. Each mobile sensor knows its physical location at all times during moving. This is a reasonable assumption as most of wireless devices (e.g., mobile phones or vehicles) are equipped with GPS or some other approximate but less burdensome localization algorithms [20].

Sensing Model. When mobile sensors move within a certain range of a threat source, the reading of the sensors will be triggered. We define the reading of a sensor S using a threshold model: $S = \mathbf{1}_{\{T(r) \geq \tau\}} = \mathbf{1}_{\{c/r^2 \geq \tau\}}$ where τ is a threshold for detection and $\mathbf{1}_{\{.\}}$ is the indicator function. That is, if the intensity $T(r)$ at the sensor location is greater than the threshold τ, the sensor reports a positive reading (i.e., detection of a threat source), otherwise the sensor reports a negative reading. In the case of multiple threat sources of the same type, the threshold model can be represented as: $S = \mathbf{1}_{\{T_{total} \geq \tau\}} = \mathbf{1}_{\{\sum_{i=1}^{N} c_i/r_i^2 \geq \tau\}}$. Moreover, as with any sensing device, the reading of mobile sensors may not be 100% accurate. Additionally, transient scenarios, e.g., a person who is walking on the street just went through a radioactive therapy, may also trigger false alarms. In this work, we do not explore how to improve the reliability of sensor readings. Instead, to live with unreliable sensor readings, we focus our work on design robust detection mechanisms with readings from multiple sensors to collaboratively detect the presence of the threat sources.

3.3 Mobile Sensing Enabled Threat Detection

Our framework employs a mobile sensing approach to detect security threats in urban environments. The readings of mobile sensors (e.g., equipped with smartphones or mounted on vehicles) will be reported to a central monitoring server along with the positions of the sensors when readings occur. The report of sensor readings will utilize the existing wireless infrastructure by sending the data over a cellular uplink or making use of the WiFi connections depending on cost/delay tradeoffs.

The detection mechanism running at the back-end server comprises three main components: *Mean Shift Procedure*, *Building Clustering Hierarchy*, and *Threat Source Identification*. The Mean Shift procedure is the core element during the detection process. It finds the clustering of the densest area in the sensing data through mode finding of a density function. The accuracy of modes detection relies on appropriate bandwidth selection in Mean Shift clustering. Since the impact range of a threat source is unknown, it is a challenging task to choose an appropriate bandwidth. To find the optimal bandwidth, we develop a technique that produces a hierarchy of clusters by repeatedly applying the Mean Shift procedure under different bandwidths. Finally, threat sources are identified by choosing the clusters with top number of sensor readings under the optimal bandwidth. This approach has the main advantage of filtering out the clusters that are formed by unreliable sensor readings and makes the detection robust. We describe these components in detail in the following sections.

4 Threat Source Detection Using Mean Shift

The Mean Shift procedure is a key component in our threat source detection framework. Detecting a threat source is equivalent to the mode detection in the

Mean Shift procedure. There have been many applications using Mean Shift such as image analysis [21], texture segmentation [22], objective tracking [23, 24] and data fusion [25]. The Mean Shift Clustering is a nonparametric clustering technique using the mode finding Mean Shift procedure, which is ideal to identify the presence of threat sources based on the readings of mobile sensors together with spatial information. Because it does not assume any prior knowledge of the number of clusters, and can handle arbitrarily shaped clusters. In particular, Mean Shift is a procedure for locating the maxima of a density function given discrete data sampled from that function [26]. We next detail the kernel density estimation to derive the density estimator and the density gradient estimation to find the modes for detecting threat sources based on the description and notation in [27].

4.1 Kernel Density Estimation

Kernel density estimation is a non-parametric way of estimating the probability density function of a random variable [28, 29]. Let $\mathbf{x}_i \in R^d, i = 1, 2, .., n$, be n independent and identically distributed d-dimensional data points. The multivariate kernel density estimator with kernel $K(\mathbf{x})$ and a symmetric positive definite $d \times d$ bandwidth matrix $\mathbf{H}$ is given by:

$$\hat{f}_{\mathbf{H}}(\mathbf{x}) = n^{-1} \sum_{i=1}^{n} K_{\mathbf{H}}(\mathbf{x} - \mathbf{x}_i), \tag{1}$$

where

$$K_{\mathbf{H}}(\mathbf{x}) = |\mathbf{H}|^{-\frac{1}{2}} K(\mathbf{H}^{-\frac{1}{2}}\mathbf{x}). \tag{2}$$

The d-variate kernel $K(\mathbf{x})$ is a bounded function with compact support satisfying the following properties:

$$\int_{R^d} K(\mathbf{x})d\mathbf{x} = 1; \quad \int_{R^d} \mathbf{x}K(\mathbf{x})d\mathbf{x} = 0;$$

$$\lim_{||\mathbf{x}|| \to \infty} ||\mathbf{x}||^d K(\mathbf{x}) = 0; \quad \int_{R^d} \mathbf{x}\mathbf{x}^T K(\mathbf{x})d\mathbf{x} = c_K I.$$

where c_K is a constant.

In practice, the bandwidth matrix $\mathbf{H}$ is chosen proportional to the identity matrix $\mathbf{H} = h^2\mathbf{I}$, with $h > 0$. Employing the bandwidth parameter h, the kernel density estimator (1) becomes

$$\hat{f}_h(\mathbf{x}) = \frac{1}{nh^d} \sum_{i=1}^{n} K\left(\frac{\mathbf{x} - \mathbf{x}_i}{h}\right), \tag{3}$$

where $K(\mathbf{x})$ is a class of radially symmetric kernels satisfying:

$$K(\mathbf{x}) = c_{k,d} k(||\mathbf{x}||^2). \tag{4}$$

The function $k(x)$ is called the profile of the kernel and $c_{k,d}$ is the normalization constant, which makes $K(\mathbf{x})$ integrate to one.

The commonly used kernels in literature are Epanechnikov Kernel, Uniform Kernel and Gaussian Kernel. In this work, we used Uniform Kernel whose profile is given by:

$$\text{Uniform Kernel:} \qquad k_U(x) = \begin{cases} 1, & 0 \le x \le 1 \\ 0, & x > 1 \end{cases} \tag{5}$$

Using the profile notation, the density estimator (3) can be written as

$$\hat{f}_{h,K}(\mathbf{x}) = \frac{c_{k,d}}{nh^d} \sum_{i=1}^{n} k\left(\left\|\frac{\mathbf{x}-\mathbf{x}_i}{h}\right\|^2\right). \tag{6}$$

The modes of the density function $f(\mathbf{x})$ are located at the zeros of the gradient function $\nabla f(\mathbf{x})$. The Mean Shift procedure can locate these modes without estimating the underlying density.

4.2 Density Gradient Estimation

The gradient of the density estimator (6) is

$$\nabla \hat{f}_{h,K}(\mathbf{x}) = \frac{2c_{k,d}}{nh^{d+2}} \sum_{i=1}^{n} (\mathbf{x}-\mathbf{x}_i) k'\left(\left\|\frac{\mathbf{x}-\mathbf{x}_i}{h}\right\|^2\right), \tag{7}$$

where $k'(x)$ is the derivative of the profile.

If we define $g(x) = -k'(x)$, and introducing $g(x)$ into (7), we have

$$\begin{aligned} \nabla \hat{f}_{h,K}(\mathbf{x}) &= \frac{2c_{k,d}}{nh^{d+2}} \sum_{i=1}^{n} (\mathbf{x}_i-\mathbf{x}) g\left(\left\|\frac{\mathbf{x}-\mathbf{x}_i}{h}\right\|^2\right) \\ &= \frac{2c_{k,d}}{nh^{d+2}} \left[\sum_{i=1}^{n} g\left(\left\|\frac{\mathbf{x}-\mathbf{x}_i}{h}\right\|^2\right)\right] \left[\frac{\sum_{i=1}^{n} \mathbf{x}_i g\left(\left\|\frac{\mathbf{x}-\mathbf{x}_i}{h}\right\|^2\right)}{\sum_{i=1}^{n} g\left(\left\|\frac{\mathbf{x}-\mathbf{x}_i}{h}\right\|^2\right)} - \mathbf{x}\right]. \end{aligned} \tag{8}$$

The first term of the above equation is proportional to the density estimate at $\mathbf{x}$ computed with kernel $G(\mathbf{x}) = c_{g,d}\, g(\|\mathbf{x}\|^2)$

$$\hat{f}_{h,G}(\mathbf{x}) = \frac{c_{g,d}}{nh^d} \sum_{i=1}^{n} g\left(\left\|\frac{\mathbf{x}-\mathbf{x}_i}{h}\right\|^2\right). \tag{9}$$

The second term

$$\mathbf{m}_{h,G}(\mathbf{x}) = \frac{\sum_{i=1}^{n} \mathbf{x}_i g\left(\left\|\frac{\mathbf{x}-\mathbf{x}_i}{h}\right\|^2\right)}{\sum_{i=1}^{n} g\left(\left\|\frac{\mathbf{x}-\mathbf{x}_i}{h}\right\|^2\right)} - \mathbf{x} \tag{10}$$

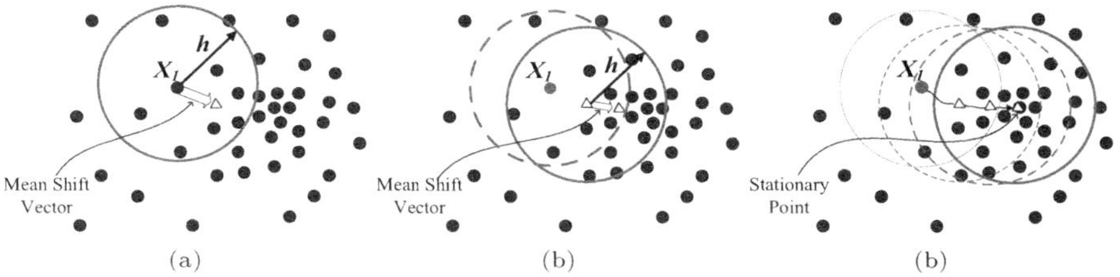

Fig. 1. Illustration of Mean Shift procedure using a uniform kernel. (a) Start from a data point $\mathbf{x}_1$, calculate the Mean Shift vector $\mathbf{m}_{h,G}(\mathbf{x}_1)$, which is the difference between $\mathbf{x}_1$ and weighted mean of data points within the window with a radius of bandwidth h. (b) Shift the window by $\mathbf{m}_{h,G}(\mathbf{x})$, recompute the Mean Shift vector $\mathbf{m}_{h,G}(\mathbf{x}_1)$. (c) Reach the stationary point in the densest area, which is the mode.

is the *Mean Shift* which is the difference between the weighted mean and $\mathbf{x}$, the center of the kernel window. The Mean Shift vector always points toward the direction of the maximum increase in the density. Therefore, the Mean Shift procedure is guaranteed to converge to a point where the gradient of density function is zero. This point is the mode (i.e., the security threat source) obtained from running Mean Shift by going through the following steps iteratively:

1. Compute the Mean Shift vector $\mathbf{m}_{h,G}(\mathbf{x})$,
2. Translate the kernel window of $G(\mathbf{x})$ by $\mathbf{m}_{h,G}(\mathbf{x})$, and re-compute the weighted mean,
3. Stop iteration if gradient is close to zero.

4.3 Cluster Formulation

For each reported reading of a mobile sensor, the Mean Shift procedure under a fixed bandwidth will find the mode (i.e. the point where the gradient of density function is zero) in the density function that the sensor reading is associated with. The mode of the density function is equivalent to the location of the densest area of the data set of all the reported sensor readings, i.e., the location of the security threat source. To form a cluster, all the sensor readings associated with the same mode belong to the same cluster. And the modes could be used to represent the cluster center. Alternately, we can define the *basin of attraction* of the mode as the set of all mobile sensor readings that converge to the same mode [27]. The sensor readings which are in the same basin of attraction is associated with the same cluster. The number of clusters is obtained by the number of modes. Figure 1 illustrates the cluster formulation for detecting security threat sources. The detailed algorithm flow is described in Algorithm 1. ClusterCenters[] is a one dimensional matrix that holds the clusters formulated under a specific fixed bandwidth.

Algorithm 1. Threat Source Detection Using Mean Shift Clustering under a Fixed Bandwidth

Let $S = (s_1, .., s_n)$ be the sensor reading dataset, h be the bandwidth;
Let ClusterCenters be the modes;
Let ClusterMembership be the cluster membership of each point;
ClusterCenters=[]; ClusterMembership=[];
$i \leftarrow 0$;
$j \leftarrow 0$;
repeat
  For the data point s_i, run Mean Shift procedure to get the mode M ;
  if (M is not in the ClusterCenters) **then**
    $j \leftarrow j + 1$
    ClusterCenters[j] $= M$;
    ClusterMembership[i] $= j$;
  else
    if (M is at the lth position of the ClusterCenters) **then**
      ClusterMembership[i] $= l$;
    end if
  end if
  $i \leftarrow i + 1$;
until i==n
return ClusterCenters;
return ClusterMembership;

5 Robust Threat Source Localization

5.1 Challenges on Bandwidth Selection

The bandwidth selection in Mean Shift directly affects the performance of the Mean Shift Clustering, and consequently affects the accuracy of detecting security threat sources. Since the impact range of the threat sources is unknown, it is hard, if not possible, to identify the significant clusters with the optimal bandwidth during Mean Shift Clustering. If the bandwidth h is estimated too large, it will produce an oversmoothed density estimate, resulting in only one mode in the estimated density even if there are multiple treat sources present. On the other hand, a too small bandwidth h will cause the density estimate to produce too many clusters, which is called undersmoothed. To address these challenges, we develop a technique that perform threat source localization by using adaptive bandwidth. Our technique is described as follows.

5.2 Building the Cluster Hierarchy

Researchers summarized four different techniques for bandwidth selection [27] including plug-in rule [30], Least Squares Cross Validation [31] and stability of the decomposition [32]. In order to achieve robust detection of security threat sources, we designed a technique to find the optimal bandwidth with significant clusters by building a hierarchy of clusters. The repeated clustering of the

Algorithm 2. Building the Cluster Hierarchy

```
Let S = (s_1, .., s_n) be the dataset, h be the bandwidth and p be the bandwidth step;
Set h= minimum non-zero distance between any two points in S;
i ← 0;
ClusterCenters{i}=S; ClusterMembership{i}=1 : n;
repeat
    (newClusterCenters, newClusterMembership) = MeanShiftClustering(S, h);
    ClusterCenters{i + 1} ← newClusterCenters;
    ClusterMembership{i + 1} ← newClusterMembership;
    i ← i + 1;
    h ← h + p;
until size(ClusterCenters{i})==1
return ClusterCenters;
return ClusterMembership;
```

data is accomplished by iteratively running Mean Shift with increasingly larger bandwidths on top of the mobile sensing data. This iterative procedure is closely related to the stability of decomposition [32] for a density shape estimate. The optimal bandwidth is taken as the center of the largest operating range of the bandwidth over which the same cluster set is obtained for the given sensing data. This means that the shapes of the estimated densities are unchanged over this operating range of the bandwidth. This technique can yield a suitable bandwidth estimate by finding all cluster centers, i.e., threat sources, over the chosen operating range.

In particular, for the initial bandwidth value h_{min} (i.e. the smallest bandwidth value), we use the minimum non-zero distance between any two points in the reported mobile sensing data set. This is the lowest level of the hierarchy consisting of individual sensing data points. In this case, each individual data point either remains as its own cluster of size one or it is merged into a larger cluster when mobile sensors have same readings (e.g., two people walked together side by side or two vehicles passed the same location within a very short time period). In the subsequent iteration of Mean Shift, the bandwidth is increased by the step p, which is the bandwidth increment. With the increased bandwidth, Mean Shift is applied to the original data points to produce a new set of clusters. We continue this process until there is only one cluster left. This is the algorithm stop criteria that we employ in our technique of building cluster hierarchy. This criteria means that the bandwidth is increased large enough so that no new cluster can be generated beyond this cluster hierarchy. Intuitively, we are repeatedly blurring the data by using a larger bandwidth. Algorithm 2 shows the procedure to build the cluster hierarchy. ClusterCenters{} is a two dimensional matrix. ClusterCenters{i} holds the clusters formulated in the ith iteration. The constructed cluster hierarchy will be used in the next section to perform multiple threat source localization.

Algorithm 3. Threat Source Localization Using Adaptive Bandwidth

Build cluster hierarchy using Algorithm 2;
Set L be the threats source candidate list;
Let k be the number of the threats source;
$i \leftarrow 0$;
repeat
    For each cluster C in ClusterCenters$\{i\}$, calculate the number of data points in C from ClusterMembership$\{i\}$;
    Add the top clusters with top k number of points into the candidate list L;
    $i \leftarrow i + 1$;
until size(ClusterCenters$\{i\}$)$==k$
For each cluster C in the candidate list L, calculate the lifetime of C;
Return the cluster centers with top k lifetime,

5.3 Localizing Multiple Threat Sources Using Adaptive Bandwidth

The challenging task is to identify the significant clusters, i.e., multiple threat sources, embedded in the constructed cluster hierarchy. We adopt the concept of the *lifetime* of a cluster to identify multiple threat sources from the built cluster hierarchy and localize their positions. The concept of the *lifetime* of a cluster is proposed in [33], which measures the range of bandwidth over which a cluster survives, i.e., the difference between the bandwidth when the cluster is formed and the bandwidth when the cluster is merged with other clusters. Specifically, given the cluster C_i formed at the ith iteration and the cluster C_{i+1} formed at the $(i+1)$th iteration, we define the cluster C_i is survived if the distance between C_i and C_{i+1} is less than the half of the bandwidth at ith iteration. And the distance between two clusters is measured as the distance between these two centers of the clusters.

By using the definition of *survive*, we can calculate the lifetime of each cluster in the cluster hierarchy. Recall that in the mobile sensing data, there can be many isolated data points due to the unreliable sensing devices and the environmental noise. Each of those isolated data points may form one cluster and the lifetime of these clusters could be large. It is thus necessary to filter out these clusters whose number of data points is small. To accurately identify the existence of multiple threat sources, when given the number of the threat source is k, our technique first chooses clusters with top k longest lifetimes at each Mean Shift Clustering iteration (i.e., with different bandwidths) and puts them into a candidate list. We then return cluster centers with the top k longest lifetimes from the constructed candidate list, the top clusters chosen from all the iterations, i.e., under all the bandwidths, as the identified security threat sources. The positions of these cluster centers are the estimated locations of these sources. The threats source localization algorithm is described in Algorithm 3.

6 Simulation Evaluation

6.1 Methodology

To validate our mechanism, we simulated the scenarios of mobile sensing with sensors mounted on city taxicabs. The simulation area is about 25 by 25 street blocks, similar to the size of downtown Manhattan in New York City. The length of each square block is 20 units, which scale to 200 feet in real distance. Since total number of taxicabs in the New York City is about 13,000, it is reasonable to assume that there are 1,500 to 2,000 taxicabs operating in this area. For illustration, we deploy 1,500 taxicabs installed with detection sensors of 5% error rates (false positive rate and false negative rate).

We first randomly generate positions of these taxicabs. We then randomly put one or two security threat sources in the study region. Based on the models described in Section 3, we can assign a probability of positive detection for each mobile sensor. We only focus on the data points with positive sensor readings. We declare a correct detection if the detected cluster center is within the impact range of the security threat source. We repeat the same simulation steps 500 times and compute the percentage of times that the proposed algorithm correctly detects true sources as the detection rate of our mechanism.

6.2 Results

Figure 2 illustrates a set of data points with positive mobile sensor readings depicted as triangles. The detected cluster center obtained from our mechanism is shown as red circles. In this simulation scenario, there is only one threat source located at (250, 250) shown as a blue diamond and the impact range of the source is 20 units (i.e. 200 feet). The red triangles represent sensors with true positive readings while the blue triangles represent false positive readings. We simulated the sensor readings with low quality (e.g. 5% error rates) together with

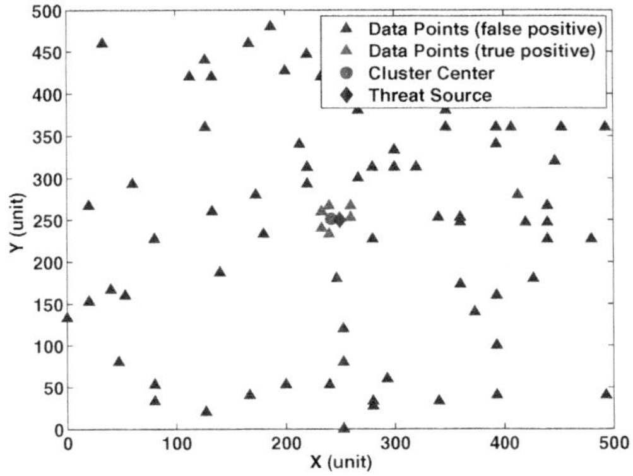

Fig. 2. An illustration of threat source detection using Mean Shift Clustering

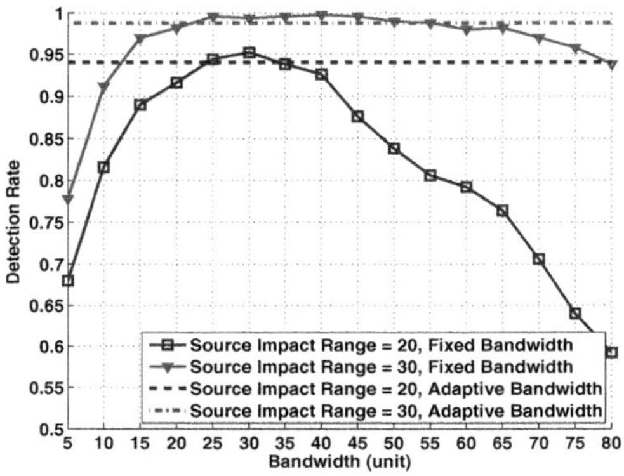

Fig. 3. One security threat source scenario: detection rate vs. bandwidth

environmental noises. As a result, a large amount of false positive readings occur. This shows the challenges in accurate threat source detection. Moveover, the cluster center detected by our mechanism using the Mean Shift Clustering with bandwidth of 30 units is located extremely close to the true threat source. This indicates the feasibility of achieving accurate detection by using our mechanism.

Single Threat Source Detection. We first study the detection rate when there is only one security threat source present with an impact range of 20 units and 30 units respectively. The results of applying Algorithm 1 with various fixed bandwidths and Algorithm 3 using adaptive bandwidth (with the bandwidth increment step $p = 5$) are displayed in Figure 3.

The two curves of detection rate obtained from Algorithm 1 under different impact ranges initially increase with bandwidths, reach maximum detection rate, and then decrease. This is because a smaller bandwidth causes an undersmooth density estimate, whereas a bigger one causes an oversmooth density estimate. Both of the estimates result in inaccurate clustering of the sensing data. Based on the curves shown in the figure, the optimal bandwidth, where we achieve the best detection rate, is 30 units and 40 units for the impact source range of 20 units and 30 units respectively. This suggests that we would choose a bandwidth larger than the source impact range. At the optimal bandwidth, the detection rate is 95.2% and 99.8% for impact source range of 20 units and 30 units respectively. In general, the detection rate of a bigger impact source range is higher than that of a smaller impact range. This is because most likely we will obtain a larger area with more true positive sensor readings to make cluster detection more accurate under a larger source impact range. Furthermore, we observed that the detection results are less sensitive to the bandwidth under larger impact source range, i.e., the detection rate under the impact range of 30 units does not drop as much after its optimal point as the rate under the impact range of 20 units.

Moreover, our Algorithm 3 returns the detection rate of 94% for the impact range of 20 units and 98.8% for the impact range of 30 units. These two detection rates are plotted horizontally in the Figure 3 to compare with the results from

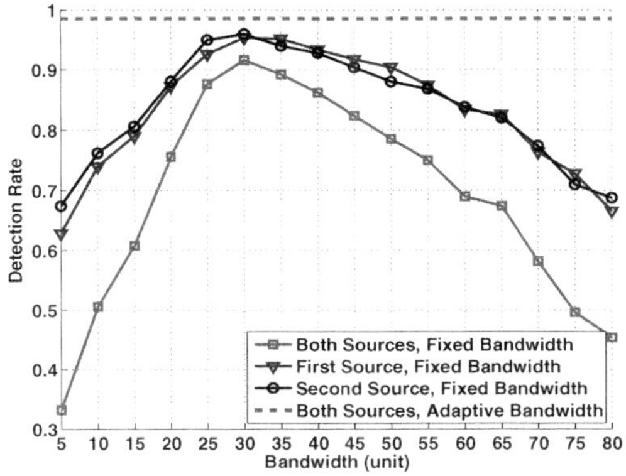

Fig. 4. Two security threat sources scenario: detection Rate under different bandwidths.

Algorithm 1 under various fixed bandwidths. The key observation is that the results of our technique is very close to that obtained under the optimal bandwidth with a fixed bandwidth. Therefore, without knowing the source impact range, our proposed method achieves similar high detection rate and is more feasible in practice than using the Mean Shift with a fixed bandwidth.

Multiple Threat Source Detection. We then move on to the scenario of two threat sources with impact range of 20 units. Figure 4 displays detection rates of correctly identifying each individual source and both under various fixed bandwidths using Algorithm 1 and Algorithm 3 respectively. Examining the curves of detection rate constructed using various fixed bandwidths, the detection rate for either one of individual source is much higher than that of both threat sources. Consistent with the single threat source case in Figure 3, the best detection rate is reached at bandwidth of 30 units under source impact range of 20 units. In addition, we attain an optimal rate of 91.6% for detecting both sources. Furthermore our proposed adaptive bandwidth method achieves a 98.6% detection rate which outperforms all the results using fixed bandwidths. This shows that our mechanism is highly effective in capturing multiple number of security threats.

We next turn to examine the localization error resulted from our threat source localization. Figure 5 shows the cumulative distribution function (CDF) of the localization error under various fixed bandwidths for the single threat source scenario. A localization error is defined as the distance between the detected cluster center to the true source location. The dashed vertical line shows cut off the CDF curves at 20 units which is the source impact range. We observed that we have better localization accuracy when the bandwidth is close to the optimal bandwidth (i.e. 30 units). This is consistent with what we observed in Figure 3 in which a bandwidth value closer to the optimal bandwidth results in a better detection rate. Additionally the median error of the optimal bandwidth is 5 units (scaled to 50 feet) whereas the 90th percentile error is 12.6 units (scaled to 126

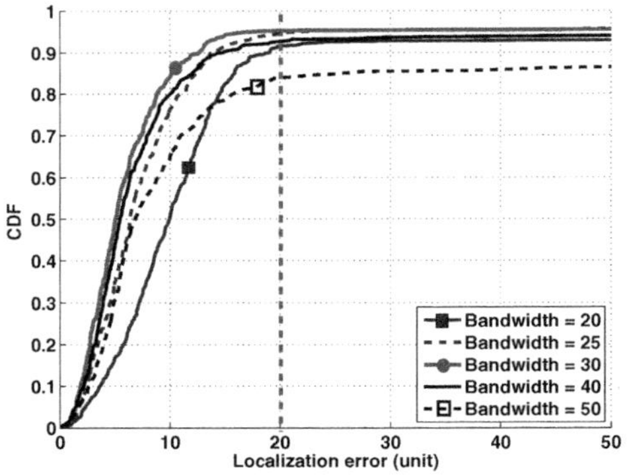

Fig. 5. Localization error under the source impact range of 20

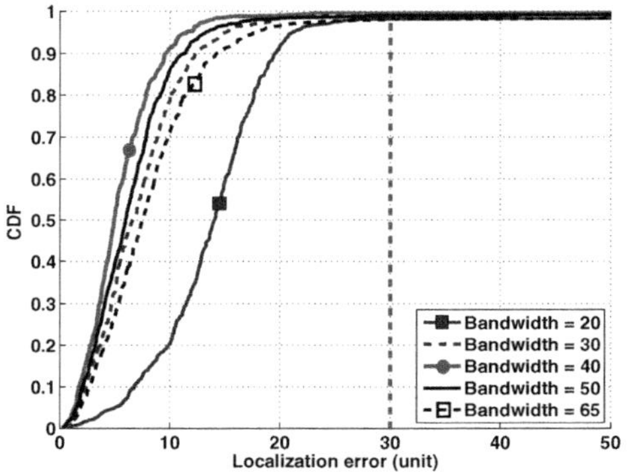

Fig. 6. Localization error under the source impact range of 30

feet). Moreover, we observed that the localization error increases sharply when we failed to detect the threat source.

A similar plot for source impact range of 30 units is depicted in Figure 6. Again, we have the similar observation as Figure 5: a bandwidth value closer to the optimal bandwidth results in a better localization result. Moreover, the localization error of impact range of 30 units is smaller than that of the impact range of 20 units under the optimal bandwidth (i.e., 40 units). Specifically, the median errors of these two cases are about the same, however, the 90th percentile error of the impact range of 30 units is only 9.5 units, which is 30 feet shorter that of the impact range of 20 units. This is due to a longer impact range.

7 Conclusion

In this paper, we proposed to use mobile sensing, which utilizes the ubiquitous nature of mobile devices, for surveillance of security threats in urban environments, e.g., environmental pollution sources or nuclear radiation materials. We proposed to detect the security threats sources using all the sensors collaboratively under the presence of unreliable sensor readings. We formulated security threat source detection as a significant cluster detection problem, in which we identify whether one or more spatial clusters exist in the area significantly from the background noise. We developed a detection mechanism grounded on Mean Shift Clustering procedure to both detect and localize the threat sources. In the proposed detection mechanism, a clustering hierarchy is built and the clusters who have the longest lifetime are returned as the threat sources. We evaluated the effectiveness of our mechanism by simulating mobile sensing using taxicabs in a metropolitan area. Our results obtained from simulations show that our detection mechanism can achieve over 90% detection rate for both single and multiple threat sources with low median localization errors, thereby strongly indicating the feasibility of detecting threat sources in urban environments using our approach.

Acknowledgments. This research was supported in part by NSF grants CNS-0954020 and CCF-1018270.

References

1. Kinkade, S., Verclas, K.: Wireless technology for social change: Trends in mobile use by NGOs. UN Foundation–Vodafone Group Foundation Partnership, United Kingdom (2008) (retrieved November 15, 2008)
2. World Health Organization (WHO): WHO Air Quality Guidelines for Particulate Matter, Ozone, Nitrogen Dioxide and Sulfur Dioxide: Global Update 2005 (2006)
3. Panofsky, W.: Nuclear proliferation risks, new and old. Issues in Science and Technology 19(4), 73–74 (2003)
4. IAEA: Trafficking in Nuclear and Radioactive Material in 2005 (2006), `http://www.iaea.org/NewsCenter/News/2006/traffickingstats2005.html`
5. Campbell, A., Eisenman, S., Lane, N., Miluzzo, E., Peterson, R.: People-centric urban sensing. In: Proceedings of the 2nd Annual International Workshop on Wireless Internet, p. 18. ACM, New York (2006)
6. Johnson, P., Kapadia, A., Kotz, D., Triandopoulos, N., Hanover, N.: People-centric urban sensing: Security challenges for the new paradigm. Technical report, Citeseer (2007)
7. Burke, J., Estrin, D., Hansen, M., Parker, A., Ramanathan, N., Reddy, S., Srivastava, M.: Participatory sensing. In: World Sensor Web Workshop, pp. 1–5 (2006)
8. Archer, D., Beauchamp, B., Mauger, G., Nelson, K., Mercer, M., Pletcher, D., Riot, V., Schek, J., Knapp, D.: Adaptable radiation monitoring system and method. US Patent App. 10/874,127 (June 21, 2004)
9. Glenn, F.: Radiation detection and measurement. John Wiley&Sons, New York (1989)
10. Morelande, M., Ristic, B., Gunatilaka, A.: Detection and parameter estimation of multiple radioactive sources. In: International Conference on Information Fusion (2007)
11. Nemzek, R., Dreicer, J., Torney, D.: Distributed sensor networks for detection of mobile radioactive sources. In: IEEE Nuclear Science Symposium Conference Record, vol. 3, pp. 1463–1467. IEEE, Los Alamitos (2004)

12. Brennan, S., Mielke, A., Torney, D., Maccabe, A.: Radiation detection with distributed sensor networks. IEEE Computer 37(8), 57–59 (2004)
13. Cheng, J., Xie, M., Chen, R., Roberts, F.: A mobile sensor network for the surveillance of nuclear materials in metropolitan areas. Technical report, DIMACS Technical Report, Rutgers University (2009)
14. Purdue University: Cell phone sensors detect radiation to thwart nuclear terrorism. ScienceDaily (January 24, 2008)
15. Glaz, J., Naus, J., Wallenstein, S.: Scan statistics. Springer, Heidelberg (2001)
16. Balakrishnan, N., Koutras, M.: Runs and scans with applications. Wiley, New York (2002)
17. Fu, J., Lou, W.: Distribution theory of runs and patterns and its applications: a finite Markov chain imbedding approach. World Scientific Pub. Co. Inc., Singapore (2003)
18. DematteIš, C., Molinari, N., Daurès, J.: Arbitrarily shaped multiple spatial cluster detection for case event data. Computational Statistics & Data Analysis 51(8), 3931–3945 (2007)
19. Wein, L., Wilkins, A., Baveja, M., Flynn, S.: Preventing the importation of illicit nuclear materials in shipping containers. Risk Analysis 26
20. Langendoen, K., Reijers, N.: Distributed localization in wireless sensor networks: a quantitative comparison. Comput. Networks 43(4), 499–518 (2003)
21. Comaniciu, D.: An algorithm for data-driven bandwidth selection. IEEE Transactions on Pattern Analysis and Machine Intelligence 25(2), 281–288 (2003)
22. Yang, X., Liu, J.: Unsupervised texture segmentation with one-step mean shift and boundary Markov random fields* 1. Pattern Recognition Letters 22(10), 1073–1081 (2001)
23. Comaniciu, D., Ramesh, V., Meer, P.: Kernel-based object tracking. IEEE Transactions on Pattern Analysis and Machine Intelligence, 564–575 (2003)
24. Peng, N., Yang, J., Liu, Z.: Mean shift blob tracking with kernel histogram filtering and hypothesis testing. Pattern Recognition Letters 26(5), 605–614 (2005)
25. Chen, H., Meer, P.: Robust fusion of uncertain information. IEEE Transactions on Systems, Man, and Cybernetics, Part B: Cybernetics 35(3), 578–586 (2005)
26. Cheng, Y.: Mean shift, mode seeking, and clustering. IEEE Transactions on Pattern Analysis and Machine Intelligence 17(8), 790–799 (1995)
27. Comaniciu, D., Meer, P.: Mean shift: A robust approach toward feature space analysis. IEEE Transactions on Pattern Analysis and Machine Intelligence 24(5), 603 (2002)
28. Parzen, E.: On estimation of a probability density function and mode. The Annals of Mathematical Statistics, 1065–1076 (1962)
29. Rosenblatt, M.: Remarks on some nonparametric estimates of a density function. The Annals of Mathematical Statistics 27(3), 832–837 (1956)
30. Sheather, S., Jones, M.: A reliable data-based bandwidth selection method for kernel density estimation. Journal of the Royal Statistical Society. Series B (Methodological) 53(3), 683–690 (1991)
31. Park, B., Marron, J.: Comparison of data-driven bandwidth selectors. Journal of the American Statistical Association 85(409), 66–72 (1990)
32. Fukunaga, K.: Introduction to statistical pattern recognition. Academic Pr., London (1990)
33. Leung, Y., Zhang, J., Xu, Z.: Clustering by scale-space filtering. IEEE Transactions on Pattern Analysis and Machine Intelligence, 1396–1410 (2000)

Detection of Fabricated CTS Packet Attacks in Wireless LANs

Xiaocheng Zou and Jing Deng

Department of Computer Science
University of North Carolina at Greensboro
Greensboro, NC 27412, USA
{x_zou,jing.deng}@uncg.edu

Abstract. IEEE 802.11 standard employs the RTS/CTS handshake procedure to avoid transmission collision and to improve network throughput. However, such an exchange may become a system vulnerability when malicious nodes send fabricated control messages such as CTS to make false claims of channel reservation. To the best of our knowledge, there exists no systematic detection technique for such fabricated control messages. In this paper, we investigate the adverse effects of such attacks on channel throughput and delivery ratio. In order to mitigate these effects, we propose an approach to detect the fabricated messages. With the help of two-hop neighborhood information, our technique enables jamming detection and allows the targeted node to send a message, which instructs neighboring nodes to ignore the fabricated control message. We perform ns-2 simulations to evaluate the benefit of our scheme.

Keywords: WLAN, Fabricated Control Message, CTS, NAV, Detection.

1 Introduction

IEEE 802.11 wireless LAN (WLAN) standard protocol was designed twenty years ago. Nowadays, this standard [1] is hugely popular in civilian, industrial and military networks. However, the inherent assumption of cooperative users can lead to critical confidentiality and trustworthiness issues, even though wireless network security has been the focus of many research [14, 19, 21].

One technique in IEEE 802.11 standard protocol is distribution coordinating function (DCF), which coordinates medium access for contending nodes. DCF is in fact the carrier sense multiple access with collision avoidance (CSMA/CA) schema which employs RTS/CTS mechanism to combat the hidden/exposed terminal problems. In this scheme, communication goes through a sequence of control/data packet dialogues: Request-To-Send (RTS) packet, Clear-To-Send (CTS) packet, Data (DATA) packet, and Acknowledgment (ACK) packet. DCF allows different nodes in the network to have fair shares of the medium usage.

In order to avoid collisions by packet transmissions from different nodes at various locations, a special field termed Network Allocation Vector (NAV) is included in RTS, CTS, and ACK packets. After receiving the NAV values on these

X. Zhang and D. Qiao (Eds.): QShine 2010, LNICST 74, pp. 105–115, 2011.

control packets, nodes can only use the channel after the NAV timer expired. While this technique works well in allowing nodes to reserve the channel, it also opens the door to malicious attackers or selfish nodes in the network to gain unfair access or prohibit other nodes from accessing the channel.

Researchers have identified several weaknesses that might be exploited by an attacker or a selfish user in the network. For example, a selfish node may choose a small interval time in the back-off procedure [9] or delaying SIFS (Short Inter Frame Spacing) interval time instead of DIFS (Distributed Inter Frame Spacing) between the process of exchanging frames [12]. This would always give the attacker itself a better chance of successful channel reservation. Similarly, it may also achieve the same goal by sending out fabricated control packets to interfere with other nodes . This is sometimes termed *intelligent jamming* [6, 8], as compared to physical jamming [17].

Compared to jamming detection [15, 18] at the physical layer, detection of intelligent jamming is more challenging. Such jammings consume less energy for the attacker while achieving a similar result - denying all other nodes' access to the channel. Due to the lack of proper data-link layer authentication techniques, any node in network could send out control packets such as RTS, CTS, and ACK. With these packets, it could dominate the channel by assigning an arbitrarily large value of NAV. Fabricated RTS attacks can be detected as nodes can sense the status of the channel for the data packet transmission with a longer carrier sensing range, or lower sensing threshold [20]. Fabricated CTS or ACK attacks are more subtle. Among others, one difficulty of detecting such attacks is that these control packets do not carry the packet sender's ID [1].

In this paper, we investigate the CTS jamming attacks and propose our solution to detect these attacks. Our assumption is that a malicious node broadcasts fabricated CTS frame specifying a certain amount of NAV duration time. We further assume that attackers cannot fake the source address of any message that they send out, with a radio-signal fingerprinting technique employed by the honest nodes. On the other hand, they can change the targeting address on the fabricated CTS messages at will. Neighbors of the malicious node are forced to be quiet for this period of time. This strategy could significantly reduce network throughput and diminish network's capacity to perform expected functions. Furthermore, we study the impact of an intelligent malicious node that can adapt to our detection technique by randomly alternating the targeting address in the CTS packets.

We design a schema called address inspection schema (AIS) to detect these attack behaviors. Our main idea is to compare the destination field on the CTS frame with the neighborhood information. If the address does not belong to the two-hop neighborhood set, then the control packet will be labeled as fabricated immediately. Otherwise, there must exist one node in the neighborhood with an ID the same as the destination field and may have sent out an RTS packet. If this node had not sent out an RTS request, it should notify other nodes to ignore the CTS message. Our technique requires some extra information: the two-hop neighborhood information that can be carried on the periodic HELLO messages.

Our paper is organized as follows: Section 2 discusses recent related works. In Section 3, our scheme is explained in detail. Simulation-based performance evaluation is presented in Section 4. In Section 5, we summarize our work and discuss future works.

2 Related Work

Security issues in wireless networks have received considerable attentions from research community in recent years. In [16], Wood and Stankovic identified vulnerabilities at each of the layers between physical layer and transport layer due to different kinds of Denial-of-Service (DoS) attacks. Due to nature of wireless communication, openness, and sharing of physical medium, it is relatively easy for malicious nodes in the network to launch jamming radio signal to disrupt normal operation of network. Xu et al. [17] examined the radio interference attack problem and categorized four jamming attack models: constant jammer, deceptive jammer, random jammer, and reactive jammer. They further designed two schemes to detect jamming attacks by employing empirical methods based on signal strength.

At MAC layer, the randomness of random access protocols (such as IEEE 802.11 medium access control) allows misbehaving or malicious nodes to gain priority to access the shared medium after deviating their behavior from normal operation [2, 7, 13]. In Raya et al.'s paper [12], they found that greedy nodes making a slight modification to some parameters defined in 802.11 standard protocol could substantially increase the chance of channel occupation. The following two categories were classified: one is greedy nodes sending out selectively scrambled frames to increase victim's contention window, which gives rise to collision occurrence on victim's side who is supposed to received RTS, CTS, and ACK packets; the other is nodes manipulating protocol parameters to increase bandwidth share by transmitting after SIFS instead of DIFS, assigning large value to NAV, and reducing back-off time. A detection mechanism was designed but its effectiveness could deteriorate if its existence is known to the attacker.

Radosavac et al. [9, 10] concentrated security issues on malicious nodes choosing not to comply to standard protocol by selecting small back-off interval in order to obtain more shares of the channel over honest and normal nodes. Through modeling observation of sequence measurements of back-off interval used by malicious node, they adopt minimax robust detection approach with objective to optimize performance for the worst-case situations. Furthermore, they presented a method to decrease the number of required samples in the minimax robust detection approach. Therefore, observing node could arrive at a decision as soon as possible. Unfortunately, these techniques only work for back-off interval maneuvering attacks.

Assigning large value to Network Allocation Vector (NAV) is another way that a malicious node could use to lower channel utilization. As pointed out in Bellardo and Savage's paper [3], attackers could fabricate certain control packets with large value in duration field in order to reserve the channel for a long period of time. This is because normal nodes who received such control packets would have to update their NAV variable and be quiet. They proposed to place a limit on the duration field in order to mitigate the effect of such attacks. This would work for attacks of changing the NAV values on RTS and ACK frames, but not for CTS frame. This is because the hidden nodes from the data sender could not overhear the RTS frame and have no way of limiting the values for NAV on a subsequent CTS frame. In this paper, we present an approach to allow even hidden nodes to distinguish unsolicited CTS frames from legitimate ones, with the help of two-hop neighbor information.

Other solutions have also been investigated. Ray et al. [11] explained the false blocking problem from RTS/CTS mechanism in IEEE 802.11, which would not only propagate to entire network, but also give rise to deadlock situations. To solve this, they presented RTS validation approach to allow nodes receiving RTS packets defer a small period of time ending at the time when corresponding DATA packet is supposed to begin, instead of deferring the longer period specified in the duration field. Zhang et al. [20] studied jamming ACK attack, which has two advantages to attacker, low energy consumption for attacker and great damage to victim. The size of ACK packet is short and it consumes small amount of energy. An Extend NAV schema (ENAV) scheme was proposed to extend the ACK transmission window from T_{ACK} to $R \cdot T_{ACK}$, which reduces the chance of collision between normal ACK packet and fabricated ACK packet. Chen et al. [4] proposed NAV validation approach to check that a subsequent packet will be received at certain time. For instance, DATA frame should be received within $RTS\_DATAHEAD\_Time$ after RTS. Similarly, ACK frame is supposed to be received within CTS_ACK_Time after CTS packet. However, malicious nodes switching between CTS and ACK packet could avoid detection.

3 Jamming Detection

In this section, we introduce a countermeasure called address inspection schema (AIS) to mitigate the effect of CTS jamming attack.

First of all, we declare several notations that will be used throughout this work. We define N_k as the neighbor set of node k. Furthermore, we use N'_k to represent the two-hop neighbor set of node k, which can be computed by the union of neighbor sets of node k's neighbor nodes. So, $N'_k = \bigcup_{j \in N_k} N_j$.

The main idea of our AIS technique is to check the targeting address carried on the CTS packets. With the help of two-hop neighborhood information, nodes can decide whether the targeting address of a CTS packet is legitimate. This is because, except in dynamic networks, all overheard CTS packets should have targeting addresses that belong to the two-hop neighborhood set. This is true for each of the neighbors of the CTS packet sender. The decision-making procedure for each node receiving or overhearing CTS packet has the following phases:

Prerequisite phase: Node k sends out HELLO message carrying N_k to all its neighbors so that other nodes can obtain their neighborhood information. This phase should be performed periodically. It is important to ensure the freshness of N'_k.

Inspection phase: Node k inspects targeting address specified in the RA (Receiver Address) field of CTS packet. One of the following scenarios may arise

- **I1:** the targeting address is k and node k has sent an RTS packet. The CTS packet is obviously legitimate. Node k proceeds with the normal operation;
- **I2:** the targeting address is k and node k has not sent an RTS packet. The CTS packet is obviously illegitimate. Node k proceeds with the Clearance phase below;
- **I3:** the targeting address is not k and it belongs to the set N'_k. The CTS packet could be legitimate. Node k proceeds with the normal operation, i.e., updating NAV;
- **I4:** the targeting address is not k and it does not belong to the set N'_k. The CTS packet is illegitimate. Node k ignores the CTS packet.

Clearance phase: In this phase, node k sends out a control packet, termed Clear Reservation (CR), to instruct neighbor nodes to ignore the channel reservation from previous CTS control packet. All nodes overhearing a CR message should ignore the CTS packet, recover the original NAV value.

In order to be able to recover the original NAV value after fabricated CTS attack detection, nodes overhearing CTS messages should not simply update their NAV values right away. Instead, they should keep a copy of the FCS of the CTS message and record the current NAV value before updating it. When a CR message is overheard, they will use these information to look for NAV value to recover.

The information carried by CR packet includes frame control, identification of previous CTS packet, source address, etc. Frame control field has the same structure as illustrated in IEEE 802.11 specification, except one new value is introduced for subtype field, CCTS, meaning clear previous CTS packet's reservation. The FCS' field is copied from the FCS field in the fabricated CTS packet. This functions as identification for the detected fabricated CTS message. A detailed CR packet format is provided in Table 1.

Table 1. Packet format of the clear reservation control packet

FIELDS	BYTE	REMARKS
frame control	2	control fields
TA	6	source address
FCS'	4	FCS of the suspected CTS packet
FCS	4	FCS of this message

An illustrative example is provided in Fig. 1. Under different attack methods, the neighboring nodes, if not all, will detect such fabricated CTS messages and clear the channel reservation.

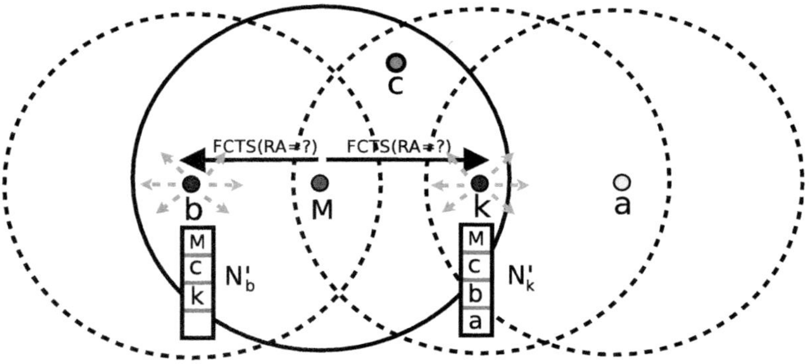

Fig. 1. A scenario illustrating fabricated CTS jamming detection. The network is consisted of five nodes, a, b, c, k, and M. Node M is a malicious node sending out CTS jamming packets randomly. If the targeting address of the fabricated CTS packet from node M is a node outside of this neighborhood, e.g., node x, nodes b, c, and k will detect the jamming and ignore the CTS packet. If node M sends a fabricated CTS targeting at node k, then node k detects it and broadcasts a CR message to notify node c (note that node b is still suffered from the attack). If node M sends a fabricated message to node a, node b detects it and ignores the CTS message (note that node k cannot detect the jamming message).

3.1 Discussions

We have the following discussions regarding to the AIS operation.

Incomplete Detection: As can be seen from the previous discussions, under some attacks, only some neighbors will detect the attack and ignore the fabricated CTS message. Other nodes will still be forced to be silent. This should not have significant impacts on the throughput recovery of the AIS scheme: with some of the nodes in the neighborhood ignoring the fabricated CTS message, they are free to send out channel request or data transmission, occupying the channel instead wasting it for idle. This beats the purpose of the attack.

Communication Overhead: There are two types of additional/revised packets that need to be transmitted: HELLO messages containing each node's neighbor list and the CR message. The HELLO messages are usually broadcast periodically even without the AIS scheme. We only modify the HELLO message to include the neighbor list of the message sender, so that the neighbors can gather information about two-hop neighbors. Note that such information may require some time to obtain.

The CR message will be sent by the node whose ID serves as the targeting address on the fabricated CTS packets. This message is only sent when the node is under attack. As we explained in the Introduction section, we assume that fabricating source address is difficult for attacker (with radio fingerprinting technique in place [5]). Only the node with ID as the targeting address on a suspected CTS message can send a CR message.

4 Performance Evaluation

4.1 Experiment Setup

In order to study the characteristics and evaluate the performance of AIS, we set up simulation experiments using NS2. The wireless transmission range is 250 meters. One node is put into the network serving as the attacker which would periodically send out CTS jamming packets.

We simulated two different types of attacking strategies for the attacker: one targets at non-existing node address, which is termed as "Blind Fabricated CTS" or "Blind FCTS"; the other targets at random node address, which is termed "Focus Fabricated CTS", or "Focus FCTS".

Then, we carried out simulation for the following four scenarios.

- **normal:** network under no FCTS attacks and the AIS;
- **Blind FCTS:** network under Blind FCTS attacks but without AIS running;
- **Blind FCTS + AIS:** network under Blind FCTS attacks and AIS is running;
- **Focus FCTS + AIS:** network under Focus FCTS attacks and AIS is running.

Unless specified otherwise, all remaining parameters used in simulations are listed in Table 2. The attack period is the duration for each FCTS packet and the attack interval is the interval between two consecutive attacks.

Table 2. Simulation Parameters

Simulation	25 sec.	Routing Protocol	AODV
Attack Start Time	8th sec.	AIS Start Time	13th sec
Attack Period	6 msec.	Attack Interval	7 msec.
CBR data rate	120Kb	CBR packet size	100 bytes

Our evaluations focus on two major metrics: throughput and delivery ratio. Throughput is defined as the total traffic transmitted in network. Throughput can be considered as the indicator of network functionality. Note that the throughput presented here is the so-called "instant throughput", which measures the instantaneous throughput, or the number of bits transmitted/received successfully in a unit time. The second major metric that we investigate is delivery ratio, defined as the number of received packets at the receiver divided by the number of transmitted packets. This represents the success ratio of actual transmission.

We first present the results of a pre-assigned network, in which a total of $N = 12$ regular nodes are placed in a field of a 500×500 meters. The attacker locates at the center of the network. As Fig. 2 shows, without attacker in network, the data transmission is stable, and overall trend of transmission stays at a horizontal level. However, when the Blind FCTS attack is introduced, throughput drops to

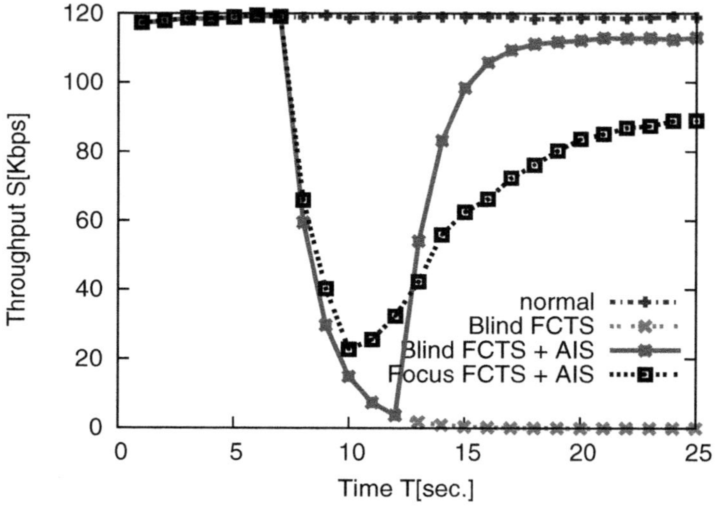

Fig. 2. Throughput performance in normal, Blind FCTS only, Blind FCTS plus AIS and Focus FCTS plus AIS networks. Jamming attack starts at 8th second. AIS kicks in at 13th second. As we can see, AIS helps network to restore most portion of original transmission when network is under attack.

almost 0, starting from the 8th second, which is the attack starting time. This is because the sender is forced to be silent after the Blind FCTS attack.

However, with the help from AIS, victim nodes would ignore illegitimate channel reservation from the attacker, this is demonstrated by two curves in Fig. 2, "Blind FCTS + AIS" and "Focus FCTS + AIS". After the AIS scheme is activated at 13th second, the throughput curves quickly climb up and approach stable throughput. For Blind FCTS attacks, AIS allows every node to detect such attacks and ignore the corresponding NAV reservations. Based on Fig. 2, the last part of "Blind FCTS + AIS" curve is very close to the curve of the normal network, showing that the network has recovered the throughput to original level.

For Focus FCTS attacks, AIS could only recover a majority of the throughput since attacker alternated targeting address randomly. This could be observed in Fig. 2, i.e., the gap in the stable throughput region between "Focus FCTS + AIS" curve and "Blind FCTS + AIS" curve. The reason for such a gap is the detection failure by the AIS scheme (such as the failed detection by node k when node M sends an FCTS message to node a). In addition, we could observe that the lowest point of "Focus FCTS + AIS" curve is around 20, which is different from that of "Blind FCTS + AIS" scenario. The protection described in IEEE 802.11 standard could explain such a phenomenon: when a node receives unexpected CTS packet targeting to itself, it will discard this packet and is free to use the channel later on.

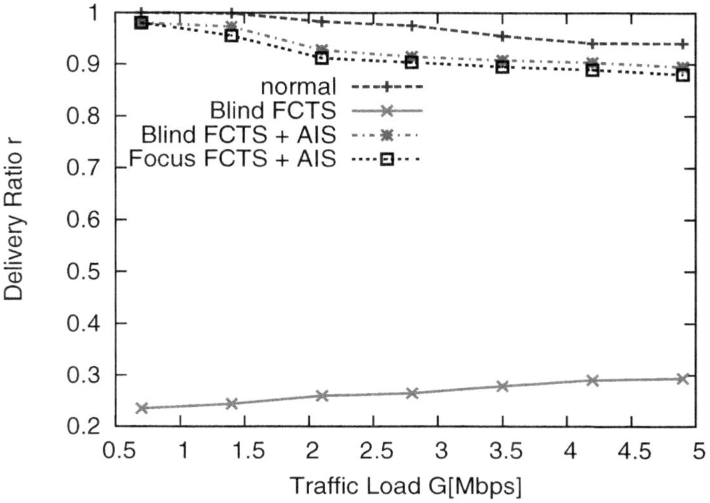

Fig. 3. Delivery ratio comparison in normal, Blind FCTS only, Blind FCTS plus AIS and Focus FCTS plus AIS networks as traffic load increased

We also simulated our scheme in a network where $N = 40$ nodes are placed randomly in a region of 1200×1200. We selected more sender-receiver pairs in order to observe the effect of attack and AIS in a network with higher traffic load.

In Fig. 3, we present delivery ratio results in normal, Blind FCTS only, Blind FCTS plus AIS and Focus FCTS plus AIS networks. The results were obtained through stable conditions, i.e., no dynamic behavior by the attacker or AIS during the observing window. In normal network, overall trend of delivery ratio stays at a high level, and drops slightly at the end, caused by the heavy traffic load. Delivery ratio in Blind FCTS only network is about 0.22 with low traffic load. Seemingly surprisingly, it rises to 0.28 as the traffic load increases. This can actually be explained by the additional pairs of communications, some of which might not be jammed by the attacker.

Networks with AIS running maintain a high delivery ratio, dropping slightly with heavy traffic load. This shows that the use of the AIS technique allows nodes to detect the FCTS attacks and are free to use the channel.

5 Conclusion

In wireless networks, MAC layer has many vulnerabilities and can suffer from different types of attacks. In this work, we have investigated the fabricated CTS attacks to the MAC scheme in wireless LANs. In this attack, an attacker sends fabricated CTS packets with large NAV values to falsely claim the use of the shared channel. We have proposed AIS to mitigate the impact of such jamming attacks. With the help of tow-hop neighborhood information, nodes could distinguish legitimate CTS packets from fabricated ones by observing the targeting address on the CTS packet. When such targeting address falls within the two-hop neighborhood of the attacker, some nodes in the network will be able to detect the attack and ignore the illegitimate claim of channel reservation. Our simulations showed that jamming attack could be easily distinguished, and a significant portion of network throughput can be recovered.

In our future work, we will investigate the jamming attack in mobile networks and evaluate the performance of our proposed scheme in such networks. An approach of delayed action can be used: only after detecting a fabricated CTS message a few times will a node ignore the NAV value from the message. This will provide extra protection for communication of mobile nodes. Theoretical analysis of the performance of our scheme will be performed as well. Furthermore, the overhead of two-hop neighborhood information will be investigated in different networks.

Acknowledgment. We would like to thank Adam Wright, Nihar Kurapati, and Rahul Reddy Maram for their constructive suggestions in the design of our scheme.

References

1. IEEE standard for wireless LAN medium access control and physical layer specifications, p. 802 (November 11, 2007)
2. Awerbuch, B., Curtmola, R., Holmer, D., Nita, C.: Odsbr: An on-demand secure byzantine resilient routing protocol for wireless ad hoc networks. ACM Transactions on Information and System Security (TISSEC) (January 2008)
3. Bellardo, J., Savage, S.: 802.11 denial-of-service attacks: Real vulnerabilities and practical solutions. In: USENIX Security Symposium, vol. 12, pp. 2–2 (2003)
4. Chen, D., Deng, J., Varshney, P.K.: Protecting wireless networks against a denial of service attack based on virtual jamming. In: ACM MobiCom 2003 Poster Session, San Diego, CA, USA (September 14-19, 2003)
5. Hall, J.: Enhancing intrusion detection in wireless networks using radio frequency fingerprinting. In: Proceedings of the 3rd IASTED International Conference on Communications, Internet and Information Technology (CIIT), Kranakis, pp. 201–206 (2004)
6. Lazos, L., Poovendran, R., Ritcey, J.A.: Analytic evaluation of target detection in heterogeneous wireless sensor networks. ACM Transactions on Sensor Networks (TOSN) (March 2009)
7. Li, M., Liu, Y.: Rendered path: Range-free localization in anisotropic sensor networks with holes. IEEE/ACM Transactions on Networking, 320–332 (February 2010)
8. Proano, A., Lazos, L.: Selective jamming attacks in wireless networks. In: 2010 IEEE International Conference on Communications (ICC) (May 2010)
9. Radosavac, S., Baras, J.S., Koutsopoulos, I.: A framework for MAC protocol misbehavior detection in wireless networks. In: Workshop on Wireless Security, pp. 33–42 (2005)

10. Radosavac, S., Crdenas, A.A., Baras, J.S., Moustakides, G.V.: Detecting IEEE 802.11 MAC layer misbehavior in ad hoc networks: Robust strategies against individual and colluding attackers. Journal of Computer Security 15, 103–128 (2007)
11. Ray, S., Carruthers, J.B., Starobinski, D.: RTS/CTS-induced congestion in ad hoc wireless lans. Ad Hoc Wireless LANs (2003)
12. Raya, M., Hubaux, J.-P., Aad, I.: DOMINO: a system to detect greedy behavior in ieee 802.11 hotspots. In: International Conference on Mobile Systems, Applications and Services, pp. 84–97 (2004)
13. Richa, A., Scheideler, C., Schmid, S., Zhang, J.: A jamming-resistant mac protocol for multi-hop wireless networks. In: Lynch, N.A., Shvartsman, A.A. (eds.) DISC 2010. LNCS, vol. 6343, pp. 179–193. Springer, Heidelberg (2010)
14. Stajano, F., Anderson, R.: The resurrecting duckling: Security issues in ad-hoc wireless networks. Computer 35, 22–26 (2002)
15. Strasser, M., Danev, B., Capkun, S.: Detection of reactive jamming in sensor networks. ACM TOSN (2010)
16. Wood, A.D., Stankovic, J.A.: Denial of service in sensor networks. Computer 35, 54–62 (2002)
17. Xu, W., Trappe, W., Zhang, Y., Wood, T.: The feasibility of launching and detecting jamming attacks in wireless networks. In: International Symposium on Mobile Ad Hoc Networking and Computing, pp. 46–57 (2005)
18. Xuan, Y., Shen, Y., Shin, I., Thai, M.T.: On trigger detection against reactive jamming attacks: A clique-independent set based approach. In: 2009 IEEE 28th International Performance Computing and Communications Conference, IPCCC (December 2009)
19. Yang, H., Luo, H., Ye, F., Lu, S., Zhang, L.: Security in mobile ad hoc networks: challenges and solutions. Wireless Communication 11, 38–47 (2004)
20. Zhang, Z., Wu, J., Deng, J., Qiu, M.: Jamming ACK attack to wireless networks and a mitigation approach. In: Proc. of IEEE Global Telecommunications Conference / Wireless Networking Symposium (GLOBECOM 2008), New Orleans, LA, USA, November 30 - December 4. ECP, vol. 950, pp. 1–5 (2008)
21. Zhou, L., Haas, Z.J.: Securing ad hoc networks. IEEE Wireless Networks 13, 24–30 (2002)

Channel Aware and Queue Aware Scheduling in LTE Uplink

Javad Hajipour[1], Amr Mohamed[2], and Victor C.M. Leung[1]

[1] University of British Columbia
[2] Qatar University
{hajipour,vleung}@ece.ubc.ca, amrm@qu.edu.qa

Abstract. Long Term Evolution (LTE) uses single carrier frequency division multiple access (SCFDMA) technique as the multiple access scheme in the uplink, due to its low peak to average power ratio (PAPR) compared to orthogonal frequency division multiple access (OFDMA). This advantage is achieved when the Resource Blocks (RBs) allocated to a user are contiguous in frequency domain. Considering this constraint we devise Channel and Queue Aware Scheduling (CQAS) algorithms to keep the users' queue sizes low and at the same time utilize the system resources efficiently. Using extensive simulations we verify the performance of these algorithms in terms of system throughput and queue size probability.

Keywords: LTE, SCFDMA, Resource Allocation, Channel and Queue Aware Scheduling.

1 Introduction

Long Term Evolution (LTE) standardization [1], proposed by the Third Generation Partnership Project (3GPP), aims at providing high speed data and multimedia services to mobile users and therefore has recently attracted a lot of attention. To achieve its goals, LTE employs orthogonal frequency division multiple access (OFDMA) as the downlink (DL) multiple access solution. OFDMA is a promising technique which provides high spectral efficiency, scalable bandwidth and robustness against multipath impairment. However, since it modulates data on multiple carriers, it leads to a high peak to average power ratio (PAPR). This makes it unattractive for the uplink (UL) where mobile handsets have power and amplifier limitations. Therefore as an alternative to OFDMA, LTE has selected single carrier frequency division multiple access (SCFDMA) for UL. SCFDMA, also known as discrete Fourier transform spread OFDMA (DFT-Spread OFDMA) can be considered as a modified version of OFDMA in which using DFT, data symbols are transformed into frequency domain before being mapped onto orthogonal subcarriers. SCFDMA can lead to lower PAPR while providing OFDMA benefits at the same time. This becomes possible when all the subcarriers allocated to a user are adjacent.

Both in DL and UL, multiple access is performed by allocating frequency and time resources in the units of Resource Blocks (RB) [2] to users. Each RB is

X. Zhang and D. Qiao (Eds.): QShine 2010, LNICST 74, pp. 116–127, 2011.

composed of several adjacent subcarriers in frequency domain and symbols in time domain. Resource allocation and scheduling play an essential role in the efficient use of network resources. Since channel conditions for different users independently vary over the time and frequency domains, channel aware resource allocation methods that utilize multiuser diversity by allocating RBs to the users with favourable channel conditions are desirable. In OFDMA, frequency domain packet scheduling (FDPS) has flexibility in allocating RBs to users, whereas in SCFDMA, RBs allocated to a user need to be contiguous in the frequency domain [2] . This constraint makes it difficult for FDPS to exploit multiuser diversity efficiently.

Recently several papers have worked on packet scheduling in SCFDMA and proposed different heuristic and suboptimal algorithms. The channel dependent scheduling algorithms in [3,4] do not consider any contiguity constraint on allocated RBs. In [5], authors have shown that applying the contiguity constraint results in about 2.7 dB decrease in PAPR. They have proposed three suboptimal algorithms which have been verified through simulations. Similar work has been performed in [6] with the proof to show the NP-hardness of the problem. Both [5] and [6] have taken into account the fact that channel conditions of users over RBs are correlated in the frequency domain, which makes it possible to cope with the contiguity constraint and at the same time utilize channels efficiently. A more general case is considered in [7], which defines a utility function to represent various scheduling policies and devises two approximation algorithms with polynomial runtime. However, how to choose the set of contiguous RBs used in the algorithms has not been explained.

Most of the previous work has considered an infinitely backlogged model, where users' queues are permanently backlogged and therefore always have data to transmit. In this paper we consider a finite-queue model for users where data packets arrive based on a random process and are buffered in users' queues before transmissions. To the best of our knowledge this is the first time that joint Channel and Queue Aware Scheduling (CQAS) is addressed in the LTE UL and the contiguity constraint for allocation of RBs in SCFDMA is considered together with finite data packets available in users' queues for transmissions. We propose different CQAS algorithms for allocation of RBs considering these two constraints. Using simulations, we investigate the performance of the proposed algorithm in terms of queue length and system throughput.

The remainder of this paper is organized as follows. Section 2 introduces the system model and CQAS. CQAS algorithms for SCFDMA are presented in section 3. Section 4 provides the performance evaluation. We conclude the paper in Section 5.

2 Preliminaries

2.1 System Model

We consider UL in a single cell of an LTE system, where a base station (BS) is located in the center and there are K active users inside the cell. Transmission bandwidth is divided into N RBs indexed by $n, n = 1...N$. In each time slot, FDPS decides about the allocation of RBs to the users. More than one RBs can be allocated to a single user with contiguity constraint applied. However each RB can be allocated to at most one user. We assume each user has a single flow of data where packets arrive randomly and are queued in a buffer before they are transmitted in the assigned RBs. We assume that BS has perfect knowledge about the queue size of users and channel condition of every user in all the RBs. Channel information can be obtained through Sounding Reference Signals (SRS) [8] which are sent by users as Channel Quality Indicator (CQI) and queue size information can be achieved based on the users' Buffer Status Reports (BSR)[9], together with the history of their scheduled rates. It is also assumed that channel state remains constant during each time slot. Based on this information and according to the scheduling algorithms, FDPS determines which user to transmit in each RB.

2.2 Channel and Queue Aware Scheduling

In wireless networks, users experience different channel conditions due to different path loss, shadowing and fading. Considering these, it is possible to use system resources efficiently. Opportunistic channel aware scheduling algorithms like Max C/I [10] allocate the channels to the user with the best condition, to obtain the maximum possible throughput in the system. However this method leads to the starvation of the users who experience deep fading in their channels. To prevent this, Proportional Fair Scheduling (PFS) [11] considers the history of the users access to the system in addition to their instantaneous achievable rates. Therefore it is possible to allocate resources to all the users whenever they have higher ratio of achievable rate to average rate. PFS is mostly considered for services that have infinitely backlogged queues. For services with finite buffer sizes it is important to maintain queue lengths in a reasonable range to keep the buffer overflow probability as low as possible. For this purpose, CQAS policies [12],[13],[14] take the length of users' queues into account as well as their channel conditions and give higher priority for transmission to the users with good channel condition or larger queue size. One of the metrics used, is MaxWeight [14]. One version of it is as follows

$$\rho_k = Q_k \min(Q_k, r_k) \tag{1}$$

where ρ_k, Q_k and r_k are, respectively, the MaxWeight metric, the queue length and achievable channel rate for user k. Other than the constraints mentioned in the previous section, UL CQAS has one other challenge. It is due to the fact that each user might have different maximum transmission power and based on the number of RBs allocated to it, different power levels will be available for transmission on each RB. This will have an effect on the modulation scheme and coding rate used for transmission on each RB given the transmissions quality objective, e.g., bit-error rate (BER). Therefore it will affect the number of bits the user can transmit on each RB and as a result the number of RBs needed to transmit a specific number of bits. In the following, we summarize these challenges

- RBs allocated to a user should be contiguous in frequency domain.
- Queues of users have a finite amount of data to send.
- Different users might have different power constraints and based on the number of RBs allocated to each user, transmission power on each RB can be different, and therefore different number of bits can be transmitted on each RB while satisfying the BER objective.

In this paper, considering these challenges, we propose resource allocation algorithms in SCFDMA and evaluate their performance.

3 CQAS in SCFDMA

In this section we propose three algorithms. In the first one we aim to calculate the number of needed RBs for each user to empty its queue and allocate the needed RBs continuously to the users that utilize them better. In the next two algorityms, we use the MaxWeight metric to allocate the RBs to users that have higher values for this metric on the corresponding RB. In all of the following algorithms it is assumed that the users use their maximum power and this power will be divided equally among all the subcarriers of allocated RBs.

3.1 Algorithm 1

This algorithm is illustrated in **Algorithm 1**. Available users are the users that have data in their queues to transmit. $P_{max,k}$ in line 6 is the maximum power of user k and $r_k(j,p)$ indicates the achievable channel rate of user k on RB j if the power used for this RB is p. The algorithm starts from the first RB and performs the following for each available user.

Algorithm 1

1: **Let** S **be the set of available users**
2: **Initialize** $n = 1$.
3: **Do:**
4: **For each** $k \in S$ **compute the following**
5: $M_k = min(N - n + 1,$ **number of RBs needed to send** Q_k **bits**$)$
6: $C_k = min(Q_k, \sum_{j=n}^{n+M_k-1} r_k(j, \frac{P_{max,k}}{M_k}))$
7: $\alpha_k = \frac{C_k}{M_k}$
8: **select** $k^* = \arg\max_k \alpha_k$
9: **Allocate** $RBs : n...n + M_{k^*} - 1$ **to user** k^*
10: $n = n + M_{k^*}$.
11: $S = S - \{k^*\}$
12: **while** $(n \leq N)$ **and** $(\sum_{k \in S} Q_k > 0)$

It first assumes that one RB will be enough and therefore all the power of the user will be given to that RB. Based on this assumption, the channel achievable rate is calculated. If this rate does not empty the queue in one slot, the number of needed consecutive RBs will be considered two and the achievable rate over two RBs will be computed assuming maximm power divided equally among them. This calculation is continued to compute the number of RBs needed to empty the queue. Then user k^* is selected according to line 8. α_k is a measure of utilization of RBs by user k. The larger this parameter is, the better the RBs are used. In other words it shows that a user has better channel conditions on its allocated RBs. After selecting k^*, its needed RBs are assigned to it and the user will be deleted from the available users list. The similar procedure will continue for the rest of RBs to allocate them to available users.

3.2 Algorithm 2

In this algorithm, illustrated in **Algorithm 2**, we consider a MaxWeight metric for user k over each RB n, as follows

$$\rho_{k,n} = Q_k \min(Q_k, r_k(n, \frac{P_{max,k}}{N})) \tag{2}$$

where Q_k is the updated queue length of user k **before** allocation of RB n, which depends on the allocation of previous RBs. Starting from first RB, RBs are assigned one by one, based on the MaxWeight metric, as follows.

For each RB n, $n = 1...N$, this metric is calculated for the available users, and the user with highest metric value is selected and assigned to RB n. Then the information about the queue sizes and the number of assigned RBs are updated and considered in the allocation of next RBs.

Since the number of RBs that are going to be allocated to each user is not known a priori, therefore for calculating achievable channel rate of a user on an

Algorithm 2

Let S , U_k **respectively be the set of available users and the set of RBs already allocated user** k

Initialize $k_0 = 0$,$n = 1$ **and** $U_k = \emptyset$ **for every user** k

Do:

  For each $k \in S$ **compute the following**

    $\rho_{k,n} = Q_k min(Q_k, r_k(n, \frac{P_{max,k}}{N}))$

  Select $k^* = \arg\max_k \ \rho_{k,n}$

  $Q_{k^*} = Q_{k^*} - min(Q_{k^*}, \sum_{j \in \{U_{k^*} + \{n\}\}} r_{k^*}(j, \frac{P_{max,k^*}}{|U_{k^*}|+1}) - \sum_{j \in U_{k^*}} r_{k^*}(j, \frac{P_{max,k^*}}{|U_{k^*}|}))$

  $U_{k^*} = U_{k^*} + \{n\}$

  If $k^* \neq k_0$ **then** $S = S - \{k_0\}, k_0 = k^*$

  If $Q_{k^*} = 0$ **then** $S = S - \{k^*\}$

  $n = n + 1$

while $(n \leq N)$**and**$(\sum_{k \in S} Q_k > 0)$

RB, we assume the allocated power by user k for each RB is $\frac{P_{max,k}}{N}$, as in (2). However after selection of the user according to line 6, the power for the RB is considered equal to $\frac{P_{max,k}}{|U_k|+1}$, where U_k is the set of RBs already assigened to user k. Based on this power, achievable channel rates are computed and queue length of selected user is updated according to line 7.

In case that the selected user's queue will be emptied after allocation of the RB, that user will be deleted from the list of available users. Also in order to comply with contiguity constraint, if the selected user for the current RB is different from the user assigned to the previous RB, the user of previous RB will be deleted from the list of available users and will not be considered in the allocation of the next RBs.

3.3 Algorithm 3

This algorithm, illustrated in **Algorithm 3**, utilizes the channel correlation in frequency domain, using similar idea as in [5] and [6]. Figure 1 demonstrates this idea.

Algorithm 3

Let S, A **and** U_k **respectively be the set of available users, the set of available RBs and the set of RBs already assigned to user** k

Calculate the metric $\rho_{k,n} = Q_k \ min(Q_k, r_k(n, \frac{P_{max,k}}{N}))$ **for each user** k, **on each RB** n **and sort them in descending order of values in the set** V

Do:

  Pick k, n **corresponding to first** $\rho_{k,n}$ **in** V

  Initialize $n_e = n, n_h = n - 1$

  Do:

    check$(k, n_e, 1, t1)$

    check$(k, n_h, -1, t2)$

  while t_1 **or** t_2

  $S = S - \{k\}$

  $V = V - \{\rho_{i,x} | i = k\}$

while $A \neq \emptyset$ **and** $(\sum_{k \in S} Q_k > 0)$

Check$(k, m, direction, result)$

  if $m \in A$ **and** $Q_k > 0$ **and** $\arg\max_i \rho_{i,x} = k$

$$Q_k = Q_k - min(Q_k, \sum_{j \in \{U_k + \{m\}\}} r_k(j, \frac{P_{max,k}}{|U_k| + 1}) - \sum_{j \in U_k} r_k(j, \frac{P_{max,k}}{|U_k|}))$$

    $U_k = U_k + \{m\}$

    $A = A - \{m\}$

    $m = m + direction$

    $result = true$

  else

    $result = false$

  end if

end check

At first, it computes MaxWeight metrics for all the users on all RBs. Then it starts from the RB that has the highest value of metric among all RBs and allocates it to the corresponding user. After updating the queue size of that user, it moves one RB towards the head of the RB sequence and if the same user has the highest metric, it is allocated to the same user. Then the similar procedure is performed one RB towards the end of the RB sequence. This is continued as long as the user has data in its queue and the metric of user on either sides are the highest among other users. Then these RBs and the user are deleted from the list of available RBs and available users.

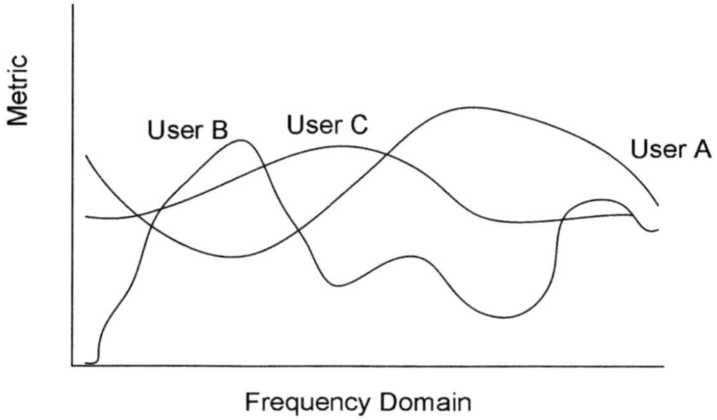

Fig. 1. Utilizing channel correlations for resource allocation

During allocation of adjacent RBs to the user, power allocation to RBs and queue size information are updated in the same way as in **Algorithm 2**. The same procedure is performed for the remaining RBs as long as there are data in queues of available users.

4 Performance Analysis

4.1 Simulation Parameters

To evaluate the performance of the proposed algorithms we have conducted Matlab simulations over 2000 time slots. Simulation parameters are shown in table 1.

Results are presented in terms of system throughput as well as average and maximum queue sizes in the system. Rates are computed, assuming a minimum mean square error (MMSE) equalizer at the receiver and using the following equation [4]:

$$r_k = \frac{BW|U_k|}{N} . \log_2(1 + \beta\gamma_k) \tag{3}$$

Table 1. Simulation Parameters

Parameter Name	Setting
Cell Radius	1000m
Min UE-BS distance	50m
Pathloss and Lognormal Shadowing	From [15]
UE speed	3 km/h
Fast Fading Model	Vehicular A [15]
System Bandwidth	3MHz
Subcarrier Spacing	15 kHz
RB size	12 subcarriers
Number of RBs:N	15
Number of Users:K	10
Time Slot Duration	1ms
User Maximum power	24dBm
Noise Power Spectral Density	-174dBm/Hz
Traffic Model	Poisson
Packet Inter Arrival Time	8ms
Packet Size	2kbit

where BW is the system bandwidth, $|U_k|$ is the number of RBs allocated to user k, β is a constant related to the BER by $\beta = \frac{-1.5}{ln(5BER)}$, and γ_k is the SNR for the data delivered with RBs in U_k defined as

$$\gamma_k = \left(\frac{1}{\frac{1}{|U_{k,sub}|}\sum_{i\in U_{k,sub}} \frac{\gamma_{k,i}}{\gamma_{k,i}+1}} - 1\right)^{-1} \tag{4}$$

where $|U_{sub,k}|$ is the number of subcarriers allocated to user k which is equal to $12|U_k|$, $\gamma_{k,i}$ is the SNR of user k on subcarrier i which is defined as

$$\gamma_{k,i} = \frac{P_k^{(sub)} H_{k,i}}{\sigma_n^2 \triangle f} \tag{5}$$

where $P_k^{(sub)}$ is the allocated power of user k to each of its assigned subcarriers, $|H_{k,i}|$ is the channel gain of user k on subcarrier i including path loss, shadowing and fast fading, σ_n^2 is the noise power per Hz and $\triangle f$ is the subcarrier spacing in Hz.

4.2 Numerical Results

In this subsection, we investigate the performance of our proposed algorithms. As a refernce for our comparisions, we use Alg3(riding peaks) in [6], by considering PF metric in **Algorithm** 3 as follows

$$\rho_{k,n} = \frac{r_k(n, \frac{P_{max,k}}{N})}{R_k} \tag{6}$$

where R_k is the average rate of user k up to current time slot.

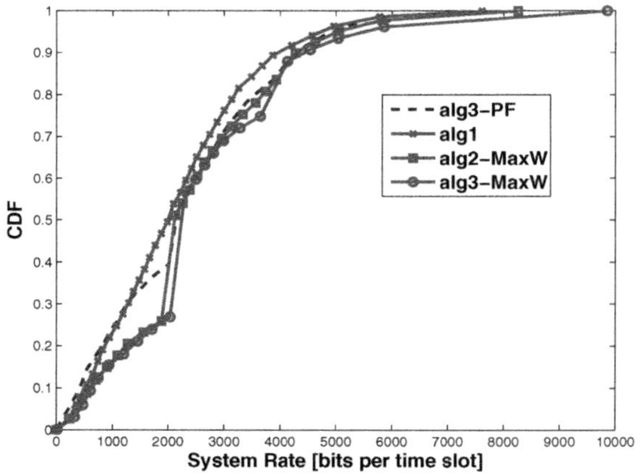

Fig. 2. Distribution of system throughput in each time slot; $K = 8$

Figure 2 shows the probability distribution of system throughput in each time slot. Reference algorithm is shown with *alg3-PF*, to distinguish it from *alg3-MaxW*, which is the **Algorithm** 3 with MaxWeight metric. Also to indicate the use of MaxWeight metric in **Algorithm** 2, it is shown as *alg2-MaxW*.

It is observed that *alg3-MaxW* has the best performance, as it has higher probability for higher throughputs than the other algorithms. *alg1* has the lowest throughput performance. The reason is that it tries to empty the queues and therefore allocates RBs as much as needed. This does not allow multiuser diversity to be utilized over several RBs as much as the other algorithms. *alg2-MaxW* utilizes multiuser diversity by allocating RBs to users that have higher MaxWeight metric. Unlike *alg3-PF*, it considers queue size of users in the allocation of RBs and therefore is able to allocate RBs to the users that have good channel and at the same time have data to transmit. However because *alg2-MaxW* does not start allocation from the RBs with highest metric, it sometimes removes the users from the available users list in the first steps of allocation. Hence it cannot utilize their probable better channel conditions on the next RBs.

Figures 3 and 4 illustrate the probability distribution of the average queue size and maximum queue size in the system, respectively.

It is observed that *alg1* has the lowest performance in terms of users' queue sizes. The reason is that although it aims at emptying the queues, it does not reach its goal as it does not allow more users to utilize RBs to transmit data. The opposite is true with *alg3-MaxW*, which utilizes system resources more efficiently by allocating channels with the best metrics to the corresponding users. Therefore it is able to allow more bits to be sent from the users queues in each time slot.

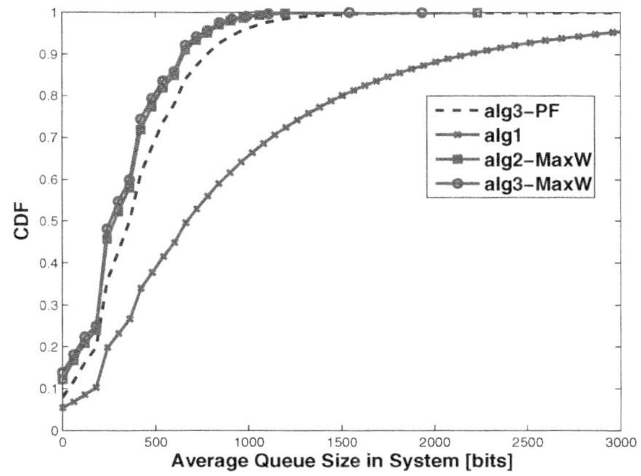

Fig. 3. Distribution of average queue size in each time slot; $K = 8$

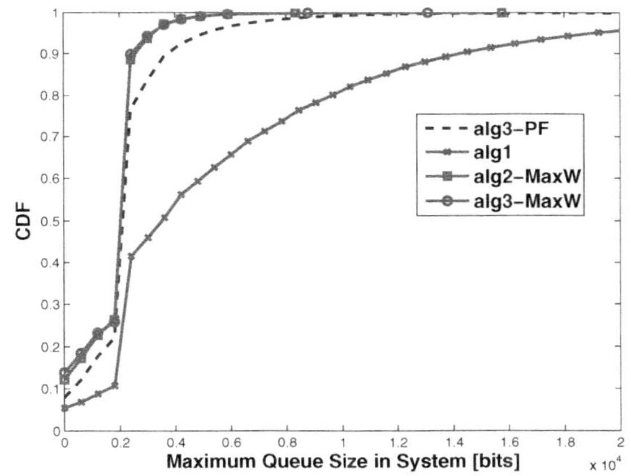

Fig. 4. Distribution of maximum queue size in each time slot; $K = 8$

Figure 5 shows the average queue size of users over time, for different number of users in different algorithms. We observe that as the number of users increases, the queue sizes of users increases rapidly with *alg1* and *alg3-PF* and cannot be kept within a reasonable range, while the average queue sizes for *alg3-MaxW* and *alg2-MaxW* has negligible increase. This shows that *alg3-MaxW* is best able to control the queue size, followed by *alg2-MaxW*, whereas *alg1* and *alg3-PF* are least able to do so.

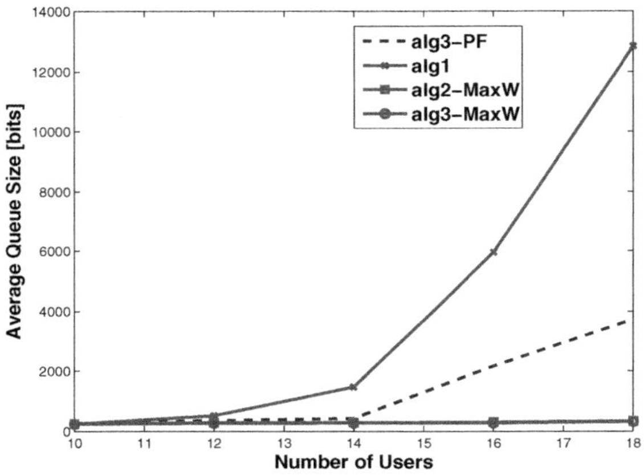

Fig. 5. Average queue size over time v.s. number of users

5 Conclusion

In this paper we have studied channel and queue aware scheduling in LTE UL. Considering the contiguity constraint in allocation of RBs to users, we have addressed the finite queue model for the first time in SCFDMA bandwidth and power allocation. We have proposed three algorithms and using simulations, we have evaluated their performance in terms of system throughput and queue sizes of users in the system. Numerical results show the best performance when using MaxWeight metric and starting allocation from the RBs with the highest metric value. In our future work we will try to improve these algorithms to provide quality of service support for different types of services.

Acknowledgement. This work is supported in part by the Qatar Telecom (QTel) Grant no. QUEX-Qtel-09/10-10 and grants from TELUS and the Natural Sciences Engineering Research Council of Canada (NSERC).

References

1. 3GPP TS36.300: Evolved Universal Terrestrial Radio Access (E-UTRA) and Evolved Universal Terrestrial Radio Access Network (E-UTRAN): Overall Description (October 2010)
2. Holma, H., Toskala, A.: LTE for UMTS-OFDMA and SC-FDMA Based Radio Access (2009)
3. Lim, J., Myung, H.G., Oh, K., Goodman, D.J.: Proportional Fair Scheduling of Uplink Single-Carrier FDMA systems. In: IEEE PIMRC (September 2006)
4. Lim, J., Myung, H.G., Oh, K., Goodman, D.J.: Channel-Dependent Scheduling of Uplink Single Carrier FDMA Systems. In: IEEE PIMRC (September 2006)

5. Angel, L., Berardinelli, G., Frattasi, S., Mogensen, P.: Channel-Aware Scheduling Algorithms for SC-FDMA in LTE Uplink. In: IEEE PIMRC (September 2008)
6. Lee, S., Pefkianakis, I., Meyerson, A., Xu, S., Lu, S.: Proportional Fair Frequency-Domain Packet Scheduling for 3GPP LTE uplink. In: IEEE INFOCOM, pp. 2611–2615 (April 2009)
7. Yang, H., Ren, F., Lin, C., Zhang, J.: Frequency-Domain Packet Scheduling for 3GPP LTE Uplink. In: IEEE INFOCOM (2010)
8. 3GPP TR 25.814: Physical layer aspects for evolved Universal Terrestrial Radio Access (UTRA) (October 2006)
9. 3GPP TS 36.321: Technical Specification Group Radio Access Network; Medium Access Control (MAC) protocol specification (Release 8) (March 2008)
10. Shariat, M., Quddus, A., Ghorashi, S., Tafazolli, R.: Scheduling as an important cross-layer operation for emerging broadband wireless systems. IEEE Communications Surveys and Tutorials 11, 74–86 (2009)
11. Bender, P., Black, P.J., Grob, M., Padovani, R., Sindhushyana, N., Viterbi, S.: CDMA/HDR: a bandwidth efficient high speed wireless data service for nomadic users. IEEE Communications Magazine 38, 70–77 (2000)
12. Tassiulas, L., Ephremides, A.: Stability properties of constrained queueing systems and scheduling policies for maximum throughput in multihop radio networks. IEEE Transactions on Automatic Control 37(12), 1936–1948 (1992)
13. Andrews, M., Kumaran, K., Ramanan, K., Stolyar, A., Whiting, P., Vijayakumar, R.: Providing quality of service over a shared wireless link. IEEE Communications Magazine 39, 150–153 (2001)
14. Andrews, M., Zhang, L.: Scheduling algorithms for multi-carrier wireless data systems. In: ACM Mobicom (2007)
15. Jeruchim, M.C., Balaban, P., Shanmugan, K.S.: Simulation of Communication Systems, 2nd edn. Kluwer Academic/Plenum, New York (2000)

Optimal Channel Pairing and Power Allocation for Multi-channel Multi-hop Relay Networks

Mahdi Hajiaghayi[1], Min Dong[2], and Ben Liang[1]

[1] Dept. of Electrical and Computer Engineering, University of Toronto, Canada
{mahdih,liang}@comm.utoronto.ca
[2] Faculty of Eng. and App. Sci., University of Ontario Institute of Technology, Canada
min.dong@uoit.ca

Abstract. We study the problem of channel pairing and power allocation in a multi-channel, multi-hop relay network to enhance the end-to-end data rate. OFDM-based relays are used as an illustrative example, and the amplify-and-forward and decode-and-forward relaying strategies are considered. Given fixed power allocation to the OFDM subcarriers, we observe that a sorted-SNR subcarrier pairing strategy is data-rate optimal, where each relay pairs its incoming and outgoing subcarriers by their SNR order. For the joint optimization of subcarrier pairing and power allocation, we show that it is optimal to separately consider the two subproblems, for both individual and total power constraints. This separation principle significantly reduces the computational complexity in finding the jointly optimal solution. We further establish the equivalence between sorting SNRs and sorting channel gains in the jointly optimal solution, which allows simple implementation of optimal subcarrier pairing at the relays. Simulation results are presented to demonstrate the performance gain of the jointly optimal solution over some suboptimal alternatives.

1 Introduction

The emerging 4G wireless systems adopt a multi-channel relaying architecture, through Orthogonal Frequency Division Multiplexing (OFDM) and the installation of wireless relay stations. As opposed to a narrow-band single-channel relay, an OFDM relay has access to multiple channels (called subcarriers). It may receive a signal from one subcarrier and transmit a processed version of the signal on a different subcarrier. This *multi-channel relaying capability* can be exploited to process the incoming signals adaptively, taking advantage of the the diverse strength of different channels. In this work, we consider the general problem of jointly optimizing, the pairing of incoming and outgoing channels in multi-channel capable relays, and power allocation to the transmitted signals on these channels, to maximize the end-to-end data rate in a *multi-hop relaying* network. Since the OFDM-based relay architecture is the best known example for multi-channel relaying, we refer to it throughout this paper for the purpose of illustration.

The concept of OFDM subcarrier pairing (SP), which maps incoming and outgoing subcarriers at the relay, was first proposed independently in [1] and [2] for a *dual-hop* (*i.e.*, single-relay) amplified-and-forward (AF) OFDM relay system. For relaying without the direct source-destination link available, [1] used integer programming to

X. Zhang and D. Qiao (Eds.): QShine 2010, LNICST 74, pp. 128–144, 2011.

find the optimal pairing that maximizes the sum SNR. Whereas from a system-design perspective, [2] proposed an SP scheme which is optimal in the noise-free case, with the assumption of uniform power allocation. These works sparked interests for more research in this area. Subsequently, [3] proved that the *sorted-SNR subcarrier pairing* (sorted-SNR SP) scheme, in the absence of the source-destination link, is optimal for both AF and decode-and-forward (DF) dual-hop OFDM relaying systems. In sorted-SNR SP, the subcarriers are paired in the order of SNR seen on each subcarrier. In [4], the authors considered the direct source-destination path to achieve full diversity and presented two suboptimal SP schemes for this case.

The related problem of optimal power allocation (PA) for a *dual-hop* OFDM system was studied by many [5,6,7] for different relay strategies and power constraints. The problem of jointly optimizing SP and PA in a dual-hop OFDM system was studied for DF relaying in [8] and [9], where direct transmission links were assumed unavailable. [8] assumed a total power constraint shared between the source and the relay, while [9] considered individual power constraints imposed on the source and the relay separately. In both cases, it was shown that the joint optimization solution can be obtained by separately optimizing SP and PA. Moreover, the SP scheme which maps the subcarriers solely based on their channel gains is the optimal pairing scheme independent of the optimal PA solution.

Similar study on SP and PA in a multi-hop setting has been scarce. The authors of [10] proposed an adaptive PA algorithm to maximize the end-to-end rate under the total power constraint. For a similar network with DF relaying, [11] studied the problem of joint power and time allocation under the long-term total power constraint to maximize the end-to-end rate. Furthermore, in [10], the idea of subcarrier pairing was mentioned for further performance enhancement in addition to PA. However, no claim was provided on the optimality of sorted SP under the influence of PA. The optimal joint SP and PA solution remained unknown.

In this paper, we present a comprehensive solution for jointly optimizing SP and PA to maximize the source-destination data rate in a multi-channel (e.g., OFDM) multi-hop relay network. The main observations from our work are summarized as follows:

- Given fixed power allocation, the sorted-SNR SP scheme is optimal in multi-hop relaying. Specifically, SP can be separated into individual pairing problems at each relay, where the relay matches the incoming subcarriers to the outgoing subcarriers according to the order of SNRs seen over the subcarriers.
- The problem of joint SP and PA optimization can be decomposed into separate problems that can be solved independently. This separation principle holds for both AF and DF relaying strategies, and for either total or individual power constraints imposed on the transmitting nodes.
- With joint SP and PA optimization, the subcarriers are optimally paired according to their channel-gain order, without the need for knowledge of power allocation on each subcarrier. This allows simple relay implementation for optimal operation.

The generalization from the dual-hop case to the multi-hop case is non-trivial. Intuitively, to maximize the end-to-end rate, the choice of SP at each relay would affect the choices of SP at other relays, which also depend on the specific power allocation scheme used. Therefore, it is not apparent that the optimal SP can be decomposed into independent pairing problems at each relay, or that SP and PA can be separately considered. Besides, we will see later that the techniques used in the dual-hop case cannot be simply extended to the multi-hop case. New techniques are required.

The rest of this paper is organized as follows. In Section 2, we present the system model and problem formulation. Given a fixed PA solution, Section 3 shows that the optimal SP scheme is one based on the sorted SNR. The joint optimization problem of PA and SP is considered in Section 4, and the separation principle between PA and SP is proven. The optimal SP and PA solution is discussed in Section 5 for multi-hop relaying under both total and individual power constraints. The simulation results are provided in Section 6, and finally a summary is given in Section 7.

2 System Model and Problem Statement

We consider an M-hop relay network where a source node communicates with a destination node via $(M-1)$ intermediate relay nodes as illustrated in Fig. 1. For broadband communication between the nodes, the frequency bandwidth is split into multiple subbands for data transmission. A practical system with such an approach is the OFDM system where the bandwidth is divided into N equal-bandwidth subcarriers. We denote by $h_{m,n}$, for $m = 1, \cdots, M$ and $n = 1, \cdots, N$, the channel on subcarrier n over hop m. The additive noise at hop m is modeled as an i.i.d. zero mean Gaussian random variable with variance σ_m^2. We define $a_{m,n} = \frac{|h_{m,n}|^2}{\sigma_m^2}$ as the *normalized* channel gain, often shortened to channel gain, over subcarrier n of hop m. We make the common assumption that the full knowledge of global channels is available at a central controller, which determines the optimal subcarrier pairing and power allocation. We further assume that the destination is out of the transmission zone of the source, and therefore, there is no direct transmission link. For M-hop relaying, a transmission from source to destination occupies M equal time slots, one for each hop. In the mth slot, $m = 1, \cdots, M$, the mth node (the source node if $m = 1$, otherwise the $(m-1)$th relay node) transmits a data block to the $(m+1)$th node (the destination node if $m = M$, otherwise the mth relay node) on each subcarrier. Our study is constrained to half-duplex transmissions, where the relay nodes cannot send and receive at the same time on the same frequency. However, the transmission of different data blocks in different hops may occur concurrently, depending on the scheduling pattern for spatial reuse of spectrum.

2.1 Relaying Strategies

In this work, we consider two types of relaying strategies: AF and DF. In AF relaying, a relay amplifies the data received from an incoming subcarrier and directly forwards it to the next node over an outgoing subcarrier. In DF relaying, a relay attempts to decode the received data from the previous node over each incoming subcarrier and forwards a version of the decoded data on an outgoing subcarrier to the next node. We consider the simple repetition-coding based DF relaying [12,13], where the relay is required to fully decode the incoming message, re-encodes it with repetition coding, and forwards it to the intended user.

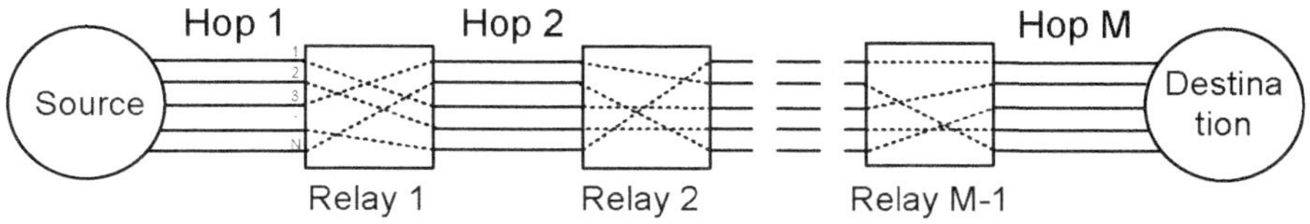

Fig. 1. Illustration of multi-hop OFDM-based network employing subcarrier pairing scheme

2.2 Subcarrier Pairing

The relay conducts SP, matching each incoming subcarrier with an outgoing subcarrier. As the channels on different subcarriers exhibit various quality, a judicious SP scheme can potentially lead to significant improvement in system spectral efficiency.

We denote path $\mathcal{P}_i = (c(1,i), \cdots, c(M,i))$, where $c(m,i)$ specifies the index of the subcarrier at hop m that belongs to path $\mathcal{P}_i$. For example, $\mathcal{P}_i = (3,4,2)$ indicates that path $\mathcal{P}_i$ consists of the third subcarrier at hop 1, the fourth subcarrier at hop 2, and the second subcarrier at hop 3. Once subcarrier pairing is determined at all the relays, the total N disjoint paths $\mathcal{P}_1, \cdots, \mathcal{P}_N$ can be identified from the source to the destination.

2.3 Power Allocation

Denote the power allocated to subcarrier n over hop m by $P_{m,n}$. The SNR obtained on this subcarrier is represented by $\gamma_{m,n} = P_{m,n}a_{m,n}$. For each path $\mathcal{P}_i$, let $\tilde{\gamma}_{m,i} \triangleq \gamma_{m,c(m,i)}$ represent the SNR seen over hop m on this path.

Let $\mathbf{P}_i = (P_{1,c(1,i)}, \cdots, P_{M,c(M,i)})$ be the PA vector for all subcarriers along path $\mathcal{P}_i$. The source-to-destination equivalent SNR of path $\mathcal{P}_i$ is denoted by $\gamma_{\text{SD}}(\mathcal{P}_i, \mathbf{P}_i)$. For AF relaying, it is given by [14],

$$\gamma_{\text{SD}}^{\text{AF}}(\mathcal{P}_i, \mathbf{P}_i) = \left(\prod_{m=1}^{M} \left(1 + \frac{1}{\tilde{\gamma}_{m,i}}\right) - 1 \right)^{-1}, \tag{1}$$

and, in Section 5, we will also use its upper bound [14],

$$\gamma_{\text{SD}}^{\text{AF}}(\mathcal{P}_i, \mathbf{P}_i) \approx \left(\sum_{m=1}^{M} \frac{1}{\tilde{\gamma}_{m,i}} \right)^{-1}, \tag{2}$$

whose approximation gap vanishes as the SNR becomes large. For DF relaying, we have

$$\gamma_{\text{SD}}^{\text{DF}}(\mathcal{P}_i, \mathbf{P}_i) = \min_{m=1,\cdots,M} \tilde{\gamma}_{m,i} \,. \tag{3}$$

We consider two types of power constraints:

Total power constraint: The power assignment $P_{m,n}$, for $m = 1 \cdots M$ and $n = 1 \cdots N$, must satisfy the following aggregated power constraint

$$\sum_{l=1}^{M} \sum_{n=1}^{N} P_{m,n} = P_t. \tag{4}$$

Individual power constraint: The power assignment $P_{m,n}$, for $n = 1, \cdots, N$, needs to satisfy the power constraint of the individual node m, *i.e.*,

$$\sum_{n=1}^{N} P_{m,n} = P_{mt}\,, \qquad m = 1, \cdots, M, \tag{5}$$

where P_{mt} denotes the maximum allowable power at node m.

2.4 Objective

Our goal is to design a jointly optimal SP and PA strategy to maximize the source-destination rate under multi-hop relaying. The source-destination rate achieved through path $\mathcal{P}_i$ is given by

$$R_{\mathrm{SD}}(\mathcal{P}_i, \mathbf{P}_i) = \frac{1}{F_s} \log_2(1 + \gamma_{\mathrm{SD}}(\mathcal{P}_i, \mathbf{P}_i)),$$

where F_s is the spatial reuse factor. In a multi-hop relaying that allows concurrent transmissions, F_s takes value between 2 and M ($F_s \geq 2$ under the half-duplex assumption). The sum rate of all paths determines the total source-destination rate of the system, *i.e.*,

$$R_t = \sum_{i=1}^{N} R_{\mathrm{SD}}(\mathcal{P}_i, \mathbf{P}_i). \tag{6}$$

It is a function of both $\{\mathcal{P}_i\}$ and $\{\mathbf{P}_i\}$, which should be jointly optimized:

$$\max_{\{\mathcal{P}_i\},\{\mathbf{P}_i\}} \quad R_t \tag{7}$$

$$\text{subject to} \quad (4) \quad \text{or} \quad (5),$$

$$\mathbf{P}_i \succeq 0, \quad i = 1, \cdots, N, \tag{8}$$

where $\succeq$ signifies the element-wise inequality.

3 Optimal Multi-hop Subcarrier Pairing under Fixed Power Allocation

In this section, we first consider the case when PA is fixed and given. In this case, the optimization problem in (7) can be re-written as

$$\max_{\{\mathcal{P}_i\}} \sum_{i=1}^{N} R_{\mathrm{SD}}(\mathcal{P}_i, \mathbf{P}_i) \tag{9}$$

and the optimal SP $\{\mathcal{P}_i^*\}$ is a function of $\{\mathbf{P}_i\}$. To simplify the notation, in this section we rewrite $R_{\mathrm{SD}}(\mathcal{P}_i)$ and $\gamma_{SD}(\mathcal{P}_i)$ and drop their dependency on $\mathbf{P}_i$ with the understanding that $\{\mathbf{P}_i\}$ is fixed. In the following, we solve (9) to obtain the optimal SP scheme under this fixed PA. We emphasize that here the generalization from the dual-hop case

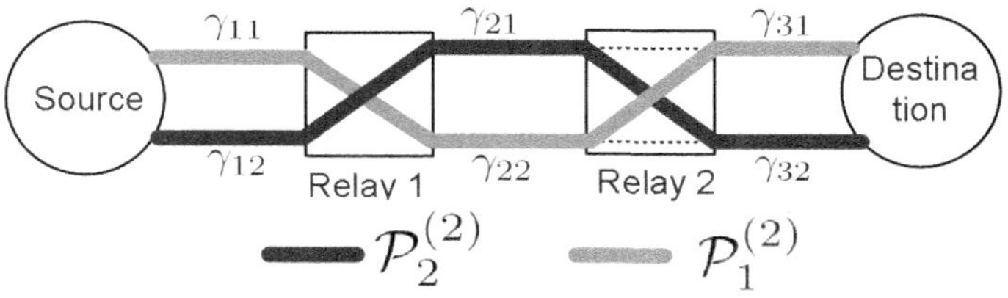

Fig. 2. Three-hop relay with two subcarriers

to the multi-hop case is non-trivial. Intuitively, there is no obvious way to decouple the sequence of pairings at all $(M-1)$ relays. Indeed, the *equivalent* incoming channel from a source to a relay and the *equivalent* outgoing channel from that relay to the destination depend on how the subcarriers are paired over multiple hops. However, we will show that the optimal SP solution over multiple hops can in fact be decomposed into $(M-1)$ independent SP problems, where the mapping of incoming and outgoing subcarriers at each relay is only based on the sorted SNR over those subcarriers, and therefore can be performed individually per hop.

In the following, we first establish the optimality of the sorted-SNR SP scheme for the case of $M=3$ and $N=2$, and then we extend the result to arbitrary M and N.

3.1 Optimal Subcarrier Pairing for Three-Hop Relaying

Two-Subcarrier Case ($N=2$). We first consider a three-hop relaying network with two subcarriers, as depicted in Fig. 2. Without loss of generality, we assume subcarrier 1 exhibits equal or larger SNR than subcarrier 2 over all the three hops, *i.e.*,

A1: $\gamma_{m,1} \geq \gamma_{m,2}$, for $m = 1, 2, 3$.

The optimal SP scheme for this case is given in Proposition 1.

Proposition 1. *For $M=3$ and $N=2$, the solution to* (9) *is the sorted-SNR SP scheme performed on each relay, i.e., $\{\mathcal{P}_i^*\} = \{(1,1,1),(2,2,2)\}$ under assumption A1.*

Proof. At relay 1, there are two ways to pair the subcarriers: (1) subcarriers 1 and 2 over hop 1 are matched with subcarriers 1 and 2 over hop 2, respectively; (2) subcarriers 1 and 2 over hop 1 are matched with subcarriers 2 and 1 over hop 2, respectively. These two ways of pairing lead to the following two sets of disjoint paths from the source to the destination: $\{\mathcal{P}_i^{(1)}\} = \{(1,1,c(3,1)),(2,2,c(3,2))\}$ and $\{\mathcal{P}_i^{(2)}\} = \{(1,2,c(3,1)),(2,1,c(3,2))\}$, where the superscript j in $\{\mathcal{P}_i^{(j)}\}$ indicates a different set of path selection.

By considering the *equivalent* subcarrier channels from the source to the second relay, using the known optimality result for dual-hop relaying, it is easy to see that $c(3,1)=1$ and $c(3,2)=2$ are optimal for $\{\mathcal{P}_i^{(1)}\}$. Furthermore, we only need to show

$$\log_2\left(1+\gamma_{\rm SD}(\mathcal{P}_1^{(1)})\right) + \log_2\left(1+\gamma_{\rm SD}\left(\mathcal{P}_2^{(1)}\right)\right) \geq$$
$$\log_2\left(1+\gamma_{\rm SD}\left(\mathcal{P}_1^{(2)}\right)\right) + \log_2\left(1+\gamma_{\rm SD}\left(\mathcal{P}_2^{(2)}\right)\right), \quad (10)$$

for the case of $c(3,1) = 1$ and $c(3,2) = 2$ for both $\{\mathcal{P}_i^{(1)}\}$ and $\{\mathcal{P}_i^{(2)}\}$, since the case of $c(3,1) = 2$ and $c(3,2) = 1$ for $\{\mathcal{P}_i^{(2)}\}$ can be similarly proven. Inequality (10) for the AF and DF relaying cases are separately proven as follows:

AF Relaying: By inserting (1) into inequality (10) we need to show

$$\begin{aligned} &\left(1+(Q_1^{(1)}-1)^{-1}\right)\left(1+(Q_2^{(1)}-1)^{-1}\right) \geq \\ &\left(1+(Q_1^{(2)}-1)^{-1}\right)\left(1+(Q_2^{(2)}-1)^{-1}\right), \end{aligned} \tag{11}$$

where

$$\begin{aligned} Q_1^{(1)} &= (1+\frac{1}{\gamma_{1,1}})(1+\frac{1}{\gamma_{2,1}})(1+\frac{1}{\gamma_{3,1}}), \\ Q_2^{(1)} &= (1+\frac{1}{\gamma_{1,2}})(1+\frac{1}{\gamma_{2,2}})(1+\frac{1}{\gamma_{3,2}}), \\ Q_1^{(2)} &= (1+\frac{1}{\gamma_{1,1}})(1+\frac{1}{\gamma_{2,2}})(1+\frac{1}{\gamma_{3,1}}), \\ Q_2^{(2)} &= (1+\frac{1}{\gamma_{1,2}})(1+\frac{1}{\gamma_{2,1}})(1+\frac{1}{\gamma_{3,2}}). \end{aligned} \tag{12}$$

The following lemma is used to prove (11).

Lemma 1. *With assumption A1, we have*

$$(Q_1^{(1)}-1)(Q_2^{(1)}-1) \leq (Q_1^{(2)}-1)(Q_2^{(2)}-1). \tag{13}$$

Proof. We omit the proof for brevity.

We proceed by considering the subtraction of the RHS from the LHS of (11),

$$\begin{aligned} &\underbrace{(Q_1^{(1)}-1)^{-1}+(Q_2^{(1)}-1)^{-1}+(Q_1^{(1)}-1)^{-1}(Q_2^{(1)}-1)^{-1}}_{A} - \\ &\left(\underbrace{(Q_1^{(2)}-1)^{-1}+(Q_2^{(2)}-1)^{-1}+(Q_1^{(2)}-1)^{-1}(Q_2^{(2)}-1)^{-1}}_{B}\right) \\ &\geq A(Q_1^{(1)}-1)(Q_2^{(1)}-1) - B(Q_1^{(2)}-1)(Q_2^{(2)}-1) \qquad (14) \\ &= Q_2^{(1)}+Q_1^{(1)}-Q_1^{(2)}-Q_2^{(2)} \\ &= (\frac{1}{\gamma_{2,2}}-\frac{1}{\gamma_{2,1}})\left[(1+\frac{1}{\gamma_{1,2}})(1+\frac{1}{\gamma_{3,2}})-(1+\frac{1}{\gamma_{1,1}})(1+\frac{1}{\gamma_{3,1}})\right] \qquad (15) \\ &\geq 0, \qquad (16) \end{aligned}$$

where the inequality in (14) holds because of (13) and the fact that $Q_i^{(j)} - 1 > 0$ for $i = 1, 2$ and $j = 1, 2, 3$.

DF Relaying: Inserting (3) into inequality (10), we need to show

$$
\begin{aligned}
&(1+\min(\gamma_{1,1},\gamma_{2,1},\gamma_{3,1}))\,(1+\min(\gamma_{1,2},\gamma_{2,2},\gamma_{3,2})) \geq \\
&(1+\min(\gamma_{1,1},\gamma_{2,2},\gamma_{3,1}))\,(1+\min(\gamma_{1,2},\gamma_{2,1},\gamma_{3,2}))\,. \qquad (17)
\end{aligned}
$$

We can consider all possible relations among $\gamma_{m,n}$, for all $m = 1,2,3$ and $n = 1,2$ subject to assumption A1. For example, when $\gamma_{1,1} \leq \gamma_{2,1} \leq \gamma_{3,1}$, $\gamma_{1,2} \leq \gamma_{2,2} \leq \gamma_{3,2}$, $\gamma_{2,2} \leq \gamma_{1,1} \leq \gamma_{3,2}$, and $\gamma_{3,2} \leq \gamma_{2,1}$, (17) yields

$$(1+\gamma_{1,1})(1+\gamma_{1,2}) \geq (1+\gamma_{2,2})(1+\gamma_{1,2}), \qquad (18)$$

which is clearly true. It is not hard to list all other relations, and (17) can be similarly verified. The details are omitted for brevity. ■

Multi-subcarrier case ($N \geq 2$). Here, we provide an argument to extend the result in Proposition 1 to a system with an arbitrary number of subcarriers.

Proposition 2. *For $M = 3$ and $N \geq 2$, the solution to* (9) *is the sorted-SNR SP scheme performed on each relay.*

Proof. Suppose the optimal pairing does not follow the pairing rule of sorted SNR. There is at least one relay (say, Relay 2) that has two pairs of incoming and outgoing subcarriers that are mis-matched according to their SNR. That is, there exist two subcarriers i_1 and i_2 over hop 2, and two subcarriers j_1 and j_2 over hop 3 that are respectively paired with each other while $\gamma_{2,i_1} < \gamma_{2,i_2}$ and $\gamma_{3,j_1} > \gamma_{3,j_2}$. Note that these two subcarrier pairs belong to two disjoint source-destination paths that can be regarded as a 2-subcarrier relay system. From Proposition 1, we know that pairing subcarriers i_1 with j_2 and i_2 with j_1 at relay 2 achieves a higher rate than the existing pairing over these two paths. Hence, by switching to this new pairing while keeping the other paths the same, we could increase the total rate. This contradicts our assumption on the optimality of a non-sorted SNR SP scheme. Hence, there is no better scheme than sorted-SNR SP to obtain the maximum sum rate. ■

3.2 Optimal Subcarrier Pairing for Multi-hop Relaying

We next extend the result in Section 3.1 to a relaying network with an arbitrary number of hops ($M \geq 3$) in the following proposition.

Proposition 3. *The solution to* (9) *is the sorted-SNR SP scheme individually performed at each relay.*

Proof. (By induction.) It is shown in Proposition 2 that the sorted-SNR SP is optimal for $M = 3$. Suppose the claim holds for $M \leq L$. Now consider $M = L + 1$ as shown in Fig. 3(a). Let $\gamma_{eq,n}$ be the nth equivalent channel SNR from the source to Relay $L - 1$, corresponding to the nth incoming subcarrier of that relay. Then, the $L + 1$-hop network can be converted to a 3-hop network, with an equivalent relay whose incoming subcarriers have SNR $\{\gamma_{eq,n}\}$ and outgoing subcarriers remain the same as those of

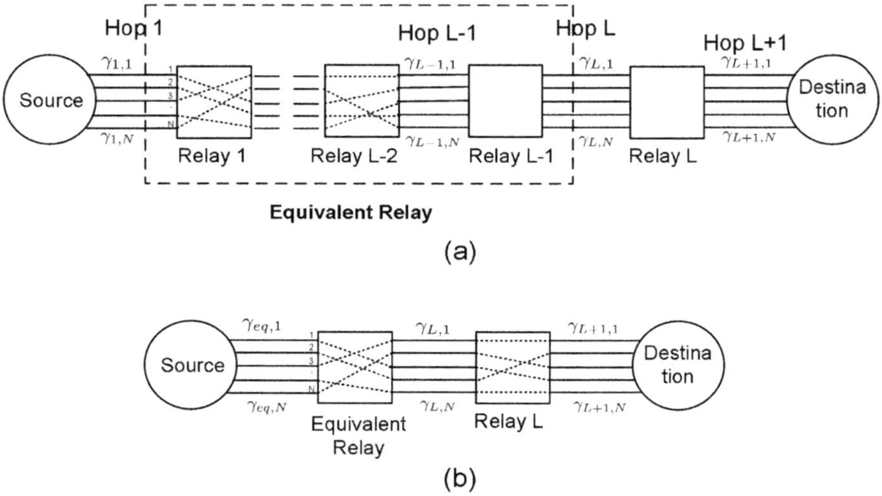

Fig. 3. Converting a $(L+1)$-hop network to a 3-hop network via an equivalent relay

Relay $L-1$, as shown in Fig. 3(b). Hence, from Propositon 2, the optimal SP is one where $\{\gamma_{eq,n}\}$ and $\{\gamma_{L,n}\}$ are sorted and paired at the equivalent relay, and $\{\gamma_{L,n}\}$ and $\{\gamma_{L+1,n}\}$ are sorted and paired at Relay L. Note that the sorted-SNR pairing at Relay L is independent of how the subcarriers are paired at the other relays.

Next, ignore Relay L and replace it by equivalent subcarriers from Relay $L-1$ to the destination. We now have a L-hop network. From the induction hypothesis, the sorted-SNR SP is optimal. In particular, the subcarriers at each of Relays $1, 2, \ldots, L-2$ are sorted by their SNR and paired. Since the equivalent SNR $\{\gamma_{eq,n}\}$ at the equivalent relay is computed by applying (1) or (3) over these sorted and paired subcarriers, it is easy to see that $\{\gamma_{L-1,n}\}$ and $\{\gamma_{eq,n}\}$ are ordered in the same way. Therefore, sorting and pairing $\{\gamma_{eq,n}\}$ and $\{\gamma_{L,n}\}$ at the equivalent relay is the same as sorting and pairing $\{\gamma_{L-1,n}\}$ and $\{\gamma_{L,n}\}$ at Relay $L-1$. ■

4 Jointly Optimal Subcarrier Pairing and Power Allocation: A Separation Principle

So far, given a fixed PA scheme, we have found that the optimal SP scheme for (9) is SNR based, which depends on the transmission power allocated to each subcarrier. We next present the solution for (7) by jointly optimizing SP and PA.

The apparent coupling of SP and PA makes a direct exhaustive search for the jointly optimal solution prohibitively complex. Instead, we will show that the joint optimization problem can be decoupled into two separate PA and SP subproblems. Specifically, we prove that the jointly optimal solution is obtained by pairing subcarriers based on the order of their *channel gains*, followed by optimal PA over the paired subcarriers.

This separation principle holds for a variety of scenarios, including AF and DF relaying under either total or individual power constraints. The solution obtained through such separate optimization bears drastically lower computational complexity in comparison with the exhaustive-search alternative.

Our argument for the separation principle is briefly summarized as follows. We first show that, at a global optimum, between two subcarriers, the subcarrier with a higher channel gain exhibits a larger SNR. This relation implies that the SNR-based ordering of subcarriers is the same as one based on channel gain. Hence, we conclude that the sorted SP scheme based on channel gain is optimal when PA is also optimized. In the following, the above argument is first verified for a multi-hop system consisting of two subcarriers ($N = 2$). We then explain how it can be extended to a system with an arbitrary number of subcarriers.

4.1 Two-Subcarrier Case ($N = 2$)

For both types of power constraint, the following proposition holds:

Proposition 4. *For $N = 2$, in the optimal PA and SP solution for* (7), *at each hop, the subcarrier with better channel gain also provides a higher received SNR, i.e., if $a_{m,i} \geq a_{m,j}$, then $\gamma^*_{m,i} \geq \gamma^*_{m,j}$, for $m = 1, \cdots, M$; $i, j \in \{1, 2\}$; and $i \neq j$.*

Proof. We prove the proposition by contradiction. Let $\mathcal{P}_1$ and $\mathcal{P}_2$ represent the two disjoint source-destination paths corresponding to the optimal SP scheme. Consider any hop m along these paths. Without loss of generality, let subcarrier 1 belong to $\mathcal{P}_2$, subcarrier 2 belong to $\mathcal{P}_1$, and $a_{m,1} \geq a_{m,2}$. Suppose at optimality $\gamma^*_{m,i} < \gamma^*_{m,j}$, i.e., $P^*_{m,2}a_{m,2} > P^*_{m,1}a_{m,1}$, where $P^*_{m,1}$ and $P^*_{m,2}$ are the power allocated to subcarriers 1 and 2, respectively. Let $P_{mt} = P^*_{m,1} + P^*_{m,2}$.

Consider the following alternate allocation of power between subcarriers 1 and 2 over hop m

$$\hat{P}_{m,1} = \frac{a_{m,2}}{a_{m,1}} P^*_{m,2}, \qquad \hat{P}_{m,2} = \frac{a_{m,1}}{a_{m,2}} P^*_{m,1}. \tag{19}$$

We further swap the two subcarriers so that subcarrier 1 belongs to path $\mathcal{P}_1$ and subcarrier 2 belongs to path $\mathcal{P}_2$. Since $\hat{P}_{m,1}a_{m,1} = P^*_{m,2}a_{m,2}$ and $\hat{P}_{m,2}a_{m,2} = P^*_{m,1}a_{m,1}$, the above procedure of power re-allocation and subcarrier swapping does not change the end-to-end rate. However, the sum power after the procedure is reduced:

$$\begin{aligned}\hat{P}_{m,1} + \hat{P}_{m,2} &= \frac{a_{m,2}}{a_{m,1}}(P_{mt} - P^*_{m,1}) + \frac{a_{m,1}}{a_{m,2}} P^*_{m,1} \\ &= \frac{a_{m,2}}{a_{m,1}} P_{mt} + \frac{(a_{m,1})^2 - (a_{m,2})^2}{a_{m,1}a_{m,2}} P^*_{m,1} \\ &< \frac{a_{m,2}}{a_{m,1}} P_{mt} + \frac{(a_{m,1})^2 - (a_{m,2})^2}{a_{m,1}a_{m,2}} \frac{a_{m,2}}{a_{m,1} + a_{m,2}} P_{mt} \\ &= P_{mt},\end{aligned} \tag{20}$$

where inequality (20) is obtained from our initial assumption that $P^*_{m,1}a_{m,1} < P^*_{m,2}a_{m,2}$, which can be rewritten as $P^*_{m,1} < \frac{a_{m,2}}{a_{m,1}+a_{m,2}} P_{mt}$, and that $a_{m,1} \geq a_{m,2}$. This contradicts our initial assumption that the original PA is globally optimal. ∎

4.2 Multi-subcarrier Case ($N > 2$)

The result of Proposition 4 can be generalized to $N \geq 2$. We have the following separation principle for jointly optimal SP and PA.

Proposition 5. *The joint optimization of PA and SP in* (7) *is equivalent to the following separate optimization problem:*

$$\max_{\{\mathbf{P}_i\}} \left\{ \max_{\{\mathcal{P}_i\}} \sum_{i=1}^{N} R_{SD}(\mathcal{P}_i, \mathbf{P}_i) \right\}.$$

Furthermore, the optimal SP is independent of $\{\mathbf{P}^*\}$ *and is performed individually at each relay based on sorted channel gain.*

Proof. A similar proof by contradiction as it was used in Section 3.1 for Proposition 2 can be applied. We omit the details here.

5 Optimal Power Allocation for Multi-hop Relaying

So far we have obtained the optimal SP at all relays. We next find the optimal PA solution for a given SP scheme.

5.1 Individual Power Constraint

Without loss of generality, we assume $a_{m,1} \geq a_{m,2} \geq \cdots \geq a_{m,N}$ for all hops $m = 1, \cdots, M$. From Proposition 5, the subcarriers with the same index are paired, and a path with the optimal SP consists of all the same subcarrier index, *i.e.*, $\mathcal{P}_i^* = (i, \cdots, i)$.

Hence, for DF relaying, the source-destination sum rate in (6) reduces to

$$R_t^{\text{DF}} = \frac{1}{F_s} \sum_{n=1}^{N} \min_{m=1,\cdots,M} \log_2(1 + P_{m,n} a_{m,n}). \tag{21}$$

Maximizing (21) over $\{P_{m,n}\}$ under individual power constraints in (5) can be cast into the following optimization problem using a set of auxiliary variables $\mathbf{r} = [r_1, \cdots, r_N]^T$:

$$\begin{aligned} &\max_{\mathbf{r},\mathbf{P}} \frac{1}{F_s} \sum_{n=1}^{N} r_n \\ &s.t. \quad i) \quad r_n \leq \log_2(1 + P_{m,n} a_{m,n}), \quad m = 1, \cdots, M, \quad n = 1, \cdots, N \\ &\qquad\; ii) \quad \sum_{n=1}^{N} P_{m,n} = P_{mt}, \quad m = 1, \cdots, M \\ &\qquad\; iii) \quad P_{m,n} \geq 0 \quad m = 1, \cdots, M, \quad n = 1, \cdots, N \end{aligned} \tag{22}$$

where $\mathbf{P} \triangleq [P_{m,n}]_{M \times N}$. Since the objective function is linear, and all the constraints in problem (22) are convex, the optimization problem in (22) is convex. We can obtain the optimal solution by employing the standard convex optimization tools, and therefore we omit the details.

Unlike DF relaying, the achievable sum rate for the case of AF relaying is not generally concave in $P_{m,n}$. Therefore, we have a non-convex optimization problem formulated as

$$\max_{\mathbf{P}} \frac{1}{F_s} \sum_{n=1}^{N} \log_2 \left(1 + \left(\prod_{m=1}^{M} \left(1 + \frac{1}{P_{m,n} a_{m,n}}\right) - 1\right)^{-1}\right) \tag{23}$$

$$s.t. \quad i) \quad \sum_{n=1}^{N} P_{m,n} = P_{mt}, \quad m = 1, \cdots, M$$

$$ii) \quad P_{m,n} \geq 0.$$

A non-convex optimization solver may be applied to produce the optimal solution, or we may resort to a suboptimal solution as follows. Instead of the exact rate formula in (23), based on (2), an upper-bound approximation can be used, which is given by

$$R_t^{\text{up}} = \frac{1}{F_s} \sum_{n=1}^{N} \log_2 \left(1 + \left(\sum_{m=1}^{M} \frac{1}{P_{m,n} a_{m,n}}\right)^{-1}\right). \tag{24}$$

Proposition 6. *R_t^{up} in* (24) *is concave with respect to* $\{P_{m,n}\}$.

Proof. The proof follows from the concavity of (2) with respect to $\{P_{m,n}\}$, which can be shown by considering its Hessian matrix. The details are omitted. ∎

Given Proposition 6, the optimization of $\{P_{m,n}\}$ to maximize R_t^{up} is a convex optimization problem, and we again may resort to the standard convex optimization tools to obtain the solution.

5.2 Total Power Constraint

With the same assumption as in the previous section, we have $a_{m,1} \geq a_{m,2} \geq \cdots, a_{m,N}$ for all hops $m = 1, \cdots, M$, and $\mathcal{P}_i^* = (i, \cdots, i)$. Let P_i be the total power allocated along path $\mathcal{P}_i^*$. We can define an equivalent channel gain, $a_{eq,i}$, corresponding to the maximum achievable end-to-end SNR by optimally allocating P_i among the subcarriers on this path:

$$a_{eq,i} = \frac{\max_{\mathbf{P}_i} \gamma_{\text{SD}}(\mathcal{P}_i, \mathbf{P}_i)}{P_i}. \tag{25}$$

For DF relaying, the equivalent channel gain corresponding to path $\mathcal{P}_i^*$ is given by [10]

$$a_{eq,i}^{DF} = \left(\sum_{m=1}^{M} \frac{1}{a_{m,i}}\right)^{-1}, \quad i = 1, \cdots, N, \tag{26}$$

where the two following facts have been applied: 1) the end-to-end rate of a path $\mathcal{P}_i^*$ is equal to the minimum rate over a subcarrier of that path; 2) to maximize the end-to-end rate on one path, the total power allocated to the path must be shared among the subcarriers on this path such that all subcarriers exhibit the same SNR. For AF relaying, the equivalent channel gain for path $\mathcal{P}_i^*$ can be expressed as

$$a_{eq,i}^{AF} = \left(\sum_{m=1}^{N} \frac{1}{\sqrt{a_{m,i}}} \right)^{-2}, \quad i = 1, \cdots, N, \tag{27}$$

where the approximation of end-to-end SNR in (2) over each path is used.

Then, power allocation among the subcarriers over all hops essentially reduces to the problem of power allocation among the paths. This problem then can be formulated as

$$\max_{P_1,\cdots,P_N} \quad \frac{1}{F_s} \sum_{i=1}^{N} \log_2 (1 + P_i a_{eq,i}) \tag{28}$$

$$s.t. \quad \sum_{i=1}^{N} P_i = P_t, \quad P_i \geq 0 \quad i = 1, \cdots, N \,.$$

This problem has the classical water-filling solution

$$P_i^* = \left[\frac{1}{\lambda F_s \ln 2} - \frac{1}{a_{eq,i}} \right]^+, \tag{29}$$

where the Lagrange multiplier λ is chosen such that the power constraint in (28) is met.

6 Simulation Results

In this section, through Monte-Carlo simulations, we first examine the performance of the optimal SP scheme when a power allocation solution is given. We then compare the performance of the joint optimal SP and PA with other suboptimal SP and PA schemes.

For both DF and AF relaying strategies, we consider a multi-hop OFDM-based cooperative network with $N = 64$ and $M = 4$, and the direct transmission between the source and destination is not available. The source, the three relays, and the destination are placed on a straight line with equal distance from each other. The distance between every two nodes is denoted by d_r. The spatial reuse factor is set to $F_s = 3$. A frequency-selective fading channel is simulated, using the L-tap filter model [15] with $L = 11$. Moreover, the pathloss exponent of 3 and the noise variance of 10^{-4} are assumed throughout the simulation.

Uniform power allocation with sum power 100W at each node ($i.e.$, $\sum_{n=1}^{N} P_{m,n} = 100, m = 1, \cdots, 4$) is assumed for both Figs. 4 and 5. Fig. 4 depicts the performance of the achievable end-to-end sum rates vs. d_r for AF relaying under the sorted-SNR SP scheme and direct pairing scheme (*i.e.*, the indices for the incoming and outgoing subcarriers are the same). We observe that a gain of more than 10% can be achieved by conducting optimal pairing compared with direct pairing. Similarly, for DF relaying,

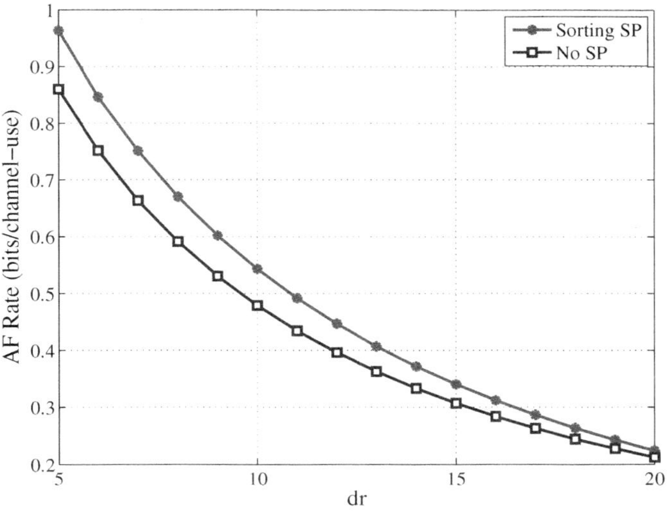

Fig. 4. Rate vs. inter-nodes distance under optimal and suboptimal SP schemes for a multi-hop AF OFDM network with $M = 4$ and $N = 64$

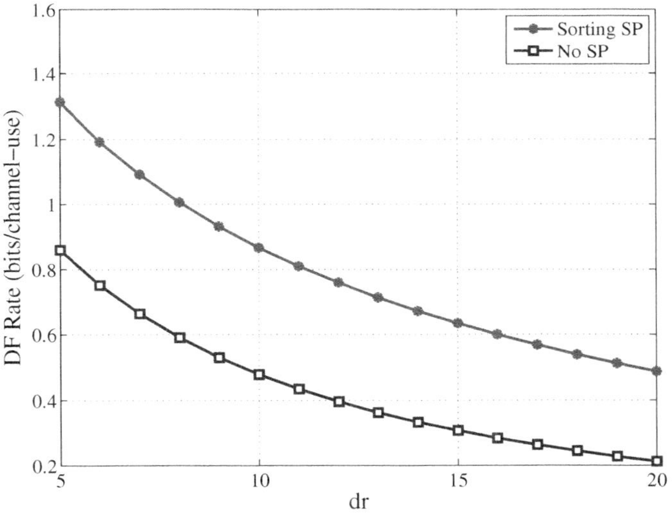

Fig. 5. Rate vs. inter-nodes distance under optimal and suboptimal SP schemes for a multi-hop DF OFDM network with $M = 4$ and $N = 64$

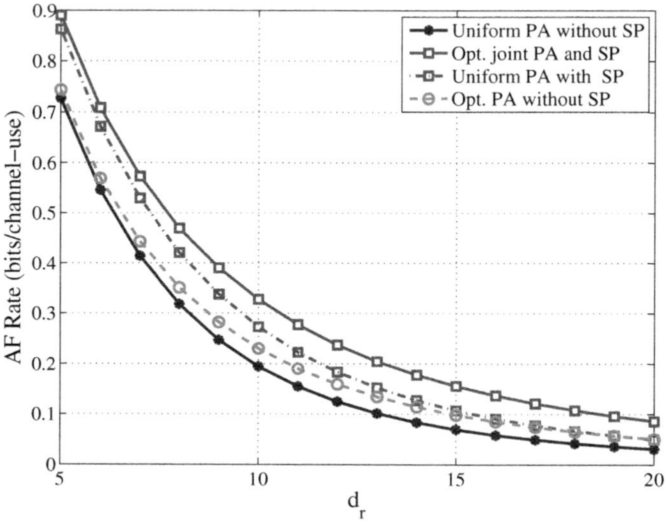

Fig. 6. Rate vs. inter-nodes distance under different SP and PA schemes with total power constraint $P_t = 100W$ for a multi-hop OFDM AF network with $M = 4$ and $N = 64$

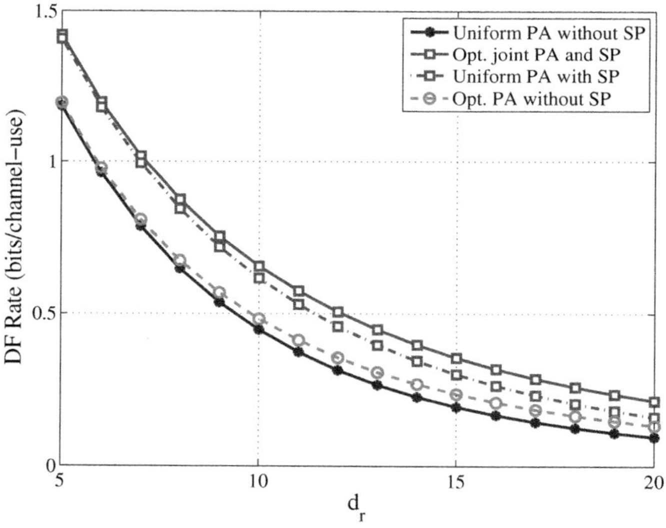

Fig. 7. Rate vs. inter-nodes distance under different SP and PA schemes with total power constraint $P_t = 100W$ for a multi-hop OFDM DF network with $M = 4$ and $N = 64$

Fig. 5 plots the achievable rates vs. d_r. We observe that the gain is even more significant, from 25% when d_r is small to over 150% when d_r is large.

Optimal power allocation leads to further performance improvement when combined with subcarrier pairing. Figs. 6 and 7 illustrate the total achievable rates vs. d_r for AF and DF relaying, respectively, for a total power constraint of 100W ($i.e.$, $\sum_{m=1}^{M} \sum_{n=1}^{N} P_{m,n} = 100$). We compare the end-to-end rate of jointly optimized SP and PA with that of suboptimal alternatives including 1) *Opt. PA without SP*, where only water-filling power allocation is applied; 2) *Uniform PA without SP*, where the total power is uniformly distributed among the N paths obtained from direct subcarrier pairing; 3) *Uniform PA with SP*, where the total power is uniformly distributed among the N paths obtained from the optimal sorted SP. We observe that, through joint optimization of PA and SP, a significant performance improvement can be obtained for both the DF and AF multi-hop relaying cases, especially when the SNR is low, e.g., when d_r is large. Similar observations are made when we increase the number of relays between the source and the destination. Those results are omitted to avoid redundancy.

7 Conclusion

In this paper, we have studied the problem of jointly optimizing spectrum and power allocation to maximize the source-to-destination data rate for a multi-channel M-hop relaying network. OFDM relays are used as example. For a fixed power allocation, we have shown that the general SP problem over multiple hops can be decomposed into $(M-1)$ independent SP problems at each relay, where the sorted-SNR scheme is optimal. We then proved that a jointly optimal solution for the SP and PA problems can be achieved by decomposing the original problem into two separate PA and SP problems solved independently. The solution obtained through the separate optimization bears considerably lower computational complexity compared with exhaustive-method alternatives. The separation principle was shown to hold for a variety of scenarios including AF and DF relaying strategies under either total or individual power constraints. For all these scenarios, the optimal SP scheme maps the subcarriers according to their channel gain order, independent of the optimal PA solution. From simulation, we observed that significant gains in data rate can be achieved from employing jointly optimal SP and PA in multi-channel multi-hop relaying, especially when the networking environment is challenging.

References

1. Hottinen, A., Heikkinen, T.: Subchannel assignment in OFDM relay nodes. In: Annual Conf. on Information Sciences and Systems, Princeton, NJ (March 2006)
2. Herdin, M.: A chunk based OFDM amplify-and-forward relaying scheme for 4g mobile radio systems. In: Proc. IEEE Int. Conf. Communications (ICC), vol. 10 (June 2006)
3. Li, Y., Wang, W., Kong, J., Hong, W., Zhang, X., Peng, M.: Power allocation and subcarrier pairing in OFDM-based relaying networks. In: Proc. IEEE Int. Conf. Communications (ICC) (May 2008)
4. Shen, Z., Wang, X., Zhang, H.: Power allocation and subcarrier pairing for OFDM-based AF cooperative diversity systems. In: Proc. IEEE Vehicular Technology Conf, VTC (April 2009)

5. Vandendorpe, L., Duran, R., Louveaux, J., Zaidi, A.: Power allocation for OFDM transmission with DF relaying. In: Proc. IEEE Int. Conf. Communications, ICC (2008)
6. Hammerstrom, I., Wittneben, A.: Power allocation schemes for amplify-and-forward MIMO-OFDM relay links. IEEE Trans. Wireless Commun. 6(8), 2798–2802 (2007)
7. Hajiaghayi, M., Dong, M., Liang, B.: Using limited feedback in power allocation design for a two-hop relay OFDM system. In: Proc. IEEE Int. Conf. Communications (ICC), vol. 2 (June 2009)
8. Wang, W., Yan, S., Yang, S.: Optimally joint subcarrier matching and power allocation in OFDM multihop system. EURASIP J. Appl. Signal Process, USA (March 2008)
9. Wang, W., Wu, R.: Capacity maximization for OFDM two-hop relay system with separate power constraints. IEEE Trans. Veh. Technol. 58(9), 4943–4954 (2009)
10. Zhang, X., Gong, Y.: Adaptive power allocation for multi-hop OFDM relaying systems. In: Int. Conf. on Infor., Commun., Signal Processing, ICICS (2007)
11. Zhang, X., Jiao, W., Tao, M.: End-to-end resource allocation in OFDM based linear multi-hop networks. In: Proc. IEEE INFOCOM (2008)
12. Cover, T., El Gamal, A.: Capacity theorems for the relay channel. IEEE Trans. Inform. Theory 25(5), 572–584 (1979)
13. Laneman, J., Wornell, G., Tse, D.: An efficient protocol for realizing cooperative diversity in wireless networks. In: Proc. IEEE Int. Symp. on Infor. Theory, ISIT (2001)
14. Hasna, M., Alouini, M.-S.: Outage probability of multihop transmission over Nakagami fading channels. IEEE Commun. Lett. 7(5), 216–218 (2003)
15. Tse, D., Viswanath, P.: Fundamentals of wireless communication. Cambridge University Press, New York (2005)

Resource Allocation with Minimum End-to-End Delay Differential Consideration in Multi-hop Cognitive Radio Networks

Yean-Fu Wen[1] and Wanjiun Liao[2]

[1] Department of MIS,
National Chiayi University, Taiwan (R.O.C)
yeanfu@mail.ncyu.edu.tw
[2] Department of Electrical Engineering,
National Taiwan University, Taiwan (R.O.C)
wjliao@cc.ee.ntu.edu.tw

Abstract. In cognitive radio networks, devices can dynamically sense, negotiate, and switch to available spectral bands so as to enhance spectrum utilization. The available spectral resource may vary with time, location, and spectral bands. This leads to many implementation difficulties, and one especially challenging one is how to fairly allocate these resources for multiple concurrent transmission flows in a multi-hop wireless environment. In this paper, we attempt to minimize the maximum end-to-end delay differential among all multi-hop flows within interference range. Flows within the same interference range may be on different routing paths with different network conditions such as hop count, network load and primary user's behavior in previous hops. Determining how to fairly allocate resources to flows within the same interference range among a disjoint set of spectral bands in terms of minimal end-to-end delay differential becomes an important issue. We consider the accumulated delays (including sensing and negotiating delay, and queuing delay) up to this hop, and the rates of channel error and primary-user interruption on different bands. We then adopt four approximation schemes to solve this problem. The simulation results show that the average end-to-end delay differential with our proposed algorithms for all flows is minimized.

Keywords: Cognitive radio networks, delay differential, negotiation, priority, resource allocation, spectral bands.

1 Introduction

Cognitive radio (CR) is the technology that allows secondary users (SUs) to use the radio spectrum unoccupied by primary users (PUs) [1] [2]. With CR, unused spectral bands in existing wireless networks can be better utilized. To avoid affecting the operation of PUs, SUs should keep sensing channels and switch to other available spectrum holes when PUs appear on the channel. The available

X. Zhang and D. Qiao (Eds.): QShine 2010, LNICST 74, pp. 145–160, 2011.

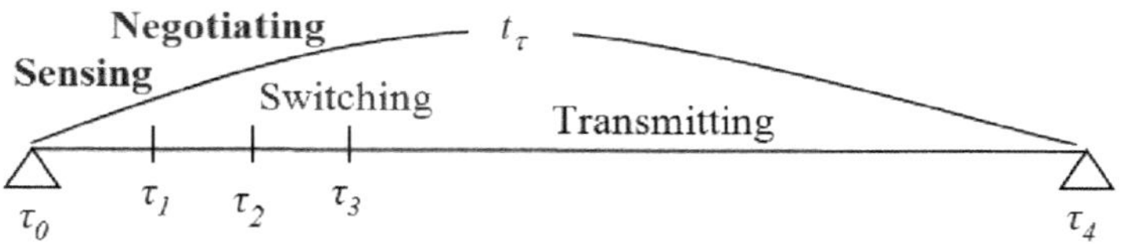

Fig. 1. A general operating cycle for a CRN

spectrum holes refer to the spectral bands such that 1) no PUs are present or currently using them, 2) SUs are free to use them, and 3) the transmission between two SUs will not affect the normal operation of PUs on neighboring links [3]. Since the presence behavior of PUs may vary with time, location, and different radio technology, the available spectral band set available for each SU may also vary. Moreover, for a certain SU node, the characteristics of the channels in the associated available spectral band set may also be different. The important factors that affect the channels include link error rate, PU interruption rate (due to its presence on the link), link length (or equivalently, transmission range and interference range), and the delay in the queue.

To send data on a channel, SUs need to search for available spectral bands to form a spectral channel. As shown in Figure 1, SUs follow an operation cycle with sensing, negotiating, switching, and transmitting [3]. In each cycle, the amount of resources used for transmission depends on: 1) sensing period ($\tau_0 - \tau_1$): the time used to sense and obtain unoccupied spectral bands; 2) negotiating period ($\tau_1 - \tau_2$): the time used to negotiate sensed free spectral bands with neighboring SUs; and 3) switching period ($\tau_2 - \tau_3$): the time used to switch to another spectral channel formed by a different subset of spectral bands. The sensing range is the area within which SUs are able to detect whether or not any PU sends signal on a spectral band. Even though a CR is able to sense in a large area (e.g., 1 km^3), the sensed free spectral bands, which are out of the transmission range (e.g., 100 m^3), may not be available for current transmissions. In addition, the time is too short to sense for longer range. Thus, we treat the effect of the sensing range as equivalent to that of the interference range. This leads to the effect that the negotiation range is equivalent to the sensing range with the same reasons, i.e., 1) out-of-range sensing result messages require longer path transmission; 2) out-of-range free spectral bands may not be available for SUs within the transmission range, and 3) out-of-range SUs which adopt the same spectral bands for transmission will not affect the SUs within the interference range. Therefore, the effects of sensing range and negotiation range in this work are limited to the interference range.

In general, the transmission range determines the interference range. A node which can reach its next hop nodes to form a set of links is initialized by the maximum transmission range, but the range of some links may be less than the maximum transmission range. The link range is determined by the required signal strength for the link nodes to communicate with each other. Accordingly, the sensing and negotiation ranges for each link are also determined.

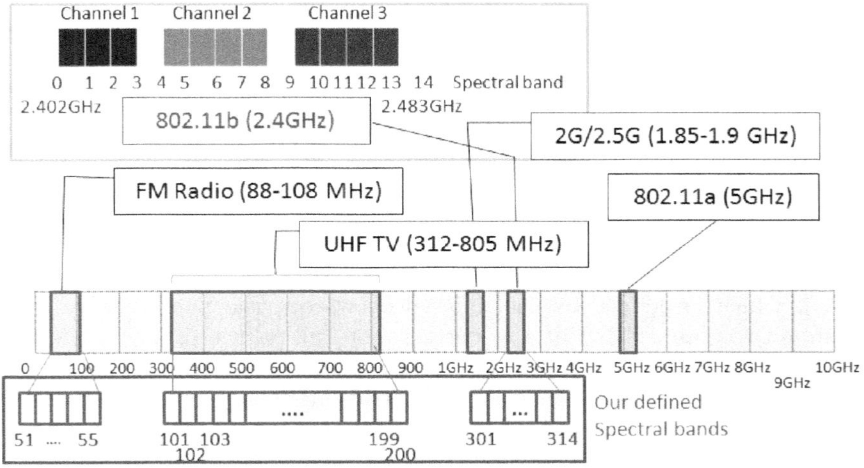

Fig. 2. The defined spectral bands marked with a serial number

The trade-off between the transmission range and the amount of available resources must be balanced. If a CR node chooses a longer link (i.e., a larger transmission range), the number of spectral bands and available duration decrease because it may need to wait for more PUs to release the spectral bands within the sensing range. In addition, the sensing range is determined by the sensing time and the capability of a CR device. The faster the sensing time, the shorter the sensing ranges. Therefore, the transmission range and the sensing range are both controlled to enlarge the amount of available spectral bands and transmission time with a reasonable sensing time.

A spectral band is the basic unit to model wireless transmission resource. The pool of spectral bands is marked with a set of numbers from 1 to M, where $|M|$ is the total number of spectral bands, as shown in Figure 2. Then, one to multiple bands are moderated as a spectral channel. For example, four spectral bands form one channel in IEEE 802.11 wireless standard [4]. When PUs in an area do not use channels 1 and 2 (which cover frequency bands $0-3$ and $4-8$, respectively), frequency bands 0 to 8 are available for use by SUs in that area. Then, each link can adopt one or more channels for transmissions.

In this paper, we will explore the resource allocation problem in multi-hop cognitive radio network with quality-of-service guarantees. We consider that each flow in the CR network has an end-to-end delay constraint. Then, we would like to design an efficient spectrum sharing mechanism among flows such that the maximum end-to-end delay for each flow is within its tolerance. We suppose that the routing path for each flow is given, and the presence behavior of PUs on a channel follows a certain distribution. The transmission time over each link on the path may be affected by many factors, including the PU presence behavior and the link condition (e.g., link error rate, link length, and wireless interference). The higher the PU interruption rate, the longer the packets must wait in the outgoing queue and the longer the packet delay. For each node, the number of flows passing through may not be identical, and the routing path (together with the accumulated delay up to this node) of each flow may also be different. Moreover, the number of hops and the link condition on the remaining path to be traversed by each flow may also be different. Our problem is then to determine an efficient resource allocation scheme to ensure the delay requirement of each flow is satisfied. We take into account the elapse time of each flow has been traversed and its remaining until the deadline in our allocation problem.

To tackle this problem, we introduce the concept of delay differential (denoted by d_ρ), which is defined as the difference between the packet delay and the maximum tolerable delay at the current hop for packet ρ. The packet delay here includes all possible delay components incurred at this hop, including the nodal delay, the queuing delay, and the transmission time. The average per-hop tolerable delay is approximated by the maximum tolerable end-to-end delay divided by the number of hops on the path that a flow traverses to reach its destination. Thus, the maximum tolerable delay at the kth hop for an n-hop path is then expressed by $D - h \times (k + 1)$, where D is the maximum end-to-end tolerable delay for this flow, and h is the per-hop tolerable delay (i.e., D/n). The smaller the delay differential, the longer the waiting time that the packet can tolerate at this hop without missing the deadline. A negative delay differential indicates that the packet may have experienced harsh path condition in previous hops and/or may have more hops to travel on the remaining path so that it should be allocated spectral bands with priority in order not to miss the deadline. A path with larger hop count tends to be more sensitive to small delay differentials. In other words, a longer path needs to be allocated resource with higher priority at each hop; otherwise, the deadline may be missed with a higher probability. This differential value is carried in each packet header, and the delay differential of each packet to the current node can be used as an indicator for scheduling. In this way, for each out-going link, the delay differential values of all flows in the interference range can be minimized.

Figure 3 shows an example of two flows with a maximum end-to-end tolerable delay of $D = 500$ unit time. Flow 1 is routed via path $\{o, a, c, e, g\}$, and Flow 2, via path $\{b, c, d, f\}$. For Flow 1, there are five nodes to traverse, and the accumulated tolerable delay at node c (i.e., via link (c, e) to node e) is $h_{1c} = 300$. For Flow 2, the accumulated tolerable delay at node c (via link (c, d) to node d) is $h_{2c} = 250$. Suppose that a packet x for Flow 1 is queued on the outgoing link (c, e) with a packet delay of 225, so its delay differential is $d_x = -75$. Similarly, for packet y of Flow 2 queued on link (c, d) with a packet delay of 275, its delay differential is $d_y = 25$. In this case, packet y is more likely to miss the deadline when it reaches the destination due to 1) having experienced more harsh path conditions in previous hops, and/or 2) having to travel more remaining hops to reach its destination. Therefore, Flow 2 should be scheduled with priority for transmission.

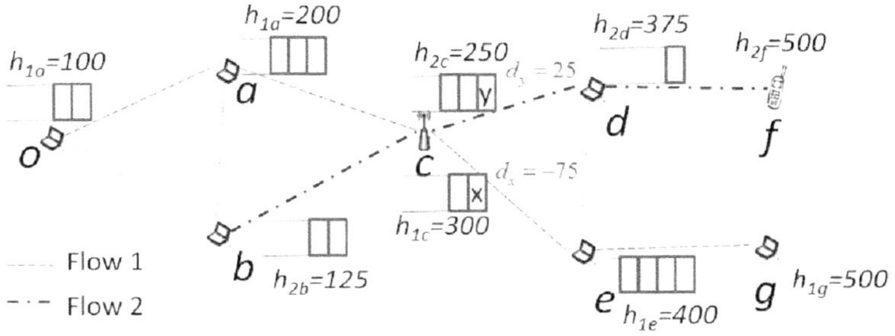

Fig. 3. An example illustrating the delay differential concept

The number of available spectral bands varies with different time and locations. The basic idea of our method is that the packet with larger delay differential has higher priority for spectral allocation, resulting in less available spectral bands for its neighbors. In other words, a transmission pair of nodes negotiating for more spectral bands can moderate higher bit-rate, which means more queued packets can be sent to the next hop in this cycle time so that the aggregated delay differential is decreased in the next hop.

The main contribution of this work is to allocate available spectral bands for CR network with minimal end-to-end delay guarantees by the following mechanisms:

- **Local resource allocation to achieve global load balancing.** We consider the effect of the sensing and negotiation range within the interference range, which is determined by the transmission range. We then introduce the concept of delay differential as the local fairness metric, and a distributed algorithm to ensure end-to-end delay fairness.
- **Four approximation schemes for proportional allocation under different CR network conditions.** The proportional resource allocation is designed according to CR network conditions, such as the queue size, delay differential, and PU interruption rate to approximate the delay differential fairness among the flows such that the relative importance of such attributes are properly requested.
- **Packet level performance evaluation to achieve flow level load balancing.** The basic unit of our model and performance evaluation is the packet. The flow transmission load balancing is achieved by packet fairness scheduling within interference range. In this way, the delay differential caused by various CR network conditions is balanced hop-by-hop for each packet to achieve end-to-end flow fairness.

The rest of the paper is organized as follows. In Section 2, the network model and problem are described. In Section 3, the proposed allocation schemes and evaluation algorithm are presented. In Section 4, the simulation results and related discussions are shown. Finally, this paper is concluded in Section 5.

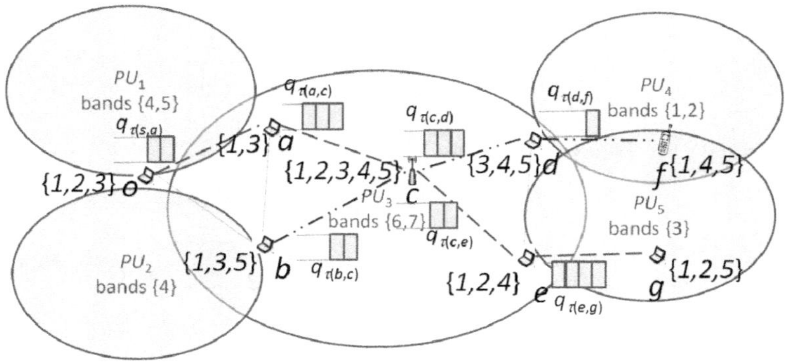

Fig. 4. The network model of our problem

2 Network Model and Problem Description

2.1 Network Architecture

The network is modeled as a connected graph $G(V, L)$, where V represents a set of nodes in the network, L is a set of links (u, v) such that nodes u and v are within the transmission range of each other. The node set V contains two subsets V_{SU} and V_{PU} for SU nodes and PU nodes, respectively. Each node $v \in V_{SU}$ is static and associated with a finite buffer for each CR and each link $(u, v) \in L$ has a capacity $c_{\tau(u,v)}$ at time τ. For each node $v \in V_{PU}$, the wireless technology employed by node v will determine its transmission range and the allocated set of spectral channels. In this work, the routing path for each flow is given so that each node has to maintain a table or cache for transmission to its next hop node. As such, the interference (negotiation) range can be obtained. The link (u, v) may not be connected if there is no available spectral band for the link. For example, Figure 4 shows an 8-node CR network with two flows over two paths $\{o, a, c, e, g\}$ and $\{b, c, d, f\}$, respectively. The transmission range of each link is based on the link length at each time slot τ. There are no available spectrum bands for link (b, c) because the set of spectral bands $\{1, 3, 5\}$ has been allocated to links (a, c) and (c, d).

2.2 Spectral Band Pool

The sensed spectral bands within a CR pool are numbered to represent a series of spectral bands. Several available spectral bands can be formed as a spectral channel. Suppose that there are $|M|$ spectral bands in a CR network. Let $M_{\tau v}$ denote the set of available spectral bands sensed and obtained by node v at time τ. Each available spectral band $m \in M_{\tau v}$, where $v \in V_{SU}$ and $\tau \in T$, can be moderated bandwidth b_m under the Signal to Noise Ratio (SNR) limitation. A spectral band m is "available" for a link (u, v) if 1) the spectral band is currently not in use by PUs located within the interference range of the link,

2) the transmission range r_m of the link using this band can cover both nodes u and v, and 3) the duration tm available for the transmission over the link is sufficiently long. Once the conditions $t_m^* = \min\{t_m\}$ and $r_m^* = \min\{r_m\}$, $\forall m \in M_{\tau v}, v \in V_{SU}, \tau \in T$, are satisfied for a subset of spectral bands, the intersection of these subsets of available spectral bands, denoted by $M_{\tau(u,v)}$, are selected for link (u, v) [3]. For example, suppose the two sets of spectral bands $M_{\tau o} = \{1, 2, 3\}$ and $M_{\tau a} = \{1, 3\}$ satisfy the transmission range and duration limitations indicated above. Then, the only common spectrum band both link nodes o and a can select for transmission at time slot is $M_{\tau(o,a)} = \{1\}$, because spectral band $\{3\}$ has been assigned to link (a, c).

2.3 Link-Based Interference Model

If a pair of nodes on link (u, v) selects a common set of available spectral bands for transmission, they have to adjust their transmission power based on the distance between nodes u and v in order to avoid interference with PUs and other SUs. With the interference consideration, the channel capacity for each transmission link is calculated based on the negotiated spectral bands, error rates, and interferences [5] [6]. Since each receiving node must reply with an ACK for a successful transmission, any sender, even those located near its receiver, cannot send or receive data on the same spectral bands at the same time. Without proper coordination, this operation may result in collisions and data retransmissions, which may degrade the performance. Therefore, the interference range should cover a two-hop distance from the sender or receiver when allocating a disjoint set of spectral bands for link (u, v). In other words, the range is equivalent to the negotiation range in which the neighboring SUs within the range should exchange and contend for resource with each other.

In contrast to the node-based interference model [7], which assumes a local resource at each intermediate node for each flow, we consider how to allocate available spectral bands within an interference range of a link in order to create simultaneous multi-flow transmissions. The model in [8] uses the link-based interference model and describes interference constraints as a conflict graph.

Consider nodes u and v for link (u, v) $(u, v \in V_{SU})$. The received power $P_r(v) = P_t(u)L(l_{(u,v)})$, where $P_t(u)$ is the sending power, and $L(l_{(u,v)})$, the degradation function of link length $l_{(u,v)}$ at node v, has to exceed a threshold so as to correctly receive a data packet from sender u. Hence, we have $SNR_{(u,v)} \geq \theta_v$, where $SNR_{(u,v)} = P_t(u)L(l_{(u,v)})/\sigma$ is the SNR of the wireless link (u, v), with σ being a constant and θ_v being the SNR threshold for a node to correctly decode a signal. The transmission range r_m of sender u $(\forall m \in M_{\tau u}, u \in V_{SU}, \tau \in T)$ is the longest distance from node v that node u's data packets can be correctly decoded and can be determined once the transmission power $P_t(u)$ of node u and θ_u are given. For a multi-hop CR network, multiple pairs of nodes may transmit data packets simultaneously. In addition to thermal noise, the transmission from node u to node v may be interfered by other concurrent transmitters. Let $K_{(u,v)}$ denote the set of concurrent transmitters within the interference range of link (u, v). The Signal to Interference Ratio (SIR) for link (u, v) defined in [9] can be expressed by (1). Namely, for node v to receive a data unit from node u correctly, the $SNR_{\tau(u,v)}$ value of link (u, v) must exceed the threshold θ_v at time τ, i.e.,

$$SIR_{\tau(u,v)} = \frac{P_t(u)L(l_{(u,v)})}{\sum_{k\in K} \left[P_t(k)\left(L(l_{(k,u)}) + L(l_{(k,v)})\right)\right] + \sigma}, \forall(u, v) \in L, \tau \in T \quad (1)$$

In this model, we assume that each node $v \in V_{SU}$ has an SNR threshold θ_v, which must be satisfied in order to have successful reception of one data unit from a transmitter. The SIR threshold β_v, provided that $\theta_v > \beta_v$, is also given so as to guarantee correct signal decoding when concurrent transmissions are performed. Based on Shannon's theorem [10], the supportable bit rate of any communication link incident to this node is guaranteed to be at least $c_u = b_m log_2(1+\beta_u)$, where b_m is the bandwidth of a spectral band m. Only when the SIR of the received signal is smaller than β_u, can the supportable bit rate of this node be assumed to be zero and hence the communication is prohibited for this wireless link.

Let $P_{max}(u)$ denote the maximum transmission power of node u. Suppose that each node u can adjust its transmission power $P_t(u)$, $0 \leq P_t(u) \leq P_{max}(u)$, such that the signal power of the receiver node v is slightly higher than $\theta_v \times \sigma$. Then, the maximum supportable bit rate of a wireless link incident to node v is given by $c_{v,\max} = B_v \log_2(1+\theta_v)$, provided that there is no interference from the neighboring nodes. The maximum interference budget B_v that a node can sustain to correctly decode the signal from the transmitter is given by $B_v = (\theta_v \times \sigma/\beta_v) - \sigma$. For a certain node k, the interference caused by another concurrent transmission from node u to node v, denoted by $I_{(u,v)k}$, can be expressed by:

$$\begin{aligned} I_{(u,v)k} &= \frac{P_t(u)\times L(l_{uk})+P_t(v)\times L(l_{vk})}{B_k} \\ &= \frac{L(l_{uk})\cdot\theta_k\cdot\beta_k+L(l_{vk})\cdot\theta_k\cdot\beta_k}{L(l_{uv})\cdot(\theta_k-\beta_k)}, \forall(u, v) \in L, k \in V \end{aligned} \quad (2)$$

Let I_{mk} denote the interference indicator for the communications performed among the set of contending nodes of node k on spectral band m at time slot τ, $\forall m \in M_{\tau(u,v)}$, $(u, v) \in L, k \in K_{(u,v)} \in V_{SU}, \tau \in T$. Binary variable $I_{mk} = 1$ if node k interferes with the transmission from node u to node v on spectral band m at time τ; otherwise, $I_{mk} = 0$. The set of nodes which renders the interference ratio $|I_{mk}| \geq 1$ is called the set of contending nodes for node v. Transmissions from a contending node will drop the supportable bit rate of node v to zero and will prohibit node k from accessing the wireless medium.

The interference model is dependent on the signal strength of a pair of transceivers. Only one pair of nodes can select spectral band m within the interference range of the considered link. The question is how to determine the interference range. The interference range of a link is not fully dependent on the distance between two nodes. Instead, it is affected by the transmission range of the selected spectral bands. Thus, given the transmission range for each intended transceiver pair, the interference range is the transmission range when node k reaches node u or v. Therefore, it can be expressed by (3).

$$\begin{cases} \frac{\left(r_m - l_{\tau(k,u)}\right)}{N} \leq I_{mk}, \\ \forall m = n, m \in M_{\tau(u,v)}, n \in M_{\tau(k,j)}, (u,v) \in L, (k,j) \in L, \tau \in T \\ \frac{\left(r_m - l_{\tau(k,v)}\right)}{N} \leq I_{mk}, \\ \forall m = n, m \in M_{\tau(u,v)}, n \in M_{\tau(k,j)}, (u,v) \in L, (k,j) \in L, \tau \in T. \end{cases} \tag{3}$$

2.4 Problem Description

In this problem, each SU has a minimum requirement of bit-rates in term of SNR (or equivalently, bit error rate (BER)), under the minimum transmission duration (t_m) and transmission range (r_m). Our goal is to minimize the maximum delay differential d_ρ among the transmitting packet $\rho \in P$ within the interference (negotiation) area at a specific cycle time $\pi \in T$. The objective function of this min-max problem is formulated by (4).

$$\min \max_{\rho} d_\rho \tag{4}$$

$$\text{where } d_\rho = \sum_{\tau \in \pi} \sum_{u \in V_{SU}} \phi_{\tau \rho u} t_\tau - \sum_{u \in V_{SU}} \phi_{\pi \rho u} h_{su}, \forall \rho \in P, \pi \in T, s \in S. \tag{5}$$

The first term in (5), i.e., $\sum_{\tau \in \pi} \sum_{u \in V_{SU}} \phi_{\tau \rho u} t_\tau$, aggregates the transmission cycle time t_τ with a decision variable $\phi_{\tau \rho u}$ to determine whether or not a packet ρ is stored on node u at time τ, because each packet can only stay in a node for each cycle time. Each cycle time includes the total transmission delay and nodal delay from the source node to the outgoing node located within the considered interference range (i.e., negotiation range). The nodal delay includes sensing, negotiating, selecting, and switching delays to periodically execute the exchange of spectral bands for each CR. The transmission delay is calculated based on the packet delay differential, queue size and allocated spectral bands. So we can aggregate the packet delay from cycle 0 to current time π with variable $\phi_{\tau \rho u}$ to indicate the number of cycle times that packet ρ has experienced. The second term $\sum_{u \in V_{SU}} \phi_{\pi \rho u} h_{su}$ obtains the maximum tolerable delay h_{su} for a packet ρ to stay at node u at the current time π for flow s with decision variable $\phi_{\tau \rho u}$. So the difference between packets delays and current maximum tolerable delay is calculated for each packet ρ as a delay differential d_ρ.

3 Solution Approach

To minimize the maximum delay differential within the interference range of the considered link by a greedy method, the available spectral bands are allocated to these packets with maximum delay differential within the current time slot with the highest priority. The problem is how much resource should be allocated to flows such that the delay differentials among the flows are balanced. In other words, we have to determine the link to which common available spectral band is assigned to optimally utilize the resources. This problem is NP-hard. Thus, we propose some approximation schemes to find a near-optimal solution.

Each node maintains a spectrum band pool in which the conditions of each spectral band are recorded. If there are no packets to be transmitted on a link between two nodes, they will cooperatively sense available spectral bands on the neighboring links. Once a node has a packet to transmit, they will sense and negotiate with neighboring nodes for available spectral bands. In this paper, we propose four schemes to determine the proportion $x_{\tau(u,v)}$ of allocated bands. The schemes to allocate available spectral bands to minimize transmission delay differential are described as follows:

- Equivalent amount of available spectral bands (EQU): In addition to the available bands which are sensed available only for a CR, but not available for others, it contends with other nodes based on the weight of spectral bands to obtain equivalent amount of available spectral bands based on the negotiation message:

$$x_{\tau(u,v)} = \frac{1}{\sum\limits_{(i,j)\in L} (I_{(u,v)i} \cdot y_{s(i,j)})}, \forall (u,v) \in L, s \in S, \tau \in T \tag{6}$$

  where $y_{s(i,j)}$ is the indicator that link (i,j) is on the path for flow $s \in S$.
- Proportional Queue-size allocation (QS): To negotiate the available spectral bands in line with the proportion of current queue size, which is denoted by $z_{\tau(u,v)}$, i.e., the queue size of outgoing link (u,v) at time τ, divided by the summation of queue sizes within the interference range, each CR selects the available bands based on the proportion:

$$x_{\tau(u,v)} = \frac{z_{\tau(u,v)}}{\sum\limits_{(i,j)\in L} I_{(u,v)k} \cdot z_{\tau(i,j)}}, \forall (u,v) \in L, k \in V_{SU}, \tau \in T \tag{7}$$

- Weighted delay-differential (WDD): In this scheme, each CR node negotiates the proportion of available bands based on the weight of accumulated delay differential of all other nodes within the interference range:

$$x_{\tau(u,v)} = \frac{\sum\limits_{\rho\in q_{\tau(u,v)}} d_\rho}{\sum\limits_{\rho\in q_{\tau(i,j)}} d_\rho I_{(u,v)i}}, \forall (u,v) \in L, (i,j) \in L, \tau \in T \tag{8}$$

  where $q_{\tau(u,v)}$ denotes the set of packets ρ stored in the queue of the outgoing link (u,v) at time τ.
- Weighted delay-differential with the rate of interruption (WDDI): Similar to the previous three schemes to obtain the proportion of available frequency bands, the main idea of this scheme is to design the proportion based on the weight of the delay differential of all queued packets discounted by the rate of PU interruption:

$$x_{\tau(u,v)} = \frac{\sum\limits_{\rho\in q_{\tau(u,v)}} d_\rho(1-\theta_{(u,v)})}{\sum\limits_{\rho\in q_{\tau(i,j)}} d_\rho I_{(u,v)i}(1-\theta_{(i,j)})}, \forall (u,v) \in L, (i,j) \in L, \tau \in T \tag{9}$$

  where $\theta_{(u,v)}$ denotes the average PU interruption rate on link (u,v).

The allocation policies described above are based on the proportion values $x_{\tau(u,v)}$. First, each CR exchanges the available spectral bands, queue size, and allocation proportion $x_{\tau(u,v)}$ between neighboring nodes within the interference range. Therefore, each link keeps the spectral bands only available for itself. Second, the extra spectral band requirements are calculated and the remaining available spectral bands are selected. Third, the selected available spectral bands that will be used by a link are announced. If there is a subset of spectral bands which are also selected by other links, the links with lower proportion value have to try other available spectral bands. The winner will then switch to the selected bands and send its packets to the next hop.

The set of sensed available spectral bands at each time slot is allocated to each transmission link based on the above proportion. Then, the link capacity is calculated. Once the link capacity is determined, the link and nodal delays are calculated to compare the average end-to-end delay differential.

4 Evaluation Results

In this section, we describe our simulation environment used to evaluate the proposed schemes.

4.1 Simulation Environment

In all simulations, SUs and three types of PUs are deployed in an area. All the nodes are within transmission range of one another. The maximum transmission range for each node is set from 1.6 to 2.5 units for SUs. The maximum transmission ranges for the three types of PUs are set to 1.0, 1.5, and 2.0 units, respectively, and these are 30 spectral bands. There are six flows, and each flow has different demands.

We consider the issues that affect our proposed model and cause different aggregate flow and delay differentials. The arrival process and service process of PUs follow Poisson and Exponential distributions with arrival rate λ and service rate μ, respectively. We consider flows with Constant Bit Rate (CBR) and the traffic demand of each flow is set to γ_s units per unit time. Each packet is queued at a finite buffer of capacity $T = 1000$ units.

4.2 Simulation Scenarios

The average end-to-end delay differentials are evaluated under four scenarios:

- Scenario A. Maximum tolerable end-to-end delay: In this case, we are concerned with the maximum tolerable end-to-end delay that will still satisfy the delay requirement for flows. As described in the previous section, many factors (e.g., different number of hops, different remaining number of hops, and queue lengths) can cause different capacity requirements. With this value, we can determine the delay differential at each hop so that the end-to-end delay differential is controlled at each hop.

Table 1. Simulation parameters

Parameter	Value	Parameter	Value
Number of SUs	100-220	Number of PUs	120
Average rate of interruption	0.2	Traffic demand	(0,1)
Transmission power level	2.0	Maximum radio range	1.6 – 2.5
Link failure rate	(0,0.2)	Sensing range	3.2-5
Number of pairs	6	Number types of network	3
Arrival rate λ of a PU	1/2	Service rate μ of a PU	2 slots
Total time slot (T)	2000	Slot time	1.0
Bandwidth / spectral bands (MHz)	15, 20,25	Negotiation time per node-to-node pair	0.01
Number of spectral bands	30	Maximum tolerable end-to-end delay	500 unit time

- Scenario B. Transmission and sensing ranges: we vary the transmission range (i.e., sensing range). Increases in the transmission range limit the number of available spectral bands for SUs, but lead to increased sensing time, and more negotiations despite further distances can be reached.
- Scenario C. Number of SUs: As in a general network, the connectivity is satisfied with the least number of SUs. The effects on the network are tested for different numbers of SUs.
- Scenario D. Traffic load of SUs: The arrival rate is controlled in order to compare how traffic loads affect the delay differentials. We increase the number of packets for each flow to test the network capacity and the effect of queue size.

4.3 Simulation Results

- Scenario A: This scenario evaluates the average end-to-end delay differential for each scheme. The amount of maximum tolerable end-to-end delay varies under each condition. As shown in Figure 5, under the condition of less tolerable delay, the delay differential is higher, so we have to adjust the sensed available resource precisely. The trend of each simulation curve for each scheme is that the longer the tolerable delay, the smaller the delay differentials. Overall, the more the features of a CR network concerned in the allocation metrics (i.e., WDDI scheme), the better the performance.
- Scenario B: This case evaluates the effect of the transmitting and sensing range on the average end-to-end delay differential, as shown in Figure 6. The trade-off between selecting a larger transmission range and the routing path with less hop count is evaluated. When the transmission range is small, the performance is improved when the range increases. However, with a larger range, more time is spent in sensing and negotiations. As a result, the performance decreases as the delay differential increases when the range becomes larger. In addition, a larger interference range results in less spectral bands available for sharing. Hence, there is a critical point to set the suitable

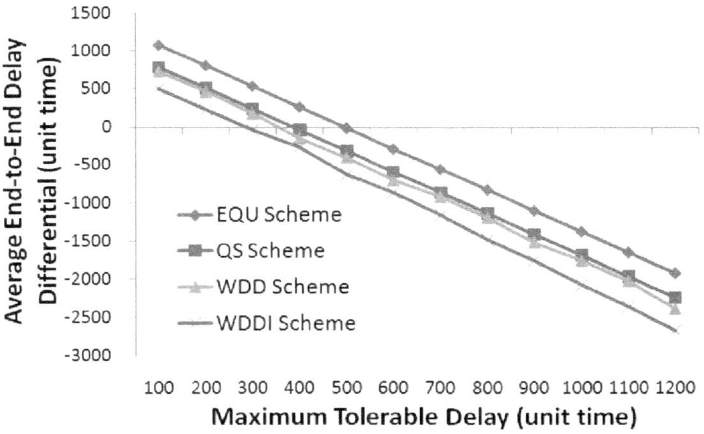

Fig. 5. The effect of the maximum tolerable delay on the average end-to-end delay differential

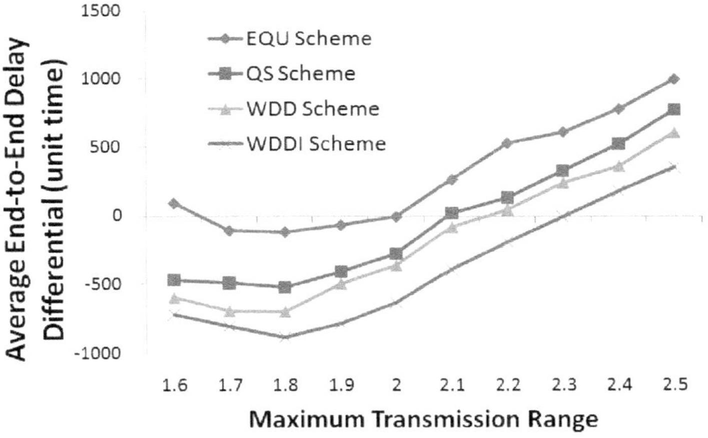

Fig. 6. The effect of transmission and sensing range on the average end-to-end delay differential

maximum transmission range for each scheme. The critical points for QS, WDD, and WDDI are smaller than that for EQU.

- Scenario C: This case evaluates the effect of the number of SUs on the average end-to-end delay differential. As shown in Figure 7, there is a trade-off between the number of negotiations and the number of cooperative sensing nodes. When the number of SUs increases with a fixed number of flows, the more the cooperative sensing nodes, the more the available frequency bands obtainable, at the expense of more negotiation overhead and increased

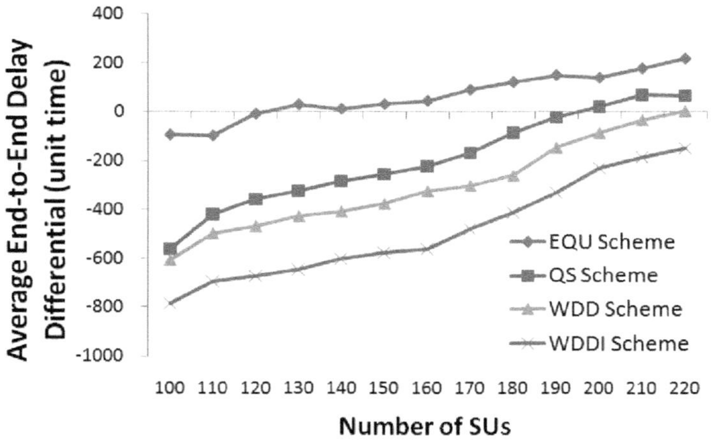

Fig. 7. The effect of the number of SUs on the average end-to-end delay differential

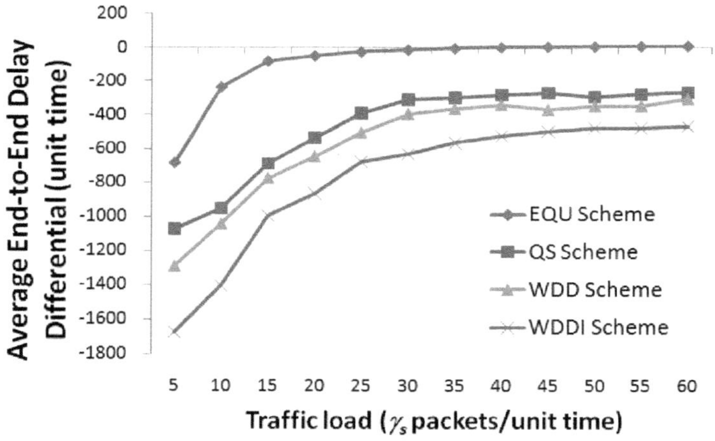

Fig. 8. The effect of traffic load on the average end-to-end delay differential

interference. Thus, with a fixed number of spectrum bands for sensing by random selection, the increased number of available bands is limited, but with longer negotiation time and more interferences, so the transmission duration is decreased. Thus, the more the duplicative sensing with larger negotiation overheads, the more the interference, which in turn causes higher average end-to-end delay differential.

– Scenario D: This case evaluates the effect of the traffic load on the average end-to-end delay differential. As shown in Figure 8, when the traffic load increases, the queue size increases, and as a result, the delay differential increases. However, the increase ratio is small when the traffic load is large. The reason is that the overdue packets will not be dropped when comparing the delay differentials among different approaches. When the capacity is limited, the amount of traffic it can handle is also limited. In other words, if the buffer is full, it blocks the transmission from the previous link. As a result, more packets will be queued at the source node, resulting in fewer packets at the receiver. Overall, the WDDI scheme outperforms the other schemes even when the traffic load is high. This is because more available spectral bands will be allocated to packets with higher delay differential, and to links with higher rates of PU interruption.

5 Conclusion

In this paper, we study the resource allocation problem for cognitive radio networks with QoS considerations. Our approach is to minimize the maximum delay differential among different flows, and to minimize the end-to-end delay, according to the sensed available spectral bands. We consider a multi-hop network with a spectral band pool and link-based interference model. Based on CR features, four allocation schemes are proposed and evaluated in terms of the average end-to-end delay differentials. Our results show that the more the resources given to packets with higher delay differentials and to links with higher rates of interruption, the better the performance. In the future, we will further consider call admission and packet dropping in our formulation.

Acknowledgment. This work was supported in part by the Excellent Research Projects of National Taiwan University, under Grant Number 97R0062-06, and in part by National Science Council (NSC), Taiwan, under Grant Number NSC99-2221-E-002-030-MY3 and Grant Number NSC 98-2221-E-415-005.

References

1. Mitola, J., Maguire, G.Q.: Cognitive Radio: Making Software Radios More Personal. IEEE Personal Communications 6(4), 13–18 (1999)
2. Chakravarthy, V., Li, X., Wu, Z., Temple, M., Garber, F., Kannan, R., Vasilakos, A.: Novel Overlay/Underlay Cognitive Radio Waveforms Using SD-SMSE Framework to Enhance Spectrum Efficiency-Part I: Theoretical Framework and Analysis in AWGN Channel. IEEE Transactions on Communications 57(12), 3794–3804 (2009)
3. Wen, Y.F., Liao, W.: On the Routing in Wireless Ad Hoc Cognitive Wireless Networks. In: 71th IEEE International Conference on Vehicular Technology, pp. 1–5. IEEE Press, Taipei (2010)
4. Wireless LAN Medium Access Control (MAC) and Physical Layer (PHY) Specifications, IEEE Standard 802.11 (1999)
5. Gupta, P., Kumar, P.: Capacity of Wireless Networks. IEEE Transactions on Information Theory 46(2), 388–404 (2000)

6. Wu, K.-D., Liao, W.: Flow Allocation in Multi-Hop Wireless Networks: A Cross-Layer Approach. IEEE Transactions on Wireless Communications 7(1), 269–276 (2008)
7. Wu, K.-D., Liao, W.: On Service Differentiation for Multimedia Traffic in Multi-Hop Wireless Networks. IEEE Transactions on Wireless Communications 8(5), 2464–2472 (2009)
8. Liu, T., Liao, W.: Interference-Aware QoS Routing for Multi-Rate Multi-Radio Multi-Channel IEEE 802.11 Wireless Mesh Networks. IEEE Transactions on Wireless Communications 8(1), 166–175 (2009)
9. Low, S.H., Lapsley, D.E.: Optimization Flow Control: Basic Algorithm and Convergence. IEEE/ACM Transactions on Networking 7(6), 861–874 (1999)
10. Shannon, C.E.: A Mathematical Theory of Communication. Bell Labs Technical Journal 27, 379–423, 623–656 (1948)
11. Caccetta, L., Kulanoot, A.: Computational Aspects of Hard Knapsack Problems. Nonlinear Analysis 47, 5547–5558 (2001)
12. Garey, M.R., Johnson, D.S.: Computers and Intractability: A Guide to the Theory of NP-Completeness. W. H. Freeman & Co., New York

Network Coding Based QoS-Provisioning MAC for Wireless Smart Metering Networks

Hang Su and Xi Zhang

Networking and Information Systems Laboratory
Department of Electrical and Computer Engineering
Texas A&M University, College Station, TX 77843, USA
su@tamu.edu, xizhang@ece.tamu.edu

Abstract. The advanced metering infrastructure (AMI) with smart metering networks, which enables the two-way information exchange between utilities and end-users, is the critically important component of the overall smart grid architecture. In this paper, we propose a network coding based QoS-provisioning MAC protocol for wireless smart metering networks, which aims at providing reliable and low-latency communications services between smart meters and coordinator of the utilities. In particular, to efficiently utilize the limited bandwidth resource, we develop our proposed MAC protocol based on the contention-free orthogonal frequency division multiple access (OFDMA) methodology. Our proposed MAC protocol incorporates the opportunistic resource allocation, cooperative grouping, and network coding technique. Under our proposed MAC protocol, the coordinator adaptively allocates the resource blocks to each smart meters based on the channel conditions between the coordinator and the smart meter and among the smart meters. Also conducted are simulations to verify and evaluate our proposed MAC protocol.

Keywords: Advanced metering infrastructure (AMI), medium access control (MAC) network coding, orthogonal frequency division multiple access (OFDMA), smart grid.

1 Introduction

Recently, the smart grid, which delivers both energy and information, has receives intensive research interests from academia, industry, and military. The motivation behind the smart grid is to improve power reliability and quality, increase the generation and transmission efficiency, and enable a wide adoption of renewable energy and electric vehicles. Due to the two-way information exchange between the utilities and the power end-users (industrial or home consumers), the smart grid enables the new load shifting and energy efficiency programs at the side of end-users that have power usage and price awareness and choice.

The novel power demand managements [1] are one of the most import applications of the smart grid, which impact the energy flow towards the end-user endpoints. With the help of smart metering networks, the end-user consumption

X. Zhang and D. Qiao (Eds.): QShine 2010, LNICST 74, pp. 161–171, 2011.

can now possibly be controlled at energy access point level. Smart meters with the features of power-line connect/disconnect allow end-users automatically to disconnect homes from the grid, in case of a security issues. The power demand managements also encourage end-users to reduce their energy consumption, especially during the power grid's peak loads. This can be achieved by implementing the smart metering networks which enable the two-way information exchange between utilities and end-users.

The smart metering networks, which consist of advanced wireless/wireline communications and networking technologies as well as new and improved solid-state metering devices, plays the critical important role in the smart grid, as shown in Fig. 1. The smart metering networks allow the utilities to not only record and analyze the end-users' utility usage, but also keep end-users informed of updated utility prices and perform remote utility management, all in a reliable and real-time manner. Furthermore, cooperating with the intelligent supervisory control and data acquisition (SCADA) system, the smart metering networks can minimize the unplanned end-user outages and ensures the optimum power distribution network configuration.

To implement the above functionalities and guarantee the efficient power management, the design criteria of the smart metering networks include low latency, reliable and robust information exchange between utility and end-users. In this paper, we propose a network coding based QoS-provisioning MAC protocol for wireless smart metering networks. The proposed MAC protocol aims at providing reliable and low-latency communications services between smart meters and coordinator of the utilities. In particular, to efficiently utilize the limited bandwidth resource, we develop our proposed MAC protocol based on the contention-free orthogonal frequency division multiple access (OFDMA) methodology. Our proposed MAC protocol incorporate the opportunistic resource allocation, cooperative grouping, and network coding technique. Under our proposed MAC protocol, the coordinator adaptively allocate the resource blocks to each smart meters based on the channel conditions between the coordinator and the smart meter and among the smart meters. Also conducted are simulations to verify and evaluate our proposed MAC protocol.

The rest of this chapter is organized as follows. Section 2 presents the related work. Section 3 describes the system model. Section 4 proposes the network coding based QoS-provisioning MAC protocol. Section 5 evaluates our scheme based on simulation experiments. The chapter concludes with Section 6.

2 The Related Works

The communications and networking technology plays the key role in the smart grids. The authors of [2] discussed the applications of communications and networking on smart grids from the industry perspective. The authors of [3] proposed a routing protocol for the AMI in smart grids based on the framework of the IPv6 routing protocol for low power and lossy networks. The authors of [4] implemented a ZigBee-based AMI which employs a multichannel frequency

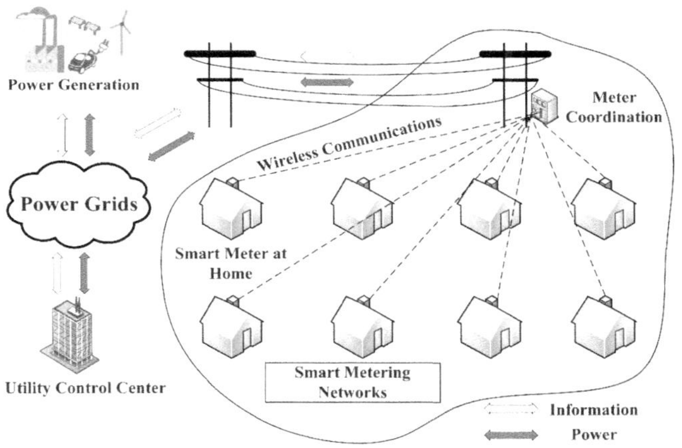

Fig. 1. Illustration of the smart metering networking in the smart grid

hopping system to tackle the co-existence problem with the WiFi and Bluetooth networks.

The network coding is a powerful tool to improve the network throughput and delay performance. The concept of network coding was first coined in [5] in the context of wired multicast communications. The conventional wireless-network coding [6, 7], which takes advantage of the broadcast nature of wireless medium to reduce the number of transmissions, were proposed to enhance the network throughput under the 2-way relay scenarios. By not only utilizing the broadcast nature of wireless medium, but also exploiting the inherent physical-layer coding capability through analogously mixing the simultaneously arrived radio waves at the relay nodes, the analog network coding [8, 9] can further increase the network throughput. However, the analog wireless network coding requires the strict synchronization among the source, destination, and relay, and thus is difficult to implement in practice. In this paper, we use only the conventional wireless network coding technique to further improve the network throughput and reliability of the smart metering networks.

3 System Model

To efficiently exploit the previous bandwidth resource, we propose to use the OFDMA as the access methodology for the wireless smart metering networks. The smart meters conduct two-way communications with the coordinator. The smart meters report the utility usage periodically to the coordinator while the coordinator periodically broadcasts utility price to the smart meters. The frequency resources is divided into $M = aN$ resource blocks, where a is a positive integer. For notational convenience, we let M be an even number. The transmission across time is assumed to be independent, and smart meters have independent channels towards the coordinator. We assume the transmission on each resource block uses the same transmit power, modulation and channel coding. The transmission bit rate of the modulated signal over one resource block is R bits/second. Let α be the channel path loss exponent, and c be some constant, then the packet success rate for a transmission of ℓ bits using a single resource block is $e^{-cd^{\alpha}}$. The transmission time in this case is ℓ/R. Alternatively, through resource block multiplexing, a transmission can use 2 resource blocks to transmit ℓ bits in $\ell/(2R)$ seconds. Using the independence assumption, we assume that the packet success rate in this case is also $e^{-cd^{\alpha}}$.

For the same source and destination node pair, the channel allows a maximum of D_0 independent resource blocks, that is the maximum diversity order. Therefore, if a smart meter transmits ℓ bits of data over $r \geq 2$ resource blocks, the increased diversity order improves the probability of success to

$$P_s(d, r) = 1 - (1 - e^{-cd^{\alpha}})^{\min(r, D_0)}. \tag{1}$$

We assume the same probability of success for transmitting L bits of data over $2r \geq 4$ resource blocks.

Let $d_{i,0}$ be the distance between smart meter i and the coordinator, and $d_{i,j}$ be the distance between smart meter i and smart meter j. We denote the probability of success for sending a packet from smarter meter i to the coordinator over a single resource block by $P_s(d_{i,0}, 1)$, and the probability of success for smarter meter j to overhear the transmission by $P_s(d_{i,j}, 1)$. We assume that these probabilities are known at the coordinator, who uses the information for group assignment and scheduling uplink transmissions.

In this paper, to guarantee the real-time information exchange between the smart meters and coordinator, we concentrate on the amount of time that is required to guarantee that all uplink transmissions are received successfully at the coordinator. The uplink transmissions are broken down into multiple transmission opportunities to guarantee reliability. This strategy to improve reliability is to provide two transmission opportunities to an uplink transmission to improve packet success rate as illustrated in Fig. 2. In particular, the time is divided into

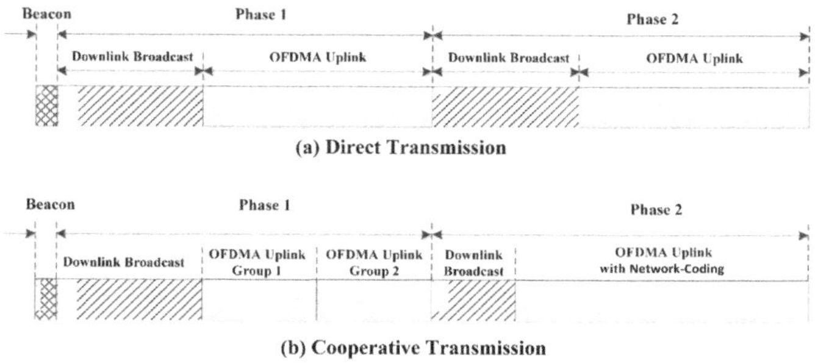

Fig. 2. The illustration of the transmission schemes of our paper

superframes each with fixed length. Each superframe starts with the beacon signal. Besides the beacon signal, the rest of the superframe is partitioned into two transmission phases, which aims at increase the reliability of message delivery. In each transmission phase, we employ the time-division duplex (TDD) for the downlink and uplink communications.

Between the two consecutive uplink transmissions, the coordinator broadcasts a downlink message to all smart meters to acknowledge packet receptions, and send resource assignment information (e.g., channel assignment etc.) for the next uplink transmission.

4 Our Proposed Transmission Schemes

4.1 The Basic Direct Transmission Scheme

We use the basic direct transmission scheme as the baseline for comparison. As shown in Fig. 2(a), in the first transmission opportunity, every smart meter transmits their data using frequency diversity order $a = M/N$. No control overhead is used. The probability of success for transmission from smart meter i is $P_s(d_{i,0}, a)$.

After each transmission opportunity, if the coordinator still fails to decode some packets from some smart meters, the coordinator requests in the feedback those failed smart meters to retransmit their packets again. To meet the time budget during the subsequent transmission opportunities, at most X, with $X \leq M$, nodes may retransmit. Let K be the number of packets that are still missing from the coordinator. If $K \geq X$ and $X = M/2$, then only $M/2$ smart meters can retransmit, and each retransmission uses 2 resource blocks, and thus the diversity order is 1. If $K < X$, the system guarantees that each smart meter has at least 2 resource blocks to transmit its data at diversity order 1. The coordinator successively allocates 2 resource blocks at a time to increase the diversity order of the node that has the worst effectively probability of success. We list the detailed algorithm as follows:

Initialization: Let E be the set of nodes with failed transmission. Let the remaining number of resource blocks be $r = M - 2K$, the effective success probability of node i be

$$P_i = \begin{cases} 1, & \text{if } i \notin E, \\ P_s(d_{i,0}, 1), & \text{otherwise,} \end{cases} \tag{2}$$

and the diversity order of node i be $D_i = I_{\{i \in E\}}$, where $I_{\{.\}}$ is the indicator function.

Step 1) Find the node with the lowest effective probability of success $\tilde{i} = \arg\min_{i:D_i<D_0} P_i$. If $D_i < D_0$ for all i, then find the retransmitting node with the lowest diversity order $\tilde{i} = \arg\min_{i \in E} D_i$.

Step 2) Assign a resource block to increase the diversity order of the transmission from $\tilde{i}$. Update $P_{\tilde{i}} := P_s(d_{\tilde{i}}, D_i + 1)$, and then $D_{\tilde{i}} := D_{\tilde{i}} + 1$, and $r := r - 2$.

Step 3) If $r > 0$, go to step 1).

Step 4) Node i retransmits data using $2D_i$ resource blocks, at diversity order D_i.

4.2 The Basic Grouping Transmission Scheme

Instead of having all smart meters to transmit simultaneously during the first transmission opportunity. As shown in Fig. 2(b), we divide the N smart meters into two groups, namely $\mathcal{A}$ and $\mathcal{B}$, with each group having exactly $N/2$ nodes. The grouping algorithm will be discussed in the next section. During the first transmission time phase, group $\mathcal{A}$ nodes first transmit their packets to the coordinator while group $\mathcal{B}$ nodes overhearing the transmissions; and then group $\mathcal{B}$ nodes transmit while group $\mathcal{A}$ nodes overhearing. Since each node is equipped with a single transceiver, nodes within the same relay group cannot overhear packets from those in the same group. Since group $\mathcal{B}$ nodes transmit their packets after group $\mathcal{A}$, each node in group $\mathcal{B}$ can include a $N/2$-bit control overhead to specify whether it has correctly received the packets from the nodes in group $\mathcal{A}$. When node i in group $\mathcal{A}$ transmits, it transmits the packet at diversity order $a = M/N$ over $2a$ resource blocks. The coordinator receives a packet successfully with probability $P_s(d_{i,0}, a)$, and a smart meter j in group $\mathcal{B}$ overhears the packet successfully with probability $P_s(d_{i,j}, a)$. In this sense, node j in group $\mathcal{B}$ transmits its packet at diversity order $a = M/N$ over $2a$ resource blocks. The coordinator receives the packet successfully with probability $P_s(d_{j,0}, a)$. Similar to the basic transmission scheme, in subsequent retransmissions, the system allocates its available resources so that each failed transmission has at least one direct or relay retransmission. Thus, the system allocates resources to improve the weakest effective probability of success. We list the detailed algorithm as follows:

Let E be the set of up to $M/2$ nodes that have failed previous transmissions and have the shortest distance to the coordinator Let $Y(i)$ be the set of nodes that successfully overhear the packet from node i. Since only the nodes in group $\mathcal{B}$ overhear the packet from the nodes in group $\mathcal{A}$, $Y(i) =$ for $i \in B$. Then, for each node $i \in E$, the system chooses a node $\tilde{j}(i) = \arg\max_{j\in\{i\}\cup Y(i)} P_s(d_{j,0}, 1)$ to retransmit the packet for node i, using 2 resource blocks and diversity order of 1.

Let $r = M - 2K$, and

$$P_i = \begin{cases} 1, & \text{if } i \notin E, \\ P_s(d_{\tilde{j}(i)}, 1), & \text{if } i \in E. \end{cases} \tag{3}$$

Also, let $D_{ij} = I_{\{i\in E\}} I_{\{j=\tilde{j}(i)\}}$ be the diversity order of the transmission from node j to relay for node i (Index ii means that node i transmits its own packet). The algorithm proceeds as follows.

Step 1) Find the node with the lowest effective probability of success $\tilde{i} = \arg\min_{i:D_{ij}<D_0\forall j} P_i$. If $D_{ij} < D_0$ for all i and j, then find the retransmitting relaying node with the lowest diversity order $\tilde{j} = \arg\min_{i\in E} \min_j D_{ij}$.

Step 2) Find the best relay for retransmitting the packet from node $\tilde{i}$.

$$\tilde{j}(\tilde{i}) = \arg \max_{j\in\{\tilde{i}\}\cup Y(\tilde{i}):D_{\tilde{i}j}<D_0} P_s(d_{j,0}, 1). \tag{4}$$

Step 3) To assign a resource block, perform

$$D_{\tilde{i}\tilde{j}(\tilde{i})} := D_{\tilde{i}\tilde{j}(\tilde{i})} + 1. \tag{5}$$

$$P_{\tilde{i}} := 1 - \prod_{j} [1 - P_s(d_{j,0}, D_{\tilde{i}j})]. \tag{6}$$

$$r := r - 2. \tag{7}$$

Step 4) If $r > 0$, go to step 1).

Step 5) Node j retransmits data from node i using $2D_{ij}$ resource blocks with diversity order of D_{ij}.

4.3 The Network Coding Based Grouping Transmission Scheme

Similar to the basic grouping transmission scheme, the network coding based grouping transmission scheme also divide the transmissions into two groups, as shown in Fig. 2(b). In the network coding based grouping transmission scheme, we employ the network coding to improve the reliability of message delivery. The coordinator assigns the resource blocks to the smart meters with good channel condition in transmission phase 2. Since the smart meters in group $\mathcal{B}$ can overhear the packets sent from the smart meters in group $\mathcal{A}$ and receive the packets sent from the coordinator to all other smart meters in transmission phase 1, as shown in Fig. 3. Based on the relay report, smart meters in group $\mathcal{B}$ will report to the coordinator on which smart meters in group $\mathcal{A}$ it can successfully overhear and whether it receives the message sent from the coordinator to the other smart meters. Thus, the coordinator is aware of the set of smart meters in group $\mathcal{B}$, denoted by $\mathcal{R}$, which successfully receive the packets sent from/to the smart meters which fail to communicate with the coordinator.

At the beginning of the transmission phase 2, the coordinator chooses the relay smart meters from those in $\mathcal{R}$ to help the smart meters which fail to transmit/receive packets to/from the coordinator. The coordinator allocates the resource blocks to the relay smart meters in the downlink slot of the transmission phase 2. The relay smart meters conduct the XOR operations to combine the

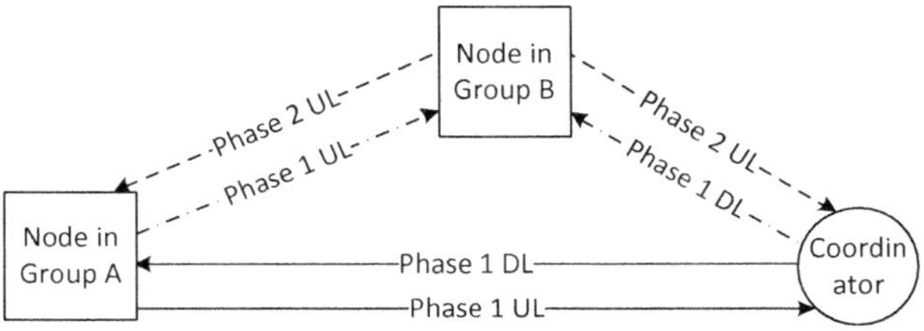

Fig. 3. The network coding aided relay

downlink and uplink packets for the smart meters which fail to transmit/receive packets, and then broadcast the combined packets by using the allocated resource blocks during the transmission phase 2, as shown in Fig.3. Upon the coordinator and the smart meters receive the combined packets, they can successfully decode the missing messages by conducting the XOR operations on the combined packets and their own packets.

The resource allocation in Phase 2 transmission is similar to the basic grouping transmission scheme. The difference between them is the selection criteria for the relay node. In the 2-way-relay network-coding transmission scheme, we need to consider the both the link between the relay and the coordinator and the link between the relay and the source node. Thus, for each smart meter $i \in E$, the system chooses the relay $\tilde{j}(i) = \arg\max_{j \in \{i\} \cup Y(i)} P_s(d_{j,i}, 1) P_s(d_{j,0}, 1)$ to retransmit the packet for node i, using 2 resource blocks and diversity order of 1. The effective probability of success, denoted by P_i, for the network coding based grouping transmission scheme can be written ad

$$P_i = \begin{cases} 1, & \text{if } i \notin E, \\ P_s(d_{\tilde{j}(i),i}, 1) P_s(d_{\tilde{j}(i),0}, 1), & \text{if } i \in E. \end{cases} \quad (8)$$

Also, let $D_{ij} = I_{\{i \in E\}} I_{\{j = \tilde{j}(i)\}}$ be the diversity order of the transmission from node j to relay for node i (Index ii means that node i transmits its own packet).

The details of the network coding based grouping transmission scheme are as follows:

Step 1) Find the node with the lowest effective probability of success $\tilde{i} = \arg\min_{i: D_{ij} < D_0 \forall j} P_i$. If $D_{ij} < D_0$ for all i and j, then find the retransmitting relaying node with the lowest diversity order $\tilde{j} = \arg\min_{i \in E} \min_j D_{ij}$.

Step 2) Find the best relay for retransmitting the packet from node $\tilde{i}$. Considering the maximum diversity order, redefine

$$\tilde{j}(\tilde{i}) = \arg \max_{j \in \{\tilde{i}\} \cup Y(\tilde{i}): D_{\tilde{i}j} < D_0} P_s(d_{j,0}, 1) P_s(d_{j,i}, 1). \quad (9)$$

Step 3) To assign a resource block, perform

$$D_{\tilde{i}\tilde{j}(\tilde{i})} := D_{\tilde{i}\tilde{j}(\tilde{i})} + 1. \quad (10)$$

$$P_{\tilde{i}} := 1 - \prod_j [1 - P_s(d_{j,0}, D_{\tilde{i}j}) P_s(d_{j,i}, D_{\tilde{i}j})]. \quad (11)$$

$$r := r - 2. \quad (12)$$

Step 4) If $r > 0$, go to step 1).

Step 5) Smart meter j combines the data from the source i and coordinator by using XOR operation, and then broadcasts the combined data using $2D_{ij}$ resource blocks with diversity order of D_{ij}.

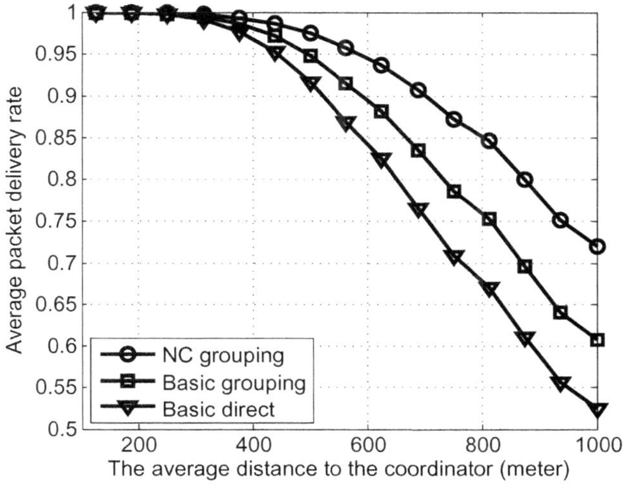

Fig. 4. The impact of the average distance to the coordination on the average packet delivery rate

4.4 The Grouping Algorithm

Since the smart meters is divided into two groups to proceed the cooperations, we develop a simple, but effective, grouping algorithm. Assume that the system only knows $P_s(d_{i,0}, 1)$ for all i. Since only the nodes in group $\mathcal{B}$ in direct transmission scheme may help relay packets of the nodes in group $\mathcal{A}$, the system may benefit from direct transmission scheme only if group $\mathcal{B}$ nodes are better performing than the nodes in group $\mathcal{A}$. Hence, the grouping algorithms simply assigns the $N/2$ smart meters with the poorest P_i to group $\mathcal{A}$, and the $N/2$ smart meters with the best P_i to group $\mathcal{B}$.

5 Performance Evaluations

To meet the delay requirement of the meter reading defined by NIST [10], we set the length of the superframe to be equal to 30 ms. The simulation results shown in this section is the average results of the 100 simulation runs.

Figure 4 shows the impact of the average distance to the coordination on the average packet delivery rate with different schemes. As the average distance to the coordination increases, the average packet delivery rate decreases. Our proposed network coding based grouping scheme decreases at the slowest speed. When the average distance to the coordination keep increasing, the advantage between our proposed network coding based grouping scheme over the other two schemes becomes larger. When the average distance to the coordinator reaches 1000 meters, the packet delivery rate of our proposed network coding based grouping scheme can still be 0.72, while the basic grouping transmission scheme

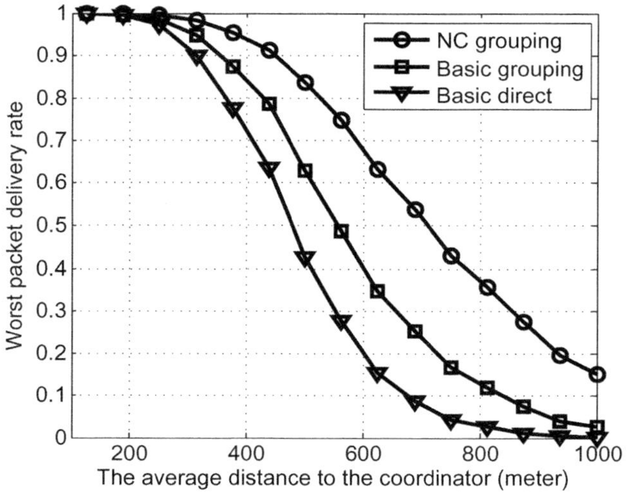

Fig. 5. The impact of the average distance to the coordination on the worst packet delivery rate among all smart meters

achieving less than 0.61 packet delivery rate and the basic direct transmission scheme having less than 0.53 packet delivery rate.

We proceed to study the worst packet delivery rate, which is defined as the worst packet delivery rate achieved among all smart meters. Generally speaking, the worst packet delivery rate is achieved by the smart meter with the worst channel condition. As shown in Fig. 5, when the average distance to the coordination is large, the worst packet delivery rates of all schemes are small. If setting the threshold for the worst packet delivery rate to be 0.9, then the maximum average distance to the coordination achieved by our proposed network coding based grouping transmission scheme is 446 meters, while the maximum average distances for the basic grouping transmission and the basic direct transmission scheme are 357 and 312 meters, respectively, which shows the superiority of our proposed network coding based grouping transmission scheme.

6 Conclusions

This paper proposed a network coding based QoS-provisioning MAC protocol for wireless smart metering networks to provide reliable and low-latency communications services between smart meters and coordinator of the utilities. Our proposed MAC protocol combine together the opportunistic resource allocation, cooperative grouping, and network coding technique. In particular, We developed our proposed MAC protocol based on the contention-free orthogonal frequency division multiple access (OFDMA) methodology, which aims at efficiently utilizing the limited bandwidth resource. Specifically, our proposed MAC protocol enables the coordinator to adaptively allocate the resource blocks to each smart meters based on the channel conditions between the coordinator and the smart meter and among the smart meters. The extensive simulation results showed that our proposed MAC protocol significantly outperforms the basic transmission protocols in terms of packet reliability.

References

1. Verschueren, T., Haerick, W., Mets, K., Develder, C., De Turck, F., Pollet, T.: Architectures for smart end-user services in the power grid. In: IEEE/IFIP Network Operations and Management Symposium Workshops, p. 316 (2010)
2. Heile, B.: Smart grids for green communications (industry perspectives). IEEE Wireless Commun. Mag. 17(3), 4–6 (2010)
3. Wang, D., Tao, J., Zhang, Z., Abouzeid, A.: Rpl based routing for advanced metering infrastructure in smart grid. In: IEEE ICC Workshop on Smart Grid Communications (2009)
4. Tung, H.Y., Tsang, K.F., Lam, K.L.: Zigbee sensor network for advanced metering infrastructure. In: Proc. Digest of Technical Papers Int Consumer Electronics (ICCE) Conf., pp. 95–96 (2010)
5. Ahlswede, R., Cai, N., Li, S.-Y., Yeung, R.: Network information flow. IEEE Trans. Inf. Theory 46(4), 1204–1216 (2000)
6. Wu, Y., Chou, P.A., Kung, S.-Y.: Information exchange in wireless networks with network coding and physical-layer broadcast. In: Proc. Conference on Information Sciences and Systems, CISS (2005)
7. Katti, S., Rahul, H., Hu, W., Katabi, D., Medard, M., Crowcroft, J.: XORs in the air: practical wireless network coding. In: Proc. ACM SIGCOMM 2006, pp. 243–254 (2006)
8. Zhang, S., Liew, S.C., Lam, P.P.: Physical-layer network coding. In: Proc. ACM MobiCom 2006, pp. 358–365 (2006)
9. Katti, S., Gollakota, S., Katabi, D.: Embracing wireless interference: analog network coding. In: Proc. ACM SIGCOMM 2007, pp. 397–408 (2007)
10. NIST. Application communication requirements matrix (2010), `http://collaborate.nist.gov/twiki-sggrid/bin/view/SmartGrid/PAP02Objective1`

A Strategy-Proof and Non-monetary Admission Control Mechanism for Wireless Access Networks

Xiaohan Kang, Juan José Jaramillo, and Lei Ying

Department of Electrical and Computer Engineering
Iowa State University, Ames IA 50011
{xkang,jjjarami,leiying}@iastate.edu

Abstract. We study admission control mechanisms for wireless access networks where (i) each user has a minimum service requirement, (ii) the capacity of the access network is limited, and (iii) the access point is not allowed to use monetary mechanisms to guarantee that users do not lie when disclosing their minimum service requirements. To guarantee truthfulness, we use auction theory to design a mechanism where users compete to be admitted into the network. We propose admission control mechanisms under which the access point intelligently allocates resources based on the announced minimum service requirements to ensure that users have no incentive to lie and the capacity constraint is fulfilled. We also prove the properties that any feasible mechanism should have.

Keywords: Auctions, truth-telling, admission control, resource allocation.

1 Introduction

Resource allocation has been one of the most important issues in the design of communication networks. Given a wireless access network, in which the users have various QoS requirement and the access point has limited resource, admission control mechanisms are vital to achieve stability, fairness and efficiency of the system. In this paper, we consider a wireless access network with multiple users and a single access point. We assume that the access point network is a public network, and is not allowed to charge users for accessing the network. We study the case when the QoS requirements are private, and users are allowed to selfishly disclose any value that would give them better service. Then the problem is to design an admission control mechanism such that the true QoS of the users can be collected without the use of any pricing scheme, and as many users are admitted as possible. A natural choice is to set up a game with the users, such that the selfish users are incentivized to tell the truth. Originated in economics theory, auction mechanisms have been found very useful in this kind of situation, since they are designed to entice selfish bidders tell the truth when allocating limited resources.

X. Zhang and D. Qiao (Eds.): QShine 2010, LNICST 74, pp. 172–187, 2011.

Various auction mechanisms have been well-studied. Myerson [1] has obtained the optimal auction mechanism in closed form mathematical expression. His result, however, only works for limited utility functions without constraint over the resource. The well-known VCG mechanism has been proved to guarantee truth-telling while achieving the social optimum [2]. The challenge of designing an effective auction mechanism in the proposed setting is the access point is not allowed to charge the users, so the auction mechanism has to be non-monetary. For example, using a simple example, it will be shown that the VCG mechanism cannot be easily adopted in our scenario due to the non-monetary requirement. Credit schemes that were developed to incentivize cooperation in wireless networks could also be adapted to guarantee users do not lie about their true requirements, but they would also require secure mechanisms to avoid tampering with the virtual money [7], [8], [9], [10], [11], [12], [13]. Recently, Hou and Kumar have proposed a bidding game between users and access point that maximizes the total utility, but in the iterative process the users are forced to bid specific values instead of bidding selfishly [4]. In this paper, we seek to design auction mechanisms that are truthful and do not use any money-based scheme for general utility functions.

Our contributions are therefore fourfold.

i. We model the admission control problem as an auction mechanism design with resource constraint, where the payment is related to the allocated service rate.
ii. We use a simple example to demonstrate that a direct adoption of VCG auction mechanism does not work in the non-monetary setting.
iii. Third, we come to two theorems that help us understand the essence of strategy-proof mechanisms. The first theorem (Theorem 2) shows the impossibility of a reasonable truth-telling mechanism to be based on probabilistic decisions, and the second theorem (Theorem 3) shows that the truthfulness of a mechanism is equivalent to the existence of a lowest winning bid regardless of one's own bid.
iv. We propose our mechanism and show that it has the desired properties and admits at least half of the optimal number of users with high probability in an asymptotic sense.

This paper is organized as follows. Section 2 gives the model of this problem and the set of assumptions. Section 3 illustrates the failure of VCG in our setting. Section 4 analyzes the problem and characterizes the properties for feasible auction mechanisms. Section 5 gives our proposed mechanism and shows feasibility and the performance bound. Section 6 concludes this paper.

2 Model

Consider a multiple access network with n users and a single access point (AP), where only one user can transmit at any given time. Let $\mathcal{N} = \{1, 2, \cdots, n\}$ be the set of all users. The service rate of the AP is assumed to be 1. Each user is assumed to have an arrival rate $\lambda_i = 1$ and a quality of service (QoS) q_i, which indicates the fraction of packages that has to be delivered successfully and is only known to user i. We refer to $t_i = 1 - q_i$ as the maximum allowed drop rate of user i and suppose that the range of t_i is $T_i = [a_i, b_i] \subset [0, 1]$. We let $T = T_1 \times T_2 \times \cdots \times T_n$ and $T_{-i} = T_1 \times T_2 \times \cdots \times T_{i-1} \times T_{i+1} \times \cdots \times T_n$ where $\times$ denotes the Cartesian product.

In order to decide which users to serve and the QoS the AP should provide, the AP sets up an auction, in which each user, or bidder, bids a requested drop rate $s_i \in T_i$, which can be different from the true drop rate t_i, and the AP decides the set of users that get admitted and assigns drop rate x_i to user i. By Myerson's revelation principle [1], we only consider direct revelation mechanisms, i.e., the mechanisms in which users submit bids in the form of drop rate values rather than in the form of any other strategies. Then the mechanism can be described by the *outcome functions* (p, x), where p and x are the set of functions

$$\begin{aligned} p(s) &= (p_\phi(s), \phi \subset \mathcal{N}) \\ x(s) &= (x_{i,\phi}(s), \phi \subset \mathcal{N}) \end{aligned}$$

with $p_\phi : T \to \mathbb{R}$ being the probability that only the subset of users ϕ of $\mathcal{N}$ get admitted and $x_{i,\phi} : T \to \mathbb{R}$ being the actually assigned drop rate for user i if the subset ϕ is admitted. For convenience we set $x_{i,\phi}(s)$ to be the assigned drop rate by mechanism (p, x) if $i \in \phi$ and $p_\phi(s) > 0$, and 1 otherwise, i.e., all packets of user i are dropped, which is equivalent to not admitting user i.

The utility function for user i, given the true drop rate t_i and the assigned drop rate x_i, is $u_i(t_i, x_i)$. It is assumed that u_i is non-increasing with x_i when $x_i \leq t_i$, and equals 0 when $x_i > t_i$.

Given mechanism (p, x), the utility for user i with true drop rate t_i, bidded drop rate s_i and others bidding s_{-i} is

$$U_i(p, x, t_i, s_i, s_{-i}) = \sum_{\phi : i \in \phi} u_i(t_i, x_{i,\phi}(s)) p_\phi(s) . \tag{1}$$

Throughout the paper we evaluate the utility of one user by examining each of the possible bid vectors of other users.

We now present some definitions we will use throughout the paper.

Definition 1 (Incentive compatibility - IC). *A mechanism (p, x) is* incentive compatible *if for any utility function, any $i \in \mathcal{N}$, any $t \in T$ and any $s_i \in T_i$, we have*

$$U_i(p, x, t_i, t_i, t_{-i}) \geq U_i(p, x, t_i, s_i, t_{-i}) \tag{2}$$

where $U(\cdot)$ is defined in (1). That is any possible true drop rate vector t is a Nash equilibrium, in which any user has no incentive to lie if all the other users bid their true drop rates.

Definition 2 (Weak-incentive compatibility - weak-IC). *A mechanism (p, x) is* weakly-incentive compatible *if for any utility function, for any $i \in \mathcal{N}$ and any $s_i \in T_i$, we have*

$$U_i(p, x, t_i, t_i, t_{-i}) \geq U_i(p, x, t_i, s_i, t_{-i}), \quad \forall t \in T^* \tag{3}$$

for some $T^* \subset T$ with $\mathcal{L}(T \backslash T^*) = 0$, where $\mathcal{L}(\cdot)$ is the Lebesgue measure.

Definition 3 (Feasibility). *A mechanism (p, x) is* feasible *if it satisfies*

i. Probability constraint (P):
For any $\phi \subset \mathcal{N}$ and any $t \in T$,

$$\sum_{\psi \subset \mathcal{N}} p_\psi(t) = 1 \quad and \quad p_\phi(t) \geq 0 . \tag{4}$$

ii. Capacity constraint (CC):
For any $i \in \mathcal{N}$, any $t \in T$, and any $\phi \subset \mathcal{N}$ with $p_\phi(t) > 0$,

$$\sum_{j \in \phi} (1 - x_{j,\phi}(t)) \leq 1 \quad and \quad x_{i,\phi}(t) \geq 0 . \tag{5}$$

iii. Individual rationality (IR):
For any $\phi \subset \mathcal{N}$ with $p_\phi(t) > 0$, any $t \in T$, and any $i \in \phi$,

$$x_{i,\phi}(t) \leq t_i . \tag{6}$$

iv. Incentive compatibility (IC) in (2).

Note that IC is also known as strategy-proofness or truthfulness.

Definition 4 (Weak-determinism - weak-D). *A mechanism (p, x) is* weakly-deterministic *if for any $t \in T$ and any $i \in \mathcal{N}$,*

$$(\forall j \in \mathcal{N}, j \neq i \Rightarrow t_j \neq t_i) \Rightarrow \sum_{\phi : i \in \phi} p_\phi(t) \in \{0, 1\} .$$

That is, for a given bid vector t, if user i is the only one who bids t_i, then either the user always gets admitted, or the user never gets admitted.

Definition 5 (Determinism - D). *A mechanism (p, x) is* deterministic *if there exists a function $\psi : T \to \mathcal{P}(\mathcal{N})$ such that for any $t \in T$,*

$$p_\phi(t) = \begin{cases} 1 & if\ \phi = \psi(t) \\ 0 & otherwise \end{cases}$$

where $\mathcal{P}(A)$ is the power set of A. Determinism implies that the winning set is always unique. We call $\psi(\cdot)$ the winning set function.

Note that for a deterministic mechanism (p, x) we can have the payment denoted as

$$x_i(t) = \begin{cases} x_{i,\psi(t)}(t) & \text{if } i \in \psi(t) \\ 1 & \text{otherwise} \end{cases} . \tag{7}$$

Definition 6 (Anonymity). *A mechanism* (p, x) *is* anonymous *if for any* $t \in T$*, any* $\pi \in \Gamma_n$*, and any* $\phi \subset \mathcal{N}$*,*

$$p(\pi(t)) = \pi(p(t))$$
$$x_\phi(\pi(t)) = \pi(x_\phi(t))$$

where Γ_n *is the set of all permutations of* n *indices. That is, the outcome of the auction does not depend on the identity of the bidders.*

Definition 7 (Monotonicity). *A mechanism* (p, x) *is* monotonic *if for any* $i \in \mathcal{N}$*, any* $t_{-i} \in T_{-i}$*, and any* $s_i, s_i' \in T_i$ *with* $s_i < s_i'$*,*

$$\sum_{\phi: i \in \phi} p_\phi(t_{-i}, s_i) \leq \sum_{\phi: i \in \phi} p_\phi(t_{-i}, s_i') \ .$$

That is, given that others' bids are fixed, a user's chance of getting admitted should not decrease when the user bids higher.

We are interested in feasible auction mechanisms that are weakly-deterministic, anonymous and monotonic. Feasibility implies that the decision is in the capacity region, no user is forced to participate, and no one has incentive to lie about his type. Weak-determinism, anonymity and monotonicity are properties we consider desirable for a fair mechanism.

3 VCG Auctions without Money

In this section we show why a simple VCG algorithm is not suitable for our problem.

3.1 VCG Auctions

In this section we temporarily turn our attention to deterministic mechanisms (p, c), where $p \colon T \to \{0, 1\}^n$ is the indicator function of users' admittance and $c \colon T \to \mathbb{R}^n$ is the payment, which can be money.

We say that a mechanism (p, c) is called a *VCG mechanism* [5] if

$$p \in \arg\max_{p'} \sum_i \theta_i(t_i) p_i'(t)$$

and for any $i \in \mathcal{N}$, there exists $h_i \colon T_{-i} \to \mathbb{R}$ such that

$$c_i(t) = h_i(t_{-i}) - \sum_{j \neq i} \theta_j(t_j) p_j(t)$$

where $\theta_i(\cdot)$ is the valuation function of user i in [5].

It has been proved that the VCG mechanism satisfies IC in (2) [5]. Usually the functions h_i's are chosen according to the *Clarke pivot rule*, that is,

$$h_i(t_{-i}) = \max_{p'} \sum_{j \neq i} \theta_j(t_j) p_j'(t)$$

and it can be shown that with Clarke pivot rule, IR is always satisfied and $h_i \geq 0$ [5]. Intuitively, user i pays an amount equal to the total damage that the user causes to the other users. Then the utility for user i given bid vector t is

$$u_i(p, c, t) = \theta_i(t_i)p_i(t) - c_i(t) .$$

Notice that the utility function is quasi-linear with payment function c [6].

3.2 Adaptation of VCG Auctions to Money-Free Scenario

In this subsection we apply a simple VCG auction mechanism in this money-free, capacity-constrained scenario and show why it is not suitable.

Let us consider the deterministic mechanism above. Let $\theta_i(t_i) = t_i$ and use the Clarke pivot payment

$$c_i(t) = \max_{p'} \sum_{j \neq i} t_j p'_j(t) - \sum_{j \neq i} t_j p_j(t)$$

then we have $0 \leq c_i(t) \leq t_i$. We consider the drop rate as the payment for those admitted users, that is,

$$x_i(t) = \begin{cases} c_i(t) & \text{if } p_i(t) = 1 \\ 1 & \text{otherwise} \end{cases}$$

then this drop rate is well defined and is no larger than t_i. The utility function for user i would be

$$u_i(p, x, t) = \begin{cases} t_i - x_i & \text{if } p_i(t) = 1 \\ 0 & \text{otherwise} \end{cases}$$

and is quasi-linear with x.

However, one may wonder if CC in (5) is still satisfied, since $x_i(t) \leq t_i$ for $p_i(t) = 1$. The answer is no. To see this, we consider the following simple example.

Example 1. Let $n = 3$ and $t = (0.6, 0.6, 0.2)$, then the optimal decision should be $\phi = \{1, 2\}$, that is, the first two users are admitted. If either of the first two users is removed, the optimal decision will be admitting the other of the two only, so the harm is 0. If user 3 is removed, nothing changes, so the harm is still 0, then we have the payment

$$c(t) = \{0, 0, 0\}$$

and the drop rate

$$x(t) = \{0, 0, 1\}$$

which violates CC because the QoS's of the first two users are then both 1, which cannot be supported by the AP simultaneously.

In summary, even though the simple VCG auction satisfies P, IR, IC, and maximizes the number of admitted users, it does not fulfill the CC constraint. Therefore the simple VCG algorithm is not a feasible mechanism for our scenario.

3.3 Further Analysis of the Failure of VCG

The failure comes from the interaction between the allocation results and the payment. As we mentioned above, to make a VCG algorithm work, the utility function has to be quasi-linear, where the payment is the linear part which can be freely chosen after allocation decision is made. This kind of behavior resembles money. Without money, however, the choice of drop rate has a feedback effect on the allocation decision, which leads to a violation on the capacity constraint.

Hence the lack of monetary payments makes VCG auction mechanisms not feasible. Thus, we must come up with better solutions in our setting.

4 Analysis of the Problem

In this section we analyze feasible strategy-proof mechanisms in non-monetary scenarios. We start with a mechanism (p^*, x^*), which is inspired by the uniform-price auction [2]. We then show the flaws of (p^*, x^*) and prove the general properties of feasible strategy-proof non-monetary auction mechanisms.

4.1 The First Mechanism

We now propose the mechanism (p^*, x^*), and prove that this mechanism satisfies the capacity constraint and weak-IC in (3), and admits at least half of the maximal possible number of users. After that we point out the flaws of the weak-IC concept.

Given bid vector $t \in T$, the mechanism (p^*, x^*) is described as follows.

Step i. Let α be a rearrangement of the indices such that $t_{\alpha(1)} \geq t_{\alpha(2)} \geq \cdots \geq t_{\alpha(n)}$. If several users bid the same, just arrange them randomly. Introduce a pseudo-bidder with $t_{n+1} = 0$ and $\alpha(n+1) = n+1$.

Step ii. Find the largest index m^* with $1 \leq m^* \leq n$ such that $m^* \cdot (1 - t_{\alpha(m^*+1)}) \leq 1$. The winning set is $\psi^*(t) = \{\alpha(1), \alpha(2), \cdots, \alpha(m^*)\}$.

Step iii. Set the payment $x_i^*(t) = x^*(t) = t_{\alpha(m^*+1)}$ if $i \in \psi^*(t)$.

The idea of the mechanism is basically that we start from the higher bidders and try to admit as many users as possible, with payment equal to the highest losing bid.

Notice that the rearrangement α and the winning set ψ^* here might not be unique if several users bid the same value, in which case both of them will be probabilistic functions instead of deterministic functions. However, when there are no two users bidding the same value, α and ψ^* are deterministic functions.

Given the bids $t \in T$ and a corresponding rearrangement α, we let

$$\sigma_i(t) = t_{\alpha(i)}, \quad 1 \leq i \leq n+1$$

and

$$\sigma(t) = (\sigma_1(t), \sigma_2(t), \cdots, \sigma_{n+1}(t))$$

where $\sigma(t)$ is the unique sorted vector of t with $\sigma_1(t) \geq \sigma_2(t) \geq \cdots \geq \sigma_{n+1}(t)$, regardless of the possible different choices of α. Then the index chosen in step ii is a function of the bids given by

$$m^*(t) = \max\{m \in \mathcal{N} | m \cdot (1 - \sigma_{m+1}(t)) \leq 1\}$$

where $m^*(t)$ is always well-defined because

$$1 \cdot (1 - t_{\alpha(2)}) \leq 1 .$$

Similarly, the payment for those admitted users is also determined by

$$x^*(t) = t_{\alpha(m^*(t)+1)} .$$

Example 2. Let the bid vector be $t = (t_1, t_2, t_3, t_4) = \{0.5, 0.6, 0.7, 0.6\}$.

Step i. The rearrangement could be $\alpha = (\alpha(1), \alpha(2), \alpha(3), \alpha(4), \alpha(5)) = (3, 2, 4, 1, 5)$. ($\alpha$ could also be $(3, 4, 2, 1, 5)$) So $t_{\alpha(1)} \geq t_{\alpha(2)} \geq t_{\alpha(3)} \geq t_{\alpha(4)} \geq t_{\alpha(5)} = 0$ and the sorted vector of t is $\sigma(t) = (0.7, 0.6, 0.6, 0.5)$.

Step ii. $2 \cdot (1 - 0.6) = 0.8 \leq 1$ and $3 \cdot (1 - 0.5) = 1.5 > 1$, so $m^*(t) = 2$ and the winning set is $\psi^*(t) = \{2, 3\}$ since $\alpha = (3, 2, 4, 1, 5)$. ($\psi^*(t)$ would be $\{3, 4\}$ if $\alpha = (3, 4, 2, 1, 5)$.)

Step iii. The payment for either of the two winners is $x^*(t) = t_{\alpha(3)} = t_4 = 0.6$. (Note that if $\alpha = (3, 4, 2, 1, 5)$, then $x^*(t)$ would still be 0.6.)

Lemma 1. *The mechanism (p^*, x^*) satisfies the P, CC, IR and weak-IC constraints.*

The proof of Lemma 1 is deferred to Appendix A.1. Note that (p^*, x^*) does not satisfy feasibility because it is not IC. To see this, just consider two users bidding the same drop rate. The chance of getting admitted is half for either user. However the chance of either user increases to 1 when he raises his bid by a small amount and the other user keeps the original bid.

We now show that (p^*, x^*) admits at least half of the maximal possible number of users.

Theorem 1 (Scalability of (p^*, x^*)). *For any true value of drop rate $t \in T$, if there exists some feasible mechanism that admits M users, then (p^*, x^*) can admit at least $\lfloor \frac{M}{2} \rfloor$ users.*

The proof of Theorem 1 is deferred to Appendix A.2.

The problem about (p^*, x^*) is that it is only weakly-IC but not IC. This means that if equal bids happen, although with low probability, users might have incentive to lie. For example, in a two-user case, if both users bid the same value, each of them would have half chance of getting admitted. But if one of them increases his bid by a small amount, he would win with the same payment and probability 1. Thus indistinguishable bids make (p^*, x^*) fail for IC.

4.2 Impossibility for Probabilistic Decisions of Equal Bids

We now show that to fulfill strict IC with some assumptions mentioned above, a weak-deterministic mechanism has to be deterministic. That is, if several users bid exactly the same value, then the only choice for guaranteeing truth-telling is to admit either all or none of them.

Theorem 2 (Impossibility). *For a mechanism (p, x) that satisfies P, IR, IC, anonymity and monotonicity, (p, x) is weakly-deterministic if and only if (p, x) is deterministic. That is, (p, x) admits either all or none of the equal bids.*

The proof of Theorem 2 is omitted due to the page limit and can be found in [16].

We should note that (p^*, x^*) is weakly-deterministic because users might be randomly admitted when bidding the same. Then by Theorem 2, to achieve IC we need to design deterministic mechanisms, that is, mechanisms with only deterministic outcomes.

4.3 Lowest Winning Bid Theorem

We further show that any deterministic mechanism with IC must be illustrated by a *lowest winning bid function.*

Definition 8 (Lowest winning bid mechanism). *A deterministic mechanism (p, x) is a* lowest winning bid mechanism *if there exists some function $z\colon T_{-i} \rightarrow T_i$ such that for any $t_{-i} \in T_{-i}$,*

$$\begin{cases} \text{if } s_i \geq z(t_{-i}) & \text{then } i \in \psi(t_{-i}, s_i) \text{ and } x_i(t_{-i}, s_i) = z(t_{-i}) \\ \text{if } s_i < z(t_{-i}) & \text{then } i \notin \psi(t_{-i}, s_i) \text{ and } x_i(t_{-i}, s_i) = 1 \end{cases}$$

where $\psi(\cdot)$ is the winning set function of (p, x) defined in the definition of determinism (Definition 5).

The function $z(\cdot)$ is called the lowest winning bid function *of (p, x).*

Theorem 3 (Lowest winning bid). *A deterministic mechanism satisfies IC if and only if it is a lowest winning bid mechanism.*

The proof of Theorem 3 is omitted due to the page limit and can be found in [16].

The lowest winning bid theorem shows us what a deterministic truth-telling mechanism should look like. Notice that no other assumptions are needed for this theorem, so it remains valid in a general setting. More importantly, this theorem gives us an efficient approach to design deterministic truth-telling mechanisms.

Although Theorem 3 does not work for (p^*, x^*) due to weak-determinism, we do have the following similar result.

Lemma 2 (Infimum winning bid function for (p^*, x^*)). *Under mechanism (p^*, x^*), knowing others' bid t_{-i}, the infimum of user i's winning bids is given by*

$$\bar{z}^*(t_{-i}) = \min\{\sigma_j(t_{-i}) | j(1 - \sigma_j(t_{-i})) \leq 1\} . \quad (8)$$

The proof of Lemma 2 is deferred to Appendix A.3.

Note that $\bar{z}^*(t_{-i})$ is not the lowest winning bid for user i because bidding this value does not guarantee winning.

5 Our Proposed Mechanism

Now we introduce the mechanism (p^{**}, x^{**}), which is a truth-telling mechanism based on the previous (p^*, x^*). We first construct (p^{**}, x^{**}) by the so-called *lifting trick*. After that we show that (p^{**}, x^{**}) is feasible, and has very close performance to (p^*, x^*) in an asymptotic sense.

5.1 Lifting Trick

Let

$$z^{**}(t_{-i}) = \begin{cases} \bar{z}^*(t_{-i}) + a(1 - \max_{\substack{j \neq i \\ t_j \neq 1}} t_j) & \text{if } t_k \neq 1 \text{ for some } k \in \mathcal{N} \\ 1 & \text{otherwise} \end{cases} \tag{9}$$

where $0 < a < 1$ is a constant. Recall that $\bar{z}^*(\cdot)$ here is the infimum winning bid function of (p^*, x^*) in (8). We note that (p^{**}, x^{**}) is a deterministic mechanism based on the lowest winning bid function z^{**} in (9) and by Theorem 3 we know that (p^{**}, x^{**}) satisfies IC. Pictorially, we lift one corner of the infimum winning bid function $\bar{z}^*$, so that bidding the new function guarantees winning.

Example 3. Take $a = 0.1$ in (9). Again, let the bid vector be $t = (t_1, t_2, t_3, t_4) = \{0.5, 0.6, 0.7, 0.6\}$. By (8) we can calculate the infimum winning bid under (p^*, x^*) for each user:

$$\begin{aligned} \bar{z}^*(t_{-1}) &= \bar{z}^*(0.6, 0.7, 0.6) = 0.6 \ , \\ \bar{z}^*(t_{-2}) &= \bar{z}^*(0.5, 0.7, 0.6) = 0.6 \ , \\ \bar{z}^*(t_{-3}) &= \bar{z}^*(0.5, 0.6, 0.6) = 0.6 \ , \\ \bar{z}^*(t_{-4}) &= \bar{z}^*(0.5, 0.6, 0.7) = 0.6 \ . \end{aligned}$$

Then the lowest winning bid under (p^{**}, x^{**}) for each user is

$$\begin{aligned} z^{**}(t_{-1}) &= \bar{z}^*(t_{-1}) + a(1 - \max_{j \neq 1} t_j) = 0.63 \ , \\ z^{**}(t_{-2}) &= \bar{z}^*(t_{-2}) + a(1 - \max_{j \neq 2} t_j) = 0.63 \ , \\ z^{**}(t_{-3}) &= \bar{z}^*(t_{-3}) + a(1 - \max_{j \neq 3} t_j) = 0.64 \ , \\ z^{**}(t_{-4}) &= \bar{z}^*(t_{-4}) + a(1 - \max_{j \neq 4} t_j) = 0.63 \ . \end{aligned}$$

Since only user 3's bid is higher than or equal to his lowest winning bid, we have the winning set $\psi^{**}(t) = \{3\}$ and $x_3^{**}(t) = 0.64$.

5.2 Properties of (p^{**}, x^{**})

Lemma 3. *(p^{**}, x^{**}) is feasible, and $0 \leq z^{**}(t_{-i}) \leq 1$ for any $t_{-i} \in T_{-i}$.*

The proof of Lemma 3 is deferred to Appendix A.4.

Now we show that (p^{**}, x^{**}) has very close performance to (p^*, x^*) in an asymptotic sense. Assume that the drop rate vector t is drawn from a distribution with joint probability density function $f: T \rightarrow \mathbb{R}^+$. Then, we have the following theorem.

Theorem 4. *If the distribution function f is bounded by K, the probability that (p^{**}, x^{**}) behaves differently from (p^*, x^*) is at most aK.*

The proof of Theorem 4 is deferred to Appendix A.5.

We notice from Theorem 4 that the probability of different behaviors between the two mechanisms is of order $O(a)$.

6 Conclusion

In this paper, we studied the problem of designing a strategy-proof non-monetary auction mechanism for wireless networks. The motivation is to let the users tell the truth when bidding their resource requirements, and to admit as many users as possible. We gave a general model for this problem and showed that due to the lack of monetary payments in this scenario, a simple adoption of VCG mechanism would violate the capacity constraint. We analyzed the problem and found some properties that any strategy-proof auction mechanism should satisfy. Finally we proposed a feasible mechanism which is truthful even with equal bids, and showed that it could admit at least half of the maximal number of users with high probability in an asymptotic sense.

As possible topic for future work, discrete pricing models might be considered rather than continuous pricing models. Also, the assumption of weak-determinism could be weakened, and more specific utility functions could be considered for better performance. Furthermore, the lower bound of the number of admitted user might be improved.

Acknowledgement. Research supported by NSF Grant CNS-0953165, and DTRA Grants HDTRA1-08-1-0016 and HDTRA1-09-1-0055.

References

1. Myerson, R.B.: Optimal Auction Design. Mathematics of Operations Research 6(1), 58–73 (1981)
2. Krishna, V.: Auction Theory. Academic Press, San Diego (2002)
3. Varian, H.R.: Position Auctions. International Journal of Industrial Organization 25, 1163–1178 (2007)
4. Hou, I.H., Kumar, P.R.: Utility Maximization for Delay Constrained QoS in Wireless. In: Proceedings of IEEE INFOCOM, pp. 1–9 (2010)

5. Nisan, N., Roughgarden, T., Tardos, E., Vazirani, V.V.: Algorithmic Game Theory. Cambridge University Press, New York (2007)
6. Maillé, P., Tuffin, B.: Why VCG Auctions Can Hardly Be Applied to the Pricing of Inter-domain and Ad Hoc Networks. In: 3rd EuroNGI Conference on Next Generation Internet Networks, pp. 36–39 (2007)
7. Anderegg, L., Eidenbenz, S.: Ad Hoc-VCG: a Truthful and Cost-Efficient Routing Protocol for Mobile Ad Hoc Networks with Selfish Agents. In: Proc. ACM 9th Annual International Conference on Mobile Computing and Networking (MobiCom 2003), San Diego, pp. 245–259 (2003)
8. Buttyán, L., Hubaux, J.P.: Enforcing Service Availability in Mobile Ad-Hoc WANs. In: Proc. International Symposium on Mobile Ad Hoc Networking & Computing (MobiHoc 2000), Boston, pp. 87–96 (2000)
9. Buttyán, L., Hubaux, J.P.: Stimulating Cooperation in Self-organizing Mobile Ad Hoc Networks. ACM/Kluwer Mobile Networks and Applications 8(5), 579–592 (2003)
10. Crowcroft, J., Gibbens, R., Kelly, F., Östring, S.: Modelling Incentives for Collaboration in Mobile Ad Hoc Networks. In: Proc. WiOpt 2003, France (2003)
11. Srinivasan, V., Nuggehalli, P., Chiasserini, C.F., Rao, R.R.: Energy Efficiency of Ad Hoc Wireless Networks with Selfish Users. In: Proc. European Wireless Conference, Italy (2002)
12. Srinivasan, V., Nuggehalli, P., Chiasserini, C.F., Rao, R.R.: Cooperation in Wireless Ad Hoc Networks. In: Proc. IEEE INFOCOM 2003, San Francisco, vol. 2, pp. 808–817 (2003)
13. Zhong, S., Chen, J., Yang, Y.R.: Sprite: a Simple, Cheat-Proof, Credit-Based System for Mobile Ad-Hoc Networks. In: Proc. IEEE INFOCOM 2003, San Francisco, vol. 3, pp. 1987–1997 (2003)
14. Goldberg, A.V., Hartline, J.D., Karlin, A.R., Wright, A., Saks, M.: Competitive Auctions. Games and Economic Behavior 55(2), 242–269 (2003)
15. Mehta, A., Vazirani, V.V.: Randomized Truthful Auctions of Digital Goods are Randomizations Over Truthful Auctions. In: Proceeding of the 5th ACM Conference on Electronic Commerce 2004, New York (2004)
16. Kang, X., Jaramillo, J.J., Ying, L.: A Strategy-Proof and Non-monetary Admission Control Mechanism for Wireless Access Networks (2010), `http://arxiv.org/abs/1010.2713`

A Appendix

A.1 Proof of Lemma 1

i. Probability constraint (P)
The probability constraint is obviously satisfied. For those t such that the winning set $\psi^*(t)$ is determined, $\sum_{\phi \subset \mathcal{N}} p_\phi(t) = p_{\psi^*(t)}(t) = 1$. For those t such that there are M possible winning sets, the probability for each of them would be $\frac{1}{M}$ and $\sum_{\phi \subset \mathcal{N}} p_\phi(t) = M \cdot \frac{1}{M} = 1$.

ii. Capacity constraint (CC)
For any $t \in T$, for any $\phi \subset \mathcal{N}$ with $p_\phi(t) > 0$,

$$\sum_{i \in \phi}(1 - x^*_{i,\phi}(t)) = m^*(t) \cdot (1 - x^*(t))$$

$$= m^*(t) \cdot \left(1 - t_{\alpha(m^*(t)+1)}\right)$$
$$\leq m^*(t) \cdot \frac{1}{m^*(t)}$$
$$= 1 \ .$$

Also $x^*_{i,\phi}(t) \geq 0$. Thus, (p^*, x^*) satisfies CC.

iii. Individual rationality (IR)
For any $t \in T$, any $\phi \subset \mathcal{N}$ with $p_\phi(t) > 0$, and any $i \in \phi$,

$$x^*_{i,\phi}(t) = x^*(t) = t_{\alpha(m^*(t)+1)} \leq t_i \ .$$

iv. Weak-incentive compatibility (weak-IC)
We only consider the set of *distinguishable* bid vectors

$$T_D = \{t \in T | t_i \neq t_j \forall i \neq j\}$$

that is, the set of bid vectors with no equal bids from any two users. For $t \in T_D$ and $i \in \mathcal{N}$, the result of user i bidding s_i would be

$$p^*_i(t_{-i}, s_i) = \begin{cases} 1 & \text{if } s_i > x^*(t) \\ \frac{1}{2} & \text{if } s_i = x^*(t) \\ 0 & \text{if } s_i < x^*(t) \end{cases}$$

with payment $x^*_i(t_{-i}, s_i) = x^*(t)$ if admitted. Note that $p^*_i(t_{-i}, s_i) = \frac{1}{2}$ when $s_i = x^*(t)$ since there is only one other user who bids $x^*(t)$.

We first consider the case of $i \in \psi^*(t)$. We then have $t_i > x^*(t)$. If $s_i > x^*(t)$, then user i still gets admitted with the same payment. If $s_i < x^*(t)$, then user i gets rejected. If $s_i = x^*(t)$, then user i either gets admitted with the same payment, or get rejected, both of which have probability 1/2. So user i cannot get better utility in the first case.

We then consider the case of $i \notin \psi^*(t)$. Now we have $t_i < x^*(t)$. If $s_i > x^*(t)$, then user i gets admitted with payment $x^*(t)$ higher than true value t_i. If $s_i < x^*(t)$, the user i still does not get admitted. If $s_i = x^*(t)$, then user i either gets admitted with payment too high to accept, or does not get admitted at all, both of which have probability 1/2. So user i cannot get better utility in the second case.

Thus, for any $t \in T_D$, no user has incentive to lie. As $T \backslash T_D$ has measure zero, we have weak-IC.

□

A.2 Proof of Theorem 1

Suppose (p^*, x^*) admits only m users, that is,

$$m^*(t) = m$$

and $(\bar{p}, \bar{x})$ has a chance of admitting at least $2m + 2$ users given bid vector t, that is,

$$\exists \phi \subset \mathcal{N}, |\phi| \geq 2m + 2, \bar{p}_\phi(t) > 0 \ .$$

Then we have

$$\begin{aligned}
\sum_{i \in \phi}(1 - \bar{x}_{i,\phi}(t)) &\geq \sum_{i \in \phi}(1 - t_i) && (10)\\
&\geq \sum_{i=1}^{2m+2} (1 - \sigma_i(t)) \\
&\geq \sum_{i=m+2}^{2m+2} (1 - \sigma_i(t)) \\
&\geq (m + 1)(1 - \sigma_{m+2}(t)) \\
&> 1 && (11)
\end{aligned}$$

where the inequality (10) comes from IR and (11) comes from the definition of the mechanism.

Thus $(\bar{p}, \bar{x})$ admits at most $2m + 1$ users. This is equivalent to the statement that if some feasible mechanism admits M users, (p^*, x^*) can at least admit $\lfloor \frac{M}{2} \rfloor$ users. □

A.3 Proof of Lemma 2

This can be checked by directly going through the process of (p^*, x^*). If $t_i > \bar{z}^*(t_{-i})$, we can see that $x^*(t) = \bar{z}^*(t_{-i}) < t_i$, so user i wins. If $t_i < \bar{z}^*(t_{-i})$, we have $x^*(t) \geq t_i$, so user i loses. If $t_i = \bar{z}^*(t_{-i})$, user i wins with some probability between 0 and 1, which depends on the number of users bidding $\bar{z}^*(t_{-i})$. □

A.4 Proof of Lemma 3

P comes from determinism. IR comes from the definition of lowest winning bid mechanisms. IC comes from Theorem 3. So only the proof of CC requires some more effort.

By the construction of $z^{**}(\cdot)$ in (9), we have for any $t_{-i} \in T_{-i}$ with $t_k \neq 1$ for some $k \in \mathcal{N}\backslash\{i\}$, $z^{**}(t_{-i}) > \bar{z}^*(t_{-i})$. Then by Lemma 2, we note that (p^{**}, x^{**}) is stricter than (p^*, x^*), that is, for any $t \in T$, if user i gets admitted in (p^{**}, x^{**}), then user i also gets admitted in (p^*, x^*). Let $\psi^{**}(t)$ be the winning set function of (p^{**}, x^{**}). Fix $t \in T$, for any ψ with $p^*_\psi(t) > 0$, we have

$$\begin{aligned}
\sum_{i\in\psi^{**}(t)} (1 - x_i^{**}(t)) &= \sum_{i\in\psi^{**}(t)} (1 - z^{**}(t_{-i})) && (12)\\
&\leq \sum_{i\in\psi^{**}(t)} (1 - \bar{z}^*(t_{-i})) && (13)\\
&\leq \sum_{i\in\psi} (1 - \bar{z}^*(t_{-i})) && (14)\\
&= \sum_{i\in\psi} (1 - x^*_{i,\psi}(t)) && (15)\\
&\leq 1 && (16)
\end{aligned}$$

where (12) comes from the definition of (p^{**}, x^{**}), (13) and (14) comes from the fact that (p^{**}, x^{**}) is stricter than (p^*, x^*), (15) comes from the definition of (p^*, x^*), (16) is because (p^*, x^*) satisfies CC.

If all bids in t_{-i} are 1, then $z^{**}(t_{-i}) = \bar{z}^*(t_{-i}) = 1$. Otherwise, for any $t_{-i} \in T_{-i}$ with some entry not equal to 1,

$$\begin{aligned}
z^{**}(t_{-i}) &= \bar{z}^*(t_{-i}) + a(1 - \max_{\substack{j\neq i\\ t_j\neq 1}} t_j)\\
&\leq \max_{\substack{j\neq i\\ t_j\neq 1}} t_j + a(1 - \max_{\substack{j\neq i\\ t_j\neq 1}} t_j)\\
&\leq \max_{\substack{j\neq i\\ t_j\neq 1}} t_j + (1 - \max_{\substack{j\neq i\\ t_j\neq 1}} t_j)\\
&\leq 1\ .
\end{aligned}$$

Since $z^{**}(t_{-i}) \geq \bar{z}^*(t_{-i}) \geq 0$, we get $0 \leq z^{**}(t_{-i}) \leq 1$. □

A.5 Proof of Theorem 4

Let $\mathbf{T}$ be the vector of random variables with values taken in T, and as usual let $\mathbf{T}_i$ and $\mathbf{T}_{-i}$ be the corresponding components. Note that $T \subset [0, 1]^n$. Then the probability that (p^{**}, x^{**}) and (p^*, x^*) have different results is

$$\begin{aligned}
D &= \Pr\left(\bar{z}^*(\mathbf{T}_{-i}) < \mathbf{T}_i < z^{**}(\mathbf{T}_{-i}) \text{ for some } i\right)\\
&\leq \sum_{i=1}^{n} \Pr\left(\bar{z}^*(\mathbf{T}_{-i}) < \mathbf{T}_i < z^{**}(\mathbf{T}_{-i})\right)
\end{aligned}$$

$$= \sum_{i=1}^{n} \int_{B_i} f(t)\mathrm{d}t$$
$$\leq \sum_{i=1}^{n} \mathcal{L}(B_i)K$$

where

$$B_i = \{t \in T | \bar{z}^*(t_{-i}) < t_i < z^{**}(t_{-i})\}$$

and $\mathcal{L}(\cdot)$ denotes the Lebesgue measure. Then

$$\begin{aligned}
\mathcal{L}(B_i) &= \int_{B_i} 1\mathrm{d}t \\
&= \int_{T_{-i}} \int_{\bar{z}^*(t_{-i})}^{z^{**}(t_{-i})} 1\mathrm{d}t_i \mathrm{d}t_{-i} \\
&\leq \int_{[0,1]^{n-1}} (z^{**}(t_{-i}) - \bar{z}^*(t_{-i}))\mathrm{d}t_{-i} \\
&= \int_{[0,1]^{n-1}} a(1 - \max_{j \neq i} t_j)\mathrm{d}t_{-i} \\
&= a - a \int_{[0,1]^{n-1}} \max_{j \neq i} t_j \mathrm{d}t_{-i} \\
&= a - a\frac{n-1}{n} \\
&= \frac{a}{n} \ .
\end{aligned}$$

Thus

$$D \leq aK \ .$$

□

Enabling Multi-packet Transmission and Reception: An Adaptive MAC Protocol for MANETs

Hui Xu[1,*], J.J. Garcia-Luna-Aceves[1], and Hamid R. Sadjadpour[2]

[1] Computer Engineering Department
[2] Electrical Engineering Department
University of California, Santa Cruz
Santa Cruz, CA 95064, USA
{xuhui,jj,hamid}@soe.ucsc.edu

Abstract. To increase network capacity, advanced physical layer (PHY) techniques have been developed to support new transmission paradigm where one node could send different packets concurrently to multiple receivers (multi-packet transmission: MPT) or receive packets concurrently from multiple senders (multi-packet reception: MPR). To exploit and support them for high performance, new type of medium access control (MAC) protocol is needed. Especially, MPT or MPR requirements are dynamic according to traffic conditions. In this paper, an adaptive MAC approach (AMPTR) is proposed to enable dynamic MPT or MPR requirements for mobile ad hoc networks (MANETs). The proposal includes two main parts: access coordination and data transmission. The access coordination process comprises channel access contention and coordination to make handshakes for multiple concurrent transmissions. Once channel access coordination is completed, multiple transmissions can then be carried out concurrently, where frame aggregation is used for network capacity and throughput improvement and accordingly block acknowledgement is employed for efficiently reporting multiple packet reception status. We evaluate the performance of AMPTR through simulations, and show that the AMPTR scheme has much higher network throughput and smaller packet delivery delay than currently widely used multiple access preventing schemes.

* This research was partially sponsored by the U.S. Army Research Laboratory under the Network Science Collaborative Technology Alliance, Agreement Number W911NF-09-0053, by the Army Research Office under agreement number W911NF-05-1-0246, by the National Science Foundation under grant CCF-0729230, and by the Baskin Chair of Computer Engineering. The views and conclusions contained in this document are those of the authors and should not be interpreted as representing the official policies, either expressed or implied, of the U.S. Army Research Laboratory or the U.S. Government. The U.S. Government is authorized to reproduce and distribute reprints for Government purposes notwithstanding any copyright notation hereon.

X. Zhang and D. Qiao (Eds.): QShine 2010, LNICST 74, pp. 188–203, 2011.

1 Introduction

In mobile ad hoc networks (MANETs), there are two directions to increase network throughput: one is to increase the number of concurrent transmissions; the other is to increase throughput for each transmission. Much work has focused on developing advanced physical layer techniques to support multiple concurrent transmissions while providing more reliable data transmission. For example, multiple-input multiple-output (MIMO) uses multiple antennas at the wireless transmitter and receiver, and exploits the spatial dimension and multi-path channel signals to enable the ability of the transmitter transmitting multiple data streams to the receiver concurrently and the receiver decoding more than one data stream with higher success probability [1,2].

When data streams aim for different receivers (multi-packet transmission: MPT) or come from different transmitters (multi-packet reception: MPR), coordinations among nodes are needed [3], and the effectiveness of multiple concurrent transmissions (such as MPT or MPR) on improving network capacity is theoretically analyzed and proved in [4]. These kinds of MPT and MPR transmission paradigms attract more interest and have not been fully exploited. There are many MAC protocols proposed to exploit the advanced physical ability of MPR and allow multiple transmitters to concurrently access medium/channel [5,6,7,3]. Particularly, [8,3] have tried to modify IEEE 802.11 DCF scheme [10] to support MPR by negotiating concurrent transmissions with extra bytes in handshake control packets (such as RTS, CTS and ACK). However, all the above proposals use fixed transmission architecture, i.e., supporting MPT or MPR while not both concurrently; in addition, they neglect traffic conditions and network layer transmission requirements. While in practice, a node may want to transmit to multiple receivers or receive multiple packets from multiple transmitters concurrently. One unexploited multiuser transmission direction is to adaptively support MPT or MPR, which is exploited in our proposal AMPTR.

In this paper, we aim to provide an adaptive MAC protocol which can provide medium access according to traffic flow requirements. When a node which obtains access permission has packets for multiple receivers, MPT will be carried out; otherwise, the node will give right to its receiver to initiate more neighbors transmitting to it concurrently, i.e., MPR will be triggered. The adaptability and compatibility of this proposal is also represented in that when MPT or MPR handshake fails, normal peer-to-peer transmission could still be carried out. Our design principle is that each node performs Carrier Sensing Medium Access (CSMA) scheme [10], and that once channel access is permitted, the maximum number of streams the channel could support will be exploited, i.e., fulfill the channel by scheduling that number of concurrent transmissions. The process includes two parts: access coordination and data transmission. The access coordination is combined with both contention and coordination, where intended transmitters follow CSMA scheme to compete channel, and use handshakes to coordinate its neighbors' behaviors. Once coordination is complete, the transmitter or multiple transmitters could access the channel concurrently and multiple transmission could be completed.

The rest of this paper is organized as follows: In Section 2, we present our adaptive MPT and MPR medium access control mechanism in detail. Then, Section 3 gives some experimental results and validates the effectiveness of the proposed scheme in greatly improving throughput and reducing end-to-end delay. We finally conclude in Section 4.

2 Adaptive Multiple Packet Transmission and Reception (AMPTR)

2.1 Scheme Overview

Our scheme includes four phases which is shown in Fig. 1. They are access contention, access coordination, data transfer and block acknowledgement. During access contention period, any node with packets in transmission queue is competing the access to channel using carrier sensing like that in 802.11 DCF; once a node gets the access to channel, transmission handshakes among multi-users which try to trigger M (the maximum number of supported concurrent transmissions) different data streams concurrently will be carried out, that is access coordination phase; after coordination is done, multiple concurrent data transmissions with frame aggregation scheme for transmitting multiple packets together can then be carried out; after that, receivers should send block acknowledgment back to transmitters to report the reception condition for each packet since frame aggregation is carried out.

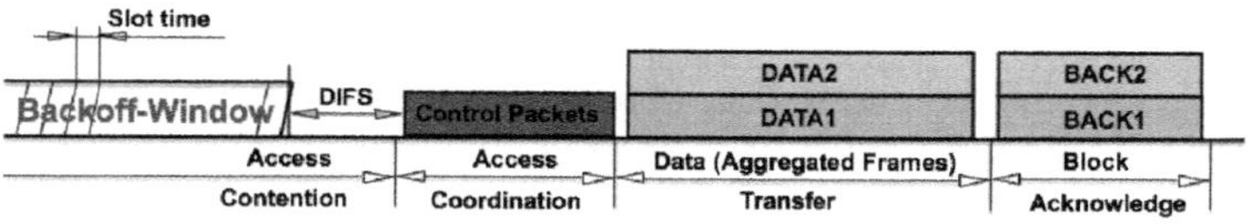

Fig. 1. The AMPTR scheme

We will present each phase of AMPTR in detail in the following subsections. To make our presentation easier, without loss of generality, we assume that $M = 2$, that is, each node can at most support two concurrent transmissions, and is half-duplex, i.e., can only transmit or receive at one time.

2.2 Access Contention

Assume that a node S has packets ready in its transmission queue, similarly to the IEEE 802.11 DCF scheme, it first listens the channel signal. When the perceived power is lower than a threshold for a period of DIFS, node S regards the channel idle. After the medium is sensed idle for a time interval (a distributed interframe space, i.e., DIFS), node S thinks the channel available. Otherwise, the node needs to defer transmission attempt with a random backoff timer. Its value is uniformly distributed in $[0, CW]$, where CW stands for contention window and is initially set to CW_{min}, and doubled after each time the transmission incurs a transmission failure and schedules a retransmission, up to its maximum value CW_{max}. The backoff timer is suspended whenever the channel becomes busy, and reactivated after the channel is sensed idle for a time interval (a distributed interframe space or DIFS) and decremented by one for each physical slot time. The node starts to make handshakes for transmission when its backoff timer reaches zero.

The big difference between AMPTR and IEEE 802.11 DCF scheme is that in AMPTR there is one special case to decrease the value of backoff timer and therefore to increase the transmission chance. When a receiver tries to trigger one more neighbor node to transmit packets to it, if a node has packet for the receiver and wins the transmission chance among multiple nodes which also have packets for the receiver and are also ready for transmissions, its backoff timer can then be decreased directly to zero.

2.3 Access Coordination

Assume that each node has two additional queues for storing unicast packets and that only packets with the same next hop destination could be stored in the same queue, that is, each node has two separate queues for two different destinations. Whenever there are arrival network layer packets, the node will try to dequeue them into the two separate queues until meet broadcast packets or unicast packets aiming for the third next hop destination.

Given that a node's backoff timer decreases step by step into zero, i.e., it wins the channel access by carrier sensing and backing off, it will again try to dequeue packets in the transmission queue into the two separate queues as many as possible, and then check whether there are unicast packets in the two separate queues. If the two queues are both empty, there must be a broadcast packet in the transmission queue, and the node can just send it out. Otherwise, handshakes before transmitting data should be carried out to coordinate the transmission behaviors among multiple users.

Our scheme is adaptive with traffic conditions and could dynamically support MPT or MPR; it is also compatible since when multiple concurrent transmissions could not be carried out, the normal point-to-point packet transmission could still be executed.

Suppose that a node S is ready for making handshakes, if its two separate queues both are not empty, i.e., there are packets ready for transmission to two different receivers, denoted by A and B, efficient handshakes among node S, A and B for MPT can then be carried out: node S first sends one Request to Send (RTS) packet to both A and B with the indication of response order for them to avoid response packet collision; when node S receives response packets of Clear to Send (CTS) from both of them successfully, it can then send packets concurrently to both A and B, i.e., MPT can then be carried out. The sketch of our scheme about how to support MPT process by making handshakes among one transmitter and two receivers is shown in Fig. 2.

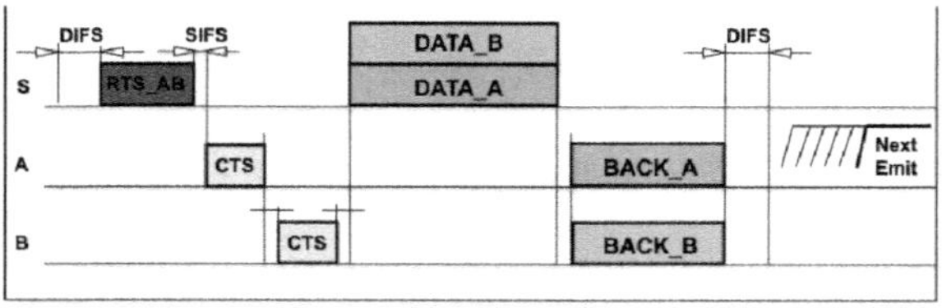

Fig. 2. The sketch of MPT process

If only one of two separate queues is not empty, i.e., there are packets ready for transmission only to one receiver of node A, node S will send a Request to Send (RTS) packet to A with the indication for the receiver A to take over the right of triggering one more transmission to it. The receiver A then broadcasts out a Ready to Receive (RTR) packet to all its neighbors, and any node U which receives the RTR packet and has packets ready for transmission to the receiver A will first make physical carrier sensing. After the channel is sensed idle for a short Interframe space (SIFS), node U will send a Reply to RTR (RRTR) packet to inform the receiver A that it is ready to be the another transmitter, and at the same time decrease its own backoff timer directly to zero for channel access. Since those nodes (excluding the first transmitter S) which have packets ready for transmission to the receiver A start physical carrier sensing for a period of SIFS, the node U which first receives RTR will send RRTR first, and at the same time other nodes will sense the ongoing RRTR transmission and stop their attempts for sending RRTR, i.e., the RRTR from node U could be successfully received by the receiver A. Finally, the receiver A could send one CTS packet to both node S and U, and the two transmitters can then send packets concurrently to the receiver A, i.e., MPR can then be carried out. The sketch of supporting MPR process by making handshakes among two transmitters and one receiver is shown in Fig. 3.

However, in MPT case it is possible that the CTS packets from receivers could not be successfully received or that only one of the CTS packets could be successfully received. For the first situation, the RTS packet could be retransmitted with the increment of Short Retransmission Counter (SRC) by one; for the second situation (e.g., the transmitter S does not receive the CTS packet from node B), a new RTS only for node B will be sent out also with the increment of

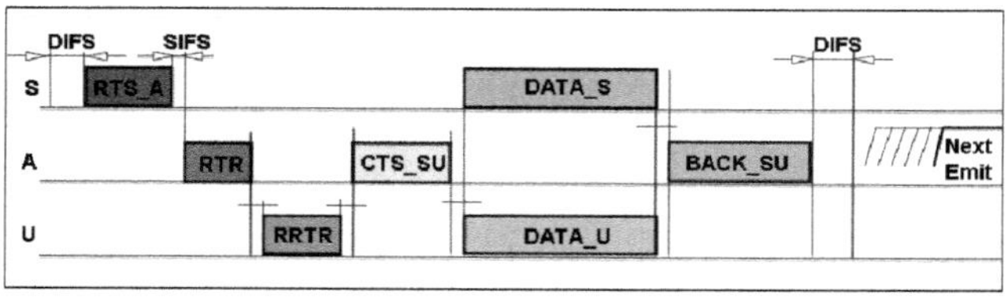

Fig. 3. The sketch of MPR process

SRC by one. The RTS packet keeps on being transmitted until one of the events occur: the handshake for MPT successes, the timer for handshake is out, and the maximum value of SRC predefined is met. When the handshake stops, if there is still only one CTS packet received, e.g., from node A, a normal point-to-point packet transmission between node S and A will be carried out, which shows the compatibility for previous transmission format of our AMPTR MAC protocol; if finally there is no CTS packet received, the transmitter S will lose the channel access for data transmission and a new random backoff timer with a doubled contention window size will be scheduled for retry with the increment of Long Retransmission Counter (LRC) by one.

In MPR case if the transmitter S does not receive the RTR packet from the receiver A, it could be regarded that node A has not successfully received the RTS packet from node S, therefore the RTS packet of node S should be retransmitted until node S receives the RTR packet, the handshake timer is out, or the retry limit is reached, i.e., the maximum value of SRC is met. If the RTR packet is still not received, data packet retransmission will be scheduled with a random backoff timer with a doubled contention window size and the LRC value will be increased by one. After the receiver A sends out the RTR control packet, if it does not receive the RRTR packet from any neighbor node U except the first transmitter S, a retransmission for the RRTR packet will be carried out until the receiver A receives a RRTR packet, the time deadline for sending the CTS packet back is met, or the retry limit is reached. If the RRTR packet is still not received, a CTS packet to the only transmitter S will be send out; otherwise, the CTS packet to both transmitter S and U will be sent out. If the first transmitter S receives a CTS packet, packets for the receiver A will be sent out with the antenna and packet information contained in the CTS packet; if node S does not receive any CTS packet, a random backoff timer with a doubled contention window size will be set for data retransmission and the LRC value will be increased by one. After sending out the RRTR packet, the second transmitter U will wait for the CTS packet. If node U does not receive any CTS packet, a random backoff timer with previous value before sending the RRTR packet will be set for channel access contention; otherwise, packets for the receiver A will be sent out with the antenna and packet information contained in the CTS packet.

Note that to make handshakes for MPT and MPR, four kinds of control packets are utilized: RTS, CTS, RTR and RRTR. Different with simple unicast control packet in previous work, to make handshakes efficiently both RTS and CTS packet should be compatible. The RTS packet in MPR and the CTS packet in MPT cases still aim to one receiver; however, the RTS packet in MPT and the CTS packet in MPR case should be sent out to multiple receivers, i.e., they should follow a multicast transmission format of single transmitter and multiple receivers. To incorporate the special requirement of transmission compatibility, the RTS and CTS packet should spare two address spaces for possible two receivers; when there is only one receiver (the RTS packet in MPR case and the CTS packet in MPT case), the second address space will be assigned a predefined number which is known as invalid address to all the nodes in network. One of the most important functions of control packets is to exchange each other's information among transmitters and receivers to get agreement on transmissions. Especially, for multiple concurrent data transmissions, there are extra information needed, e.g., for virtual MIMO-based concurrent transmissions, antenna and pilot information is needed. Therefore, in RTS packet a transmitter needs to include its antenna information for each of its receivers (one or two receivers). Based on those information, receivers can decide antenna weights for data transfer which should also be included in CTS packets. In addition, since we plan to use frame aggregation scheme to increase network throughput, transmitters need to indicate in RTS packets for each receiver how many packets will be aggregated and sent; accordingly receivers have to include the permitted packet number in CTS packets. The maximum number of packets which could be aggregated for one time data transmission could be predefined. In summary, the control packet formats of RTS and CTS for supporting multiple concurrent transmissions are respectively shown in Fig. 4 and Fig. 5.

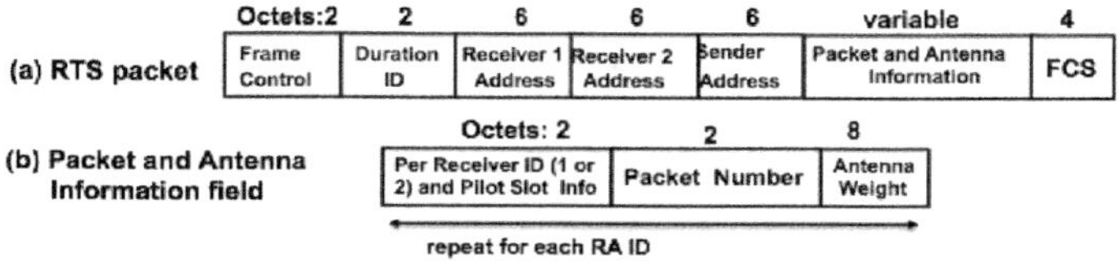

Fig. 4. Format of RTS packet

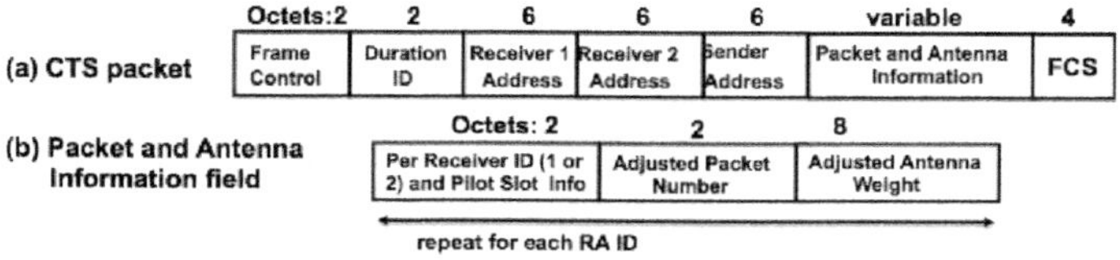

Fig. 5. Format of CTS packet

Note that in MPR case, the RTR packet is broadcasted to all its neighbors by the receiver to announce that more transmissions to it are needed. Therefore, the number of more permitted transmissions should be included in the packet, and in our example with the maximum number of concurrent transmissions $M = 2$, one more transmission to the receiver should be triggered, i.e., the number included in the RTR packet should be one. The node which wins the right for the one more transmission sends a unicast RRTR packet to the receiver to declare that it is ready for transmission, and similarly its antenna and packet information should also be included. In summary, the packet formats of RTR and RRTR are respectively shown in Fig. 6 and Fig. 7.

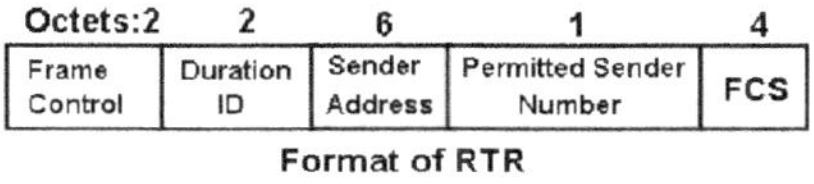

Fig. 6. Format of RTR packet

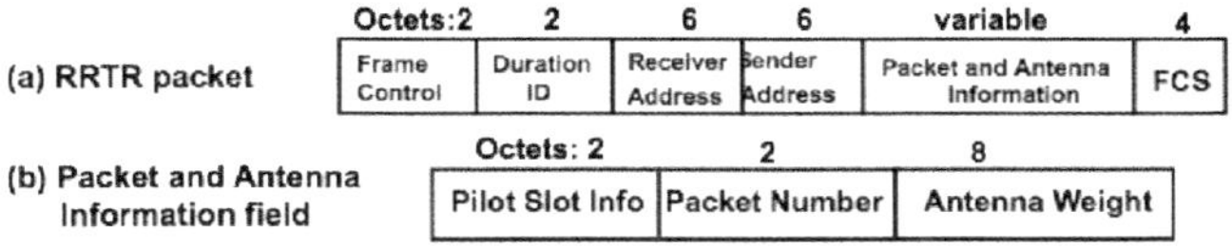

Fig. 7. Format of RRTR packet

2.4 Data Transfer

As we have presented in subsection 2.3, every node has two separate queues for storing packets aiming to two different neighbor nodes. Since every node in the queue has the same next hop destination, it provides the possibility to send them all together to increase network throughput. Therefore, we include in the RTS or RRTR packets the number of packets transmitters want to transfer, which is determined by the number of packets in the queue and the maximum number of packets permitted in one transmission (predefined in the network settings). The receivers will return back the number of permitted packets in transmissions based on the situations of transmitters and receivers, and the transmission conditions, such as the number of concurrent transmissions, the number of packets to be transmitted in each concurrent transmission, channel conditions, and so on.

After handshake is done, all or part of the packets in those queues could be sent out together. To do that, in our AMPTR protocol multiple MAC layer packets are designed to be concatenated into a large MAC layer packet, i.e., frame aggregation. For each MAC layer packet, it contains a MAC header and MAC data unit coming from network layer. When it enters into PHY layer for packet transmission, the PHY header and the frame check sequence (FCS) are then appended. In a large aggregated MAC packet, the multiple MAC packets which come from network layer and are attached with their own MAC headers are combined together to form the MAC data unit; each of the multiple MAC packets is assigned a fixed size space to be easily accessed by receivers for later packet retrieve. This aggregated packet format is shown in Fig. 8, and the case that there is only one data packet for transmission could be regarded as a special case with the number of aggregated packet as one.

When receivers successfully receive the aggregated data packet, with the fixed space size for each aggregated packet they could extract each MAC packet out. By checking parameters of those extracted packets (such as traffic ID, packet ID,

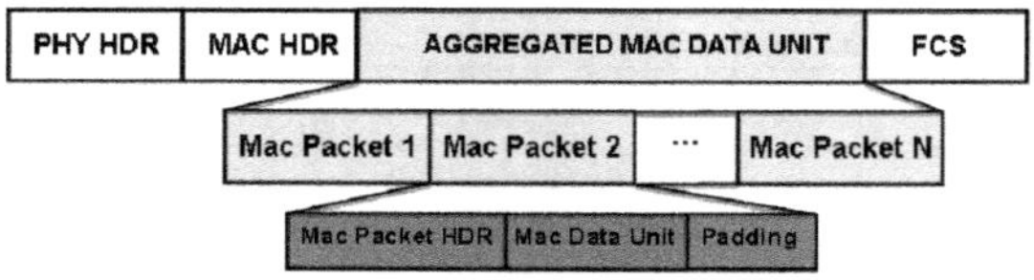

Fig. 8. Format of aggregated MAC packet

packet sequence number and so on), we could record which packets have been successfully received.

2.5 Block Acknowledgement

As shown in Fig. 1, after those aggregated packets are transmitted to receivers, transmitters expect to receive positive acknowledgement (ACK) packets from receivers. To provide efficient acknowledgements for multiple aggregated packets, we choose to add some bytes (namely, Block ACK Bitmap) in the ACK packet and use each bit of it to map the reception status of corresponding aggregated packet, namely aggregated block acknowledgement, i.e., BACK.

Our AMPTR scheme aims to support acknowledgements for all kinds of transmissions dynamically (MPT, MPR and normal point-to-point transmission). In MPT transmission process, as shown in Fig. 2, the transmitter S is expected to receive two BACKs from both node A and B. Since after data transmission transmitters and receivers know each other well to have concurrent transmissions, those two BACKs from node A and B could be concurrently transmitted to node S and then could be successfully received. In MPR transmission process, as shown in Fig. 3, both transmitters S and U are waiting for acknowledgements from node A. That is, after a time interval of Short Inter Frame Space (SIFS), node A will send one BACK aiming to both S and U. For this case, we propose that node A can include all the acknowledgements to two transmitters in one packet, and receiver ID could be included to differentiate them. Compared to the aggregated data packet the BACK packet is still pretty small even for two aggregated data packets.

To satisfy our special and dynamic acknowledgement requirement, the BACK packet should include two address spaces for possible two receivers. Especially, the transmitter of the BACK packet needs to indicate the reception status of any aggregated and transmitted packet coming from certain traffic flow of certain receiver, i.e., traffic ID (TID), packet ID, receiver ID and packet sequence number should be included in the BACK packet. To achieve above functions, the format of BACK packet is designed as Fig. 9, where BA represents a block acknowledgement. When the BACK packet only has one receiver, the left address space can then be assigned a predefined number known as invalid address.

If a BACK packet is not received within an interval of two round trip packet transmission, the sender assumes that the data packet transmission has experienced a transmission failure, and then contends channel access for retransmission

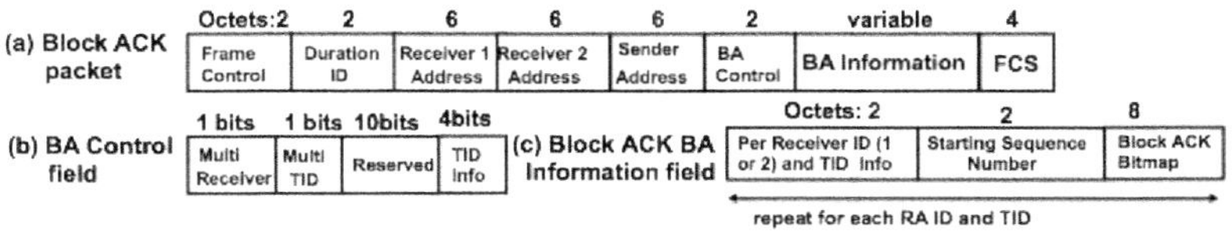

Fig. 9. Format of the Block ACK frame

with a random backoff timer with a doubled contention window size with the increment of Long Retransmission Counter (LRC) by one. If the sender can successfully receive a BACK packet, and extract for itself which aggregated packets are correctly transmitted. Those correctly received data packets will be removed from the sending queue and check transmission queue for next round of channel access contention.

3 Performance Evaluation

We evaluate the performance of AMPTR with and without frame aggregation scheme against that of IEEE 802.11 DCF through discrete-event simulations conducted in Qualnet v4.0 [11]. For our simulations, we used two network scenarios, namely a connected mesh and multi-hop topologies. In connected mesh scenarios, nodes are divided into clusters within which every node is within transmission range of one another, and traffics are generated within each cluster, i.e., destinations can be reached within one hop. In multi-hop scenarios, mobile nodes are randomly deployed and a packet may travel several hops before it reaches the destination.

As performance metrics, we used aggregate throughput for the whole network, average packet delivery ratio and average packet delivery delay, and MPTR/MPT ratio. In the connected mesh (single hop) scenario, packet delivery is carried out in one hop and packet delivery delay is calculated as the difference between the time a packet arrives at the queue and the time a packet gets transmitted successfully. In multi-hop scenarios, the evaluation metric of aggregate throughput is defined as aggregated traffic "end-to-end" throughput, that is, in the whole network how many bits per second are successfully sent out to reach destinations which may be multiple hops away from sources; similarly, end-to-end multi-hop packet delivery ratio and multi-hop end-to-end packet delivery delay are used for evaluation. In AMPTR protocol, when traffic flows are not high enough to allow MPT or MPR handshake to success, the conventional point-to-point transmission will be carried out. The MPTR ratio is the percentage of MPT and MPR transmissions occupied over all the data transmissions through the whole simulation. The MPT ratio plus MPR ratio equals to the MPTR ratio.

3.1 Simulation Setup

In our simulations, 36 nodes are deployed over a square area with $1000 \times 1000\, m^2$, the data rate for channel is 54 $Mbps$ and the transmission range is set to 250 m.

Fix-size (1024 bytes) data packets generated from CBR sources are continuously sent out. We use omni-directional antennas, and TwoRayGround propagation model. Each data point in graphs is an average of 10 different simulation runs. Each run is conducted with a random seed with a time duration of 10 minutes.

In connected mesh (single hop) scenarios, nodes are static, and placed in scheduled positions to form clusters. 36, 72, 108, 144 or 180 concurrent flows are simulated, one node can simultaneously generate several traffic flows to multiple destinations and multiple sources can simultaneously generate traffic flows for one same destination. Those kinds of traffic conditions enhance the probability of MPT or MPR processes.

In multi-hop network scenarios, nodes are randomly placed and move according to the random way-point model with speeds varying between 0 and 5 m/s and with no pause. Multiple concurrent flows are also simulated, while sources and destinations are randomly selected such that a node may be the source for multiple destinations and a node may be a destination for multiple sources.

For both single and multiple hop scenarios, approximately, half of the nodes are sources and half are destinations, and all the traffic flows start at the same time and use the same traffic rate for one network configuration. The traffic rate varies from low to high, that is, traffic flows with 2, 5, 10, 20, 25, 33 and 50 packets per second (the inverse of traffic data rate) are individually simulated.

3.2 Simulation Results

We consider the advantages of our adaptive proposal for MPT and MPR in improving network capacity and reducing waiting and service time for packet transmissions, in terms of network throughput and packet delivery delay respectively, under various number of traffic flows, or various traffic data rate. We also observe the packet delivery ratio (PDR) and MPTR/MPT ratio, which reflect behaviors of MAC protocols. To be comprehensive, we make simulations in both single hop and multi-hop scenarios.

Fig. 10 (a) presents a comparison of simulation results on aggregate network throughput for single hop scenarios under various number of traffic flows. AMPTR with frame aggregation performs significantly better than IEEE 802.11 DCF and yields almost double times the throughput of 802.11 DCF, and even AMPTR without frame aggregation can still achieve around 30 percent higher throughput than that of 802.11 DCF. The performance gain of our AMPTR over 802.11 DCF is mainly due to two aspects: the ability of nodes in AMPTR transmitting or receiving multiple data streams concurrently, and the ability of nodes aggregating multiple packets for the same destination and sending them altogether in one time. Since in our simulation each node could concurrently transmit or receive two different data streams, the network throughput of AMPTR without frame aggregation is expected two times that of 802.11 DCF. The reason why the practical results are not as good as what we have expected is as follows. First, the handshake phase of AMPTR takes place among multiple nodes (may be more than two) and it takes more time and control messages to complete than that of 802.11 DCF. Secondly, the handshake for MPT or MPR may not

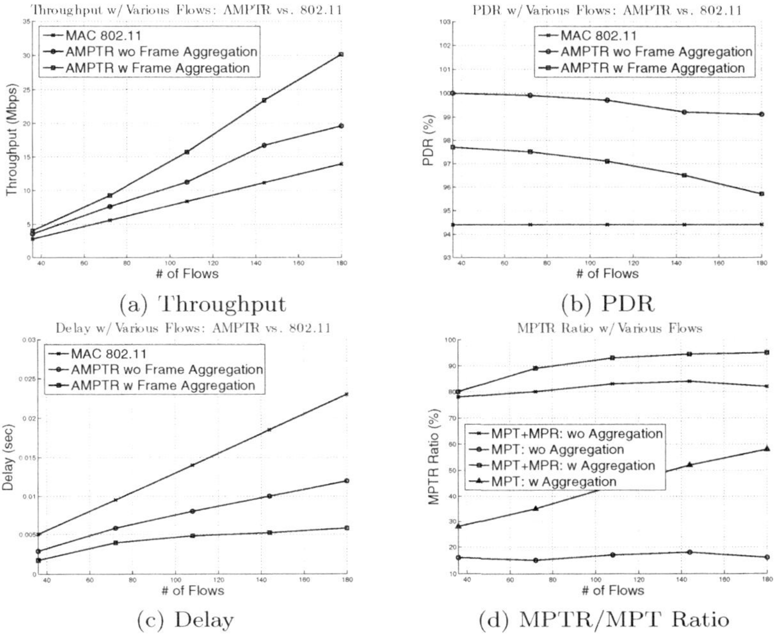

Fig. 10. Protocol Performance vs Traffic Flow Number in Single Hop Scenarios

success, and at that time normal point-to-point will take place, which has spent longer time to achieve transmission agreement than 802.11 DCF does. Thirdly, the number of packets for the same destination varies and depends on traffic conditions, and sometimes frame aggregation may not be carried out when there is only one packet ready for transmission at that time.

From Fig. 11 (a) we observe that in multi-hop scenarios AMPTR performs slightly worse than it in single hop scenarios. That is because in multi-hop scenarios the aggregate throughput is for traffics which may need multiple hops to reach destinations. Also because the traffic packet delivery should overcome more challenges to reach destinations multiple hops away, the PDR in multi-hop scenarios (Fig. 11 (b)) is much less than that in single hop scenarios (Fig. 10 (b)). The MPTR/MPT ratios for both scenarios are presented in Fig. 10 (d) and Fig. 11 (d), respectively. We observe that the MPTR ratio in multi-hop scenarios is smaller than that in single hop scenarios, and accordingly its throughput is also lower than that in single hop scenarios.

Similarly for packet delivery delay, since each node could concurrently transmit two different data streams, and especially multiple packets could be aggregated and sent out in one time, the packet waiting for service time could be greatly reduced which in turn causes the great reduction in packet delivery delay. Our expectation has been validated in Figs. 10 (c) and 11 (c) for both single and multi-hop scenarios. In single hop scenarios, from low to high traffic flows

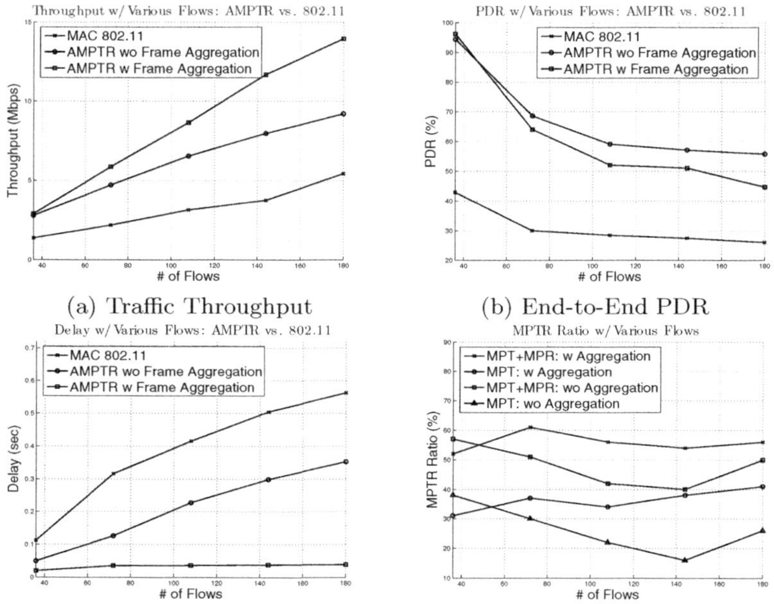

(a) Traffic Throughput (b) End-to-End PDR

(c) End-to-End Delay (d) MPTR/MPT Ratio

Fig. 11. Protocol Performance vs Traffic Flow Number in Multi-hop Scenarios

the delay of AMPTR with frame aggregation can be only around 1/3 to 1/4 of that of 802.11 DCF; even the performance of AMPTR without fame aggregation is worse than that of AMPTR with frame aggregation, it is only 60 percent of the delay of 802.11 DCF. In multi-hop scenarios, the delay is not for one single hop, but for multiple hops from a source to a destination, i.e., end-to-end delay; the delays for both 802.11 and AMPTR without aggregation increase by around 20 times, while that with frame aggregation only increases by around 6 times. The advantage of frame aggregation is shown as it can reduce many times handshakes into one; considering the possible retransmission for each hop, the number of handshakes reduced will be affected by the delivery hop number, the number of aggregated frames plus retransmission times. The more handshakes it can reduce, the smaller the packet delivery delay will be. However, from Figs. 10 (b) and 11 (b) we notice that the frame aggregation also has disadvantages. Because the packet size is increased, the transmission failure probability is also increased which in turn caused the lower PDR than that of AMPTR without frame aggregation, even though AMPTR with frame aggregation still has higher PDR than 802.11 DCF.

We also take a look at the protocol performance under various traffic data rate, i.e., varying the number of arrival packets per second, in Figs. 12 and 13 with the number of traffic flows of 36*2. In single hop scenarios, it is obvious and interesting that as the number of arrival packets increases, the MPTR/MPT

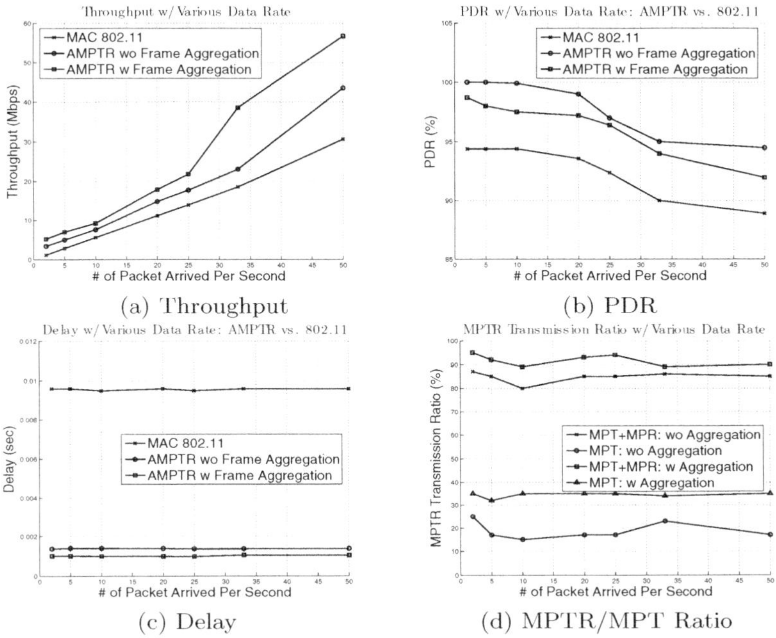

Fig. 12. Protocol Performance vs Arrival Packet Number Per Second in Single Hop Scenarios

ratio keeps the same, as shown in Fig. 12 (d). It is heuristic that data rate does not change the traffic pattern. However, since there are more packets sent out per second and for AMPTR with frame aggregation more packets for possible aggregation, the throughput for all the protocols increases and especially AMPTR with frame aggregation achieve larger increment ratio than other protocols (shown in Fig. 12 (a)). Note that in Fig. 12 (c) as data rate increases the packet delivery delay basically keeps the same, which may occur when the data rate is much smaller than the packet delivery delay. For example, the maximum data rate in our simulations is $1/50 = 0.02$, while all the packet delivery delay is less than 0.01. That is, nodes could send packets out before new packets come. While traffic data rate increases, the PDR decreases slightly (Fig. 12 (b)), which may be caused by the channel access schedule effected by increasing data rate. However, multi-hop scenarios have more complicated situations (Fig. 13). From Figs. 13 (b), 13 (a) and 13 (c), we observe that as the number of arrival packets increases, the PDR of AMPTR decreases lighter than that of 802.11 DCF, and that the throughput of AMPTR increases while that of 802.11 DCF has no obvious increment, and that the delay of 802.11 DCF increases greatly while that of AMPTR increases very slightly. All those observations demonstrate the stability and advantage of our proposal in multi-hop scenarios as traffic packet rate increases.

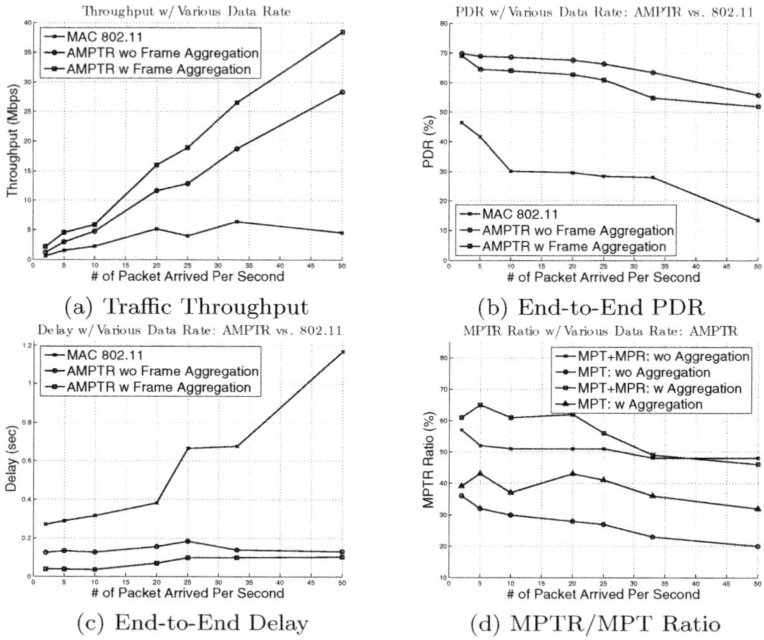

Fig. 13. Protocol Performance vs Arrival Packet Number Per Second in Multi-hop Scenarios

In summary, our AMPTR proposal can greatly improve network capacity and reduce packet transmission waiting and service time for ont only single hop networks but also random deployed mobile multi-hop topologies.

4 Conclusions

We presented an adaptive MAC protocol which exploits the advanced physical layer MPT and MPR ability, by supporting nodes which access channel to transmit multiple packets concurrently or receive multiple packets concurrently. We implemented our proposal in Qualnet and demonstrated its advantages in improving network throughput and reducing packet delivery delay by comparing its performances with those of 802.11 DCF.

References

1. Wang, Z., Ji, M., Sadjadpour, H., Garcia-Luna-Aceves, J.J.: Cooperation-Multiuser Diversity Tradeoff in Wireless Cellular Networks. In: IEEE Globecom 2009 Wireless Networking Symposium, USA, pp. 1–6. IEEE Press, Honolulu (2009)
2. Paulraj, A., Nabar, R., Gore, D.: Introduction to Space-Time Wireless Communications. Cambridge University Press, Cambridge (2003)

3. Mirkovic, J., Zhao, J., Deneneer, D.: A MAC Protocol with Multi-User MIMO Support for Ad-Hoc WLANs. In: IEEE 18th International Symposium on Personal, Indoor and Mobile Radio Communications, pp. 1–5. IEEE Press, Athens (2007)
4. Sadjadpour, H., Wang, Z., Garcia-Luna-Aceves, J.J.: The Capacity of Wireless Ad Hoc Networks with Multi-Packet Reception. IEEE Transactions on Communications 58(2), 600–610 (2010)
5. Jin, H., Jung, B.C., Hwang, H.Y., Sung, D.K.: A MIMO-Based Collision Mitigation Scheme in Uplink WLANs. IEEE Communications Letters 12(6), 417–419 (2008)
6. Ke, B.W., Zhang, Y.J., Liew, S.C.: Media Access Control with Spatial Correlation for MIMO Ad Hoc Networks. In: IEEE International Conference on Communications, pp. 3660–3665. IEEE Press, Glasgow (2007)
7. Park, M., Choi, S.-H., Nettles, S.M.: Cross-layer MAC Design for Wireless Networks Using MIMO. In: IEEE Global Telecommunications Conference, pp. 2870–2874. IEEE Press, St. Louis (2005)
8. Mirkovic, J., Orfanos, G., Reumerman, H.-J., Deneneer, D.: A MAC Protocol for MIMO Based IEEE 802.11 Wireless Local Area Networks. In: IEEE Wireless Communications and Networking Conference, pp. 2131–2136. IEEE Press, Hong Kong (2007)
9. Mirkovic, J., Zhao, J., Deneneer, D.: A MAC Protocol with Multi-User MIMO Support for Ad-Hoc WLANs. In: IEEE 18th International Symposium on Personal, Indoor and Mobile Radio Communications, pp. 1–5. IEEE Press, Athens (2007)
10. IEEE 802.11 Working Group: Part 11: Wireless LAN Medium Access Control (MAC) and Physical Layer (PHY) Specifications: Amendment 4: Enhancements for Higher Throughput. IEEE-SA Standards Board (2008)
11. Scalable Network Technologies: Qualnet Simulator v. 4.0

Admission Control and QoS Provisioning in Multi-service MDA for IEEE 802.11s-Based Wireless Mesh Networks

Jihene Rezgui, Abdelhakim Hafid, and Michel Gendreau

Network Research Laboratory
University of Montreal, Canada
{rezguiji,ahafid}@iro.umontreal.ca,
{michel.gendreau}@cirrelt.ca

Abstract. In this paper, we design an efficient scheduler for Mesh Deterministic Access (MDA) in IEEE 802.11s-based Wireless Mesh Networks (WMNs), called Enhanced Differentiated MDA (ED-MDA). ED-MDA is combined with an Efficient Admission Control algorithm, called EAC, that guarantees QoS for different service classes and provides higher capacity in WMNs. EAC processes both new and HandOver (HO) calls to maintain a balance between two conflicting requirements: maximizing the resource utilization and minimizing the dropping rate. To establish priority between new and HO calls, especially to avoid the forced termination of an ongoing call, ED-MDA coupled with EAC reserves the minimum amount of necessary resources while maintaining an acceptable HO call dropping rate and high resource utilization. Particularly, our model provides an efficient adaptive adjustment of the Contention Free Period (CFP) duration to make efficient use of the scarce wireless resources while supporting different services with different QoS requirements, such as delay. Simulations show that ED-MDA together with EAC outperforms existing schemes.

Keywords: Wireless mesh networks, muti-services, admission control, medium access control and quality of service.

1 Introduction

Wireless Mesh Networks (WMNs) have recently emerged as a promising technology for the next-generation wireless networks [1]. A WMN consists of two types of nodes: Mesh Clients (MCs) and Mesh Routers (MRs). The MRs form a wireless mesh backbone infrastructure that forwards most of the traffic between MCs and Internet gateways. In general, MRs have very little mobility and operate just like stationary routers, except for being connected by wireless links using the very popular IEEE 802.11 WLAN standard or other wireless technologies, such as WiMax. Using more than one radio interface in each MR allows better channel diversity resulting in less interference and therefore more throughput and capacity. However, this only improves the performance for best effort traffic [2] since supporting Quality of Service (QoS) for real-time traffic in WMNs remains an open challenge. For example, real-time connections (RTC)

X. Zhang and D. Qiao (Eds.): QShine 2010, LNICST 74, pp. 204–221, 2011.

require low delays and low packet losses, which are difficult to guarantee because of the random contention used by the traditional CSMA/CA MAC in IEEE 802.11-based WMNs. Nevertheless, some studies solve this problem by enforcing contention-based MAC with complex admission control schemes [3].

In this paper, we are concerned with QoS provisioning as well as the mobility management in multi-service IEEE 802.11s WMNs. Thus, the allocation and efficient use of limited wireless resources are to be studied. The challenge is, to support multimedia applications in order to meet users's expectations in terms of delay and packet losses while maintaining a reasonable high use of radio resources. Generally, WMNs consist of MRs whose small transmission range may result in frequent handoffs of mobile MCs. This inherent property of the mobile MC may often lead to high packet delays and loss rates [4]. Thus, a suitable admission control protocol must be designed to avoid the forced termination of an ongoing call so that a mobile MC could be able to freely move across the WMN while maintaining its ongoing communications uninterrupted. In multiple radio WMNs, it has been proved that there are still channels that may interfere even if the highest quality channel assignment heuristics [5] are adopted. The concept of clique [3] is used in this paper to identify the conflicting links that need to be carefully managed to limit contentions. Recall that a clique is defined as a subset of logical links that interfere with each other.

Mesh Deterministic Access (MDA) aims to provide stringent medium control access (MAC) delay guarantees for real-time services such as voice over IP (VoIP) which is a condition that can hardly be satisfied in classical IEEE 802.11 standard. The MDA scheme studied in [6-9] extends the typical IEEE 802.11 medium instantaneous reservation procedure, also known as the virtual carrier sensing (V-CS), to a more advanced reservation procedure using scheduled MDA OPportunities (MDAOPs) within two-hop neighborhood. MDAOPs are first negotiated between neighboring mesh nodes by exchanging broadcast setup messages, then, MDAOPs reservations are performed in multiples of time-slot unit, during the Delivery Traffic Indication Message (DTIM) periodic interval. To limit the message broadcast signaling overhead, MDA-related messages are sent only within two-hop neighborhood. IEEE 802.11 V-CS [6] is performed by a four way handshake procedure in which request-to-send (RTS), clear-to-send (CTS), data and acknowledgment packets are exchanged between two communicating nodes, while a network-allocation vector (NAV) is set by the other nodes in the physical sensing range (P-CS). V-CS works flawlessly in single hop wireless networks but can cause severe interference in wireless links that are multiple hops away and share the same channel within an overlapped transmission or interference range. Therefore, in multi-hop wireless networks, such as WMNs, MDA, with less multi-hop interference, was adopted in the IEEE 802.11s draft amendment [6] to provide stringent delay bounds. Although, this adopted scheme of MDA has reduced, to a certain extent, the delay bounds it lacks the concept of differentiating frames with different priorities. Basically, MDA is supposed to provide a channel access with equal probabilities for all stations contending for the channel deterministic access in a distributed manner. However, equal access probabilities are not desirable among stations with different priority frames.

Hiertz et al. [6] studied the MAC in IEEE 802.11s by first describing the mandatory Enhanced Distributed Channel Access (EDCA) and then the optional feature MDA. In opposition to EDCA, MDA is particularly designed to support multi-hop

WMNs and is based on scheduled medium access. In addition, we note that QoS support in IEEE 802.11s is not sufficient to provide lower delays and losses required by multimedia applications. This being said, there exists a number of contributions [10-12] that propose techniques in the context of IEEE 802.11 to improve voice QoS and network capacity. However, these contributions neither take into account real and no real time traffic at the same time nor differentiate between new and HO calls.

We note that interferences outside the two-hop neighborhood can still occur with MDA; this may degrade WMN performance. The contributions in [8-9] address the problem of interferences outside the two-hop neighborhood in MDA. Cicconetti et al. [8] use dynamic relocation (called, MDA-DR) of conflicting MDA time-slots that are two hops away from each requesting node. However, MDA-DR limits its performance improvement to throughput which makes it suitable only for data applications such as file transfer and Web browsing and not for delay sensitive applications such as voice. Furthermore, it lacks an admission control scheme that limits packet losses. In [9], Medium Access through REservation (MARE) is proposed as a generalization of the RTS/CTS V-CS procedure to multi-hop wireless networks. In this scheme, rather than using the excessive beaconing broadcast of each node in MDA, MARE uses a more elaborated RTS/CTS-like scheme that is able to consistently notify the appropriate group of neighboring nodes to which a given set of time-slots has been reserved. However, no scheduling and admission control was developed for MARE which limits its efficiency as well as its QoS guarantees for voice traffic. It is worth noting that several research studies focused on defining admission control mechanisms in wireless networks to differentiate between HO and new calls [13] or between different types of traffic with different QoS requirements [14] and [15]. However, current MDA or improved MDA schemes ([8] and [9]) do not support differentiation between different classes of services and also between HO and new calls.

Our contributions, in this paper, can be summarized as follows: (1) We propose a novel scheduler algorithm for per-flow MDA, called ED-MDA, which takes into account traffic differentiation; (2) We elaborate an adaptive adjustment of CFP (Contention Free Period) for each service class by prioritizing RTC in order to provide QoS guarantees in terms of delay and blocking probability that does not exceed predefined thresholds; (3) We propose a new admission control algorithm, called EAC, in conjunction with ED-MDA. EAC reserves the minimum amount of necessary resources (i.e., time-slots) to maintain an acceptable HO call dropping rate and provide high resource utilization; it also establishes priority between new and HO calls to avoid interrupted communication for highly mobile users; and (4) We present a centralized static algorithm (Alg. I) which searches for an initial feasible Upper Bound of the number of Time-Slots for each service class based on Erlang-b formula [16], called UBTS. Thus, starting with this good initial solution rather than a randomly generated one, ED-MDA coupled with EAC provides a better final adjustment of CFP/time-slots while improving the initial solution according to the state of the blocking probability constraint (Section 4).

The remainder of the paper is organized as follows. Section 2 presents our notations, assumptions, and network model. Section 3 proposes an analytical formulation of the differentiated MDA (ED-MDA). Section 4 describes our admission control algorithm EAC. Section 5 evaluates the proposed solution via simulations. Finally, Section 6 concludes this paper.

2 Overview of ED-MDA

In this section, we present a network model for WMNs. In particular, we first define key concepts, namely, transmission and interferences cliques, and then present the assumptions/notations used in the rest of this paper. We also present an overview of the proposed scheme ED-MDA.

2. 1 Assumptions and Notations

We consider a multi-hop WMN as illustrated in Fig. 1. MRs (e.g., MR_3 and MR_{17}) aggregate and forward traffic from/to MCs. These MRs communicate with each other to form a multi-hop wireless backbone network. This backbone network forwards traffic to/from the gateway access points (e.g., P_1 and P_2) so that a packet traverses multiple hops (MRs) to reach a destination. We model the backbone of a WMN as an undirected graph called the connectivity graph G = (V, E), where V represents the set of mesh nodes, and E represents the set of edges between nodes (connections). Among these nodes, we notice the subset P such that, $P \subset V$ to be the gateway access points that connect clients to the Internet. In the rest of this paper, MRs and gateway access points are collectively referred to as mesh nodes. Finally, $\forall$ (u, v) $\in$ V, an edge e = (u, v) $\in$ E if the distance between u and v, which is denoted by d(u, v), is smaller than the transmission range (i.e., $d(u, v) \le r$), where r represents the radio transmission range of nodes u and v.

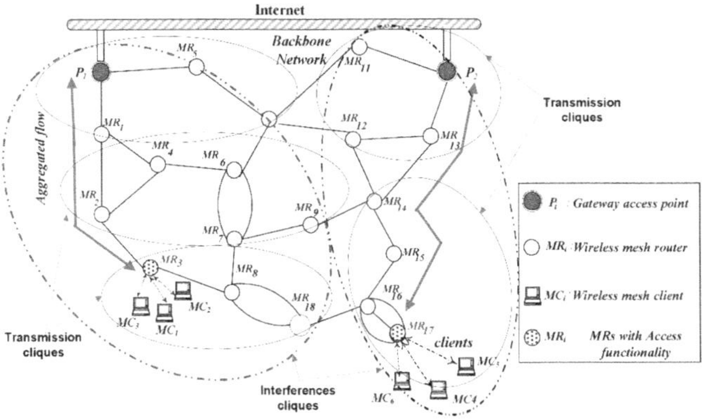

Fig. 1. A typical WMN where transmission cliques vs. interferences cliques are shown, topology of mesh routers mutually interfering within cliques

Generally, we can define in any topological network two types of cliques (see Fig. 1): A transmission clique and an interference clique. Transmission cliques are defined as sets of nodes, sharing the same channel and having a pairwise distance smaller than or equal to the radio transmission range of nodes in the pair (i.e., $d(u, v) \le r$). Interference cliques are defined as sets of nodes that use the same channel and have a pairwise distance in the interval $]r, R]$, where R is the interference range. Interference cliques are used to identify nodes in the Carrier Sense Range (CSR). These interference cliques are briefly called cliques in the rest of the paper. Fig. 1 shows a sample MR-WMN that consists of two interference cliques and a number of transmission cliques inside each of the interference clique. Routers MR_{14}, MR_{15}, MR_{16} and MR_{17} compose the transmission clique where all routers links are in transmission range of each other. The logical topology is usually built using a Dijkstra shortest path based routing algorithm with the number of hops as the routing metric which is suitable for delay-sensitive RTC.

We note that the Erlang-b formula [16] describes the probability of call lost/blocked in a network (all time-slots are busy) until a time-slot becomes free. In Table 1, we assume that $i \in [1..4]$, $j \in [1..2]$ where i represents the service class and j is equal to 1 if it is an HO call, 2 otherwise (i.e., new call).

Table 1. Notations

$AIFS_{AC[i.j]}$	Arbitrary Inter Frame Space of radio for service class $AC[i.j]$
τ	Time-slot duration
$L_{AC[i.j]}$.	Packet size (including PHY and above) for service class $AC[i.j]$
O	MDAOP offset for a flow in a clique
π	MDAOP periodicity for a flow in a clique
δ	MDAOP duration for a flow in a clique
σ	DTIM utilization at time t
α	Fraction of the DTIM period reserved for MDA
T	DTIM period
C	IEEE 802.11 transmission rate in a clique
$T_{AC[i.j]}$	Upper bound of the number of time-slots for service class $AC[i.j]$
CFP	Contention Free Period to serve *RTC*
CP	Contention period to serve no RTC (*NRTC*)
E	Total offered load in Erlang-b
$B_{AC[i.1]}$	Dropping probability threshold of *HO* voice (i=1)/video (i=2) calls
$B_{AC[i.2]}$	Blocking probability threshold of *new* voice (i=1)/video (i=2) calls
$A_{AC[i.j]}$	Total number of *HO* (j=1) *or New* (j=2) [voice (i=1)/video (i=2)] calls admitted
$R_{AC[i.j]}$	Total number of *HO* (j=1) *or New* (j=2) [voice (i=1)/video (i=2)] calls rejected
$D_{AC[i.j]}$	The required delay tolerance in a clique for service class $AC[i.j]$
$D^{\max}_{AC[i.j]}$	The maximal delay for maximum number of hops m in a path for service class $AC[i.j]$

2.2 MDA without Traffic Differentiation

In MDA, the time between consecutive DTIM beacon frames is divided into time-slots of length $32\mu s$. The periodic broadcast of beacon frames to all radios in a clique allows the synchronization of these DTIM intervals. Initially, MRs reserve the wireless medium for MDAOPs, which are reserved as multiples of time-slots during a given Contention Free Period (CFP) of a maximum access fraction (MAF=αT) of the DTIM interval T (see Fig.2-(a)). The remaining part of the interval as illustrated in Fig. 2-(a) is the contention period (CP) used for throughput-sensitive data applications. Current MDA does not support traffic differentiation and has same behavior for all service classes in the network.

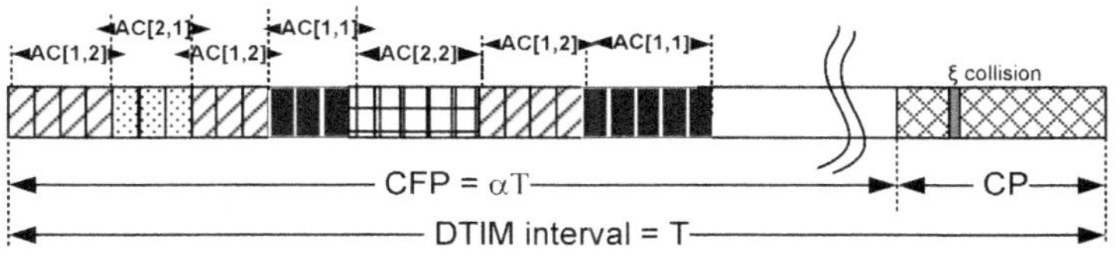

(a) Per-flow MDAOP schedules for MR_3 and MR_{17} RTC illustrated in Fig.1 over **MDA**

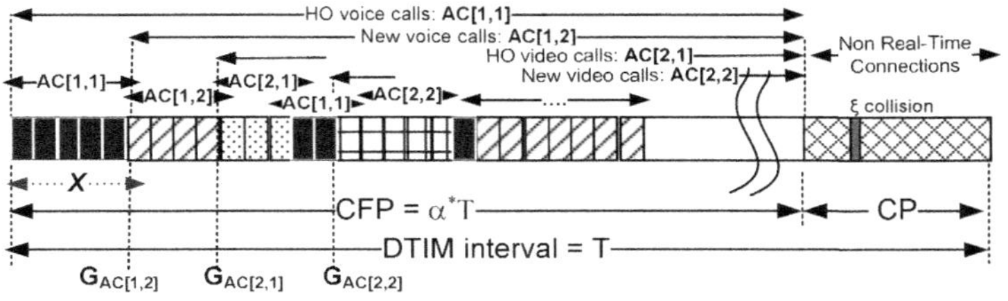

(b) Per-flow MDAOP schedules for MR_3 and MR_{17} RTC illustrated in Fig.1 over **ED-MDA** where X is the reserved guard to AC [1,1]

Fig. 2. (a) MDA per-flow schedule; and (b) ED-MDA per-flow schedule

2.3 MDA with Traffic Differentiation: ED-MDA

In this Section, we define the service classes supported by ED-MDA (see Tables 2-3). Then, we describe the details of ED-MDA functionality in the presence of multi-service traffic and HO calls (see Fig. 2-(b)); particularly, we will show how HO calls are prioritized over new calls to maintain high resources utilization.

Table 2. QoS requirement: 4 service classes

Class-id	**Delay sensitivity**	**Throughput sensitivity**	**Example**
AC[1]	Very high ($\leq$ 150ms)	No	Voice
AC[2]	High ($\leq$ 300ms)	High	Videoconference
AC[3]	Medium	Medium	Streaming
AC[4]	No	No	Best effort

Service classes

Generally, we can identify four types of classes related to the traffic nature either real time connections or not (see Table 2).

a) Voice traffic (real-time connections: RTC)

Typically, voice traffic is delay-sensitive, but tolerates some frame losses. Acceptable upper bound limit of one way voice delay according to ITU-T G.114 must be smaller than 150ms [17]. Besides, voice packet loss (%) according to ITU-T P.862 (Perceptual Evaluation of Speech Quality, PESQ) should be kept below 10% [18]. Empirically, this value is found to reach 6.79% if the maximum allowed delay is 120ms [18].

b) Video traffic (real-time connections: RTC)

Video traffic is delay-sensitive, but it tolerates some frame losses. In fact, the acceptable upper bound limit of one way video delay according to ITU-T H.261 must be smaller than 300ms [17]. Additionally, video packet loss (%) should be kept below 12%. Typically, a video traffic source generates frames at a constant rate over its active period. A popular example of RTC is a videoconference connection.

c) Video traffic (no real-time connections: NRTC)

In this kind of video traffic, the requirement is to reserve enough time-slots in each of the MDAOP periodicity sub-intervals in order to transmit 1 video packet (one downstream) of size $L_{\text{streaming}}$over a clique having a transmission rate of C. This service class is considered as unidirectional calls when compared to voice/videoconference connections, the latter being bidirectional calls. An example of NRTC is video streaming.

d) Data traffic (no real-time connections: NRTC)

Data traffic is delay tolerable, but data packet loss (%) should be kept below 10%.

Recall that NRTCs traffic (AC [3] and AC [4]) in the proposed ED-MDA will be treated during the contention period (CP) and will not be studied in the scope of this paper. Indeed, we are only interested in supporting RTCs (AC [1] and AC [2]) that require low delays, low packet losses, and traffic differentiation support which are difficult to guarantee with both IEEE.802.11 protocol [2] and current MDA [6].

2) Priority setup in the presence of handover/new Calls

In the ED-MDA model we are studying, we consider a WMN consisting of several cliques. A clique serves heterogeneous users who require different service classes in the form of new calls generated in the clique or HO calls coming from adjacent cliques. According to the requirements of each service class in terms of blocking and delay constraints, it seems natural to consider in the DTIM interval this setup:

$$\Pr(AC[1]) > \Pr(AC[2]) > \Pr(AC[3]) > \Pr(AC[4]) \tag{1}$$

where Pr(x) denotes the priority of service class x.

Since we take HO calls into account and the no-delay-sensitive service classes are treated in CP duration in our proposed ED-MDA scheme (see Fig. 2-(b)), the problem under investigation is transformed and will be subject to different set of priority rules. The newly setup priority and QoS requirements as stated in (1) and Table 2 during the CFP period will become, in our model as follows:

Table 3. QoS requirement: 4 new service classes

Sub-class-id	**Delay sensitivity**	**Throughput sensitivity**	**Service class**
AC[1.1]	Very high ($\leq$ 150ms)	No	Handoff voice calls
AC[1.2]	Very high ($\leq$ 150ms)	No	New voice calls
AC[2.1]	High ($\leq$ 300ms)	High	Handoff video calls
AC[2.2]	High ($\leq$ 300ms)	High	New video calls

$$\Pr(AC[1.1]) > \Pr(AC[1.2]) > \Pr(AC[2.1]) > \Pr(AC[2.2]) \tag{2}$$

where $\Pr(AC[i.j]$ denotes the priority of service class i, $i \in [1..2]$ and the sub-class-id j denotes whether the call is a HO or new call with $j \in [1..2]$.

We assume that each call request (i.e., call) of service class $AC[i.j]$ includes the corresponding blocking probability and delay thresholds, $B_{AC[i.j]}$ and $D_{AC[i.j]}$ for $i \in [1..2]$ and $j \in [1..2]$, respectively. In addition, the call request contains a field called the relative blocking probability, $RB_{AC[i.j]}$ (see (3)), for accumulated blocking probability in WMN. $RB_{AC[i.j]}$ is updated at each intermediate WMN node and it is defined to be the total number of admitted calls per the total number of both admitted and rejected calls in WMN (computed according to the service class). Fig. 3 shows the main fields of the call request. $RB_{AC[i.j]}$ will be compared to the $B_{AC[i.j]}$ threshold according to each service class in our proposed admission control algorithm (see Section 4) and it is expressed as shown by (3).

$$RB_{AC[i.j]} = \frac{A_{AC[i.j]}}{A_{AC[i.j]} + R_{AC[i.j]}}, i \in [1..2], j \in [1..2] \quad RB_{AC[i.j]} \tag{3}$$

Source	Destination	**AC [i.j]**	$B_{AC[i,j]}$	$D_{AC[i,j]}$	$RB_{AC[i,j]}$

Fig. 3. Main fields of the call request packet in IEEE 802.11s (ED-MDA)

3 The Real Time Connections Analysis in the ED-MDA Per-Flow Scheme

In this section, we analytically derive the MDAOP duration in number of time-slots $\delta^k_{AC[i.j]}$ and the MDAOP periodicity $\pi^k_{AC[i.j]}$ parameters in ED-MDA scheme in order to support QoS of RTC.

3.1 Bounds Conditions for CFP Adaptive Adjustment

In this sub-section, we introduce our model along with the underlying assumptions, so as to statistically describe the characterization of the traffic being studied and the guard computation of service classes in ED-MDA.

3.1.1 Traffic Characterization

We assume that our model follows the $M/M/1$ queuing model, the call arrival is described by a Poisson process and the inter-arrival time is exponentially distributed. We denote by E, $E=\lambda h$, the total amount of traffic offered in erlangs where λ is the call arrival rate and h denotes the call-holding time. The Erlang-b formula [16] is expressed by the probability: $P_b^{AC[i.j]} = \psi_b[E, T_{AC[i.j]}]$; this means that a new arriving call is blocked (meaning that all time-slots are busy) until a time-slot becomes free. Since the $\psi_b[E, T_{AC[i.j]}]$ formula assumes that the number of resources is known in advance (which is not the case in our model), we bound the Erlang-b to find an upper limit of the time-slots according to each service class. In our model, blocking threshold is considered to be known, while the number of time-slots is unknown. This can be better illustrated in UBTS (see ALG. I) where the probability of blocking is bounded by the required blocking probability $B_{AC[i.j]}$ and 1. $T_{AC[i.j]}$ is the upper bound of the number of resources (time-slots) and it is determined based on Erlang-b computation (see Alg. I).

ALG. I: COMPUTING THE NUMBER OF TIME-SLOTS FOR EACH SERVICE CLASS: UBTS

Input: $AC[i.j]$: service class, E: total traffic, $B_{AC[i.j]}$: blocking probability threshold, CFP, τ : time-slot duration, T:DTIM interval.
Output: $T_{AC[i.j]}$: Time-slots upper bound for each service class .

Hypothesis: **No resources (k=0) condition** $\psi_b[E,0]=1$;
we assume that $\psi_b[E,k-1] > \psi_b[E,k]$ for any k resources;
Step0: **Bounded recurrence form of Erlang-b**

$$B_{AC[i.j]} \leq P_b^{AC[i.j]} = \psi_b[E, T_{AC[i.j]}] = \frac{E^{T_{AC[i.j]}} \Big/ T_{AC[i.j]}!}{\sum_{k=0}^{T_{AC[i.j]}} E^k \Big/ k!} \leq 1 ;$$

For fixed k, we can express $P_b^{AC[i.j]}(k)$ as follows:

$$B_{AC[i.j]} \leq P_b^{AC[i.j]}(k) = \frac{E \times \psi_b[E,k-1]}{E \times \psi_b[E,k-1] + k} \leq 1 ;$$

then $$B_{AC[i.j]} \leq P_b^{AC[i.j]}(k) = \frac{E \times P_b^{AC[i.j]}(k-1)}{E \times P_b^{AC[i.j]}(k-1) + k} \leq 1 ;$$

Step1: **Bounds conditions for *k***

We determine $$0 \leq k = P_b^{AC[i.j]}(k-1)[\frac{E}{P_b^{AC[i.j]}(k)} - E] \leq E \times P_b^{AC[i.j]}(k-1)[\frac{1}{B_{AC[i.j]}} - 1]$$

Step2: **Time-slots upper bound according to each service class**

if ($i == 1$ *and* $j == 1$) **then** $T_{AC[1.1]} = \left\lceil \frac{CFP}{\tau} \right\rceil = \left\lceil \frac{\alpha^* T}{\tau} \right\rceil$;

else for $k\{1,2,\ldots, T_{AC[i.j]}\}, T_{AC[i.j]} = \left\lceil E \times P_b^{AC[i.j]}(k-1)[\frac{1}{B_{AC[i.j]}} - 1] \right\rceil$;

where $(i=2, j \in [1.2]) \cup (i=1, j=2)$

where $B_{AC[i.j]}$ is a given blocking probability threshold for each service class; for example for voice call, $B_{AC[1.2]}$ is equal to 6%. This constraint is used to guarantee a smaller rate of blocking probability inside the network in order to meet users's expectations and to subsequently increase the operator's benefits.

By periodically applying steps (0-2) in UBTS, ED-MDA can compute an adequate upper bound of the number of time-slots to be reserved for each service class. Although this reservation scheme guarantees the finding of suitable time-slots, it may cause interferences among time-slots that correspond to different users. Thus, a guard interval mechanism must be set to ensure that distinct transmissions belonging to different service classes do not interfere with each other.

3.1.2 Guard Computation

The class with highest priority in our case corresponds to HO voice calls; these calls could use up the CFP duration for voice packet transmission in the absence of other service classes, as shown in Fig.2 (b). Additionally, a HO voice call takes up a minimum portion of the whole DTIM ($CFP - G_{AC[1.2]}$ denoted by X) in order to transmit its packets without sharing them with other classes as shown in Fig.2-(b). ED-MDA manages the priorities among new and HO calls according to inequality (2); it operates under a time-slot reservation scheme expressed by the adjustment of CFP to maintain the HO priority (see Fig.2 (b)) for video or voice calls.

The parameters $G_{AC[1.2]}, G_{AC[2.1]}, G_{AC[2.2]}$ are the respective guard intervals (computed according to service class). The guard interval threshold in ED-MDA is defined to be the maximum number of slots, during DTIM interval, that traffic of a given service class can use. The purpose of using guard intervals is to give higher priority to handoff calls over new calls and to protect against interferences among service classes. This is expressed in the following equations:

$$G_{AC[1.1]} = [0, \lceil CFP = \alpha^{*} T \rceil] \tag{4}$$

$$G_{AC[1.2]} = [\lceil X \rceil + 1, \frac{T_{AC[1.2]}}{K_{AC[1.2]}}] \tag{5}$$

where $K_{AC[1.2]}$ is the number of slots for voice

$$G_{AC[2.j]} = [\frac{T_{AC[i.2]}}{K_{AC[i.2]}} + 1, \frac{T_{AC[2.j]}}{K_{AC[2.j]}}] \tag{6}$$

where $K_{AC[2.j]}$ is the number of slots for video. If $i = 1$ then $j=1$ and if $i=2$ then $j=2$ in (6).

3.2 ED-MDA Formulation

In per-flow ED-MDA scheduling, $2\delta^{k}_{AC[i.j]}$ is the number of time-slots reserved for each service classes (see Eq. 7) in each of the $\pi^{k}_{AC[i.j]}$ (see Eq. 8) sub-intervals that satisfies a hard constraint on a maximal delay $D^{\max}_{AC[i.j]}$ for maximum number of hops m in a path. In fact, we need to reserve enough time-slots during the sub-intervals for transmitting 2 voice or video conference (RTC) packets of size $L_{AC[i.j]}$, one upstream and the other downstream (bidirectional calls) over a clique having a transmission rate C. We note that this transmission occurs after duration $AIFS_{AC[i.j]}$. To prevent exceeding the one-hop delay, the periodicity $\pi^{k}_{AC[i.j]}$ in the ED-MDA reservation request has to be sufficiently upper and lower bounded by: $\pi^{k}_{AC[i.j]} \geq T / D^{\max}_{AC[i.j]}$. For the sake of simplicity, we consider a uniform distribution of $D^{\max}_{AC[i.j]}$ over interfering links even though a better repartition may take into account the non uniformity of traffic load over these links. Thus, the MDAOP duration (Eq. 7) and periodicity (Eq.8) are expressed as follows:

$$2\delta^{k}_{AC[i.j]} = 2\left\lceil \frac{AIFS_{AC[i.j]} + \frac{L_{AC[i.j]}}{C}}{\tau} \right\rceil \quad k \in N \text{ and} \tag{7}$$

$$\pi^{k}_{AC[i.j]} = \frac{DTIM}{D^{\max}_{AC[i.j]}} = \frac{T}{D^{\max}_{AC[i.j]}} \quad \text{where} \tag{8}$$

$$D^{\max}_{AC[i.j]} = \frac{D_{AC[i.j]}}{m} \tag{9}$$

The maximal delay is denoted by $D^{\max}_{AC[i.j]}$, i.e., the hard constraint on maximal delay for maximum number of hops m in a path and $D_{AC[i.j]}$ is the required delay by the service class $AC[i.j]$.

4 Admission Control Scheme EAC

The Call Admission Control (CAC) is an important task to ensure QoS requirements in multi-service WMNs. Therefore, in this study, ED-MDA is combined with a CAC for multimedia service networks, called EAC. The main objectives of EAC are: (1) differentiating between traffic with different QoS requirements; (2) providing an efficient adjustment of the CFP period to maintain high utilization of radio resources; and (3) providing higher priority to HO calls.

4.1 CFP Values' Update Based on EAC Operation

To guarantee QoS to RTC in terms of delay and blocking probability, we need an efficient fine-tuning of the CFP duration according to service classes. EAC defines three variables: (1) α: holds the current fraction of the DTIM period reserved for MDA; (2) β: a multiplicative factor which is used to update CFP; and (3) VCS (Violation Constraint State): assumes 0 when the state of the required blocking threshold is violated; otherwise assumes 1. VCS-v is defined as a vector of VCSs (e.g., VCS_v="01").

The first variable is common in IEEE 802.11s whereas the latter two variables are EAC specific.

The parameter β is a multiplicative factor used to update the value of α and subsequently CFP according to the service class and VCS. A similar multiplicative factor (equal to 2) is implicitly defined in the original IEEE 802.11 binary exponential backoff; Contention Window (CW) is doubled upon each transmission failure. However, β in EAC can be either assigned statically, i.e. before runtime and remains constant or dynamically (adaptive).

EAC first stores the most recent S states of VCSs upon each call request where S indicates the length of VCS_v. For example, in the case of S=2 and VCS_v is equal to "10" means that VCS is equal to 1 followed by VCS is equal to 0. EAC re-adjusts α accordingly to the rules shown in Alg.II. Besides, EAC checks the value of VCS_v; if the latter is equal to "11" and the current VCS is equal to 1 then α is divided by β. In other words, a slow decrease of α /CFP is preferred in this case since a QoS violation (for at least two states) is unlikely to be encountered. In the case where VCS_v is equal to "00" indicating a QoS violation and that the current VCS is equal to 1, α will revert to its current value. Furthermore, EAC multiplies the value of α by β when the current VCS is equal to 0

4.2 Time-Slots Reservation over ED-MDA by EAC Algorithm

EAC algorithm performs the necessary verification of the incoming traffic in terms of the DTIM utilization and decides whether the DTIM parameters (e.g., fractions of the DTIM parameter) should be modified according to the service class. If a violation in the blocking probability constraint ($B_{AC[i.j]}$) is encountered during a period T , EAC adjusts all the DTIM parameters in a way that may lead to a better utilization of the available resources. To realize DTIM parameters variation, we assume the existence of an online measurement mechanism that measures DTIM parameters variations and computes a suitable time interval, denoted by T , to capture periods with minimal DTIM parameters variation.

The algorithm starts by looking for RTC traffic (Step 0-1 of Alg. II); once found, EAC proceeds with the initialization phase (see step 2 of Alg. II). We apply UBTS algorithm to search for initial feasible CFP adjustments (i.e., computes adequately the upper bound of the number of time-slots for each $AC[i.j]$). EAC initializes this feasible solution rather than a randomly generated one. Then, according to our ED-MDA classification (see Table 3), EAC periodically checks whether DTIM exceeds a certain threshold ($\frac{CFP}{\tau}$) as well as the ($\frac{\alpha_{AC[i.j]}T}{\tau}$) threshold for the handoff voice $AC[1.1]$ and for the others service classes, respectively. Based on the above, it then rejects or accepts the incoming calls (see Alg. II, line-7).

In case of a rejected call, EAC increments the rejection parameter (Alg. II, line-8) that can be used later for eventual adjustment of CFP parameters. Otherwise, (a call is accepted), EAC recalculates the relative blocking parameter $RB_{AC[i.j]}$ (see Eq.3) and compares it to the given threshold blocking parameter; if $RB_{AC[i.j]}$ exceeds the threshold ($B_{AC[i.j]}$), then a violation of the constraint is detected; in this case, EAC re-adjusts the parameters accordingly as depicted in line-13 of Alg. II. Otherwise, it uses the information from line-11 of Alg. II.

In fact, EAC makes use of blocking probability statistics, VCS and VCS_v that are collected during the period of time T, to dynamically trigger or not CFP adjustment at the end of the time period (see Section 4.1).

ALG. II: EFFICIENT ADMISSION CONTROL: EAC

Input: Call request: k ($AC[i.j]$, $B_{AC[i.j]}, D_{AC[i.j]}$) in clique q_j where $B_{AC[i.j]}$ and $D_{AC[i.j]}$ are the blocking probability and the delay required for the service class AC[i.j], respectively.
Output: Admission decision: accept or reject and CFP adjustment

1 Receive the call request from user with QoS requirements ($B_{AC[i.j]}, D_{AC[i.j]}$)
2 Step 0: **Throughput-sensitive traffic**
if ($k \in NRTC$) **then** the call is served in CP duration;
3 Step 1: **Delay and throughput-sensitive traffic**
if ($k \in RTC$) {
4 Step 2: **Initialization**
if ($k \in AC[1.1]$) **then** i=Voice, j=HO;
if ($k \in AC[1.2]$) **then** i=Voice, j=New;
if ($k \in AC[2.1]$) **then** i=Video, j=HO;
if ($k \in AC[2.2]$) **then** i=Video, j=New;
5 Step 3: Upper bound **Time-slots computation**
Call UBTS /* Initial feasible CFP adjustment based time-slots computation for each $AC[i.j]$ */
6 Step 4: **DTIM utilization test**
$U = \delta_{AC[i.j]} + \sigma$;
7 **if** ($U > (\frac{CFP}{\tau})$ && $U > \frac{\alpha_{AC[i.j]}T}{\tau}$)
8 **then** $R_{AC[i.j]}{++}$; exit; // The call k is rejected
9 $A_{AC[i.j]}{++}$; // The call k is admitted in CFP duration
10 $\sigma = \sigma + \delta_{AC[i.j]}$; // Update of DTIM utilization
11 Step 5: **CFP adjustment**
if ($RB_{AC[i.j]} < B_{AC[i.j]}$) **then** VCS =1;
// No violation of the constraint: currently state
12 **then if** (VCS_v=="00")
$\alpha_{AC[i.j]} = \alpha_{AC[i.j]}$; **else** $\alpha_{AC[i.j]} = \alpha_{AC[i.j]} / \beta$;
13 **else if** (VCS==0) **then** $\alpha_{AC[i.j]} = \alpha_{AC[i.j]} \times \beta$;
// Violation of the constraint: currently state
Repeat Step 4 Every period of time T }
14 Update of the number of time-slots for each $AC[i.j]$;

5 Simulation Results

In this section, we conduct a simulation study using ns-2 to evaluate and compare the performance of ED-MDA with other existing schemes. We evaluate several metrics: 1) end-to-end delay; 2) packet loss rate; and 3) call blocking/dropping probability. Note that the end-to-end delay is the sum of the access delays (the queuing delay, which is negligible in low load, and the contention delay) experienced in hops along the path's call from the MC to the gateway.

5.1 Simulation Configurations

The WMN topology used in simulations is arranged as a regular grid of 5x5 802.11 stations acting as MRs. More precisely, the node spacing is about 100 m and the position of each node is deviating from the regular grid by choosing a random perturbation of an angle in [0, 2π] and a radius in [0 m, 25 m]. This kind of disturbance has been adopted in several existing WMN performance studies (e.g., [3, 8]). We notice that in tests with MDA, all MRs apply the DCF and its MAC parameters after the scheduling of MDAOPs. In the first scenario (scenario 1), simulations are performed under combination of two service classes (voice and video calls) while in the second scenario (scenario 2) four service classes (HO/new voice and HO/new video calls) are considered as described in sub-section 3.3.2. In the simulation results, we compare our proposed ED-MDA performance to: (1) WLAN 802.11 CSMA/CA for DCF [2]; (2) WLAN 802.11s MDA [8]; (3) WLAN 802.11s EDCA [6]; and (4) An improved version of MDA that considers dynamic relocation [8] of reserved MDAOPs two-hop away; we call this scheme MDA-DR.

5.2 Results Analysis

MAC Access Method Analysis—**DCF/MDA/MDA-DR/EDCA/ED-MDA Using Scenario 1:** As presented in Section 1, the IEEE 802.11s draft 2.0 allows an optional contention-free MDA access method besides the well-known CSMA/CA contention-based access DCF method. We study the performance of these access methods when transmitting data on the channel of each clique compared to our proposed scheme, ED-MDA, which is based on MDA while taking into account multi-service traffic. Moreover, ED-MDA prioritizes HO calls over new calls; this feature will be simulated in scenario 2.

Fig. 4 presents the average end-to-end delay experienced by network voice calls when using the various simulated schemes. In the case of low offered load (1 to 4

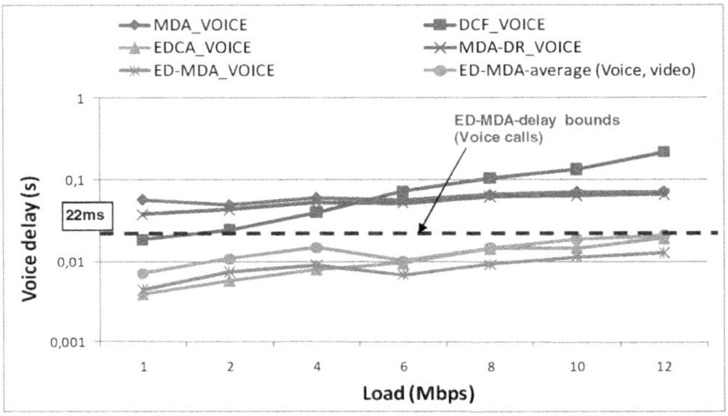

Fig. 4. Delay for voice calls using different access methods

Mbps), where collisions are very rare, the DCF method provides lower delays than MDA and MDA-DR. We note that DCF, MDA and MDA-DR do not grant higher priority to voice calls compared with video calls; this explains their poor performance with respect to voice delay. Besides, due to the scheduling of MDAOPs although the time-slots are available in a low offered load condition, the average delay is higher with MDA and MDA-DR compared to DCF. In highly offered load situation (e.g., 10 Mbps), the average delay with MDA or MDA-DR does not exceed 224 ms; it is bounded by the DTIM interval, which is equal to 32ms, multiplied by the maximum number of hops, which is equal to 7 in our topology. Fig. 4 shows that EDCA and ED-MDA schemes present lower delays, which is expected since these two schemes consider traffic differentiation and give priority to voice calls.

In the case of low offered load, the methods based on MDA wait for longer periods of time before transmitting in specific reserved contiguous time-slots even when collisions are rare and time-slots are available. However, in high offered load situation the delay in these schemes is bounded as explained above. Nevertheless, the delay provided by DCF and EDCA increases boundlessly with the increase of the offered load. We observe that the link delay with ED-MDA experienced by voice packets never exceeds the hard constraint on the maximal delay of $D_{AC[1]}/7=150/7=21.42$ ms.

Fig. 5 shows that ED-MDA delay, experienced by video packets never exceeds the hard constraint on the maximal delay of $D_{AC[2]}/7=300/7=42.85$ ms.

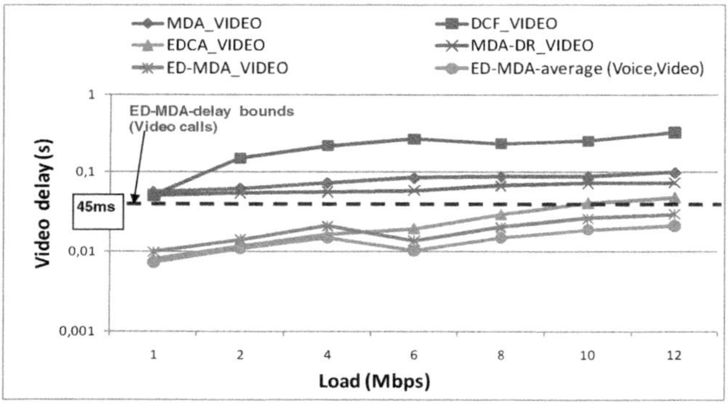

Fig. 5. Delay for video calls using different access methods

MAC Access Method Analysis—**MDA/ED-MDA Using Scenario 2:** in addition to video and voice calls, we take into account new and HO calls in this scenario.

In order to study the impact of the reserved guard $CFP\text{-}G_{AC[1,2]}$ on the HO voice calls dropping probability (see Fig.6), we simulated the network behavior with the same parameters as in ED-MDA (called W.ED-MDA) and without ED-MDA (called W/O.ED-MDA). Also, similar tests are done to investigate the impact of the guard threshold for HO video calls. We note that the admission control policy of EAC, with guard thresholds, offers HO calls dropping rate much lower than without guard thresholds (i.e., MDA). This dropping rate is 2 to 12 times lower for an offered load ranging from 2 to 10 Mbps. Indeed, up to 2 Mbps, all HO voice or video calls with ED-MDA are admitted but over 4Mbps, the dropping probability ranges from

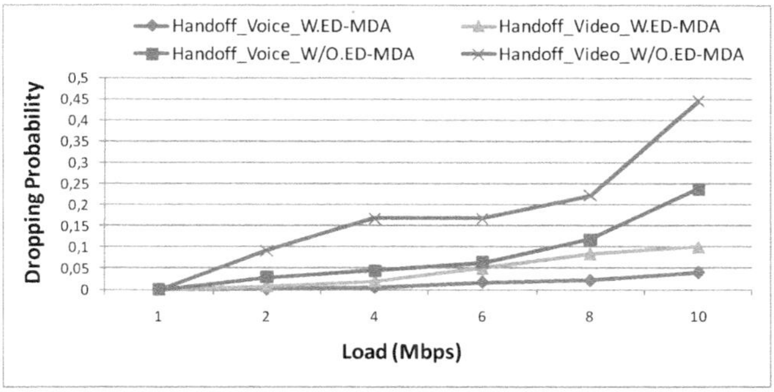

Fig. 6. HO calls dropping probability with and without ED-MDA

0.4% to 2% for HO voice and HO video, respectively, which is very acceptable for users. However, when offered load is equal to 10 Mbps, the dropping probability does not exceed 4% which is still acceptable for voice traffic.

Fig. 7 shows that new connections blocking probability increases with the offered load. We observe that for offered loads varying between 1 and 8Mbps using ED-MDA scheme, we get an acceptable blocking rate (0 to 5% for new voice calls and 0 to 9% for new video calls); beyond these values, the blocking rate for video calls exceeds 11% when the offered load exceeds 8Mbps. Fig. 7 also shows that the lowest blocking rate is offered when using W/O.ED-MDA; this is expected since the new and HO calls are treated similarly; as long as time-slots are available, these calls are admitted. However, when a guard threshold is used for new calls using ED-MDA, a part of the CFP (i.e., a number of time-slots) is reserved exclusively for HO voice calls and the guard threshold for HO video calls is higher than the guard threshold for new video calls; the new calls have less chance to be admitted which causes the increase of call blocking probability.

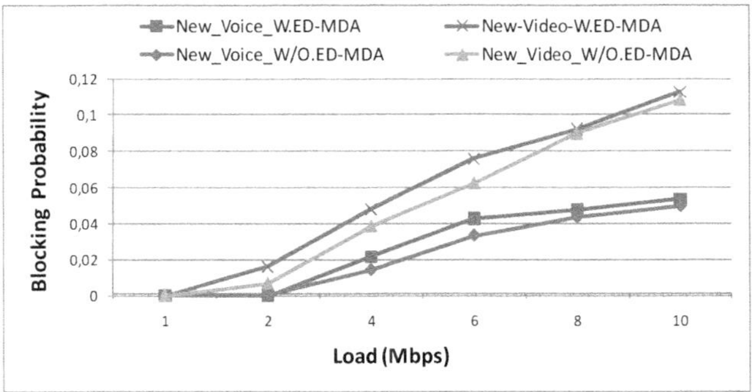

Fig. 7. New calls blocking probability with and without ED-MDA

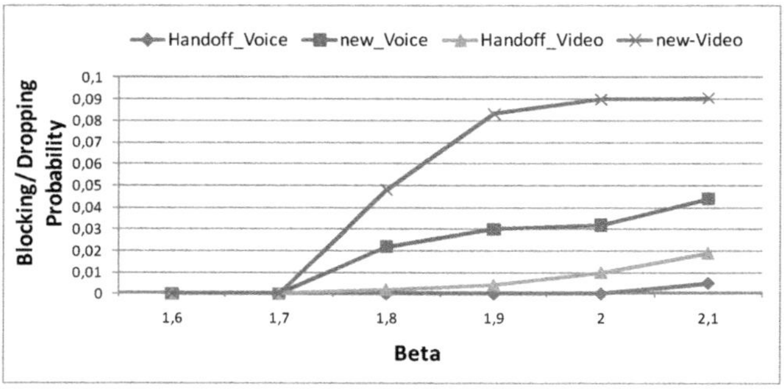

Fig. 8. Impact of β on the HO and new calls Blocking/Dropping probability for an offered load equal to 10 Mbps with ED-MDA (S=2)

Fig. 8 shows the blocking and dropping probability while varying β. We observe that the dropping of handoff calls is much lower (starting from β equal to 1.8) than new calls regardless of the nature of the call (voice or video). An interrupted communication is a very frustrating phenomenon that may happen to MCs. Thus, our proposed admission control protocol EAC coupled with ED-MDA avoids the forced termination of an ongoing call at the expense of slightly higher blocking of new calls. This is a price worth paying to provide low delays and loss rates while satisfying QoS requirements of multi-service WMNs.

6 Conclusion

In this paper, we have shown that MDA that carefully reserves dimensioned CFP according to each service class (ED-MDA) provides better voice/video QoS in terms of packet loss, delay and dropping rate. Furthermore, we proposed an admission control algorithm EAC that takes the HO and multiple service classes into account. Our proposed scheme ED-MDA combined with EAC is based on the guard interval principle which reserves necessary time-slots (i.e., resources) to meet and grants high priority to HO calls. Compared to MDA mechanism without guard interval, the simulation results show that our proposed solution outperforms the current MDA and many others existing schemes regarding the management of HO calls. Indeed, it offers a lower dropping probability and a lower delay for voice/video HO calls at the cost of a slightly higher blocking rate for new calls.

References

[1] Akyildiz, I.F., Wang, X., Wang, W.: Wireless mesh networks: a survey. Computer Networks 47(4), 445–487 (2005)

[2] Ergin, M., Gruteser, M., Luo, L., Raychaudhuri, D., Liu, H.: Available bandwidth estimation and admission control for QoS routing in wireless mesh networks. Computer Communications 31, 1301–1317 (2008)

[3] Rezgui, J., Hafid, A., Gendreau, M.: Distributed Admission Control in Wireless Mesh Networks: Models, Algorithms, and Evaluation. IEEE Transactions on Vehicular Technology 59(3), 1459–1473 (2010)

[4] Rezgui, J., Hafid, A., Gendreau, M., Rong, B.: Congestion-aware Clique-based Handoff in Wireless Mesh Networks. In: Proc of IEEE MSN, pp. 388–395 (2009)

[5] Raniwala, A., Gopalan, K., Chiueh, T.: Centralized Channel Assignment and Routing Algorithms for Multichannel Wireless Mesh Networks. ACM of Mobile Computing and Communications Review 8, 50–65 (2004)

[6] Hiertz, G.R., Max, S., Zang, Y., Junge, T., Denteneer, D.: IEEE 802.11s MAC fundamentals. In: Proc. of IEEE MeshTech, pp. 1–8 (2007)

[7] Hiertz, G.R., Max, S., Junge, T., Denteneert, D., Berlemann, L.: IEEE 802.11s-Mesh deterministic access. In: Proc. of EW, pp. 1–8 (2008)

[8] Cicconetti, C., Lenzini, L., Mingozzi, E.: Scheduling and dynamic relocation for IEEE 802.11s mesh deterministic access. In: Proc. of IEEE Secon, pp. 19–27 (2008)

[9] Gallardo, J., Makrakis, D., Mouftah, H.: MARE: an Efficient Reservation-Based MAC Protocol for IEEE 802.11 s Mesh Networks. In: Proc. of IEEE MESH, pp. 97–102 (2009)

[10] Niculescu, D., Ganguly, S., Kim, K., Izmailov, R.: Performance of VoIP in a 802.11 wireless mesh network. In: Proc. of IEEE Infocom, pp. 1–11 (2006)

[11] Wei, H., Kim, K., Kashyap, A., Ganguly, S.: On admission of VoIP calls over wireless mesh network. In: Proc. of IEEE ICC, pp. 1990–1995 (2006)

[12] Bayer, N., Xu, B., Rakocevic, V., Habermann, J.: Application-aware scheduling for VoIP in Wireless Mesh Networks. Journal of Computer Networks 54(2), 257–277 (2010)

[13] Zhang, Y., Liu, D.: An adaptive algorithm for call admission control in wireless networks. In: Proc. of IEEE Globecom, pp. 3628–3632 (2001)

[14] Wei, Y., Lin, C., Ren, F., Raad, R., Dutkiewicz, E.: Dynamic handoff scheme in differentiated QoS wireless multimedia networks. Computer and Communications Journal 27(1), 1001–1011 (2004)

[15] Belbekkouche, A., Rezgui, J., Hafid, A.: QoS provisioning for Wireless Mesh and Optical Burst Switching Convergence. In: Proc. of IEEE WCNC, pp. 1–6 (2010)

[16] Kleinrock, L.: Queueing Systems. Theory, vol. 1, 448 pages. Wiley Interscience, New York (1975)

[17] Siegel, E.D.: Designing Quality of Service Solutions for the Enterprise. John Wiley & Sons, Inc., Chichester (1999) ISBN: 0471333131

[18] Shim, C., Xie, L., Zhang, B., Sloane, C.J.: How delay and packet loss impact voice quality in Voip. Qovia, Inc. (2003), `http://www.qovia.com`

Empirical Analysis of Local Round Trip Time for Wireless Traffic Differentiation

Guangzhi Qu and Michael M. Nefcy

Computer Science and Engineering Department, Oakland University,
Rochester, MI, 48309, USA
{gqu,mmnefcy}@oakland.edu

Abstract. This paper focuses on how to differentiate wireless traffic from wired peer by using the temporal TCP characteristics of SYN, FIN, and ACK local round trip times (LRTT) found in all TCP sessions. With these session-based temporal characteristics, traffic from wireless and wired nodes can be differentiated by exploiting the fundamental differences between Ethernet and 802.11b/g/n. The effort of this paper is then on analyzing the resulting empirical LRTT data extensively and designing several algorithms for effective wireless host discovery. Most algorithms are light-weight, with little memory overhead, and can be easily implemented on commodity hardware. Ultimately, SYN, FIN, and ACK LRTTs can be compared against each other to discover wireless hosts regardless of network speeds.

Keywords: Wireless Network, Temporal Analysis, Local Round Trip Time.

1 Introduction

Network security and resource management are vital components to the productivity of any modern day business network. Whether the network is a small home office or a large university, significant time and effort must be devoted towards protecting computers and services against threats. This all but demands that network administrators keep a close watch on their networks by planning hardware and software resources to carefully keep unauthorized users out. Such a picture becomes further complicated as more and more innovative technologies become available for business adoption, ultimately creating further network vulnerabilities. How successfully network administrators maintain security in the face of old and emerging threats depends largely on the deployment of firewalls, intrusion detection systems, and other network defense tools.

One particular piece of technology that network administrators know they need to keep a close eye on is 802.11 wireless networking. Wireless networking has seen tremendous growth and widespread business adoption in the past decade. In fact, commodity 802.11 devices have become nearly ubiquitous today, emerging as an easy, convenient solution that can already exceed megabit Ethernet speeds [1]. Challenge in differentiating the wireless network traffic from the traditional Ethernet traffic draws huge interests and attention to the system administration.

For example, take the simplified network illustrated in Fig. 1. The local area network has both wired (LAN) hosts and wireless (WLAN) hosts present. These hosts

X. Zhang and D. Qiao (Eds.): QShine 2010, LNICST 74, pp. 222–237, 2011.

are potentially very diverse, with some being servers and workstations, and still others as PDAs and phones. The hosts have different operating systems and hardware processing speeds, as well as different networking technologies. For instance, an internal host could use wired networking based on either 10/100/1000 BaseT gigabit IEEE 802.3ab Ethernet or 10/100 BaseT IEEE 802.3u megabit Ethernet (referred to as either gigabit or megabit, respectively). Alternatively, a host could use wireless networking based on IEEE 802.11b, 802.11g, or 802.11n wireless LAN (referred to as 802.11b, 802.11g, or 802.11n respectively). Further, the network infrastructure can have similar diversity as well. Typically, such infrastructures are wired 10/100/1000 BaseT IEEE 802.3ab gigabit Ethernet infrastructures or 10/100 BaseT IEEE 802.3u megabit Ethernet infrastructures (referred to as either gig-backbones or meg-backbones for simplicity's sake). Lastly, multiple wireless access points are also present, with similar diversities as the rest of the wireless hosts.

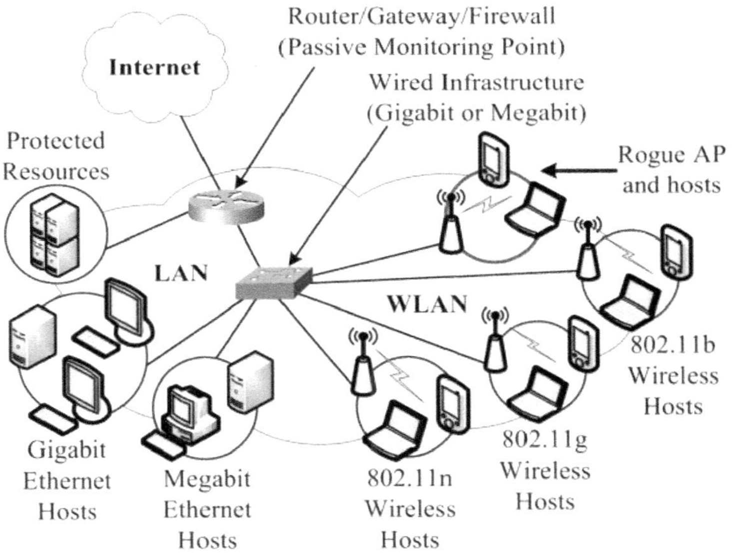

Fig. 1. A Typical Network Topology

In order to manage this heterogeneous network environment, the system administrators have to understand the network traffic. The question is how we can classify the network traffics in real-time? More specifically, how we can discover wireless hosts? What measurable metrics differentiate wireless from wired host traffic properties? And what algorithms can decide on such metrics in a scalable, real-time fashion with low overhead requirements?

To answer these questions robustly, this work takes the approach of using local round trip time (LRTT) metrics and some simple algorithms to discover wireless hosts in a heterogeneous, real-world, small office networking environment. Previous research has demonstrated that LRTT is an effective metric for wireless host differentiation from the rest of network traffic [3, 4]. The specific definition for LRTT used here is the time any TCP/IP session packet pair takes to traverses the local side of a LAN between the gateway/router and the host itself. This study's novelty is similar to Watkins et al. [4]: it uses only packet pairs found within TCP/IP sessions.

Several simple, stateless algorithms are used to decide on the developed LRTT metrics. These just use empirically-derived thresholds to classify wireless traffic in a straight-forward fashion. More complicated learning algorithms will be evaluated in later research, but are presently excluded from analysis.

2 Analysis of LRTT Metrics

An in depth analysis of local round trip time metrics is provided in this section, demonstrating how they reveal wireless hosts on different host and network infrastructure technologies. This research also differs from previous work done by Watkins et al [4] and others in that a much more robust analysis of LRTTs collected within TCP/IP sessions is performed. Analysis of these separate metrics illustrates how Ethernet hosts (gigabit and 100-megabit) are separated from 802.11 wireless hosts (802.11b, 802.11g, and 802.11n) on different network infrastructure hardware generations (10/100/1000 BaseT gigabit and 10/100 BaseT megabit). The following analysis starts with the settings and assumptions for LRTTs. From them, different results for Ethernet and 802.11 wireless networking are derived using a high-level approach. Further, how these results are affected by network infrastructure changes is presented as well.

2.1 Settings for LRTT Analysis

Consider two hosts communicating with each other, as depicted in Fig. 2. An external host communicating with an internal network host first sends an incoming packet to the local gateway. The incoming packet is then propagated on the network infrastructure, represented by link L_1. If the internal host uses wired networking, it receives the incoming packet directly from L_1. Otherwise, if the host uses wireless networking, it must receive the incoming packet from a radio-frequency link between a wireless access point and the host, represented by link L_1. Once the packet is received, the internal host sends a responding outgoing packet back to the gateway via L_1 (for the wired host) or L_2+ L_1 (for the wireless host). Finally, the gateway relays the outgoing packet back out to the external host.

Due to the diversity of networking technologies, the two communicating hosts may have slightly different settings. The result is the three primary variants shown in Fig. 2 (based on the host's networking technology). In the top scenario, the internal host uses gigabit or megabit. In the middle scenario, the internal host uses 802.11b or 802.11g wireless LAN. And finally, in the bottom scenario, the internal host uses IEEE 802.11n wireless LAN. These different network technologies each translate to different properties. For instance, the link speeds can vary, allowing the internal host to communicate at maximum rates of 1000mbps, 100mbps, 11mbps, 54mbps, or 600mbps for gigabit, megabit, 802.11b, 802.11g, or 802.11n respectively. Further, Ethernet hosts can transmit and receive packets simultaneously (full-duplex), whereas 802.11b and 802.11g hosts can either only transmit or receive at once (half-duplex). Unlike 802.11b or 802.11g, the newer 802.11n also allows full-duplex just like Ethernet (although not as guaranteed).

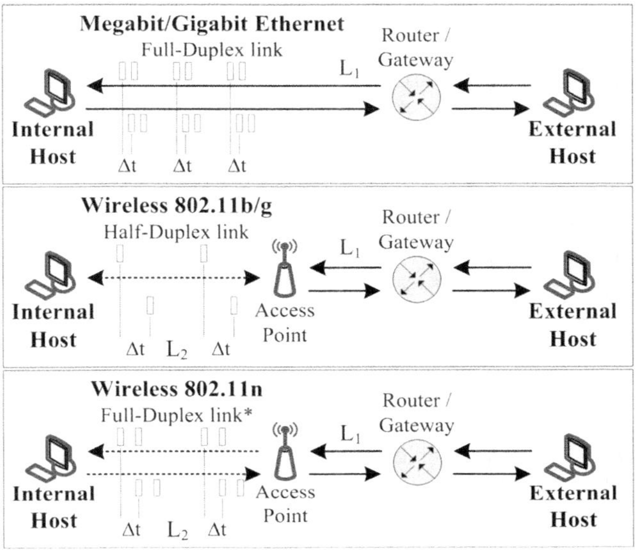

Fig. 2. The settings for analysis of LRTT metrics

Both the incoming and outgoing packets must pass over the network infrastructure, or backbone, represented by link L_1. This infrastructure can have similar diversity as well, and is typically either a gigabit-backbone or megabit-backbone. The network infrastructure limits the overall speed of the packets, as well as the ultimate link speed of the internal hosts. Additionally, the path of L_1 may have multiple chained links, where the packets must be propagated through a series of backbone switches.

As packets traverse the infrastructure, they pass through a central router or gateway. This gateway is the monitoring point, where all TCP/IP packet headers of network traffic are continuously, passively captured. From this traffic, unique local hosts are identified and their TCP/IP connection streams are analyzed. If a packet is incoming, its header is stored in a single, large, temporary queue for finding LRTTs (an M/D/1 queuing model). Conversely, if a packet is an outgoing response, its header is matched against its corresponding incoming packet from the queue, creating an incoming-outgoing-packet-pair. The arrival times of the packets in the packet-pair form a time difference, which is called the LRTT.

2.2 Assumptions of LRTT Analysis

There are two main types of assumptions made, the first involve time and the second involves TCP/IP itself. Though covered in more detail below, essentially this approach assumes LRTTs will only be affected by network factors and can readily be derived from TCP/IP session streams.

Time Assumptions. Pairs of corresponding incoming and outgoing packets are used to take time measurements for each unique internal network host. A core assumption of these metrics is that packet-pair LRTTs will only vary with a significant difference based on the host's network technology and packet size.

Consider a packet-pair travelling on the network setting shown in Fig. 2. The LRTT of this packet-pair is influenced by various factors of the path the packets travel on. From the incoming packet (P_{in}) arriving to the outgoing packet (P_{out}) leaving, these factors are: (1) The processing (proc), queuing (queue), and transmission (trans) of P_{in} at the gateway; (2) The processing , queuing, transmission and propagation (prop) of P_{in} along the network backbone (L_1) to the internal host; (3) The processing, queuing, and transmission of P_{out} by the internal host; (4) Finally, the processing queuing, transmission, and propagation of P_{out} back along the infrastructure (L_1) to the gateway. Note that if the internal host is wireless, both P_{in} and P_{out} must also pass through the wireless access point and its time delays as well (L_2). Each factor adds a specific Δt_i to the total LRTT for P_{in} and P_{out}. Further, each Δt_i is specific for a given packet size. If P_{in} and P_{out} arrive at times t_{in} and t_{out}, then all Δt_i factors for the packet-pair can be stated as ΔT.

Ultimately, the dependence of each time delay factor on the size of each packet and each device is not as confounding as it seems. For a fixed packet size, some Δt_i factors will remain mostly constant between network technologies. Further, any Δt_i that remains constant is not scientifically significant and can thus be ignored (it may event drop out of the equation if both P_{in} and P_{out} are the same size). For instance, most hosts have CPU speeds fast enough to render host processing and queuing times insignificant. The same goes for the gateway itself, especially since all incoming packets experience the same gateway time delays for a given packet size. Therefore, the only factors that remain are infrastructure related delays and host transmission delays, as seen in the final equation for ΔT. And thus, for a given P_{in} and P_{out}, only the variability that remains significant are changes in the host's network technology and changes in the infrastructure network technology.

Packet size cannot be ignored, nor can it be dismissed as an easily calculation based on transmission speed alone. True, a faster network can transmit a same-size packet in less time than a slower one, but this is not always the case. In wireless networks, larger packet time delays are dominated by transmission rates, whereas smaller packets are dominated by processing, queuing, and propagation rates. The effects of large vs. small packets on time delay have been studied on wireless LANs [2, 3, 5], but also appear in wired networking too (similarly due to the timing of MAC protocols, but with opposite results; see Section 6). In effect, the overall rate of transmission and propagation depends on the per packet time spent in the link protocol vs. spent in the data itself. Thus, if the link-time to data-time ratio is low, then smaller packets transmit faster due to the high overall time efficiency (low time overhead per frame). But if the ratio of link-time to data-time is high, then larger packets transmit faster due to the low overall time efficiency (high time overhead per frame). In addition, larger packets are take longer to transmit, longer to propagate, and are more affected by congestion and interference. Also, larger packets need more network process and queuing time for a given host or infrastructure as well. In summation, a nonlinear time dependency exists between packet size and total time delay for each network technology variant which can't be explained by link speed alone. Thus packet-pairs of different sizes need to be kept separate in order to achieve the most accurate measuring of ΔT.

Stated another way, within each host network technology type, within each network infrastructure technology type, the total time delay will remain constant for a fixed packet-pair size. Any other variables simply don't have the right variance to be significant. Again, the ultimate goal is to show the host's network medium type. So if the host processing takes too long, the resulting ¢T won't reveal the underlying technology and would be discarded anyway. Furthermore, with larger packets showing transmission effects the most, and smaller packets showing the medium effects the most, this approach of taking both into account will scale with faster networking speeds.

TCP/IP Assumptions. Producing pairs of sequential, corresponding packets relies on several TCP/IP assumptions. First, both the external and internal hosts use TCP/IP for most, if not all, of their communications. This guarantees that packets from these hosts will appear in the monitoring queue of the central gateway. Second, the most basic TCP/IP standards for managing a connection need to be followed. This means that hosts need to follow the 3-way establishing handshake, the 2-way continuing transfer, and the 4-way termination handshake, as depicted in Fig. 3. Either host can initiate each of those three general pieces of the TCP / IP session, leading to the six possible permutations shown. Each permutation has specific packet sequences, which must follow standards. Each packet, besides having IP addresses and port numbers, also has identifiable TCP/IP flags, acknowledgement numbers, and sequence numbers. The packet is matched to which sequence it belongs to based on such information. Note that not all packets in a sequence are usable. In Fig. 3, the unusable packets are shown in black, whereas the usable packets are shown in color (with red denoting incoming packets, and blue denoting outgoing packets). Only the packet-pairs which form incoming and then outgoing communications can be used, and they are given specific packet case numbers in Fig. 3. Packet-pairs form case-pairs of either 0-1, 2-3, 4-5, 6-7, and 8-9. Further, these case-pairs fall into three different categories based on their TCP/IP roles: 3-way establishing SYN-pairs, 2-way continuing ACK-pairs, and 4-way terminating FIN-pairs. Each set of SYN-, ACK-, or FIN-packet pairs form their own local round trip time categories, which are referred to as $LRTT_{FIN}$, $LRTT_{SYN}$, and $LRTT_{ACK}$, respectively. Each LRTT category is assumed to have packet-pairs of distinct, relatively constant sizes (based on the total size of both packets in a packet-pair). Further, each packet-pair is assumed to be directly sequential in the overall packet stream – i.e. no other packets are transmitted between the incoming packet "stimulus" and the outgoing packet "response." In totality, these assumptions guarantee that matching packet pairs form effective, fully passive "pings" of a local host, estimating the local host's LRTT latencies for three different packet-pair class sizes.

All assumptions made here hold well by default, or can be enforced through filtering rules. In order to communicate, hosts need to follow basic TCP/IP standards, so these assumptions are well met most of the time. Any packets not meeting the packet cases expected are simply discarded. Any packet-pairs that are not directly sequential (with no other packets being transmitted from the host in between) are also discarded. Such rules further ensure that the assumptions hold. As discussed in Section 6, the three classes of LRTTs to, in fact, fall into mostly constant sizes. The median pair-pair sizes for FIN-, SYN-, and ACK-pairs are 114, 122, and 1574 bytes respectively (see Table 1). Of final note is that the assumption on packet-pair sizes was not enforced in this specific approach, as only ACK-pairs had noteworthy size variation.

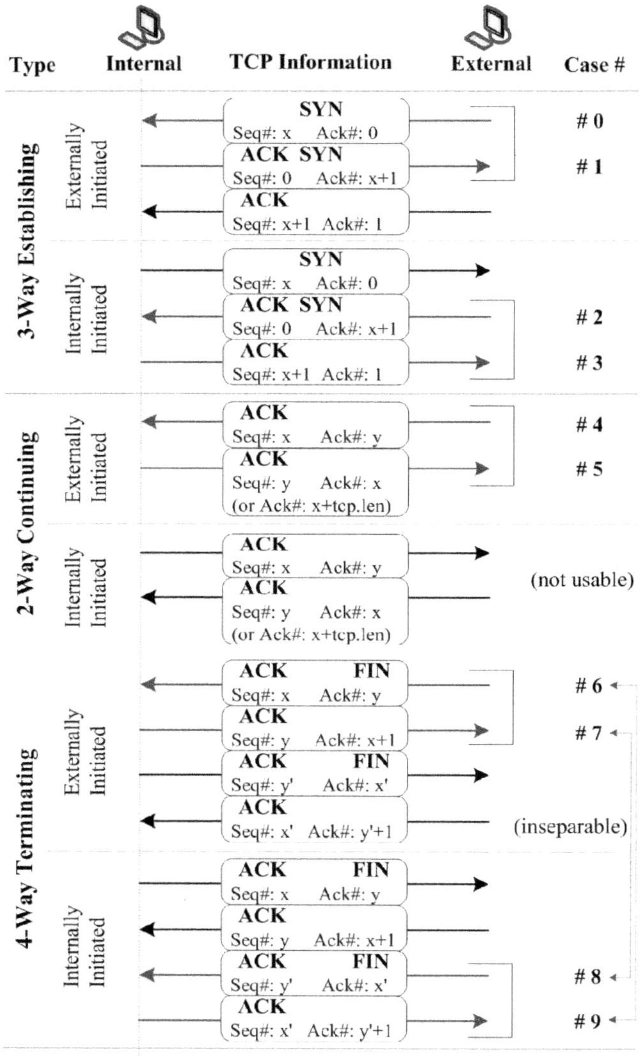

Fig. 3. Assumptions for LRTT metrics

Overall, this approach utilizes small amounts of overall network traffic. Less than 10% of the network's data stream is used. Further, of that utilized traffic, less than 10% is made up of FIN- and SYN-pairs. This is significant, since this approach places more importance is such the smaller packet pairs. Thus a large weight is placed on by far the smallest overall percentages of packets, which delays detection times to an average of one second. Aside from this disadvantage, these packet-pair rules nevertheless create three powerful indices of LRTTs, which can be used to effectively differentiate wired from wireless host activity.

2.3 Ethernet Analysis

A closer inspection of local round trip time in the Ethernet host setting is provided in this subsection. As illustrated in the top scenario of Fig. 3, an Ethernet host typically uses full-duplex communications with link rates of 1000mbps or 100mbps for either gigabit or megabit network interfaces, respectively. Full-duplex communications allow for efficient, long distance, simultaneous transmitting and receiving of packets. Today, practically all Ethernet hosts use full-duplex networking on switch-based infrastructures, making the use of Ethernet's half-duplex CSMA/CD MAC protocol a rare enough event to ignore. Lastly, the noteworthy differences between megabit and gigabit hosts are that the later has a faster link rate, faster processing, and lower overall latency.

Network infrastructure complications aside, this all has several implications for Ethernet host LRTTs. First, full-duplex communications allow for very low noise and congestion. Second, the efficiency of the Ethernet MAC layer makes inter-packet spaces and time-overhead per packet very small (low time overhead per frame). Third, since hosts can send whenever they want, packets have very small queue waiting times before they're transmitted. All these factors mean Ethernet time delays are very short to begin with, even without considering the very fast link rates Ethernet provides. Thus Ethernet LRTTs will be the smallest in general, with gigabit hosts showing smaller LRTTs than megabit hosts (Equation (1)). Further, since the medium has high overall time efficiency, smaller packets will transmit faster within a given Ethernet type (Equation (2)).

Short packet queues on network interfaces and high time efficiency mean that the overall variance in packet time delays remain low. Stated another way, the faster an interface can process its packet queues, and the less link-layer negotiating that takes place, the more consistently a given packet size is transmitted. Thus gigabit hosts are expected to have smaller LRTT variability than megabit ones due to their faster processing and transmission rates (Equation (1)). Similarly, since smaller packets can be processed faster and spend less time in transmission per frame, their variability should be less than larger Ethernet packet transmissions (Equation (2)). Let $f^1(.)$ be a function that calculates the mean value of the input, $f^2(.)$ for median and $f^3(.)$ for variance of LRTT categories (FIN, SYN, or ACK). In summation, these LRTT derivations in Ethernet are as follows:

$$f^{1,2,3}(LRTT_x(Gigabit)) < f^{1,2,3}(LRTT_x(Megabit)) \,. \tag{1}$$

$$f^{1,2,3}(LRTT_{FIN}(Eth)) < f^{1,2,3}(LRTT_{SYN}(Eth)) < f^{1,2,3}(LRTT_{ACK}(Eth)) \,. \tag{2}$$

2.4 Wireless Analysis

For Now a closer analysis of local round trip times associated with wireless hosts is given. Referencing the middle and bottom views in Fig. 2, wireless networking technology encompasses more diversity than Ethernet does. Despite this, most wireless technology shares many of the same traits. For instance, 802.11b, 802.11g, and 802.11n all use the CSMA/CA algorithm in their MAC protocols, as well as random back-off and wait DCFs when collisions occur. Further, each technology often shares the same channels, causing interference and contention due to their inherent shared medium nature.

There are also many notable differences between wireless network technologies too. The link-rate speeds of 802.11b, 802.11g, and 802.11n are typically 11mbps, 54 mbps, and 300+ mbps. These correspond to actual data-rate speeds of approximately 5 mbps, 22 mbps, and 94 mbps respectively (see Table XX). The previous generations of wireless technology, 802.11b & 802.11g, use half-duplex communications, which often leads to significant network throughput reductions with only a handful of users. The latest generation of wireless technology, 802.11n, uses almost full-duplex with the help of MIMO and SDM. Other differences are that 802.11b/g uses the 2.4GHz spectrum, with 20Mhz channel bands, where as 802.11n has the option of using the 5GHz spectrum, with 40Mhz channel bands. By using an effective Greenfield spectrum and larger channels, 802.11n achieves significantly less interference and faster throughput. Other advances of 802.11n include MAC-layer packet aggregation with block level acknowledgments, and up to 4x4 spatial antenna streams. All of this allows 802.11n have link rates of over 600mbps, which may achieve effective gigabit Ethernet performance for the average user.

Again ignoring network infrastructure complications for now, all these factors elicit differences in wireless LRTT behavior. Half-duplex communications, random back-off and wait DCFs, 3-way handshakes, and other MAC-layer properties make wireless networking have high time-overhead per frame. This is even more pronounced in 802.11n, which, even though it uses MIMO/full-duplex, still uses the DIFS and SIFS spaces of wireless MAC frames. With high inefficiency per frame, this situation becomes the opposite of Ethernet. Here, the larger packet-pairs will be dominated by transmission delays. Thus the faster wireless medium is, the faster it will transmit larger ACK-pairs (Equation (3)). Furthermore, the greater time inefficiencies will manifest the most in the smallest packets, giving FIN- and SYN-pairs the largest delays (Equation (4)).

Similarly, wireless hosts will have to wait and synchronize carefully before sending, leading to larger queue time delays on top of larger processing and propagation delays. The faster the wireless link, the faster queues can be serviced, leading to smaller variability for the larger packets (which are dominated by transmission rate) (Equation (3)). This is due to larger frames needing to be carefully controlled to ensure fair throughput, which thus reduces $LRTT_{ACK}$ variability. At the same time, the smallest packets (already experiencing large medium delays) will show the largest variation in total time delay since they will often have to wait the longest relative to their size (Equation (4)).

Lastly, due to the high time efficiency of Ethernet and the low time efficiency of 802.11b/g/n, the smallest packets experience opposing effects on LRTTs. Wireless hosts see the most delays with the smallest packets, while Ethernet hosts see the least (Equation (5)). Moreover as wireless hosts see the most variance with the smallest packets, Ethernet hosts see the least (Equation) as well. These derivations are summarized below:

$$f^{1,2,3}(LRTT_{ACK}(w_n))<f^{1,2,3}(LRTT_{ACK}(w_g))<f^{1,2,3}(LRTT_{ACK}(w_b))\ . \tag{3}$$

$$f^{1,2,3}(LRTT_{ACK}(w_x))<f^{1,2,3}(LRTT_{SYN}(w_x))<f^{1,2,3}(LRTT_{FIN}(w_x))\ . \tag{4}$$

$$f^{1,2,3}(LRTT_{FIN,SYN}(Eth))<f^{1,2,3}(LRTT_{FIN,\ SYN}(w_x))\ . \tag{5}$$

The differences between these equations and those for Ethernet are due to the ambiguity surrounding smaller packet sizes, since smaller packets are dominated by the network medium in wireless networking. However, using the link rates as approximations for the data rates of each host network technology type, equations (1) and (3) can still be combined to a general result. Since the largest packets will be dominated by transmission delays in both Ethernet and wireless, and since gigabit, megabit, 802.11n, 802.11g, and 802.11b are decreasing in link rates, then:

$$f^{1,2}(LRTT_{ACK}(w_{b,11m}))>f^{1,2}(LRTT_{ACK}(w_{g,54m}))>$$
$$f^{1,2}(LRTT_{ACK}(w_{n,300m}, Megabit))> f^{1,2}(LRTT_{ACK}(Gigabit))\ . \tag{6}$$

$$f^{1,2}(LRTT_{ACK}(w_{b,11m}))>f^{1,2}(LRTT_{ACK}(w_{g,54m}))>$$
$$f^{1,2}(LRTT_{ACK}(Megabit))> f^{1,2}(LRTT_{ACK}(W_{n,600m}, Gigabit))\ . \tag{7}$$

Note that megabit is comparable to 300mbps 802.11n due to the high wireless link overhead yielding nearly equivalent actual data rates. Thus assuming an 802.11n link rate of 300mbps, megabit and 802.11n cannot be intuitively ordered (Equation (6)). Similarly, assuming an 802.11n link rate of 600mbps, data rates for gigabit and 802.11n are ambiguous, so they cannot be intuitively sorted either (Equation (7)). As 802.11n becomes faster, its data rates may become indistinguishable from those of effective gigabit data rates. However, since no 600mbps 802.11n technology was available for evaluation, equation (7) will have to be verified at a later date.

2.5 Network Infrastructure Analysis

This subsection analyzes how network infrastructure influences the network time delays of internal hosts. Principally, this research only considers gigabit and megabit twisted pair Ethernet networking backbones. The differences between these two are an important consideration for maintaining RWAP detection accuracy on both current and legacy networks. Also, 802.11n wireless hosts currently achieve 300mbps link rates, making them effectively indistinguishable from megabit and gigabit Ethernet on megabit-backbones. Finally, since most previous studies were done using only megabit infrastructure, providing analysis of both is critical for comparing studies.

Gigabit and megabit network backbones alter LRTTs due to various factors. The link rate (i.e. transmission) differences have the largest effects, but disparities in processing and queuing times also exist. Since gigabit and megabit networking have link rates of 1000mbps and 100mbps respectively, packets on gigabit-backbones experience smaller transmission delays. In addition, gigabit-backbones often have faster, more efficient processing and queuing rates than megabit-backbones as well.

Thus on gigabit-backbones, hosts will experience smaller, less distorted LRTTs, making wireless host differentiation easier. Conversely, on megabit-backbones, hosts experience more latency and distortion of their medium's time characteristics, leading to larger LRTTs.

Moreover, gigabit Ethernet and 802.11n wireless hosts must accept lower data rates while on megabit-backbones. This means gigabit and 802.11n LRTTs will be no smaller than similar megabit hosts. In addition, since larger packets are dominated by transmission rates, the large ACK-pair LRTTs of gigabit, megabit, and 802.11n will be very similar on megabit-backbones. This may invalidate equations (6) and (7), with respect to such hosts. Finally, megabit-backbones are optimized for megabit hosts by default, so LRTTs may be smaller specifically for megabit hosts. In contrast, gigabit hosts will forced into a legacy mode, possibly leading to larger LRTTs than megabit hosts.

Overall, the ultimate goal is to create LRTT metrics and rules of host behavior that are agnostic with respect to network infrastructure speeds. Since only the faster gigabit, megabit, and 802.11n hosts are affected by infrastructure changes, a megabit-backbone essentially squeezes LRTT measurements together, distorting results, and making wireless host discovery less accurate. However, this loss of accuracy will be shown to be minor (but measurable).

So consider the worst case scenario: very fast, contention-free 802.11n hosts share a megabit-backbone with highly congested gigabit and megabit hosts. The high Ethernet congestion increases dropped and retransmitted packets, while contention free wireless hosts conversely see fewer delays. In this situation, the differentiation of wireless hosts will be more problematic, since both absolute and relative LRTT trends are affected. However, such extreme conditions would show only a slight decrease in the high MAC time efficiency of Ethernet. Further, retransmitted packets would experience delays too large to be useful and thus would be ignored anyway. And the already low MAC time efficiency of wireless would stay the same regardless. Therefore the overall increased delays in Ethernet LRTTs would be negligible. Lastly, the more normal scenario sees Ethernet as free from contention while wireless quickly degrades with just small numbers of hosts. So under normal conditions, wireless host behaviors increase their separation, aiding in wireless detection. Thus, in either extreme or normal conditions, wireless host are still differentiated.

3 Evaluation

This section presents the evaluation environment, the evaluation procedure, and the collected network traffic data sets.

To achieve real world results, a small, extremely diverse test environment was created following the setting shown in Fig. 1. The evaluation procedure is specified as follows. First, the network infrastructure was constructed as either gigabit or megabit Ethernet. Second, network traffic was captured on the router's LAN interface. Third, traffic captures were fed into a packet analysis program for convenient packet summary generation. Fourth, packet summary outputs were sent through a tracking program that continuously classified hosts using wireless host detection algorithms.

Due to the page limitation, we show only the cumulative distribution functions of how the behavior of 3 different packet-pair sizes from 5 different host types change over 2 different network backbone technologies. Figures 4-6 show the cumulative distribution functions (CDFs) of FIN-, SYN-, and ACK-LRTTs for Gig-Backbone data. Similarly, Figures 7-8 show the corresponding CDFs for Meg-Backbone data.

As an example, consider Fig. 4. This figure shows CDFs of FIN-pair local round trip times. Solid lines represent Gig-Backbone data, and each color represents a specific host network technology type. So the solid red line is the cumulative distribution of gigabit host LRTTs on the gigabit network infrastructure, etc. Since the graph is logarithmic on the horizontal axis, roughly 95% of all LRTTs are $1*10^{-3}$ seconds or less. In fact, most Ethernet traffic in general falls before $1*10^{-3}$s and is clearly distinct from wireless traffic. More importantly, despite the different properties of 802.11n, 802.11g, and 802.11b wireless types, all wireless $LRTT_{FIN}$ CDFs are virtually identical. The separate grouping of Ethernet and wireless technologies confirms results from other research showing that the smallest packets show affects from the transmission medium the best [34].

Compare the Gig-Backbone trends from Fig. 4 to the Meg-Backbone ones of Fig. 7. The Meg-Backbone data shows that $LRTT_{FIN}$ CDFs for wireless host types are similar to those from the gigabit infrastructure. However, gigabit host LRTTs increase (shift to the right) and become more variable (smaller slope). This is most likely due to gigabit hosts operating in legacy modes when on megabit infrastructure. Thus LRTTs from gigabit hosts become slightly more like those from 802.11n hosts when on megabit infrastructure. Interestingly enough, megabit hosts show the opposite: their LRTTs are slightly smaller and less variable in the Meg-Backbone dataset, increasing separation from wireless hosts. This behavior is seen repeatedly in later figures as well.

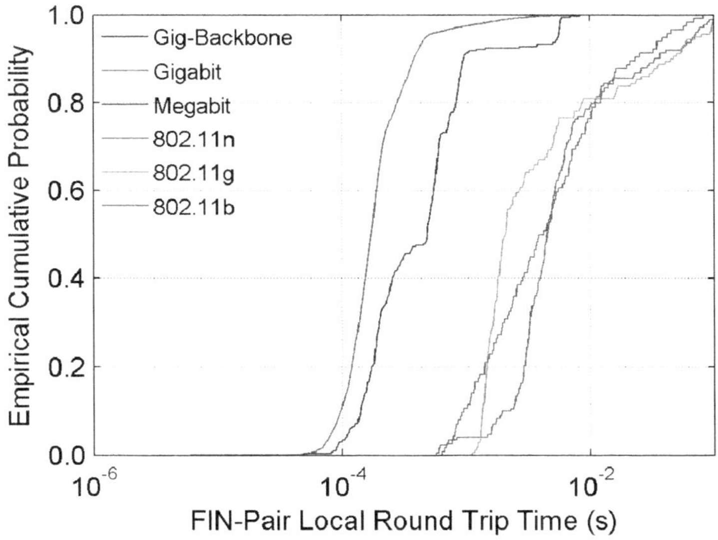

Fig. 4. CDF of Gigabit-Backbone FIN-LRTTs per host network technology type

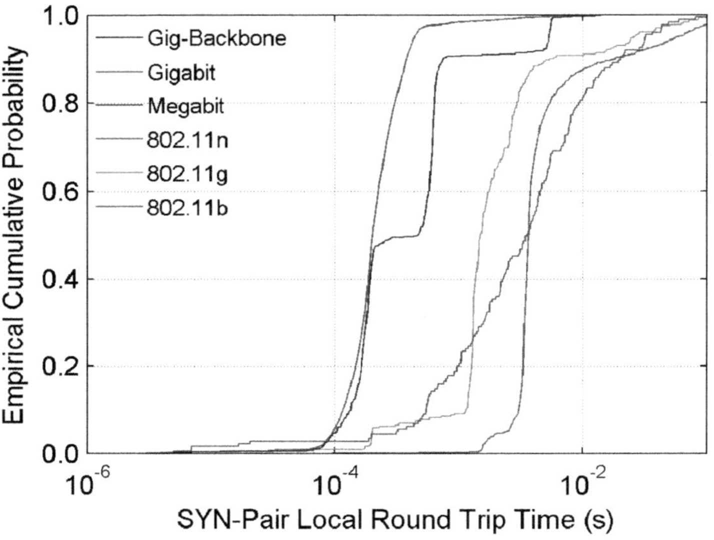

Fig. 5. CDF of Gigabit-Backbone SYN-LRTTs per host network technology type

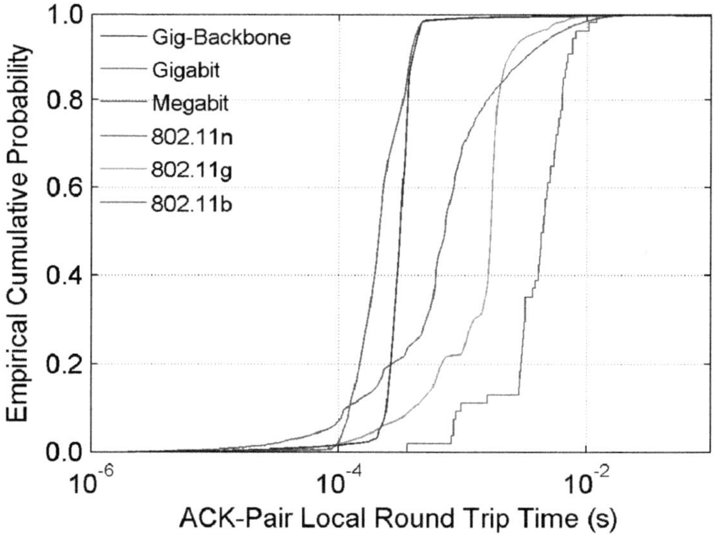

Fig. 6. CDF graphs of Gigabit-Backbone ACK-LRTTs per host network technology type

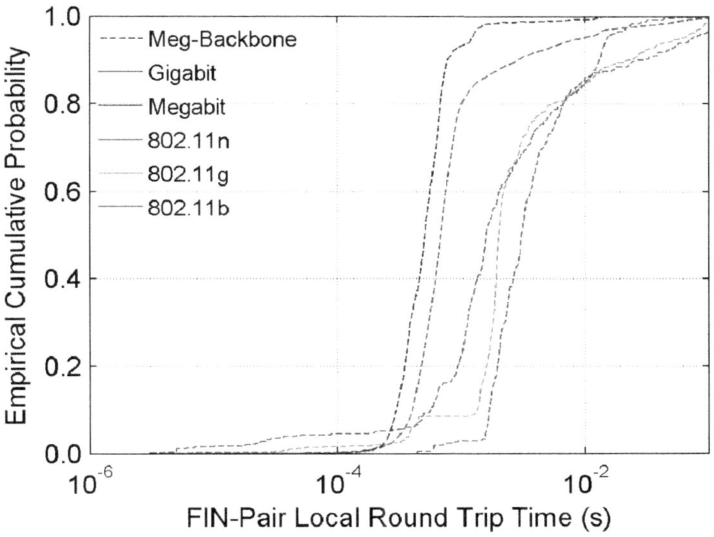

Fig. 7. CDF of Megabit-Backbone FIN-LRTTs per host network technology type

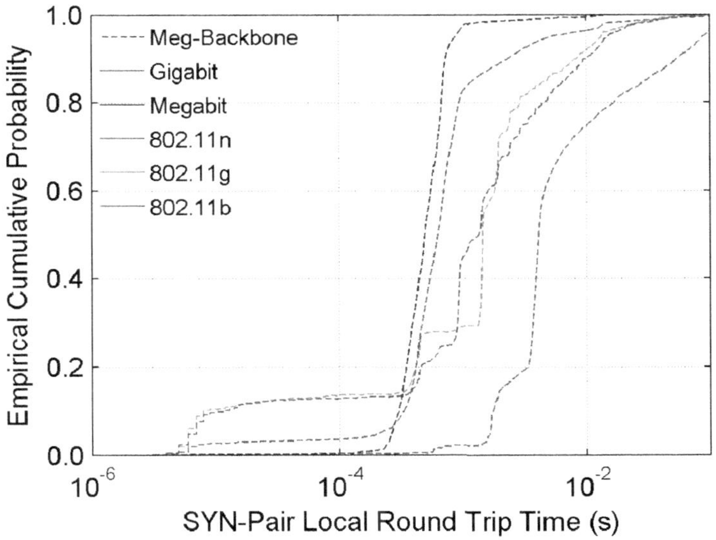

Fig. 8. CDF of Megabit-Backbone SYN-LRTTs per host network technology type

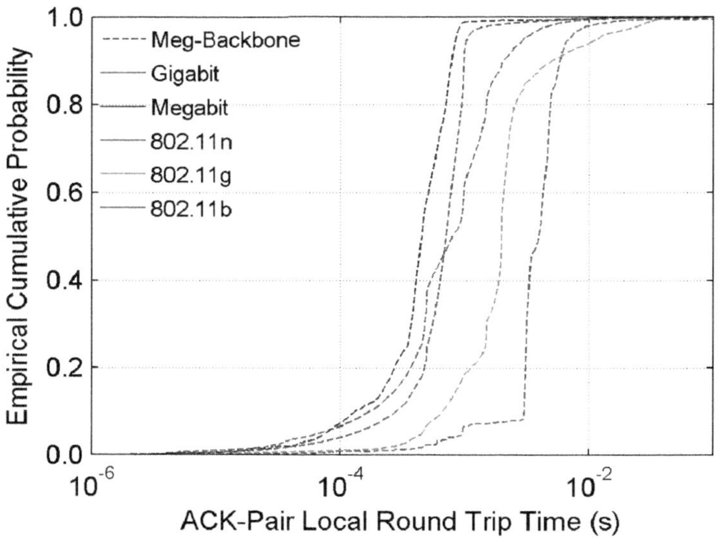

Fig. 9. CDF of Megabit-Backbone ACK-LRTTs per host network technology type

4 Conclusions

The analysis covered in this work reveals that TCP session based LRTTs can discover wireless hosts regardless of host network technology or network infrastructure type. Individual trends illustrate wireless $LRTT_{ACK}$ delays being higher in 802.11b/g wireless hosts than in Ethernet hosts. These ACK-pairs also become smaller as link rates increase. Further, the variance of wireless $LRTT_{ACK}$ delays will also become smaller as link rates increase. Yet no matter how fast a wireless host is, its smaller packet-pair $LRTT_{FIN,SYN}$ delays will remain large and highly variable. Relatively, this means that wireless $LRTT_{ACK}$ and $LRTT_{FIN,SYN}$ values will show greater separation as wireless technology advances. Thus, simple LRTT cut-offs classify slower wireless hosts, and increased separation of LRTT categories classify 802.11n hosts.

As for infrastructure, gigabit-backbones allow further separations of 802.11n metrics, increasing detection as wireless technology improves. Gigabit-backbones also show better detection overall by distorting LRTTs the least, while megabit-backbones confound wireless host separation. The worst case scenario sees the similar "squeezing" of host LRTTs together as in megabit-backbones, but wireless LRTTs are still distinct enough to stand out. So regardless of network backbone technologies and packet traffic conditions, wireless host differentiation is maintained.

Furthermore, these patterns in LRTT differences hold over different network gigabit- and megabit-backbone infrastructures, regardless of traffic conditions. Lastly, these derivations use only the characteristic link-layer timing differences of Ethernet and wireless, which are as unlikely to change as they are to be faked. Thus this approach to wireless host discovery remains robust against spoofing and wireless technology advances.

References

1. Cisco. 802.11n: The Next Generation of Wireless Performance. (2009), http://www.cisco.com/en/US/prod/collateral/wireless/ps5678/ps6973/ps8382/prod_white_paper0900aecd806b8ce7_ns767_Networking_Solutions_White_Paper.html
2. Cheng, L., Marsic, I.: Fuzzy Reasoning for Wireless Awareness. International Journal of Wireless Information Networks 8(1), 15–26 (2001)
3. Mano, C.D., et al.: RIPPS: Rogue Identifying Packet Payload Slicer Detecting Unauthorized Wireless Hosts Through Network Traffic Conditioning. ACM Trans. Inf. Syst. Secur. 11(2), 1–23 (2008)
4. Watkins, L., Beyah, R., Corbett, C.: A Passive Approach to Rogue Access Point Detection. In: Global Telecommunications Conference, GLOBECOM 2007. IEEE, Los Alamitos (2007)
5. Mano, C.: Defending against malicious rogue system threats. University of Notre Dame (2006)

On the Design of Access Network Packet Consolidation Protocol of CDMA 1xEV-DO Systems

Venkatesh Ramaswamy, Pradeepkumar Mani, and Olufemi Adeyemi

Airvana Network Solutions,
19 Alpha Road, Chelmsford, MA 01824
{vramaswamy,pmani,fadeyemi}@airvana.com

Abstract. We seek to formulate an optimal packet consolidation policy to transmit messages in the forward link control channel of CDMA 1xEV-DO air-interface standard such that the policy consumes minimum number of physical layer transmission slots. The consolidation policy is constrained by a given set of rate and delay constraints. We present an analysis of the mean message delay in the system and formulate the process as an generalized optimization problem to determine the optimal value for a variety of system-wide objectives, subject to various rate and delay constraints. We verify our analytical results through simulations, and present numerical examples to illustrate the design principles.

Keywords: 1xEV-DO, Control Channel, Packet Consolidation Protocol, Performance Analysis.

1 Introduction

The Control Channel (CC) in a 1xEV-DO [1,2] network is the signaling plane in the network that carries signaling messages from the Access Network (AN) to the Access Terminal (AT). The messages are transmitted to the AT by packing them in Synchronous capsules or Asynchronous capsules. The synchronous capsules are scheduled for transmission at fixed instances in time (once every 256 slots, 1 slot = 1.667 msec), where as the asynchronous capsules are scheduled as and when needed [4,5].

The synchronous capsule can hold up to eight 1xEV-DO medium access control (MAC) packets (each of length 1024 bits) if the synchronous capsule CC rate is chosen to be 76.8 Kb/s or up to four MAC packets if the synchronous capsule CC rate is 38.4 Kb/s. The asynchronous capsule is exactly one MAC packet in length regardless of the CC rate; but the length of each MAC packet varies based on the packet transmission format chosen from the following[1]:

[1] interpret the tuple $< X, Y, Z >$ as X = length of the physical layer packet in bits, Y = number of physical layer slots used for transmission, Z = length of preamble in chips; the transmission rate offered by a particular format is simply X/Y expressed in bits/slot or $3X/5Y$ bits/sec.

X. Zhang and D. Qiao (Eds.): QShine 2010, LNICST 74, pp. 238–250, 2011.

- For synchronous capsule, the transmission format is $< 1024, 16, 1024 >$ for 38.4 Kb/s or $< 1024, 8, 512 >$ for 76.8 Kb/s
- For the asynchronous capsule, one of the following transmission formats can be used: $< 128, 4, 1024 >$, $< 256, 4, 1024 >$, $< 512, 4, 1024 >$, $< 1024, 16, 1024 >$, or $< 1024, 8, 1024 >$. All packet formats can be used when the synchronous capsule CC rate is 76.8 Kb/s, while only 1, 2, and 4 can be used for a synchronous capsule CC rate of 38.4 Kb/s.

It must be noted that for the transmission format chosen, if there are not enough CC bits to fill up an entire MAC packet, then the remaining bits are padded as zeros. These padding bits constitute wasted transmission effort. For ease of exposition, for the remainder of this paper, we will only use a synchronous capsule CC rate of 76.8 Kb/s.

Each of the message types has a priority value assigned to it, which can be used to control its resource access. For example, the priority values can be used to determine the order in which the messages are packed into the synchronous or asynchronous capsules. While all message types can be packed into the synchronous capsule, some messages are "synchronous-only" and cannot be packed in an asynchronous capsule. Some message types that are common to both synchronous and asynchronous capsules are the *Traffic Channel Assignment* message, and the *Access Channel Acknowledgement* message while *Page* message, and *Sector Parameters* message constitute some of the "synchronous-only" or "exclusive" messages.

The synchronous capsule is periodically scheduled, and hence provides very little scope for implementation of a scheduling/consolidation policy to achieve some objective while satisfying certain constraints. On the other hand, the asynchronous capsule can be formed and scheduled in a flexible manner, and is a very valuable tool in enforcing a particular consolidation policy.

We are particularly interested in a consolidation policy that minimizes the number of slots consumed by the CC capsules for a given arrival process to the queues. We seek to minimize the number of slots consumed because the MAC packets carrying control channel capsules are Time-Division Multiplexed (TDM) with the MAC packets carrying user data from the Forward Traffic Channel (FTC). While the CC capsules use a transmission format such that a maximum rate of 76.8 Kb/s is supported, the FTC packets use transmission formats that support up to 3.1 Mb/s. As with traditional networking systems, the control plane (CC) data has higher priority in resource access (physical layer slots) when compared to data plane (FTC) data. Hence, sufficient care must be exercised when scheduling packets because an erroneous scheduling policy could adversely impact the FTC traffic.

In the absence of any delay constraints for a particular message type, the optimal scheduling strategy is one that schedules transmission of an asynchronous capsule after accumulating enough bits to fill up the packet format that offers the maximum CC rate. In the presence of delay constraints, it may not be acceptable to delay transmission of certain messages until enough bits are accumulated. Frequent and unnecessary scheduling of the asynchronous capsules to meet delay constraints can lead to significant reduction in available transmission slots for FTC traffic, thus impacting overall network data throughput. Hence, we seek a policy that chooses the optimal frequency of forming the asynchronous capsules based on the arrival process and given delay constraints.

This paper has four sections. In Section 2, we describe our model, and formulate the problem of designing an optimal packet consolidation policy as a generalized optimization problem. We also provide computational formulas needed to solve the problem. We dedicate Section 3 to validating the correctness of our model through simulations, and presenting numerical results obtained from our model. In Section 4, we draw conclusions and provide pointers for future work.

2 Analysis

In this section, we model the control channel packet consolidation protocol at the Access Network. Figure 1 depicts the CC capsule creation process. For ease of representation, the exclusive queues are consolidated and shown as one queue (colored red). There are k common queues (colored blue) with the arrival to the ith queue modeled as a Poisson process with parameter λ_i. The queue i has a higher priority than queue j if $i < j$. The aggregate arrival to the exclusive queues are also modeled as a Poisson process with rate λ_E. Synchronous capsules are formed at the end of each control channel cycle. Asynchronous capsules are formed every τ seconds termed as the *asynchronous capsule cycle*. Note that τ is a design parameter. Synchronous capsules are formed by first packing messages from the exclusive queues, and then depending on the availability of space in the capsule, messages from the asynchronous capsules are added. The queues are drained always in the order of their priority—higher priority queues are drained before draining the lower priority queues. Asynchronous capsules are formed independent of the synchronous capsule, and are formed every τ seconds from messages in the common queue. Again the higher priority queues are drained before draining the lower priority queues. The maximum size of a synchronous capsule is denoted by μ_E messages, whereas that of an asynchronous capsule is given by μ_C messages. The offset between the beginning of a CC cycle and the first asynchronous capsule in a CC cycle is denoted by x. The maximum expected waiting time of the ith common queue message is denoted by the $d^i_{\max}$. Let n be the asynchronous capsule formation frequency, given by T/τ. That is, there will be an average of T/τ asynchronous capsules in a CC cycle. The number of physical layer slots used for transmitting the asynchronous capsules in a CC cycle is denoted by U and can be represented simply by $nS(\mu_C)$, where $S(\mu_C)$ is the number of slots used by a packet of size μ_C. We refer to U as the forward link *control channel utilization*.

We consider the forward link control channel utilization as the performance parameter in designing the system. Our objective is to find the optimal way of creating the control channel packets that minimizes the forward link control

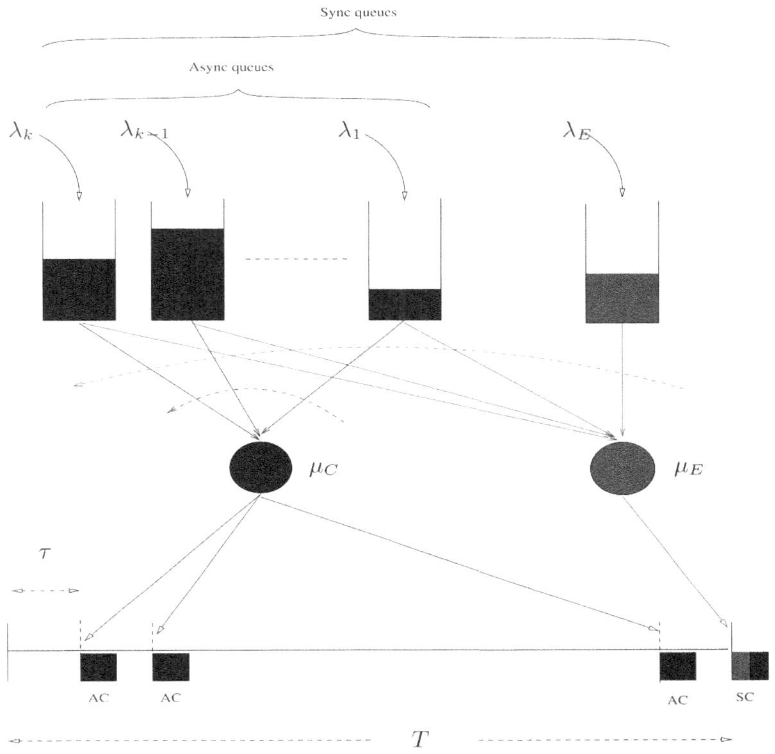

Fig. 1. Illustration of the control channel packet consolidation procedure at the access network

channel utilization under the constraint that expected delay of each type of message is within a maximum acceptable expected delay for that message. In 1xEV-DO, control channel is time division multiplexed with the traffic channel in the forward link. If the control channel transmission occupied more slots to transmit, it means that there will be less number of slots available for user data traffic—resulting in a low forward link user throughput. If the asynchronous capsules are formed frequently, then they are likely to have more padding bits. This means that some of the forward link slots are used to transmit padding bits, wasting air-link resources. On the other hand, if asynchronous capsules are formed less frequently, then some of the messages in the common queues might miss their delay deadline, and will have to be re-transmitted. Therefore, the rate at which asynchronous capsule are created must be such that forward link CC utilization is minimized while meeting the deadline of the messages.

The optimum policy is the solution of the following optimization problem, Problem P1.

Problem P1

Minimize $U = n\ S(\mu_C)$,

subject to

$$(128 - \mu_C)(256 - \mu_C)(512 - \mu_C)(1024 - \mu_C) = 0, \quad (1)$$

$$(S(\mu_C) - 4)(S(\mu_C) - 8) = 0, \quad (2)$$

$$S(\mu_C) \geq \frac{8\mu_c}{1024}, \quad (3)$$

$$S(\mu_C) \geq 4, \quad (4)$$

$$d^i_{\text{avg}}(n) \leq d^i_{\max} \qquad i = 1 \cdots k, \quad (5)$$

$$n \geq \left\lceil \frac{(\lambda_E + \lambda_C)T - \mu_E}{\mu_C} \right\rceil, \quad (6)$$

$$n \geq 1. \quad (7)$$

The size of the asynchronous capsule can only be 128, 256, 512 or 1024 messages. Constraint (1) enforces this. For asynchronous capsule size of 128, 256 or 512 messages, four physical layer transmission slots are required, whereas a capsule size of 1024 messages requires 8 physical layer transmission slots. Constraints (2), (3) and (4) capture these requirements. The average delay experienced by messages in each queue, denoted by $d^i_{\text{avg}}(n)$ for the ith queue must be less than a maximum average delay for the ith queue, $d^i_{\max}$. This constraint is given in (5). Constraint (6) guarantees system stability—the average number of arrivals in a CC cycle must be less than the maximum number of messages that can be drained in a CC cycle, and finally constraint (7) enforces that there be at least one asynchronous capsule formed in a CC cycle.

In order to solve the above optimization problem, we need to derive a computational formula for obtaining the expected delay suffered by an arbitrary message in each common queue. We consider the ith queue, and refer an arbitrary message arriving at the queue as tagged message. Let the arrival time of this tagged message be t sec after the beginning of the last CC cycle. The idea exploited to compute the delay experienced by the tagged message is simple—the waiting time of the tagged message is equal to the time it takes to drain all the high priority message that are already in the system. Note that this may also include messages that arrived after t. For stability, we assume that in the beginning of a CC cycle, the average number of messages held over from previous CC cycles is zero. For the purpose of analysis, we also assume that a message arriving in a control channel cycle will be removed from the queue using one of the asynchronous capsules or using the synchronous capsule before the end of the cycle. Note that when the load in control channel is high, messages may only be removed in the subsequent control channel cycles. This assumption is not very realistic, and can make our delay analysis less accurate at high operating loads. We compute the average delay experienced by a tagged message for the case when the tagged customer arrives after the first asynchronous capsule is created in a CC ($t > x$) and for the case when it arrives before the first asynchronous capsule is created ($t \leq x$).

2.1 Case I: $t > x$

The components needed to compute the expected delay are depicted in Figure 2. We start by computing the expected number of messages with higher priority that arrived before the tagged message that are in the queue by time t. This can be determined by computing the number of messages left in the system during three intervals which include:

- at time x, there will be an average of $\sum_{j=1}^{i} \lambda_j x$ messages with higher priority than the tagged message, and right after x, there will be $\left[\sum_{j=1}^{i} \lambda_j x - \mu_C\right]^+$ messages left in the system[2]. Let this be B_1;
- between x and $t - a$, there would be $\lfloor n(t-x)/T \rfloor$ asynchronous capsule cycles, and each of them would leave $\left[\sum_{j=1}^{i} \lambda_j T/n - \mu_C\right]^+$ messages, totaling to $\lfloor n(t-x)/T \rfloor \left[\sum_{j=1}^{i} \lambda_j T/n - \mu_C\right]^+$ messages; Denote this by B_2;
- during the time period a, there would be an average of $\sum_{j=1}^{i} \lambda_j a$ arrivals, represented by B_3.

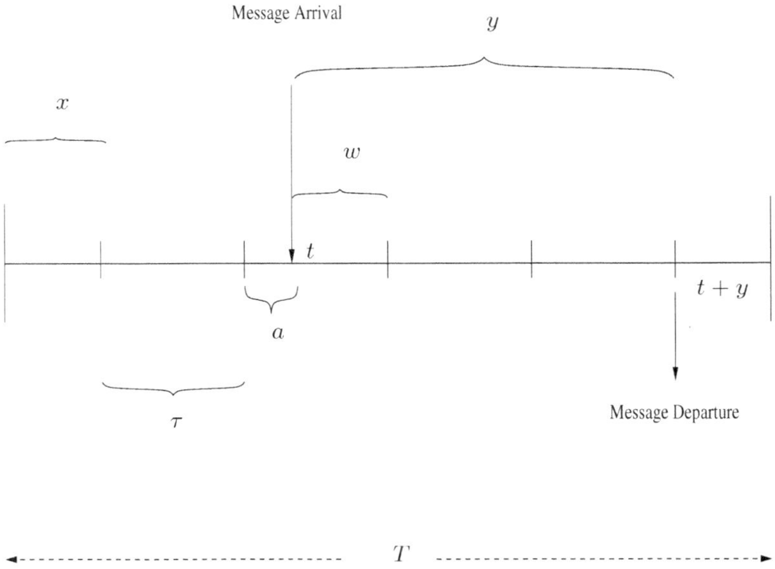

Fig. 2. Diagram illustrating various components of message delay when $t > x$

[2] The notation $[x]^+$ denotes $\max\{0, x\}$.

The total number of messages that arrived before the tagged message that needs to be removed before removing the tagged message is then obtained by adding the number of messages during each of the above three intervals. Now let y be the delay experienced by the tagged message. Then $y\sum_{j=1}^{i-1}\lambda_j$ messages arrived, on an average, when the tagged message was waiting in the queue. The expected number of messages that needs to be removed after t and before removing the tagged messages is then $B_1+B_2+B_3+y\sum_{j=1}^{i-1}\lambda_j$, which requires $\lfloor(B_1+B_2+B_3+y\sum_{j=1}^{i-1}\lambda_j)/\mu_C\rfloor$ asynchronous capsule cycles after the next asynchronous capsule. After t, the next asynchronous capsule cycle happens after w seconds (see Figure 2) given by $(\lfloor n(t-x)/T\rfloor+1)T/n-(t-x)$. Therefore, it would require $w+T/n\lfloor(B_1+B_2+B_3+y\sum_{j=1}^{i-1}\lambda_j)/\mu_C\rfloor$ sec after t to remove all the messages that need to be removed before removing the tagged message. Therefore the expected delay experienced by a tagged message arriving to queue i at time t, represented by $d^i_{\{t>x\}}(n,t,x)$ is the solution to the following equation

$$d^i_{\{t>x\}}(n,t,x)=\Bigg\{y:y-w + \frac{T}{b}\left\lfloor\frac{B_1+B_2+B_3+y\sum_{j=1}^{i-1}\lambda_j}{\mu_C}\right\rfloor=0\Bigg\}.$$

Note that the above equation can yield a set of values for $d^i_{\{t>x\}}(n,t,x)$—we only need to consider the minimum positive solution. Substituting all the values in the above equation, we get (8). Because of our assumption that messages will be removed before the beginning of the next CC cycle, we can write the expected delay as $\min{(T-t), d^i_{\{t>x\}}(n,t,x)}$. We now turn our attention to the computation of the delay when $t\le x$.

2.2 Case II: $t \le x$

Figure 3 portrays this scenario. The average number of messages that are of higher priority than the tagged message remaining after time x, denoted by B_4, is simply $\left[\sum_{j=1}^{i-1}\lambda_j x+\lambda_i t-\mu_c\right]^+$. Let y be the average delay experienced by the tagged message, then the number of messages that are of higher priority than the tagged message arriving after x that needs to be removed before servicing the tagged message is $(t+y-x)\sum_{j=1}^{i-1}\lambda_j$. Let this be B_5. It is obvious that after x, we need $\lfloor(B_4+B_5)/\mu_c\rfloor$ asynchronous capsule cycles to remove all the messages with higher priority than the tagged message. Therefore the average total delay will be $x-t+\lfloor(B_4+B_5)/\mu_c\rfloor T/n$, where $x-t$ represents the time to next asynchronous capsule after t. The expected delay for this case denoted by $d^i_{\{t\le x\}}(n,t,x)$ can be computed as a solution to (9). As in the previous case, we select the minimum positive solution of the equation. Combining the delay for both the cases and assuming that messages are not carried over to the next control channel cycle, we can represent the delay the tagged message as given in (10).

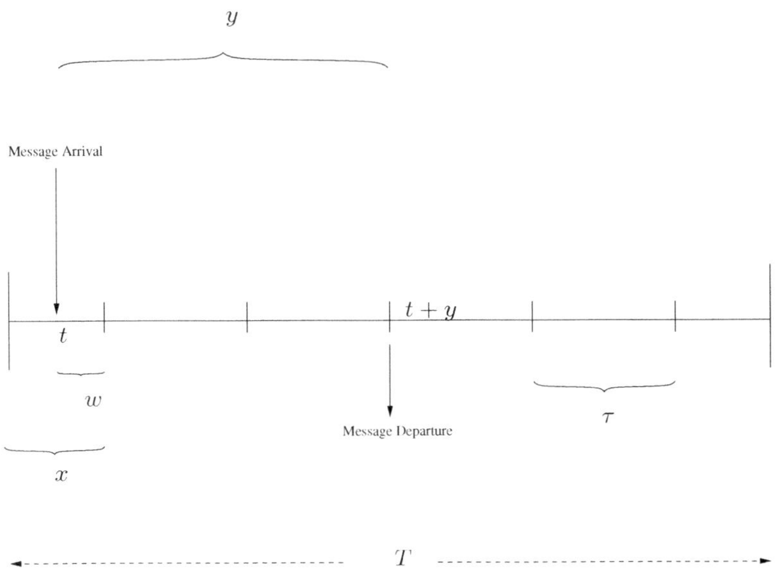

Fig. 3. Diagram illustrating various components of message delay when $t \leq x$

We assumed in our model that all messages arriving during a CC cycle will be removed by the end of that cycle. This assumption is not always true, especially when the CC load is high, but will let us work on a CC cycle in isolation. This makes the computation of delay much easier. We shall show later by simulation that errors introduced by this assumption is negligible for normal operating parameters.

$$\begin{aligned} d^i_{\{t>x\}}(n,t,x) = & \min_{y\geq 0}\Bigg\{ y : y - \left(\left\lfloor \frac{n(t-x)}{T} \right\rfloor + 1\right)\frac{T}{n} - (t-x) \\ & + \left\lceil \frac{\left[\sum_{j=1}^{i}\lambda_j x - \mu_c\right]^+ + \left[\sum_{j=1}^{i}\lambda_j \frac{T}{n} - \mu_c\right]^+ \lfloor\frac{n(t-x)}{T}\rfloor + \left[t - \left(x + \lfloor\frac{n(t-x)}{T}\rfloor\frac{T}{n}\right)\right]\sum_{j=1}^{i}\lambda_j + y\sum_{j=1}^{i-1}\lambda_j}{\mu_c} \right\rceil \frac{T}{n} = 0 \Bigg\}. \end{aligned} \tag{8}$$

$$d^i_{\{t\leq x\}}(n,t,x) = \min_{y\geq 0}\left\{ y : y - \left((x-t) + \left\lceil \frac{\left[\sum_{j=1}^{i-1}\lambda_j x + \lambda_i t - \mu_c\right]^+ + (t+y-x)\sum_{j=1}^{i-1}\lambda_j}{\mu_c} \right\rceil \frac{T}{n} \right) = 0 \right\}. \tag{9}$$

$$d^i(n,t,x) = \begin{cases} \min\left\{(T-t), d^i_{\{t>x\}}(n,t,x)\right\}, & t > x \\ \min\left\{(T-t), d^i_{\{t\leq x\}}(n,t,x)\right\}, & t \leq x \end{cases} \tag{10}$$

The message arrivals to all the queues are assumed to be Poisson. If there is at least one arrival in a CC cycle, then because of the stationary and independent increment properties of the Poisson process, we know that the arrival time of that message will be uniformly distributed in the CC cycle [3]. Therefore the arrival time, t of the tagged message is uniformly distributed in $(0, T)$. The offset between the CC cycle and the asynchronous capsule cycle is also uniformly distributed in $(0, \tau)$. Therefore, averaging over t and offset x gives the expected delay as

$$d^i_{\text{avg}}(n) = \frac{1}{T\tau} \int_0^{\tau} \int_0^{T} d^i(n,t,x)\; dt\; dx. \tag{11}$$

3 Numerical Examples

In this section, we check the reasonableness of the model using simulations, and present a limited set of numerical results obtained from the model. For validation, we consider a system where there are two common queues and one exclusive queue. We assume the following parameters for simulation. The arrival process to highest priority common queue is Poisson with rate $\lambda_1 = 0.3$ messages/slot,

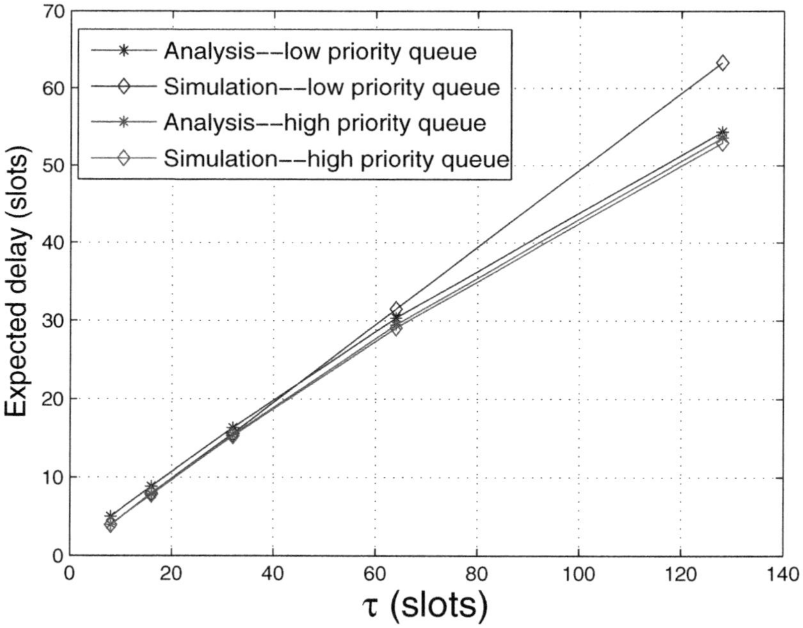

Fig. 4. Delay experienced by high priority and low priority messages for different values of τ—analysis and simulation. The following parameters are used: $\lambda_1 = 0.3$ msgs/slot (high priority), $\lambda_2 = 0.1$ msgs/slot (low priority), $\lambda_E = 0.1$ msgs/slot, $\mu_E = 128$ msgs, and $\mu_C = 64$ msgs.

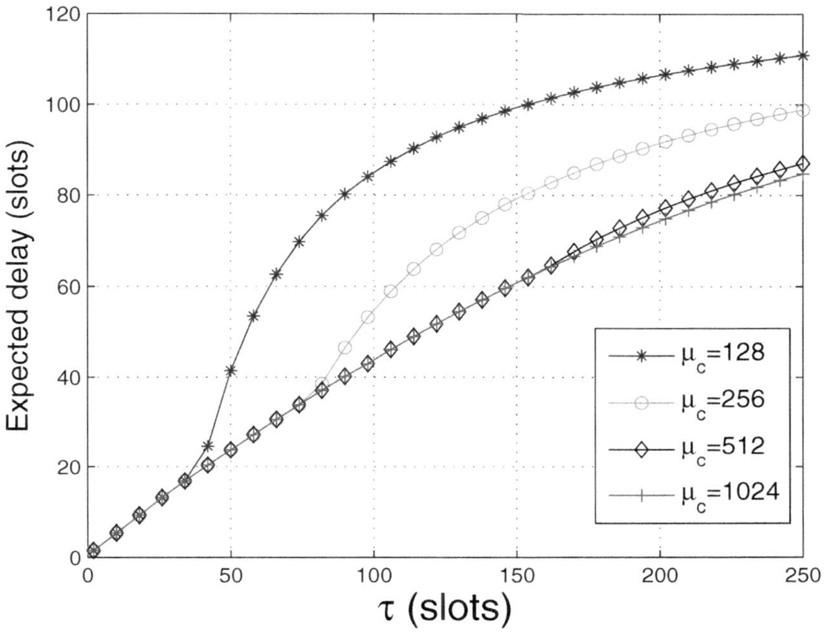

Fig. 5. Delay experienced by high priority messages as a function of τ for all allowed asynchronous capsule sizes. The arrival rate of high priority messages is assumed to be 0.4 msgs/slot, and that of exclusive queue is assumed to be 0.1 msgs/slot. $\mu_E = 128$ msgs.

and that of the lower priority queue is Poisson with rate $\lambda_2 = 0.1$ messages/slot. The arrivals to the exclusive queue is also assumed to be Poisson with rate $\lambda_E = 0.1$ messages/slot. The synchronous and asynchronous capsule sizes are assumed to be 128 messages and 64 messages, respectively. All the messages, arriving to both common and exclusive queues are assumed to be of equal length. Figure 4, shows the results from simulation and analysis. We can see that the model matches with the simulation for the high priority queue. However for the low priority queue, the model matches well with the simulation when the asynchronous capsule cycle (τ) is less than 64 slots ($\approx$106 ms). The difference between the model and the simulation is expected due to our assumption that all the messages arriving in a CC cycle departs by the end of the cycle. In reality, there will be cycles when messages are carried over to the next or subsequent cycles, and when τ is large, the carried over messages will be delayed in multiples of τ resulting in a much higher delay than the case when message can leave the same CC cycle it arrived. Therefore, at higher loads and larger asynchronous capsule cycles, the model predicts a lower delay than the actual delay. In most practical systems, the average delay of messages of interest is well within 100 ms, and therefore our approximate model works well within the operating region.

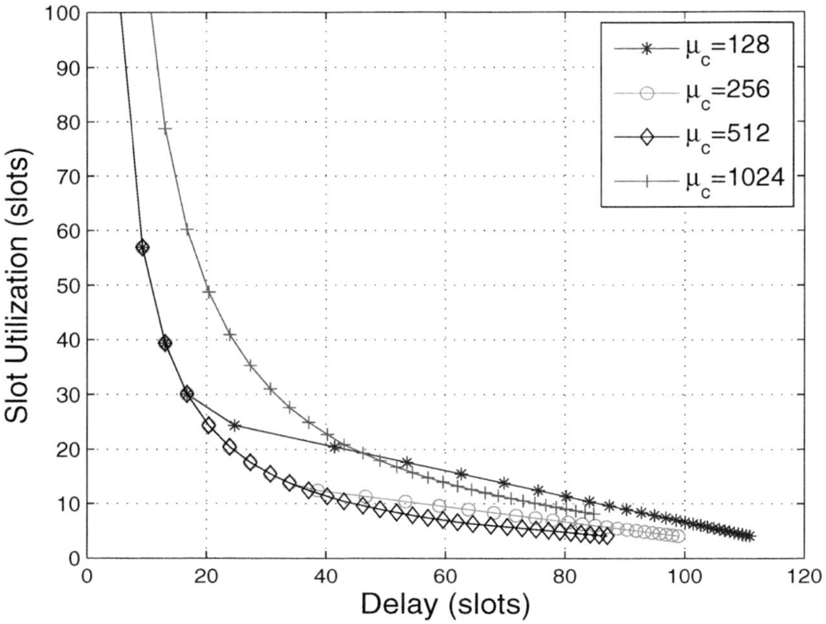

Fig. 6. Forward link physical layer slot utilization as a function of the experienced delay of high priority messages for different values of asynchronous capsule sizes. The parameters used are same as in Figure 5.

In the next set of results, we assume that only the highest priority common queue has the delay constraint. The arrival rate of the highest priority common queue is assumed as $\lambda_1 = 0.4$ messages per CC slot. The aggregate arrival rate for the exclusive queue is fixed to be $\lambda_E = 0.1$ msgs/slot, and the synchronous capsules are assumed to be of size 128 messages. Figure 5 plots delay as a function of τ for different values of asynchronous capsule size. We use (11) to compute the expected delay for different values of μ_C. From the plots, we can see that increasing the asynchronous capsule size gives diminishing returns on the expected delay of the highest priority messages.

Now we wish to find the optimum asynchronous capsule formation policy satisfying the delay constraints of the highest priority queue and other constrains given in Problem P1. In Figure 6, we plot forward link control channel slot utilization versus the delay experienced by the highest priority messages for different values of asynchronous capsule sizes. The optimum policy is then the selection of asynchronous capsule size that minimizes the slot utilization for each value of maximum acceptable expected delay of the highest priority messages. Once the capsule size is selected, we can select τ that satisfies the delay constraint for the selected μ_C using Figure 5.

In the above example, suppose that we want to find the optimum value of asynchronous capsule size and asynchronous capsule cycle that minimizes the forward link CC utilization, while achieving a delay constraint of 40 slots ($\approx$ 66 ms) for the highest priority message. Then, from Figure 6, we know that asynchronous capsule size of 512 messages achieves the minimum slot utilization, and from Figure 5, we know that to meet the 40 slots delay constraint of the highest priority message using a asynchronous capsule size of 512 messages, we need to form asynchronous capsules approximately every 90 slots.

In certain cases, the access network may not be able to transmit in all the available asynchronous capsule sizes. For example, if the feasible asynchronous capsule sizes are 128 and 512 messages, then depending on the delay requirements, different asynchronous capsule sizes will yield optimum performance. In our example, when maximum expected delay is less than 45 slots, choosing μ_C as 128 yields the minimum forward link CC utilization. However, for delay requirements for more than 45 slots, it is better to opt for μ_C as 512.

We note in passing that the procedure developed can be used to determine other performance measures. For instance, it is straight forward to determine the FL traffic channel utilization for a given value of asynchronous capsule cycle and asynchronous capsule size.

4 Concluding Remarks

In this work, we studied the optimal design of control channel packet consolidation protocol at the access network in a 1xEV-DO system. The design discussed herein allows for optimal formation of asynchronous capsules such that forward link control channel utilization is minimized while satisfying the delay constraints of all signaling messages. We first described the operation of packet consolidation protocol, and then formulated the problem of designing an optimal policy as a generalized optimization problem. We presented a simple model for computing the average delay of messages needed to solve the optimization problem. We validated our model through limited set of simulation results, and finally presented some numerical examples to illustrate the design principles. Our hope is that the guidelines presented here provides cellular operators useful order-of-magnitude estimates for the parameters of interests.

This work can be extended in a number of interesting ways. The delay model presented uses some simplifying assumptions; a more accurate delay model can be worked out. It appears worthwhile to develop the delay distribution of messages. Currently, we consider the average delay as an optimization constraint. An interesting extension would be to include constraints that guarantee that the probability that the delay exceeds a certain threshold is less than a maximum acceptable value. Finally, comparing the results from the model to that of a real 1xEV-DO network would be a valuable next step.

References

1. 3GPP2 C.S0024-A v1.0, CDMA 2000 High Data Rate Packet Data Air Interface Specification, www.3gpp2.org
2. Bhushan, N., Lott, C., Black, P., Attar, R., Jou, Y.-C., Fan, M., Ghosh, D., Au, J.: CDMA2000 1xEV-DO revision a: a physical layer and MAC layer overview. IEEE Communications Magazine 44(2) (2006)
3. Gallager, R.: Discrete stochastic processes, 1st edn. Springer, Heidelberg (1995)
4. Gopal, T.: EVDO Rev. A control channel bandwidth analysis for paging. In: Proc. of IEEE Wireless Communications and Networking Conference, pp. 3262–3267 (2007)
5. Ramaswamy, V., Chung, J.: Performance analysis of the quick idle state protocol of CDMA 1xEV-DO Rev. B systems. To Appear in Proc. of IEEE Global Telecommunications Conference (2010)

Performance Analysis of Cloud Computing Centers

Hamzeh Khazaei[1], Jelena Mišić[2], and Vojislav B. Mišić[2]

[1] University of Manitoba, Winnipeg, Manitoba, Canada
hamzehk@cs.umanitoba.ca
http://www.cs.umanitoba.ca/~hamzehk
[2] Ryerson University, Toronto, Ontario, Canada

Abstract. Cloud computing is a computing paradigm in which different computing resources, including infrastructure, hardware platforms, and software applications, are made accessible to remote users as services. Successful provision of infrastructure-as-a-service (IaaS) and, consequently, widespread adoption of cloud computing necessitates accurate performance evaluation that allows service providers to dimension their resources in order to fulfil the service level agreements with their customers. In this paper, we describe an analytical model for performance evaluation of cloud server farms, and demonstrate the manner in which important performance indicators such as request waiting time and server utilization may be assessed with sufficient accuracy.

Keywords: cloud computing, performance analysis, M/G/m queuing system, response time.

1 Introduction

Significant innovations in virtualization and distributed computing, as well as improved access to high-speed Internet, have accelerated interest in cloud computing [15]. Cloud computing is a general term for system architectures that involves delivering hosted services over the Internet. These services are broadly divided into three categories: Infrastructure-as-a-Service (IaaS), which includes equipment such as hardware, storage, servers, and networking components are made accessible over the Internet); Platform-as-a-Service (PaaS), which includes computing platforms—hardware with operating systems, virtualized servers, and the like; and Software-as-a-Service (SaaS), which includes sofware applications and other hosted services [11]. A cloud service differs from traditional hosting in three principal aspects. First, it is provided on demand, typically by the minute or the hour; second, it is elastic since the user can have as much or as little of a service as they want at any given time; and third, the service is fully managed by the provider – user needs little more than computer and Internet access. Cloud customers pay only for the services they use by means of a customized service level agreement (SLA), which is a contract negotiated and agreed between a customer and a service provider: the service provider is required to execute service

X. Zhang and D. Qiao (Eds.): QShine 2010, LNICST 74, pp. 251–264, 2011.

requests from a customer within negotiated quality of service(QoS) requirements for a given price.

Due to dynamic nature of cloud environments, diversity of user's requests and time dependency of load, providing expected quality of service while avoiding over-provisioning is not a simple task [17]. To ensure that the QoS perceived by end clients is acceptable, the providers must exploit techniques and mechanisms that guarantee a minimum level of QoS. Although QoS has multiple aspects such as response time, throughput, availability, reliability, and security, the primary aspect of QoS considered in this work is related to response time [16].

Cloud computing has been the focus of much research in both academia and industry, however, implementation-related issues have received much more attention than performance-related ones; in this paper we describe an analytical model for evaluating the performance of cloud server farms and verify its accuracy with numerical calculations and simulations. we assume that any request goes through a *facility node* and then leaves the center. A facility node may contain different computing resources such as web servers, database servers, and others, as shown in Fig. 1. We consider the time a request spends in one of those facility node as the response time; response time does not follow any specific distribution. Our model is flexible in terms of cloud center size and service time of customer requests; We model the cloud environment as an $M/G/m$ queuing system which indicates that inter-arrival time of requests is exponentially distributed, the service time is generally distributed and the number of facility nodes is m. Also, due to the the nature of cloud environment (i.e., it is a service provider with potentially many customers), we pose no restrictions on the number of facility nodes. These two characteristics, general service time and large number of nodes, have not been adequately addressed in previous research.

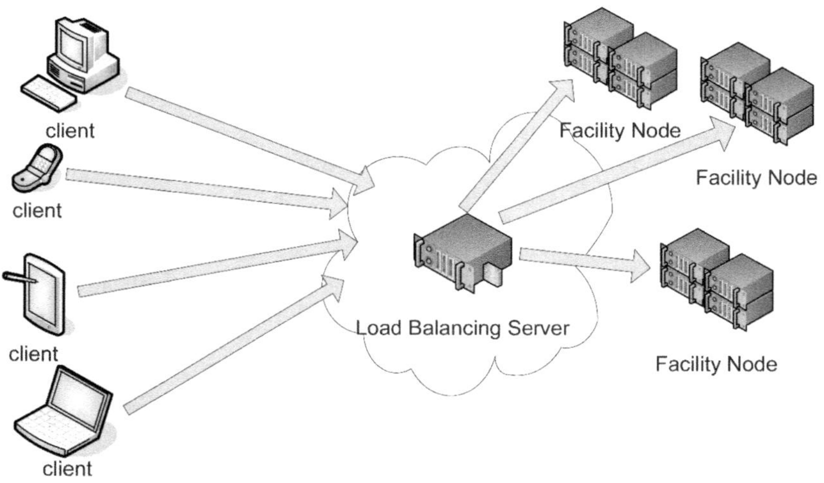

Fig. 1. Cloud clients and service provider

The remainder of the paper is organized as follows: Section 2 gives a brief overview of related work on cloud performance evaluation and on performance characterization of $M/G/m$ queueing systems. We introduce our analytical model in Section 3, and present performance results obtained with it in Section 4, using discrete event simulation to validate them. Discussion of our approach and outlook to future research activities complete the paper.

2 Related Work

As mentioned above, most of the research related to cloud computing has dealt with implementation issues, while performance-related issues have received much less attention. For example, [20] studied the response time in terms of various metrics, such as the overhead of acquiring and realizing the virtual computing resources, and other virtualization and network communication overhead. To address these issues, they have designed and implemented C-Meter, a portable, extensible, and easy-to-use framework for generating and submitting test workloads to computing clouds.

In [18], the cloud center was modeled as an $M/M/m/N$ queuing system, which was been used to compute the distribution of response time. Inter-arrival and service times were both assumed to be exponentially distributed, and the system had a finite buffer of size N. The response time was broken down into waiting, service, and execution periods, assuming that all three periods are independent which is unrealistic, based on their own argument.

In [17], the authors consider a cloud center which is modelled as the classic open network; they obtained the distribution of response time based on assumption that inter-arrival time and service time are both exponential. Using the distribution of response time, they found the relationship among the maximal number of tasks, the minimal service resources and the highest level of services.

Theoretical analyses have mostly relied on extensive research in performance evaluation of $M/G/m$ queuing systems, as outlined in [2,5,7,8,9,19]. As solutions for mean response time and queue length in $M/G/m$ systems can't be obtained in closed form, suitable approximations were sought. However, most of these provide reasonably accurate estimates of mean service time only when number of servers is comparatively small, (say, less than twenty or so), but fail for large number of servers [1,3,13,14]. Approximation errors are particularly pronounced when the offered load ρ is small, and/or when both the number of servers m and the coefficient of variation of the arrival process for service requests, CV, are large. As a result, these results are not directly applicable to performance analysis of cloud computing server farms where the number of servers is huge and service request arrival distribution is not generally known.

3 The Analytical Model

We model a cloud server farm as a $M/G/m$ queuing system which indicates that the inter-arrival time of requests is exponentially distributed, the service times of customers' requests are independent and identically distributed random variables with a general distribution whose service rate is μ; both μ and CV, the coefficient of variation defined as standard deviation divided by the mean, are finite.

A $M/G/m$ queuing system may be considered as a Markov process which can be analysed by applying the embedded Markov chain technique. Embedded Markov Chain techique requires selection of Markov points in which the state of the system is observed. Therefore we monitor the number of the tasks in the system (both in service and queued) at the moments immediately before the task request arrival. If we consider the system at Markov points and number these instances 0, 1, 2, ..., then we get a Markov chain [4]. Here, the system under consideration contains m servers, which render service in order of task request arrivals.

Task requests arrival process is Poisson. Task request interarrival time A is exponentially distributed with rate to $\frac{1}{\lambda}$. We will denote its Cumulative Distribution Function (CDF) as $A(x) = Prob[A < x]$ and its probability density function (pdf) as $a(x) = \lambda e^{-\lambda x}$. Laplace Stieltjes Transform (LST) of interarrival time is $A^*(s) = \int_0^\infty e^{-sx} a(x)dx = \frac{\lambda}{\lambda+s}$.

Task service times are identically and independently distributed according to a general distribution B, with a mean service time equal to $\bar{b} = \frac{1}{\mu}$. The CDF of the service time is $B(x) = Prob\,[B < x]$, and its pdf is $b(x)$. The LST of service time is $B^*(s) = \int_0^\infty e^{-sx} b(x)dx$.

Residual task service time is time from the random point in task execution till the task completion. We will denote it as B_+. This time is necessary for our model since it represents time distribtion between task arrival z and departure of the task which was in service when task arrival z occured. It can be shown as well that probability distrubtion of elapsed service time (between start of the task execution and next arrival of task request B_- has the same probability distribtion [12].

The LST of residual and elapsed task service times can be calculated in [12] as

$$B_+^*(s) = B_-^*(s) = \frac{1 - B^*(s)}{s\bar{b}} \tag{1}$$

The offered load may be defined as

$$\rho \triangleq \frac{\lambda}{m\mu} \tag{2}$$

For practical reasons, we assume that the system never enters saturation, which means that any request submitted to the center will get access to the required facility node after a finite queuing time. Furthermore, we also assume each task is serviced by a single server (i.e., there are no batch arrivals), and we do not distinguish between installation (setup), actual task execution, and finalization components of the service time; these assumptions will be relaxed in our future work.

3.1 The Markov Chain

We are looking at the system at the moments of task request arrivals – these points are selected as Markov points. A given Markov chain has a steady-state solution if it is ergodic. Based on conditions for ergodicity [4] and the above-mentioned assumptions, it is easy to prove that our Markov Chain is ergodic. Then, using the steady-state solution, we can extract the distribution of number of tasks in the system as well as the response time.

Let A_n and A_{n+1} indicate the moment of nth and $(n+1)^{th}$ arrivals to the system, respectively, while q_n and q_{n+1} indicate the number of tasks found in the system immediately before these arrivals; this is schematically shown in Fig. 2. If v_{n+1} indicates the number of tasks which are serviced and depart from the system between A_n and A_{n+1}, the following holds:

$$q_{n+1} = q_n - v_{n+1} + 1 \tag{3}$$

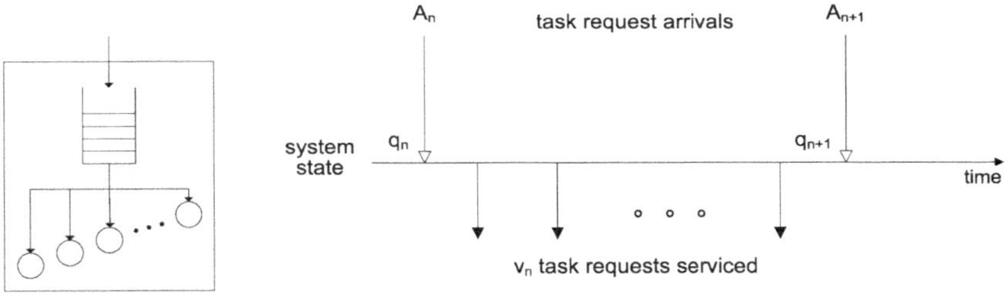

Fig. 2. Embedded Markov points

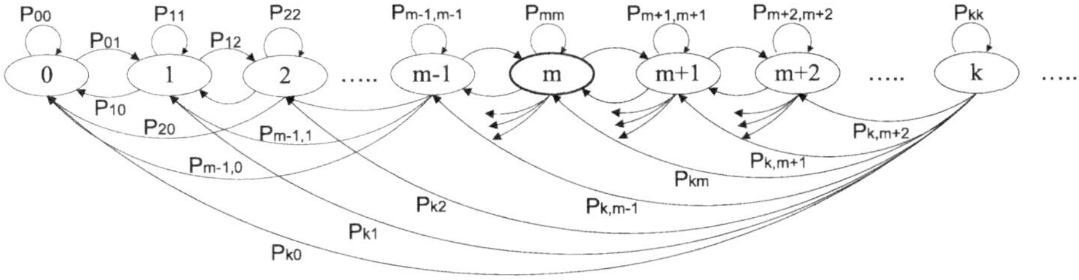

Fig. 3. State-transition-probability diagram for the $M/G/m$ embedded Markov chain

We need to calculate the transition probabilities associated with this Markov chain, defined as

$$p_{ij} \triangleq Prob\left[q_{n+1} = j | q_n = i\right] \tag{4}$$

i.e., the probability that $i+1-j$ customers are served during the interval between two successive task request arrivals. Obviously for $j > i+1$

$$p_{ij} = 0 \tag{5}$$

since there are at most $i+1$ tasks present between the arrival of A_n and A_{n+1}. The Markov state-transition-probability diagram as in Fig. 3, where states are numbered according to the number of tasks currently in the system (i.e those in service and those awaiting service). For clarity, some transitions are not fully drown, esp. those originating from states above m. We have also highlighted the state m because the transition probabilities are different for states on the left and right hand side of this state (i.e., below and above m).

3.2 Departure Probabilities

Due to ergodicity of the Markov chain, an equilibrium probability distribution will exist for the number of tasks present at the arrival instants; so we define

$$\pi_k = \lim_{n \to +\infty} Prob\,[q_n = k] \tag{6}$$

From [12], the direct method of solution for this equilibrium distribution requires that we solve the following system of linear equations:

$$\pi = \pi \mathbf{P} \tag{7}$$

where $\pi = [\pi_0, \pi_1, \pi_2, \ldots]$, and $\mathbf{P}$ is the matrix whose elements are one-step transition probabilities p_{ij}.

To find the elements of the transition probability matrix, we need to count the number of tasks departing from the system in between two successive arrivals. Consider the behaviour of the system, as shown in Fig. 4. Each server has zero or more departures during the time between two successive task request arrivals (the inter-arrival time). Let us focus on an arbitrary server, which (without loss of generality) could be the server number 1. For a task to finish and depart from the system during the inter-arrival time, its remaining duration (residual service time defined in (1)) must be shorter than the task inter-arrival time. This probability will be denoted as P_x, and it can be calculated as

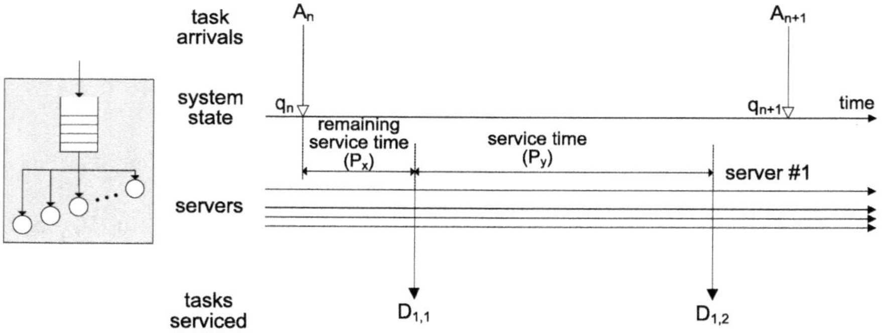

Fig. 4. System behaviour in between two arrivals

$$P_x = Prob\,[A > B_+] = \int_{x=0}^{\infty} P\{A > B_+ | B_+ = x\} P\{B_+ = x\}$$
$$= \int_0^{\infty} e^{-\lambda x} dB_+(x) = B_+^*(\lambda) \tag{8}$$

Physically this result presents probability of no task arrivals during residual task service time.

In the case when arriving task can be accommodated immediately by an idle server (and therefore queue length is zero) we have to evaluate the probability that such task will depart before next task arrival. We will denote this probability as P_y and calculate it as:

$$P_y = Prob\,[A > B] = \int_{x=0}^{\infty} P\{A > B | B = x\} P\{B_+ = x\}$$
$$= \int_0^{\infty} e^{-\lambda x} dB(x) = B^*(\lambda) \tag{9}$$

However, if queue is non-empty upon task arrival following situation may happen. If between two successive new task arrivals a completed task departs from a server, that server will take a new task from the non-empty queue. That task may be completed as well before the next task arrival and if the queue is still non-empty new task may be executed, and so on until either queue gets empty or new task arrives. Therefore probability of $k > 0$ job departures from a single server, given that there are enough jobs in the queue can be derived from expressions (8) and (9) as:

$$P_{z,k} = B_+^*(\lambda)(B^*(\lambda))^{k-1} \tag{10}$$

note that $P_{z,1} = P_x$.

Using these values we are able to compute the transition probabilities matrix.

3.3 Transition Matrix

Based on our Markov chain, we may identify four different regions of operation for which different conditions hold; these regions are schematically shown in Fig. 5, where the numbers on horizontal and vertical axes correspond to the number of tasks in the system immediately before a task request arrival (i) and immediately upon the next task request arrival (j), respectively.

Regarding the region labelled 1, we already know from Eq. 5 that $p_{ij} = 0$ for $i + 1 < j$.

In region 2, no tasks are waiting in the queue, hence $i < m$ and $j \leq m$. In between the two successive request arrivals, $i + 1 - j$ tasks will complete their service. For all transitions located on the left side of state m in Fig. 3, the probability of having $i + 1 - j$ departures is

$$p_{ij} = \binom{i}{i-j} P_x^{i-j} (1 - P_x)^j P_y + \binom{i}{i+1-j} P_x^{i+1-j} (1 - P_x)^{j-1} (1 - P_y) \quad \text{for } i < m, j \leq m \tag{11}$$

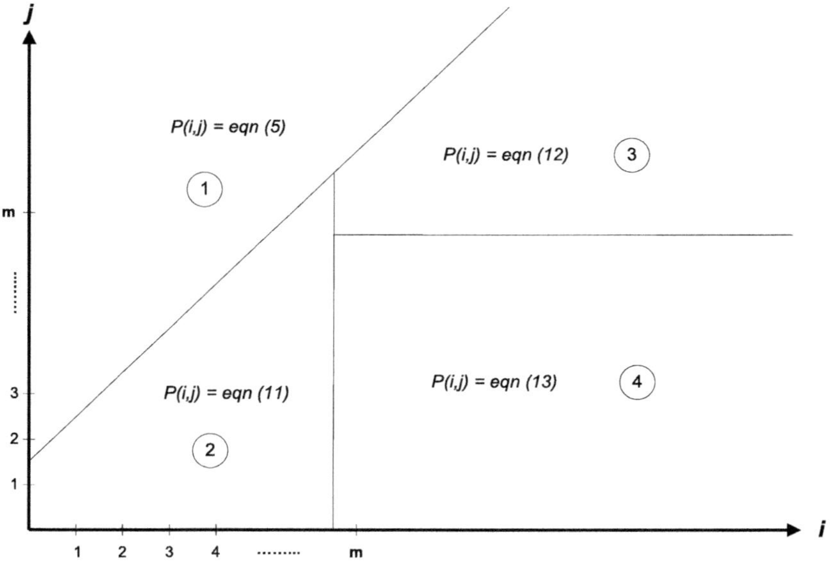

Fig. 5. Range of validity for p_{ij} equations

Region 3 corresponds to the case where all servers are busy throughout the inter-arrival time, i.e., $i, j \geq m$. In this case all transitions remain to the right of state m in Fig. 3, and state transition probabilities can be calculated as

$$p_{ij} = \sum_{s=\phi}^{\sigma} \binom{m}{s} P_x^s (1 - P_x)^{m-s} P_{z,2}^{i+1-j-s} (1 - P_{z,2})^s \qquad \text{for } i, j \geq m \tag{12}$$

In the last expression, the summation bounds are $\sigma = min\,[i + 1 - j, m]$ and $\phi = min\,[i + 1 - j, 1]$.

Finally, region 4, in which $i \geq m$ and $j \leq m$, describes the situation where the first arrival (A_n) finds all servers busy and a total of $i - m$ tasks waiting in the queue, which it joins; while at the time of the next arrival (A_{n+1}) there are exactly j tasks in the system, all of which are in service. The transition probabilities for this region are

$$p_{ij} = \sum_{s=1}^{\sigma} \binom{m}{s} P_x^s (1 - P_x)^{m-s} \binom{\eta}{\alpha} P_{z,2}^{\psi} (1 - P_{z,2})^{\zeta} \beta \qquad \text{for } i \geq m, j < m \tag{13}$$

where we used the following notation:

$$\begin{aligned} \sigma &= min\,[m, i+1-j] \\ \eta &= min\,[s, i+1-m] \\ \alpha &= min\,[s, i+1-j-s] \\ \psi &= max\,[0, i+1-j-s] \\ \zeta &= max\,[0, j-m+s] \\ \beta &= \begin{cases} 1 & \text{if } \psi \leq i+1-m \\ 0 & \text{otherwise} \end{cases} \end{aligned} \tag{14}$$

4 Numerical Validation

The steady-state balance equations outlined above can't be solved in closed form, hence we must resort to a numerical solution. To obtain the steady-state probabilities $\pi = [\pi_0, \pi_1, \pi_2, ...]$, as well as the mean number of tasks in the system (in service and in the queue) and the mean response time, we have used the probability generating functions (PGFs) for the number of tasks in the system:

$$P(z) = \sum_{k=0}^{\infty} \pi_z z^k \tag{15}$$

and solved the resulting system of equations using Maple 13 from Maplesoft, Inc. [6]. Since the PGF is an infinite series, it must be truncated for numerical solution; we have set the number of equations to twice the number of servers, which allows us to achieve satisfactory accuracy (as will be explained below), plus the necessary balance equation

$$\sum_{i=0}^{i=2m} \pi_i = 1. \tag{16}$$

the mean number of tasks in the system is, then, obtained as

$$E[QS] = P^{'}(1) \tag{17}$$

while the mean response time is obtained using Little's law as

$$E[RT] = E[QS]/\lambda \tag{18}$$

We have assumed that the task request arrivals follow the gamma distribution with different values for shape and scale parameters; however, our model may accommodate other distributions without any changes. Then, we have performed two experiments with variable task request arrival rate and coefficient of variation CV (which can be adjusted in the gamma distribution independently of the arrival rate).

To validate the analytical solutions we have also built a discrete even simulator of the cloud server farm using object-oriented Petri net-based simulation engine Artifex by RSoftDesign, Inc. [10].

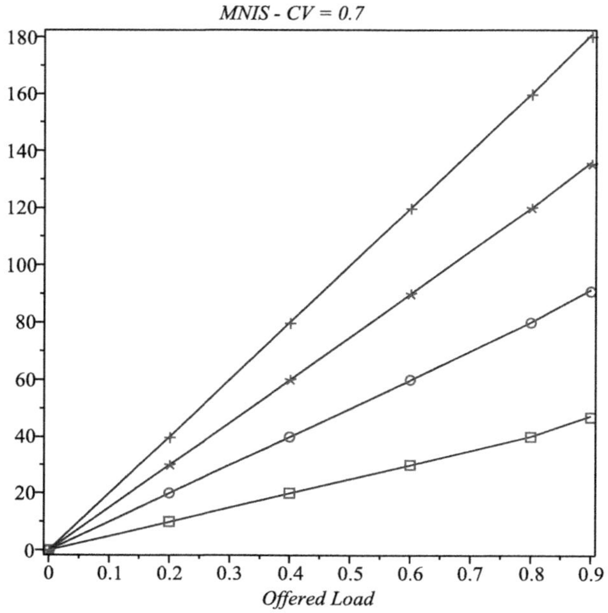

(a) $CV = 0.7$.

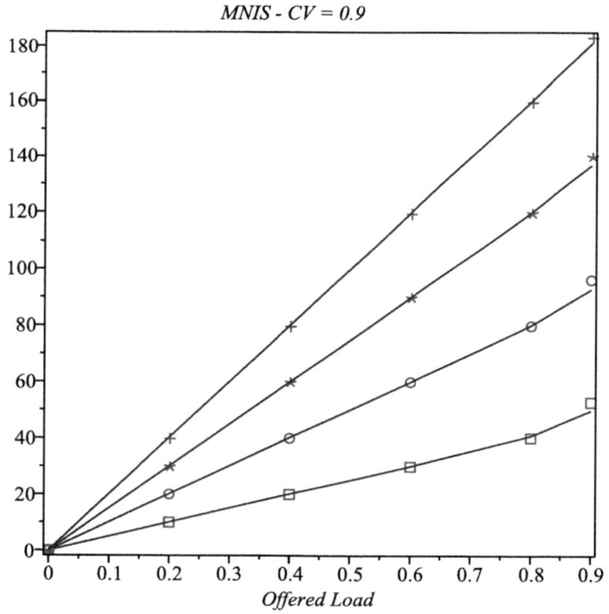

(b) $V = 0.9$.

Fig. 6. Mean number of tasks in the system: $m = 50$ (denoted with squares), 100 (circles), 150 (asterisks), and 200 (crosses)

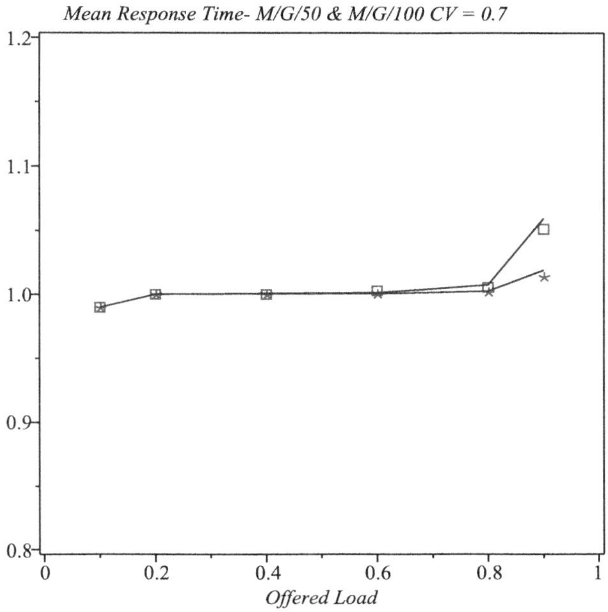

(a) Results for $CV = 0.7$, $m = 50$ and 100 servers.

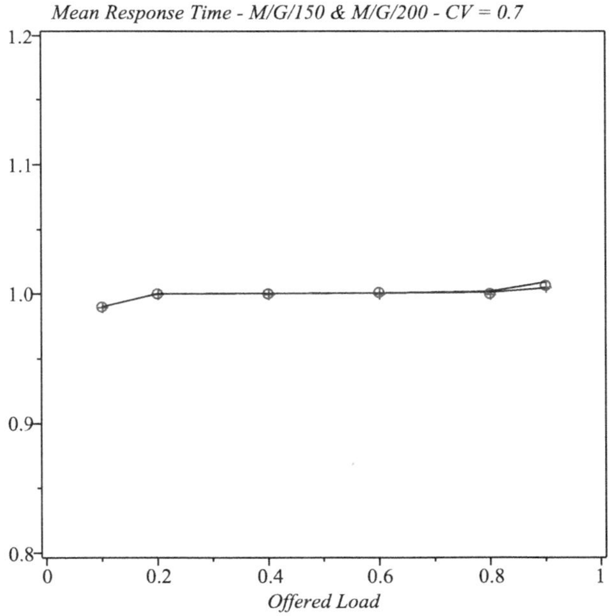

(b) Results for $CV = 0.7$, $m = 150$ and 200 servers.

Fig. 7. Mean response time $CV = 0.7$, $m = 50$ (denoted with squares), 100 (asterisks), 150 (circles), and 200 (crosses)

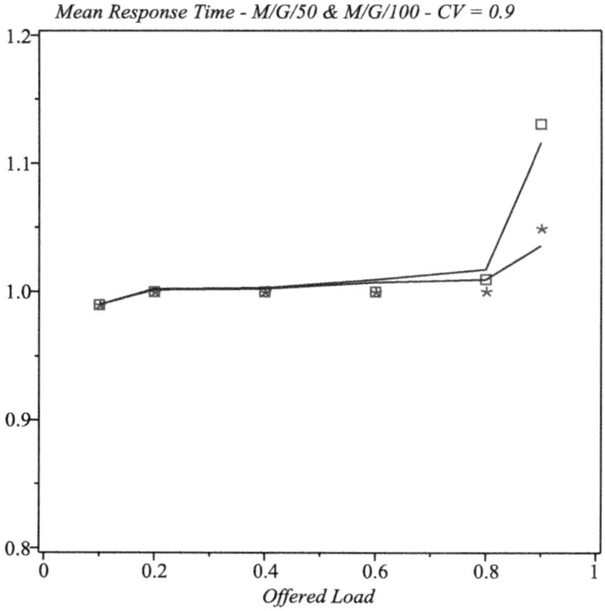

(a) Results for $CV = 0.9$, $m = 50$ and 100 servers.

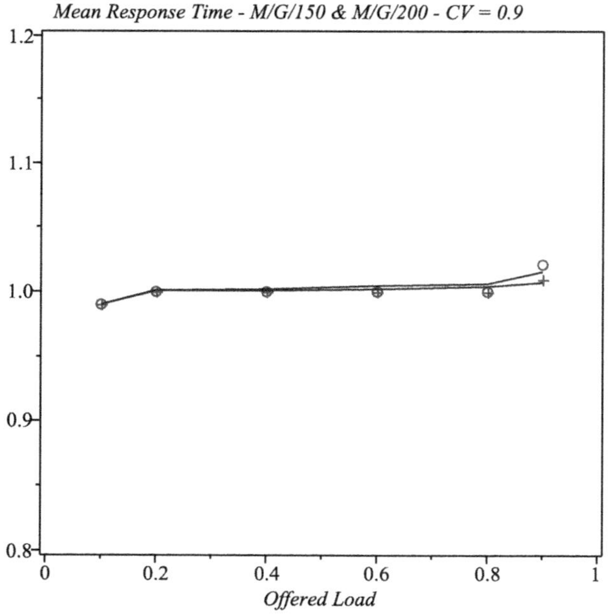

(b) Results for $CV = 0.9$, $m = 150$ and 200 servers.

Fig. 8. Mean response time for $CV = 0.9$, $m = 50$ (denoted with squares), 100 (asterisks), 150 (circles), and 200 (crosses)

The diagrams in Fig. 6 show analytical and simulation results (shown as lines and symbols, respectively) for mean number of tasks in the system as functions of the offered load ρ, under different number of servers. Two different values of the coefficient of variation, $CV = 0.7$ and 0.9, were used; the corresponding results are shown in Figs. 6(a) and 6(b). As can be seen, the results obtained by solving the analytical model agree very well with those obtained by simulation.

The diagrams in Fig. 8 show the mean response time, again for the same range of input variables and for the same values of the coefficient of variation. As above, solid lines correspond to analytical solutions, while different symbols correspond to different number of servers. As could be expected, the response time is fairly steady up to the offered load of around $\rho = 0.8$, when it begins to increase rapidly. However, the agreement between the analytical solutions and simulation results is still very good, which confirms the validity of our modelling approach.

5 Conclusions

Performance evaluation of server farms is an important aspect of cloud computing which is of crucial interest for both cloud providers and cloud customers. In this paper we have proposed an analytical model for performance evaluation of a cloud computing center. Due to the nature of the cloud environment, we assumed general service time for requests as well as large number of servers; in the other words, our model is flexible in terms of scalability and diversity of service time. We have further conducted numerical experiments and simulation to validate our model. Numerical and simulation results showed that the proposed method provided a quite accurate computation of the mean number of tasks in the system and mean response time.

In future work we plan to extend our model for burst arrivals of requests or a kind of task including several subtasks; we are also going to examine other types of distributions as service time which are more realistic in cloud computing area, e.g. Log-Normal distribution. Looking in to the facility node and breaking down the response time into several components such as setup, execution, return and clean up time will be another dimension of extension. We will address all these issues in our future work.

References

1. Boxma, O.J., Cohen, J.W., Huffel, N.: Approximations of the mean waiting time in an $M/G/s$ queueing system. Operations Research 27, 1115–1127 (1979)
2. Hokstad, P.: Approximations for the $M/G/m$ queues. Operations Research 26, 510–523 (1978)
3. Kimura, T.: Diffusion approximation for an $M/G/m$ queue. Operations Research 31, 304–321 (1983)
4. Kleinrock, L.: Queueing Systems. Theory, vol. 1. Wiley-Interscience, Hoboken (1975)

5. Ma, B.N.W., Mark, J.W.: Approximation of the mean queue length of an $M/G/c$ queueing system. Operations Research 43, 158–165 (1998)
6. Maplesoft, Inc. Maple 13. Waterloo, ON, Canada (2009)
7. Miyazawa, M.: Approximation of the queue-length distribution of an $M/GI/s$ queue by the basic equations. J. Applied Probability 23, 443–458 (1986)
8. Nozaki, S.A., Ross, S.M.: Approximations in finite-capacity multi-server queues with poisson arrivals. J. Applied Probability 15, 826–834 (1978)
9. Page, E.: Tables of waiting times for $M/M/n$, $M/D/n$ and $D/M/n$ and their use to give approximate waiting times in more general queues. J. Operational Research Society 33, 453–473 (1982)
10. RSoft Design. Artifex v.4.4.2. RSoft Design Group, Inc., San Jose, CA (2003)
11. searchcloudcomputing.techtarget.com. Cloud computing definition (2010), `http://searchcloudcomputing.techtarget.com/sDefinition/0,,sid201_gci128 7881,00.html`
12. Takagi, H.: Queuing Analysis. Vacation and Priority Systems, part 1, vol. 1. Elsevier Science Publisher B.V., Amsterdam (1991)
13. Takahashi, Y.: An approximation formula for the mean waiting time of an $M/G/c$ queue. J. Operational Research Society 20, 150–163 (1977)
14. Tijms, H.C., Hoorn, M.H.V., Federgru, A.: Approximations for the steady-state probabilities in the $M/G/c$ queue. Advances in Applied Probability 13, 186–206 (1981)
15. Vaquero, L., Rodero-Merino, L., Caceres, J., Lindner, M.: A break in the clouds: towards a cloud definition. ACM SIGCOMM Computer Communication Review, 39(1) (2009)
16. Wang, L., Laszewski, G.V., Younge, A., He, X., Kunze, M., Tao, J., Fu, C.: Cloud computing: a perspective study. New Generation Computing 28, 137–146 (2010)
17. Xiong, K., Perros, H.: Service performance and analysis in cloud computing, Los Alamitos, CA, USA, pp. 693–700 (2009)
18. Yang, B., Tan, F., Dai, Y., Guo, S.: Performance evaluation of cloud service considering fault recovery. In: Jaatun, M.G., Zhao, G., Rong, C. (eds.) Cloud Computing. LNCS, vol. 5931, pp. 571–576. Springer, Heidelberg (2009)
19. Yao, D.D.: Refining the diffusion approximation for the $M/G/m$ queue. Operations Research 33, 1266–1277 (1985)
20. Yigitbasi, N., Iosup, A., Epema, D., Ostermann, S.: C-meter: A framework for performance analysis of computing clouds. In: CCGRID 2009: Proceedings of the 2009 9th IEEE/ACM International Symposium on Cluster Computing and the Grid, Washington, DC, USA, pp. 472–477 (2009)

Efficient Stream Processing in the Cloud

Dung Vu[1], Vana Kalogeraki[2], and Yannis Drougas[3]

[1] Department of Computer Science and Engineering,
University of California-Riverside, USA
dungv@cs.ucr.edu
[2] Department of Informatics, Athens University of Economics and Business, Greece
vana@aueb.gr
[3] Environmental Systems Research Institute, Redlands, USA
drougas@esri.com

Abstract. In the recent years, many emerging on-line data analysis applications require real-time delivery of the streaming data while dealing with unpredictable increase in the volume of data. In this paper we propose a novel approach for efficient stream processing of bursts in the Cloud. Our approach uses two queues to schedule requests pending execution. When bursts occur, incoming requests that exceed maximum processing capacity of the node, instead of being dropped, are diverted to a secondary queue. Requests in the secondary queue are concurrently scheduled with the primary queue, so that they can be immediately executed whenever the node has any processing power unused as the results of burst fluctuations. With this mechanism, processing power of nodes is fully utilized and the bursts are efficiently accommodated. Our experimental results illustrate the efficiency of our approach.

Keywords: Stream Processing, Peer-to-Peer, Distributed Systems.

1 Introduction

Over the years, we have experienced the proliferation of distributed stream processing systems that deal with large volume and high-rate data streams. A number of stream processing systems have been developed, including Aurora [2], STREAM [3], TelegraphCQ [1] and Cougar [4]. These systems are characterized by continuous, large-volume, high-rate data streams that are generated by geographically distributed sources and processed concurrently and asynchronously by one or more processing components (*e.g.,* filtering operations aggregation operators, or top-K querying) to perform various tasks such as IP network traffic monitoring and analysis for detecting DoS attacks, location tracking, text mining, financial data analysis, multimedia delivery, and outlier detection in sensor networks [3,1]. More recently, the *cloud computing* model promotes the development of an infrastructure comprising large groups of servers that enables the sharing of computational, storage and network resources rather than having dedicated servers and personal commodity machines to run the applications.

X. Zhang and D. Qiao (Eds.): QShine 2010, LNICST 74, pp. 265–281, 2011.

Such an infrastructure has important flexibility, scalability and economical advantages. This model has been applied successfully by a number of companies such as Amazon, IBM and Google [15] [16].

In such an environment, burst management becomes an important challenge to provide real-time delivery of the streaming data while dealing with the sharing of the computing resources and the unpredictable increase in the volume of data. Data streams that arrive in bursts may create overloads at the processing and networking resources, causing losses of critical data and severely affecting the application performance. The problem is challenging because of the sharing of the same computing and network resources by multiple competing applications, combined with the short duration and unpredictability of the occurrence of the bursts. In [5] the authors report, that, to guarantee the application response time for an increase from 90% to 100% of bursts in storage systems, the corresponding resource capacity needs to increase from 4 to 7 times. Even a small burst increase from 99.9% to 100%, which is only accounted for the last 0.1% of the bursts, the capacity needs to increase by a factor of 2.4 times. As a result, attempts to accommodate 100% of bursts is very expensive and sometimes inpractical.

We believe that careful scheduling of the execution of the data streams on the system resources can maximize the probability that the timing requirements of the applications are met and furthermore reduce resource costs. Scheduling policies that decide the processing ordering of the data streams on the system resources where bursts occur, can greatly assist in compensating for delays. However, scheduling distributed streaming applications brings additional challenges. The scheduling algorithm must be distributed since the streaming applications invoke components concurrently and asynchronously on multiple nodes and the occurrence of a burst is not confined into a single node, rather it affects multiple nodes running components of the streaming applications. Furthermore, the scheduling algorithm should adapt dynamically to changes in the application behavior and maintain accurate resource measurements, since the burst produced by one application can cause queuing delays experienced in the execution of other distributed streaming applications in the system.

In this paper we address the problem of scheduling distributed data streams to efficiently accommodate bursts in distributed stream processing systems. Our main idea is to dynamically control the execution of the applications and delay the processing of bursty data streams when resource capacities are exceeded. Our approach is as follows: we implement a primary and a secondary queue at the Scheduler component of each node, to store data streams pending execution. Requests in the primary queue are scheduled to meet their timing requirements while the ones in the secondary queue are served with best effort. When the rate of the incoming requests exceeds a node's processing capacity and the Scheduler's primary queue is full, requests of the excess rate, instead of being dropped, are diverted to the secondary queue. When there is space in the primary queue due to temporarily reduced rates, requests from the secondary queue are brought back to the primary queue and scheduled for execution. The advantage of our approach, is that, by selectively delaying the execution of the bursty applications,

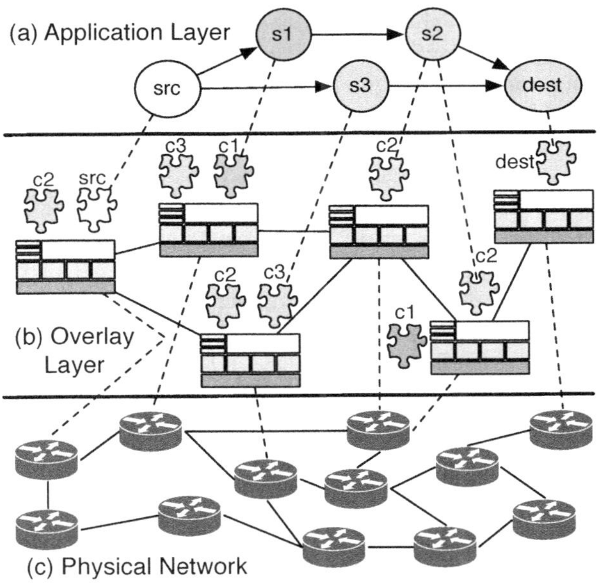

Fig. 1. The architecture of our distributed stream processing system

we can effectively reduce the effects of the bursts. With this mechanism, the effect of the bursts is confined only to the bursty applications while missed application deadlines are reduced to a minimum. Furthermore, by considering queuing delays when scheduling the requests on the system resources, applications that are projected to miss their deadlines are dropped early on. Detailed experimental results of our proposed mechanism over our distributed stream processing system Synergy [6] demonstrate the efficiency and benefits of our approach.

2 System Model and Problem Formulation

2.1 Scheduling Model

Each application app_q is represented as an acyclic graph that consists of a sequence of the services to be invoked. The services are distributed across multiple peers in the system; the distributed stream processing application is executed collaboratively by the peers of the system that host the corresponding services. The execution of an application in the system is triggered by a user request; this will trigger the instantiation of all the services that comprise the application. The instantiation of a service on a node is called *component*. A component operates on collections of tuples generated by a data source, called *data units* (or application data units - ADUs). Examples of data units are sequences of picture or audio frames (for example, in a multimedia application), or sets of measured values (for example, in a sensor data analysis application). The size of a data unit depends on the application. Upon the issue of a user request, our system discovers and instantiates the appropriate components on the system nodes to perform the processing required by the application.

The streaming applications are aperiodic and their arrival times is not known a priori. Each application app_q is characterized by the following parameters: (i) Deadline D_q is the time interval, starting from the time the user submits the request, within which the application must complete execution. (ii) Rate r_q represents the delivery rate of the data units of the application as requested by the user. (iii) Projected Execution Time $Proj\_Exec_q$ is the estimated amount of time required for the distributed streaming application to complete. The projected execution time of the application includes the processing times of the components comprising the application, the queuing times at the Schedulers' queues and the corresponding communication times. The difference between the $Deadline_q$ and the $Proj\_Exec_q$ is called the $Laxity_q$ of the application and represents a measure of urgency of the application. Upon reception of a data unit at a node, the data unit is inserted into the Scheduler's queue waiting to be processed. The order imposed on the data units on the Scheduler's queue depend on the scheduling algorithm implemented in the system.

Each component is characterized by the following parameters: (i) Processing time τ_{c_i} is the processing time required for component c_i to execute locally on the node and (ii) σ_{c_i} is the queuing time on the Scheduler's queue. We denote as (iii) r_{c_i} the incoming rate of the component c_i and (iv) sel^{c_i} be the selectivity of component c_i; this represents the ratio of output rate to input rate for the component. (v) Finally, we let $u_j^{c_i}$ represent the resource requirements of component c_i for resource j (where resource j can represent CPU, memory or bandwidth). Note that the resource requirements and component selectivity characteristics can be provided by the user prior to application execution or acquired through profiling at run-time with low-overhead.

We have implemented our approach in Synergy, our peer-to-peer stream processing system [6]. Synergy is middleware whose goal is to support the execution of distributed stream processing applications with QoS requirements. Each node in Synergy consists of the following modules: (i) a discovery module, that runs over the Pastry DHT, which is responsible for discovering components at run-time with low-overhead, (ii) a composition module is responsible for running a component composition algorithm to dynamically select components to configure and instantiate distributed stream processing applications, (iii) a monitoring module that is responsible for maintaining current resource availability, and, (iv) a routing module that routes data streams between Synergy nodes. We have extended Synergy with a scheduling component that implements our scheduling approach (described in this paper). Synergy's architecture is shown in Figure 1.

2.2 Problem Formulation

The objective of our Scheduling approach is as follows: Given a number of Q applications submitted into the system, our goal is to minimize the number of applications that miss their deadlines. We present our problem formulation below:

The system must satisfy the following conditions: node and link capacity constraints, flow conservation constraint, and deadline constraint.

To satisfy the node and link capacity constraints, the sum of the utilization $u_j^{c_i}$ of resource j $(1 \leq j \leq J)$ made by all components c_i in a node n must not exceed the total resource availability U_j^n of resource j on that node as follows:

$$\forall n \in \mathcal{N}, \sum_{c_i \in n} r_{c_i} \cdot u_j^{c_i} \leq U_j^n, 1 \leq j \leq J \tag{1}$$

where r_{c_i} and $u_j^{c_i}$ are the assigned rate and unit resource needs for component c_i respectively.

To satisfy the flow conservation constraint, the output rate of a component c_i is computed based on the input rate of the component and the *selectivity* of the component sel^{c_i}. The selectivity of each component depends on the service of the component. Let $\mathcal{O}(c_i)$ be the set of downstream components of component c_i. Thus, the flow conservation constraint is given by:

$$\forall c_i, \sum_{c_j \in \mathcal{O}(c_i)} r_{c_j} = sel^{c_i} \cdot r_{c_i} \tag{2}$$

To satisfy the deadline constraint, the application q must finish execution before its deadline $\mathcal{D}_q$. Let $\mathcal{C}_j^q$ be a subset of components for application q. The execution time of the application $\mathcal{C}_j^q$ is the sum of the processing times τ_{c_i} of each component c_i invoked by the application, the queuing time σ_{c_i} experienced by each component c_i waiting at the Scheduler's queue for other components in the same node to execute, and the corresponding communication times δ_{c_i} between two adjacent components c_i and c_{i+1}. Since the end-to-end execution time of the application must be smaller than its deadline, the deadline constraint for the application q is:

$$\forall q \in \mathcal{Q}\ , \forall \mathcal{C}_j^q \in \mathcal{C}^q,$$

$$\sum_{c_i \in \mathcal{C}_j^q} \tau_{c_i} + \sum_{c_i \in \mathcal{C}_j^q} \delta_{c_i} + \sum_{c_i \in \mathcal{C}_j^q} \sigma_{c_i} \leq \mathcal{D}_q \tag{3}$$

Let r_{c_i} be processing rate of component c_i, then the processing time τ_{c_i} of component c_i is: $\tau_{c_i} = \frac{1}{r_{c_i}}$. Then, the deadline constraint becomes as follows:

$$\forall q \in \mathcal{Q}, \forall \mathcal{C}_j^q \in \mathcal{C}^q,$$

$$\sum_{c_i \in \mathcal{C}_j^q} \left(\frac{1}{r_{c_i}} + \delta_{c_i} + \sigma_{c_i} \right) \leq \mathcal{D}_q \tag{4}$$

Our goal is to find the excess load that exceeds the capacity of the primary scheduling queue and divert this load to a secondary queue, so that we meet the node capacity and also minimize the effect on the currently scheduled data units as follows:

- Find excess load.
- Delay the execution of the excess load. Whatever excess load exists,
- We schedule it later using the least-laxity scheduling algorithm so that we minimize the deadline misses.

Let $r_{c_i}^{Max}$ be the maximum input rate that component c_i can admit without overloading, and $r_{c_i}^{Burst}$ be the burst rate that exceeds component c_i's maximum allowed input rate. Then, $r_{c_i}^{Excess} = r_{c_i}^{Burst} - r_{c_i}^{Max}$ is the excess rate of component c_i. The number of missed data units of a component is directly related to its excess rate. In addition, the excess rate of a component has direct impact on drops of other components in the node. This will be further discussed in section 2.3. Our scheduling algorithm aims to reduce data unit drops by minimizing the effects of excess rate of individual components.

2.3 Laxity-Based Scheduling

In this section, we discuss our Least-Laxity Scheduling (LLS) scheduling technique under bursty input rates. LLS has been successfully employed in distributed real-time systems such as in [7]. In least-laxity scheduling, each data unit is associated with a laxity value that represents a measure of urgency for the data units; these will be ordered in the scheduler's queue based on their laxity values. We compute the laxity value L_q of an application q as the difference between the deadline and the end-to-end projected execution time of the application:

$$L_q = Deadline_q - Proj\_Exec_q \tag{5}$$

The application with the smallest laxity value has the highest priority in the system. The laxity value for each data unit of the application is computed initially at the sources and is adjusted as it gets propagated in the distributed system, based on actual processing and network conditions. The purpose of this ordering of the data units is to allow for compensation for delays that were introduced at previous nodes. Especially in the presence of multiple distributed stream applications, the introduction of a new application may cause existing applications to experience higher delays due to queuing. If a data unit is delayed at the node's queue, its laxity value will diminish and thus its priority will increase. Streaming applications with negative laxity values are estimated to miss their deadlines, thus they are dropped from the queue. This approach allows us to implement a distributed scheduling algorithm that executes across multiple nodes and dynamically adapts to the current load conditions.

In least laxity scheduling, two data units belonging to different streaming applications arriving at the same node and having the same laxity value will be treated the same. When two data units have the same laxity value, their relative order in the queue is their arrival order. We finally note that out-of-order transmission of data units of the same application may potentially happen in the case that a packet of a flow was delayed enough locally at a queue that its laxity value became smaller than the laxity value of an earlier data unit from the same application. Data units with smaller laxity values are still treated as more urgent. When data units of the same flow arrive at the destination, they can be post-ordered based on their id-numbers.

3 Our Two-Queue Scheduling Approach

In this section we present the operation of our approach. Our solution accommodates bursts by dividing the data units to be processed into two separate queues: A *primary* queue Q_1 and a *secondary* queue Q_2. Data units in Q_1 are scheduled based on their timing requirements. The ones in Q_2 are only guaranteed best-effort service: Data units in Q_2 will only be serviced when there are resources available on the node. This way we avoid over-penalizing the rest of the data units that are already scheduled for execution: Should there be enough idle time after the burst, they will also be processed.

The primary queue Q_1 is characterized by its predefined maximum size, $\max(Q_1)$. Data units in Q_1 are scheduled to be executed by their deadline. On the contrary, data units are put in Q_2 when the system is considered to be overloaded (Q_1 is full). Data units in Q_2 are executed in a best-effort manner. This means that they will be processed on the earliest slack of the system. $\max(Q_1)$ is defined based on the average processing time and the average laxity of the the data units on the node. More formally:

$$\max(Q_1) = \frac{1}{|C_n|} \sum_{i \in n} \frac{1}{r_{c_i}} \cdot \frac{1}{\tau_{c_i}}$$

where C_n is the set of components running on node n and τ_{c_i} is the average processing time of component c_i. The computation of $\max(Q_1)$ is based on the same logic as in [5]. The average input rate r_{c_i} of each component c_i is constantly monitored during system operation. The average processing time τ_{c_i} for a component c_i is extracted through profiling.

Under stable conditions, when there is no burst, the combined rates of all streaming applications do not exceed the system capacity. In this case, the ADUs are processed and propagated only through primary queues from source to the destination. No ADUs enter the secondary queue.

When a burst occurs, this means that the combined incoming rates of all streaming applications exceeds system capacity. As a result, the primary queues of overloaded nodes have become full. The excess load is then diverted to the secondary queues and kept there in least laxity order. If at any time during execution the laxity values become negative, the corresponding ADUs are dropped.

This way, incoming requests with excessive rates will be kept at the secondary queue Q_2 to wait for a chance to run while their deadlines are not yet missed. The primary queue Q_1 is never empty as long as the secondary still has requests. This is, because, requests in the secondary queue Q_2 are moved back to the primary queue as soon as one or more slots are available. New incoming requests still have a chance to be directed to the primary queue.

ADUs in the primary queue have a higher probability of meeting their response time requirements. Therefore, they have lesser risk of deadline miss. Scheduling for this queue would be chosen to best support the secondary queue. The secondary queue however always uses LLS, since data units are more likely to miss their deadlines. After the burst occurs, the data units stored in Q_2 are pushed back to Q_1 and processed in a timely manner.

4 Performance Evaluation

4.1 Experimental Setup

We have implemented our approach as a scheduler component in our Synergy distributed stream processing system [6] and evaluated its performance. Synergy runs on top of the FreePastry library [8], an open source implementation of the Pastry DHT which is used for component discovery and collecting statistics. Our system is deployed on a 10/100 LAN network of Debian Linux 2.6.20 workstations consisting of Intel Pentium 4 2.66GHz and Intel Xeon 3.06GHz processors, whose main memory varied from 1GB to 2GB of RAM. Our system is written in Java and was developed with Eclipse using Java 1.6.0. We used the timing function provided by the JVM 1.6.0 with time granularity of $1msec$, which is adequate for our experiments. We run a series of experiments to evaluate the performance and demonstrate the working of our approach.

4.2 Experimental Results

In the first set of experiments our goal was to evaluate the performance of our approach under different burst intensities. To model the fluctuating nature of bursts, we designed a burst pattern that has two bursts with a period of normal rate after each burst. Our experimental system employs 6 unique services instantiated as unique components on the processing nodes. Each component is replicated and available at multiple nodes. Each experiment instantiates concurrently 11 applications. Each application submits a unique service request structured as a DAG; each service request invokes 4 to 6 service components which are located on different nodes. To generate the service request, the source component of an application dispatches a stream of data units which is 250 bytes long. The normal rate required by each service request ranges from 8 Kpbs to 45 Kbps, which is, 80% of the maximum request service rate, as can be obtained through profiling. We run a series of experiments with different burst intensities from no bursts 0% (normal rate) to 100% burst. Results are averaged over 5 runs with 90% confident interval where possible.

To demonstrate the working and benefits of our approach we have implemented two additional scheduling approaches as follows: (a) **First Come First Served (FCFS)**: orders the data units on the scheduler's queue based on the order in which they arrive at the node, (b) **Earliest Deadline First (EDF)**:

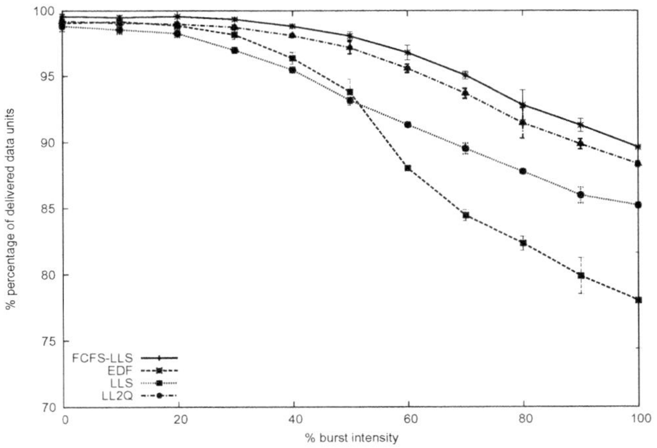

Fig. 2. Percentage of delivered data units on time, under various burst intensities

uses the deadline of the data units to decide the ordering; the data unit with the smallest deadline is ordered first. To demonstrate the advantages of our two-queue approach, we have compared the following strategies: **FCFS-LLS** uses the FCFS scheduler for the primary queue and the LLS scheduler for the secondary queue, while the **LLS-LLS (LL2Q)** strategy uses the LLS scheduler for both queues. We have also compared our two-queue scheduling approach with an approach that uses a single queue for scheduling. In particular, the **LLS** approach uses a single queue where the data units are ordered based on the least laxity scheduling algorithm, while the **EDF** scheduling approach uses a single queue where the data units are ordered based on the EDF scheduling.

On-time Delivery: In this experiment we measure the percentage of data units that are delivered on time as a function of burst intensity. Our approach that employs two-queues for scheduling leads to over 90% of data units that are delivered, as shown in figure 2. As the figure indicates, especially the FCFS-LLS scheduling has the highest percentage. An interesting observation that we notice is that the two-queue approach LLS-LLS performs better than single queue LLS. The reason for this is that the single queue LLS experiences a large number of context switching by which data units are frequently switched in the queue for the incoming data units which have smaller laxities. The longer the single queue, the worse the problem is, and as a result this significantly affects the system performance. In the two-queue approach, the problem is less since the primary queue's size is smaller, just enough to keep requests that their response time are guaranteed. When FCFS is used as the primary queue's order, there is no problem of context switching.

Dropped Data Units: The percentage of dropped data units as the results of deadline missing is shown in figure 3. As the figure indicates, both two-queue scheduling approaches perform better than any single-queue approach, especially with higher burst intensities.

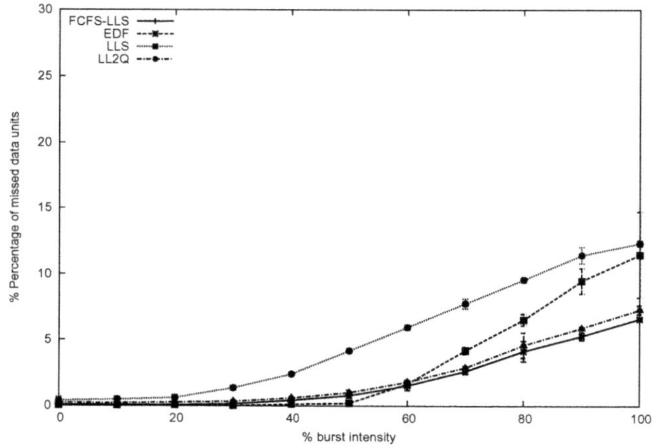

Fig. 3. Percentage of data units with missed deadlines, under various burst intensities

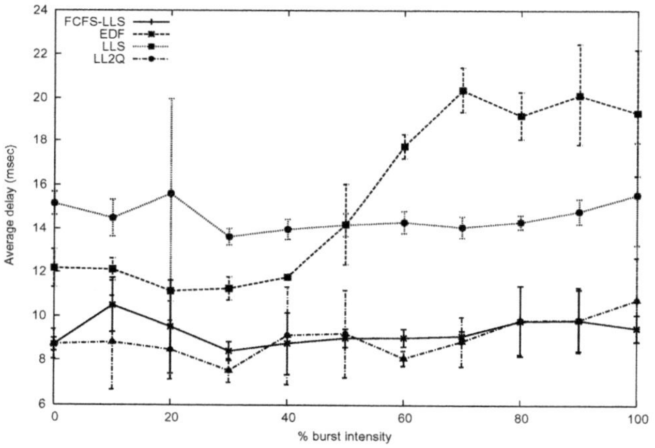

Fig. 4. Average end-to-end delay, in msec, under various burst intensities

Average End-to-End Delay: In the next experiment we measure the end-to-end delay experienced by the data units in the system. Figure 4 shows the average end-to-end delay of data units with different scheduling approaches. Both two-queue scheduling approaches decrease the average delay and are not affected by the burst intensity. With LLS the data units with the least laxity are given top priority. If they are deliverable, it means that their laxities are still positive and their delays are small accordingly. The two-queue approach with LLS even enhances the delay. Since data units are in the primary queue, with small size and given a top priority, will not wait long for processing. When they are processed, their delays are still small. This is one of interesting features of the two-queue approach.

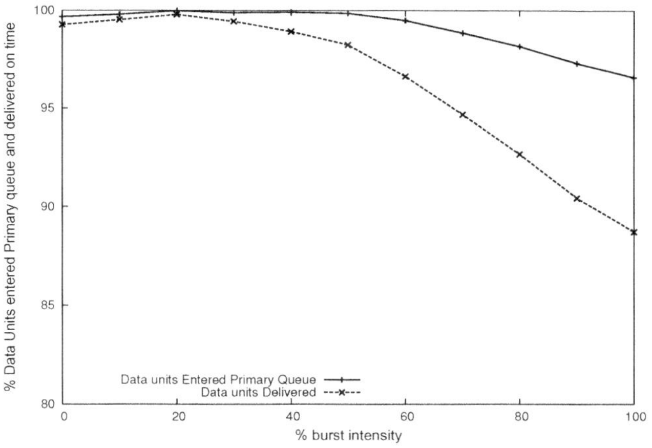

Fig. 5. Percentage of data units inserted and delivered on-time in the primary queue

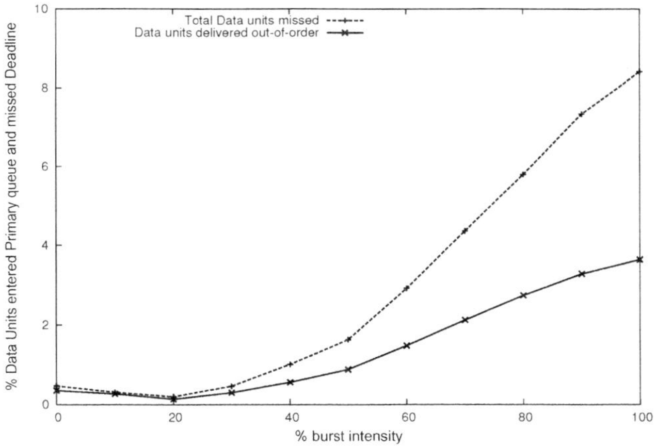

Fig. 6. Percentage of data units of the primary queue with missed deadlines and the corresponding percentage of data units delivered out of order

Performance of the two queues approach: In this set of experiments we illustrate the working of our approach by showing separately the behavior of each of the queues. Figure 5 shows the percentage of the data units of the primary queue that enter the primary queue and those that are delivered on-time. Note that there is some percentage of data units of the primary queue that are not delivered on time. As the burst intensity increases the percentage of data unit delivery drops. There are two reasons for this: As figure 6 shows, the out-of-order delivery accounts for about 50% of the misses. Furthermore, some data units that enter the primary queue and are then delivered, they could come from the secondary queues in the upstream nodes. The problem of out-of-order delivery could be easily mitigated by post-ordering at the destination based on

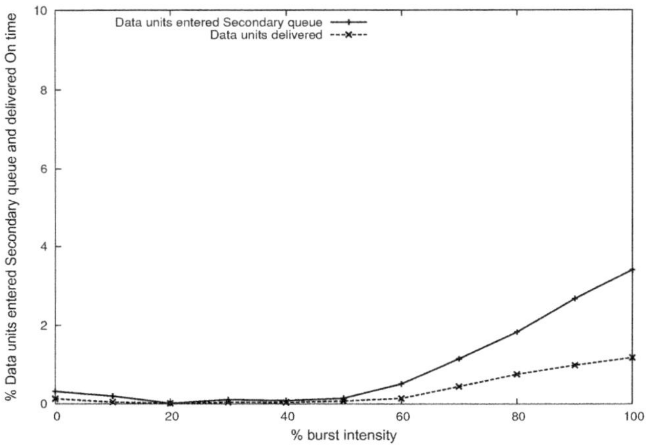

Fig. 7. Percentage of data units inserted and delivered on-time in the secondary queue

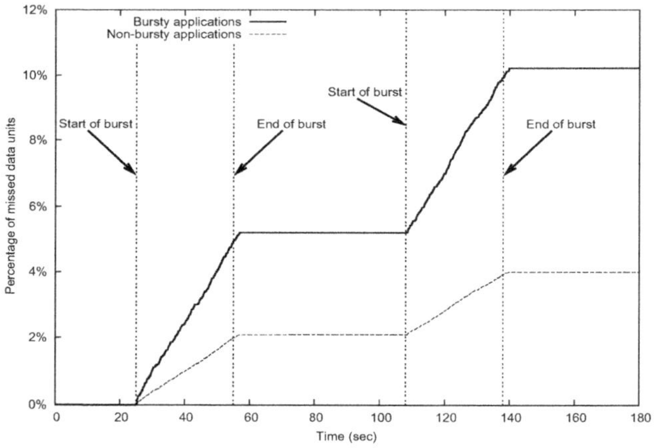

Fig. 8. EDF: Percentage of missed data units for bursty and non-bursty applications

their ID-numbers. Figure 7 shows about 4% of workload are diverted to the secondary queue, and half of this load are on-time deliverable.

Mitigating effects of bursty applications: In the next set of experiments we explored how the approach handles a mix of bursty and non-bursty applications. Using the same double peak burst pattern as in the first experiment, however, 9 applications, accounted for 70% of total workloads, have 100% burst intensity while the remaining applications, or 30% of total workloads have no bursts. This set of experiments demonstrates how our system handles both bursty and non-bursty applications.

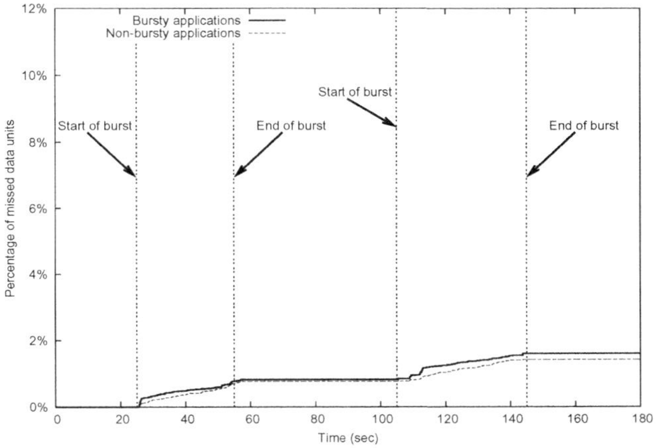

Fig. 9. FCFS-LLS: Percentage of missed data units for bursty and non-bursty applications

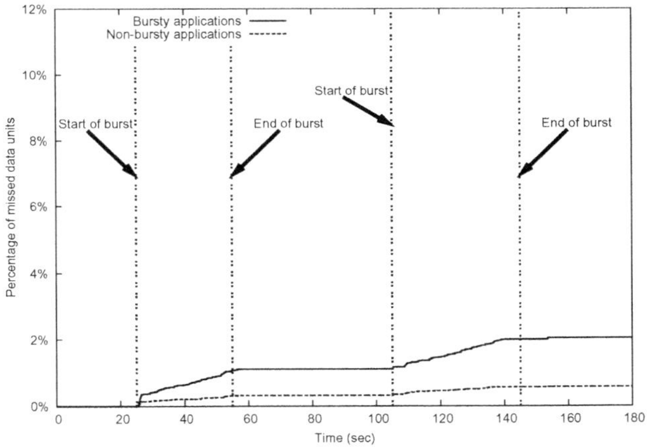

Fig. 10. LL2Q: Percentage of missed data units for bursty and non-bursty applications

Figures 8, 9, 10, and 11 show the percentage of data unit misses of bursty and non-bursty applications on different scheduling approaches as a function of time. Figure 8 and 11 show non-bursty applications with single queue scheduling, both EDF and LSS, are affected by the burst. This results into an increased number of data unit misses as soon as the bursts occurs. Figure 9 and 10, on the other hand, show data unit misses are low with FCFS-LLS and LL2Q (LLS-LLS) scheduling. With an appropriate size, the primary queue does not give bursty applications the chance to greedily occupy the processing queue due to their high rate. If the queue is large, as in a single queue scheduling, the situation will get worse with bursts. Data units from bursts application will overwhelm the queue and

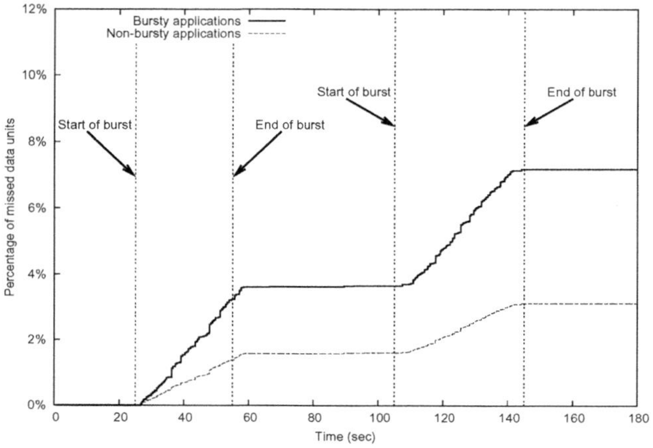

Fig. 11. LLS: Percentage of missed data units for bursty and non-bursty applications

data units of none-burst application will be affected by a high miss rate together with burst application since the system is unable to handle. This observation is significant for critical application that need to maintain a required throughput and not affected by bursts.

5 Related Work

Distributed stream processing systems have become increasingly popular in recent years for the development of applications that are characterized by high-rate, large-volume data streams. Examples include Aurora [2], STREAM [3], TelegraphCQ [1] and Cougar [4].

We have developed Synergy, a distributed stream processing system in [6] and have investigated various challenges related to dynamic rate allocation in [17], decentralized media streaming and transcoding [10]. Following are recent works (including ours) that propose solutions to accommodate bursts.

In [11], the authors propose a two-tie distributed control algorithm with the goal of maximizing the entire system weighed throughput, achieve low end-to-end latency, and stabilize the system coping with bursty workload. In the first tie, Lagrange multipliers are used to maximize the resource utilization. In the second tie, CPU and flow control algorithms are used to adjust input rates with feedback from downstream components. With this mechanism, any exceed input rate will be dropped. In our approach, in contrast, the exceed input rates instead of being dropped, are diverted to a secondary queue and re-scheduled to be processed later with best effort. To avoid context switching overhead and decrease memory cache miss, in [11] a batch scheduling is employed to process several data units at a time, however, this scheduling strategy does not address bursts.

In [12], the authors propose centralized and distributed load shredding approach to address bursts in distributed stream processing systems. A number of shredding plans are generated in advance for certain load conditions, so that the system can react to overload fast and in a lightweight manner. Under this approach, any input rates that exceed the maximum allowed rate in pre-configured shredding plans will be dropped. Our approach, with two-queue scheduling, is able to claim back data units that would otherwise be dropped, like in this approach.

In [13], the authors propose a multi-parametric programming approach to maximize the system utility in response to changing workloads. The approach consists of an off-line and an on-line version. The off-line version transforms the utility optimization problem into a linear function of CPU utilization, while the on-line version produces optimal solutions based on workload variation in polynomial time. This approach also shreds input rates that exceed the rate establised by their solution.

Our previous work on burst management [14] presents a solution that accommodates bursts by re-distributing the load among the processing nodes of the system. Furthermore, our previous work does not differentiate between the applications, and treats all of them the same. In this paper we present a totally different approach to accommodate bursts using a two-queue scheduling approach. In particular, this paper focuses on maximizing the potential of each individual application (and of the system overall). The method we present considers the urgencies of the applications and their resource requirements to decide the scheduling ordering of the data streams on the system resources. This maximizes the probability that the deadlines of the most urgent applications (and thus most important applications) will be met.

In [5], the authors propose two queues to handle bursty workload for storage systems. Our work, on the other hand, studies the effect of employing a secondary queue for distributed stream processing systems which, to the best of our knowledge, has not been proposed before. This is a much more difficult problem because: (1) stream processing applications are distributed and thus the occurrence of a burst is not confined in one node, rather it can affect remote nodes that run components of the same applications, and (2) distributed stream processing applications have end-to-end deadlines which are affected by all nodes running components of the applications. Our approach aims to meet the end-to-end deadlines of the applications. Furthermore, in order to deal with the unique requirements of the distributed stream processing systems, we have made additional contributions: we have implemented several different scheduling algorithms and we have studied the behavior of the two-queue structure under these different scheduling strategies. In particular, we have implemented and compared the following scheduling algorithms: least laxity scheduling, earliest deadline first and first come first served.

6 Conclusions

In this paper, we have investigated the problem of accommodating unpredicted data bursts in streaming applications deployed over cloud computing infrastructures. Our approach takes into account the fluctuation nature of bursts and their effects on existing applications, and employs two scheduling queues with different scheduling policies. When bursts occur, any excess workload is diverted to a secondary queue, where requests can be processed whenever the node has any unused processing power. Our experimental results on our Synergy distributed stream processing system show that our two-queue approach efficiently reduces data unit drops, improves end-to-end delay compared to other popular single queue approaches, and significantly mitigates the effects of bursts on non-bursty applications.

Acknowledgements. This research has been supported by the European Union through the Marie-Curie RTD (IRG-231038) Project and by AUEB through a PEVE2 Project.

References

1. Chandrasekaran, S., Cooper, O., Deshpande, A., Franklin, M.J., Hellerstein, J.M., Hong, W., Krishnamurthy, S., Madden, S., Raman, V., Reiss, F., Shah, M.: TelegraphCQ: Continuous Dataflow Processing for an Uncertain World. In: CIDR, Asilomar, CA (January 2003)
2. Tatbul, N., Çetintemel, U., Zdonik, S.B., Cherniack, M., Stonebraker, M.: Load Shedding in a Data Stream Manager. In: VLDB 2003, Berlin, Germany, pp. 309–320 (2003)
3. Arasu, A., Babcock, B., Babu, S., Cieslewicz, J., Datar, M., Ito, K., Motwani, R., Srivastava, U., Widom, J.: STREAM: The Stanford Data Stream Management System (March 2005)
4. Madden, S., Gehrke, J.: Query Processing in Sensor Networks. IEEE Pervasive Computing, vol 3(1) (March 2004)
5. Lu, L., Varman, P., Doshi, K.: Graduated QoS by Decomposing Bursts: Don't Let the Tail Wag Your Server. In: ICDCS 2009, Montreal, QC, Canada, pp. 12–21 (June 2009)
6. Repantis, T., Gu, X., Kalogeraki, V.: Synergy: Sharing-Aware Component Composition for Distributed Stream Processing Systems. In: van Steen, M., Henning, M. (eds.) Middleware 2006. LNCS, vol. 4290, pp. 322–341. Springer, Heidelberg (2006)
7. Kalogeraki, V., Melliar-Smith, P.M., Moser, L.E.: Dynamic Scheduling of Distributed Method Invocations. In: IEEE Real-Time Systems Symposium (RTSS), Orlando, FL (December 2000)
8. FreePastry (2006), `http://freepastry.org/FreePastry`
9. Drougas, Y., Kalogeraki, V.: RASC: Dynamic Rate Allocation for Distributed Stream Processing Applications. In: International Parallel and Distributed Processing Symposium (IPDPS), Long Beach, CA (March 2007)
10. Chen, F., Kalogeraki, V.: RUBEN: A Technique for Scheduling Multimedia Applications in Overlay Networks. In: Globecom 2004, Dalas, TX (November 2004)

11. Amini, L., Jain, N., Sehgal, A., Silber, J., Verscheure, O.: Adaptive Control of Extreme-scale Stream Processing Systems. In: ICDCS 2006, Lisboa, Portugal (2006)
12. Tatbul, N., Çetintemel, U., Zdonik, S.: Staying FIT: Efficient Load Shedding Techniques for Distributed Stream Processing. In: VLDB 2007, Vienna, Austria, pp. 159–170 (September 2007)
13. Chen, Y., Lu, C., Koutsoukos, X.: Optimal Discrete Rate Adaptation for Distributed Real-Time Systems. In: Real Time Systems Symposium (RTSS), Tucson, AZ (December 2007)
14. Drougas, Y., Kalogeraki, V.: Accommodating Bursts in Distributed Stream Processing Systems. In: 23rd International Parallel and Distributed Processing Symposium (IPDPS), Rome, Italy (May 2009)
15. Amazon Elastic Computer Cloud (Amazon EC2), `http://aws.amazon.com/ec2/`
16. IBM Cloud Computing, `http://www.ibm.com/ibm/cloud/`
17. Drougas, Y., Kalogeraki, V.: RASC: Dynamic Rate Allocation for Distributed Stream Processing Applications. In: International Parallel and Distributed Processing Symposium (IPDPS), Long Beach, CA (March 2007)

Deterministic Algorithm for Coded Cooperative Data Exchange*

Alex Sprintson[1], Parastoo Sadeghi[2], Graham Booker[1], and Salim El Rouayheb[3]

[1] Texas A&M University, College Station, Texas, USA
{spalex,gbooker}@tamu.edu
[2] Australian National University, ACT, Canberra, Australia
parastoo.sadeghi@anu.edu.au
[3] University of California at Berkeley, Berkeley, California USA
salim@eecs.berkeley.edu

Abstract. We consider the problem of cooperative data exchange in a group of wireless clients. In this problem each client initially holds a subset of packets and needs to obtain all packets held by other clients. Each client can broadcast its own packets or a combinations thereof to other clients via an error-free broadcast channel. Assuming that clients know which packets are available to other clients, our goal is to minimize the total number of transmissions needed to satisfy the demands of all clients. We present a deterministic algorithm that computes an optimal solution to this problem in polynomial time.

1 Introduction

In this paper, we consider the problem of cooperative data exchange between a group of wireless clients that share a common lossless broadcast channel. In this problem, a set of n packets $X = \{x_1, \ldots, x_n\}$ needs to be delivered to k clients. Each client initially holds a subset X_i of packets in X and needs to obtain all packets held by other clients. Our goal is to design a communication scheme that enables all clients to obtain all packets with the minimum number of transmissions.

For example, consider the instance of the problem shown in Fig. 1(a) where there are three wireless clients that need to obtain three packets $x_1, x_2, x_3 \in \mathrm{GF}(2^m)$. Initially, the clients hold packets $\{x_2, x_3\}$, $\{x_1, x_3\}$ and $\{x_1, x_2\}$, respectively, i.e., each client is missing one packet. A simple cooperative scheme consists of three uncoded transmissions. However, this is not an optimal solution since the clients can send coded packets which satisfy demands of multiple clients. The number of transmissions for this example can be decreased to two by letting the first client broadcast $x_2 + x_3$ and the second client broadcast x_1 (see Fig. 1(b)).

In this paper, we present an algorithm that finds, in polynomial time, the optimal solution for the cooperative data exchange problem. In particular, the algorithm finds an encoding scheme that achieves the minimum number of transmissions over a small finite field.

* The work of Alex Sprintson was supported by NSF grant CNS-0954153. The work of Parastoo Sadeghi was supported under Australian Research Councils Discovery Projects funding scheme (project no. DP0984950).

X. Zhang and D. Qiao (Eds.): QShine 2010, LNICST 74, pp. 282–289, 2011.

Fig. 1. Coded data exchange among three clients

Related Work

Cooperative communication at the physical, network and application layers has been the subject of extensive research in the past few years. Physical-layer user cooperation in the form of signal relaying has been shown to result in higher data rates, extended coverage, and robustness to link outages [9, 13, 15]. Network coding [1, 12] is another powerful technique that has been proposed to enhance achievable data rates, as well as other aspects, through packet-level encoding at intermediate nodes [11].

Direct information exchange problems were recently considered in [6] and [7] where it was assumed that the packets available at different network nodes follow a certain random distribution.

A closely-related coding problem with "subset" side information is the *Index Coding* problem [2, 3, 4, 5, 8] which was originally motivated by satellite broadcast applications with cashing clients. However, the Index Coding setup is centralized and non-cooperative with a single transmitter server holding all packets and passive clients having different demands. A related problem of set reconciliation between two or more similar sets was studied in [14].

When the clients have only information about certain neighboring nodes, and can communicate to a restricted number of them, our setting is related to that of gossip algorithms studied in the literature (see e.g., [16]).

In our previous work [17] we have presented an efficient randomized algorithm for the distributed data exchange problem and established several bounds on the minimum number of transmissions. In this paper, we extend our results by presenting an efficient deterministic algorithm for this problem.

2 Model

Consider a set of n packets $X = \{x_1, \ldots, x_n\}$ to be delivered to k clients belonging to the set $C = \{c_1, \ldots, c_k\}$. The packets are elements of a finite alphabet which will be assumed to be a finite field $\mathbb{F}_q$ throughout this paper. At the beginning, each client knows a subset of the packets denoted by $X_i \subseteq X$, while the clients collectively know all the packets in X, i.e., $\cup_{c_i \in C} X_i = X$. We denote by $\overline{X}_i = X \setminus X_i$ the set of packets missing at client c_i. We refer to X_i as the *has* set of client c_i and to $\overline{X}_i$ as its *wants* set. We assume that each client knows the indices of packets that are available to other clients.[1] Without loss of generality, we assume that each packet in X is needed by at least one client.

The clients exchange packets over a lossless broadcast channel with the purpose of making all packets in X available to all clients. The data is transferred in communication rounds, such that at round i one of the clients, say c_{t_i}, broadcasts a packet $p_i \in \mathbb{F}_q$ to the rest of the clients in C. Packet p_i may be one of the packets in X_{t_i}, or a combination of packets in X_{t_i} and the packets $\{p_1, \ldots, p_{i-1}\}$ previously transmitted over the channel. Our goal is to devise a scheme that enables each client $c_i \in C$ to obtain all packets in $\overline{X}_i$ while minimizing the total number of transmissions. We refer to the minimum number of transmissions required to satisfy all clients as OPT.

Our scheme uses linear coding over the field $\mathbb{F}_q$. As discussed in [17], linear codes are sufficient to achieve the minimum number of transmission in our problem. With linear coding, any packet p_i transmitted by the algorithm is a linear combination of the original packets in X, i.e.,

$$p_i = \sum_{x_j \in X} \gamma_i^j x_j,$$

where $\gamma_i^j \in \mathbb{F}_q$ are the *encoding coefficients* of p_i. We refer to the vector $\gamma_i = [\gamma_i^1, \gamma_i^2, \ldots, \gamma_i^n]$ as the *encoding vector* of p_i. The i-th *unit* encoding vector that corresponds to the original packet x_i is denoted by $u_i = [u_i^1, u_i^2, \ldots, u_i^n]$, where $u_i^i = 1$ and $u_i^j = 0$ for $i \neq j$. We also denote by U_i the set of unit vectors that correspond to packets in X_i.

A client c_i is said to have a *unique* packet x_j if $x_j \in X_i$ and $x_j \notin X_\ell$ for all $\ell \neq i$. A unique packet can be broadcast by the client holding it without any penalty in terms of optimality. Thus, without loss of generality, we assume that there are no unique packets in the system. Also, without loss of generality, we assume that all k clients initially have distinct packet sets.

3 Deterministic Algorithm

In this section, we present a deterministic algorithm for the data exchange problem. For clarity, we describe and analyze the behavior of the algorithm in terms of encoding vectors, rather than the original packets. That is, instead of saying that a packet $p_i = \sum_{x_j \in X} \gamma_i^j x_j$ has been transmitted, we say that we transmit the corresponding encoding vector $\gamma_i = [\gamma_i^1, \gamma_i^2, \ldots, \gamma_i^n]$.

3.1 Algorithm Description

Our algorithm operates over a finite field $\mathbb{F}_q$. The size q of $\mathbb{F}_q$ must be larger than $2k$, where k is the number of clients. For a client $c_j \in C$ we define by $\Gamma(c_j) = \mathtt{span}(U_j)$, i.e., $\Gamma(c_j)$ is the set of all possible encoding vectors in $\mathbb{F}_q^n$ that can be generated by client c_j. Then, each vector $\gamma_i \in \Gamma(c_j)$ can be written as

$$\gamma_i = \sum_{u_g \in U_j} \gamma_i^g u_g,$$

[1] This can be achieved by exchanging packet indices at the beginning of data exchange. The indices can also be piggybacked on the data packets to reduce overhead.

where each γ_i^g is an element of $\mathbb{F}_q$.

The algorithm executes in iterations. In each iteration we identify a client that will be transmitting at that round. The key idea is that at each round i we only determine the client c_{t_i} that will transmit a packet at that round, but not the encoding coefficients of the packet that will be transmitted. The encoding coefficients of each transmitted packets will be determined at the last stage of the algorithm.

More specifically, for each client $c_j \in C$ we maintain a counter b_j that specifies the number of the packets that will be transmitted by that client. Initially, $b_j = 0$ for all $c_j \in C$. Once we have determined that client c_{t_i} is transmitting a packet a round i, we increment the corresponding counter b_{t_i}. We denote by $B_i = (b_1, b_2, \ldots, b_k)$ the vector that specifies the number of transmissions made by each client $c_j \in C$ at iteration i. We refer to B_i as a *counting vector*.

Definition 1. *We say that a set of vectors Γ fits $B_i = (b_1, b_2, \ldots, b_k)$ if Γ can be partitioned into k disjoint subsets $\Gamma^1, \Gamma^2, \ldots, \Gamma^k$, such that for each $\Gamma^j, 1 \leq j \leq k$ it holds that:*

1. $|\Gamma^j| = b_j$;
2. $\Gamma^j \subseteq \Gamma(c_j)$.

We also denote by $\mathcal{M}(B_i)$ the collection of all sets of vectors that fit B_i.

That is, the set Γ that fits B_i is a union of b_1 vectors from $\Gamma(c_1)$, b_2 vectors from $\Gamma(c_2)$, $\ldots$, and b_k vectors from $\Gamma(c_k)$.

Definition 2. *Let B_i be a counting vector and let U_j be the set of encoding vectors available to client j. Let $\mathcal{M}(B_i)$ the collection of all sets of encoding vectors that fit B_i. Then, we define* $\texttt{Maxrank}(B_i, U_j)$ *as follows:*

$$\texttt{Maxrank}(B_i, U_j) = \max_{\Gamma \in \mathcal{M}(B_i)} \text{rank}(\Gamma \cup U_j).$$

For given B_i and U_j, the value of $\texttt{Maxrank}(B_i, U_j)$ can be efficiently computed in polynomial time. First, note, that there exits a set $\Gamma' \in \mathcal{M}(B_i)$ that maximizes the value of $\text{rank}(\Gamma \cup U_j)$ that only contains unit vectors u_g (corresponding to packets in X). Thus, we can compute $\texttt{Maxrank}(B_i, U_j)$ by constructing a bipartite graph $G(V_1, V_2, E)$, as depicted in Fig. 2. The nodes of V_1 correspond to packets in $X = \{x_1, \ldots, x_n\}$. For each client $c_g \in C$, set V_2 contains b_g nodes, each node is connected to all nodes in V_1 that correspond to packets in X_g. In addition, set V_2 contains $|X_j| = |U_j|$ nodes, each node is connected to a corresponding packet in X_j. Note that each node in V_2 corresponds to a linear combination of packets in X, and our goal is to maximize the number of independent linear combinations. It is easy to verify that the value of $\texttt{Maxrank}(B_i, U_j)$ is equal to the maximum size of a matching in $G(V_1, V_2, E)$.

The formal description of the algorithm, referred to as *Deterministic Data Exchange (DDE)*, appears on Fig. 3.

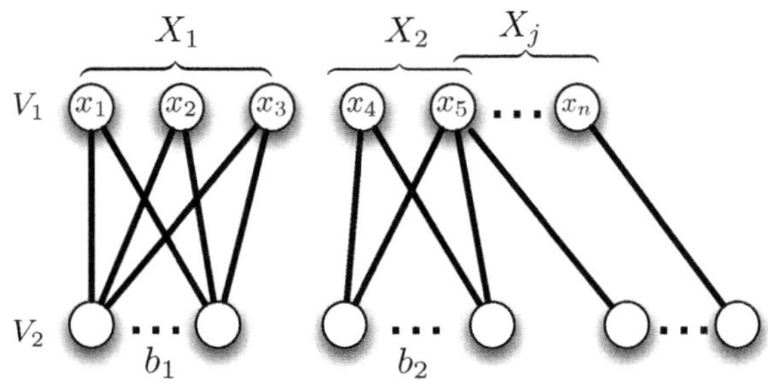

Fig. 2. The auxiliary graph $G = (V_1, V_2, E)$ used for computing the value of $\mathtt{Maxrank}(B_i, U_j)$

Algorithm DDE $(C, \{U_j, c_j \in C, \mathbb{F}_q\})$

***For** each $c_j \in C$ **do**:*
  $b_j \leftarrow 0$
enddo
$B_0 \leftarrow (b_1, b_2, \ldots, b_k)$
$i \leftarrow 1$
***while** there exists a client $c_j \in C$ for which it holds that*
  $\mathtt{Maxrank}(B_{i-1}, U_j) < n$ ***do***
  Let c_{t_i} be a client for which $\mathtt{Maxrank}(B_{i-1}, U_{t_i})$ is maximum.
  $b_{t_i} \leftarrow b_{t_i} + 1$
  $B_i = (b_1, b_2, \ldots, b_k)$
  $i \leftarrow i + 1$
endwhile
$\hat{i} \leftarrow i - 1$
Find a vector set $\hat{\Gamma} \in \mathcal{M}(B_{\hat{i}})$ that satisfies $\mathrm{rank}(\hat{\Gamma} \cup U_j) = n$ *for all $c_j \in C$*
***return** $\hat{i}$ encoding vectors $\gamma_1, \ldots, \gamma_{\hat{i}}$ that correspond to elements in $\hat{\Gamma}$*

Fig. 3. Algorithm DDE

3.2 Algorithm Analysis

We proceed to analyze the correctness of the algorithm. Consider an iteration i of the algorithm. Recall that the vector $B_i = (b_1, b_2, \ldots, b_k)$ specifies the number transmissions made by each client $c_l \in C$ during iterations $1, \ldots, i$. Recall also that the collection $\mathcal{M}(B_i)$ includes all possible sets of encoding vectors that fit B_i.

Let OPT_i be the minimum number of additional rounds (i.e., starting from round $i+1$) necessary to satisfy requests of all clients in C, provided that during iterations $1, \ldots, i$ the number of transmissions made by each client is consistent with B_i. We define OPT_0 as the optimal solution to the problem at hand, i.e., $OPT_0 = OPT$. Also, let $L_i = (l_1, l_2, \ldots, l_k)$ the number of additional transmissions that need to be done by the clients $c_j \in C$ to achieve optimum OPT_i. We refer to L_i as a *forward counting vector* at iteration i. Note that $OPT_i = \sum_{j=1}^{k} l_j$. Note also that the set L_i must satisfy that $\mathtt{Maxrank}(B_i + L_i, U_j) = n$ for each client $c_j \in C$, where $B_i + L_i = (b_1 + l_1, b_2 + l_2, \ldots, b_k + l_k)$.

In order to prove the optimality of the algorithm it is sufficient to show that for each iteration $i = 1, 2, \ldots$ it holds $OPT_i = OPT_{i-1} - 1$.

Theorem 1. *For each iteration i of the algorithm it holds that* $OPT_i = OPT_{i-1} - 1$

Proof: Consider iteration i of the algorithm. Let c_{t_i} be the client selected at that iteration. Let $B_{i-1} = (b_1, b_2, \ldots, b_k)$ be the counting vector and $L_{i-1} = (l_1, l_2, \ldots, l_k)$ be the forward counting vector at iteration $i - 1$.

First, consider the case in which $l_{t_i} > 0$. Let B_i be a vector formed from B_{i-1} by incrementing b_{t_i} by one. Let L_i be a vector formed from L_{i-1} be decrementing l_{t_i} by one. Note that $B_i + L_i = B_{i-1} + L_{i-1}$. Thus, after iteration i we need $OPT_{i-1} - 1$ transmissions to satisfy $\texttt{Maxrank}(B_{\hat{i}}, U_j) = n$ for each client $c_j \in C$. Hence it holds that $OPT_i = OPT_{i-1} - 1$.

Next, suppose that $l_{t_i} = 0$. Note that for each client $c_j \in C$ it holds that $\texttt{Maxrank}(B_{i-1} + L_{i-1}, U_j) = n$. Then, there exist vector set $Q_{i-1} \in \mathcal{M}(L_{i-1})$ and $\Gamma_{i-1} \in \mathcal{M}(B_{i-1})$ that satisfy, for each client $c_j \in C$,

$$\text{rank}(\Gamma_{i-1} \cup Q_{i-1} \cup U_j) = n. \tag{1}$$

Also, the definition of Maxrank implies that for each client $c_j \in C$ there exists a vector set Γ_{i-1} that satisfies

$$\text{rank}(\Gamma_{i-1} \cup U_j) = \texttt{Maxrank}(B_{i-1}, U_j). \tag{2}$$

By using the standard network coding techniques (see e.g., [12]) it can be shown that there exist sets of vectors $\Gamma_{i-1} \in \mathcal{M}(B_{i-1})$ and $Q_{i-1} \in \mathcal{M}(L_{i-1})$ that satisfy the conditions of both Equations (1) and (2) for all clients $c_j \in C$, provided that the field size q is larger than the number of constraints ($2k$).

Since client c_{t_i} has the largest value of $\texttt{Maxrank}(B_{i-1}, U_j)$ among all clients in $c_j \in C$, it holds that

$$n - \texttt{Maxrank}(B_{i-1}, U_{t_i}) = n - \text{rank}(\Gamma_{i-1} \cup U_{t_i}) \le \sum_{j=1}^{k} l_j - 1 = OPT_{i-1} - 1.$$

This implies that there exists at least one vector $v \in Q_{i-1}$ such that the set $\Gamma_{i-1} \cup \{Q_{i-1} \setminus \{v\}\} \cup U_{t_i}$ is of rank n.

Let v be such a vector and let c_{i^*} be a client for which it holds that $v \in \Gamma(c_{i^*})$. We denote by $\tilde{Q}_{i-1} = Q_{i-1} \setminus \{v\}$. Note that for each client $c_j \in C \setminus \{c_{t_i}\}$ it holds the rank of vector set $S_j = \Gamma_{i-1} \cup \tilde{Q}_{i-1} \cup U_j$ is at least $n - 1$. Let C' be as subset of $C \setminus \{c_{t_i}\}$ such that for each $c_j \in C'$ it holds that $\text{rank}(S_j) = n - 1$.

Let c_j be a client in C' and let ζ_j be the normal vector to the span of S_j. Note that ζ_j is non-zero according to the definition of C'. Note that ζ_j can be written as

$$\zeta_j = \sum_{u_g \in U_{t_i}} \beta_g u_g + \sum_{u_g \in \overline{U}_{t_i}} \beta_g u_g,$$

where $\overline{U}_{t_i}$ is the set of unit encoding vectors that correspond to $\overline{X}_{t_i} = X \setminus X_{t_i}$.

Lemma 1. *There exists* $u_g \in U_{t_i}$ *such that* $\beta_g \neq 0$.

Proof: Suppose that it is not the case. Then, ζ_j can be expressed as $\zeta_j = \sum_{u_g \in \overline{U}_{t_i}} \beta_g u_g$. Then, ζ_j is a normal to $\mathtt{span}(U_{t_i})$. Since ζ_j is a normal to $\mathtt{span}(S_j)$ it is also normal to $\mathtt{span}(\Gamma_{i-1} \cup \tilde{Q}_{i-1})$. Thus, ζ_j is a normal to $\mathtt{span}(\Gamma_{i-1} \cup \tilde{Q}_{i-1} \cup U_{t_i})$ which contradicts the fact that $\mathrm{rank}\{\Gamma_{i-1} \cup \tilde{Q}_{i-1} \cup U_{t_i}\} = n$. ■

Let $\hat{\gamma}$ be a projection of ζ_j to $\mathtt{span}(U_{t_i})$, i.e., $\hat{\gamma} = \sum_{u_g \in U_{t_i}} \beta_g u_g$ and let $\Gamma_i = \Gamma_{i-1} \cup \{\hat{\gamma}\}$. Note that Lemma 1 implies that $\langle \hat{\gamma}, \zeta_j \rangle \neq 0$.

Lemma 2. *For each client $c_j \in C'$ it holds that*

$$\mathrm{rank}\{S_j \cup \{\hat{\gamma}\}\} = \mathrm{rank}\{\Gamma_i \cup \tilde{Q}_{i-1} \cup U_j\} = n.$$

Proof: By way of contradiction, suppose that $\mathrm{rank}\{S_j \cup \{\hat{\gamma}\}\} = n - 1$. Then, vector $\hat{\gamma}$ belongs to $\mathtt{span}(S_j)$. However, this contradicts the fact that $\langle \hat{\gamma}, \zeta_j \rangle \neq 0$. ■

Lemma 2 implies that after iteration i the demands of all clients in C' are satisfied by transmitting vectors in $\tilde{Q}_{i-1}$, i.e., with $OPT_{i-1} - 1$ additional transmissions. The same holds for other clients in $C \setminus C'$, because for each client $c_j \in C \setminus C'$ it holds that $\mathrm{rank}\{S_j = \Gamma_{i-1} \cup \tilde{Q}_{i-1} \cup U_j\} = n$. Hence $OPT_i = OPT_{i-1} - 1$ and the lemma follows. ■

Lemma 3. *At Step 11 of Algorithm* DDE *it is possible to find a vector set $\hat{\Gamma} \in \mathcal{M}(B_{\hat{i}})$ that satisfies* $\mathrm{rank}(\hat{\Gamma} \cup U_j) = n$ *for all $c_j \in C$.*

Proof: At the end of iteration $\hat{i}$ of the algorithm, for each client $c_j \in C$ it holds that $\mathtt{Maxrank}(B_{\hat{i}}, U_j) = n$. The definition of $\mathtt{Maxrank}$ implies that for each client $c_j \in C$ there exists a vector set $\Gamma^j \in \mathcal{M}(B_{\hat{i}})$ such that $\mathrm{rank}(\Gamma^j \cup U_j) = n$. By using standard network coding techniques (see e.g., [10, 12]) we can find a set of vectors $\hat{\Gamma} \in \mathcal{M}(B_{\hat{i}})$ that satisfy the demands of all clients, provided that the field size q is larger than the number of constraints (k). ■

We summarize our results with the following theorem:

Theorem 2. *Algorithm* DDE *computes, in polynomial time, the optimal solution to the data exchange problem.*

Proof: Follows from Theorem 1 and the fact that OPT_0 is equal to the optimal solution OPT. ■

The computational complexity of the algorithm is comparable to that of the standard network coding algorithms such as due to Jaggi et al. [10].

4 Conclusions

We studied the problem of direct information exchange between a group of wireless clients with the goal to minimize the total number of transmissions between clients. We presented deterministic algorithm that provides an optimal solution in polynomial time.

There are many open problems for future research. One direction is to explore the two related issues of providing incentives to guarantee continued cooperation between clients and fairness to clients in terms of transmission load during data exchange. An additional interesting aspect of data exchange which can be considered is the energy cost associated with each transmission. This is of particular importance in networks with heterogeneous terminals that have different power consumption and battery options.

References

1. Ahlswede, R., Cai, N., Li, S.-Y.R., Yeung, R.W.: Network Information Flow. IEEE Trans. Inf. Theory 46(4), 1204–1216 (2000)
2. Alon, N., Hassidim, A., Lubetzky, E., Stav, U., Weinstein, A.: Broadcasting with Side Information. In: Proc. of the 49th Annual IEEE Symposium on Foundations of Computer Science (FOCS), pp. 823–832 (2008)
3. Bar-Yossef, Z., Birk, Y., Jayram, T.S., Kol, T.: Index Coding With Side Information. In: Proc. of the 47th Annual IEEE Symposium on Foundations of Computer Science (FOCS), pp. 197–206 (2006)
4. Birk, Y., Kol, T.: Coding-On-Demand by an Informed Source (ISCOD) for Efficient Broadcast of Different Supplemental Data to Caching Clients. IEEE Trans. Inf. Theory 52(6), 2825–2830 (2006)
5. Blasiak, A., Kleinberg, R., Lubetzky, E.: Index Coding via Linear Programming. IEEE Trans. Inf. Theory (April 2010) (submitted for publication), `http://arxiv.org/abs/1004.1379`
6. Courtade, T., Wesel, R.: Efficient Universal Recovery in Broadcast Networks. In: 48th Annual Allerton Conference on Communication, Control, and Computing, Monticello, IL, USA (September-October 2010)
7. Courtade, T., Wesel, R.D.: A Deterministic Approach to Rate-Compatible Fountain Communication. In: Information Theory and Applications Workshop (ITA), San Diego, CA, USA (January/February 2010)
8. El Rouayheb, S., Sprintson, A., Georghiades, C.: On the Index Coding Problem and Its Relation to Network Coding and Matroid Theory. IEEE Transactions on Information Theory 56(7), 3187–3195 (2010)
9. Hong, Y.-W., Huang, W.-J., Chiu, F.-H., Kuo, C.-C.J.: Cooperative Communications in Resource-Constrained Wireless Networks. IEEE Signal Process. Mag. 24(3), 47–57 (2007)
10. Jaggi, S., Sanders, P., Chou, P.A., Effros, M., Egner, S., Jain, K., Tolhuizen, L.: Polynomial Time Algorithms for Multicast Network Code Construction. IEEE Trans. Inf. Theory 51(6), 1973–1982 (2005)
11. Katti, S., Rahul, H., Hu, W., Katabi, D., Médard, M., Crowcroft, J.: XORs in The Air: Practical Wireless Network Coding. In: Proc. of ACM SIGCOMM, Pisa, Italy (2006)
12. Koetter, R., Médard, M.: An Algebraic Approach to Network Coding. IEEE/ACM Trans. on Networking 11(5), 782–795 (2003)
13. Laneman, J.N., Tse, D.N.C., Wornell, G.W.: Cooperative Diversity in Wireless Networks: Efficient Protocols and Outage Behavior. IEEE Trans. Inf. Theory 50(12), 3062–3080 (2004)
14. Minsky, Y., Trachtenberg, A., Zippel, R.: Set Reconciliation with Nearly Optimal Communication Complexity. IEEE Trans. Inf. Theory 49(9), 2213–2218 (2003)
15. Sendonaris, A., Erkip, E., Aazhang, B.: User Cooperation Diversity–Part I: System Description. IEEE Trans. Commun. 51(11), 1927–1938 (2003)
16. Shah, D.: Gossip algorithms. Foundations and Trends in Networking 3(1) (2007)
17. Sprintson, A., Sadeghi, P., Booker, G., El Rouayheb, S.: A Randomized Algorithm and Performance Bounds for Coded Cooperative Data Exchange. In: Proceedings of 2010 IEEE International Symposium on Information Theory (ISIT 2010), Austin, Texas, USA (June 2010)

Dynamically Provisioned Priority-Aware Algorithms in Shared Mesh Optical Networks

Alireza Nafarieh[1], Shyamala Sivakumar[2], William Phillips[1], and William Robertson[1]

[1] Dalhousie University, Internetworking Engineering, Halifax, Canada
{ali.nafarieh,william.phillips,bill.robertson}@dal.ca
[2] Saint Mary's University, Information Systems and Science, Halifax, Canada
ssivakumar@smu.ca

Abstract. The paper introduces two novel algorithms, one each for static and dynamic traffic types, to improve the availability of high priority connection requests over shared mesh optical networks. The proposed algorithms are a complementary study to the previous work. The paper also proposes a new metric, maximum path availability, by which the proposed algorithms improve the performance of high priority requests by reducing blocking probability, increasing availability satisfaction rate, and with better resource utilization using dynamic negotiation of service level agreement parameters between a customer and service providers. In a multi-homed network topology, the introduced algorithms along with the proposed metric can help the customers to modify, refine and further process the final connection requests to better comply with service providers' network capacity. The simulation results show improvements on preserving the high priority class of traffic for both static and dynamic traffic types compared to other protection schemes in shared mesh optical networks.

Keywords: Shared mesh optical networks, priority-aware algorithm, dynamic service level agreement negotiation, static maximum path availability algorithm, dynamic maximum path availability algorithm, and maximum path availability metric.

1 Introduction

Typically in a shared-mesh path restoration scheme over WDM networks, the service providers' goal is to provide a reliable connection with minimum resource allocation. This paper explores the need for having a dynamic mechanism to propagate and refine the requirements requested by service level agreement (SLA) contracts, and develops the algorithms that are capable of handling and working in a dynamic environment. The paper discusses how the proposed algorithms benefit from dynamic SLA parameters propagation mechanisms to affect network performance and to develop new priority-aware algorithms which focus on SLA parameters including link availability. The paper introduces two priority-aware algorithms one for static and the other for dynamic traffic types for different priority levels. In a multi-homed network topology, the link availability information can be communicated using dynamic SLA negotiation mechanisms [1] and [2]. The customer side of the network is exposed to

X. Zhang and D. Qiao (Eds.): QShine 2010, LNICST 74, pp. 290–306, 2011.

SLA information from all the service providers to which it is connected. The customer has the choice to pick the service provider that is the most suitable for satisfying the requested connection. It is shown that the proposed algorithms benefit from dynamic SLA negotiation between customer and service providers to reduce the blocking rate of high priority traffic streams, increase the availability satisfaction rate of connection requests, improve resource utilization including wavelength usage, and preserve high priority requests.

Two different types of traffic, static and dynamic, are considered in this paper and for each type of traffic, an algorithm is proposed. The characteristics of static traffic are known *a priori* and in this paper are considered as a matrix of requests with known parameters including source, destination, requested availability, and requested level of priority. In contrast, dynamic traffic has no knowledge of subsequent connection requests and requests are processed in the order in which they are received. Both traffic types, static and dynamic, are classified to three different priority traffic classes, Gold, Silver and Bronze.

The paper focuses on the priority-aware algorithms rather than the dynamic SLA negotiation mechanism, and it is assumed that there is an automatic mechanism for SLA parameters negotiation between service providers and customers. Examples of the protocols used for dynamic SLA negotiation have been discussed in [3]. Although some RFC standards such as [4] and [5] have discussed the extensions to OSPF-TE opaque link state advertisements (LSAs), none of them supports and propagates SLA parameters. As a brief explanation of how the negotiation can be accomplished, the dynamic mechanism through which some vital SLA parameters are negotiated is assumed to be done by defining new extensions to OSPF-TE opaque LSAs [6]. It is assumed that a dynamic SLA negotiation mechanism can be implemented and the proposed algorithms can benefit from this facility to improve network performance.

The paper is organized as follows: in Section 2, the related work concerning priority-aware algorithms is discussed. Section 3 introduces two novel priority-aware algorithms for static and dynamic traffic types. In Section 4, simulation environment and performance analyses of the proposed algorithms are presented. Section 5 presents conclusions and future work.

2 Related Work

In previous work [7], a priority-aware pre-provisioning algorithm has been proposed based on the SLA-parameters negotiation for shared-mesh WDM networks. The pre-provisioning algorithm in [7] shows performance improvements over conventional shared-mesh protection [8] and [9] and priority-aware algorithms [10] and [11] with respect to availability satisfaction rate for high-priority connection requests. The simulation results in [7] have shown that the static pre-provisioning algorithm preserves the high-priority connection requests better than the lower priority connection requests. However, the nature of SLA negotiation in the pre-provisioning algorithm is static as the pre-provisioned requests are calculated based on the initial link availability of the network. That is, although the pre-provisioning algorithm in [7] benefits from a static SLA negotiation, there is no dynamic mechanism to inform the customer about the changes in the availability of the requested paths.

The conventional shared mesh algorithm presented in standards [8] and [9] takes advantage of constraint-based shortest path algorithm for path calculation. Although link availability is one of the most prominent parameters in SLA parameters negotiations, the algorithm in [8] and [9] does not consider the availability of the links as a constraint in the path calculations.

The priority-aware algorithms in [10] and [11] have presented a new insight to the definition of the path availability with respect to traffic priority level. In [10] the algorithm has been applied to a static traffic type and in [11] to a dynamic traffic type. The network performance improvements for both algorithms have been proven through both mathematical model and heuristics. Although the algorithms presented in [10] and [11] are priority-aware algorithms and they improve the blocking rate and availability satisfaction rate of shared mesh optical networks, they have no knowledge of what happens in the network dynamically in terms of availability of the requested paths based on the current status of the network resources.

The algorithm discussed in [12] has considered SLA parameters as important factors to guarantee customers' reliability. The new cost function definitions for both primary and backup paths' calculations in [12] and [13] have enabled the algorithm to introduce a novel case of protection, partial link-disjoint protection, and to increase the availability satisfaction rate and reduce restoration time of shared mesh WDM networks. Although the algorithm in [12] is an SLA-based algorithm, it does not benefit from a dynamic SLA communication mechanism to inform the customer about available paths which meet the availability requirements.

In contrast to the above mentioned related and previous work, the proposed work focuses on algorithms which take advantage of a dynamic mechanism for propagating requested paths' availabilities. The proposed algorithms use a dynamic environment to better serve high priority traffic.

3 Paper's Contribution

The main contribution of this paper can be summarized in three parts. In the first part, an algorithm is proposed by which new metric called maximum path availability (MPA) is introduced. The proposed metric plays an important role in performance improvements of the algorithms presented in the second and the third parts of the contribution. In fact the proposed metric is the basis of the introduced algorithms in this paper.

The MPA algorithm calculates the highest path availability offered by the service providers for any pairs of source and destination in a dynamic manner. That is, the algorithm updates the MPA matrix after any connection changes, either a connection request or a connection release. In fact, the algorithm is a generalized mechanism of the algorithm presented in [7]. In the algorithm presented in [7], it is assumed that the network has no traffic load when the initial maximum path availability is calculated and the parameter is calculated once forever, and is not updated over the network operation. In the previous work [7], the algorithm results in a parameter called initial maximum availability (IMPA) which is a special case of the MPA algorithm presented in this paper. The MPA algorithm in this paper updates the MPA matrix dynamically. Moreover, it is assumed that this dynamically updated metric is propagated through the network by a dynamic SLA negotiation mechanism and all the nodes have a unique and updated picture of the network in terms of MPA matrix.

As of the second and third parts of the contribution, two algorithms have been introduced taking advantage of the proposed metric presented in the first part of the contribution. The algorithms try to improve the performance of the high priority requests with regard to blocking probability and resource utilization. In this case, two different algorithms for static and dynamic types of the traffic are proposed, the static maximum path availability (SMPA) and the dynamic maximum path availability (DMPA) algorithms. To analyze the SMPA and DMPA algorithms' performance, two different simulation environments are developed and evaluated. The modules and the algorithms building the static and dynamic maximum path availability algorithms are discussed in Section 4 in detail, and the simulation environments are presented in Section 5.

4 Proposed Algorithms

4.1 Static Maximum Path Availability Algorithm

For static traffic type, it is assumed that the connection requests are known a priori. Each request is characterized by source, s, destination, d, requested availability for the nth connection, A_{rn}, and requested traffic priority level, p. The nth request can be shown as $C_n\ (s,\ d,\ A_{rn},\ p)$. The block diagram of SMPA algorithm is presented in Fig. 1. First, the requests are prioritized to find the highest priority request which meets the path availability requirements. The requests are stored in form of a connection request matrix (CRM), and are sorted in descending order of the requests' priority. The first request in the CRM whose availability is met is sent for the further processing. Secondly, the routing and wavelength assignment (RWA) module is applied to the request to find the most optimum paths. The wavelength and graph update module updates the link-wavelength status and the graph topology respectively. In a multi-homed network in which the customer can be served by several service providers as shown in Fig. 2, the maximum path availability algorithm module calculates dynamically the highest path availability offered by the service providers for any pairs of source and destination.

The steps by which the connections prioritizing module of the SMPA algorithm affects the connection request matrix is shown in Fig. 3. Three levels of priorities, $p=1,\ 2,\ 3$, have been defined as Gold, Silver, and Bronze respectively. The requests belonging to each level of priority are stored in a corresponding set; S_p. Based on the prioritizing module presented in Fig. 3, the requests, C_{np}, with the highest priority are served first. As long as there are requests in the highest level of priority set which can meet the path availability requirements, for instance S_{Gold}, SMPA will not process the lower priority set, S_{Silver} for example. Those requests which meet the requirements, i.e. the requested availability of the requests (A_{rCn}) is lower than the offered availability (A_{oCn}) are kept in another set associated with the same priority level, S_{pp}, and are sent to service providers to be established. Those requests which do not meet the requirements requested in SLA are left for possible future opportunity and the next request in the same set will be processed. If the set regarding a certain level of priority

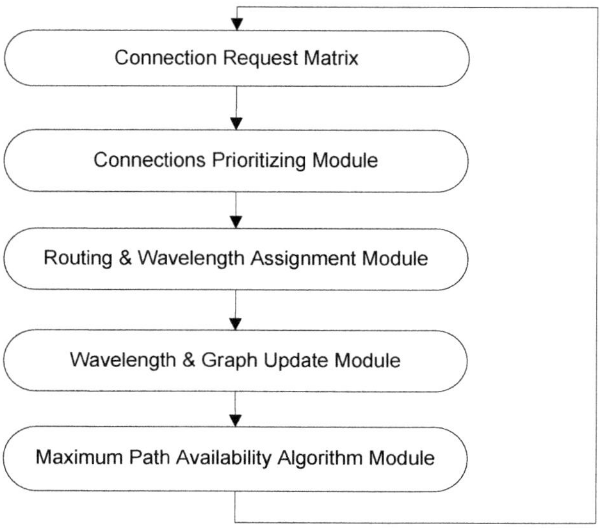

Fig. 1. SMPA algorithm block diagram

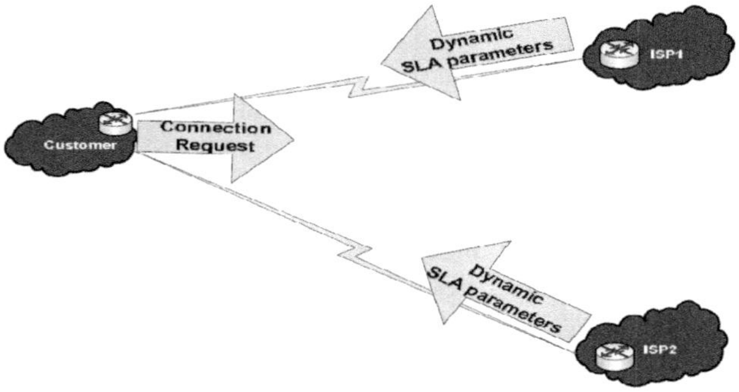

Fig. 2. Multi-homed network topology

is empty or the existing requests cannot be served, the next lower priority level set is processed. The output of the prioritizing module saved in S_{pp} is a refined request which will be processed by RWA module to be established as a connection. The connection requests in S_{pp} which are sent to RWA module for further processing are permanently removed from S_{pp} set. The RWA and wavelength update modules have been discussed in [7] in greater detail. The routing scheme which is used in the RWA module to determine the primary and backup paths is adaptive routing [14].

Before a path computation algorithm, Dijkstra's, is applied to the prioritized request to find the primary path, the cost of the links of the graph is modified by cost function presented in (1). Based on the cost function, if there is no bandwidth available on the link, the link is removed from the graph; otherwise the cost of the link is a function of the link availability. The way that the cost function is calculated in a logarithmic basis has been discussed in [15] in detail.

$$C_{P(i,j)} = \begin{cases} \infty & \omega_{ij} = 0 \\ -ln(a_{(i,j)}) & \omega_{ij} > 0 \end{cases} \tag{1}$$

where $C_{p(i,j)}$ is the cost of the link between nodes i and j for a primary path , $a_{(i,j)}$ is the availability associated with link between nodes i and j, ω_{ij} is the number of free wavelengths on the link between the nodes i and j.

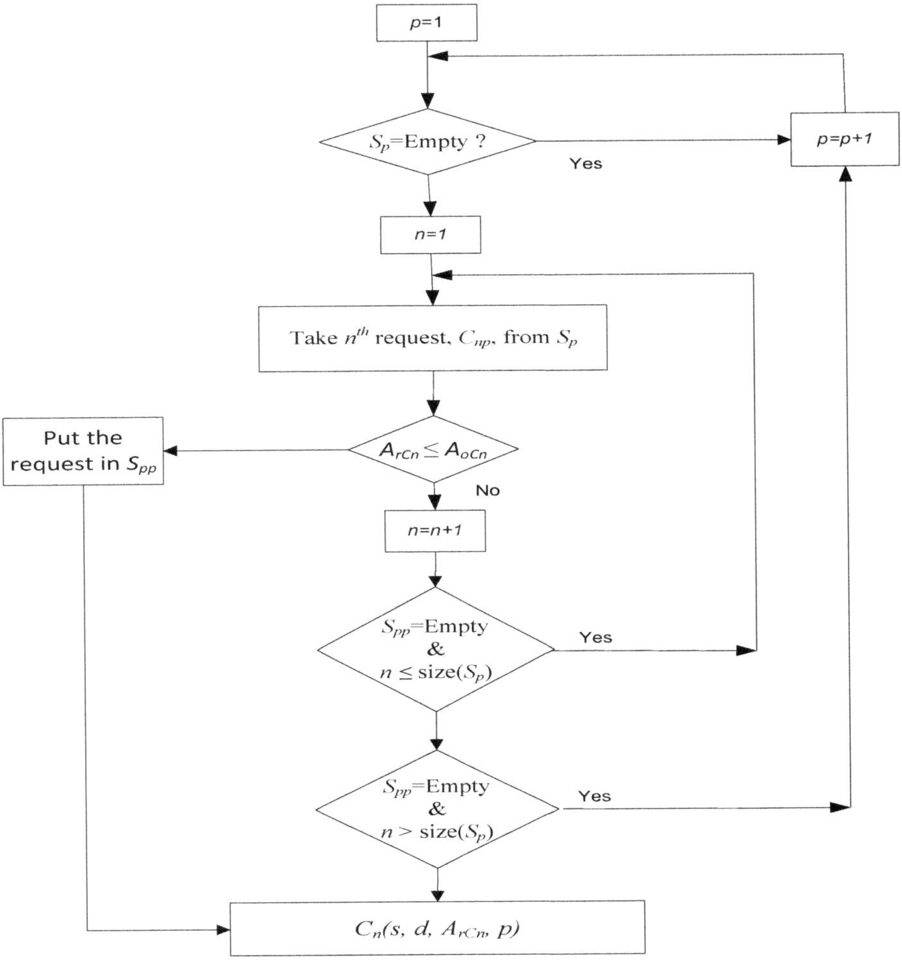

Fig. 3. Connections prioritizing module of SMPA algorithm

Following this step the wavelengths are assigned to the path based on a per-link basis, and the wavelength usage matrix is updated in this step. If the path computation finds no way from s to d, the request is said to be blocked. The wavelength assignment of the primary paths follows the First-Fit (FF) algorithm [14]. In the FF technique, the wavelengths are numbered and the lowest numbered free wavelengths are selected.

Before the RWA module calculates the backup path, the cost of the links of the graph are changed one more time based on (2) [7]. The cost function based on the logarithmic formula has been discussed in detail in [15].

$$C_{b(i,j)} = \begin{cases} \infty & \omega_{ij} = 0 \\ -ln(a_{(i,j)}) * \omega_{ij} & \omega_{rsvd} > \omega_B \\ -ln(a_{(i,j)}) * \omega_{ij} + 1 & \omega_{rsvd} \leq \omega_B \end{cases} \tag{2}$$

where $C_{b(i,j)}$ is the cost of the link between nodes i and j for a backup path, ω_{rsvd} is the number of the reserved wavelengths on the link for the shared backup paths, and ω_B is the number of required wavelengths if the primary path fails.

As (2) denotes, to setup a backup path, the algorithm looks for the paths with the highest available bandwidth and lowest number of shared paths on each link. The algorithm checks if the backup path can share any wavelength considering link-disjointness constraint. Then, it follows FF technique [14] to allocate a wavelength to the links forming the path.

After calculating the primary path and finding totally link-disjoint primary-backup paths pair, the wavelength and graph update module modifies the graph topology matrix by removing the links forming the primary path. The graph topology matrix is an m x m matrix of zeros and ones showing which nodes are connected to each other.

The MPA module of SMPA algorithm is an algorithm by itself. The module is responsible for calculating the highest path availability offered by the service providers for any certain source and destination pairs. The parameters advertised in the dynamic SLA mechanism is the availability of the links forming the graph. However, proper SLA negotiation needs the information about availability of all possible paths for any pair of source and destination. This is accomplished by applying the MPA module. After cost modification, the MPA module calculates the best possible availability offered by the service providers. MPA algorithm uses (3) and (4) to calculate the path availability of the primary and backup paths respectively.

$$A_{pC_n} = \prod_{(i,j) \in C_{n(primary-path)}} a_{(i,j)} \tag{3}$$

$$A_{bC_n} = \prod_{(i,j) \in C_{n(\,backup-path)}} a_{(i,j)} \tag{4}$$

where A_{pCn} and A_{bCn} are the availability of the primary and backup paths for the nth connection request respectively.

Equation (5) results in the maximum offered path availability between any requested pair of source and destination using path availabilities for primary and backup paths. If the value of $MPA_{(s,d)}$ is zero, this means the network has no capacity at that time for serving the request and the request is not sent to the service providers and considered blocked.

$$MPA_{(s,d)} = A_{pC_n} + A_{bC_n} - A_{pC_n} \cdot A_{bC_n} \tag{5}$$

where A_{pCn} and A_{bCn} are the path availabilities of the primary and backup paths respectively, $MPA_{(s,d)}$ is the maximum offered path availability for a source-destination pair in the nth connection request.

Table 1 shows the pseudo code describing the algorithm used in MPA module. The MPA algorithm calculates an m x m matrix, the MPA matrix, for a network topology of m nodes.

Table 1. MPA calculation algorithm

1. If $s=d$, $MPA_{(s,d)}=0$ Else source=s & destination=d for all values $s\epsilon\{1,2,3,\ldots,m\}$, $d\epsilon\{1,2,3,\ldots,m\}$
2. Modify cost of the links of the graph through (1) for all values of i & $j=1,2,\ldots,m$
3. Run *Dijkstra's* algorithm to calculate the primary path for the given source, destination, and the pre-calculated cost function in 3
4. If no primary path found, $MPA_{(s,d)}=0$ Else go to 5
5. For all links forming primary path, link-wavelength usage matrix is updated and saved as a new matrix
6. If any elements of new link-wavelength matrix are zero, same elements on the link-availability matrix becomes zero. The modified link-availability matrix is saved in a new matrix
7. Repeat steps 2&3 with the new link-availability matrix to find the backup path
8. Calculate the path availabilities through (3) & (4) for all links forming primary and backup paths
9. Compute $MPA_{(s,d)}$ for a specific pair of source-destination in the nth connection request through joint-availability function [12] presented in(5)
10. Repeat steps 1-9 for all values $s\epsilon\{1,2,3,\ldots,m\}$, $d\epsilon\{1,2,3,\ldots,m\}$ to build entire MPA matrix

4.2 Dynamic Maximum Path Availability Algorithm

In dynamic traffic, only one request is processed at a time in the order they are received, and the algorithm has no knowledge of the next request. After each request is processed, the graph topology and wavelength usage matrices are updated. Each request is established, blocked, or buffered for further process. The nth connection request is in form of C_n (s, d, A_r, p) with the requested parameters source, s, destination, d, requested availability, A_{rn}, and the requested priority level, p, respectively. Since each established connection changes the link-wavelength usage matrix and consequently it may change the graph topology matrix, after processing any request, the cost matrix of the entire network is updated through (1).

After the cost modification, the MPA module calculates the best possible availability offered by the service providers. As described in Section 3.1, the MPA module uses (3), (4), and (5) to calculate the path availability of the primary and backup paths and the maximum offered path availability between any requested pair of source and destination. If the value of $MPA_{(s,d)}$ is zero, this means the network has no capacity at that time for serving the request and the request is not sent to the service providers and considered blocked.

In this paper, the requests which are refused to be sent to service providers are considered blocked. Typically, in the real world scenario, such calculations take place in the customer premises and reduce the overhead of the control plane in transport networks which is one of the advantages of the proposed algorithm.

As shown in Fig. 4, if the requested availability is lower than the offered one, this means the requirements requested by the customer can be met, then the original request will be sent to the service provider for further processing including routing, wavelength assignment, wavelength usage update, and graph topology modification modules, otherwise the best availability offer from service providers will replace the requested one and the modified request is now sent to the proper service provider which is capable of fulfilling the request. The DMPA algorithm considers two availability threshold parameters at each level of priority for the customer, the lower bound availability threshold (A_{thpLB}) and the higher bound availability threshold (A_{thpHB}). Using these threshold parameters, a customer will be able to decide whether it accepts the offered parameters for a certain level of priority. The threshold parameters entirely depend on the customer and are different for any level of priorities. Some statistical research in [16] shows that the presented numerical values for the improved average availability of the different topologies can be used as threshold levels. The results presented in [16] can be a good resource for choosing thresholds that are close enough to the real world parameters.

The offered availability should be in the range of pre-defined threshold availabilities, as it is shown in (6). If (6) is fulfilled the request is modified by new parameters and is sent to the RWA module for further processing, otherwise it is blocked. The inequality (6) can be written as (7). Equation (8) presents $WMPA_{(s,d)}$ as weighted maximum path availability for a pair of source and destination, s and d. In (9), WA_{thp}, weighted availability thresholds of LB and HB for a specific level of priority, p, is defined. Then (7) can be summarized as (10).

$$A_{thpLB} \leq MPA_{(s,d)} \leq A_{thpHB} \tag{6}$$

$$1 \leq \frac{MPA_{(s,d)}}{A_{thpLB}} \leq \frac{A_{thpHB}}{A_{thpLB}} \tag{7}$$

$$WMPA_{(s,d)} = \frac{MPA_{(s,d)}}{A_{thpLB}} \tag{8}$$

$$WA_{thp} = \frac{A_{thpHB}}{A_{thpLB}} \tag{9}$$

$$1 \leq WMPA_{(s,d)} \leq WA_{thp} \tag{10}$$

Since the connections are determined with and treated based on their priority levels, p, A_{thpHB} can be considered 1 for all priority level traffic. With this assumption, for all values of s and d, $MPA_{(s,d)} \leq 1$.

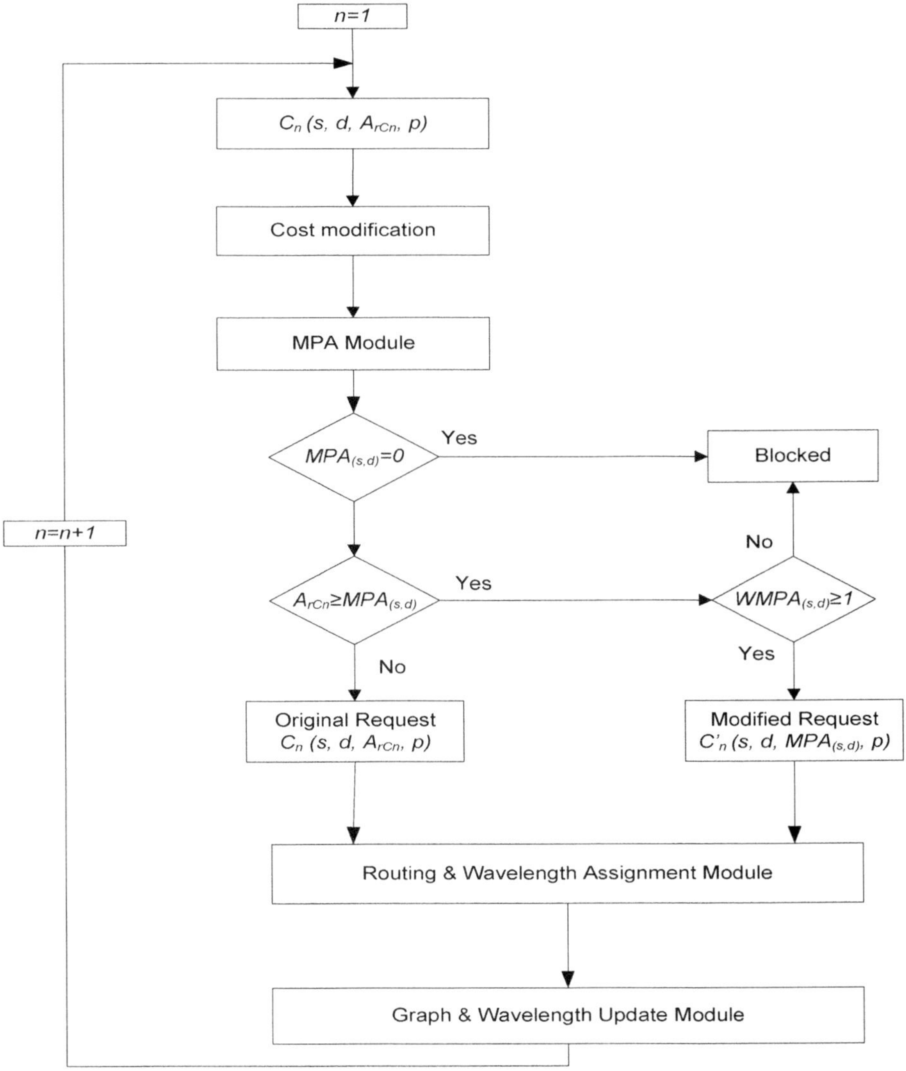

Fig. 4. DMPA algorithm flowchart

Then (10) can be simplified as (11). In fact, (11) is the final constraint for the customer to either accept or refuse the service provider's offer. If the traffic belongs to the class of Gold services and the weighted offered availability requirement presented in (11) is not met, the request is either refused or buffered for further processing. However, in this paper, for the sake of simplicity, it is assumed that no request is buffered; they are either established or blocked.

$$WMPA_{(s,d)} \geq 1 \tag{11}$$

The MPA, RWA, and graph and wavelength update modules follow the same algorithms described and discussed for the SMPA algorithm in Section 3.1. The MPA module is responsible to calculate $MPA_{(s,d)}$ for any pair of source and destination of nth connection request requested in $C_n(s, d, A_r, p)$ for a network topology of m nodes. The RWA module calculates the primary and backup paths for the requested connection. The graph and wavelength update module modifies the topology and wavelength usage matrices dynamically based on the current status of the network.

5 Performance Analysis

5.1 Static Traffic Analysis

The performance in terms of blocking probability (BP) of different classes of traffic, Gold, Silver, and Bronze for static traffic case is analyzed as BP-G, BP-S, and BP-B respectively. In addition, average number of allocated wavelengths per connection for fulfilling the connection requirement (AWPC), and the percentage of high-priority provisioned requests (HPPR) are investigated by SMPA in this paper. Several protection schemes are studied in the static traffic analysis: no protection (NP), standard shared path protection (SSPP) [8] and [9], priority-aware algorithm (PAA) [10], static pre-provisioning algorithm (SPA) [7], static maximum path availability algorithm (SMPA), and SMPA and SPA algorithm from previous work together (SMPA+SPA). All existing schemes are compared with two proposed cases; SMPA and SMPA+SPA algorithms.

Connection availability requests are uniformly distributed between three classes of traffic: Gold class with the availability of 0.9999, Silver class with the availability of 0.9990, and Bronze class with no availability significance. In addition, it is assumed that the primary and the backup paths in any protection scheme are totally disjoint.

Table 2 shows that SMPA algorithm improves the blocking rate of Gold requests more than 40% in comparison to the other protection schemes. The SMPA algorithm also brings 11-16% improvement for different protection schemes. Applying SPA algorithm on top of the SMPA brings 18% more improvement on preserving Gold requests, but it increases the blocking rate of Silver requests. However, the blocking rate of the SMPA+SPA algorithm for Silver class of traffic is still comparable with other existing algorithms.

This great improvement in decreasing blocking rate for Gold class of traffic does not degrade the resource utilization. Fig. 5 shows that although no improvement in bandwidth allocation is seen in either SMPA or SMPA+SPA algorithms, the average number of allocated wavelengths per connection is almost the same for different priority-aware algorithms. However, no-protection (NP) scheme has the minimum amount of bandwidth consumption which is obviously because of considering no protection paths as backup for primary paths. In addition, since standard shared mesh protection method, SSPP, does not count the link availabilities into account as a constraint in the path calculation, it has the maximum bandwidth consumption among all studied protection schemes.

Fig. 6 shows an increase on the percentage of high priority provisioned requests including Gold and Silver which are served by either SMPA or SMPA+SPA algorithms. It is indicated that 52% of high-priority requests which were blocked in other protection schemes are now provisioned by the proposed mechanism. SMPA+SPA algorithm barely improves the total number of provisioned high priority requests in comparison to SMPA algorithm, and it works well just for Gold requests. However, as Fig. 6 shows, the SMPA+SPA algorithm still has 12-28% improvement in number of provisioned high-priority requests comparing to the other existing algorithms. Since SMPA algorithm works based on the traffic priority, the Bronze traffic gets the minimum attention. From the business point of view, applying the proposed algorithm brings more money to service providers since serving the quality of service based traffic is more lucrative than low priority traffic.

The total number of Gold, Silver, and Bronze requests among all possible connection requests is dictated by the network topology presented in Fig. 7. This paper uses the same network topology used in other existing algorithms in [7], [10], [11], and [12] for the sake of consistency and fair results comparison.

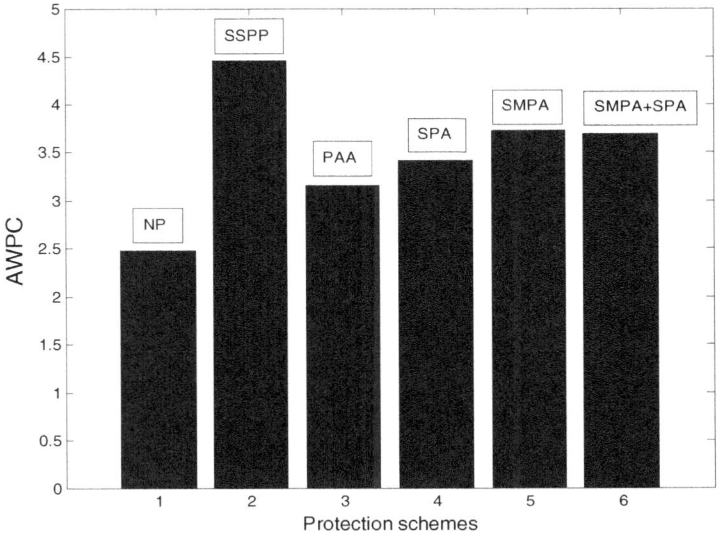

Fig. 5. Average number of allocated wavelengths per connection with respect to different protection schemes

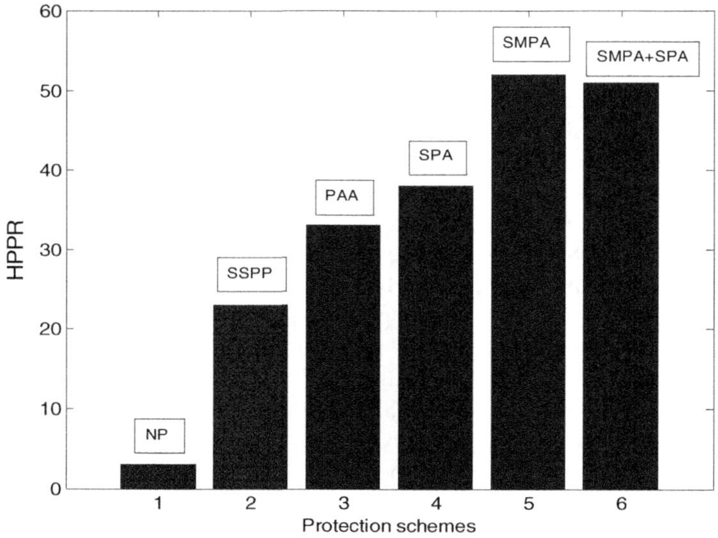

Fig. 6. Percentage of high-priority requests provisioned by different protection schemes

Table 2. Blocking rate percentage comparison for several protection schemes and algorithms

	NP	SSPP	PAA	SPA	SMPA	SMPA+SPA
BP-G	100	89	81	65.5	47	29
BP-S	97	66	61	64	50	66
BP-B	32	68	26	35	62	65

5.2 Dynamic Traffic Analysis

The lightpaths in the dynamic traffic pattern are requested dynamically with randomly generated availability requests so that the algorithm has no knowledge about the coming request. The simulation environment is similar to the previous work [7] environment. The links have wavelength conversion capability with 8 wavelengths per each link. The link availabilities are uniformly distributed between 0.99 and 0.9995. Connection availability requests are uniformly distributed between 0.99 and 1.00. A Poisson process with arrival rate of β is considered for the arrival process of connection requests. The holding time of the connections follows an exponential distribution with the mean value of μ=1. No waiting queue has been considered for this process. The primary and the backup paths are considered totally disjoint. The total number of connection requests over entire simulation period is 10^5. The topology selected for the simulation is NSFNet shown in Fig. 7 with 14 nodes and 21 bidirectional fiber connections of the same physical distance.

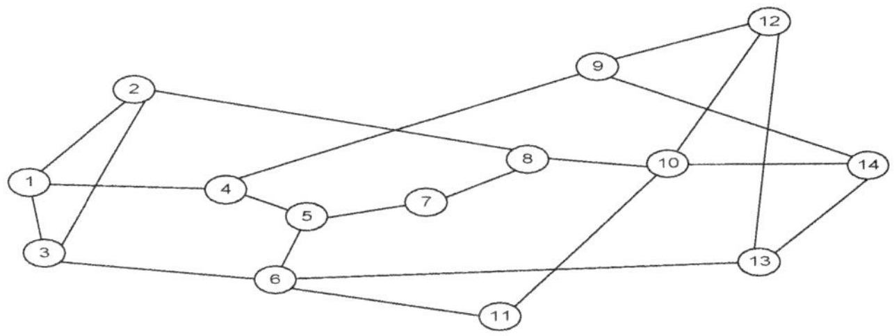

Fig. 7. Network topology, NSFNet

Based on practical values for different protection schemes and several network topologies presented in [16], for the simulation purposes A_{thpLB} has been considered 0.9997 and 0.9988 for Gold and Silver traffic respectively. However, theses marginal numbers can vary depending on the customer needs.

In dynamic traffic analysis, the availability satisfaction ratio (ASR), the blocking probability (BP), and the average number of link-wavelength per connection request (AWPC) of DMPA algorithm is compared with other existing algorithms for dynamic traffic case. ASR represents the percentage of provisioned connections whose availability requirements are met over all provisioned connections. BP denotes the percentage of blocked connection requests over all arriving requests. AWPC shows the average number of the wavelengths allocated for each connection. The performance of the proposed algorithm for the dynamic traffic is compared with the schemes in which there are either no automatic SLA negotiations (SSPP) [8] and [9] or just static negotiation (SPA) [7]. In the previous work [7], the performance of SPA has been compared with SSPP and PAA algorithms [10] and [11].

As Fig. 8 Show, the blocking rate of DMPA algorithm is improved in comparison to two other algorithms, SSPP and SPA schemes. Results in Fig. 8 shows a 47% decrease in connection blocking probability of DMPA algorithm in comparison to SSPP and SPA algorithms.

Simulation results in Fig. 9 shows a 15% increase in ASR performance of the DMPA algorithm compared to SSPP and SPA algorithms. Fig. 8 and 9 show that although the previous work [7] could improve ASR compared to the SSPP scheme, it could not improve the BP. In contrast, the proposed algorithm, the DMPA algorithm, improves both ASR and BP.

Fig. 10 shows how much better the DMPA algorithm saves the network resources in terms of number of assigned wavelengths per connection when compared with the SSPP and SPA algorithms. The average AWPC for DMPA is around 4.0 wavelengths while it is 4.6 for SPA and 4.75 for SSPP. Since the number of connection requests is large, the DMPA algorithm saves a huge amount of network resources. In addition, Fig. 10 shows that the performance of SPA in terms of AWPC is degraded for small values of offered loads. However, the DMPA has a good performance for both small and large offered loads.

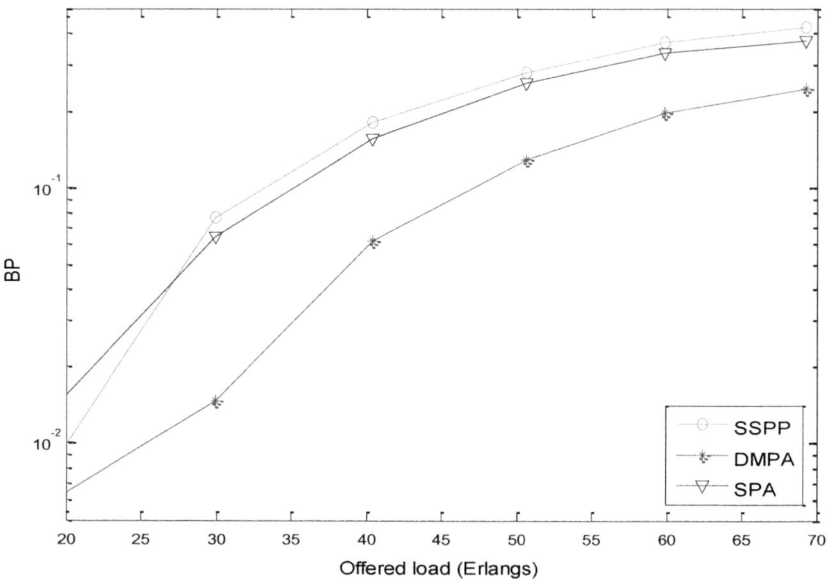

Fig. 8. Blocking rate performance of DMPA algorithm

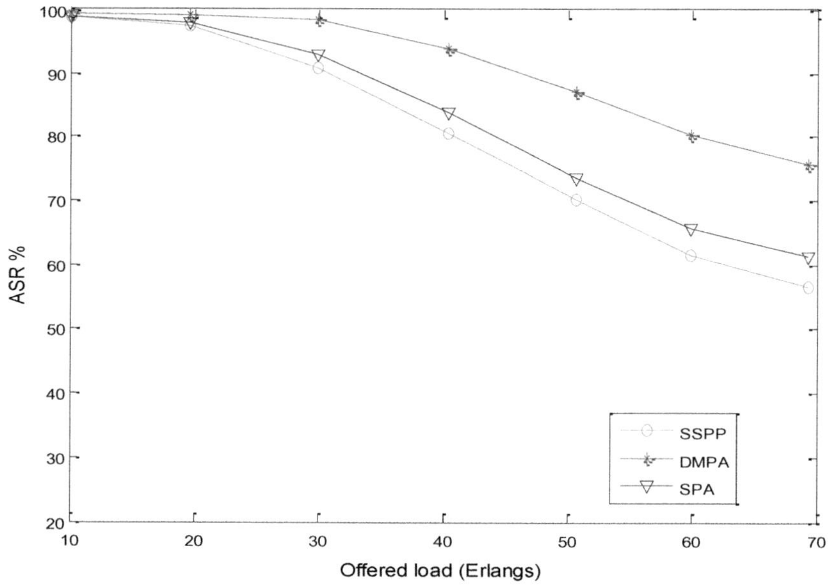

Fig. 9. Availability satisfaction rate performance of DMPA algorithm

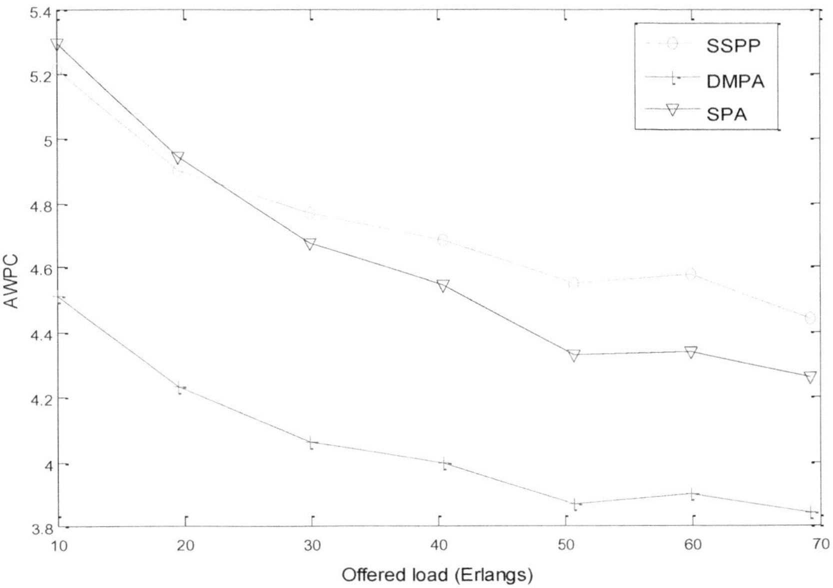

Fig. 10. Average number of allocated wavelengths per connection

6 Conclusion

This paper is a complementary study to the previous work presented in [7]. In this paper, two priority-aware algorithms have been introduced for shared mesh survivable WDM networks. The SMPA and the DMPA algorithms have been introduced to improve the network performance for static and dynamic traffic types respectively. The proposed algorithms take advantages of a dynamic negotiation mechanism of SLA parameters which can help customers to have a better picture of the entire network with respect to path availabilities. This fact can help customers to modify and process their requests before they send them out to the service providers for the further processing. The paper has also proposed a new metric, maximum path availability, by which the proposed algorithms improve the performance of high priority connection requests The algorithms reduce the blocking probability of the high priority requests, increase availability satisfaction rate, better preserve high priority connection requests, and reduce the average number of allocated wavelengths per connection. In addition, the algorithms also reduce the calculation steps done by service providers.

The performance analysis in both types of traffic, static and dynamic, shows lower blocking probability and resource consumption, higher availability satisfaction rate and resource utilization, and more preserved high priority traffic for the customer. In addition, the proposed algorithms can provide more benefits and fewer calculations and decision processes for service providers.

An automatic mechanism for SLA parameters negotiation between service providers and customers by defining new extensions to OSPF-TE opaque LSAs will be presented in future work. In addition, a buffered request service module in the DMPA algorithm which can benefit from certain algorithms to accommodate potentially blocked high priority connection requests is a potential subject for future research studies.

References

1. Nguyen, T.M.T., Boukhatem, N., Doudane, Y.G., Pujolle, G.: COPS-SLS: A service level negotiation protocol for the internet. IEEE Communications Magazine, 158–165 (2002)
2. Green, L., Mirchandani, V., Cergol, I., Verchere, D.: Design of a Dynamic SLA Negotiation Protocol for Grids, GridNets (2007)
3. Sarangan, V., Chen, J.C.: Comparative study of protocols for dynamic service negotiation in the next generation internet. IEEE Communications Magazine, 151–156 (2006)
4. Katz, D., Kompella, K., Yeung, D.: Traffic Engineering (TE) Extensions to OSPF Version 2. IETF RFC 3630 (2003)
5. Kompella, K., Rekhter, Y.: OSPF Extensions in Support of Generalized Multi-Protocol Label Switching (GMPLS). IETF RFC 4203 (2005)
6. Yin, Y., Kuo, G.S.: An Improved OSPF-TE in GMPLS-Based Optical Networks. In: International Conference on High Performance Switching and Routing, pp. 241–245 (2005)
7. Nafarieh, A., Phillips, W., Robertson, W., Sivakumar, S.: Statically Pre-provisioned Priority-aware Algorithm for Shared-Mesh Optical Networks. In: International Conference on Transparent Optical Networks, pp. 1–4 (2010)
8. Lang, J., Rajagopalan, B., Papadimitriou, D.: GMPLS recovery functional specification. IETF RFC4426 (2006)
9. Mannie, E., Papadimitriou, D.: Recovery (Protection and Restoration) Terminology for GMPLS. IETF RFC4427 (2006)
10. Fawaz, W., Sawah, T., Rjeily, C.: Priority-aware optical shared protection: An offline evaluation study. Journal of Computer Applications 32, 1677–1684 (2009)
11. Fawaz, W., Chen, K., Pujolle, G.: Priority-enabled optical shared protection: An online efficiency evaluation study. Journal of Computer Applications 30, 3690–3697 (2007)
12. He, R., Lin, B., Li, L.: Dynamic service-level-agreement aware shared-path protection in WDM mesh networks. Journal of Computer Application 30, 429–444 (2007)
13. Lin, R., Li, L.: A New Network Availability Algorithm for WDM Optical Networks. In: 5th International Conference on Computer and Information Technology, pp. 480–484 (2005)
14. Zang, H., Jue, J.P., Mukherjee, B.: A Review of routing and wavelength assignment approaches for wavelength-routed optical WDM networks. Optical Networks, 47–60 (2000)
15. Zhang, J., Zhu, K., Zang, H., Matloff, N., Mukherjee, B.: Availability-aware provisioning strategies for differentiated protection services in wavelength-convertible WDM mesh networks. IEEE ACM Transactions on Networking 15(5), 1177–1190 (2007)
16. Segovia, J., Calle, E., Vila, P.: Availability analysis of GMPLS connections based on physical network topology. In: International Conference on Optical Networking Design and Modeling, pp. 1–6 (2008)

NAT-ZigBee: NAT-Based Address Assignment for Scalable ZigBee Networks

Yuan-Yao Shih[1], Yu-Kai Huang[2],
Ai-Chun Pang[3], Pang-Feng Liu[4], and Wei-Ni Chu[4]

[1] Research Center for Information Technology Innovation,
Academia Sinica
ckm@citi.sinica.edu.tw
[2] Graduate Institute of Networking and Multimedia,
National Taiwan University
d94944009@csie.ntu.edu.tw
[3] Graduate Institute of Networking and Multimedia,
Department of Computer Science and Information Engineering,
National Taiwan University
acpang@csie.ntu.edu.tw
[4] Department of Computer Science and Information Engineering,
National Taiwan University
{pangfeng,r95074}@csie.ntu.edu.tw

Abstract. ZigBee has been considered as one of the promising communication protocols for low-rate wireless personal area networks. Among the well-known ZigBee topologies, ZigBee cluster-tree is especially suitable for wireless sensor applications with its supporting of power-saving operation and light-weight routing. Due to the restriction of configuration parameters, some devices could be prohibited to join a ZigBee cluster-tree network and become *orphans*. The orphan problem leads to the difficulty in smoothly increasing the network coverage or device density. In this paper, we propose an NAT-based address assignment framework for a scalable ZigBee network which locally resolves the orphan problem without the reconfiguration of a whole network, while the light-weight ZigBee tree routing is maintained. The experimental results show that the proposed framework can significantly reduce the number of orphan devices compared to the original ZigBee address assignment scheme.

1 Introduction

Recent advances in wireless communications and micro-electromechanical technologies have strong impacts on the development of wireless sensor networks (WSNs) [2]. IEEE 802.15.4 is a promising standard for wireless sensor applications with special considerations on energy efficiency and communication overheads [1]. Based on the physical (PHY) and medium access control (MAC) layers of IEEE 802.15.4, the ZigBee alliance defines the specification for upper-layer (including network and application layers) protocols. Among the well-known ZigBee topologies, ZigBee cluster-tree is especially suitable for low-power and

X. Zhang and D. Qiao (Eds.): QShine 2010, LNICST 74, pp. 307–315, 2011.

low-cost WSNs because it supports multi-hop communication with power-saving operation, and requires few resources for routing.

For a ZigBee cluster-tree network, some system configuration parameters must be determined before its topology formation. The parameters include the maximum number of children of a router (Cm), the maximum number of child routers of a router (Rm), and the depth of the network (Lm). Under the topology configuration, an accordingly distributed address assignment policy is executed for routing purpose. The topology configuration and address assignment enable a light-weight routing protocol for ZigBee devices, but limit the overall system capacity. In ZigBee, a device can successfully join a network if it associates with a proper parent device (i.e., router) and is assigned a network address. Due to the restriction of Cm, Rm, and Lm, some devices with a physical connection to a router could fail to join the network and become unaddressed. This problem is referred to as the orphan problem.

This paper aims at the proposing of flexible address assignment for a ZigBee cluster-tree network so that the system capacity can be effectively increased without violating the ZigBee specification. In the past years, notable research results have been presented for address assignment in wireless ad-hoc and sensor networks. In wireless ad hoc networks, some automatic address assignment approaches based on the concept of IP networks were presented to avoid human interventions [13] [15] [14]. The methodologies proposed in [8,6,7] ensure address uniqueness for node mobility. The reduction of address space for energy efficiency in WSNs were also investigated [3][11][17]. Unfortunately, the results of these works could not be directly applied to ZigBee networks due to its unique system characteristics.

Some alternative addressing configurations which might violate the ZigBee specification were proposed by Yen, et al. [16] as an effort to reduce the possibility of unaddressed devices. On the other hand, the work done by Pan et al. [10] is one of the pioneering studies to address the orphan problem in a ZigBee cluster tree network. They proved the NP-Completeness of the orphan problem, and proposed a better network formation to reduce the number of orphan devices. Pan et al. [9] also proposed ZigBee addressing and routing schemes for a special network topology.

Even though the existing approaches dealt with the orphan problem to increase system capacity for ZigBee networks, the scalability issue has not been well addressed in these works. In a constructed WSN, the information of some interest areas could be required for further investigation. Then new sensor devices should be deployed in an existing ZigBee network to increase the network coverage or device density in the areas. However, the newly deployed devices could be unaddressed due to their geographical locations such that the expected performance is hardly achieved. In this paper, we propose an NAT (Network Address Translation) based address assignment framework, called "NAT-ZigBee" for a scalable ZigBee cluster tree network. Once a newly deployed device fails to join the network, one of its neighboring devices is upgraded to or replaced with an NAT router to assign a private address and perform the address translation for the orphan device, while the remaining devices are not affected. In NAT-ZigBee, the existing ZigBee tree routing is not violated, and the orphan problem can be locally addressed without the reconfiguration of a whole system. The capability of our NAT-ZigBee is evaluated by extensive simulations.

The remainder of this paper is organized as follows. Section 2 overviews the ZigBee system architecture and motivation. Section 3 describes the proposed NAT-ZigBee framework for a scalable ZigBee network, and the performance of our proposed NAT-ZigBee is also evaluated by extensive simulation results in this section. Finally, Section 4 concludes this paper.

2 Zibgee Overview and Motivation

The ZigBee alliance aims at the development of a reliable, cost-effective, and low-power wireless network [5]. It specifies the network-layer and application-layer protocols upon the IEEE 802.15.4 standard [1] which defines the PHY and MAC for low-rate wireless personal area networks. This section overviews the ZigBee system architecture, and elaborates on the motivation of this work.

There are three types of ZigBee devices, and they are coordinator, router and end device [5]. The ZigBee coordinator is the most complicated device in a network, and provides the initialization, maintenance, and controlling functions for the network. The ZigBee router has a forwarding capability to route sensed data to a sink device. The data forwarding is not needed for a ZigBee end device, and its hardware requirement is minimal. A ZigBee device has two kinds of address: IEEE address and network (NWK) address. The 64-bit IEEE address is designed as a unique identifier for a device, and the 16-bit network address is used for routing. Also, a ZigBee device can support up to 240 endpoints, and each endpoint is uniquely associated an application object. The endpoint in ZigBee is analogous to a port in TCP/IP.

Each device in ZigBee maintains a neighbor table which records IEEE/network address mapping, relationship (e.g., parent/child), and link-state information of the neighbors residing in the coverage area of the device. The neighbor table is used for network association when a device joins the network or tries to connect to a new parent device, and is updated every time the device receives the packets from its neighbors. According to the ZigBee specification [5], if multiple devices with the same network address exist in the neighbor table, all the related entries are invalid such that the routing operations could fail.

ZigBee supports three kinds of network topology: star, cluster-tree, and mesh. In a star network, multiple ZigBee end devices directly connect to a ZigBee coordinator, and communications between any two end devices should be done through the coordinator. For cluster-tree and mesh networks, devices can communicate with each other in a multi-hop fashion. The cluster-tree topology is especially suitable for low-power and low-cost wireless sensor networks because of its power-saving operation and light-weight routing. A superframe structure is supported in ZigBee cluster tree topology such that the power-saving operation provided in IEEE 802.15.4 MAC layer can be utilized to achieve energy efficiency. Also, a distributed address assignment scheme defined in ZigBee enables a light-weight tree routing protocol without routing information maintenance.

In a ZigBee cluster-tree network, the topology is formulated by using the following three parameters, i.e., maximum number of children of a router (Cm), maximum number of child routers of a router (Rm), and depth of the network (Lm). Based on the parameters, a parent device can locally determine a unique 16-bit network address to its child devices. The distributed address assignment procedure is described below. The ZigBee coordinator is the root of the cluster-tree network, and normally has the network address 0. Each router is pre-allocated a finite sub-block of the address space for its children. To enable the pre-allocated concept, each router of depth d calculates a $Cskip(d)$ to decide the number of addresses allocated to every router-capable child device. $Cskip(d)$ is calculated as follows.

$$Cskip(d) = \begin{cases} 1 + Cm \cdot (Lm - d - 1), & \text{if } Rm = 1, \\ 1 + \dfrac{Cm \cdot (1 - Rm^{Lm-d-1})}{1 - Rm}, & \text{Otherwise.} \end{cases}$$

Then the distributed address assignment can be executed accordingly with $Cskip(d)$. For each parent device with address A_{parent} of depth d, the network addresses A_k for its kth router-capable child and A_n for its nth end device child are defined as follows.

$$\begin{aligned} A_k &= A_{parent} + 1 + Cskip(d) \cdot (k - 1) \\ A_n &= A_{parent} + Cskip(d) \cdot Rm + n. \end{aligned} \tag{1}$$

A ZigBee device can successfully join a network if it connects to a parent device and is assigned a network address by the parent device. A parent device shall not accept children's association once the acceptance violates the parameter settings of Cm, Rm, or Lm. Therefore, some devices might be prohibited to join the network, and become orphans even if they have strong physical connections to the network. In a WSN, the information of some interest areas could be required for further investigation after the network is constructed. Since the reconfiguration of a whole network is not cost-efficient or even impossible, new sensor devices should be deployed in an existing network to increase the network coverage or device density in some interest areas. However, under the ZigBee addressing assignment, the newly deployed devices could be unaddressed due to their geographical locations. Consequently, it is difficult for a ZigBee network to smoothly scale up and achieve expected performance.

One possible approach to solve this problem is to establish a new network in the unaddressed area and install a ZigBee gateway to connect the new network to the original network. The new network will be supervised by an additional coordinator, and the coordinator can be either elected from the existing devices in the unaddressed area or served by a new powerful device. An inter-coordinator protocol is also needed to synchronize the system information between the networks. Obviously, this approach would significantly increase the deployment cost for a scaling network. On the other hand, the orphan problem can be solved by a stochastic addressing scheme proposed in the enhanced ZigBee version called ZigBee PRO [4]. ZigBee PRO randomly assigns network addresses to devices and sorts out conflicts by continuously monitoring. However, the continuing messages exchanges occurring from monitoring and resolving for address conflict could cause significant energy consumption. Since ZigBee PRO is not standardized and compatible with legacy ZigBee specification, the system upgrading requires reconfiguration and installation for a whole network.

3 NAT-Based Address Assignment for a Scalable ZigBee Network

3.1 NAT-ZigBee: An NAT-Based Address Assignment Framework

In this subsection, we propose an NAT-based address assignment framework, called NAT-ZigBee, for a scalable ZigBee cluster-tree network. This solution is motivated from the Network Address Translation (NAT) developed by IETF [12], which solves the IP address shortage in IPv4 to allow private networks with internal addresses to connect public networks.

In NAT-ZigBee, the orphans are assigned private addresses and one of the orphan's neighboring devices is upgraded to or replaced with performs address translations. The mapping table for address translations can be constructed as an extension of the existing neighbor table described in Section 2. Therefore, these original orphans can join the network to carry on their sensing tasks without significant upgrading cost and message exchanges. The proposed approach preserves existing ZigBee tree routing, and locally addresses the orphan problem without the reconfiguration of a whole system. Consequently, a ZigBee cluster-tree network can smoothly scale up.

Network Address and Port Translation (NAPT) is one of the NAT techniques to transparently map a set of private network addresses and corresponding TCP/UDP ports to a small set of public network addresses and ports. In the private network, the hosts are uniquely assigned private addresses for identification. The NAT router provides address translation service by maintaining mapping entries of private address/port and public address/port. As mentioned in Section 2, the endpoint in ZigBee acts as a port in TCP/IP to identify different applications. By combining endpoints with network addresses, the NAT approach is basically applicable in ZigBee networks. Each NAT-ZigBee router transparently assigns the private address and translates the network address/endpoint for its NAT children, so the other devices are not affected.

Our NAT-ZigBee framework follows a multi-level NAT structure where a NAT router could have one or more NAT routers as its child, such that more orphans can be accommodated to further increase network scalability. Figure 1 shows an example of our NAT-ZigBee architecture. Each top-level NAT router establishes a respective private network, and data packets can be transparently forwarded to destinations with address translations level by level.

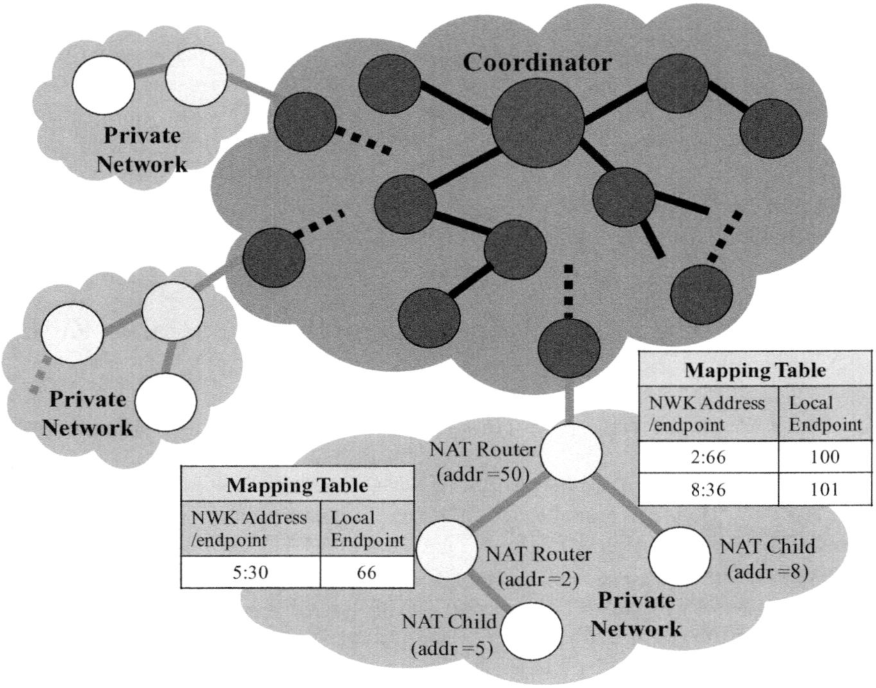

Fig. 1. The system architecture of NAT-ZigBee

3.2 Performance Evaluation

In this subsection, the capability of the proposed NAT-ZigBee is demonstrated. In the simulation model, the formation of cluster-tree topology is based on ZigBee specification [5], and the transmission range of each device is 20 meters.

In this experiment, 200 sensor devices are randomly distributed in a square terrain, and the ZigBee coordinator is placed in the center of the square. Figure 2 shows the orphan number under different network parameters. The original cluster-tree address assignment scheme specified in the ZigBee standard is used for comparison baseline. The performance of the proposed NAT-ZigBee is not compared with that in [10] since they are complementary works and can work well with each other. In the experiment, the performance of NAT-ZigBee is evaluated with the private network address spaces 5, 10, and 20.

Figure 2 (a) shows the effects of terrain length on the orphan number. It is observed that the occurrence of orphans increases as the terrain increases, since the sparse topology reduces the associations in the network. In this figure, the proposed NAT-ZigBee significantly decreases the number of orphans even with a extremely small private address space (i.e., 5). Moreover, this figure shows that the orphan problem seldom occurs when the private address is greater than 10 which is still a small portion against the whole address space.

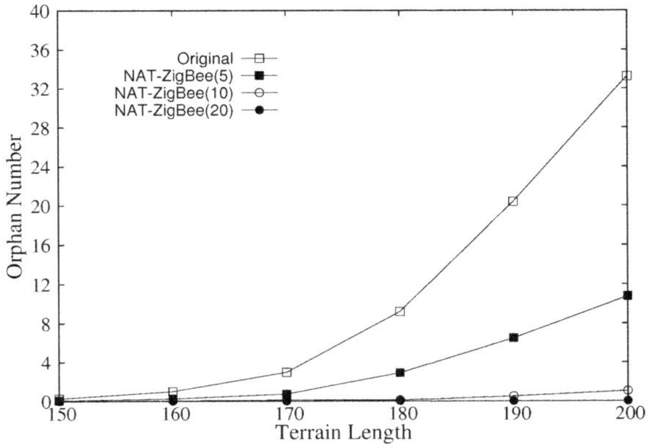

(a)

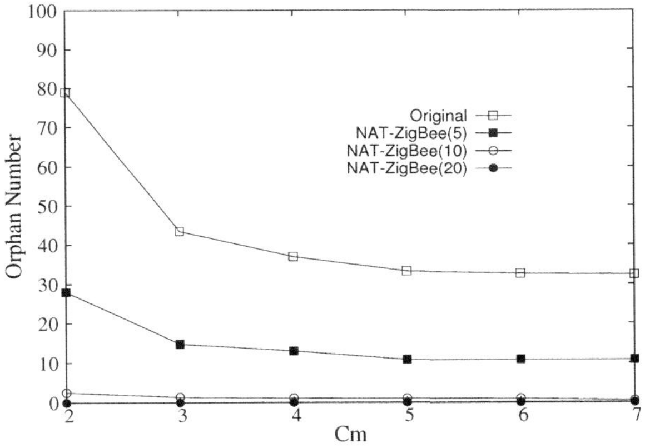

(b)

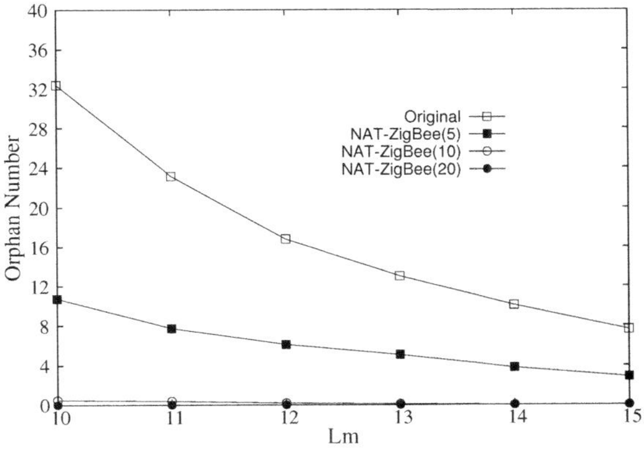

(c)

Fig. 2. The performance of the proposed NAT-ZigBee

Figure 2 (b) and (c) shows the effects of Cm and Lm on the orphan number in a 200×200 terrain. Without loss of generality, Rm is set to be equal to Cm in the experiment. Also, Figure 2 (b) adopts the setting of $Lm = 10$ and Figure 2 (c) adopts the setting of $Cm = 5$. Since the network formulation can be more easily achieved by large configuration parameters, the curves decreases as Cm and Lm increase. However, Figure 2 (b) indicates that the orphan problem cannot be eliminated even under a large Cm due to the topology restriction. Specifically, these figures show that NAT-ZigBee has a significant performance improvement even under a small Cm and Lm. Therefore, with NAT-ZigBee, the network can operate well even with inappropriate parameters settings.

4 Conclusion

ZigBee supports low-rate wireless communications for short-range data transmissions. The ZigBee cluster-tree topology is suitable for wireless sensor applications due to its power-saving operation and light-weight routing. However, the orphan problem in a ZigBee cluster-tree network leads to the difficulty in system scalability. In this paper, we proposed an NAT-based address assignment framework, called NAT-ZigBee, for a scalable ZigBee network. NAT-ZigBee locally resolves the orphan problem without the reconfiguration of a whole network while the light-weight ZigBee tree routing is maintained. The experimental results indicated that NAT-ZigBee can significantly reduce the number of orphan devices compared to the original ZigBee address assignment scheme.

Acknowledgements. This work was supported in part by the Excellent Research Projects of National Taiwan University under Contracts 99R80304 and 99R80300, by the National Science Council under Contract NSC98-2221-E-002-200-MY3, and by Institute for Information Industry.

References

1. 802.15.4-2003 IEEE Standard for Information Technology-Part 15.4: wireless medium access control (MAC) and physical layer (PHY) specifications for low-rate wireless personal area networks (LR-WPANs) (2003)
2. Akyildiz, I.F., Sankarasubramaniam, Y., Su, W., Cayirci, E.: A Survey on Sensor Networks. IEEE Comm. Magazine 40, 102–114 (2002)
3. Ali, M., Uzmi, Z.A.: An Energy-efficient Node Address Naming Scheme for Wireless Sensor Networks. In: IEEE International Networking and Communications Conference, INCC (2004)
4. ZigBee Alliance. ZigBee-Pro Stack (January 2008)
5. ZigBee Alliance. ZigBee Specifications (October 2007)
6. Bernardos, C., Calderon, M., Moustafa, H.: Survey of IP Address Autoconfiguration Mechanisms for MANETs. Internet Draft (October 2007)
7. Mohsin, M., Prakash, R.: IP Address Assignment in a Mobile Ad Hoc Network. In: MILCOM (November 2002)

8. Nesargi, S., Prakash, R.: MANETconf: Configuration of Hosts in a Mobile Ad Hoc Network. In: IEEE INFOCOM (April 2002)
9. Pan, M.-S., Fang, H.-W., Liu, Y.-C., Tseng, Y.-C.: Address Assignment and Routing Schemes for ZigBee-based Long-Thin Wireless Sensor Network. In: IEEE Vehicular Technology Conference, VTC (2008)
10. Pan, M.-S., Tseng, Y.-C.: The Orphan Problem in ZigBee-based Wireless Sensor Networks. In: ACM/IEEE International Symposium on Modeling, Analysis and Simulation of Wireless and Mobile Systems, MSWiM (October 2007)
11. Schurgers, C., Kulkarni, G., Srivastava, M.B.: Distributed On-demand Address Assignment in Wireless Sensor Networks. IEEE Transaction on Parallel and Distributed System 13(10), 1056–1065 (2002)
12. Srisuresh, P., Holdrege, M.: IP Network Address Translator (NAT) Terminology and Considerations. RFC 2663 (August 1999)
13. Sun, Y., Belding-Royer, E.M.: A Study of Dynamic Addressing Techniques in Mobile Ad Hoc Networks. Wireless Communications and Mobile Computing 4(3), 315–329 (2004)
14. Weniger, K.: PACMAN: Passive Autoconfiguration for Mobile Ad Hoc Networks. IEEE Journal on Selected Areas in Communications (JSAC) (March 2005)
15. Weniger, K.: Passive Duplicate Address Detection in Mobile Ad Hoc Networks. In: IEEE Wireless Communications and Networking Conference (WCNC), New Orleans, USA (March 2004)
16. Yen, L.-H., Tai, W.-T.: Flexible Address Configurations for Tree-Based ZigBee/IEEE 802.15.4 Wireless Networks. In: International Conference on Advanced Information Networking and Applications (2008)
17. Zhou, H., Mutka, M.W., Ni, L.M.: Reactive ID assignment for sensor networks. In: IEEE Mobile Adhoc and Sensor Systems Conference, MASS (2005)

Interference Aware Routing and Load Balancing in Wireless Sensor and Actuator Networks

Shekar Nethi and Riku Jäntti

Department of Communications and Networking, P.O. Box 13000,
School of Science and Technology, Aalto University, Finland
{shekar.nethi,riku.jantti}@aalto.fi
http://www.aalto.fi//

Abstract. In wireless sensor and actuator network (WSANs) there usually exists critical data, such as alarms and emergent event reports, that require short time delivery with high reliability but data volume is relatively small. At the same time, there exists larger volume of bulk sensor data which typically is delay tolerant and should be transmitted in the most energy efficient manner. On the 2.4 GHz ISM-band, the coexistence of the WSANs with WLANs imposes great challenge for designing reliable WSAN protocols. In this paper, we propose new sink initiated routing protocol that supports the two above mentioned traffic classes and is aware of the interference situation of the network. Critical data is transmitted using most interference resilient and reliable paths that minimize the expected transmission delay while the bulk data is transmitted in the most energy efficient manner by applying load balancing over several paths. The use of load balancing aims to even out the energy consumption of the intermediate nodes and thus maximize the operation time of the network. Our routing protocol estimates the link reliability based on the strength and activity of the interference as well as the path loss of the link and is able to adapt the routes based on changes in the propagation conditions and WLAN interference.

Keywords: WSANs, Interference Aware Routing, Load Balancing, Multipath routing.

1 Introduction

Wireless Sensors and Actuator Networks (WSANs) enable designing of novel and intriguing applications that can be used to address numerous industrial, environmental, societal and economical challenges. In WSANs, low-power sensors collect information about the physical world which is then transmitted wirelessly to actuators. Based on this reported information, actuators perform desired actions. Often in WSANs, there is no need for real-time communication, however, in real-time process control it is essential that the measurements and other critical data are delivered with minimum delay and maximum reliability. Recent years has seen growing industrial interest towards using WSANs for real-time applications, such as process control. Traditional control theory is based on the

X. Zhang and D. Qiao (Eds.): QShine 2010, LNICST 74, pp. 316–327, 2011.

assumption of deterministic measurements with fixed delays. There are several approved and reliable solutions exist for wired communications. However, these solutions are not feasible for WSANs due to the random delays introduced by the underlying wireless medium access protocols.

Another challenging issue is the coexistence of high power wideband wireless local area networks (WLAN) and low power wireless sensor and actuator networks (see Fig. 1). In case of IEEE 802.11b and g transmitters, one can find three IEEE 802.15.4 channels that are orthogonal to the WLAN channels. That is, they only experience adjacent channel interference which is at least 30 dB lower than the interference on the signal band. For IEEE 802.11n, the situation gets worse and there is only single IEEE 802.15.4 channel that only experience adjacent channel interference. The general conclusion is that coexistence on the same band is possible if there is enough spatial separation between the systems. Increasing transmission power could be an exciting approach against WLAN interference; however this introduces self interference within 802.15.4 networks. Also it is not trivial to make an optimal power control scheme, since the actual transmission power does not scale linearly with the power settings available on the devices. Another alternative is to use multipath routing, i.e. if there is enough spatial separation between relay nodes, then each next-hop node would experience different channel interference. Hence, multipath routing to some extent inherits resilience against WLAN interference.

Also, a common feature among WSANs applications is that the data will be gathered to a sink for the decision making process and hence, it is feasible to assume that a GW node exists in all WSAN applications. Some examples applications are Greenhouses, Condition Monitoring, Crane Control System and Energy Production System. Therefore, communication protocols should take into account these special properties of WSANs, such as delay-constrained, reliability and network topology, i.e. convergence of data towards the GW.

In this paper, we introduce a real time networking protocol called Sink Initiated Routing Protocol (SIRP). SIRP is designed especially for time-constrained wireless sensor and actuator networks (WSANs). SIRP combines interference aware routing and load-balancing techniques, to achieve robustness and improve

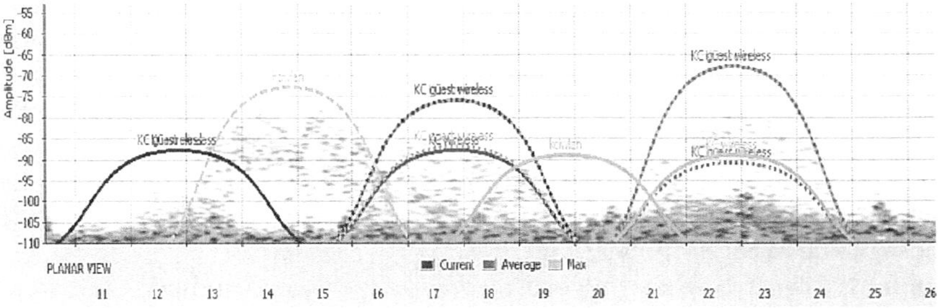

Fig. 1. Demonstration of spectrum measurement results in an industrial warehouse. The numbers on the x-axis denote IEEE 802.15.4 channels and measured power levels are represented on the y-axis.

energy efficiency of the network. Also our protocol enables global time synchronization by utilizing periodic initialization packets. Since distances between neighboring nodes are generally small in WSANs, one way time synchronization can be exploited, such as [1]. This way time synchronization hierarchy and network topology will be updated every round and thus, the impact of node failures will last only until the end of that particular round. The performance of SIRP will be compared to other existing solutions and using simulations we show that SIRP outperforms alternative designs. The rest of the paper is organized as follows. In Section II we briefly review related work and the proposed SIRP will be introduced in Section III. Simulation results are presented in Section IV. Finally, Section V concludes the paper.

2 Related Work

Due to convergent nature of data flow in many WSANs applications, the communication is predominantly from sensors to sink nodes. Standard routing solutions for other types of networks (e.g. ZigBee, MANET) cannot be employed directly to WSANs, ZigBee which is based on Adhoc On-demand Distance Vector Routing (AODV) routing protocol [5] is mainly intended to discover a single route between a source and destination node. AODV is not an optimal solution for WSANs, since it introduces higher power consumption and requires significant routing overhead to accomplish many to one communication.

Heterogeneous nature of WSANs is a good reason to employ hierarchical routing, thus, more complicated data processing operations can be transferred to cluster heads. Due to data aggregation and fusion in the cluster heads, the number of transmitted messages in the WSAN can be significantly reduced, and, consequently, the energy efficiency increased. As a representative of hierarchical routing methods in WSANs, we consider the Ripple-Zone (RZ) routing scheme [2], the introduced hierarchical routing algorithm uses the concept of a Ripple-Zone (RZ) around each actuator. Sensors are assigned to different ripples based on their distances in number of hops from the actuator. In each ripple, some sensors are chosen to perform as masters based on the Topology Discovery Algorithm (TDA) previously proposed by the authors. Each master collects data from the sensors in its zone and then transmits data to a master in the next ripple that is closer to the actuator. The Ripple-Zone (RZ) routing scheme was compared with LEACH [3], ZRP [4] and flat-topology based schemes in terms of energy efficiency. The authors have proven that their protocol is energy efficient, reliable and scalable, moreover, it can adapt to changing network topology by employing the local link failure repair method. However, it would be interesting to know, how the scheme performs in terms of latency, since it is a crucial point in most of the WSAN applications.

Similarly, the tree-based algorithm has a few shortcomings 1)Existence of bottleneck, especially nodes closer to the sink 2)They are more susceptible to node failure. SIRP is a multipath routing protocol; hence we implement load-balancing algorithm which minimizing these bottle necks. There have been developed a number of multipath routing protocols, Adhoc On-demand Multipath

Distance Vector Routing (AOMDV) [6] that is an extension to the AODV protocol for computing multiple loop-free and link-disjoint paths. In our simulation results we show that AOMDV is expensive with respect to bandwidth and power consumption.

Therefore, the solution presented in this paper addresses the requirements of a real-time control system and takes into consideration the challenges of the industrial environment in the system design. Even though the solution addresses different aspects of the complex problem, it still remains simple and easy to implement on top of IEEE 802.15.4 compliant platforms.

3 Protocol Description

In wireless sensor and actuator network there usually exists critical data that require short time delivery but data volume is relatively low (such as fire alarm, emergent event report). These data must be treated separately since they are time constrained. Therefore routing protocol should establish interference resilient links to support such communication requirements. Hence our protocol support two types of routes minimum delay Routes and a backup interference resilient route in case of critical data. SIRP achieves the following features:

1. Actuator/sensor communication and actuator/sensor communication.
2. Establishes multiple minimum delay routes for transfer of critical data.
3. Dynamic interference aware routing algorithm for successful operation in challenging environment.
4. Protocol simplicity.
5. Scalability
6. Increased network life using load balancing.

3.1 Routing Metric

In this section, we present the routing criteria to establish minimum delay routes. In SIRP, each node estimates the channel access time (CAT), waiting time of a given packet before transmission begins. CAT is a measure of the MAC layer congestion in the local neighborhood of the node and it is counted for each transmission attempt separately. That is, it does not consider retransmission delays due to unreliable transmission - link reliability is already considered in the interference part of the routing metric. In actual implementation, this task could be done by implementing a timer function, which computes the average delay of a packet before the transmission occurs. More details on route establishment will be covered in sub-section 3.4.

An estimated delay of each *(i,j)* link can be found by Little's Law.

$$\hat{T} = \frac{\hat{N}}{\hat{\lambda}} \tag{1}$$

where $\hat{\lambda}$ is given by Eq. 2 and $\hat{N}$ the expected number of packets in queue,

$$\hat{\lambda}(t+1) = (1-\alpha)\hat{\lambda}(t) + \alpha x(t), \tag{2}$$

with $\alpha \in (0,1), x(t) = \begin{cases} 0, & \text{no arrival} \\ 1, & \text{arrival} \end{cases}$

3.2 Estimating Strength and Activity of the Interferer

Fig. 2 clearly illustrates that 'activity' and 'strength' of the WLAN interferer has significant effect on the packet delivery of 802.15.4 nodes. Our goal is to offer interference resilience routes in case of critical data transmission. We assume that using existing energy detection capabilities of 802.15.4 nodes, it is possible to determine the *strength* of a WLAN interfere. This assumption is reasonable since there already exists various methodology to reliably estimate interference level, see [11]-[12]. During network initialization, each node will distributes this information to all its neighbors. Nodes are then able to evaluate the corresponding link *SINR*.

Periodic channel sensing is done to estimate the activity of the interferer using the following equation $\frac{\tau_{busy}^{i}}{T}$, where τ_{busy}^{i} is the time fraction that channel is sensed busy for node i, and T is the normalized overall sensing time. Based on the recently estimated activity and strength of the interferer, a transmitting node computes the drop probability of the link from the Fig. 2.

3.3 Load Balancing for Bulk Data

Multipath routing is a good means to provide resilience by transmitting multiple packet copies over multiple paths, and to distribute energy utilization among nodes by load balancing. An advantageous feature of WSANs that encourages

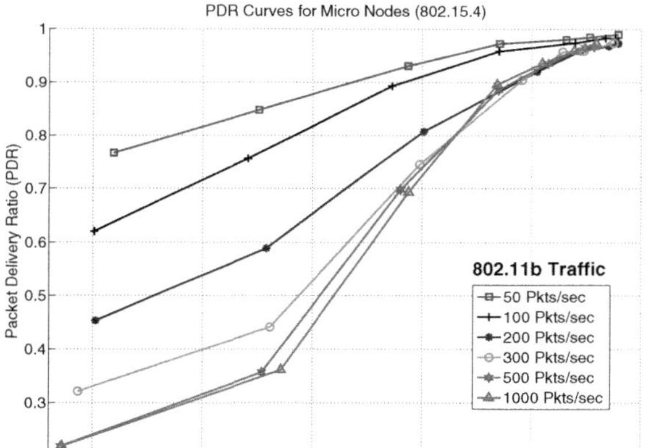

Fig. 2. Experimental packet delivery curves for micro nodes (802.15.4 compatible) under co-existence with 802.11b WLAN interferer

employing multipath routing is high node density due to which there exist many paths with dissimilar link quality as that of the primary path. Load balancing scheme for WSANs is considered in SIRP. The main idea is to reduce utilization of network resources by splitting the data packet into k packets and sending these packets on multiple paths. The number of packets sent on each nexthop node depends on the corresponding link *SINR*, recent activity of the channel and residual battery energy of the prospective node.

3.4 Sink Initiated Routing Protocol

This section present in detail our proposed routing scheme, which integrates all functionalities describe in sub-section 3.13.2 & 3.3.

1. In SIRP actuator or gateway initiates the routing procedure by sending a broadcast messages. The proposed algorithm runs locally on each node and moves on as a *wave* from the GW. The broadcast message for $node_i$ consists of the following fields.
   - Hop count: Number of hops to the gateway.
   - Interference Level (dB) at $node_i$: We assume that using energy detection capabilities, devices are able to estimate the strength of the WLAN interferer. Channel is scanned periodically and the interference level is updated.
   - T is the accumulated delay information carried by the incoming broadcast message. τ_i is the expected delay at $node_i$. Therefore, the total accumulated delay from $node_i$ to the gateway is computed as $(T + \tau_i)$.
   - Energy left in the battery of $node_i$
   - Sequence number of the broadcast wave, on every new broadcast the gateway increments the sequence number. This serves as an indication to all receiving nodes that route entries corresponding to older sequence number should be purged.
   - Packet Time stamping is done to achieve global time synchronization, which is essential for innumerable WSAN applications, for example industrial automation [7].
2. Up on receiving broadcast message, each relay node should add the new route entry only:
   - if the accumulated delay time is less than the maximum tolerable delay defined by the application,
   - if the accumulated delay is greater than the node will simply drop the broadcast message, since the delay requirement does not satisfy the required quality of service.

   Further, each nodes should now be able to compute the corresponding *SINR* of the link, this information is stored along with other routing information. In our simulations, we have fixed the maximum number of routes entries for a node.

3. Instead of rebroadcasting these packets immediately, a node chooses to defer its transmission for a random period. The defer time is a function of number of hops from the gateway, that is, higher the number longer is the defer time. This approach reduces the number of routing messages re-broadcasted considerably, thus avoiding collision of routing packets and consequently increasing the probability of route establishment. It also allows nodes further in the network to receive multiple broadcasts from all its precedors. For each intermediate node that has received a new broadcast message (that is, the sequence number is greater), we control the route entry procedure to accommodate multiple routes. If an intermediate $node_j$ receives another copy of broadcast message from the different relay node it will check if its accumulated delay of the forward path is larger than its application requirement,
4. Upon completion of defer timer, intermediate nodes will re-broadcast the message with their own interference level, battery residue value and accumulated delay T. This procedure continues until all the nodes have establishes forward path to the sink.
   (a) As said earlier each node scans the channel periodically and updates signal level and activity of the interferer. Now if a node has some critical data to forward,
      – A node simply selects the highest *SINR* link and transmit the packet, thus maximizing the reliability of the transmission.
   (b) If a node has some bulk data to forward:
      – Depending on the activity of interferer and link SINR, a node computes drop probability for each link in its routing table. It then does load balancing over a reduced set of links, which assures reliability under current interference activity.
5. Reverse path are then established when data packets from the sensors are sent back to the sink. Each intermediate node on receiving data packets makes a route entry to the source of the date packet. That is when the data packet reaches the sink it would have established a reverse path to the source of the packet i.e. sensor.
6. The frequency of routing initialization phase is controlled by the sink node. That is depending on the network performance and dynamics of the channel a sink node could increase or decrease the route initialization period. This approach complies well with the centralized architecture of wireless sensors and actuator networks.
7. In order to provide additional resilience against link failure, overhearing of messages and subsequent route updates are allowed. Finally, if a link failure is detected a route error message is broadcast. Intermediate nodes with active route information to the sink will quickly respond to these messages.

4 Simulation Results

This section presents the performance of the SIRP scheme in wireless sensors and actuators networks. The impact of the WLAN interference is taken into account. SIRP is compared with the conventional AODV, AOMDV routing protocols. The results for load balancing are also presented.

4.1 Simulation Environment

The implementation is done in ns-2 simulator [8]. This simulator is easily available public domain tool and is considered a de facto standard for implementation and verification of wireless networks. Since there is no support for co-simulation of WLAN and WSANs in ns-2, we have developed a trace based WLAN interference model. The interference model is integrated into WSANs simulations to emulate co-existence scenario. In this approach, we initially configure interference traffic and run simulations for 802.11b networks. The output trace file captures all the distances, timing and packet information of the simulation. This information is then used during WSANs simulations. For instance, when a 802.15.4 compatible node sends a packet, we check from the trace file, if any WLAN transmission occurred in that interval, based on the distance between the interferer and the destination node and the pdrcurves, we evaluate if the current packet should be dropped or not. Similarly, channel sensing and interference level are also integrated. One limitation in this approach is that WLAN network is independent of 802.15.4 network. That is 802.15.4 nodes are invisible to WLAN.

4.2 Resilience against Interference

Simulation is carried out with 50 nodes randomly distributed over an area of 1000x1000m2. Nodes are moving according to the random waypoint model. The traffic generation for 802.15.4 nodes is fixed at 8Pkts/sec. First, we present results obtained for the interference case, Fig. 3a & 3b show packet delivery ratio and average end to end delay plotted versus different WLAN traffic conditions. For both low and high interference scenario, SIRP consistently shows a 10 % increase in packet delivery over single path AODV and multi path AOMDV, whereas significant increase is seen at low interference traffic (approx. 25 % on AOMDV and approx. 50 % for AODV). In case of average end-to-end delay SIRP clearly outperforms other protocols maintaining a min delay of 0.2sec. These results show us the effectiveness of using minimum delay routing metric combined with interference avoidance algorithm.

With load-balanced methods, nodes in a network are involved into network activity with an more evenly distributed manner. This gives fairness in term of energy consumption. Fig. 3c & 3d. shows the mean battery residue and standard deviation of AODV and AOMDV and SIRP for different WLAN traffic.

4.3 Scalability

In this simulation, nodes are randomly scattered in an area of 50x50m2. We can see from Fig. 4 that SIRP reduces the overall amount of collisions using random defer time, better connectivity is achieved which also improves the overall packet delivery. Unlike AODV & AOMDV, SIRP is a sink initiated routing protocol, which means, the number of route initializations is limited by the number of sinks in the network. This approach is well suited for WSANs as it considerably reduces the overall control overhead. In case of AODV & AOMDV, each source would have to initiate a RREQ mechanism. Hence increasing the overall control messages and subsequently packet collisions.

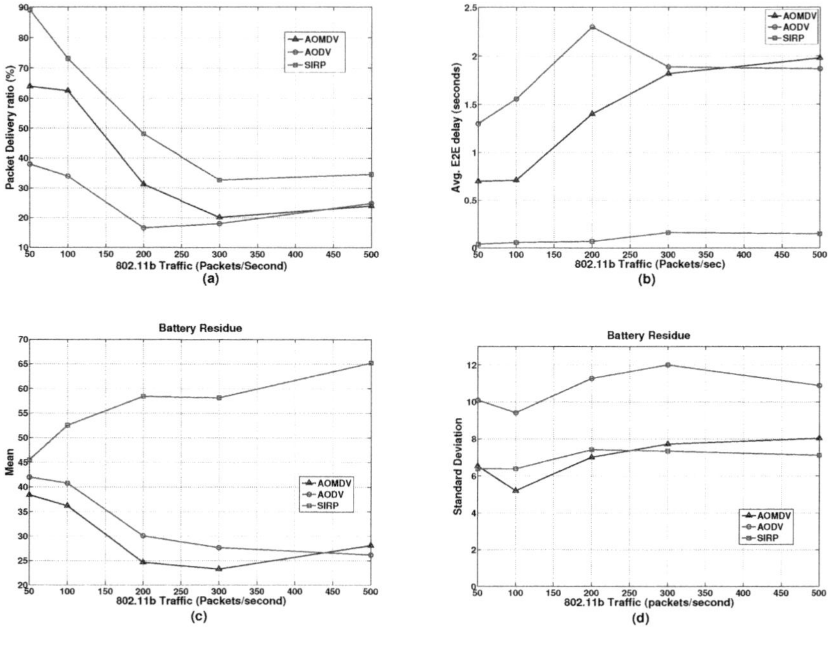

Fig. 3. Simulation Results a) Packet Delivery Ratio b) Average End-to-End Delay c) Mean Battery Residue and d) Standard Deviation

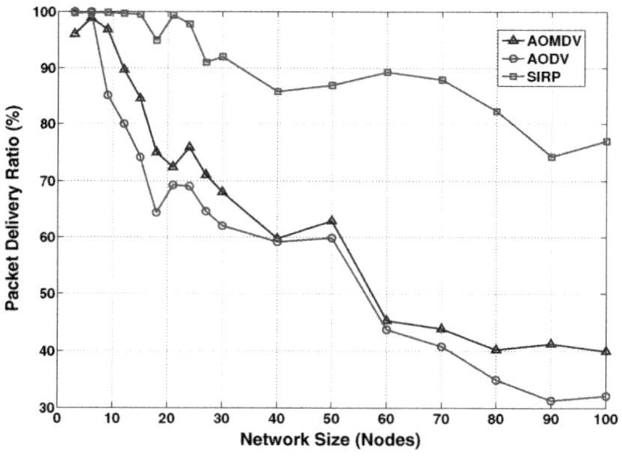

Fig. 4. Scalability for WSANs

4.4 Case Study: Wireless Air Conditioning System (WACS)

Wireless Air Conditioning System (WACS) maintains the cooling of the industrial hall at the desired temperature (21°C). Here we have a wireless sensor grid network overlaid on the industrial hall; the measurements are taken at a

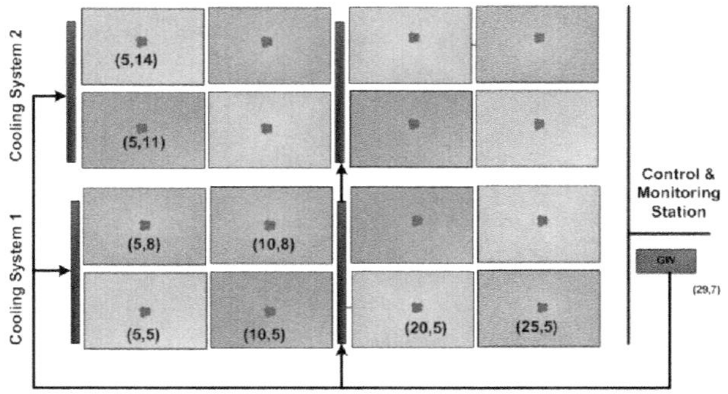

Fig. 5. Wireless Networked Controlled Application

rate of 1/min. The measurements are sent to the central control system access point which is right of the industrial hall. This access point is connected to the wired building automation system that can handle intelligent control tasks and performs cooling operation. The heat sources causing the warming of inside temperature are the lathe machines and the crane control system. Our physical model for ACU considers heat transfer between adjacent areas. We modelled an realistic wireless networked control application in PiccSIM [10]. The objective is to maintain the temperature of an industrial hall at an desired value (21°C). The industrial hall is divided in to 4 segments, each segment represents 4 sensor nodes wirelessly communicating measured temperature value to the command control. Hence depending on the packet delivery of that node the temperature estimate are computed and the subsequent control (heating/cooling) is performed.

The results in Table 1 indicate that SIRP achieves minimum control cost over AODV and AOMDV for ricean k factor equal to 0, which corresponds to maximum drops in the channel. However when the network size is small and minimum drops are seen (for k=20), there would not be any link failures, in such cases, AODV seems to have slightly better cost function than multipath SIRP.

Table 1. Control Cost (integral of squared error (ISE) between desired and actual temperature) of each cells for ricean k factor 0 & 20 respectively

	AODV	AOMDV	SIRP
k=0	13.7059	15.354	12.7822
k=20	10.3857	14.6046	12.459

The numbers of sensors are significantly higher than the sinks. This means, in case of AODV, AOMDV each sensor initiates a routing mechanism, that is more control overhead for AODV and AOMDV, therefore significantly higher energy consumption. Fig. 6c presents normalized routing overhead for AODV, AOMDV and SIRP. The convergence of all the links to the sink makes SIRP, a practical approach for wireless sensors and actuators networks. SIRP improves reliability

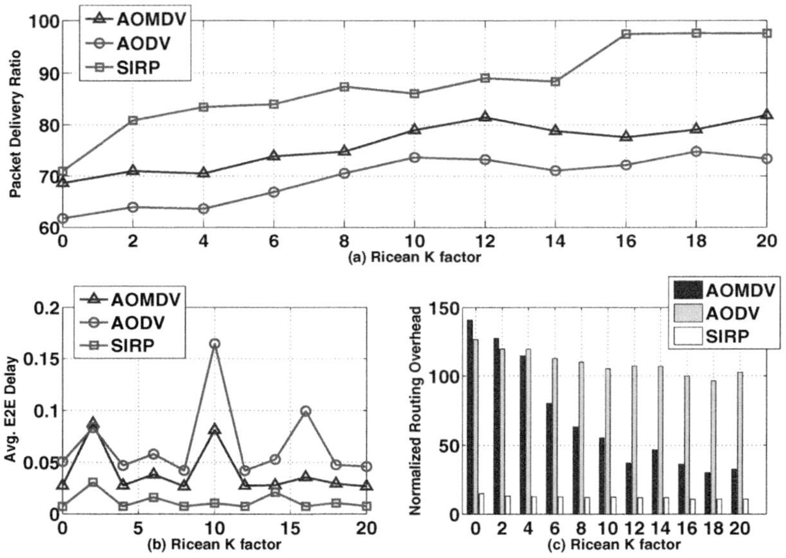

Fig. 6. Packet delivery ratio, Average end to end delay and Routing overhead for AODV, AOMDV and SIRP protocols

(Fig. 6a), minimizes average end to end delay (Fig. 6b) and is an energy efficient routing protocol (Fig. 6c).

5 Conclusions

In this paper we have presented a novel routing protocol, called Sink Initiated Routing Protocol, which is designed particularly for Wireless Sensor and Actuator Networks (WSANs). The proposed protocol achieves higher throughput while offering low packet transmission delays. Scalability is an inherent property due to the localized nature of operation. Frequency of routing initialization phase is adapted to the channel conditions and is used to deliver time synchronization and routing information which further enables rapid recovery from topology changes. The protocol is also robust and performs well under interference which is of significant importance for WSANs operating on unlicensed frequency bands. The protocol has been implemented on ns-2 and we showed a comprehensive set of simulation results, including a real-world industrial application scenario, to confirm that SIRP is suitable for WSANs.

Acknowledgments. This research work is supported by TEKES (Finnish Funding Agency for Technology and Innovation) as part of the Wireless Sensor and Actuator Networks for Measurement and Control (WiSA-II) program.

References

1. Mahmood, A., Jäntti, R.:Time synchronization accuracy in real-time wireless sensor networks. In: 2009 IEEE 9th Malaysia International Conference on Communications (MICC), pp. 652–657 (December 15-17, 2009)
2. Fei, H., Xiaojun, C., Kumar, S., Sankar, K.: Trustworthiness in Wireless Sensor and Actuator Networks: Towards Low-Complexity Reliability and Security. In: IEEE Globecom, November 28-December 2, vol. 3, pp. 1696–1700 (2005)
3. Heinzelman, W.R., Chandrakasan, A., Balakrishnan, H.: Energy-efficient communication protocol for wireless microsensor networks. In: Proceedings of the 33rd Annual Hawaii International Conference on System Sciences, January 4-7, vol. 2 (2000)
4. Haas, Z.J.: A new routing protocol for the reconfigurable wireless networks. In: IEEE 6th International Conference on Universal Personal, October 12-16, vol. 2, pp. 562–566 (1997)
5. Perkins, C.E., Royer, E.M.: Ad-hoc on-demand distance vector routing. In: Proceedings of Second IEEE Workshop on Mobile Computing Systems and Applications, WMCSA 1999, February 25-26, pp. 90–100 (1999)
6. Marina, M.K., Das S.R.: On-demand multipath distance vector routing in ad hoc networks. In: Ninth International Conference on Network Protocols, 2001, pp. 14–23 (2001)
7. Gungor, V.C., Hancke, G.P.: Industrial Wireless Sensor Networks: Challenges, Design Principles, and Technical Approaches. IEEE Transactions on Industrial Electronics 56, 4256–4258 (2009)
8. Network Simulator - ns-2, http://www.isi.edu/nsnam/ns/
9. Nethi, S., Pohjola, M., Erikkson, L., Jäntti, R.: Simulation case studies of wireless networked control systems. In: Proceedings of the 2nd ACM Workshop on Performance Monitoring and Measurement of Heterogeneous Wireless and Wired Networks (October 22-22, 2007)
10. Nethi, S., Pohjola, M., Erikkson, L., Jäntti, R.: Platform for Emulating Networked Control Systems in Laboratory Environments. In: Proc. IEEE International Symposium on a World of Wireless, Mobile and Multimedia Networks (IEEE WoWMoM 2007), Helsinki, Finland (June 18-21, 2007)
11. Razvan, M.E., Andread, T.: Minimising the effect of WiFi interference in 802.15.4 wireless sensor networks. International Journal of Sensor Networks 3(1), 43–54 (2007)
12. Mahmood, A., Jäntti, R.: A Decision Theoretic Approach for Channel Ranking in Unlicensed Bands. Journal in Wireless Networks (2010)

Sensor Distribution on Coverage in Sensor Networks

Miao Peng[1], Yang Xiao[1,*], Hui Chen[2], Qi Hao[3],
Athanasios V. Vasilakos[4], and Jie Wu[5]

[1] Dept. of Computer Science, University of Alabama, Tuscaloosa, AL 35487
[2] Dept. of Math and Computer Science, Virginia State University, Petersburg, VA 23806
[3] Electrical and computer engineering, Tuscaloosa, AL 35487
[4] Dept. of Computer and Telecom. Engineering, Univ. of Western Macedonia, Greece
[5] Dept. of Computer & Information Science, Temple University, Philadelphia, PA 19122
mpeng@crimson.ua.edu, {yangxiao,huichen}@ieee.org,
qh@eng.ua.edu, vasilako@ath.forthnet.gr, jiewu@temple.edu

Abstract. In this paper, we study the impacts of sensor node distributions on network coverage. We first show the impacts on network coverage by adopting different sensor node distributions through both analytical and simulation studies. Then, we adopt a distribution-free approach to study network coverage, in which no assumption of probability distribution of sensor node locations is needed. The proposed approach has yielded good estimations of network coverage.

Keywords: Coverage, sensor network, distribution.

1 Introduction

In most previous work concerning network coverage problems where sensors are deployed randomly, researchers assume the spatial distributions of sensor nodes are known when evaluating their proposed algorithms or protocols. Major disadvantages of such an analysis method include: 1) it is very difficult to choose an accurate sensor location distribution; 2) inaccurate distribution assumption will result in poor analysis of protocols or algorithms; and 3) changes in sensor distributions may lead to variations in system performance sometimes even invalidate the whole analysis.

Motivated by this intuition, we propose a network coverage analysis approach in which no assumption on sensor location distribution is required. Thus, the approach is in effect a distribution-free approach. The approach is suitable to solve network coverage problem concerning a great number of sensors, which are deployed randomly.

We summarize the contribution of the paper as follows, 1) we evaluated the effects of sensor location distribution via both analytical modeling and computer simulations, and have concluded that accurate sensor location distribution is important to assessment of sensor networks where a great number of sensors are randomly deployed; 2) we then propose a distribution-free sensor network modeling approach, which uses a non-parametric statistical approach; 3) we verify the approach by using our previous work in [10] as an example.

[*] Corresponding author.

X. Zhang and D. Qiao (Eds.): QShine 2010, LNICST 74, pp. 328–344, 2011.

2 Related Work

A sensor network may contain a large number of simple sensor nodes. Sensor nodes are often powered by batteries, and hence have to operate on limited energy budgets. Furthermore, it is difficult to replace batteries in the sensors deployed in inaccessible or inhospitable environments. Thus, many research efforts have been spent on the energy conservation of sensor nodes to extend sensor network life time [13]. The network lifetime is defined as the time between the initialization of the network and the first case of battery exhaustion among sensor nodes. Extending the network lifetime has been extensively researched [1-3]. A lot of protocols keep a subset of sensor nodes vigilant for sensing and communication tasks while putting the others in power-save mode [4]. On the other hand, energy efficiency should not be achieved at the cost of reduced network coverage and connectivity. Thus, the network coverage and connectivity have also been considered simultaneously in some researches [5-8]. There are many related papers in sensor networks [21-109].

In [9], the authors studied a network with sensor nodes deployed strictly in grids. Plenty of work focus on sensor networks where sensor locations follow a Poisson point process and sensors are uniformly distributed in sensing fields, e.g., [19] and [20]. In [18], barrier coverage problems are studied when sensors are distributed along the line with random offsets due to wind and other environmental factors. In [8], the authors investigate energy efficiency in more general sensor networks where the sensor nodes are deployed randomly. In [10], the authors study a randomized scheduling algorithm where sensors are uniformly distributed. The paper [14] proposes a worst and average case algorithm for coverage calculation from a perspective of computational geometry where no sensor location distribution is required. Nevertheless, little work has been done where no prior knowledge of sensor node location distribution is required.

This paper studies the impact of sensor location distributions on network coverage and provides a distribution-free approach in which no assumption on sensor location distribution is required and sensor locations can be in any distribution. To the best of our knowledge, no literature is found to apply distribution-free approach to sensor network coverage problems.

3 Coverage Intensity

3.1 Coverage Intensity

Assume n sensors are randomly deployed to form a wireless sensor network to cover a field, which we refer to as the sensing field. The sensor network runs a randomized scheduling algorithm. The randomized scheduling algorithm is given as follows. Let S denote the set of all the n sensor nodes. Let S be divided into k disjoint subsets S_j $(j = 1, 2, ..., k)$ and each sensor node is randomly assigned to one of these subsets. At any time, only one subset of sensor nodes are active and the rest of sensor nodes are inactive. The objective is to extend the network life time and maintain satisfactory coverage. We measure the coverage using coverage intensity.

Network coverage intensity is the ratio of the time when a point in the field of the sensor network is covered by at least one active sensor node to the total time. We model the sensor node deployment field as a two-dimensional Cartesian coordination system. The field ranges from *0* to X and *0* to Y on X- and Y-axis respectively. Assume that the sensing area of a sensor is the area of a circle and the sensing range of sensors is R, the radius of the circle. Let $f(x,y)$ denote the probability density function of sensor node locations. Actual deployment of sensor nodes may be unknown, and $f(x,y)$ can be any distribution. Let $P(g,h)$ denote the probability that a given point (g,h) is covered by at least one sensor node. We have

$$P(g,h) = \iint_{(x-g)^2+(y-h)^2 \le R^2} f(x,y)dxdy \tag{1}$$

Since n sensors are divided into k disjoint subsets, which take turns to wake up and perform sensing tasks while the rest of the subsets are in power-save mode. Then the probability that point (g,h) is covered by an active sensor can be written as

$$C(g,h) = 1 - [1 - P(g,h)/k]^n \tag{2}$$

Coverage intensity is the detection metrics for the whole network. Note that point (g,h) is randomly chosen from the sensing field. Thus, the network coverage intensity for the network is

$$C_n = E(C(g,h)) \tag{3}$$

It is worth noting that in the above discussion, no assumption on sensor location distribution is given, and the sensor location distribution can be any distribution, which can even be a distribution which has no explicit form.

The above derivation does not consider edge effect. Since the whole sensing field must have boundaries, a coverage area of a sensor node may not be completely inside the whole sensing field, which we refer to as the edge effect. The computer simulations in Section V show that the error rate between the simulation and analytical results is very small and can be neglected when the number of sensors is large.

3.2 Uniform Distribution

Assume that sensors are uniformly deployed in the sensing field. This case is studied in detail in [10]. For comparison purpose, we reformulate the coverage intensity using the result obtained in previous subsection. Sensor location (g,h) follows a two-dimensional uniform distribution, namely $f(x,y) = 1/(XY)$. Plug this into equations (1)-(3), we can obtain the network coverage intensity for the two dimensional uniform distribution.

$$P^U(g,h) = \iint_{(x-g)^2+(y-h)^2 \le R^2} \frac{1}{XY} dxdy = \frac{\pi R^2}{XY} \tag{4}$$

$$C^{U}(g,h)=1-[1-\frac{\pi R^{2}}{kXY}]^{n} \tag{5}$$

$$\begin{aligned} C_{n}^{U} &= E(C(g,h))=\int_{0}^{Y}\int_{0}^{X}\frac{1}{XY}\left\{1-[1-\frac{\pi R^{2}}{kXY}]^{n}\right\}dxdy \\ &=1-[1-\frac{\pi R^{2}}{kXY}]^{n} \end{aligned} \tag{6}$$

where we use superscript U to indicate that sensor locations follow a two-dimensional uniform distribution.

3.3 Two-Dimensional Gaussian Distribution

Assume that sensor nodes deployed in the sensing field follow a two-dimensional Gaussian distribution. The probability density function of the two-dimensional Gaussian distribution is given as

$$f(x,y)=\frac{1}{2\pi\sigma^{2}}e^{-[(x-X/2)^{2}+(y-Y/2)^{2}]/2\sigma^{2}}$$

Plugging this into (1), we have

$$P^{G}(g,h)=\iint_{(x-g)^{2}+(y-h)^{2}\le R^{2}}\frac{1}{2\pi\sigma^{2}}e^{-[(x-X/2)^{2}+(y-Y/2)^{2}]/2\sigma^{2}}dxdy$$

where subscript G indicates that sensor locations follow a two-dimensional Gaussian distribution.

Let $x'=x-g$ and $y'=y-h$,

$$P^{G}(g,h)=\iint_{x'^{2}+y'^{2}\le R^{2}}\frac{1}{2\pi\sigma^{2}}e^{-[(x'+g-X/2)^{2}+(y'+h-Y/2)^{2}]/2\sigma^{2}}dx'dy'$$

Let $x'=l\sin\theta$, $y'=l\cos\theta$, and $|J|=|\frac{\partial(x',y')}{\partial(l,\theta)}|=l$,

$$\begin{aligned} P^{G}(g,h) &= \int_{0}^{R}\int_{0}^{2\pi}\frac{1}{2\pi\sigma^{2}}e^{-[(l\sin\theta+g-X/2)^{2}+(l\cos\theta+h-Y/2)^{2}]/2\sigma^{2}}|J|\,dld\theta \\ &= \int_{0}^{R}\int_{0}^{2\pi}\frac{1}{2\pi\sigma^{2}}e^{-[(l\sin\theta+g-X/2)^{2}+(l\cos\theta+h-Y/2)^{2}]/2\sigma^{2}}\,ldld\theta \end{aligned} \tag{7}$$

Plug (8) into (2) and (3), we have,

$$C^{G}(g,h)=1-[1-P^{G}(g,h)/k]^{n} \tag{8}$$

$$C_n^G = E(C^G(g,h)) \tag{9}$$

3.4 GU Distribution

In this part, we assume that known sensors location distribution is the one along the x-axis, where sensor locations follow a Gaussian distribution with a mean of $X/2$, and along the y-axis, where sensor locations follow a uniform distribution with a mean of $Y/2$. For simplicity, we name this two-dimensional distribution as a GU distribution. Similar to the above, we need to calculate the probability $P(g,h)$ to obtain coverage intensity under a GU distribution. Thus, we have

$$P^{GU}(g,h) = \iint_{(x-g)^2+(y-h)^2 \le R^2} f(x)f(y)dxdy$$

where $f(x) = \frac{1}{\sqrt{2\pi}\sigma_x} e^{-\frac{(x-X/2)^2}{2\sigma_x^2}}$ and $f(y) = \frac{1}{Y}$. Note that superscript GU indicates that sensor locations follow a GU distribution.

Following the similar steps in previous subsection, we have

$$P^{GU}(g,h) = \int_0^R \int_0^{2\pi} \frac{1}{\sqrt{2\pi}\sigma_x} e^{-\frac{(l\sin\theta+g-X/2)^2}{2\sigma_x^2}} \frac{1}{Y} ldld\theta \tag{10}$$

$$C^{GU}(g,h) = 1-[1-P^{GU}(g,h)/k]^n \tag{11}$$

$$C_n^{GU} = E(C^{GU}(g,h)) \tag{12}$$

4 Distribution-Free Approach

In this section, we introduce the distribution-free approach for estimating coverage intensity. The approach uses a non-parametric statistical method [11], [16]. It does not require that the sensor location distribution to be known. Instead, it requires the locations of a few sensors among the deployed sensors.

There are many studies regarding sensor node localization. Common localization approaches rely on a few sensor anchor or beacon nodes whose locations are known in advance, for example, via GPS signals. Thus, we can have a few sensors whose locations can be accurately determined. Due to random factors in real world, such as wind, sensor location distributions are impossible to be exactly the same as assumed distributions. Since inaccurate knowledge on sensor location distributions can yield misleading or invalid network coverage estimations, we propose a distribution-free approach to estimate network coverage intensity. The approach is not based on an assumed distribution. Instead, it is based on the locations of a sample of sensor nodes whose locations are known.

In the rest of this section, we first present how we infer sensor location distribution from the locations of a sample of sensor nodes using a non-parametric statistical method, called Kernel-density estimation [11], [16]. Next, we describe the distribution-free method.

4.1 Infer Sensor Location Distribution from Locations of Sample Sensor Nodes

Denote the locations of randomly selected sample nodes as (X_i, Y_i), $i = 1, 2, \ldots, N$, where N is the sample size. The probability density at any point (x, y) can be estimated using the locations of the sample of sensor nodes, i.e.,

$$\hat{f}_h(x, y) = \frac{1}{N h_x h_y} \sum_{i=1}^{N} K\left(\frac{x - X_i}{h_x}, \frac{y - Y_i}{h_y} \right) \tag{13}$$

where $K(\square)$ is some kernel, and h_x and h_y are smoothing factors or window-width. It is quite often that $K(\square)$ is taken to be a standard Gaussian function with mean 0 and variance 1, i.e.,

$$K(u, v) = \frac{1}{2\pi} e^{-\frac{1}{2}(u^2 + v^2)} \tag{14}$$

Plugging (14) into (13), we get,

$$\begin{aligned} \hat{f}_h(x, y) &= \frac{1}{N h_x h_y} \sum_{i=1}^{N} K\left(\frac{x - X_i}{h_x}, \frac{y - Y_i}{h_y} \right) \\ &= \frac{1}{N h_x h_y} \sum_{i=1}^{N} \frac{1}{2\pi} e^{-\frac{1}{2}\left(\frac{(x - X_i)^2}{h_x^2} + \frac{(y - Y_i)^2}{h_y^2} \right)} \end{aligned} \tag{15}$$

Note that 1) window-width h_x and h_y indirectly control the variance of the Gaussian function, and 2) probability density functions to be estimated can be multi-modal [16] and by no means have to be Gaussian even though the kernel is a Gaussian function.

Choices of N, h, and $K(\square)$ are the factors determining the efficiency and effectiveness of the estimation of the probability density.

4.2 Distribution-Free Coverage Intensity Estimation

The approach has four steps, 1) obtaining the locations of the sample sensor nodes; 2) analyzing the locations and obtaining the window-width (h_x and h_y); 3) approximating sensor location distribution using Kernel-density estimation; 4) based on the Kernel-density estimation, calculating the coverage intensity.

Though N and $K(\square)$ are also factors related to the efficiency and effectiveness of the approach, they are determined empirically before sensor deployment in this paper. The above four steps are carried out after sensor deployment without using any assumed sensor location distribution.

The coverage intensity is calculated as follows. Replacing $f(x, y)$ in (1) by (13), we get,

$$\begin{aligned} P^{DF}(g,h) &= \iint_{(x-g)^2+(y-h)^2\le R^2} \hat{f}_h(x,y)dxdy \\ &= \iint_{(x-g)^2+(y-h)^2\le R^2} \frac{1}{Nh_xh_y}\sum_{i=1}^{N} K\left(\frac{x-X_i}{h_x},\frac{y-Y_i}{h_y}\right)dxdy \end{aligned} \quad (16)$$

where superscript DF indicates we are using the distribution-free approach. Plugging (16) into (2) and (3), we have,

$$C^{DF}(g,h) = 1-[1-P^{DF}(g,h)/k]^n \quad (17)$$

$$C_n^{DF} = E(C^{DF}(g,h)) \quad (18)$$

5 Simulation Verification

In this section, we perform computer simulations to verify our analytical model presented in Section III. We developed a discrete event simulation program in C++. In our program, there are scheduling events, intrusion starting events, and intrusion departure events. The program is capable of loading any sensor deployment configuration. In our simulations below, sensor nodes are deployed randomly in the sensing field. The purposes of this section are 1) to demonstrate that the analytical model in Section III is accurate; 2) the edge effect is neglectable. For coping with limited space, we show only the results for GU distributions for the first purpose. For the second purpose, we show only the results for the two-dimensional uniform distributions.

In this section, the standard deviation (σ_x) of Gaussian distribution along the x-axis is *20*, the number of deployed sensor nodes (n) is *1000*, the size of the whole sensing field is *10000*, the sensing area of each sensor is 30, and the number of subsets is 4, unless otherwise stated.

Fig. 1 shows the network coverage intensity *vs.* the number of sensor nodes with both analytical and simulation results. The figure shows that the analytical results match exactly with the simulation results. In addition, the network coverage intensity increases as the number of sensor nodes increases, and when the number of disjointed subsets (k) increases, the network coverage intensity becomes smaller.

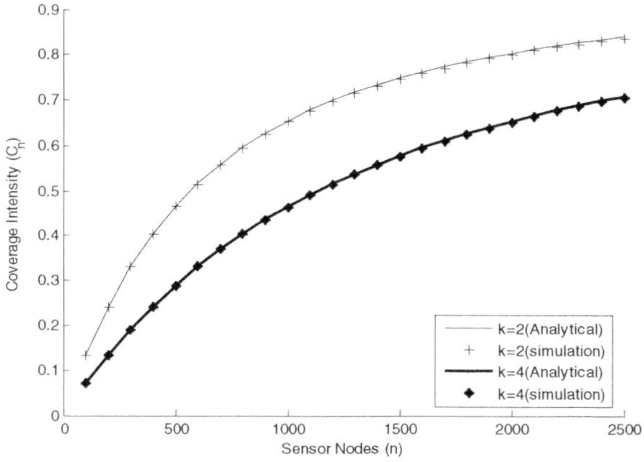

Fig. 1. Coverage Intensity *vs.* n

6 Impacts of Sensor Location Distribution on Network Coverage Estimation

In this section, we show the impacts of inaccurate sensor location distribution on network coverage estimation. Intuitively, the discrepancy between actual and estimated network coverage would occur when the knowledge of the sensor location distribution is inaccurate. We intend to demonstrate that the discrepancy is so great that the inaccurate sensor location distributions may in effect render the network coverage estimation worthless and misleading. This section is organized as follows. 1) We compare the calculated coverage intensity when sensor locations follow two-dimensional uniform and two-dimensional Gaussian distributions respectively. This case can be interpreted as that actual sensor location distribution is a two-dimensional Gaussian distribution; however, we assume the distribution is a two-dimensional uniform distribution; or vice versa. 2) Similarly, we next compare the calculated coverage intensity between two-dimensional uniform and GU distributions.

The coverage intensity for uniform distributions is calculated using equation (6), that for two-dimensional Gaussian distributions using equation (9), and that for GU distributions using equation (12). We choose $X = 100$, $Y = 100$, and $R = 3$ unless otherwise stated.

6.1 Two-Dimensional Gaussian and Uniform Distributions

Fig. 2 shows the coverage intensity *vs.* the number of sensor nodes (n) for both two-dimensional Gaussian distributions and Uniform distributions.

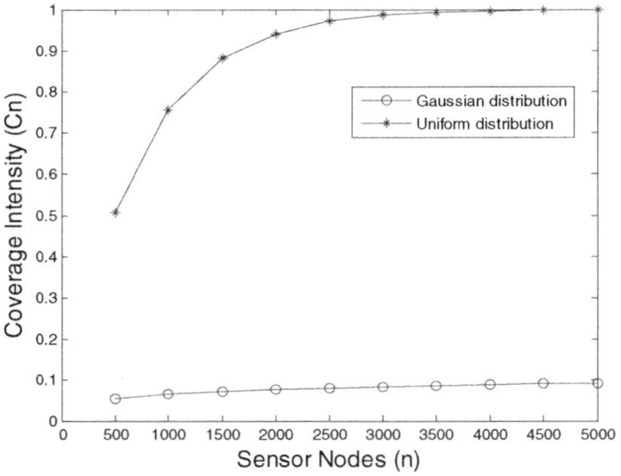

Fig. 2. C_n *vs.* n ($k = 2, \sigma = 5$)

7 Example and Evaluation of Distribution-Free

In this section, we are to demonstrate how to apply the distribution-free approach to estimate network coverage intensity. As discussed in Section IV, three factors affect the effectiveness and efficiency of the approach. The three factors are kernel $K(\square)$, sample size N, and windows-width h_x and h_y. Literature has shown that Gaussian function is a good choice to estimate probability density for continuous random variables using Kernel-density estimation method [16]. Note that probability density functions to be estimated can be multi-modal and by no means have to be Gaussian even though the kernel is a Gaussian function. Nevertheless, we have to determine sample size and windows-width beforehand. In subsection VII.A, we present some discussion on the sample size and the window-width. In subsection VII.B, we present a complete example of the distribution-free approach, and compare the result obtained from the distribution-free approach with that obtained from actual distribution.

7.1 Sample Size

Larger number of sample sensor nodes lead to better estimation of network coverage. Large sample can be obtained by deploying large number of anchor or beacon sensor nodes, or determining accurate locations of large number of sensor nodes, which is either expensive or difficult to achieve. However, when too few sample sensor nodes are chosen, the network coverage estimation can be inaccurate. In this paper, we use a simple method to determine the sample size. The method requires a few number of field experiments,

1. Deploy n sensors in a sensing field via a desirable vehicle, e.g., an aircraft or a rocket. Obtain the locations of all the sensors. The sensors are treated as a population, and we calculated the mean and the variance of the locations of the sensors. Denote the population mean and the population variance as $\overline{Y}$ and S^2 respectively.

2. Randomly select a small number of sensors. The sensors constitute a sample. Obtain their locations. Calculate the mean and the variance of the locations. Denote the sample mean and the sample variance as $\overline{y}$ and s^2 respectively.
3. Calculate the error between the sample mean and the population mean, and denote it as $r = \left(\overline{y} - \overline{Y}\right)/\overline{Y}$.
4. As suggested in [12], the proper sample size is estimated as $n = n_0/\left(1 + n_0/N\right)$ where $n_0 = \left[\left(u_{\alpha/2}S\right)/\left(r\overline{Y}\right)\right]^2$ and $u_{\alpha/2}$ is the value of the vertical boundary for the area of $\alpha/2$ in the right tail of the standard normal distribution.

Repeat the above steps for a few times to converge to the desired sample size.

7.2 Example and Evaluation of Distribution-Free Approach

Step 1:Obtain Locations of Sample Sensors

Before the sensor node deployment, according to the number of sensor nodes deployed in sensor network, we decide the number of sample sensor nodes, and randomly select the sample sensor nodes and equip them with proper components such as GPS receivers to become anchor or beacon nodes. Second, after random deployment, the locations of the sample sensor nodes are obtained via a sensor localization protocol. The locations of the sample sensors are (X_i, Y_i), $i = 1, 2, \ldots N$, where N is the sample size.

Step 2:Window-Width (h)

Many numerical methods have been developed to find h, and they mostly minimize the so-called Mean Integrated Squared Error [16]. In our experiment, we use a fast and accurate bivariate kernel density estimator as in [16] to obtain the values of window-width (h_x and h_y). For example, we obtain the bivariate window-width as $(h_x, h_y) = (3.88, 16.71)$.

Step 3:Distribution Estimation

Based on the sample location coordinates from step 1 and bivariate window-width from step 2, the density function can be calculated using equation (15) since we use Gaussian function as the kernel.

Step 4: System Performance Evaluation

In this step, we can use the estimated density function to calculate the network coverage intensity using equations (16)-(18). Fig. 3 shows the estimation results.

Fig. 3 shows the network coverage intensity *vs.* the number of sensor nodes for Uniform distribution, GU distribution and the Estimated GU distribution, where the standard deviation of Gaussian distribution along the x-axis is *5* and the number of

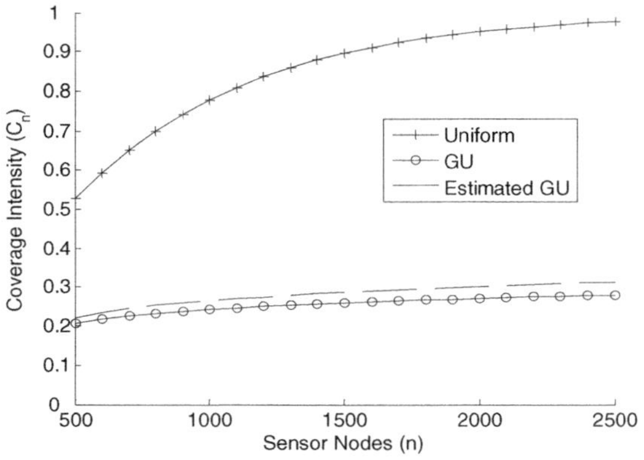

Fig. 3. Estimation performance (size of sample=50)

disjointed subsets is *2*. In the experiment, the size of the whole sensing field is *10000*; the sensing area of each sensor is 30. In Fig.3, in sensor network, the number of whole deployed sensors varies from *500* to *2500*, but we only use *50* sample sensors to estimate the distribution through the kernel density estimation method.

8 Conclusion

Network coverage problems are important to wireless sensor networks. Previous works are based on assumed probability density functions that govern the distribution of sensor nodes in the sensing field. However, the actual distribution of sensor nodes may be very different from the assumed one. Our analytical and simulation studies show that when a different assumption is used, the introduced error on the network coverage metrics can be very large and cannot be neglected.

In this paper, we first reformulate the network coverage intensity using general probability distribution. In other words, we do not assume that the sensor location distributions are known. We verified the formulization using computer simulations, which show that the analytical results and computer simulations match exactly.

Most importantly, we proposed a distribution-free approach for estimating network coverage intensity. In our proposed method, no assumption on sensor location distribution is required. Instead, we take a small sample of the actual deployment, and carry on a statistical analysis to capture the distribution function of the deployment. In practice, this small set of sample can be the sensor nodes equipped with GPS receivers, and thus their locations are known. Furthermore, we used the kernel density estimator to estimate the deployment distribution. Based on the obtained knowledge, the network coverage metrics can be calculated.

The results show that a small sample of sensor nodes yields fairly good estimates on the distribution used. In particular, compared with the case that a different assumption (the uniform distribution) than actual sensor location distribution (GU distribution) is used, the distribution-free approach yields far better results.

Acknowledgement. This work is supported in part by the US National Science Foundation (NSF) under the grant numbers CCF-0829827, CNS-0716211, and CNS-0737325.

References

1. Sankar, A., Liu, Z.: Maximum Lifetime Routing in Wireless Ad Hoc Networks. In: Proc. IEEE Infocom, pp. 1089–1097 (2006)
2. Chang, J.H., Tassiulas, L.: Maximum Lifetime Routing in Wireless Sensor Networks. IEEE/ACM Transactions on Networking 12(4), 609–619 (2004)
3. Madan, R., Luo, Z. Q., Lall, S.: A Distributed Algorithm with Linear Convergence for Maximum Lifetime Routing in Wireless Sensor Networks. In: Proc. of the Allerton Conference on Communication, Control and Computing (2005)
4. Abrams, Z., Goel, A., Plotkin, S.: Set k-cover Algorithms for Energy Efficient Monitoring in Wireless Sensor Networks. In: Proc. of IPSN (2004)
5. Liu, C., Wu, K., Xiao, Y., Sun, B.: Random Coverage with Guaranteed Connectivity: Joint Scheduling for Wireless Sensor Networks. IEEE Transactions on Parallel and Distributed Systems 17(6), 562–575 (2006)
6. Gupta, H., Das, S., Gu, Q.: Connected Sensor Cover: Self-organization of Sensor Networks for Efficient Query Execution. In: Proc. of Mobihoc (2003)
7. Wang, X., Xing, G., Zhang, Y., Lu, C., Pless, R., Gill, C.: Integrated Coverage and Connectivity Configuration in Wireless Sensor Networks. In: Proc. of Sensys (2003)
8. Zhang, H., Hou, J.: Maintaining Coverage and Connectivity in Large Sensor Networks. In: Proc. of WTASA (2004)
9. Shakkottai, S., Srikant, R., Shroff, N.: Unreliable Sensor Grids: Coverage, Connectivity and Diameter. In: Proc. of INFOCOM (2003)
10. Xiao, Y., Chen, H., Wu, K., Sun, B., Liu, C.: Modeling Detection Metrics in Randomized Scheduling Algorithm in Wireless Sensor Networks. In: Proc. of WCNC (2007)
11. Hastie, T., Tibshirani, R., Friedman, J.: The Elements of Statistical Learning - Data Mining, Inference, and Prediction. Springer Series in Statistics (2001)
12. Chaudhuri, A., Stenger, H.: Survey Sampling: Theory and Methods, 2nd edn. CRC, Boca Raton (2005)
13. Wang, L., Xiao, Y.: A Survey of Energy-Efficient Scheduling Mechanisms in Sensor Networks. Mobile Networks and Applications (MONET) 11(5), 723–740 (2006)
14. Meguerdichian, S., Koushanfar, F., Potkonjak, M., Srivastava, M.B.: Coverage Problem in Wireless Ad-hoc Sensor Networks. In: Proc. of INFOCOM (2001)
15. Savvides, A., Koushanfar, F., Potkonjak, M., Srivastava, M.B.: Location Discovery in Ad-hoc Wireless Sensor Networks. UCLA EE and CS Departments
16. Botev, Z. I.: Nonparametric Density Estimation via Diffusion Mixing. Postgraduate Series, Department of Mathematics, The University of Queensland (2007), `http://espace.library.uq.edu.au/eserv/UQ:120006/diffusion_estimator.pdf`, `http://www.mathworks.com/matlabcentral/fileexchange/authors/27236`
17. Cardei, M., Wu, J.: Coverage in Wireless Sensor Networks. In: Ilyas, M., Magboub, I. (eds.) Handbook of Sensor Networks. CRC Press, Boca Raton (2004)

18. Saipulla, A., Westphal, C., Liu, B., Wang, J.: Barrier Coverage of Line-Based Deployed Wireless Sensor Networks. In: Proc. of IEEE INFOCOM 2009, pp. 127–135 (2009)
19. Bandyopadhyay, S., Coyle, E.: An Energy Efficient Hierarchical Clustering Algorithm for Wireless Sensor Networks. In: Proc. of IEEE INFOCOM 2003, pp. 1713–1723 (2003)
20. Wan, P.J., Yi, C.W.: Coverage by Randomly Deployed Wireless Sensor Networks. IEEE Transactions on Information Theory 52(6), 2658–2669 (2006)
21. Hadi, K., Krishna, C.M.: Management of target-tracking sensor networks. International Journal of Sensor Networks 8(2), 109–121 (2010)
22. Li, M., Chen, H.: Editorial. International Journal of Sensor Networks 8(2), 63–64 (2010)
23. Schooler, E.M., Livadas, C., Kim, J., Gandhi, P., Passera, P.R., Chandrashekar, J., Orrin, S., Koyabe, M., El-Moussa, F., Dabibi, G.D.: Collaborative defence as a pervasive service: architectural insights and validation methodologies of a trial deployment. International Journal of Sensor Networks 8(2), 65–76 (2010)
24. Jardak, C., Rerkrai, K., Kovacevic, A., Riihijarvi, J., Mahonen, P.: Design of large-scale agricultural wireless sensor networks: email from the vineyard. International Journal of Sensor Networks 8(2), 77–88 (2010)
25. Alazzawe, A., Wijesekera, D., Dantu, R.: A testbed for large mobile social computing experiments. International Journal of Sensor Networks 8(2), 89–97 (2010)
26. Liang, X., Chen, M., Xiao, Y., Balasingham, I., Leung, V.C.M.: MRL-CC: a novel cooperative communication protocol for QoS provisioning in wireless sensor networks. International Journal of Sensor Networks 8(2), 98–108 (2010)
27. Xia, X., Liang, Q.: Latency-aware and energy efficiency tradeoffs for wireless sensor networks. International Journal of Sensor Networks 8(1), 1–7 (2010)
28. Singh, K., Muthukkumarasamy, V.: Key establishment protocols using environmental and physiological data in wireless sensor networks. International Journal of Sensor Networks 8(1), 8–26 (2010)
29. Stabellini, L., Zander, J.: Energy-efficient detection of intermittent interference in wireless sensor networks. International Journal of Sensor Networks 8(1), 27–40 (2010)
30. Ham, M.I., Rodriguez, M.A.: A boundary approximation algorithm for distributed sensor networks. International Journal of Sensor Networks 8(1), 41–46 (2010)
31. Ansari, J., Zhang, X., Mahonen, P.: Multi-radio medium access control protocol for wireless sensor networks. International Journal of Sensor Networks 8(1), 47–61 (2010)
32. Hyodo, K., Wakamiya, N., Nakaguchi, E., Murata, M., Kubo, Y., Yanagihara, K.: Reaction-diffusion based autonomous control of wireless sensor networks. International Journal of Sensor Networks 7(4), 189–198 (2010)
33. Wang, D.: Clustering mesh-like wireless sensor networks with an energy-efficient scheme. International Journal of Sensor Networks 7(4), 199–206 (2010)
34. Chen, Y., Nasser, N., El Salti, T., Zhang, H.: A multipath QoS routing protocol in wireless sensor networks. International Journal of Sensor Networks 7(4), 207–216 (2010)
35. Al-Rousan, M., Landolsi, T., Kanakri, W.M.: Energy consumption considerations in dynamic wireless sensor networks with nodes and base stations mobility. International Journal of Sensor Networks 7(4), 217–227 (2010)
36. Kam, M., Leng, G.: On the power law relationship of the critical transmitting range and the number of nodes of ad hoc networks. International Journal of Sensor Networks 7(4), 228–235 (2010)

37. Yang, Y., Cardei, M.: Delay-constrained energy-efficient routing in heterogeneous wireless sensor networks. International Journal of Sensor Networks 7(4), 236–247 (2010)
38. Shi, W., Tang, X.: Editorial. International Journal of Sensor Networks 7(3), 125–126 (2010)
39. Han, Q., Hakkarinen, D., Boonma, P., Suzuki, J.: Quality-aware sensor data collection. International Journal of Sensor Networks 7(3), 127–140 (2010)
40. Zhang, Y., Meratnia, N., Havinga, P.J.M.: Ensuring high sensor data quality through use of online outlier detection techniques. International Journal of Sensor Networks 7(3), 141–151 (2010)
41. Bigrigg, M.W., Matthews, H.S., Garrett Jr., J.H.: Fault perturbations in building sensor network data streams. International Journal of Sensor Networks 7(3), 152–161 (2010)
42. Zhao, W., Liang, Y.: A systematic probabilistic approach to energy-efficient and robust data collections in wireless sensor networks. International Journal of Sensor Networks 7(3), 162–175 (2010)
43. Toledo, A., Wang, X.: Efficient multipath in wireless networks using network coding over braided meshes. International Journal of Sensor Networks 7(3), 176–188 (2010)
44. Sen, A., Das, N., Murthy, S.: Coverage and connected coverage problems for sensors embedded in a temperature-sensitive environment. International Journal of Sensor Networks 7(1/2), 106–123 (2010)
45. Tsiligiridis, T., Douligeris, C.: Editorial. International Journal of Sensor Networks 7(1/2), 1–3 (2010)
46. Zink, M., Lyons, E., Westbrook, D., Kurose, J., Pepyne, D.L.: Closed-loop architecture for distributed collaborative adaptive sensing of the atmosphere: meteorological command and control. International Journal of Sensor Networks 7(1/2), 4–18 (2010)
47. Lim, H., Iqbal, M., Wang, W., Yao, Y.: The National Weather Sensor Grid: a large-scale cyber-sensor infrastructure for environmental monitoring. International Journal of Sensor Networks 7(1/2), 19–36 (2010)
48. Bagchi, S.: A distributed algorithm for energy-aware clustering in WSN. International Journal of Sensor Networks 7(1/2), 37–43 (2010)
49. Li, S., Wang, X.: Source nodes localisation algorithm for large-scale wireless sensor networks using self-organising isometric mapping. International Journal of Sensor Networks 7(1/2), 44–52 (2010)
50. Terzis, A., Musaloiu-E, R., Cogan, J., Szlavecz, K., Szalay, A., Gray, J., Ozer, S., Liang, C., Gupchup, J., Burns, R.: Wireless sensor networks for soil science. International Journal of Sensor Networks 7(1/2), 53–70 (2010)
51. Liu, J., Jiang, X., Horiguchi, S., Lee, T.: Analysis of random sleep scheme for wireless sensor networks. International Journal of Sensor Networks 7(1/2), 71–84 (2010)
52. Lehsaini, M., Guyennet, H., Feham, M.: An efficient cluster-based self-organisation algorithm for wireless sensor networks. International Journal of Sensor Networks 7(1/2), 85–94 (2010)
53. Lai, Y., Chen, Y., Chen, H.: Continuous monitoring of global events in sensor networks. International Journal of Sensor Networks 7(1/2), 95–105 (2010)
54. Lin, J., Xie, L., Xiao, W.: Target tracking in wireless sensor networks using compressed Kalman filter. International Journal of Sensor Networks 6(3/4), 251–262 (2009)
55. Chen, H.: Editorial. International Journal of Sensor Networks 6(3/4), 129–130 (2009)

56. Tennina, S., Renzo, M.D., Graziosi, F., Santucci, F.: ESD: a novel optimisation algorithm for positioning estimation of WSNs in GPS-denied environments – from simulation to experimentation. International Journal of Sensor Networks 6(3/4), 131–156 (2009)
57. Bagci, F., Kluge, F., Ungerer, T., Bagherzadeh, N.: Optimisations for LocSens – an indoor location tracking system using wireless sensors. International Journal of Sensor Networks 6(3/4), 157–166 (2009)
58. Lee, J., Yao, K.: Exploiting low complexity motion for ad-hoc localisation. International Journal of Sensor Networks 6(3/4), 167–179 (2009)
59. Guo, S., Guo, M., Leung, V.C.M.: A message complexity oriented design of distributed algorithm for long-lived multicasting in wireless sensor networks. International Journal of Sensor Networks 6(3/4), 180–190 (2009)
60. Ni, C., Hsiang, T., Tygar, J.D.: A power-preserving broadcast protocol for wireless sensor networks. International Journal of Sensor Networks 6(3/4), 191–198 (2009)
61. Pedrosa, L.D., Melo, P., Rocha, R.M., Neves, R.: A flexible approach to WSN development and deployment. International Journal of Sensor Networks 6(3/4), 199–211 (2009)
62. Shuaib, A.H., Aghvami, A.H.: Dynamic topology control for the IEEE 802.15.4 network. International Journal of Sensor Networks 6(3/4), 212–223 (2009)
63. Wang, K., Jacob, J., Tang, L., Huang, Y.: Transmission error analysis and avoidance for IEEE 802.15.4 wireless sensors on rotating structures. International Journal of Sensor Networks 6(3/4), 224–233 (2009)
64. Li, X., Liu, X., Zhao, H., Jiang, N., Parashar, M.: ASGrid: autonomic management of hybrid sensor grid systems and applications. International Journal of Sensor Networks 6(3/4), 234–250 (2009)
65. Matrouk, K., Landfeldt, B.: Prolonging the system lifetime and equalising the energy for heterogeneous sensor networks using RETT protocol. International Journal of Sensor Networks 6(2), 65–77 (2009)
66. Comeau, F., Sivakumar, S.C., Robertson, W., Phillips, W.: Energy conservation in clustered wireless sensor networks. International Journal of Sensor Networks 6(2), 78–88 (2009)
67. Sundaresan, S., Koren, I., Koren, Z., Krishna, C.M.: Event-driven adaptive duty-cycling in sensor networks. International Journal of Sensor Networks 6(2), 89–100 (2009)
68. Kawai, T., Wakamiya, N., Murata, M.: Design and evaluation of a wireless sensor network architecture for urgent information transmission. International Journal of Sensor Networks 6(2), 101–114 (2009)
69. Zhou, S., Wu, M., Shu, W.: Improving mobile target detection on randomly deployed sensor networks. International Journal of Sensor Networks 6(2), 115–128 (2009)
70. Chiang, M., Byrd, G.T.: Adaptive aggregation tree transformation for energy-efficient query processing in sensor networks. International Journal of Sensor Networks 6(1), 51–64 (2009)
71. Li, Y., Mandoiu, I., Zelikovsky, A.: Editorial. International Journal of Sensor Networks 6(1), 1–2 (2009)
72. Su, W., Lim, T.L.: Cross-layer design and optimisation for wireless sensor networks. International Journal of Sensor Networks 6(1), 3–12 (2009)
73. Das, A.N., Popa, D.O., Ballal, P.M., Lewis, F.L.: Data-logging and supervisory control in wireless sensor networks. International Journal of Sensor Networks 6(1), 13–27 (2009)

74. Popa, D.O., Mysorewala, M.F., Lewis, F.L.: Deployment algorithms and indoor experimental vehicles for studying mobile wireless sensor networks. International Journal of Sensor Networks 6(1), 28–43 (2009)
75. Stanford, J., Tongngam, S.: Approximation algorithm for maximum lifetime in wireless sensor networks with data aggregation. International Journal of Sensor Networks 6(1), 44–50 (2009)
76. Bein, W.W., Bein, D., Malladi, S.: Reliability and fault tolerance of coverage models for sensor networks. International Journal of Sensor Networks 5(4), 199–209 (2009)
77. Zhang, Y., Xiao, Y., Bales, K.L.: Primate social systems, scent-marking and their applications in mobile and static sensor networks. International Journal of Sensor Networks 5(4), 210–222 (2009)
78. Pilakkat, R., Jacob, L.: A cross-layer design for congestion control in UWB-based wireless sensor networks. International Journal of Sensor Networks 5(4), 223–235 (2009)
79. Reddy, A.M.V., Kumar, A.V.U.P., Janakiram, D., Kumar, G.A.: Wireless sensor network operating systems: a survey. International Journal of Sensor Networks 5(4), 236–255 (2009)
80. Su, I.-F., Lee, C., Ke, C.: Radius reconfiguration for energy conservation in sensor networks. International Journal of Sensor Networks 5(4), 256–267 (2009)
81. Liu, J., Hong, X.: An online energy-efficient routing protocol with traffic load prospects in wireless sensor networks. International Journal of Sensor Networks 5(3), 185–197 (2009)
82. Yang, K., Li, J., Marshall, A., Ma, Y.: Editorial. International Journal of Sensor Networks 5(3), 127–128 (2009)
83. Chen, M., Kwon, T., Mao, S., Leung, V.C.M.: Spatial-Temporal relation-based Energy-Efficient Reliable routing protocol in wireless sensor networks. Journal: International Journal of Sensor Networks 5(3), 129–141 (2009)
84. Jeong, W., Nof, S.Y.: Design of timeout-based wireless microsensor network protocols: energy and latency considerations. International Journal of Sensor Networks 5(3), 142–152 (2009)
85. Li, X., Hunter, D.K.: Distributed coordinate-free algorithm for full sensing coverage. International Journal of Sensor Networks 5(3), 153–163 (2009)
86. Gasparri, A., Panzieri, S., Pascucci, F., Ulivi, G.: An Interlaced Extended Kalman Filter for sensor networks localisation. International Journal of Sensor Networks 5(3), 164–172 (2009)
87. Bi, Y., Sun, L., Li, N.: BoSS: a moving strategy for mobile sinks in wireless sensor networks. International Journal of Sensor Networks 5(3), 173–184 (2009)
88. Tan, L., Ge, F., Li, J., Kato, J.: HCEP: a hybrid cluster-based energy-efficient protocol for wireless sensor networks. International Journal of Sensor Networks 5(2), 67–78 (2009)
89. Takata, M., Bandai, M., Watanabe, T.: RI-DMAC: a receiver-initiated directional MAC protocol for deafness problem. International Journal of Sensor Networks 5(2), 79–89 (2009)
90. Guizani, S., Hamam, H., Du, X., Chen, H.: Ad hoc systems backboned by fibres: limitation and solutions. International Journal of Sensor Networks 5(2), 90–97 (2009)
91. Liu, J., Xiao, Y., Hao, Q., Ghaboosi, K.: Bio-inspired visual attention in agile sensing for target detection. International Journal of Sensor Networks 5(2), 98–111 (2009)
92. Fayed, M., Mouftah, H.T.: Localised convex hulls to identify boundary nodes in sensor networks. International Journal of Sensor Networks 5(2), 112–125 (2009)

93. Yeh, L., Wang, Y., Tseng, Y.: iPower: an energy conservation system for intelligent buildings by wireless sensor networks. International Journal of Sensor Networks 5(1), 1–10 (2009)
94. Wang, M., Cao, J., Liu, M., Chen, B., Xu, Y., Li, J.: Design and implementation of distributed algorithms for WSN-based structural health monitoring. International Journal of Sensor Networks 5(1), 11–21 (2009)
95. Doorn, B., Kavelaars, W., Langendoen, K.: A prototype low-cost wakeup radio for the 868 MHz band. International Journal of Sensor Networks 5(1), 22–32 (2009)
96. Iyengar, R., Kar, K., Banerjee, S.: Low-coordination wake-up algorithms for multiple connected-covered topologies in sensor nets. International Journal of Sensor Networks 5(1), 33–47 (2009)
97. Chen, M., Mao, S., Xiao, Y., Li, M., Leung, V.C.M.: IPSA: a novel architecture design for integrating IP and sensor networks. International Journal of Sensor Networks 5(1), 48–57 (2009)
98. Iyer, J.V., Yu, H., Kim, H., Kim, E., Yum, K., Mah, P.: Assuring K-coverage in the presence of mobility and wear-out failures in wireless sensor networks. International Journal of Sensor Networks 5(1), 58–65 (2009)
99. Zhang, J., Lok, T.M.: Cooperative protocols for multiple-source multiple-relay wireless networks. International Journal of Sensor Networks 4(4), 209–219 (2008)
100. Wang, G., Zhang, L., Cao, J.: Hole-shadowing routing in large-scale MANETs. International Journal of Sensor Networks 4(4), 220–229 (2008)
101. Yan, H., Li, J., Sun, G., Guizani, S., Chen, H.: A novel power control MAC protocol for mobile ad hoc networks. International Journal of Sensor Networks 4(4), 230–237 (2008)
102. Renesse, R.D., Khengar, P., Friderikos, V., Aghvami, A.H.: Quality of service adaptation in mobile ad hoc networks. International Journal of Sensor Networks 4(4), 238–249 (2008)
103. El-Hajj, W., Kountanis, D., Al-Fuqaha, A., Guizani, S.: A fuzzy-based virtual backbone routing for large-scale MANETs. International Journal of Sensor Networks 4(4), 250–259 (2008)
104. Gavalas, D., Pantziou, G., Konstantopoulos, C., Mamalis, B.: ABP: a low-cost, energy-efficient clustering algorithm for relatively static and quasi-static MANETs. International Journal of Sensor Networks 4(4), 260–269 (2008)
105. Brahim, G., Khan, B., Al-Fuqaha, A., Guizani, M.: Weak many vs. strong few: reducing BER through packet duplication in power-budgeted wireless connections. International Journal of Sensor Networks 4(3), 145–154 (2008)
106. Krishnamurthy, V., Sazonov, E.: Reservation-based protocol for monitoring applications using IEEE 802.15.4 sensor networks. International Journal of Sensor Networks 4(3), 155–171 (2008)
107. Chang, C.G., Snyder, W.E., Wang, C.: Secure target localisation in sensor networks using relaxation labelling. International Journal of Sensor Networks 4(3), 172–184 (2008)
108. Zeng, P., Zang, C., Yu, H.: Investigating upper bounds on lifetime for target tracking sensor networks. International Journal of Sensor Networks 4(3), 185–193 (2008)
109. Ferrari, G., Cappelletti, F., Raheli, R.: A simple performance analysis of RFID networks with binary tree collision arbitration. International Journal of Sensor Networks 4(3), 194–208 (2008)

Sensor-Aided Navigation in GPS-Denied Environments

Ka Yang, Daji Qiao, and Wensheng Zhang

Iowa State University, Ames, IA, 50010, USA
{yangka,daji,wzhang}@iastate.edu

Abstract. In this work we propose a novel approach to navigate users in GPS-denied environments with the help of sensors. The basic idea is to deploy wireless sensors over the field of interest and use the change of signal strength from multiple sensors to navigate users to the destination. To deal with RSS irregularity in a practical environment, an adaptive scheme is proposed to decide the scheme parameters online. We evaluated the proposed scheme through simulations and experiments. Results show that the proposed scheme navigates users to the destination successfully and efficiently with low movement overhead.

Keywords: Navigation, GPS-denied, Sensor, Received Signal Strength (RSS).

1 Introduction

Global Positioning System (GPS) based navigation has been widely used in both civil and military applications. To use GPS, a receiver should receive radio signal from at least four satellites in the outer space [10] simultaneously. Thus in situations where the satellite signals cannot reach the receiver, such as sand storm or blizzard, GPS may not function properly. Moreover, since the anti-satellite techniques have become more mature [9], it is likely that the satellite positioning systems may be disabled by adversaries during war time. Therefore, it is highly desirable to have advanced navigation schemes that work under GPS-denied environments.

Many navigation systems and methods have been devised to conduct navigation in GPS-denied environments [11]. We roughly classify them as *location based systems* and *direction based systems*. For location based systems, beacons are deployed at known locations. A user calculates its position based on beacon locations and signal information such as TDoA (Time Difference of Arrival), AoA (Angle of Arrival), etc. Then a user can find a path based on its location and the destination location. One drawback of such systems is that they require the location information for beacons and measurement of distance. For direction based systems, the user finds direction towards landmarks by detecting the direction of the arriving signal, which typically requires special directional antennas.

X. Zhang and D. Qiao (Eds.): QShine 2010, LNICST 74, pp. 345–361, 2011.

In this work, we propose to navigate a user to the destination in GPS-denied environments using only the information of signal strength, with low-cost sensor devices. Here we use "sensor" as a generic term to represent wireless devices that are low-cost and suitable for large-area deployment. The measurement of signal strength, usually known as received signal strength (RSS), is readily available on commercial wireless devices. Our work is motivated by the experimental observation that RSS and distance usually has a monotone relation. Therefore, the change of RSS can be used as an indicator of the distance relation. We show in [12] that ideally, based on the change of RSS, a user can be navigated to a destination efficiently.

However, in practice, the signal strength fluctuates and cannot reliably reflect distance. Based on our in-depth experimental observations, we propose an adaptive navigation scheme based on RSS from multiple sensors, which has the following features: i) It adaptively decides the scheme parameters according to the environment to deal with RSS irregularity while maintaining a low movement overhead. ii) It navigates the user to the destination solely based on the comparison of RSS along user's movement, which incurs very low computational overhead. iii) It only requires low-cost omni-directional wireless devices.

We have implemented a test bed system using 49 commercial sensor motes to evaluate the proposed navigation scheme. Experimental results show that the scheme can navigate a user to a destination successfully and efficiently with low movement overhead.

The rest of the paper is organized as follows. Section 2 reviews the related work. Section 3 explains the system components and the motivations. Sections 4 describes the proposed navigation scheme in detail. Section 5 presents the implementation of our test bed system and reports the experimental results. Section 6 concludes the paper.

2 Related Work

Using wireless sensors for navigation purposes has been proposed in recent years [5,1,4,6]. Previous works on navigation with sensor network focus on providing user the "safest" path to the destination. To move along the path, the user needs either location information (as in [5,1]) or the directions of signals from sensors (as in [4]).

The problem of navigation has been investigated extensively in the area of robotics. Typical navigation methods for robots utilize location, range and/or bearing information from landmarks [2]. In [6], the authors propose a range-only SLAM using wireless sensor networks. The algorithm uses a predetermined RSSI-distance model to measure distance and uses an extended Kalman Filter to deal with measurement errors from the sensed RSSI. RSSI-distance model is typically considered to be environment dependent.

The RSS-distance monotonicity has been reported and used in [13]. Based on the RSS-distance monotonicity, the authors propose to use signature distance to improve the performance of existing connectivity based localization algorithms.

Different from our work, in [13], the RSS-distance relation is assumed to be monotonic from the perspective of the receiver. This requires the transmitters to have approximately the same configuration. In our work, the RSS-distance relation is based on each transmitter-receiver pair, which does not need specific calibration in practice. The RSS-distance monotonicity for current commercial wireless devices has also been validated in previous works [8,7,3].

3 Observations and Motivations

The goal of this work is to design a scheme that can navigate a user to the destination using only the RSS information, with low-cost sensor devices and low user movement overhead. In this section, we first explain the components of our proposed sensor-aided navigation system. Then, we present experimental and simulation results on RSS irregularity, a preliminary navigation scheme, and how it performs in the presence of RSS irregularity, which motivate us to design an advanced multi-sensor navigation scheme that will be discussed in Sections 4.

3.1 System Components

In our proposed sensor-aided navigation system, a large number of sensors are deployed in the field of interest. The sensors have limited energy and computing capabilities. Each sensor has a wireless interface that can communicate with nearby sensors and sample the signal strength of the received packets. All sensors are equipped with a simple omni-directional antenna. Upon probed by the user, the sensors send back *navigation beacons* to guide the user's movement.

After deployment, a destination sensor is set by, e.g., detecting a certain event of interest. Then, through network-wise flooding, sensors form a *navigation tree* rooted at the destination sensor. Each sensor is assigned a *navigation hop count* towards the destination, as shown in Fig. 3(i).

Each user, e.g., a person or a robot that needs to be navigated to the destination, carries a user sensor which is the same as the sensors deployed in the field. The user moves by steps. In each step, the user moves towards a certain direction for a certain step length. At both ends of each step, the user sensor probes nearby sensors, samples the RSS of the received navigation beacons, and makes the decision about the next step based on the collected information.

3.2 A Preliminary Navigation Scheme

Under the ideal situation when the RSS-distance relation is strictly monotonic, we show in [12] that a user can be navigated to the destination efficiently using a naive one-sensor navigation scheme (denoted as NoS) which is based on RSS information from only one sensor. Unfortunately, RSS is notorious for its irregularity and unreliability in real-world scenarios. A strict monotonic RSS-distance relation does not always hold in practice. As a result, NoS scheme performs poorly in certain practical scenarios [12].

To deal with RSS-distance irregularity, it is naturally an attractive idea to use multiple sensors to guide user's movement. To do this, we use a metric called *increasing RSS ratio* (IRR, denoted as ρ) which is defined as the percentage of sensors whose average RSS measured by the user increases at the end of a step. We conduct experiments to study the relation between ρ and user's moving direction. Experimental setup is shown in Fig. 1, where 10 sensors (all of which transmit at -24 dBm) are placed indoor along a straight line with 15 cm apart between adjacent sensors. The user sensor is placed at 80 cm away from the center of them. In each experiment, the user moves in a different direction (α) with a different step length (L). The user collects 100 RSS readings from each sensor to determine ρ. We repeat each experiment 100 times and plot the ECDF (Empirical Cumulative Distribution Function) of ρ in Fig. 2. We have a few observations from the experiments.

- ρ is affected by the moving direction α. As α decreases, ρ increases. For example, Fig. 2(i) reads that, if the observed ρ is higher than 60%, it is most likely that α is less than or equal to 45 degree, while there is a small possibility that α is 90 degree.
- It is interesting to see that, when α is 180 degree, we still observe non-zero ρ values in certain scenarios, meaning that even when the user moves away from all the beacon sensors, some of the beacon sensors' average RSS indeed increase. This is due mainly to the high RSS irregularity in environments with obstacles.
- ρ is affected by the step length L. As shown in Fig. 2, a larger L results in more separation between ECDF curves of ρ for different α values. As a result, the relation between ρ and α becomes less ambiguous. For example, Fig. 2(ii) reads that, if the observed ρ is higher than 60%, α is less than or equal to 45 degree with 100% probability.

Based on the above observations, we propose a basic multi-sensor navigation scheme, denoted as BmS, which uses ρ as an indicator of moving direction. It consists of the following steps.

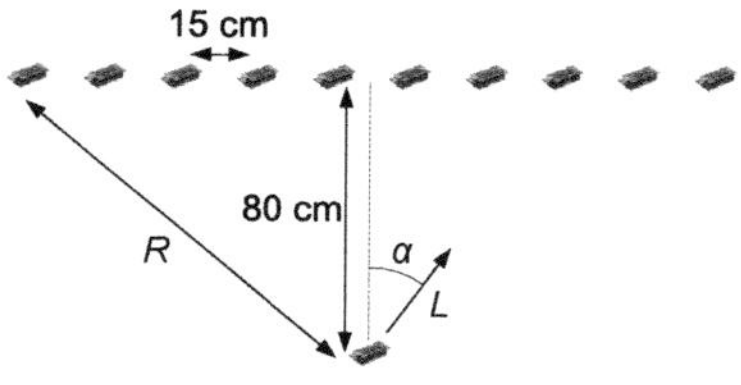

(i)

(ii)

Fig. 1. Experiments for testing multiple sensors' RSS. (i) α denotes the angle between the receiver's moving direction and the perpendicular bisector of the transmitters. R is the sensor's communication range, which is about one meter. (ii) Experimental setup.

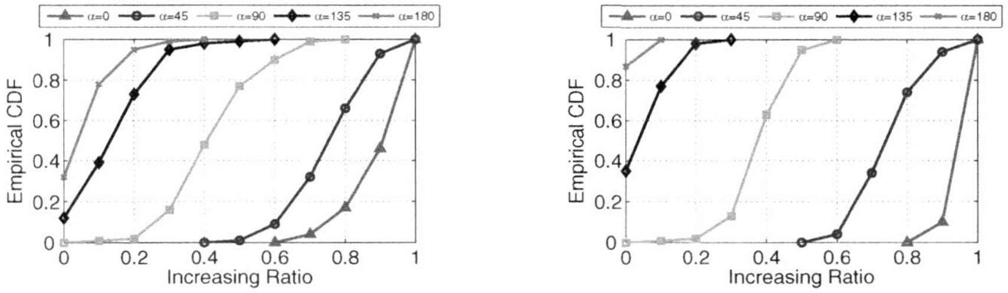

(i) $L = 10$ cm (ii) $L = 25$ cm

Fig. 2. ECDF of ρ for different α and L

- *Pick beacon sensors*: The user sensor collects the hop count information of the sensors within its communication range and selects *beacon sensors* to move towards. The beacon sensors are these that have the smallest hop count in the user sensor's communication range, which are potentially closest to the destination. For example, in Fig. 3(i), at the starting point, the user sensor overhears sensors with a hop count of 6 or 7. All the sensors with a hop count of 6 are selected as beacon sensors. Then the user takes the first step by moving towards a random direction.
- *Compare RSS*: After each step, the user sensor calculates ρ for beacon sensors at both ends of the step. If ρ is greater than a predefined threshold ρ_{th}, the user will take the next step in the same direction as the previous step. Otherwise, the user will make a clockwise turn by $(-1)^i \times (i+1) \times 90°$ for the next step, where i is the number of previous consecutive turns. An example is shown in Fig. 3(ii).
- *Switch beacon sensors*: After each step, if the user sensor overhears sensors with a hop count smaller than the current beacon sensors, it will select new beacon sensors as described in the first step.

3.3 Observations from Simulations

We define *stretching factor* as the metric to quantitatively evaluate the performance of a navigation scheme, which is the ratio of the straight-line distance between the user and the destination, to the actual distance traversed by the user to reach the destination. Clearly, the closer the stretching factor is to one, the better the navigation scheme performs. We evaluate the performance of BmS using simulations. We use the following formula to generate the RSS samples:

$$\mathrm{RSS}_d = \mathrm{RSS}_0 - 10 \times \beta \times \log(d/d_0) + X, \quad (1)$$

where RSS_0 is the reference RSS at a short distance of d_0. d is the distance between the transmitter and the receiver. β is the path loss factor. X is the zero-mean Gaussian noise, i.e., $X \sim N(0, \sigma^2)$. We evaluate BmS scheme under the RSS noise level of $\sigma^2 = 12$ (dBm)^2 that approximately corresponds to the sample

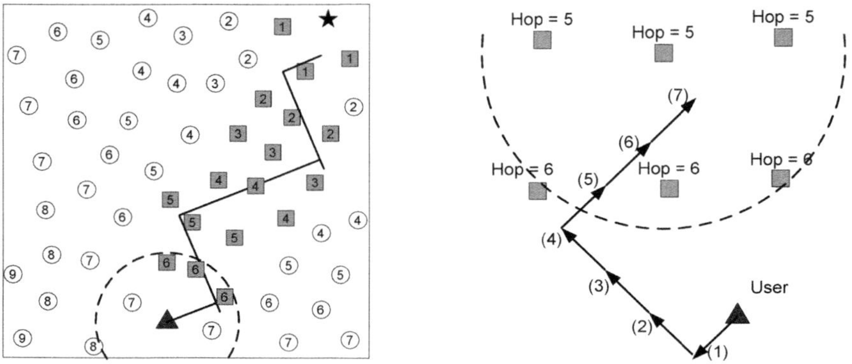

(i) (ii)

Fig. 3. The basic multi-sensor navigation scheme (BmS). In this example, ρ_{th} is set to 0.5. (i) An example network and a sample trace of a BmS user. Star: destination sensor; triangle: user sensor; squares: beacon sensors; the dashed circle represents the communication range of the user sensor. (ii) A zoom-in view of the user trace. After step 5, the user sensor switches to three new beacon sensors.

parameter	value
number of sensors	500
area size	500m × 500m
sensor TX power	0 dBm
antenna sensitivity	-94 dBm
path loss factor	3
RSS sample size	100

(i)

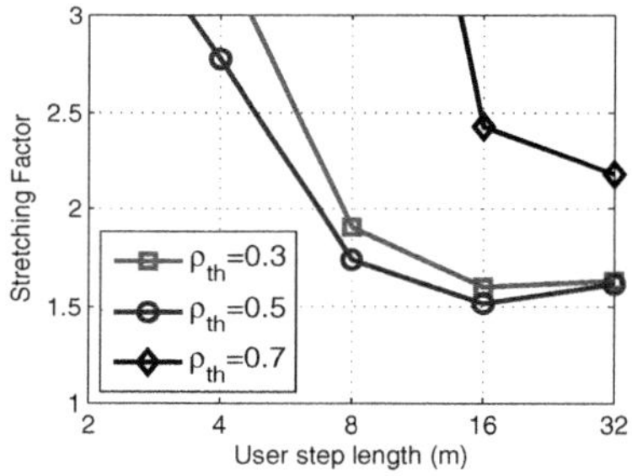

(ii)

Fig. 4. (i) Simulation parameters. (ii) Simulation results: the stretching factor of the BmS scheme under high noise level of $\sigma^2 = 12$ (dBm)^2. Each point is averaged over 50 runs.

variance observed in our indoor experiments [12]. The simulation parameters are listed in Fig. 4(i).

From the results plotted in Fig. 4(ii), we observe interesting tradeoffs in setting proper L and ρ_{th}. On one hand, as L increases, the RSS-distance monotonicity becomes more reliable, which helps to make more proper movement decisions and reduce the stretching factor. However, as L increases, if the user moves in a wrong direction, a larger movement overhead is incurred. So there is a tradeoff for choosing a proper step length. For example, when $\rho_{th} = 0.5$, the optimal step length that yields the smallest stretching factor is 16 meters.

On the other hand, for a fixed L, there is also a corresponding optimal ρ_{th}. For example, when $L = 16$ meters, $\rho_{th} = 0.5$ yields the better performance (stretching factor around 1.5), while other ρ_{th} values result in a larger stretching factor. The rationale behind this is that, if ρ_{th} is set too low, the range of α which results in $\rho \geqslant \rho_{th}$ may be too large. As a result, the user may make wrong decisions and move towards wrong directions. However, if ρ_{th} is set too high, the range of α which results in $\rho \geqslant \rho_{th}$ may be too small. Since BmS scheme makes turns in units of 90 degrees, it may not be able to find a proper direction that satisfies the ρ_{th} requirement. In fact, from the experiments we can see that user's turning angle (denote as ϕ) is another parameter that affects the performance of the navigation scheme.

Motivated by the observations on RSS irregularity and the performances of BmS schemes, we propose an advanced multi-sensor navigation scheme (denoted as AmS) and the details of AmS scheme will be presented in the next section. AmS is based on BmS but with more sophisticated control on setting the following three parameters: decision threshold for the increasing RSS ratio (ρ_{th}), the step length (L) and the turning angle (ϕ).

4 The Proposed AmS Navigation Scheme

The advanced navigation scheme proposed in this work consists of two components: *the navigation network* and *the user navigation scheme*. The navigation network serves as a base for the navigation scheme. The goal of constructing the navigation network is to let each sensor obtain its navigation hop count towards the destination, which generally reflects its distance to the destination. Due to space limitation, we do not present how the navigation network is constructed but only explain the user navigation scheme in this paper. Please refer to [12] for details about construction of the navigation network.

4.1 User Step Length (L)

As discussed in Section 3.2, the user step length should be long enough to deal with RSS irregularity but should be kept small to reduce the movement overhead. In this section we describe an online probing-based heuristic for the user to find a proper step length.

Searching for a proper user step length is initiated at the beginning of the navigation process, or triggered by change of the environment, which will be explained in Section 4.2. The flowchart of how the user step length is decided is shown in Fig. 5. The heuristic is based on the fact that the average IRR decreases as the corresponding moving direction deviates from the destination direction. Therefore, if we let the user make probing steps in different directions, by checking whether the relation of the IRRs are consistence with the direction relation, we can verify whether the user step length is proper. We explain the proposed heuristic using the example shown in Fig. 6.

The searching process starts with initial step length L_{init} and initial turning angle $\phi = 90°$. L_{init} could be the minimal step length the user can take in

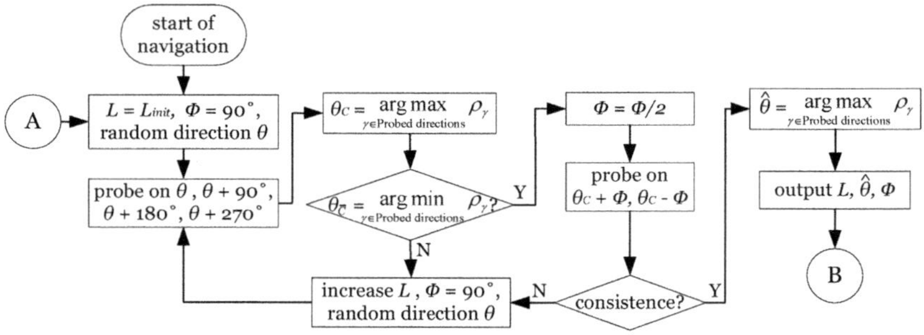

Fig. 5. The flowchart of deciding a proper L. The consistency conditions are explained in Fig. 6.

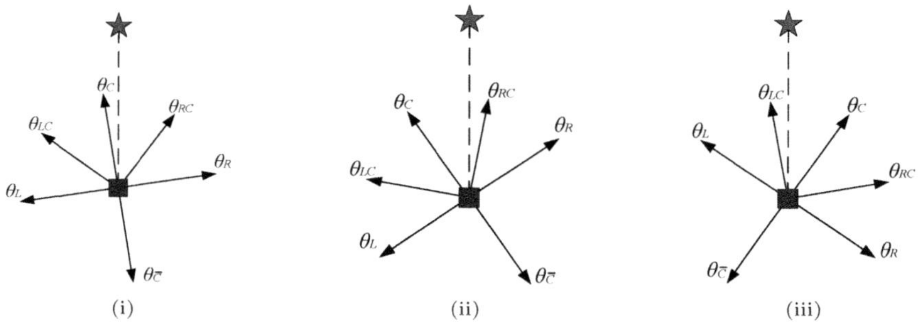

Fig. 6. Consistency conditions: (i) $\rho_{\theta_C} > \rho_{\theta_{LC}} > \rho_{\theta_L}$, and $\rho_{\theta_C} > \rho_{\theta_{RC}} > \rho_{\theta_R}$; (ii) $\rho_{\theta_{RC}} > \rho_{\theta_C} > \rho_{\theta_{LC}} > \rho_{\theta_L}$; (iii) $\rho_{\theta_{LC}} > \rho_{\theta_C} > \rho_{\theta_{RC}} > \rho_{\theta_R}$. Star indicates the location of the destination.

practice. Firstly, the user selects a random direction, say, θ and takes a step towards direction θ, $\theta + 180°$, $\theta + 90°$ and $\theta + 270°$ respectively. We denote the direction that results in the highest IRR as θ_C. Denote $\theta_C + 90°$, $\theta_C + 270°$, $\theta_C + 180°$ as θ_R, θ_L, $\theta_{\overline{C}}$ respectively (see Fig. 6). Apparently, if the step length is long enough, the IRR towards the direction of $\theta_{\overline{C}}$ should have the lowest IRR. If this condition is satisfied, the user moves on to the secondary probing stage. Otherwise, we consider the current step length to be shorter than necessary and the process restarts with a longer step length.

In the secondary probing stage, the user reduces ϕ to half of the previous ϕ (i.e., 45°) and continues to probe two directions around θ_C (i.e., $\theta_{RC} = \theta_C + \phi$ and $\theta_{LC} = \theta_C$ - ϕ), as shown in Fig. 6. Then it checks the consistency between the IRRs and the moving directions. Based on the direction of θ_C, there might be three different consistency conditions: namely, (i) $\rho_{\theta_C} > \rho_{\theta_{LC}} > \rho_{\theta_L}$ and $\rho_{\theta_C} > \rho_{\theta_{RC}} > \rho_{\theta_R}$; (ii) $\rho_{\theta_{RC}} > \rho_{\theta_C} > \rho_{\theta_{LC}} > \rho_{\theta_L}$; (iii) $\rho_{\theta_{LC}} > \rho_{\theta_C} > \rho_{\theta_{RC}} > \rho_{\theta_R}$. If any of the three conditions is satisfied, the current step length is output as the desired step length and the direction with the highest IRR is set as the moving direction for the next step, denoted as $\hat{\theta}$. For example, in Figs. 6(i)-(iii), $\hat{\theta}$ is set to be θ_C, θ_{RC} and θ_{LC} respectively.

4.2 Decision Threshold for IRR (ρ_{th}) and Turning Angle (ϕ)

In this section, we describe how a user decides ρ_{th} and ϕ. The flowchart of this process is shown in Fig. 7.

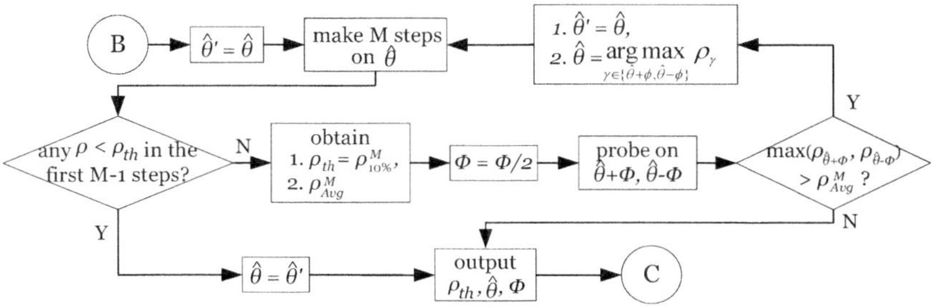

Fig. 7. The flowchart of deciding ρ_{th} and ϕ

Recall that ρ_{th} is used to decide whether the moving direction has deviated from the desired range of moving directions. To decide ρ_{th}, we let the user take M steps towards the direction of $\hat{\theta}$. We rank the M IRRs in the increasing order and get its 10th percentile (denoted as $\rho^M_{10\%}$) value. We set $\rho_{th} = \rho^M_{10\%}$. Statistically, if M is large, steps in the current moving direction has a probability of about 90% that the resulting IRR is greater than ρ_{th}. On the other hand, we want to keep M small so that if the current direction is not good, the user can react quickly to reduce the movement overhead. In our scheme, we use $M = 11$ and $\rho^M_{10\%}$ will be the second smallest IRR. Besides, the user also calculates the average of the M IRRs, denoted as ρ^M_{avg}, which will be used for the following steps.

After the step length is decided, it is possible that the output $\hat{\theta}$ is apart from the destination direction and we need to check whether there is a better moving direction. To do this, we let the user probe two directions around $\hat{\theta}$ (i.e., $\hat{\theta} + \phi$ and $\hat{\theta}$ - ϕ) with a smaller ϕ that is half of the previous ϕ. If any of the probed directions is better than the current direction, its expected IRR should be greater than ρ^M_{avg}. Therefore, if the user gets $\rho > \rho^M_{avg}$ from any of the probed directions, it will update the moving direction to be the direction which has greater IRR. Otherwise, the user stops probing and continues along the current direction of $\hat{\theta}$.

If the user changes the direction of $\hat{\theta}$, to avoid the situation where the user mistakenly chooses a bad direction, we let the user make another M steps towards the new direction of $\hat{\theta}$. If any of the first $M-1$ steps yields an IRR smaller than ρ_{th}, the new direction is considered improper and the user will change back to the previous $\hat{\theta}$. Otherwise, the user calculates ρ^M_{avg} and $\rho^M_{10\%}$, which becomes the new ρ_{th}. The user repeats this procedure, with smaller ϕ each time, until it cannot get a better direction.

4.3 Runtime Adaptation

The turning angle ϕ is used when a user encounters $\rho < \rho_{th}$. If $\hat{\theta}$ deviates from the desired range of moving direction, the user expects to find the correct direction from its current close-by directions. Thus we let the user probe two directions around θ (i.e., $\hat{\theta} + \phi$ and $\hat{\theta}$ - ϕ). To deal with outlier ρ, after the two probes, the user also makes a step towards the direction of $\hat{\theta}$. Then the user gets three IRRs, namely, $\rho_{\hat{\theta}-\phi}$, $\rho_{\hat{\theta}+\phi}$, and $\rho_{\hat{\theta}}$. Because at least one of the three probed directions is expected to be in the range of direction corresponding to ρ_{th}, without change of environmental noise, the max of $\rho_{\hat{\theta}-\phi}$, $\rho_{\hat{\theta}+\phi}$, and $\rho_{\hat{\theta}}$ should be greater than ρ_{th} with high probability. Thereby the user can continue to move towards the direction with max ρ.

In case the environment changes, e.g., the RSS noise level increases, it will render the current L too short to deal with RSS irregularity. Specifically, larger noise will increase the probability that $\rho < \rho_{th}$. Therefore, after the aforementioned probing process, the max of $\rho_{\hat{\theta}-\phi}$, $\rho_{\hat{\theta}+\phi}$, and $\rho_{\hat{\theta}}$ may still be smaller than ρ_{th}. To deal with such outlier, we let the user repeat the probing one more time. If it still cannot get $\rho > \rho_{th}$, it indicates that the environment may have changed and the user should restart from the step length searching process to find a proper L and ρ_{th}. Fig. 8 shows the flowchart of deciding whether to start a new search.

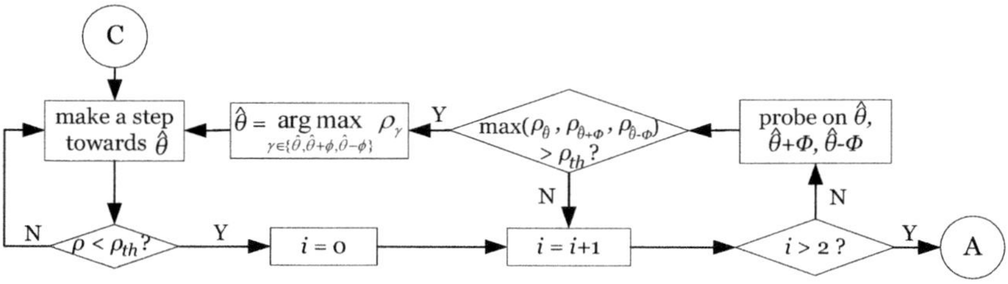

Fig. 8. The flowchart of deciding whether to start a new search

4.4 Example Trace

Fig. 9 shows an example trace of the proposed navigation scheme from simulation, whose setup is the same as described in Section 3.2. The user starts from step length of 4 meters and increases the step length by 2 meters each time. The trace shows the first 50 steps in one simulation run. For each step, we plot ρ, ρ_{th}, L and ϕ.

During steps 1∼6 and 7∼10, the user probes in step length of 4 meters and 6 meters. But the increasing RSS ratios do not pass the consistence check. The user then probes in step length of 8 meters and pass the consistence check after step 16. Then the user probes on the direction with maximum increasing RSS ratio until step 27, where the user get $\rho_{th} = 0.5$. Then the user reduces ϕ and probes around the current moving direction in steps 28 and 29. But none of the increasing RSS ratio is greater than the average increasing RSS ratio in the

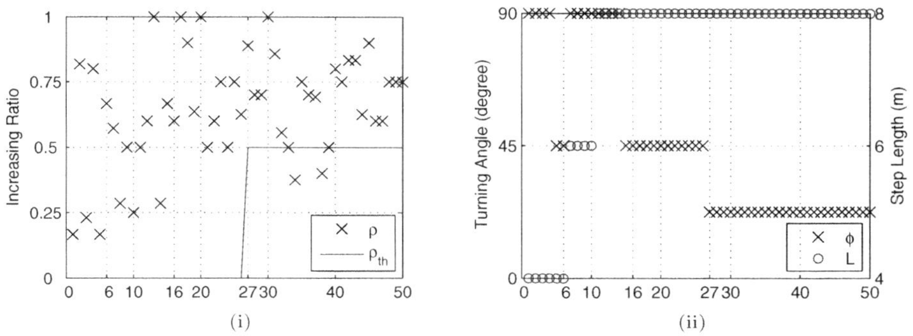

Fig. 9. Example trace of the proposed scheme from simulation. The X-axis shows the step index. (i) plots instant ρ and ρ_{th}. (ii) plots L and ϕ.

previous 11 steps. Therefore, the user stops searching for L and ρ_{th} at step 30. At step 34, the increasing RSS ratio goes under ρ_{th}, then the user probes around the current moving direction in steps 35, 36, and 37. Then it changes its direction to that of step 35 because it has the highest ρ. We can see that the user probes again in steps 39∼41 because the increasing RSS ratio drops below ρ_{th} again. Because the user stops probing for the third time, the user does not restart the searching process.

5 Performance Evaluation

We have conducted both experiments and simulations to evaluate the effectiveness of the proposed AmS scheme. In this section, we report some selected results from our experiments and simulations.

5.1 Experiments

We have implemented a small-scale test bed system consisting of 24 MicaZ motes and 25 TelosB motes. We implemented the AmS scheme on a TelosB mote. For comparison purpose, we also implemented the NoS scheme [12] and a revised NoS scheme, in which a user makes turns by 45 degree in a way same as the AmS scheme. In this section, we report some selected results of the experiments conducted using the test bed system. The implementation details of the test bed system are omitted due to space limitation. Please refer to our technical report [12] for details.

We conducted the experiments in an office of size 7 m × 4 m. The office consisted of tables, chairs and shelves which should cause strong multi-path effect. During the experiments, there were people randomly moving in the office, which caused the sensors' signal strength to be more irregular. We used this configuration to represent the worst-case scenario where there are many static and moving obstacles. The navigation network were placed on a table of size

3.2 m × 2.5 m and were deployed in two ways, namely, regular deployment and random deployment (see Fig. 10). In the regular deployment, sensors were deployed in a 7×7 grid with grid size of 0.4 m × 0.4 m. In the random deployment, we also used the same 7×7 grid, but each deploying location randomly uniformly lay in a disk centered in the grid intersection with a radius of 0.2 m. Each mote was mounted to a height of about 15 cm. Navigation sensors transmit at -24 dBm and -23 dBm for TelosB motes and MicaZ motes respectively, so that they may have a similar communication range.

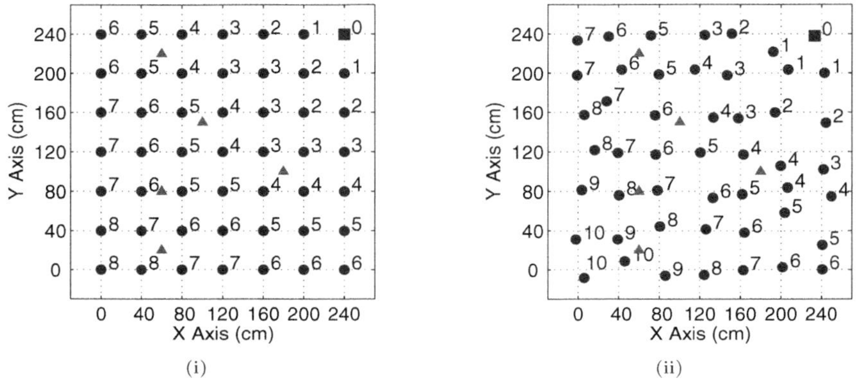

Fig. 10. The network deployment. (i) and (ii) show the network layout. Square: destination; dots: navigation sensors; triangles: the starting points for step length test in experiment I. The number to the right of each dot is the sensor's navigation hop count.

For each deployment, we ran the network formation process only once. The destination sits at the top-right corner as shown by the square marker in Fig. 10. The navigation hop count for each sensor is also shown in the figure.

Due to the limited scale of the test bed, it is infeasible to run the complete proposed scheme within the test bed. Specifically, as shown in Fig. 7, to obtain ρ_{th}, the user needs to probe M steps after deciding L, which does not always fit in the test bed. In order to evaluate the effectiveness of the proposed navigation scheme, we conducted the following two experiments.

Experiment I: user step length. In the first experiment, we only ran the step length searching heuristic to obtain a proper user step length under the test environment. In each run, the user sensor started from a random location with a random initial moving direction. There were a total of 5 starting locations, as shown by the triangles in Fig. 10. For each location the user ran the heuristic for 5 times, with different random initial directions. The initial step length was set to be 10 cm and the increase unit was 5 cm. Fig. 11(i) shows the results of the 25 runs in both deployments. Under the test environment, the searching heuristic returns the step length of around 25 cm in most times. Therefore, we used step length 25 cm for the rest of the experiments.

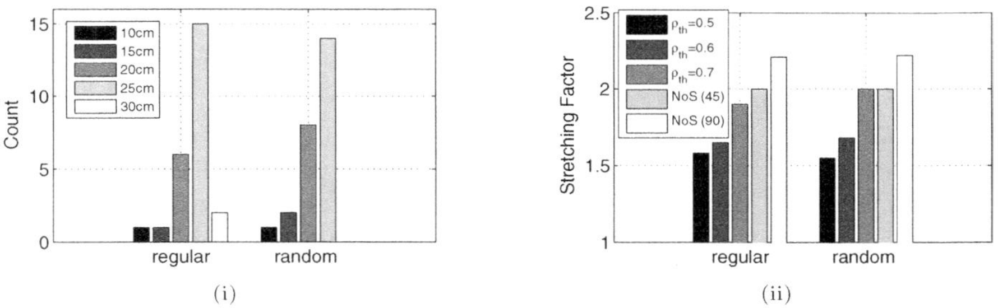

Fig. 11. (i) The results of user step length searching from 25 runs. Y axis is the count of time that the searching heuristic returns the corresponding result. (ii) Average stretching factors for different schemes in different networks.

Experiment II: stretching factor. The purpose of our scheme is to navigate users to the destination successfully and efficiently with low movement overhead. The goal of this experiment is to show that our proposed scheme incurs reasonably low overhead in practice. Due to the limited scale of the test bed, we skipped the process that decides ρ_{th} online. Instead, we ran the navigation scheme with fixed L and ρ_{th}. The scheme was slightly modified accordingly: i) in the L-deciding phase as shown in Fig. 5, the consistence check is skipped (or always returns "Y") and the user direction θ is set as the direction with maximum ρ as usual; ii) in the ρ_{th}-deciding phase as shown in Fig. 7, we set $M=1$ because ρ_{th} is decided in advance; iii) φ was always set to be 45 degree as a result of ii).

We used L obtained from experiment I (i.e., 25 cm) and different ρ_{th} (namely, 0.5, 0.6, 0.7), and for each combination we ran the scheme for 10 times. In each run the user sensor starts from a random location near the boundaries of the network. Each run terminates when the user is within a distance of 0.4 meter to the destination. The average stretching factors of different schemes are shown in Fig. 11(ii). The results using NoS are also presented for comparison. We can see that in the best case, i.e., when $L = 25$ cm and $\rho_{th} = 0.5$, the proposed scheme achieves an average stretching factor of about 1.6. Its overhead is about half of the overhead incurred in NoS. Moreover, in the proposed navigation scheme, the first 4 probing steps always let the user return to the starting location. In the experiments, these 4 steps typically contribute to about 0.45 in the stretching factor. We conjecture that the overhead brought by the searching process can be reduced if the scale of the network increases, which is evidenced by our simulation results in [12]. If we exclude the first 4 steps, the average stretching factors in the best case is around 1.15.

5.2 Simulations

From the experimental results we can see that our proposed scheme successfully finds a proper step length and can complete the navigation with reasonably low overhead with proper L and ρ_{th}. But due to the scale limitation, we did not evaluate our searching heuristic for deciding ρ_{th}. In this section, we evaluate the

AmS scheme through simulations with large-scale networks to demonstrate the adaptiveness of the AmS scheme.

Simulation Setup. We generate RSS samples in the same way as explained in Section 3.2. We use different σ^2 to represent environments with different noise levels. Note that in the simulation we did not use any of the explicit RSS-distance model information. In the simulation, the area of interest is a 2-D plane of size 1 km $\times$ 1 km. 2000 sensors are randomly uniformly deployed. The path loss factor and the antenna parameters are the same as listed in Fig. 4(i). The results are averaged over 50 simulation runs.

Adaptiveness. Our proposed navigation scheme decides the user step length L and the decision threshold for increasing RSS ratio ρ_{th} online in a heuristic manner. Ideally, the scheme should eventually return L and ρ_{th} that result in optimal performance, i.e., smallest stretching factor. Because it is hard to derive theoretically optimal L and ρ_{th}, we firstly run simulations to find the optimal combination. Specifically, we simulated the scheme with different fixed L and ρ_{th}, and the combinations resulting in the smallest stretching factor are considered "optimal". To use fixed L and ρ_{th} and compare performance with the adaptive scheme, we modify the adaptive navigation scheme in the same way as in our test bed experiment II described in Section 5.1. We ran simulations under different noise levels and selected results are shown in Fig. 12.

From Fig. 12 we can see that the optimal combination of L and ρ_{th} exists and it changes as the noise level changes. For example, when $\sigma^2 = 0$ (in Fig. 12(i)), the optimal (L, ρ_{th}) = (4 m, 0.8); when $\sigma^2 = 4$ (dBm)^2 (in Fig. 12(ii)), the optimal (L, ρ_{th}) = (6 m, 0.7). Moreover, there might be more than one optimal combinations. For example, when σ^2=12 (dBm)^2 (in Fig. 12(iv)), the optimal (L, ρ_{th}) = (8 m, 0.5) or (10 m, 0.6). We can also see that the optimal L and ρ_{th} under one noise level could perform much worse than these under a different noise level. For example, (L, ρ_{th}) = (8 m, 0.5) and (10 m, 0.6) yield the best performance when $\sigma^2 = 12$ (dBm)^2. However, when $\sigma^2 = 4$ (dBm)^2, they both incur more than twice of the overhead incurred by the optimal. This highlights the necessity of an adaptive navigation scheme in an unknown environment.

We plot the optimal stretching factors and stretching factors under different noise levels using AmS scheme in Fig. 13. We can see that our scheme performs closely to the optimal. On the other hand, fixed L and ρ_{th} may perform much worse than the optimal under certain environments. The overhead from AmS is a little higher than the optimal because: (i) the scheme has to search for proper (L, ρ_{th}) at the beginning which increases the overhead; (ii) due to the randomness of RSS, the scheme may not always be able to find the optimal (L, ρ_{th}). Fig. 14 shows an example distribution of the searching results from AmS. We can see that in most times the searching heuristic returns L and ρ_{th} close to the optimal ones. This proves that our heuristic in finding the L and ρ_{th} is effective.

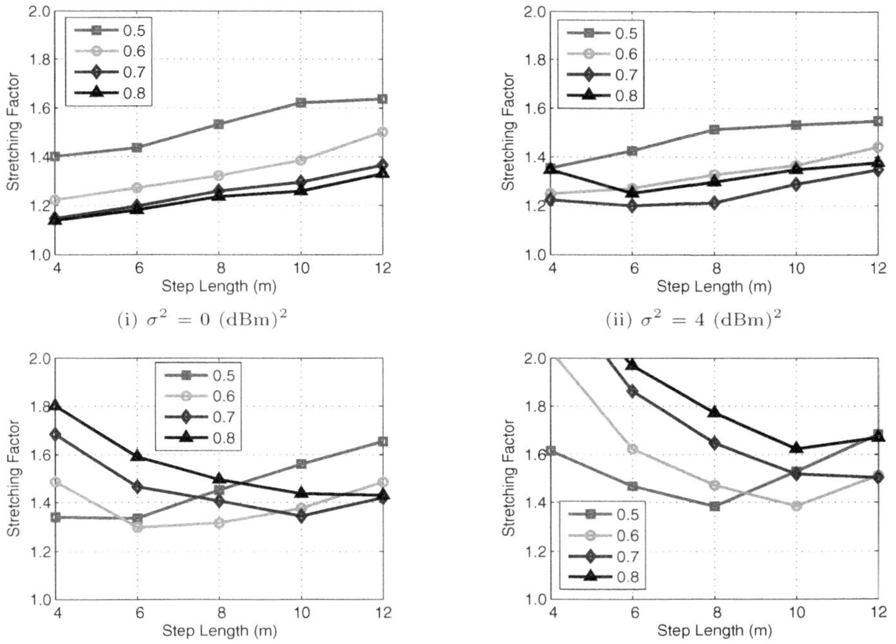

(i) $\sigma^2 = 0$ (dBm)^2 (ii) $\sigma^2 = 4$ (dBm)^2

(iii) $\sigma^2 = 8$ (dBm)^2 (iv) $\sigma^2 = 12$ (dBm)^2

Fig. 12. The stretching factor with fixed L and ρ_{th} under different noise level σ^2

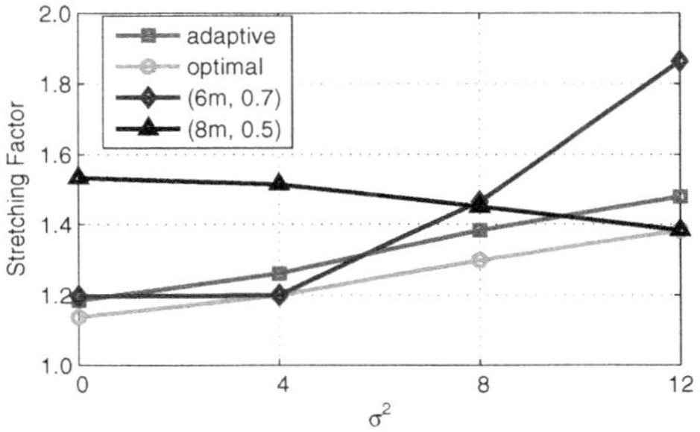

Fig. 13. Stretching factors under our adaptive navigation scheme. For comparison purpose, we also plot stretching factors under two fixed L and ρ_{th}, namely, (L, ρ_{th}) = (6 m, 0.7) and (L, ρ_{th}) = (8 m, 0.5). For simplicity, we do not plot all (L, ρ_{th}). Readers can get the performance from Fig. 12 for any (L, ρ_{th}) if interested. Clearly, the performance using fixed L and ρ_{th} varies drastically in different environments.

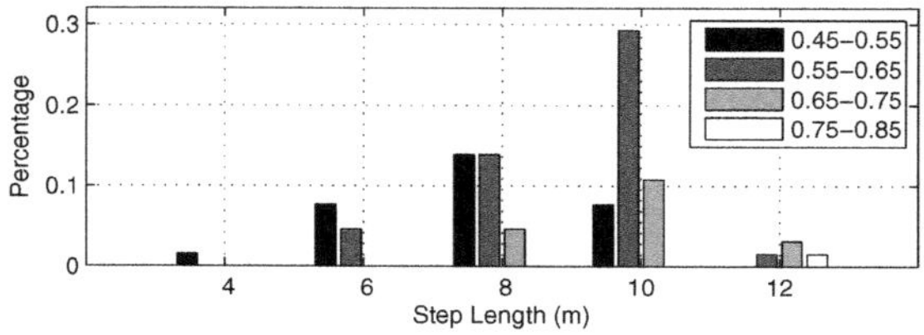

Fig. 14. We ran the L and ρ_{th} searching process for 50 times, where $\sigma^2 = 12$ (dBm)^2. This figure plots the distribution of the resulting (L, ρ_{th}). From Fig. 12(iv), we can see that the optmal (L, ρ_{th}) is (8m, 0.5) or (10m, 0.6). The distribution of (L, ρ_{th}) from our searching heuristic concentrates around these values.

6 Conclusions

In this work we investigate the problem of navigating a user to the destination using only RSS information. Our approach is based on the experimental observation that the RSS-distance relation is approximately monotonic. To deal with RSS irregularity, we propose to use increasing RSS ratio from multiple sensors as an indicator of user's moving direction. Using increasing RSS ratio, we propose an adaptive navigation scheme to decide the user step length, the decision threshold for increasing RSS ratio, and the turning angle, according to the environmental RSS noise level. We conduct both experiments and simulations to evaluate the proposed scheme. Results show that our proposed scheme can navigate users to the destination successfully and efficiently with low movement overhead under various environmental noise levels.

Acknowledgments. The research reported in this paper was supported in part by the Information Infrastructure Institute (iCube) of Iowa State University and the National Science Foundation under Grants CNS 0716744 and CNS 0831874.

References

1. Buragohain, C., Agrawal, D., Suri, S.: Distributed navigation algorithms for sensor networks. In: Proc. of INFOCOM 2006 (2006)
2. DeSouza, G.N., Kak, A.C.: Vision for mobile robot navigation: A survey. IEEE Transactions on Pattern Analysis and Machine Intellegence 2(24) (February 2002)
3. Holland, M.M., Aures, R.G., Heinzelman, W.B.: Experimental investigation of radio performance in wireless sensor networks. In: Proc. of SECON 2006 (2006)
4. Li, M., Liu, Y., Wang, J., Yang, Z.: Sensor network navigation without locations. In: Proc. of INFOCOM 2009 (2009)
5. Li, Q., Rosa, M.D., Rus, D.: Distributed algorithms for guiding navigation across a sensor network. In: Proc. of MobiCom 2003 (2003)
6. Menegatti, E., Zanella, A., Zilli, S., Zorzi, F., Pagello, E.: Range-only slam with a mobile robot and a wireless sensor networks. In: Proc. of ICRA 2009 (2009)
7. Polastre, J., Szewczyk, R., Culler, D.: Telos: Enabling ultra-low power wireless research. In: Proc. of IPSN 2005 (2005)

8. Su, W., Alzaghal, M.: Channel propagation characteristics of wireless micaz sensor nodes. Ad Hoc Networks 7(6) (August 2009)
9. Wikipedia. Anti-satellite weapon
10. Wikipedia. Global positioning system
11. Wikipedia. Navigation
12. Yang, K., Qiao, D., Zhang, W.: Sensor aided navigation in gps-denied environment. Technical Report (2010)
13. Zhong, Z., He, T.: Achieving range-free localization beyond connectivity. In: Proc. of SenSys 2009 (2009)

Time and Power Scheduling in a Wireless Network with Network Coding and Bidirectional Relaying

Maryam Mohseni and Dongmei Zhao

Department of Electrical and Computer Engineering
McMaster University
Hamilton, Ontario, L8S 4K1
dzhao@mcmaster.ca

Abstract. Using network coding in a wireless network can potentially improve the network throughput. On the other hand, it increases the complexity of resource allocations as the quality of one transmission is often affected by the link conditions of the transmitter node to multiple destination nodes. In this paper we consider a wireless network using network coding, where one relay node is available for each bidirectional link. Both digital and analog network coding strategies, referred to as DNC and ANC, respectively, are considered. All transmissions share the same frequency channel, and therefore simultaneous transmissions cause interference to each other. For ANC, transmission power of the nodes at different time slots can be co-related. We study link transmission scheduling, and in specific, the transmission time and power for each node so that the overall network throughput is maximized. Our results indicate that the two types of network coding strategies can outperform each other depending on the network conditions.

Keywords: Digital network coding, analog network coding, scheduling, power distribution.

1 Introduction

The notion of network coding (NC) was first introduced in [1] in the context of wired multicast networks. It is a technique for intermediate nodes to combine the received packets from multiple links and forward to subsequent nodes. Using NC can reduce the amount of transmitted data and potentially improve the network throughput. Significant efforts have been put to design different strategies for applying network coding in wireless networks [2]-[5]. Based on whether messages are decoded at the relay nodes, wireless network coding (WNC) can be implemented in two different ways. In the conventional WNC, multiple signals received at a relay node should be first decoded and then mixed together before being forwarded to subsequent nodes. Alternatively, the relay node can simply amplify and forward the analog signals, and this type of physical layer network coding (PNC) is also referred to as analog network coding (ANC) [6][7].

X. Zhang and D. Qiao (Eds.): QShine 2010, LNICST 74, pp. 362–374, 2011.

In contrast, the conventional WNC that requires the relay node to decode the received signals is referred to as digital network coding (DNC) in this paper.

By exploiting the broadcast nature of the wireless channel, WNC may not only improve the network throughput [8], but also combat transmission errors, save transmission power, and improve routing efficiency. While WNC has a great potential to improve the performance of wireless networks, applying WNC increases the complexity in managing the radio resources. In particular, transmission power allocations become different in a network using WNC from that in a traditional wireless network, as one message often needs to reach multiple destinations. In order for a mixed message to reach all the destinations, the transmission power of the relay station has to consider the worst link condition. This can negatively affect the network performance. The relationship between transmission power and symbol error rate is studied in [9]. The advantage of using network coding, such as throughput improvement, can be reduced significantly in a fading channel [10]. The problem of power management using WNC becomes particularly challenging in interference-limited networks, such as UWB and CDMA-based networks. However, little work has been done along this direction.

Network coding is closely related to packet transmission scheduling, which is important in order to coordinate the resource allocations and optimize the network performance. Different opportunistic scheduling schemes have been proposed for wireless networks using NC by taking advantage of the random channel conditions so that packets that are mixed in the same transmission have similar channel conditions to their respective destinations. In [11] the scheduling problem is studied jointly with channel and power allocations for broadcast traffic in an OFDMA-based WiMAX network using random network coding. In [12] joint scheduling and automatic modulation and coding is studied for a network having one transmitter and multiple destination nodes. Joint coding, routing and scheduling is studied in [13] for a wireless mesh network with orthogonal channels. Although determining the transmission time and power is one of the main aspects for scheduling, not very much work has been done on this topic for interference-limited networks with network coding.

In this paper we study time and power allocations for an interference-limited network with WNC. Both DNC and ANC are considered. All the transmissions in the network share the same frequency channel, and simultaneous transmissions can interfere with each other. Furthermore, for a network using ANC, transmissions at different time slots are co-related. An optimization problem is first formulated for the network using DNC and ANC, respectively. The objective is to maximize the system throughput, subject the transmission power and time constraints. Heuristic scheduling schemes are then proposed. The remainder of the paper is organized as follows. In Section 2 we describe the system that this work is based on. The scheduling problems for networks with DNC and ANC, respectively, are formulated and heuristic schemes are proposed in Sections 3 and 4. Numerical results are demonstrated in Section 5 to show the performance of the scheduling schemes. Section 6 concludes the paper.

2 System Description

We consider a network with M bidirectional links, indexed by $m = 1, 2, \ldots, M$. For each link, there is a relay node (R-node) that forwards data packets between the two end nodes, referred to as source node (or S-node) and destination node (or D-node), respectively. For preliminary study, we consider that the R-node for each pair of the S-node and D-node is fixed. Both the S-node and the D-node always have packets to transmit.

We use $x = s, d, r$ to represent the type of the transmitting node with s for the S-node, d for the D-node, and r for the R-node, and use $t = 1, 2, \ldots, T$ to represent the time slots. Define a set of binary variables $A_{x_m,t}$'s with $A_{x_m,t} = 1$ denoting that node x of link m transmits at time slot t and $A_{x_m,t} = 0$ denoting that the node does not transmit at the time slot. Let $P_{x_m,t}$ represent the transmission power of node x of link m, and $G_{x_m,y_n,t}$ represent the link gain between node x of link m and node y of link n at time slot t.

For DNC, we consider a simple XOR operation for the R-node to combine the packets from the S-node and D-node. It takes three time slots for the S-node and D-node of a given link to exchange one pair of packets, i.e., one packet is transmitted in each direction. In the first time slot, the S-node transmits packet k_s to the R-node; and in the second time slot, the D-node transmits packet k_d to the R-node. The two packets are decoded at the R-node, which transmits $k_d \oplus k_s$ to both the S-node and D-node in the third time slot. Upon receiving the XORed packet, the S-node recovers k_d and the D-node recovers k_s.

When using ANC, it takes two time slots for the S-node and D-node of a given link to exchange one packet in each direction. In the first time slot, the S-node transmits packet k_s and the D-node transmits packet k_d simultaneously to the R-node. The R-node does not try to decode the packets, but amplifies the mixed analog signals and forwards to both the S-node and D-node in the second time slot. Upon receiving the forwarded signal, the S-node recovers k_d and the D-node recovers k_s.

We consider that a single frequency channel is shared by all the links, and therefore simultaneous transmissions interfere with each other. The number of transmitting nodes at any given time should be limited, and the transmission power of each node should be carefully distributed, so that data can be correctly recovered at the destinations. When decoding is required, the signal-to-interference-plus-noise ratio (SINR) of the desired signal at the receiver should be above a certain threshold, γ. For ANC, transmission power at the current time slot is dependent on the transmission power and link conditions at previous time slots.

3 Scheduling for a Network Using DNC

Consider a typical bidirectional link m. For DNC, the SINR at a given time slot only depends on the transmission power and link conditions at the current time slot. If the S-node transmits at time slot t, its transmission power should satisfy the following condition in order for the R-node to correctly decode its transmitted packet k_s:

$$\frac{P_{s_m,t}G_{s_m,r_m,t}}{I_{r_m,t}+P_n} \geq \gamma, \tag{1}$$

where P_n is the power of the background noise, and $I_{r_m,t}$ is the co-channel interference that the R-node experiences from all other transmissions at time t and is given by

$$I_{r_m,t} = \textstyle\sum_{\text{all } z_n \neq s_m} P_{z_n,t}G_{z_n,r_m,t}. \tag{2}$$

Similarly, if the D-node of link m transmits at time slot t, its transmission power should satisfy the following condition in order for the R-node to correctly decode its transmitted packet k_d.

$$\frac{P_{d_m,t}G_{d_m,r_m,t}}{I_{r_m,t}+P_n} \geq \gamma, \tag{3}$$

where

$$I_{r_m,t} = \textstyle\sum_{\text{all } z_n \neq d_m} P_{z_n,t}G_{z_n,r_m,t}. \tag{4}$$

For the R-node, its transmission power should satisfy the following two conditions in order for the S-node to correctly decode packet k_d and for the D-node to correctly decode packet k_s:

$$\frac{P_{r_m,t}G_{r_m,s_m,t}}{I_{s_m,t}+P_n} \geq \gamma, \tag{5}$$

$$\frac{P_{r_m,t}G_{r_m,d_m,t}}{I_{d_m,t}+P_n} \geq \gamma, \tag{6}$$

where $I_{s_m,t}$ and $I_{d_m,t}$, respectively, represent the co-channel interference that the S-node and D-node experience from all other transmissions at time t, and their expressions can be obtained similarly to (4) as

$$I_{s_m,t} = \textstyle\sum_{\text{all } z_n \neq r_m} P_{z_n,t}G_{z_n,s_m,t}, \tag{7}$$

$$I_{d_m,t} = \textstyle\sum_{\text{all } z_n \neq r_m} P_{z_n,t}G_{z_n,d_m,t}. \tag{8}$$

Combining (1), (3), (5) and (6), and using x_m to represent the transmitter node and y_n the receiver node, we have

$$\frac{P_{x_m,t}G_{x_m,y_m,t}}{I_{y_m,t}+P_n} \geq \gamma, \tag{9}$$

where

$$I_{y_m,t} = \textstyle\sum_{\text{all } z_n \neq x_m} P_{z_n,t}G_{z_n,y_m,t}. \tag{10}$$

In (9), when $x_m = s_m$ or $x_m = d_m$, $y_m = r_m$; and when $x_m = r_m$, node $y_m = s_m$ or $y_m = d_m$.

In addition to transmission power, transmission time of the nodes for the same link should also be constrained. For each link, at most one node can transmit at any given time. That is,

$$A_{s_m,t} + A_{r_m,t} + A_{d_m,t} \leq 1. \tag{11}$$

The R-node should transmit later than the S-node and D-node of the same link. That is,

$$\sum_{\tau=1}^{t} A_{r_m,\tau} \leq \sum_{\tau=1}^{t} A_{x_m,\tau} \tag{12}$$

should be true for all $t > 0$, where $x = s$ or d.

Let C_m represent the total number of packets successfully transmitted in each direction of the bidirectional link m until time T. We have

$$C_m = \sum_{t=1}^{T} A_{r_m,t}. \tag{13}$$

Define C_m/T as the average throughput in number of packets per time slot in one direction of link m, and our objective is to maximize $\sum_m C_m/T$.

By putting together all the above constraints, we can formulate the optimization problem as follows:

$$\text{P1: } \max \sum_m C_m/T \tag{14}$$

$$\text{s.t. } \frac{A_{x_m,t}P_{x_m,t}G_{x_m,y_m,t}}{I_{y_m,t}+P_n} \geq \gamma A_{x_m,t}, \text{for all } m,\ x, \text{ and } t \tag{15}$$

$$I_{y_m,t} = \sum_{\text{all } z_n \neq x_m} A_{z_n,t}P_{z_n,t}G_{z_n,y_m,t} \tag{16}$$

$$C_m = \sum_{t=1}^{T} A_{r_m,t}, \text{ for all } m \tag{17}$$

$$A_{s_m,t} + A_{r_m,t} + A_{d_m,t} \leq 1, \text{ for all } m \text{ and } t \tag{18}$$

$$\sum_{\tau=1}^{t} A_{r_m,\tau} \leq \sum_{\tau=1}^{t} A_{s_m,\tau}, \text{ for all } m \text{ and } t \tag{19}$$

$$\sum_{\tau=1}^{t} A_{r_m,\tau} \leq \sum_{\tau=1}^{t} A_{d_m,\tau}, \text{ for all } m \text{ and } t \tag{20}$$

$$A_{x_m,t} \in \{0,1\}, \text{ for all } m,\ x, \text{ and } t \tag{21}$$

$$P_{x_m,t} \leq P_{\max}, \text{ for all } m\ ,\ x, \text{ and } t. \tag{22}$$

There are two sets of unknown variables in the above problem, $A_{x_m,t}$'s and $P_{x_m,t}$'s. The problem is a non-linear mixed-integer problem. Finding solutions to this optimization problem is not efficient. Moreover, implementing the solutions is impossible since the scheduling decisions at all the time slots are jointly optimized and it requires future information in order to make the current scheduling. Below we design a heuristic scheme to schedule the node transmissions.

We define S_m, D_m and R_m, respectively, as the number of time slots that the S-, D-, and R-nodes of link m have successfully transmitted till the current time slot. The scheduling process is given in Algorithm 1, where $\mathcal{A}_t$ is a set of the nodes that are allowed to transmit at time slot t. At the beginning of each time slot, the values of S_m, D_m and R_m are used to decide which nodes of each link can possibly be scheduled to transmit. Exchanging one pair of the packets in each bidirectional link requires one transmission from each of the S-, D-, and R-nodes. According to the time constraints, all the three nodes of the same link transmit at different time slots. For link m, if both $S_m = D_m$ and $S_m > R_m$, then the R-node of the link should be scheduled to transmit next. This is shown in Lines 6-7 of Algorithm 1. If $S_m = D_m = R_m$, either the S- or the D-node can be scheduled to transmit next, and we arbitrarily choose the S-node to transmit. This is shown in Lines 9-10. If $S_m > D_m$, then the D-node should be scheduled to transmit next. This is shown in Lines 12-13. Note that Lines 6, 9 and 12 list all possible relations among S_m, D_m and R_m. The R-node always transmits later than the S- and D-nodes of the same link, and therefore $R_m \le S_m$ and $R_m \le D_m$. Although there is no order between transmissions of the S- and D-nodes for the same link, we choose to have the S-node transmit earlier than the D-node in the scheduling, and therefore $D_m \le S_m$.

Whether the nodes with $A_{x_m,t} = 1$ can transmit depends on the required SINR, which is achieved by transmission power distribution. Given $A_{x_m,t}$'s at time slot t, the minimum required transmission power of all the nodes can be found from the following optimization problem:

$$\text{P2:} \qquad \min \max\nolimits_{x_m \in \mathcal{A}_t} P_{x_m,t} \tag{23}$$

$$\text{s.t.} \quad \frac{P_{x_m,t} G_{x_m,y_m,t}}{\sum_{z_n \in \mathcal{A}_t,\, z_n \neq x_m} P_{z_n,t} G_{z_n,y_m} + P_n} \ge \gamma, \text{ all } x_m \in \mathcal{A}_t. \tag{24}$$

If the power level for any node exceeds $P_{\max}$, the node is not allowed to transmit at the current time, i.e., the corresponding $A_{x_m,t}$ is reset to zero. This is shown in Lines 19-21. Alternatively, we can remove only the node with the highest power level from $\mathcal{A}_t$, and solve problem P2 again. The process is repeated until all the remaining nodes can have their transmission power below $P_{\max}$. However, this is not considered in this paper due to the high complexity.

Algorithm 1. scheduling when using DNC

1: Initialize $S_m = D_m = R_m = 0$ for all m.
2: Initialize $C_m = 0$ for all m.
3: Initialize $A_{x_m,t} = 0$ for all m, x and t and $\mathcal{A}_t = \emptyset$.
4: **for** $t = 1 : T$ **do**
5:   **for** $m = 1 : M$ **do**
6:     **if** $S_m = D_m$ and $S_m > R_m$ **then**
7:       Set $A_{r_m,t} = 1$ and $\mathcal{A}_t = \mathcal{A}_t \cup \{r_m\}$.
8:     **else**
9:       **if** $S_m = D_m$ and $S_m = R_m$ **then**
10:         Set $A_{s_m,t} = 1$ and $\mathcal{A}_t = \mathcal{A}_t \cup \{s_m\}$.
11:       **else**
12:         **if** $S_m > D_m$ **then**
13:           $A_{d_m,t} = 1$ and $\mathcal{A}_t = \mathcal{A}_t \cup \{d_m\}$
14:         **end if**
15:       **end if**
16:     **end if**
17:   **end for**
18:   Find $P_{x_m,t}$'s by solving problem P2.
19:   **while** There exist m and x so that $P_{x_m,t} > P_{\max}$ **do**
20:     Reset $A_{x_m,t} = 0$ and $\mathcal{A}_t = \mathcal{A}_t \setminus \{x_m\}$
21:   **end while**
22:   Find $P_{x_m,t}$'s by solving problem P2.
23:   Update S_m, D_m and R_m.
24:   **for** $m = 1 : M$ **do**

```
        C_m = C_m + A_{r_m,t}.
    end for
end for
```

4 Scheduling for a Network Using ANC

As in Section 3, we consider a bidirectional link m. When using ANC, transmissions at different time slots are dependent on each other. This is because the R-node amplifies signals and interference received from previous time slots. At time slot τ, packet k_s transmitted by the S-node reaches the R-node with the received power $P_{s_m,\tau}G_{s_m,r_m,\tau}$, and packet k_d transmitted by the D-node reaches the R-node with the received power $P_{d_m,\tau}G_{d_m,r_m,\tau}$. At the same time, the interference level that other simultaneous transmissions cause at the R-node is given by

$$I_{r_m,\tau} = \sum\nolimits_{\text{all } z_n \neq s_m, z_n \neq d_m} A_{z_n,\tau}P_{z_n,\tau}G_{z_n,r_m,\tau} \tag{25}$$

At time slot t, $t > \tau$, the mixed signals, which includes the desired signals, the interference received from time slot τ, as well as noise, are amplified with amplification factor $\beta_{m,t}$ by the R-node. The amplified desired signal includes $\beta_{m,t}P_{s_m,\tau}G_{s_m,r_m,\tau}$ for packet k_s, $\beta_{m,t}P_{d_m,\tau}G_{d_m,r_m,\tau}$ for packet k_d, and the amplified interference and noise component is $\beta_{m,t}(I_{r_m,\tau} + P_n)$.

The above amplified mixed signal reaches both the S-node and the D-node of link m. At the S-node, the received power becomes $\beta_{m,t}P_{s_m,\tau}G_{s_m,r_m,\tau}G_{r_m,s_m,t}$ for packet k_s, $\beta_{m,t}P_{d_m,\tau}G_{d_m,r_m,\tau}G_{r_m,s_m,t}$ for packet k_d, $\beta_{m,t}(I_{r_m,\tau}+P_n)G_{r_m,s_m,t}$ for the interference and noise. Meanwhile, other nodes that transmit at time t cause interference at the S-node, and the interference level is given by

$$I_{s_m,t} = \sum\nolimits_{\text{all } z_n \neq r_m} A_{z_n,t}P_{z_n,t}G_{z_n,s_m,t}. \tag{26}$$

Assume $\beta_{m,t}$ is known to all the nodes, $G_{s_m,r_m,t} = G_{r_m,s_m,t}$ and $G_{s_m,r_m,\tau} = G_{r_m,s_m,\tau}$ are known to the S-node, the S-node can remove the component of k_s from the received signal. It should recover packet k_d from the remaining signal with the desired signal power $\beta_{m,t}P_{d_m,\tau}G_{d_m,r_m,\tau}G_{r_m,s_m,t}$ and all other components as interference. In order for the S-node to correctly decode packet k_d, the following condition should be satisfied:

$$\frac{A_{r_m,t}\beta_{m,t}P_{d_m,\tau}G_{d_m,r_m,\tau}G_{r_m,s_m,t}}{\beta_{m,t}G_{r_m,s_m,t}(P_n + I_{r_m,\tau}) + I_{s_m,t} + P_n} \geq \gamma A_{r_m,t} \tag{27}$$

where P_n in the last term in the denominator is the local background noise power at the S-node.

Similarly, the SINR condition for the D-node to decode packet k_s is given by

$$\frac{A_{r_m,t}\beta_{m,t}P_{s_m,\tau}G_{s_m,r_m,\tau}G_{r_m,d_m,t}}{\beta_{m,t}G_{r_m,d_m,t}(I_{r_m,\tau} + P_n) + I_{d_m,t} + P_n} \geq \gamma A_{r_m,t}, \tag{28}$$

where $I_{d_m,t}$ is given by

$$I_{d_m,t} = \sum\nolimits_{\text{all } z_n \neq r_m} A_{z_n,t}P_{z_n,t}G_{z_n,d_m,t}. \tag{29}$$

Define $A_{x_m,t}$ in the same way as in the previous section, the optimization problem P3 can be formulated, where (36) gives the total transmission power of the R-node at time t, and (37)-(41) together specify the time constraints. Among the time constraints, (37) specifies the most recent time slot when the S-node and D-node transmit, (38) specifies that the S-node and D-node of the same link should transmit at the same time, (39) specifies that the S-node and R-node cannot transmit at the same time, and (40) indicates that the R-node transmits after the S-node for the same link. In problem P3, the unknown variables are $A_{x_m,t}$'s, $P_{x_m,t}$'s, and $\beta_{m,t}$'s. The problem is non-linear and non-convex, and cannot be solved efficiently. Below we design a heuristic scheduling scheme.

$$\text{P3: } \max \sum_m C_m/T \tag{30}$$

$$\text{s.t.} \quad \frac{A_{r_m,t}\beta_{m,t}P_{d_m,\tau}G_{d_m,r_m,\tau}G_{r_m,s_m,t}}{I_{s_m,t}+P_n+\beta_{m,t}G_{r_m,s_m,t}(P_n+I_{r_m,\tau})} \geq \gamma A_{r_m,t} \tag{31}$$

$$\frac{A_{r_m,t}\beta_{m,t}P_{s_m,\tau}G_{s_m,r_m,\tau}G_{r_m,d_m,t}}{I_{d_m,t}+P_n+\beta_{m,t}G_{r_m,d_m,t}(P_n+I_{r_m,\tau})} \geq \gamma A_{r_m,t} \tag{32}$$

$$I_{r_m,\tau} = \textstyle\sum_{\text{all } z_n \neq s_m, z_n \neq d_m} A_{z_n,\tau}P_{z_n,\tau}G_{z_n,r_m,\tau} \tag{33}$$

$$I_{s_m,t} = \textstyle\sum_{\text{all } z_n \neq r_m} A_{z_n,t}P_{z_n,t}G_{z_n,s_m,t} \tag{34}$$

$$I_{d_m,t} = \textstyle\sum_{\text{all } z_n \neq r_m} A_{z_n,t}P_{z_n,t}G_{z_n,d_m,t} \tag{35}$$

$$P_{r_m,t} = \beta_{m,t}(P_{s_m,\tau}G_{s_m,r_m,\tau} + P_{d_m,\tau}G_{d_m,r_m,\tau} + P_n + I_{r_m,\tau}) \tag{36}$$

$$\tau = \max\{t_1 < t : A_{d_m,t_1} = 1\} \tag{37}$$

$$A_{s_m,t} = A_{d_m,t}, \text{ for all } m \text{ and } t \tag{38}$$

$$A_{s_m,t} + A_{r_m,t} \leq 1, \text{ for all } m \text{ and } t \tag{39}$$

$$\textstyle\sum_{\tau=1}^{t} A_{r_m,\tau} \leq \sum_{\tau=1}^{t} A_{s_m,\tau}, \text{ for all } m \tag{40}$$

$$A_{x_m,t} \in \{0,1\}, \text{ for all } m, x, \text{ and } t \tag{41}$$

$$P_{x_m,t} \leq P_{\max}, \text{ for all } m, x, \text{ and } t \tag{42}$$

$$C_m = \textstyle\sum_{t=1}^{T} A_{r_m,t}, \text{ for all } m \tag{43}$$

Define S_m and R_m as the number of time slots that the S-node and R-node of link m have transmitted till the current time slot. Algorithm 2 gives the scheduling scheme for a network using ANC. The time constraints for the scheduling is that for each link, the S-node and D-node should transmit at the same time slot, and they should transmit at an earlier time slot than the R-node. Therefore, $S_m \geq R_m$. If $S_m = R_m$, the S-node and D-node should transmit next; if $S_m > R_m$, the R-node should transmit next. This is shown in Lines 6-12.

The scheme then finds transmission power for all the nodes in $\mathcal{A}_t$. The transmission power of any S-nodes and D-nodes at the current time slot is dependent on the transmission power of the R-node at a future time slot. Without any future information available, setting the transmission power for the S- and D-nodes involves two contradictory effects. For a given link, allowing the S- and D-nodes to transmit at high power can potentially reduce the required transmission power of the R-node at a future time slot. On the other hand, this increases the interference to other nodes at the current time slot. Given the transmission power of the S-nodes and D-nodes, the scheme finds the minimum required transmission power for all the R-nodes in $\mathcal{A}_t$ by solving problem P4 in (44). In the solution, if the required power for any of the R-nodes exceeds $P_{\max}$, either some nodes (can be S-, D-, or R-nodes) are removed from the current time slot, or the transmission power of some S-nodes and D-nodes is reduced. Therefore, different criteria can be used depending the objective. As a preliminary study, we fix the transmission power of the S- and D-nodes to $P_{\max}$ and reset $A_{r_m,t} = 0$ for all R-nodes with $P_{r_m,t} > P_{\max}$. Results of using other criteria will be reported in future work.

Algorithm 2. scheduling when using ANC

Initialize $S_m = R_m = 0$ for all m.
Initialize $C_m = 0$ for all m.
$A_{x_m,t} = 0$ for all m, x and t and $\mathcal{A}_t = \emptyset$.
for $t = 1 : T$ **do**
  for $m = 1 : M$ **do**
    if $S_m = R_m$ **then**
      $A_{s_m,t} = 1$, $A_{d_m,t} = 1$, and $\mathcal{A}_t = \mathcal{A}_t \cup \{s_m, d_m\}$
    else
      if $S_m > R_m$ **then**
        $A_{r_m,t} = 1$ and $\mathcal{A}_t = \mathcal{A}_t \cup \{r_m\}$
      end if
    end if
  end for
  if There exists r so that $A_{r_m,t} = 1$ **then**
    Find $P_{r_m,t}$'s by solving problem P4.
    while There exists m so that $P_{r_m,t} > P_{\max}$ **do**
      $A_{r_m,t} = 0$ and $\mathcal{A}_t = \mathcal{A}_t \setminus \{r_m\}$
    end while
  end if
  if There exists r so that $A_{r_m,t} = 1$ **then**
    Find $P_{r_m,t}$'s by solving problem P4.
  end if
  Update S_m and R_m for all m.
  for $m = 1 : M$ **do**
    $C_m = C_m + A_{r_m,t}$.
  end for
end for

$$\text{P4: } \min \max_{r_m \in \mathcal{A}_t} P_{r_m,t} \tag{44}$$

$$\text{s.t.} \quad \frac{\beta_{m,t} P_{d_m,\tau} G_{d_m,r_m,\tau} G_{r_m,s_m,t}}{I_{s_m,t} + P_n + \beta_{m,t} G_{r_m,s_m,t}(P_n + I_{r_m,\tau})} \geq \gamma \tag{45}$$

$$\frac{\beta_{m,t} P_{s_m,\tau} G_{s_m,r_m,\tau} G_{r_m,d_m,t}}{I_{d_m,t} + P_n + \beta_{m,t} G_{r_m,d_m,t}(P_n + I_{r_m,\tau})} \geq \gamma \tag{46}$$

$$I_{s_m,t} = \sum\nolimits_{\text{all } z_n \in \mathcal{A}_t, z_n \neq r_m} P_{z_n,t} G_{z_n,s_m,t} \tag{47}$$
$$I_{d_m,t} = \sum\nolimits_{\text{all } z_n \in \mathcal{A}_t, z_n \neq r_m} P_{z_n,t} G_{z_n,d_m,t} \tag{48}$$
$$P_{r_m,t} = \beta_{m,t}(P_{s_m,\tau} G_{s_m,r_m,t} + P_{d_m,\tau} G_{d_m,r_m,t} + P_n + I_{r_m,\tau}) \tag{49}$$
$$\tau = \max\{t_1 < t : A_{d_m,t_1} = 1\} \tag{50}$$

5 Numerical Results

We consider a network with M bidirectional links, where $3M$ nodes are located in a $3 \times M$ grid as shown in Fig. 1. For each link, the R-node is located at the middle point between the S-node and the D-node. The distance between the R-node and the S-node of the same link is denoted as w_1, and the distance between the S-nodes of two neighboring links is denoted as w_2. The link gain between any two nodes x_m and y_n includes both distance-based path loss and log-normally distributed shadowing and is given by $G_{x_m,y_n,t} = d_{x_m,y_n}^{-\alpha} \times 10^{-X}$, where X is Gaussian distributed with zero mean and standard deviation of 1. The default values are $M = 3$, $w_1 = 100$ m, $w_2 = 500$ m, $P_n = 10^{(-6)}$W, $\alpha = 2$, and $P_{\max} = 1$W. The throughput results for $T = 350$ time slots are plotted in Figs. 2 and 3 for the network using DNC, and in Fig. 4 for the network using ANC. Analytically, it takes three time slots to exchange one pair of packets between the S-node and the D-node of a given bidirectional link when using DNC. Therefore, the maximum throughput in one direction for each link is 1/3 packets per time slot. With M bidirectional links, the maximum unidirectional throughput is $M/3$ packets per time slot, or 1 packet per time slot when $M = 3$. For ANC, it takes two time slots to exchange one pair of packets in each bidirectional link. Therefore, the maximum unidirectional throughput is 1/2 packets per time slot, or 1.5 packets per time slot when $M = 3$.

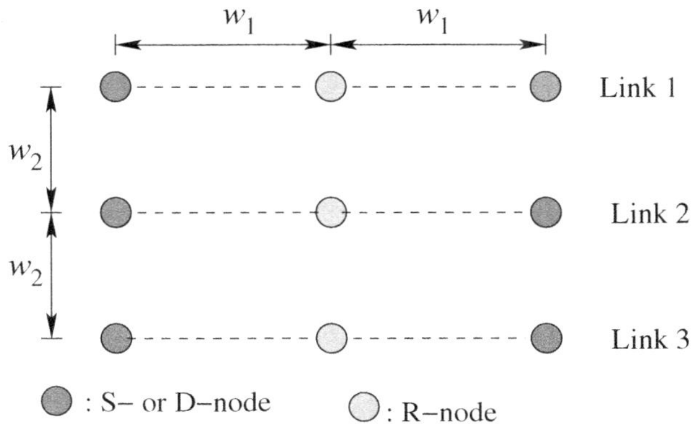

Fig. 1. Grid topology with $M = 3$

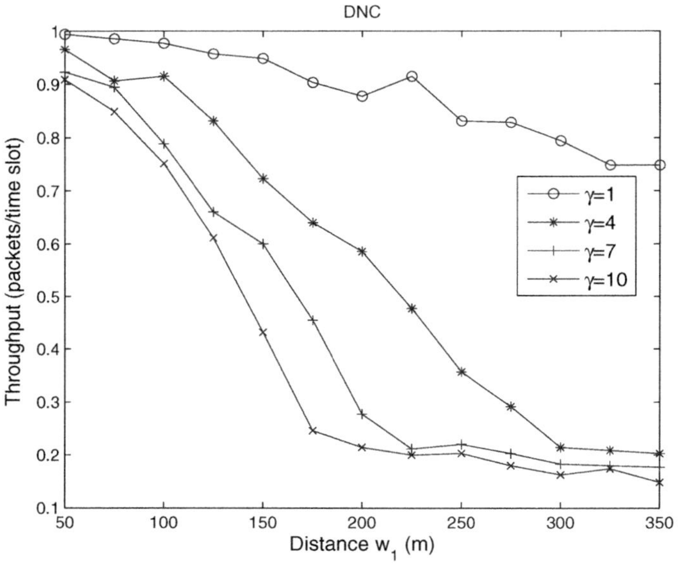

Fig. 2. Throughput versus w_1 for DNC

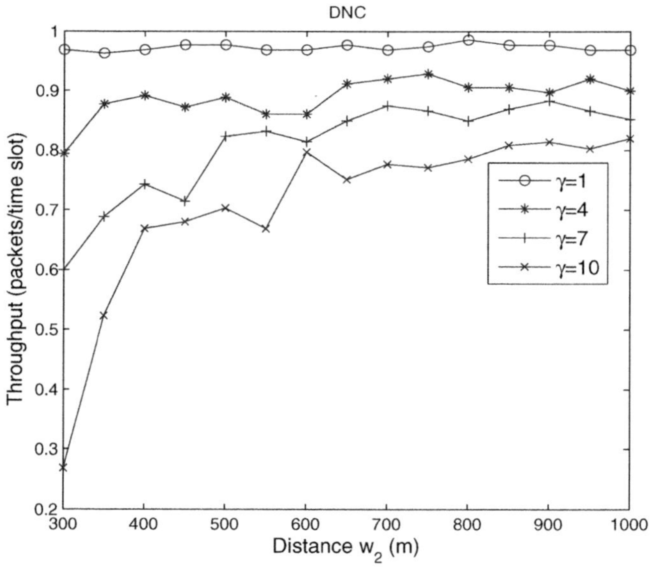

Fig. 3. Throughput versus w_2 for DNC

Fig. 2 shows that the transmission throughput decreases with w_1. This is due to the increased path loss, which increases the required transmission power. When w_1 is relatively small, the throughput is close to the upper limit, which is one packet per time slot. This indicates that the proposed scheduling scheme is efficient. On the other hand, as w_1 increases, the transmission power of the nodes increases, which increases the mutual interference and makes it more difficult to satisfy the SINR requirements. Therefore, throughput decreases as w_1 increases, and the decrease is much faster for larger γ values.

As w_2 increases, the distance between nodes of different links increases, which reduces the mutual interference between the links. Therefore, throughput increases with w_2 as shown in Fig. 3. When w_2 is relatively small, increasing w_2 can significantly increase the overall throughput. After w_2 is larger than a certain value, the effect of mutual interference on throughput performance becomes minor and further increasing w_2 does not very much increase the throughput, which is not limited by w_1. Fig. 3 also shows that the throughput is very close to the upper bound when the required SINR is very low.

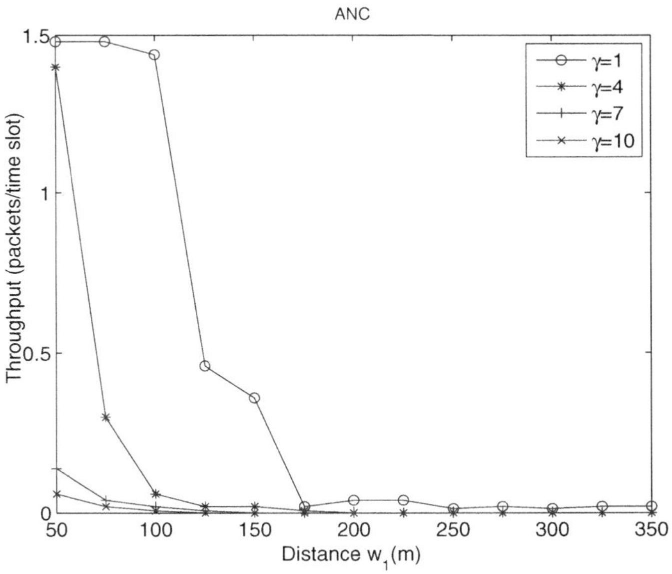

Fig. 4. Throughput versus w_1 for ANC

For ANC, the maximum throughput is achieved when both w_1 and γ are small as shown in Fig. 4. When w_1 or γ increases, the throughput can decrease very abruptly. Comparing Figs. 4 and 2 we can find that when both the required SINR and w_1 are small, using ANC can achieve much higher throughput than using DNC. On the other hand, when the required SINR is relatively high or w_1 is relatively large, using DNC is a better choice. This is basically because that when using ANC, interference and noise is also amplified by the R-node and this deteriorates the transmission quality and makes it much more difficult to satisfy relatively high SINR requirement in poor channel conditions.

6 Conclusions

We have studied transmission time and power scheduling for a network with bidirectional links and network coding. Heuristic schemes have been proposed for the network using digital network coding and analog network coding. In general, the scheduling scheme for a system using ANC is more complicated than that using DNC because transmissions at different time slots can be dependent on each other. When the SINR requirement is relatively low and channel conditions are good, using ANC can achieve higher throughput than using DNC. In other conditions, using DNC can achieve higher throughput than using ANC.

References

1. Ahlswede, R., Cai, N., Li, S.-Y., Yeung, R.: Network Information Flow. IEEE Transactions on Information Theory 46(4), 1204–1216 (2000)
2. Wu, Y., Chou, P.A., Kung, S.-Y.: Information Exchange in Wireless Networks with Network Coding and Physical-Layer Broadcast. In: Proc. CISS (2005)
3. Katti, S., Rahul, H., Hu, W., Katabi, D., Medard, M., Crowcroft, J.: Xors in The Air: Practical Wireless Network Coding. IEEE/ACM Transactions on Networking 16(3), 497–510 (2008)
4. Chachulski, S., Jennings, M., Katti, S., Katabi, D.: Trading Structure for Randomness in Wireless Opportunistic Routing. In: Proc. ACM SIGCOMM (2007)
5. Katti, S., Katabi, D.: MIXIT: The Network Meets The Real Channel. In: Proc. ACM HotNets (2007)
6. Zhang, S., Liew, S., Lam, P.: Physical-Layer Network Coding. In: Proc. MobiCom (2006)
7. Katti, S., Gollakota, S., Katabi, D.: Embracing Wireless Interference: Analog Network Coding. In: Proc. 13th ACM SIGCOMM, pp. 397–408 (2007)
8. Sagduyu, Y.E., Ephremides, A.: Joint Scheduling and Wireless Network Coding. IEEE/ACM Transactions on Networking 16(3), 497–510 (2005)
9. Lai, H.-Q., Ibrahim, A., Liu, K.J.R.: Wireless Network Cocast: Location-Aware Cooperative Communications with Linear Network Coding. IEEE Transactions on Wireless Communications 8(7), 3844–3854 (2009)
10. Li, W., Li, J., Fan, P.: Network Coding for Two-way Relaying Networks over Rayleigh Fading Channels. IEEE Transactions on Vehicular Technology (2010), available on web
11. Jin, J., Li, B.: Cooperative Multicast Scheduling with Random Network Coding in WiMAX. In: Proc. 17th International Workshop on Quality of Service Digital, IWQoS (2009)
12. Gong, S.-L., Kim, B.-G., Lee, J.-W.: Opportunistic Scheduling and Adaptive Modulation in Wireless Networks with Network Coding. In: Proc. IEEE Vehicular Technology Conference Spring 2009 (VTC Spring)
13. Zhang, X., Su, H.: Network-Coding-Based Scheduling and Routing Schemes for Service-Oriented Wireless Mesh Networks. IEEE Wireless Communications 16(4), 40–46 (2009)

Distributed Scheduling for Advance Bandwidth Reservation in High-Performance Networks

Qishi Wu and Yunyue Lin

University of Memphis, Memphis, TN 38152, USA
{qishiwu,ylin1}@memphis.edu

Abstract. High-performance networks are capable of provisioning dedicated channels through circuit/lambda-switching or MPLS/GMPLS techniques to support large-scale data transfer. These dedicated links are typically shared by multiple users through advance resource reservations, resulting in varying bandwidth availability in future time periods. Most previous efforts were focused on centralized bandwidth scheduling to improve the utilization of network resources and meet the transport requirements of application users. These centralized scheduling schemes imply the use of a central control plane, posing significant reliability and scalability challenges as the network size rapidly grows. We propose distributed algorithms for path computation and bandwidth scheduling in response to four basic bandwidth reservation requests: (i) fixed bandwidth in a fixed slot, (ii) highest bandwidth in a fixed slot, (iii) first slot with fixed bandwidth and duration, and (iv) all slots with fixed bandwidth and duration. These algorithms are developed through a rigorous extension of the classical breadth first search and Bellman-Ford algorithms to a completely distributed manner, and their performances are evaluated and analyzed through extensive simulations.

Keywords: distributed scheduling, bandwidth reservation, high-performance networks.

1 Introduction

1.1 Background

A number of large-scale applications in various science, engineering and business domains are generating colossal amounts of data, on the order of terabytes currently and petabytes or even exabytes in the near future, which must be transferred over a long geographical distance for remote operations. High-performance networks that are capable of provisioning dedicated channels have proved to be a promising solution to large data transfer and several network projects are currently underway to develop such capabilities, including UltraScience Net (USN) [17], Circuit-switched High-speed End-to-End Transport ArcHitecture (CHEETAH) [7], Dynamic Resource Allocation via GMPLS Optical Networks (DRAGON) [1], Japanese Gigabit Network II [2], Bandwidth on Demand (BoD) on Geant2 network [3], On-demand Secure Circuits and Advance Reservation

X. Zhang and D. Qiao (Eds.): QShine 2010, LNICST 74, pp. 375–391, 2011.

System (OSCARS) [4] of DOE Energy Sciences Network (ESnet), Hybrid Optical and Packet Infrastructure (HOPI) [5], and Bandwidth Brokers [21]. Such dedicated channels are a part of the capabilities envisioned for Global Environment for Network Innovations (GENI) project [6].

The deployments of high-performance networks are expected to increase significantly and proliferate into both public and dedicated network infrastructures across the globe in the coming years. An evidence of this trend in production networks is reflected by Internet2 offering on-demand circuits and Multiple Protocol Label Switching (MPLS) tunnels. MPLS improves the forwarding speed of IP routers by adopting a key concept from the world of virtual-circuit networks: a fixed-length label. MPLS is often referred to as layer-2.5, which adds a small MPLS header between the layer-2 header and the layer-3 header in a link-layer frame. Since modern optical networks have reached a very high transfer rate (at 40 Gbit/s and beyond), the true advantages of MPLS do not lie in the potential increase in switching speeds, but rather in the new traffic management capabilities that MPLS enables. OSCARS uses MPLS and Resource Reservation Protocol (RSVP) to create virtual circuits or Label Switched Paths (LSPs), while the management and operation of end-to-end virtual circuits within the network are done at the layer-3 network level. OSCARS supports advance reservation, but its underlying path computation limits connections over links returned by traceroute; hence, it does not explore all available bandwidths inside the network.

The dedicated links in high-performance networks are typically shared by multiple users through various advance reservation techniques, resulting in varying bandwidth availability in future time periods. Most previous scheduling efforts were focused on centralized advance bandwidth reservation in high-performance networks that employ a central control plane [15,16,9]. Such centralized schemes are suited for small-scale networks, but pose significant reliability and scalability challenges as the network size increases, which calls for distributed solutions for large-scale networks. To the best of our knowledge, there are very few studies on distributed advance bandwidth reservation. In this paper, we formulate four basic advance bandwidth scheduling problems and propose distributed algorithms for path computation and bandwidth scheduling. These algorithms are developed through a rigorous extension of the classical breadth first search (BFS) and Bellman-Ford algorithms to a completely distributed manner and their performances are evaluated and analyzed through extensive simulations.

1.2 Related Work

As high-speed dedicated networks are increasingly developed and deployed, many scheduling algorithms have been designed for advance bandwidth reservation. We provide below a brief survey of such efforts.

The four basic bandwidth scheduling problems discussed in this paper were first introduced in [18] and were later investigated in [19] with a detailed description on the solution to each of these problems in the centralized scheme. Guerin *et al.* studied these basic scheduling problems with several extensions to increase the flexibility of services [13]. The scheduling algorithm proposed by

Cohen *et al.* considers the flexibility of transfer start time and the capability of path switching between different paths during the transfer to improve network utilization [8]. Grimmell *et al.* formulated a dynamic quickest path problem, which deals with the transmission of a message from a source to a destination with the minimum end-to-end delay over a network with propagation delays and dynamic bandwidth constrains on the links [12]. Veeraraghavan *et al.* transferred files with varying bandwidths in different time slots in a simple case where the path is pre-specified [20]. Ganguly *et al.* generalized the problems of finding an optimal path in a graph with varying bandwidths to minimize the total transfer time, and also proposed approaches to find the minimum number of path switchings to transfer a file in a specified number of time slots [10]. Gorinsky *et al.* proposed a Virtual Finish Time First algorithm to schedule incoming files in a preemptive manner to minimize the total transfer end time over a given dedicated channel [11]. Most recently, Lin *et al.* proposed four instant scheduling problems under different constraints (fixed or variable) on both path and bandwidth [16]. Considering the path switching delay, the scheduling problem using either fixed or variable paths with variable bandwidths are proved to be NP-complete [15].

Most of the aforementioned efforts were focused on centralized bandwidth scheduling in dedicated networks managed by a central control plane, where a global repository storing all bandwidth reservations on all network links has to be maintained and updated. Such a centralized management scheme is not adequate for large-scale networks due to its reliability and scalability issues, which motivate us to develop distributed solutions.

The rest of the paper is organized as follows. We formulate four basic advance bandwidth scheduling problems in Section 2. The distributed algorithms for these scheduling problems are presented in Section 3. The scheduling performance is evaluated in Section 4.

2 Advance Bandwidth Scheduling Problems

We consider a generic control plane to support advance bandwidth reservation of dedicated channels in high-performance networks [18]. The distributed bandwidth scheduler runs on each switch in circuit/lambda-switching networks or on each router in MPLS-enabled networks, and computes routing paths in a distributed manner based on the available bandwidth of adjacent links. In MPLS-enabled networks, a Label Switched Path (LSP) is created by the signaling daemon at the start time of a bandwidth reservation. Each router along the computed path receives a path setup request via Resource Reservation Protocol (RSVP) and commits bandwidth to create the LSP. At the end time of a bandwidth reservation, the signaling daemon tears down the corresponding LSP.

We represent the topology of a dedicated network as a directed graph $G = (V, E)$ with n nodes and m links, where each link $l \in E$ maintains a list of residual bandwidths specified as segmented constant functions of time. An example of the available bandwidth over time for a link is shown in Fig. 1. We use a 3-tuple of time-bandwidth (TB) $(t_l[i], t_l[i+1], b_l[i])$ to represent the residual bandwidth

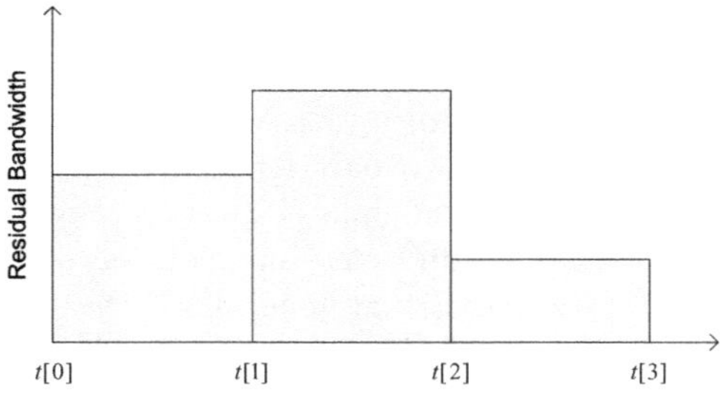

Fig. 1. A time-bandwidth example of a link

of link l at time slot $[t_l[i], t_l[i+1]]$, $i = 0, 1, 2, \ldots, T_l - 1$, where T_l is the total number of time slots of link l. The current time point is denoted by $t_l[0]$, and the future time point is denoted by $t_l[i]$ $(i > 1)$. We set $t_l[T_l] = +\infty$, which indicates that there is no bandwidth reservation on link l after time point $t_l[T_l - 1]$ and therefore $b_l[T_l - 1]$ is the initial bandwidth of link l. Let T be the maximum number of time slots on a TB list of link $l \in E$. Each node $v \in V$ only knows its neighbor nodes, and maintains the TB lists for all outgoing links from itself to its neighbor nodes.

Based on different data transport constraints and application requirements, we formulate four basic advance bandwidth scheduling problems: Given a graph $G = (E, V)$ with a time-bandwidth list TB for each link $l \in E$, source v_s and destination v_d,

- Fixed-Bandwidth: compute a path from v_s to v_d with a fixed bandwidth β in a specified time slot $[t_s, t_e]$.
- Highest-Bandwidth: compute a path from v_s to v_d with the highest available bandwidth in a specified time slot $[t_s, t_e]$.
- First-Slot: compute the earliest start time of a path from v_s to v_d with a fixed bandwidth β for a specified duration t_d.
- All-Slots: compute all start time slots of all paths from v_s to v_d with a fixed bandwidth β for a specified duration t_d.

Note that the solution to the First-Slot scheduling problem is the earliest start time, while the solution to the All-Slots scheduling problem is a union of all feasible start times. If t is a feasible start time in the solution to All-Slots, the computed path has bandwidth β from time point t to $t + t_d$.

3 Distributed Scheduling Algorithms

We propose an optimal bandwidth scheduling algorithm in a distributed manner for each of these problem. The proposed algorithms are based on the BFS and Bellman-Ford algorithms and are different from the existing link-state and distance-vector routing protocols: a node makes a routing decision based on its local TB lists and connectivity information, and only broadcasts its own information to its neighbor nodes. Although link-state routing protocols are easy to implement, the periodical broadcasting of node connectivity and link TB lists incurs a significant amount of overhead. Furthermore, if the changes in node connectivity and link bandwidth are not promptly updated, the network may operate with inaccurate information.

We use two types of routing messages between nodes for distributed path exploration for advance bandwidth scheduling: (i) bandwidth reservation message, and (ii) acknowledgment (ACK) message. The bandwidth scheduler running on every node incorporates the scheduling algorithms and handles the routing messages during path exploration. Node and link states can be updated by simple periodic message exchanges between neighbor nodes. Every node broadcasts a HELLO message to its neighbor nodes. Upon the receival of a HELLO message, a node simply replies with an ACK message to let the neighbor node know that the link between them is active.

3.1 Fixed-Bandwidth

Given a fixed-bandwidth (FB) reservation request r_i^{FB}, the fixed-bandwidth scheduling problem is to compute a dedicated channel from source node v_s to destination node v_d with specified bandwidth β in time slot $[t_s, t_e]$. The source node v_s receives r_i^{FB} from an end user, and initiates path exploration by broadcasting r_i^{FB} to its neighbor nodes. When an intermediate node receives r_i^{FB} from one of its neighbor nodes, it checks the TB lists of its outgoing links and determines whether r_i^{FB} can be scheduled on these links. An example of bandwidth reservation process is shown in Fig. 2. After receiving r_i^{FB} from node v_1, node v_2 checks the TB lists of three outgoing links (v_2, v_3), (v_2, v_4), and (v_2, v_5). If r_i^{FB} is feasible only on links (v_2, v_3) and (v_2, v_5), node v_2 sends r_i^{FB} to nodes v_3 and v_5. Once r_i^{FB} reaches the destination node v_d, node v_d replies with a positive acknowledgment message, which is echoed all the way back to the source node.

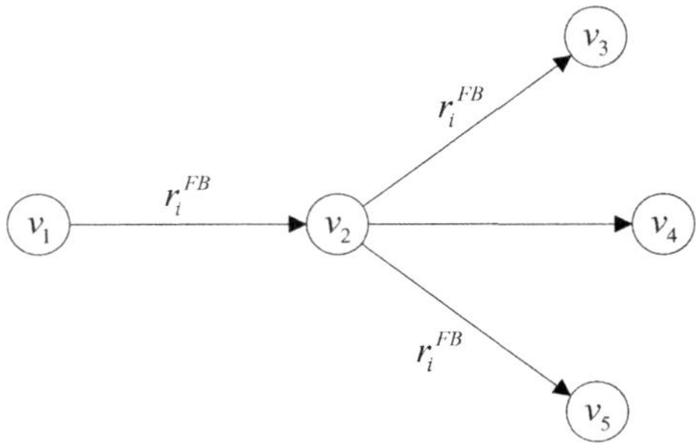

Fig. 2. An example of the bandwidth reservation process for the fixed-bandwidth problem

Algorithm 1. Scheduling algorithm for the fixed-bandwidth problem

1: Create a job queue Q to store all reservation requests.
2: Wait for routing messages.
3: **if** a fixed-bandwidth reservation request r_i^{FB} is received from its neighbor u **then**
4:     Add u to V_i^{pre}.
5:     **if** r_i^{FB} is not in Q **then**
6:         Add r_i^{FB} to Q and mark r_i^{FB} as "pending".
7:         **if** the current node is the destination of r_i^{FB} **then**
8:             Send a positive acknowledgment of r_i^{FB} to u.
9:         **else**
10:             Compute the neighbor node set S_i (excluding u) such that r_i^{FB} can be scheduled on each link between the current node and any neighbor node in S_i. If $S_i \neq \emptyset$, broadcast r_i^{FB} to all nodes in S_i; otherwise, send a negative acknowledgment of r_i^{FB} to u and mark r_i^{FB} as "failed". Initialize $n_i = |S_i|$.
11:     **else if** r_i^{FB} is marked as "failed" in Q **then**
12:         Send a negative acknowledgment of r_i^{FB} to u.
13:     Return to line 2.
14: **if** an acknowledgment of request r_i^{FB} is received from its neighbor node u and u is in S_i **then**
15:     Remove u from S_i.
16:     **if** the acknowledgment is positive **then**
17:         Allocate the bandwidth on the link between the current node and node u for r_i^{FB}. Mark r_i^{FB} as "successful". Send a positive acknowledgment of r_i^{FB} to the first node that is added to V_i^{pre}.
18:     **else**
19:         $n_i = n_i - 1$.
20:         **if** $n_i \leq 0$ **then**
21:             Mark r_i^{FB} as "failed", and send a negative acknowledgment of r_i^{FB} to all nodes in V_i^{pre}.
22:     Return to line 2.

The algorithm details for the fixed-bandwidth scheduling problem are provided in Algorithm 1. Each node maintains a job queue Q that stores bandwidth reservation requests. When a bandwidth reservation request arrives, Q is dynamically updated and the request state is changed. A bandwidth reservation request in Q is in one of three states: "pending", "successful", and "failed". The scheduling daemon waits for control messages and processes bandwidth reservation messages in lines 3-13 and acknowledgment messages in lines 14-22.

- When the current node receives a fixed-bandwidth reservation request r_i^{FB} from its neighbor node u, the algorithm first adds u to the node set V_i^{pre} that stores all the previous nodes from which r_i^{FB} is received and is used for sending back the acknowledgment. The algorithm then checks whether r_i^{FB} is in Q. If r_i^{FB} is not in Q, the algorithm adds r_i^{FB} to Q, marks r_i^{FB} as "pending", and replies with a positive acknowledgment if r_i^{FB} is destined to itself; otherwise, the algorithm computes the potential qualified neighbor node set S_i. A neighbor node is qualified if r_i^{FB} can be successfully scheduled on the link between the current node and that neighbor node based on its time-bandwidth list. If S_i is not empty, the current node broadcasts r_i^{FB} to all the nodes in S_i, which is an expansion performed in BFS. If S_i is an empty set, which means that there does not exist any qualified neighbor node, the current node replies with a negative acknowledgment to u where r_i^{FB} is received from and marks r_i^{FB} in Q as "failed". In the case that r_i^{FB} is in Q but r_i^{FB} is already marked as "failed", the current node sends a negative acknowledgment of r_i^{FB} to u since there does not exist any feasible path that passes the current node to satisfy r_i^{FB}.

– When the current node receives an acknowledgment of r_i^{FB} from its neighbor u and u is in S_i, the algorithm removes u from S_i to avoid receiving duplicate acknowledgments from the same neighbor node and checks whether the acknowledgment is positive or negative. If the acknowledgment is positive, the algorithm allocates the bandwidth on the link between the current node and u for r_i^{FB} and sends a positive acknowledgment of r_i^{FB} to the first node that is added to V_i^{pre}. Otherwise, the algorithm decreases n_i by 1. Note that n_i is initialized to be $|S_i|$ in line 10 and is used to count the number of negative acknowledgments received from neighbor nodes in S_i. If n_i reaches 0, which indicates that the current node receives negative acknowledgments from all the nodes in S_i and there does not exist a qualified link of the current node, the algorithm marks r_i^{FB} as "failed" and sends a negative acknowledgment of r_i^{FB} to all the nodes in V_i^{pre}. An example of the algorithm processing acknowledgment messages is shown in Fig. 3, where the solid line represents the fixed-bandwidth reservation request and the dashed line represents the acknowledgment. The current node is v_3 that receives r_i^{FB} from both v_1 and v_2, and broadcasts it to v_4 and v_5. In this example, $V_i^{pre} = \{v_1, v_2\}$, $S_i = \{v_4, v_5\}$. The positive acknowledgment process is shown in Fig. 3 (a): once v_3 receives a positive acknowledgment of r_i^{FB} from one node in S_i (v_5), it sends the acknowledgment to the node that is firstly added to V_i^{pre} (v_1). The negative acknowledgment process is shown in Fig. 3 (b): only when v_3 receives negative acknowledgments of r_i^{FB} from all the nodes in S_i, v_3 broadcasts the negative acknowledgment to all the nodes in V_i^{pre}. The path exploration process for r_i^{FB} is terminated when the source node of r_i^{FB} receives an acknowledgment.

Performance Tuning. Algorithm 1 is simple and scalable, but some extra work is needed to improve its performance. A deadlock may occur during the acknowledgment message processing, as shown in Fig. 4, where there is a cycle of r_i^{FB} among v_2, v_3 and v_4, but v_2 only sends r_i^{FB} to v_3 once. With the qualified neighbor node set $S_i = \{v_2, v_5\}$ for r_i^{FB}, v_4 receives a negative acknowledgment from v_5 and is waiting for the acknowledgment from v_2 before sending any acknowledgments to v_3. However, v_2 is waiting for the acknowledgement from v_3 and v_3 is waiting for the acknowledgment from v_4. Therefore, there is a deadlock among v_2, v_3 and v_4. To address this issue, we can encode a set of nodes that a bandwidth reservation request have traversed. When the algorithm computes S_i for r_i^{FB}, S_i only includes the qualified neighbor nodes that are not in the set

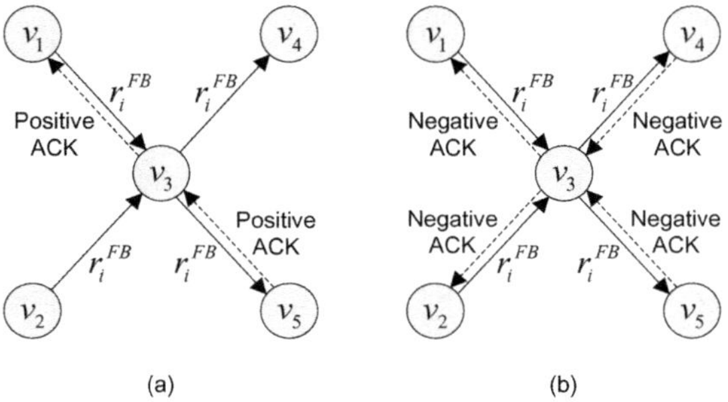

Fig. 3. An example of the acknowledging process for the fixed-bandwidth problem

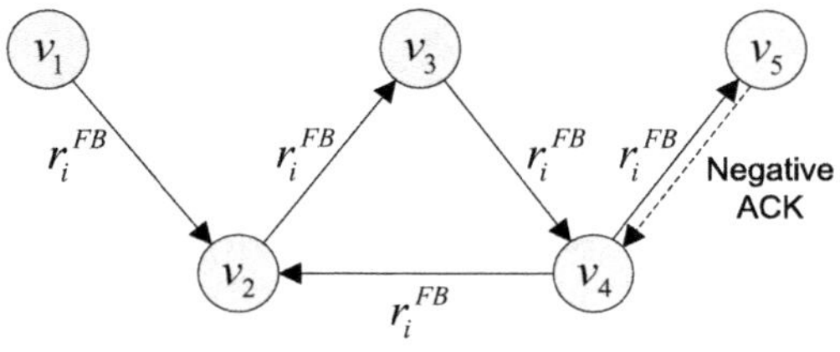

Fig. 4. An example of deadlock in Algorithm 1

of nodes that r_i^{FB} have traversed. Therefore, there is no bandwidth reservation request from v_4 to v_2 in the above example and the deadlock is avoided.

In a special case where a neighbor node u in S_i breaks down right after the current node broadcasted r_i^{FB}, and the current node receives negative acknowledgments from all the nodes in S_i except u, n_i never reaches 0 in line 23 of Algorithm 1. If there does not exist a feasible path to satisfy r_i^{FB}, the source node of r_i^{FB} may never receive a negative acknowledgment. The solution is that detecting the breakdown of a neighbor node u in S_i should be equivalent to receiving a negative acknowledgment from u for the pending bandwidth reservation request r_i^{FB}. Also, all the scheduled bandwidth reservations using u must be canceled. In this case, the current node sends a CANCEL message along the path for each scheduled bandwidth reservation request using the current node and u, and the reserved bandwidths on the corresponding links will be released. Once the source node of a bandwidth reservation request receives a CANCEL message, the source node initiates another path exploration process to find a new path. The handling of node failures can also be applied to the rest algorithms.

Algorithm Analysis. Algorithm 1 exhibits several salient features.

(i) Loop free: The job queue that maintains all incoming bandwidth reservation requests and the verification condition in line 5 ensure that each node broadcasts a bandwidth reservation request at most once. Hence, there is no loop for a bandwidth reservation request. Furthermore, the condition of whether u is in S_i in line 17 and the update of S_i in line 18 ensure that each node receives at most one acknowledgment from a neighbor node in S_i. Hence, there is no loop for an acknowledgment.

(ii) Fault tolerant: Any node and link failures can be detected by periodical HELLO messages exchanged between nodes. Hence, any node failure does not affect the path exploration process if there still exists a feasible path.

(iii) Time efficient: The runtime complexity of this algorithm is $O(T \cdot m)$ in the worst case. Unlike most distributed routing algorithms where each node must wait for a constant time period to collect all messages from its neighbor nodes, this algorithm processes each incoming routing message immediately to speed up path exploration. In the worst case, the algorithm involves $O(m)$ message communications in the entire network.

3.2 Highest-Bandwidth

Given a highest-bandwidth reservation request r_i^{HB}, the highest-bandwidth scheduling problem is to compute a dedicated channel from source node v_s to destination node v_d with the highest available bandwidth during time slot $[t_s, t_e]$. This problem can be solved by extending Dijkstra's shortest path algorithm in the centralized scheme. We propose a distributed solution based on Bellman-Ford algorithm to this problem. The source node v_s receives r_i^{HB} from an end user, initializes the highest bandwidth of r_i^{HB} to be infinity, and initiates the path exploration process by broadcasting r_i^{HB} to its neighbor nodes. The neighbor nodes compute their highest bandwidth according to the incoming r_i^{HB}, and broadcast their results only if the bandwidth value is increased. Note that the highest available bandwidth of the entire path is determined by the bottleneck bandwidth of all component links in the specified time slot. Hence, the highest bandwidth of r_i^{HB} on each node is dynamically updated during the path exploration process.

The algorithm details for the highest-bandwidth scheduling problem are provided in Algorithm 2. Let $BW_i(v_s, v_{cur})$ denote the highest bandwidth of the path found so far from source node v_s to the current node v_{cur} for r_i^{HB} in Q, and $BW_i'(v_s, v_{cur})$ denote that for the incoming r_i^{HB}. The algorithm waits for control messages, and processes bandwidth reservation messages in lines 3-17 and acknowledgment messages in lines 18-20. When the current node receives a highest-bandwidth reservation request r_i^{HB} from its neighbor node u, the algorithm checks whether r_i^{HB} is in Q; if not, the algorithm adds r_i^{HB} to Q. If the highest bandwidth of the incoming r_i^{HB} is larger than that of r_i^{HB} in Q (i.e. $BW_i'(v_s, v_{cur}) > BW_i(v_s, v_{cur})$), the algorithm updates the highest bandwidth of r_i^{HB} in Q. The algorithm computes the highest bandwidth of the path found so far from source node v_s to every neighbor node v by calculating $BW_i(v_s, v) = min\{BW_i(v_s, v_{cur}), BW_i(v_{cur}, v)\}$, where $BW_i(v_{cur}, v)$ is the highest bandwidth of the link (v_{cur}, v) during the time slot specified in r_i^{HB}. The algorithm encodes $BW_i(v_s, v)$ in r_i^{HB} and sends r_i^{HB} to v. If the highest bandwidth of r_i^{HB} in Q does not increase (i.e. $BW_i'(v_s, v_{cur}) \leq BW_i(v_s, v_{cur})$), the algorithm returns to line 2 directly to avoid message broadcasting. If the

Algorithm 2. Scheduling algorithm for the highest-bandwidth problem

1: Create a job queue Q to store all reservation requests.
2: Wait for routing messages.
3: **if** a highest-bandwidth reservation request r_i^{HB} is received from its neighbor node u **then**
4:   **if** r_i^{HB} is destined to the current node **then**
5:     Restart a timer for r_i^{HB}.
6:   **if** r_i^{HB} is not in Q **then**
7:     Add r_i^{HB} to Q.
8:   **else if** $BW_i'(v_s, v_{cur}) > BW_i(v_s, v_{cur})$ **then**
9:     $BW_i(v_s, v_{cur}) = BW_i'(v_s, v_{cur})$
10:   **else**
11:     Return to line 2.
12:   Set $v_i^{pre} = u$.
13:   **if** r_i^{HB} is not destined to the current node **then**
14:     Compute the neighbor node set S_i (excluding u).
15:     **for all** $v \in S_i$ **do**
16:       $BW_i(v_s, v) = min\{BW_i(v_s, v_{cur}), BW_i(v_{cur}, v)\}$. Encode $BW_i(v_s, v)$ in r_i^{HB} and send r_i^{HB} to v.
17:   Return to line 2.
18: **if** an acknowledgment of request r_i^{HB} is received from its neighbor node u **then**
19:   Allocate the bandwidth on the link between the current node and node u for r_i^{HB}. Forward acknowledgment of r_i^{HB} to node v_i^{pre}.
20:   Return to line 2.

current node is the destination node of r_i^{HB}, the algorithm restarts a timer for r_i^{HB}. This timer is used by the destination node to acknowledge the granting of the request since the destination node does not know when the path exploration process reaches an equilibrium. If the destination node does not receive any updated r_i^{HB} from its neighbor nodes for a period of time, it is very likely that the path exploration process for r_i^{HB} has reached an equilibrium. Once the timer for r_i^{HB} expires, the destination node determines the highest bandwidth of the entire path and sends an acknowledgment of r_i^{HB} that carries the highest bandwidth to v_i^{pre}, which is the best neighbor node on the widest path from the source node to the current node. When the current node receives an acknowledgment of r_i^{HB} from its neighbor node u, it allocates the bandwidth on the link between the current code and node u, and forwards the acknowledgment of r_i^{HB} to node v_i^{pre}. This backtracking process continues until the source node of r_i^{HB} receives the acknowledgment.

Performance Tuning. The time cost of the path exploration process is affected by the timer on the destination node, which needs to be carefully decided according to the network size and link delay. Let $DELAY$ denote the average delay of a message communication between two adjacent nodes, which includes the processing delay on two end nodes and the link delay between them. A node can estimate $DELAY$ by measuring the round trip time of a message between itself and its neighbor nodes. Since a bandwidth reservation message traverses at least 1 hop and at most $n-1$ hops from v_s to v_d, the difference between the arrival times of any two request messages is at most $(n-2)\cdot DELAY$, which could be used to set the timer on v_d.

Algorithm Analysis. Algorithm 2 also exhibits similar features as Algorithm 1.

(i) Loop free: A node broadcasts r_i^{HB} to its neighbor nodes only when the highest bandwidth of r_i^{HB} increases. When a node broadcasts r_i^{HB} to its neighbors, the highest bandwidth of r_i^{HB} does not increase during the path exploration process. Hence, there is no loop for a bandwidth reservation request. Since node v_i^{pre} is set only when the highest bandwidth of r_i^{HB} increases and the acknowledgment is sent to v_i^{pre}, there is no loop for an acknowledgment, either.

(ii) Fault tolerant: Since each node makes a local decision and acts as an autonomous system, a node or link failure would not affect the path exploration process. If one node on the computed path for r_i^{HB} breaks down after an equilibrium is achieved but before the acknowledgment of r_i^{HB} is forwarded, the source node of r_i^{HB} will never receive the acknowledgment. This problem can be solved as follows. The failure of a node can be detected by its neighbor nodes by periodical HELLO message exchanges between them. Each neighbor node then sends a negative acknowledgment of r_i^{HB} to its v_i^{pre}. The source node of r_i^{HB} eventually receives the negative acknowledgment of r_i^{HB} and may initiate another path exploration process for r_i^{HB}.

(iii) Time efficient: The runtime complexity of this algorithm is $O(m\cdot T)$ in the worst case. Except for the destination node, all other nodes process each incoming routing message immediately. In the worst case, the algorithm requires $O(n^3)$ message broadcasting as the distributed Bellman-Ford algorithm.

3.3 First-Slot and All-Slots

Given a first-slot or all-slots bandwidth reservation request, r_i^{FS} or r_i^{AS}, the first-slot or all-slots bandwidth scheduling problem is to compute the time slot with the earliest start time or all possible time slots of a dedicated channel from v_s to v_d with a fixed bandwidth β for duration t_d. Obviously, first-slot is a special case of all-slots, and the solution to all-slots can be applied to first-slot. We propose a distributed algorithm based on Bellman-Ford algorithm for these two problems.

We first define a list of start time slots $[t_i, t_{i+1}]$ for each link $l\in E$, denoted as $ST(l)$. For any time point t during a start time slot $[t_i, t_{i+1}]$, i.e. $t\in[t_i, t_{i+1}]$, link l has available bandwidth of β from time point t to time point $t+t_d$. The time slots on ST are disjoint and arranged in an ascending order. The ST list of a link can be constructed from its TB list in $O(T)$ time, and the ST list of a path can be constructed by combining the ST lists of all component links. Let $ST(v_s, v)$ denote the union of the ST lists of all paths from source node v_s to node v. Hence, $ST(v_s, v_d)$ contains all start time slots of all paths from v_s to v_d with bandwidth β for duration t_d. Let $\bigoplus$ and $\bigotimes$ denote the point-wise merging and intersection operations of the time slots in two ST lists, respectively. We have $ST(l)\bigoplus\emptyset = ST(l)$, $ST(l)\bigoplus\Re^+ = \Re^+$, $ST(l)\bigotimes\emptyset = \emptyset$,

Algorithm 3. Scheduling algorithm for the all-slots problem

1: Create a job queue Q to store all reservation requests.
2: Wait for routing messages.
3: **if** an all-slots bandwidth reservation request r_i^{AS} is received from its neighbor node u **then**
4:     **if** r_i^{AS} is destined to the current node **then**
5:         Restart a timer for r_i^{AS}.
6:     **if** r_i^{AS} is not in Q **then**
7:         Add r_i^{AS} to Q. Set $v_i^{pre} = u$.
8:     **else if** $ST_i'(v_s, v_{cur}) \not\subseteq ST_i(v_s, v_{cur})$ **then**
9:         $ST_i(v_s, v_{cur}) = ST_i(v_s, v_{cur}) \bigoplus ST_i'(v_s, v_{cur})$.
10:     **else**
11:         Return to line 2.
12:     **if** r_i^{AS} is not destined to the current node **then**
13:         Compute the neighbor node set S_i (excluding u).
14:         **for all** $v \in S_i$ **do**
15:             $ST_i(v_s, v) = ST_i(v_s, v_{cur}) \bigotimes ST_i(v_{cur}, v)$. Encode $ST_i(v_s, v)$ in r_i^{AS} and send r_i^{AS} to v.
16:     Return to line 2.
17: **if** an acknowledgment of request r_i^{AS} is received from its neighbor node u **then**
18:     Forward acknowledgment of r_i^{AS} to node v_i^{pre}.
19:     Return to line 2.

and $ST(l) \bigotimes \Re^+ = ST(l)$, where $\emptyset$ is the empty time slot and $\Re^+$ is the infinite time slot of non-negative real values.

The algorithm details for the all-slots bandwidth scheduling problem are provided in Algorithm 3. The source node v_s receives r_i^{AS} from an end user, initializes the ST list of r_i^{AS} to be $ST(v_s, v_s) = \Re^+$, and initiates the path exploration process by broadcasting r_i^{AS} to its neighbor nodes. Let $ST_i(v_s, v_{cur})$ denote the list of start time slots of the paths found so far from source node v_s to the current node v_{cur} for r_i^{AS} in Q, and $ST_i'(v_s, v_{cur})$ denote that for the incoming r_i^{AS}. The algorithm is modified from Algorithm 2 by replacing the bandwidth operation with the ST list operation. If the ST list of the incoming r_i^{AS} is not a subset of the ST list of r_i^{AS} in Q (i.e. $ST_i'(v_s, v_{cur}) \not\subseteq ST_i(v_s, v_{cur})$), the algorithm updates the ST list of r_i^{AS} in Q (i.e. $ST_i(v_s, v_{cur}) = ST_i(v_s, v_{cur}) \bigoplus ST_i'(v_s, v_{cur})$). Here, the relationship $\not\subseteq$ of two ST lists holds if at least one time slot in $ST_i'(v_s, v_{cur})$ does not belong to any time slots on $ST_i(v_s, v_{cur})$. Due to the monotonicity property of $\bigoplus$ operation, once start time slots are placed on $ST_i(v_s, v_{cur})$, they will not be removed. The algorithm then computes the start time slots of all paths found so far from source node v_s to every neighbor node v by calculating $ST_i(v_s, v) = ST_i(v_s, v_{cur}) \bigotimes ST_i(v_{cur}, v)$, where $ST_i(v_{cur}, v)$ is the ST list of link (v_{cur}, v) for r_i^{AS}. The algorithm encodes $ST_i(v_s, v)$ in r_i^{AS} and sends r_i^{AS} to v. If the current node is the destination node of r_i^{AS}, the algorithm restarts a timer for r_i^{AS}. Once the timer for r_i^{AS} expires, the destination node sends an acknowledgment of r_i^{AS} that carries all start time slots on $ST_i(v_s, v_d)$ to v_i^{pre}.

For the first-slot problem, the earliest start time is the lower boundary of the first time slot on the returned ST list. For the all-slots problem, the end user at the source node may choose one or multiple start times from the returned ST list. Once the start time t for a feasible path is decided, we can apply Algorithm 1 for the fixed-bandwidth problem to perform the actual path computation and bandwidth scheduling with $t_s = t$ and $t_e = t + t_d$.

The runtime complexity of Algorithm 3 is $O(m)$ in terms of $\bigoplus$ and $\bigotimes$ operations. Since the complexities of $\bigoplus$ and $\bigotimes$ operations are determined by the length of the ST list, which is at most $m \cdot T$ in the algorithm, the complexities of $\bigoplus$ and $\bigotimes$ operations are of $O(m \cdot T)$. Therefore, the algorithm complexity is of $O(m^2 \cdot T)$ in the worst case. Due to the similarity in the algorithm structure, the performance tuning and algorithm analysis for Algorithm 2 are applicable to Algorithm 3.

4 Performance Evaluation

We perform simulation-based evaluations for the proposed distributed scheduling algorithms. For performance comparison, we also design and implement a simple greedy algorithm. In the simulations, each simulated network is randomly generated with an arbitrary network topology with 50 nodes and 200 links, and the TB list of each link is also randomly generated with residual bandwidths ranging from 0.2 Gbps to 10 Gbps in each time slot with an identical length of 1 second. The residual bandwidths follow a normal distribution:

$$b_l[i] = 0.2 + 10 \cdot (1 - e^{-\frac{1}{2}(3x)^2}), \tag{1}$$

where x is a random variable within the range of [0,1]. There are 600 time slots in the TB list of each link.

4.1 Experimental Results for Algorithm 1

We conduct performance comparison between Algorithm 1 and the traceroute-based method for the fixed-bandwidth problem using various simulated networks. Note that traceroute is implemented in OSCARS to find the shortest path within ESnet that MPLS LSP traverses [14]. Once the entire path controlled by OSCARS is obtained, each link on the path is then checked for available bandwidth.

Fixed-bandwidth is a decision problem and the satisfiability of a fixed-bandwidth request is determined by the availability of the network resources. Algorithm 1 is an optimal algorithm that is able to find a feasible solution when there exists one. We randomly generate 200 network instances of different topologies, in each of which, we randomly generate a series of fixed-bandwidth requests with requested bandwidth β ranging from 0.24 Gbps to 2.4 Gbps at an interval of 0.24 Gbps. The duration of a request $t_e - t_s$ is constrained within the range of [1, 10]. We run Algorithm 1 and traceroute on these fixed-bandwidth requests and plot a series of acceptance rates in response to different β values in Fig. 5. The acceptance rate is defined as the ratio of successfully scheduled requests

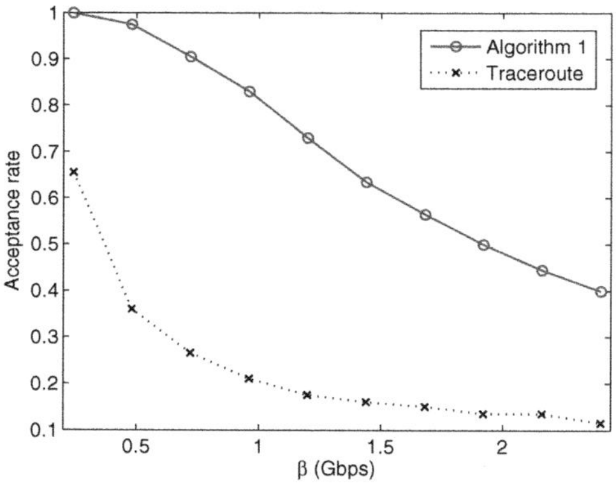

Fig. 5. Acceptance rates of Algorithm 1 and traceroute for the fixed-bandwidth problem

and the total 200 submitted requests. We observe that Algorithm 1 exhibits superior performance over the traceroute-based method. Since the requests with larger β values require more network resources, the acceptance rate decreases as β increases.

4.2 Experimental Results for Algorithm 2

We compare the performance of Algorithm 2 with that of a greedy algorithm for the highest-bandwidth problem. In the greedy algorithm, a node always chooses one neighbor node whose link has the highest available bandwidth in the specified time slot. In each of 200 randomly generated network instances, we generate a series of highest-bandwidth requests with duration $t_d = t_e - t_s$ ranging from 1 to 10 seconds at an interval of 1 second. We run Algorithm 2 and the greedy algorithm on these highest-bandwidth requests and plot the average and standard deviation of the highest available bandwidth in response to different t_d values in Fig. 6. We observe that Algorithm 2 outperforms the greedy approach in all the cases we studied. We also observe that the average highest available bandwidth decreases as t_d increases.

4.3 Experimental Results for Algorithm 3

Algorithm 3 is designed for both the first-slot and all-slots problems. We first compare the performance of Algorithm 3 with that of a greedy method for the first-slot problem. In the greedy algorithm, a node always chooses one neighbor node such that the earliest start time of the path from the source node to the neighbor node for that request is minimized. We randomly generate a series of first-slot requests with $t_d = 5$ seconds and requested bandwidth β ranging from 0.24 Gbps to 2.4 Gbps at an interval of 0.24 Gbps. We plot the average and

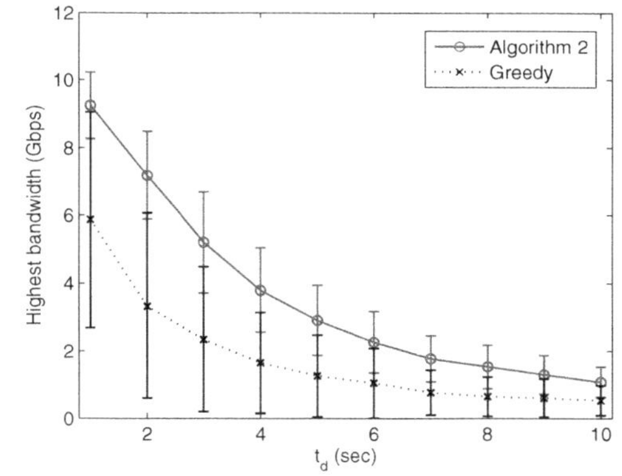

Fig. 6. The highest bandwidths (mean and standard deviation) of Algorithm 2 and the greedy method for the highest-bandwidth problem

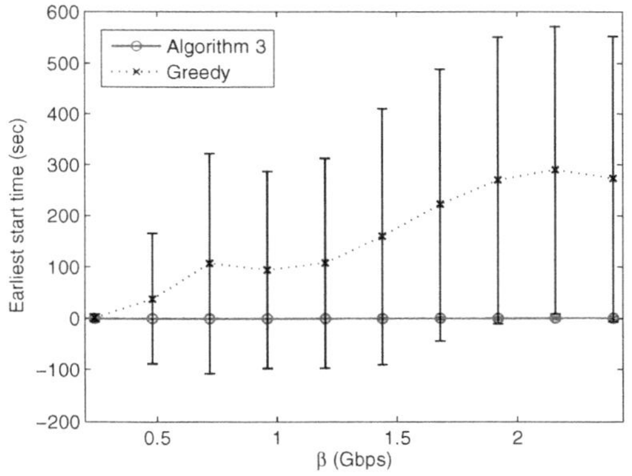

Fig. 7. The earliest start time (mean and standard deviation) of Algorithm 3 and the greedy method for the first-slot problem

standard deviation of the earliest start time in response to different β values in Fig. 7. In most of the cases, the earliest start time computed by Algorithm 3 is 0 second. The largest earliest start time is 600 seconds since we assume that there is no bandwidth reservation on each link after 600 seconds. We observe that the average of 200 earliest start times computed by Algorithm 3 is much less than that computed by the greedy method.

We also compare the performance of Algorithm 3 with that of a greedy method for the all-slots problem. The objective function in the all-slots problem is the total length of start times. In the greedy method, a node always chooses one neighbor node such that the total length of start times of the path from the source node to the neighbor node for that request is maximized. The simulation

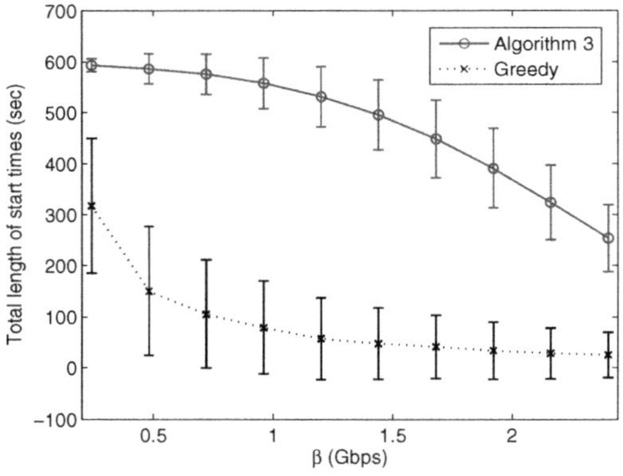

Fig. 8. The total length of start times (mean and standard deviation) of Algorithm 3 and the greedy method for the all-slots problem

settings for the all-slots problem are the same as those for the first-slot problem. We plot the average and standard deviation of the total length of start times in response to different β values in Fig. 8, and observe that the performance superiority of Algorithm 3 is dominant in all the cases.

5 Conclusion

We formulated four basic bandwidth scheduling problems in high-performance networks that support advance bandwidth reservations. We proposed a distributed algorithm for each of these problems. These algorithms are based on a rigorous extension of the classical breadth first search and Bellman-Ford algorithms. The extensive experimental results in a large set of simulated networks demonstrate the performance superiority of these algorithms in comparison with greedy approaches.

References

1. DRAGON: Dynamic Resource Allocation via GMPLS Optical Networks, `http://dragon.maxgigapop.net`
2. JGN II: Advanced Network Testbed for Research and Development, `http://www.jgn.nict.go.jp`
3. Geant2, `http://www.geant2.net`
4. OSCARS: On-demand Secure Circuits and Advance Reservation System, `http://www.es.net/oscars`
5. HOPI: Hybrid Optical and Packet Infrastructure, `http://networks.internet2.edu/hopi`
6. GENI: Global Environment for Network Innivations, `http://www.geni.net`

7. CHEETAH: Circuit-switched High-speed End-to-End Transport ArcHitecture, `http://www.ece.virginia.edu/cheetah`
8. Cohen, R., Fazlollahi, N., Starobinski, D.: Graded channel reservation with path switching in ultra high capacity networks. In: Proc. of Broadnets, San Jose, CA (2006)
9. Cohen, R., Fazlollahi, N., Starobinski, D.: Path switching and grading algorithms for advance channel reservation architectures. IEEE/ACM Transactions on Networking 17(5), 1684–1695 (2009)
10. Ganguly, S., Sen, A., Xue, G., Hao, B., Shen, B.H.: Optimal routing for fast transfer of bulk data files in time-varying networks. In: Proc. of IEEE Int. Conf. on Communications (2004)
11. Gorinsky, S., Rao, N.S.V.: Dedicated channels as an optimal network support for effective transfer of massive data. In: INFOCOM 2006 Workshop on High-Speed Networks (2006)
12. Grimmell, W.C., Rao, N.S.V.: On source-based route computation for quickest paths under dynamic bandwidth constraints. Int. J. on Foundations of Computer Science 14(3), 503–523 (2003)
13. Guerin, R.A., Orda, A.: Networks with advance reservations: the routing perspective. In: Proc. of the 19th IEEE INFOCOM (2000)
14. Guok, C., Robertson, D., Thompson, M., Lee, J., Tierney, B., Johnston, W.: Intra and interdomain circuit provisioning using the oscars reservation system. In: Proc. of the BROADNETS, San Jose, CA, October 1-5, pp. 1–8 (2006)
15. Lin, Y., Wu, Q.: Path computation with variable bandwidth for bulk data transfer in high-performance networks. In: Proceedings of INFOCOM HSN Workshop, Rio de Janeiro, Brazil (April 24, 2009)
16. Lin, Y., Wu, Q., Rao, N.S.V., Zhu, M.: On design of scheduling algorithms for advance bandwidth reservation in dedicated networks. In: The 2008 INFOCOM High-Speed Networks Workshop, Phoenix, Arizona (April 13, 2008)
17. Rao, N.S.V., Wing, W.R., Carter, S.M., Wu, Q.: Ultrascience net: Network testbed for large-scale science applications. IEEE Communications Magazine 43(11), s12–s17 (2005), An expanded version `www.csm.ornl.gov/ultranet`
18. Rao, N.S.V., Wu, Q., Carter, S.M., Wing, W.R., Ghosal, D., Banerjee, A., Mukherjee, B.: Control plane for advance bandwidth scheduling in ultra high-speed networks. In: INFOCOM 2006 Workshop on Terabits Networks (2006)
19. Sahni, S., Rao, N.S.V., Ranka, S., Li, Y., Jung, E., Kamath, N.: Bandwidth scheduling and path computation algorithms for connection-oriented networks. In: Proc. of Int. Conf. on Networking (2007)
20. Veeraraghavan, M., Lee, H., Chong, E.K.P., Li, H.: A varying-bandwidth list scheduling heuristic for file transfers. In: Proc. of IEEE Int. Conf. on Communications (2004)
21. Zhang, Z.L., Duan, Z., Hou, Y.T.: Decoupling QoS control from core routers: A novel bandwidth broker architecture for scalable support of guaranteed services. In: Proc. of ACM SIGCOMM (2000)

Proportionally Fair Selective Cooperation for Cellular Networks: Algorithm, Simulation and Analysis*

Erwu Liu[1], Qinqing Zhang[2], and Kin K. Leung[1]

[1] Department of Electrical and Electronic Engineering, Imperial College, London, UK
{erwu.liu,kin.leung}@imperial.ac.uk
[2] Applied Physics Laboratory, Johns Hopkins University, USA
Department of Computer Science, Johns Hopkins University, USA
qinqing.zhang@jhuapl.edu

Abstract. We consider the problem of 2-node cooperation for cellular networks. In the considered system, a node and a cooperative relay are selected in a proportionally fair (PF) manner to transmit to the base station (BS), which uses the maximal ratio combining (MRC) for reception. The proposed scheme integrates the PF criterion in selective cooperation to maximize the overall utility of all 2-node cooperative transmissions, and is called *proportionally fair selective cooperation* (PFSC). This technique does not require distributed space-time coding algorithms and its simplicity allows for immediate implementation in existing hardware.

In this research, we further provide a mathematical framework to analyze the performance of PFSC under a Rayleigh flat fading channel. We also present an analytical expression for quantifying the throughput gain of a PFSC-enabled cellular network over a traditional PFS-enabled cellular network without node cooperation. Using simulations, we show that our model is very accurate.

To our knowledge, it is the first time that a closed-form expression is obtained for the throughput of 2-node cooperative communication in a cellular network with the PF constraint.

Keywords: Cooperative communications, proportional fairness.

1 Introduction

In this work, we propose and analyze a practical cooperative scheme for cellular networks. The rationale for the extraordinary interest in exploiting cooperative techniques in wireless networks is very clear, as demonstrated by the high volume of publications in recent years [1, 2, 3, 4, 5, 6, 7, 8, 9, 10, 11, 12]. Cooperative diversity [4] is a new form of diversity through distributed transmission and processing with node collaboration. Transmit cooperation has nodes exchanging

* This work was supported, in part, by Johns Hopkins University, Applied Physics Laboratory's internal research and development funds.

X. Zhang and D. Qiao (Eds.): QShine 2010, LNICST 74, pp. 392–405, 2011.

each other's messages, sharing their antennas, and creating multiple paths to transmit the information. Receive cooperation has nodes forwarding information about their observations for decoding. A system with both transmit and receive cooperation is similar to a multiple-input multiple-output (MIMO) system in a networked manner. Therefore it is sometimes called a distributed MIMO or network MIMO [13]. In a network where each node is equipped with a single omnidirectional antenna, cooperative diversity can achieve similar gains from a MIMO system where each node is equipped with multiple antennas.

Researchers have proposed various network MIMO schemes for distributed wireless networks [13, 14]. However, there are several limitations that hurdle the potential implementation of a network MIMO in a practical system compared to the conventional MIMO system. For example, synchronization among distributed antennas is much harder than a conventional MIMO transmitter. Additional resources such as bandwidth, power, time and complex code design are also required to enable the cooperation. Furthermore, antenna power allocation cannot be done as that in a conventional MIMO system, *etc.* However, these limitations and implementation complexity are significantly reduced in time-division-multiple-access (TDMA) cellular networks. Consequently, there are growing interests in cooperative diversity schemes for cellular networks. In this direction, this paper considers a practical 2-node cooperative scheme which requires minimum cooperation among the nodes for TMDA cellular networks and focuses on the throughput performance.

In a 2-node cooperative scenario for TDMA cellular networks, a Node m together with a Node r are opportunistically selected by the base station (BS) to transmit at next slot in a collaborative manner. Once the pair (m, r) is scheduled, Node m will transmit and Node r will function as a decode-and-forward (DF) relay [4]. The BS will store the signal received from m and use it together with the signal received from r for successful decoding. Without node cooperation, the BS will only decode using the signal from m. With this scheme, better performance can be achieved as the BS will decode using the signals from two branches.

The BS can use various metrics to determine which (m, r) pair will be scheduled to transmit. For example, to achieve the full spatial diversity order, selective relaying [6] can be extended to this 2-node cooperative scenario, *i.e.*, only the best (m, r) pair that provides the highest instantaneous capacity to the BS is selected at next slot. One can verify that, while this greedy policy provides the highest throughput performance of the whole system, it results in highly unfair resource utilization and will starve those nodes far away from the BS. Obviously, the performance of the considered system not only depends on node cooperation, it also depends on the metric used to select the (m, r) pair. Originally from Kelly's work [15], the proportionally fair scheduling (PFS) algorithm [16, 17, 18, 19, 20, 21] has spurred the development of a large number of network utility maximization (NUM) algorithms since 1997 [22, 23, 24, 25, 26], and is implemented in current 3G networks [27] as the most-cited NUM method. Being a promising scheme for fair resource allocation, PFS has shown excellent balance between throughput and fairness via multi-user diversity and game-theoretic equilibrium. In light of this, we apply the proportionally fair (PF) criterion in selecting the pair (m, r) for cooperative communication in cellular networks and the scheduling algorithm is called *proportionally fair selective cooperation* (PFSC).

The paper is structured as follows. In Section 2, we first describe the assumptions made in our study of the problem and the resulting problem formulations, we then propose the PFSC algorithm that maximizes the overall utility of a cooperative cellular network. After that, analytical results of PFSC are presented in Section 3. Finally, in Section 4, we present simulation results to validate the theoretical model and throughput performance of PFSC, followed by the conclusion in Section 5.

2 System Model

For an N-node TDMA cellular network, let's consider the problem where these N nodes wishing to transmit data to the BS. We assume that each node can function as a DF relay as needed and single-input-single-output (SISO) is used for communications.

2.1 Problem Formulation

As shown in Fig. 1, unlike the BS in a traditional cellular network where only one node is selected for transmission at a time, the BS in the considered network schedules a pair of nodes (m, r) at each slot. Once the pair (m, r) is selected, Node m will transmit and Node r will function as a DF relay. If Node r successfully decodes the signal received from Node m, it will relay the signal to the BS. The BS will decode using the signals received from nodes m and r.

We use Θ to denote the set of nodes in the network and $\Omega = \{(m,r)|m,r \in \Theta, m \neq r\}$ to denote the set of all possible node pairs. Obviously, for an N-node cellular network, there are ${}_NP_2 = N \times (N-1)$ possible (m, r) pairs. We use $m \rightarrow r$, $(n, r) \rightarrow BS$ to denote the non-cooperative link and the cooperative link, respectively. According to Fig. 1, a transmission over a non-cooperative link together with a transmission over the corresponding cooperative link compose a relay-aided 2-node transmission. Let $R_{m,r}^{NC}[t+1]$, $R_{m,r}^{C}[t+1]$ be the instantaneous capacities of the non-cooperative link $m \rightarrow r$ and the cooperative link $(m, r) \rightarrow BS$, and $R_{m,r}[t+1]$ be the instantaneous capacity of the corresponding 2-node transmission, in next time slot t+1. $R_{m,r}[t]$'s k-point moving average throughput (*i.e.*, short-term average throughput) up to time t is denoted by $\mu_{m,r}[t]$.

Because of the superior performance of the PFS algorithm, we apply the PF criterion [15,21] in the cooperative cellular network shown in Fig. 1. Similar to the traditional PFS problem in a cellular network without cooperation, for a 2-node cooperative cellular network with PF constraint, the objective is to maximize the overall logarithmic utility of all 2-node transmission. This is formulated as follows

$$\max \sum_{(m,r)\in\Omega} \ln\left(\mu_{m,r}(t)\right) . \tag{1}$$

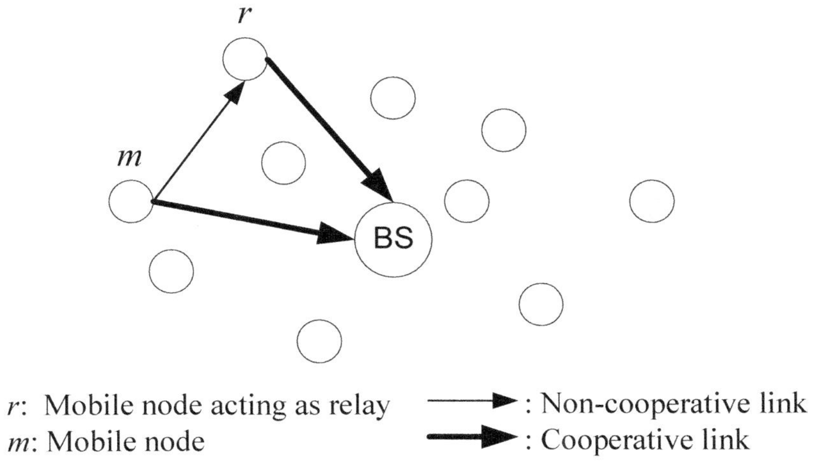

Fig. 1. 2-node cooperative cellular network

given,

$$I_{m,r}[t+1] = \begin{cases} 1 & \text{if node pair } (m,r) \in \Omega \text{ is scheduled} \\ 0 & \text{else} \end{cases} \tag{2}$$

$$\mu_{m,r}[t+1] = \left(1 - \frac{1}{k}\right) \mu_{m,r}[t] + I_{m,r}[t+1] \times \frac{R_{m,r}[t+1]}{k}. \tag{3}$$

$$R_{m,r}[t+1] = \min\left(R^{NC}_{m,r}[t+1], R^{C}_{m,r}[t+1]\right). \tag{4}$$

(2) indicates that at each time slot, only one node and one accompanying relay are selected for cooperative transmission at next time slot. The min operation in (4) accounts for the successful decoding required at both Node r and the BS.

2.2 Proportionally Fair Selective Cooperation (PFSC)

For the above optimization problem, we propose the following solution, namely the *proportionally fair selective cooperation* (PFSC) algorithm.

As pointed out earlier, it has been proved that PFS maximizes the overall logarithmic utility of a cellular network by scheduling the Node j having the maximum PF metric defined as the ratio of the instantaneous rate to its short-term average throughput tracked by an exponential moving average [15]. Similarly, PFSC maximizes the overall logarithmic utility of a 2-node cooperative cellular network by scheduling the node pair (m,r) having the maximum PFSC metric $\frac{R_{m,r}[t+1]}{\mu_{m,r}[t]}$. In other words, the proposed PFSC algorithm is the optimal solution of (1).

Algorithm 1. Proportionally Fair Selective Cooperation (PFSC) Algorithm

/* k, network size N, node set Θ, various link capacity at each slot */
Input: $k, N, \Theta, R^{NC}_{m,r}, R^{NC}_{m,BS}$ $(\forall m, r \in \Theta, m \neq r)$;

/* node pair (m^*, r^*) scheduled to transmit with data rate R_{m^*,r^*} at each slot */
Output: (m^*, r^*, R_{m^*,r^*})

/* 2-node set */
$\Omega = \{(m, r) | m, r \in \Theta, m \neq r\}$;

/* current slot */
$t = 0$;

/* initialize throughput to small value */
$\mu_{m,r}[0] = 0.000001 (\forall m, r \in \Omega)$;

for $t \geq 0$ **do**
    for $m, r \in \Omega$ **do**
        /* instantaneous capacity of 2-node transmission */
        $R_{m,r}[t+1] = \min\left(R^{NC}_{m,r}[t+1], R^{C}_{m,r}[t+1]\right)$;
        /* calculate PF metric */
        $\mathbf{M}_{m,r}[t+1] = R_{m,r}[t+1] / \mu_{m,r}[t]$;
    end
    /* the node pair (m^*, r^*) with the maximum PF metric will be scheduled */
    $(m^*, r^*) = \arg\max_{m,r} \mathbf{M}_{m,r}[t+1]$;

    /* update throughputs for all pairs */
    for $m, r \in \Omega$ **do**
        $\mu_{m,r}[t+1] = (1 - \frac{1}{k}) \mu_{m,r}[t]$;
    end
    /* update throughput of (m^*, r^*) correctly to reflect it is scheduled */
    $\mu_{m,r}[t+1] = \mu_{m,r}[t+1] + R_{m,r}[t+1]/k$;

    /* tell m^*, r^* to transmit with data rate $R_{m^*,r^*}[t+1]$. r^* acts as a DF relay */
    Output $(m^*, r^*, R_{m^*,r^*}[t+1])$;

    /* proceeds to next slot */
    $t = t + 1$;
end

To implement PFSC, the BS needs to know the channel state information (CSI) of both the $m \to BS$ and $m \to r$ links. Relay needs to know the CSI of the m to relay link, which will be sent to the BS. We assume that the network operates in a slow fading scenario, so that channel estimation is possible and CSI is available at both the BS and relays.

The differences between PFSC and PFS are: in an N-node cellular network, with PFS the BS will schedule N nodes while with PFSC the BS will schedule $N \times (N-1)$ "nodes" (*i.e.,* node pairs); with PFS, the instantaneous capacity of Node m is estimated by the link quality of $m \to BS$, while with PFSC, the instantaneous capacity of node pair (m, r) should be estimated by the link quality of both $m \to r$ and $(m, r) \to BS$.

3 Analysis of PFSC over a Rayleigh Flat Fading Channel

In this section, we analyze the proposed PFSC algorithm over Rayleigh flat fading channels, and provide closed-form theoretical results. For the analysis, we use the following assumptions:

- Both the non-cooperative link and the cooperative link are Rayleigh flat faded;
- The fading between any different links (no matter they are non-cooperative links or cooperative links) is mutually independent but not necessarily identical distributed;
- Channel fading keeps constant over each slot, and varies independently from slot to slot;
- Maximal ratio combining (MRC) is used at the BS to combine the signals from the source and relay nodes.

To include the distance-dependent path loss, we use the model seen in [4], *i.e.*,, the instantaneous SNR is the multiplication of the average SNR (determined by path loss and shadowing[1]) with an randomly fading variable. This results in a single random variable that jointly models path loss and fading. Consequently, the instantaneous SNR of a Rayleigh fading link is modeled as an exponentially distributed random variable with an average SNR (in dB) given by

$$\overline{SNR_d} = \overline{SNR_{d_0}} - \alpha \times 10 \log_{10}(d/d_0). \tag{5}$$

where α is path loss exponent, d denotes link distance, and $\overline{SNR_{d_0}}$ is the average SNR at reference distance d_0.

Refer to Fig. 1, we use $SNR^{NC}_{m,r}$, $SNR^{C}_{m,r}$ to denote the instantaneous SNRs of the non-cooperative link $m \to r$ and the cooperative link $(m, r) \to BS$, respectively.

We use $\overline{SNR^{NC}_{m,r}}$, $\overline{SNR^{NC}_{m,BS}}$, $\overline{SNR^{NC}_{r,BS}}$ to denote the mean SNRs of the non-cooperative links $m \to r$, $m \to BS$, and $r \to BS$, respectively.

[1] Without loss of generality, shadowing is not considered here.

Due to Rayleigh fading, the instantaneous SNR of a non-cooperative link is an exponentially distributed random variable with a probability density function (pdf) given by

$$\rho_{i,j}^{NC}(x) = \frac{1}{\overline{SNR_{i,j}^{NC}}} \times e^{-x/\overline{SNR_{i,j}^{NC}}}, \forall(i,j) \in \Theta, \text{or } j = BS. \tag{6}$$

With MRC, the combined SNR is the sum of the SNRs of each individual diversity branch. The pdf of the instantaneous SNR of a cooperative link is then given by

$$\rho_{m,r}^{C}(x) = \frac{1}{\overline{SNR_{m,BS}^{NC}} + \overline{SNR_{r,BS}^{NC}}} \times e^{-x/\left(\overline{SNR_{m,r}^{NC}} + \overline{SNR_{r,BS}^{NC}}\right)}. \tag{7}$$

Unless otherwise specified, system bandwidth is normalized to 1 Hz. Using Shannon formula, we can write the capacities of the non-cooperative link and the cooperative link as follows

$$R_{m,r}^{NC} = \log_2\left(1 + SNR_{m,r}^{NC}\right). \tag{8}$$

$$R_{m,r}^{C} = \log_2\left(1 + SNR_{m,r}^{C}\right). \tag{9}$$

Substituting (8) and (9) into (4), we have

$$\begin{aligned} R_{m,r}[t+1] &= \min\left(R_{m,r}^{NC}[t+1], R_{m,r}^{C}[t+1]\right) \\ &= \log_2\left(1 + \min\left(SNR_{m,r}^{NC}[t+1], SNR_{m,r}^{C}[t+1]\right)\right). \end{aligned} \tag{10}$$

We know that the distribution of the minimum of multiple independently exponential random variables is also an exponential distributed random variable, *i.e.*, for n independently exponential distributed random variables $g_1, g_2, \ldots, g_n$ with parameters $\lambda_1, \lambda_2, \ldots, \lambda_n$, $g = \min\{g_1, g_2, \ldots, g_n\}$ is also exponentially distributed, with parameter

$$\lambda = \lambda_1 + \lambda_2 + \ldots + \lambda_n. \tag{11}$$

We then have the capacity of the 2-node transmission,

$$R_{m,r}[t+1] = \log_2\left(1 + SNR_{m,r}[t+1]\right). \tag{12}$$

where $SNR_{m,r}$ is an exponentially distributed random variable with a pdf given by

$$\rho_{m,r}(x) = \left(\frac{1}{\overline{SNR_{m,BS}^{NC}} + \overline{SNR_{r,BS}^{NC}}} + \frac{1}{\overline{SNR_{m,r}^{NC}}}\right) \times e^{-\left(\frac{x}{\overline{SNR_{m,BS}^{NC}} + \overline{SNR_{r,BS}^{NC}}} + \frac{x}{\overline{SNR_{m,r}^{NC}}}\right)}. \tag{13}$$

(12) indicates that $SNR_{m,r}$ represents the effective SNR of the 2-node transmission $\{m \rightarrow r, (m,r) \rightarrow BS\}$. By (13), we obtain the mean value of $SNR_{m,r}$

$$\overline{SNR_{m,r}} = \frac{1}{\left(\frac{1}{\overline{SNR_{m,BS}^{NC}} + \overline{SNR_{r,BS}^{NC}}} + \frac{1}{\overline{SNR_{m,r}^{NC}}}\right)}. \tag{14}$$

[28] and [29] have shown that, in Rayleigh or Ricean fading networks, the Shannon capacity $R = \log_2[1+SNR]$ can be accurately approximated by Normal distribution. Specially, for a SISO link over a Rayleigh flat fading channel where SNR is exponential distributed, the Normal distribution is characterized by a mean and variance given as follows

$$E[R] = \int_0^\infty e^{-x} \times \log_2\left(1 + \overline{SNR} \times x\right) dx. \tag{15}$$

$$\sigma^2 = \int_0^\infty e^{-x} \times \left(\log_2\left(1 + \overline{SNR} \times x\right)\right)^2 dx - \left(E[R]\right)^2. \tag{16}$$

According to the above discussion, we have the following theorem.

Theorem 1. *In a Rayleigh flat fading cellular network with 2-node cooperation, the capacity of a 2-node transmission can be characterized by Normal distribution.*

Specifically, with (12)-(16), we obtain the Normal distribution that characterizes the capacity of the 2-node transmission $\{m \rightarrow r, (m,r) \rightarrow BS\}$

$$E[R_{m,r}] = \int_0^\infty e^{-x} dx \times \log_2\left(1 + \frac{x}{\left(\frac{1}{\overline{SNR_{m,BS}^{NC}} + \overline{SNR_{r,BS}^{NC}}} + \frac{1}{\overline{SNR_{m,r}^{NC}}}\right)}\right). \tag{17}$$

$$\sigma_{m,r}^2 = \int_0^\infty e^{-x} dx \times \left(\log_2\left(1 + \frac{x}{\left(\frac{1}{\overline{SNR_{m,BS}^{NC}} + \overline{SNR_{r,BS}^{NC}}} + \frac{1}{\overline{SNR_{m,r}^{SR}}}\right)}\right)\right)^2 - \left(E[R_{m,r}]\right)^2. \tag{18}$$

According to our previous research [30, 21], we have the following lemma.

Lemma 1 (PFS Throughput). *For a PFS-enabled N-node cellular network, if the instantaneous capacity is a Normally distributed random variable, the long-term average throughput of Node i is given by*

$$E[\mu_i] = \frac{E[R_i]}{N} \times \left(1 - [\phi(-M_i)]^N\right) + \int_{-M_i}^\infty y\sigma_i\rho(y) \times [\phi(y)]^{N-1} dy. \tag{19}$$

where R_i and μ_i are the instantaneous capacity and throughput of Node i, $E[R_i]$ and σ_i denote the statistical average and standard deviation of R_i, $M_i = E[R_i]/\mu_i$, $\rho(\cdot)$ and $\phi(\cdot)$ are the pdf and cdf of zero mean, unit variance standard normal distribution.

According to Theorem 1, we know that in a Rayleigh flat fading environment, the capacity of a node pair in a PFSC-enabled cellular network can be modeled by Normal distribution. With Theorem 1 and Lemma 1, we have the following important theoretical theorem for PFSC.

Theorem 2. *For a Rayleigh fading PFSC-enabled cellular network, the long-term average throughput of Node m is given by*

$$\begin{aligned} E[\mu_m] &= \sum_{r\in\Theta, r\neq m} E[\mu_{m,r}] \\ &= \sum_{r\in\Theta, r\neq m} \frac{E[R_{m,r}]}{N(N-1)} \times \left(1 - [\phi(-M_{m,r})]^{N(N-1)}\right) \\ &\quad + \sum_{r\in\Theta, r\neq m} \int_{-M_{m,r}}^{\infty} y\sigma_{m,r}\rho(y) \times [\phi(y)]^{N(N-1)-1}\, dy. \end{aligned} \tag{20}$$

where $R_{m,r}$ and $\mu_{m,r}$ are the instantaneous capacity and throughput of node pair (m, r), $E[R_{m,r}]$ and $\sigma_{m,r}$ denote the statistical average and standard deviation of $R_{m,r}$, $M_{m,r} = E[R_{m,r}]/\mu_{m,r}$.

Proof. There are ${}_NP_2 = N(N-1)$ possible node pairs in an N-node cellular network. In the beginning of this section, we have proved that in a Rayleigh fading cellular network with 2-node cooperation, the capacity $E[\mu_{m,r}]$ of a node pair (m, r) can be modeled as a Normally distributed random variable (Theorem 1). Viewing a pair in a PFSC-enabled cellular network as a node in a PFS-enabled cellular network, and applying Lemma 1, we have

$$\begin{aligned} E[\mu_{m,r}] &= \frac{E[R_{m,r}]}{N(N-1)} \times \left(1 - [\phi(-M_{m,r})]^{N(N-1)}\right) \\ &\quad + \int_{-M_{m,r}}^{\infty} y\sigma_{m,r}\rho(y) \times [\phi(y)]^{N(N-1)-1}\, dy. \end{aligned} \tag{21}$$

where $E[R_{m,r}]$ and $\sigma_{m,r}$ are given by (17) and (18).

Since $E[\mu_m] = \sum_{r\in\Theta, r\neq m} E[\mu_{m,r}]$, Theorem 2 follows immediately.

With Theorem 2 and Lemma 1, we have the following corollary.

Corollary 1. *For Node m in a Rayleigh fading cellular network, the throughput gain of PFSC over PFS is*

$$G_m = \frac{\sum\limits_{r\in\Theta, r\neq m} \begin{pmatrix} \frac{E[R_{m,r}]}{N(N-1)} \times \left(1 - [\phi(-M_{m,r})]^{N(N-1)}\right) \\ + \int_{-M_{m,r}}^{\infty} y\sigma_{m,r}\rho(y) \times [\phi(y)]^{N(N-1)-1}\, dy \end{pmatrix}}{\left(\frac{E[R_m]}{N} \times \left(1 - [\phi(-M_m)]^{N}\right) + \int_{-M_m}^{\infty} y\sigma_m\rho(y) \times [\phi(y)]^{N-1}\, dy\right)}. \tag{22}$$

where $R_{m,r}$ and $\mu_{m,r}$ are the instantaneous capacity and throughput of node pair (m,r), $E[R_{m,r}]$ and $\sigma_{m,r}$ denote the statistical average and standard deviation of $R_{m,r}$, $M_{m,r}=E[R_{m,r}]/\mu_{m,r}$.

Remark 1. *Numerical calculations show that $(\phi(-M_m))^N \ll 1$, $(\phi(-M_{m,r}))^{N(N-1)} \ll 1$, $\int_{-M_m}^{\infty} y\rho(y)\,[\phi(y)]^{N-1}dy \approx \int_0^{\infty} y\rho(y)\,[\phi(y)]^{N-1}dy$, $\int_{-M_{m,r}}^{\infty} y\rho(y)[\phi(y)]^{N(N-1)-1}dy \approx \int_0^{\infty} y\rho(y)[\phi(y)]^{N(N-1)-1}dy$, so Theorem 2 and Corollary 1 reduce to the following forms,*

$$E[\mu_{m,r}] = \frac{E[R_{m,r}]}{N(N-1)} + \sigma_{m,r}\int_0^{\infty} y\rho(y)\,[\phi(y)]^{N(N-1)}\,dy. \tag{23}$$

$$G_m = \frac{\sum\limits_{r\in\Theta,r\neq m}\left(\frac{E[R_{m,r}]}{N(N-1)} + \sigma_{m,r}\int_0^{\infty} y\rho(y)\,[\phi(y)]^{N(N-1)-1}\,dy\right)}{\left(\frac{E[R_m]}{N} + \sigma_m\int_0^{\infty} y\rho(y)\,[\phi(y)]^{N-1}\,dy\right)}. \tag{24}$$

Corollary 1 provides a closed-form expression for quantifying the throughput gain with 2-node cooperation for cellular networks with PF constraint.

4 Numerical and Simulation Results

We first performed numerical experiment to evaluate the throughput performance of PFSC algorithm over a Rayleigh flat fading channel.

4.1 Numerical Experiment

In the numerical experiment, we use the following settings: As shown in Fig. 2, 10 nodes $n_1 \sim n_{10}$ are placed in an area of 1.2 km× 1.6 km. Path loss exponent $\alpha = 2.5 \sim 4.5$ for various environment, reference distance $d_0 = 100$ m, reference SNR at d_0 is $SNR_{d_0} = 30$ dB.

Equation (5) is used to calculate the SNR (in dB) at distance d. After that, (15) and (16) are used to calculate the mean and standard deviation of the capacity for each node; (14), (17) and (18) are used to calculate the mean and standard deviation of the capacity for each (m,r) pair. Finally, we use (20) or (22) to evaluate the performance of PFSC.

We plot in Fig. 3 the throughput gain G_m of PFSC over PFS for each Node m under various path loss scenarios. We can see that with node cooperation, PFSC significantly improves performance. Fig. 3 also shows that the PFSC performance increases with path loss exponent and bad-channel nodes n_1, n_2, n_3, n_{10} benefit more from PFSC than good-channel nodes $n_4 \sim n_9$. These indicate that PFSC will perform better in relatively bad communication environment.

In Fig. 3, we notice Node n_5 has relatively small gain. This is due to two reasons: 1). Node n_5 locates near to the BS and already has relatively good channel condition. This makes relaying less efficient; 2). All possible DF relays are far from Node n_5. This means that the received SNRs may not be high enough to allow successful decoding at the DF relays, which in turn reduces the relaying gain;

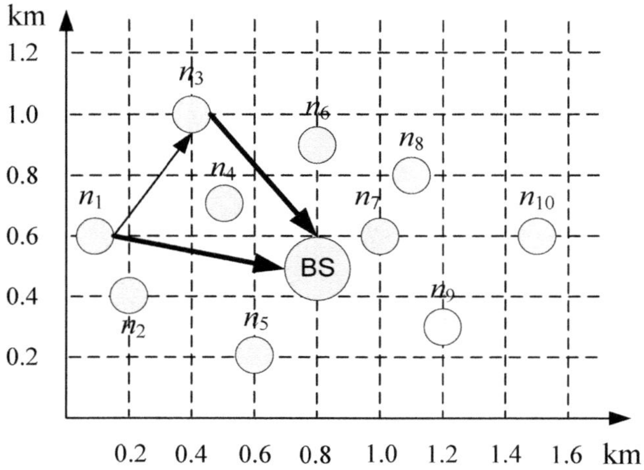

Fig. 2. Network Topology: a 10-node Cellular Network

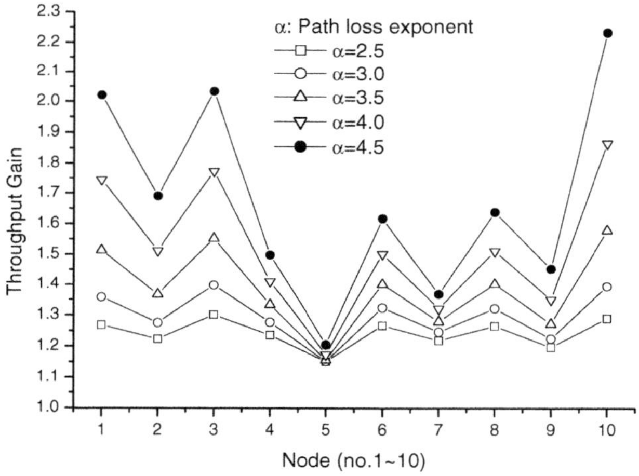

Fig. 3. PFSC v.s. PFS: a throughput comparison

4.2 Simulation

We now evaluate the accuracy of our theoretical models by simulation. We use the same network topology as shown in Fig. 2. Each link undergoes independent Rayleigh flat fading. The PFSC algorithm presented in Subsection 2.2 is used in simulation. Path loss exponent is 3.0. Moving average factor $k = 500$. The simulation runs for 8000 slots. Fig. 4 depicts the curves from simulation and theoretical results for nodes n_1, n_7, and n_9. Node n_1 is the farthest node from

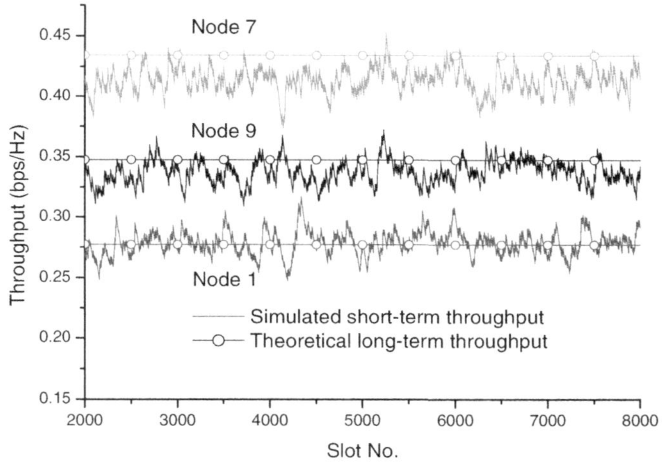

Fig. 4. PFSC for the worst/medium/best node: Simulation v.s. Analysis

the BS and Node n_7 is the nearest one. Nodes n_1, n_9, and n_7 represent the worst-, the medium-, and the best-channel nodes in a cellular network.

The validity of our theoretical results is illustrated in Fig. 4 where the solid lines are the short-term throughput from simulation and the circle ones are the long-term throughput from analysis. We can see that our theoretical analysis matches well with simulations for all three nodes with $\sim$98% accuracy.

5 Conclusion

2-node selective cooperation in cellular networks is a practical cooperation method that does not require complex code design to benefit from the spatial diversity. Adding proportional fairness into 2-node selective cooperation, we propose the proportionally fair selective cooperation (PFSC) algorithm for cellular networks. This technique integrates well-known proportional fairness with traditional MRC reception and thus allows for immediate implementation in existing hardware.

To facilitate researches, we further analyzed PFSC and presented a framework to quantify the performance of PFSC. Numerical results validates that PFSC significantly improves the throughput performance, and the simulation shows the accuracy of our analysis.

The increased performance of PFSC over PFS comes with additional overhead: To implement PFSC, the BS needs to collect the CSI of both m to the BS and m to relay links. Relay needs to know the CSI of the m to relay link, which is sent to the BS via a control/feedback channel. Even with this overhead, PFSC is still a promising solution for efficient data communications in a cooperative cellular network, due to its simplicity and superior performance in terms of throughput.

References

1. Sendonaris, A., Erkip, E., Aazhang, B.: User cooperation diversity. part i. system description. IEEE Transactions on Communications 51(11), 1927–1938 (2003)
2. Sendonaris, A., Erkip, E., Aazhang, B.: User cooperation diversity. part ii. implementation aspects and performance analysis. IEEE Transactions on Communications 51(11), 1939–1948 (2003)
3. Ng, T.C.-Y., Yu, W.: Joint optimization of relay strategies and resource allocations in cooperative cellular networks. IEEE Journal on Selected Areas in Communications 25(2), 328–339 (2007)
4. Laneman, J.N., Tse, D.N.C., Wornell, G.W.: Cooperative diversity in wireless networks: Efficient protocols and outage behavior. IEEE Transactions on Information Theory 50(12), 3062–3080 (2004)
5. Aktas, E., Evans, J., Hanly, S.: Distributed decoding in a cellular multiple-access channel. IEEE Transactions on Wireless Communications 7(1), 241–250 (2008)
6. Bletsas, A., Khisti, A., Reed, D.P., Lippman, A.: A simple cooperative diversity method based on network path selection. IEEE Journal on Selected Areas in Communications 24(3), 659–672 (2006)
7. Sung, Y., Misra, S., Tong, L., Ephremides, A.: Cooperative routing for distributed detection in large sensor networks. IEEE Journal on Selected Areas in Communications 25(2), 471–483 (2007)
8. Huang, J., Han, Z., Chiang, M., Poor, H.V.: Auction-based resource allocation for cooperative communications. IEEE Journal on Selected Areas in Communications 26(7), 1226–1237 (2008)
9. Simeone, O., Somekh, O., Kramer, G., Poor, H.V., Shamai, S.: Throughput of cellular systems with conferencing mobiles and cooperative base stations. EURASIP Journal on Wireless Communications and Networking 2008, 1–14 (2008)
10. Chen, W., Dai, L., Ben Letaief, K., Cao, Z.: A unified cross-layer framework for resource allocation in cooperative networks. IEEE Transactions on Wireless Communications 7(8), 3000–3012 (2008)
11. Papadogiannis, A., Hardouin, E., Gesbert, D.: Decentralising multi-cell cooperative processing on the downlink: a novel robust framework. EURASIP Journal on Wireless Communications and Networking (August 2009)
12. Liu, E., Zhang, Q., Leung, K.K.: Theoretical analysis of selective relaying, cooperative multi-hop networks with fairness constraints. In: Proc. IEEE International Conference on Communications ICC 2009, pp. 1–5 (June 2009)
13. Coso, A.D., Spagnolini, U., Ibars, C.: Cooperative distributed MIMO channels in wireless sensor networks. IEEE Journal on Selected Areas in Communications 25(2), 402–414 (2007)
14. Rong, Y., Hua, Y.: Optimal power schedule for distributed MIMO links. IEEE Transactions on Wireless Communications 7(8), 2896–2900 (2008)
15. Kelly, F.: Charging and rate control for elastic traffic. European Transactions on Telecommunications 8, 33–37 (1997)
16. Kushner, H.J., Whiting, P.A.: Convergence of proportional-fair sharing algorithms under general conditions. IEEE Transactions on Wireless Communications 3(4), 1250–1259 (2004)
17. Borst, S.: User-level performance of channel-aware scheduling algorithms in wireless data networks. IEEE/ACM Transactions on Networking 13(3), 636–647 (2005)

18. Bu, T., Li, L., Ramjee, R.: Generalized proportional fair scheduling in third generation wireless data networks. In: Proc. INFOCOM 2006, the 25th Joint Conference of the IEEE Computer and Communications Societies, pp. 1–12. IEEE, Los Alamitos (2006)
19. Caire, G., Muller, R.R., Knopp, R.: Hard fairness versus proportional fairness in wireless communications: The single-cell case. IEEE Transactions on Information Theory 53(4), 1366–1385 (2007)
20. Li, L., Pal, M., Yang, Y.R.: Proportional fairness in multi-rate wireless lans. In: Proc. of the 27th Joint Conference of the IEEE Computer and Communications Societies, INFOCOM 2008, pp. 1004–1012. IEEE, Los Alamitos (2008)
21. Liu, E., Leung, K.K.: Expected throughput of the proportional fair scheduling over Rayleigh fading channels. IEEE Communications Letters 14(6), 515–517 (2010)
22. Bianchi, G., Campbell, A.T.: A programmable mac framework for utility-based adaptive quality of service support. IEEE Journal on Selected Areas in Communications 18(2), 244–255 (2000)
23. Shi, L., Liu, C., Liu, B.: Network utility maximization for triple-play services. Computer Communications 31(10), 2257–2269 (2008)
24. Xu, D., Li, Y., Chiang, M., Calderbank, A.: Elastic service availability: utility framework and optimal provisioning. IEEE Journal on Selected Areas in Communications 26(6), 55–65 (2008)
25. Zhang, X., Tao, M., Ng, C.: Utility-based wireless resource allocation for variable rate transmission. IEEE Transactions on Wireless Communications 7(9), 3292–3296 (2008)
26. Gao, Q., Zhang, J., Hanly, S.: Cross-layer rate control in wireless networks with lossy links: Leaky-pipe flow, effective network utility maximization and hop-by-hop algorithms. IEEE Transactions on Wireless Communications 8(6), 3068–3076 (2009)
27. Jalali, A., Padovani, R., Pankaj, R.: Data throughput of CDMA-HDR a high efficiency-high data rate personal communication wireless system. In: Proc. IEEE 51st VTC 2000-Spring Tokyo Vehicular Technology, vol. 3, pp. 1854–1858 (January 2000)
28. Smith, P.J., Roy, S., Shafi, M.: Capacity of MIMO systems with semicorrelated flat fading. IEEE Transactions on Information Theory 49(10), 2781–2788 (2003)
29. McKay, M.R., Smith, P.J., Suraweera, H.A., Collings, I.B.: On the mutual information distribution of OFDM-based spatial multiplexing: Exact variance and outage approximation. IEEE Transactions on Information Theory 54(7), 3260–3278 (2008)
30. Liu, E., Zhang, Q., Leung, K.K.: Resource allocation for frequency-selective fading, multi-carrier systems. In: Proc. IEEE International Conference on Communications ICC 2009 (June 2009)

A Multi-objective Optimization Approach for Designing Multihop Cellular Networks

Souha Bannour, Abdelhakim Hafid, and Mariam Tagmouti

Network Research Laboratory,
University of Montreal, Canada
{bannours,ahafid,tagmoutm}@iro.umontreal.ca

Abstract. A proper design of multi-hop cellular networks (MCNs) is a key step before its deployment. It helps determining where to install the nodes and how to configure their interfaces while guaranteeing full user coverage and satisfying traffic and QoS requirements with minimum cost. Few proposals can be found in the open literature that deals with the MCN design problem. Furthermore, these proposals assume the existence of a physical topology where the locations of the nodes are fixed. In this paper, we consider the design of MCNs assuming unfixed topologies (i.e., locations of nodes are not known a priori). We start with proposing a new multi-objective optimization model for designing MCNs. This model simultaneously optimizes two conflicting objectives, namely network deployment cost and throughput while guaranteeing users' full coverage and the requirements of providers (expected amount of traffic/users and QoS). To resolve the optimization problem, we start with an exact resolution using CPLEX, and then we develop a fast and simple greedy algorithm.

Keywords: Multihop, Cellular Networks, Design problem, Multi-objective optimization.

1 Introduction

In the last decades, several applications have emerged and have been taking more space and importance in our lives, such as multimedia applications. Indeed, the exponential growth of mobile telephony, among others, has created a huge need for new services. These services have specific requirements in terms of bandwidth and Quality of Service (QoS). Thus, wireless networks, especially cellular networks, should be designed to support these requirements. In cellular networks, data rates noticeably vary depending on the positions of users in the cell due to signal fluctuations and interferences. In addition, the quality of the received signal is much more affected by path attenuation, mainly in the case of non-line-of-sight. The capacity and the coverage are two major challenges in cellular networks. So a new architecture that solves these problems will certainly be of great interest. Multihop Cellular Networks (MCNs) has been proposed as an attractive solution [1]. This architecture consists of using relay stations (RS) that work as intermediate nodes of communication to receive and transmit data to the destination. In some cases, when necessary and when conditions permit, users themselves may act as relays. Otherwise,

X. Zhang and D. Qiao (Eds.): QShine 2010, LNICST 74, pp. 406–420, 2011.

relaying is performed by dedicated equipments that are part of the network infrastructure. Thus, the communication can be carried over multiple hops instead of a single hop between a mobile station (MS) and a base station (BS). The MCNs emerged mainly in order to improve the performance and cells capacity of cellular networks. Each MS has 2 interfaces: a 3G interface to communicate directly with BSs and a WiFi interface to communicate with RSs; each RS has a 3G interface to communicate with BSs and a WiFi interface to communicate with MSs and neighbouring RSs. It has been shown [1] that MCNs reduce interference and provide better coverage and higher throughput.

Major research efforts have been focused on developing design solutions for cellular networks and WLANs. However, we can't apply these solutions because these networks strongly differ from MCNs (e.g., 1-hop communication vs. multiple-hop communication). The design of a MCN basically involves choosing the best installation location (given a set of candidate locations), the type of network nodes to install (Base Station "BS" or Relay Station "RS") and their number. It involves also deciding the channels or codes to be assigned and the power to be applied to nodes' interfaces while at the same time providing guaranteed coverage to users and adequate connectivity with minimum cost.

Most of related work on the performance improvement of MCNs assumes a fixed topology. In fact, in the open literature, we came across several contributions that compute the throughput for a given MCN topology and node locations [7- 9]. It has been proved that the number of deployed RSs has an impact on the performance of the network. In this context, the impact of the cell radius and the number of relays in the system performance was studied and evaluated in [2]. A number of existing contributions assume that the relay stations are pre-installed and try to optimize some parameters, such as traffic load [12] and delay [13]. Contributions in [3-5] studied the most convenient locations of BSs in a cellular environment without the use of RSs. So et al. [6] investigated relay placement with the assumption that the BSs are pre-installed (positions known a priori). More specifically, in the context of a single BS (i.e., cell), they proposed an approach to compute a minimum number and the best positions of relays, to use in a micro cell, in order to maintain the pre-specified uplink and downlink demands of the end users. However, this work did not consider the constraints of interferences.

The cost of deployment and throughput are the most important criteria in MCNs design. To provide better throughput, more nodes need to be deployed. However, the more nodes we use, the more expensive the deployment will be. Therefore, a single objective optimization model cannot reflect the true nature of the problem. In this paper, we define an optimization model to decide the most convenient locations for BSs and RSs simultaneously. Its objective is to minimize the network deployment cost, to maximize the network throughput, and to guarantee clients' coverage. Our model allows the design of MCNs from scratch (e.g. a new network deployment or a new geographic area) or the expansion of existing cellular network (e.g. to increase the capacity). In opposition to existing schemes, we address several issues (e.g., multi-channel, placement of BSs, placement of RSs, and interferences) at the same time by exploiting the relationship between them.

Since the problem is NP-hard, exact solutions, using for example CPLEX, are not feasible for realistic size networks. Thus, we propose a simple heuristic to produce a feasible solution (a solution that satisfies all the constraints). Our model takes into account physical interference to make it more realistic. . The complexity of the studied problem makes the model hard and non-linear. After linearization, we solved the problem using CPLEX. Then, we proposed a greedy heuristic for larger instances. It aims at placing RSs and BSs in the most appropriate position sites to guarantee wireless connectivity and users' coverage at a minimum cost.

The key contributions of the paper can be summarized as follows: (1) a novel multi-objective optimization model that optimizes two conflicting objectives: cost and throughput. It can be used for both deployment of new MCN and expansion of an existing MCN; and (2) a heuristic algorithm to solve the model for real-size networks.

The paper is organised as follows. In Section II, we present the proposed mathematical formulation of the MCNs design problem. In Section III we describe the approach used to solve the proposed model. Section IV summarizes our preliminary experimental results. Finally, Section V concludes the paper and presents future works.

2 Problem Definition and Formulation

2.1 Problem Description

A MCN consists of a number of BSs and RSs. To relay communications between MSs and BSs, RSs are linked to BSs through wireless links; they are also interlinked through wireless links. However, BSs are interlinked using wired connections. RSs have the same structure as BSs, but they are smaller and less costly since they provide far less functionalities than BSs. The MCN can use 2 different types of RSs: fixed or mobile.

In our model, we consider a wireless network with fixed RSs, multiple channels and multiple heterogeneous radios. Each node (BS, RS or MS) has 2 interfaces: a 3G (UMTS) and a WiFi interface to use according to the available resources and the distance between the sender and the receiver. We use WiFi radio if the distance and the capacity allow it; otherwise, we use the 3G interface. Each node has multiple wireless channels (12 channels in IEEE 802.11a) which are orthogonal to each other, and multiple orthogonal codes for 3G interface (W-CDMA: Wideband Code Division Multiple Access), to allow simultaneous communications.

Let S be the set of potential sites (PS) (positions where BSs or RSs can be installed); and P the set of positions of traffic concentration in the area of study called traffic spots (TS). The design problem aims at: (a) selecting a subset $N \subseteq S$ of PSs where nodes (BSs or RSs) should be installed. This means that these nodes cover the considered TSs, and their capacities can satisfy clients' request; (b) selecting a subset $B \subseteq S$ of BSs among PSs, where the connectivity is assured and all traffic generated by TSs can find a way to reach one of these BSs; and (c) maintaining the cardinalities of N and B small enough to satisfy the financial and the performance' requirements of the network planner.

2.2 Terminology

Before going any further, we describe the terminology that will be used to define our model (see Table 1).

Table 1. Inputs

Variables	Description
c_i^b	The installation cost of BS_i , $i \in S$
c_i^r	The installation cost of RS_i , $i \in S$
$d(i, j)$	The distance between two nodes i and j
TRWifi	The maximum transmission range of WiFi interface
TR3G	The maximum transmission range of 3G interface
α_{ij}	Coverage matrix. α_{ij} = 1 if 2 devices are installed in the positions PS_j and PS_i and can communicate with each other (given the devices maximum transmission range), or if 1 device is installed in the position PS_j and can communicate with a TS_i; 0 otherwise. $i \in P \cup S, j \in S$
α_{ij}^{Wifi}	Coverage matrix with respect to WiFi. α_{ij}^{Wifi} = 1 if 2 devices are installed in the positions PS_j and PS_i and can communicate with each other (given the devices maximum transmission range of WiFi signal), or if 1 device is installed in the position PS_j and can communicate with a TS_i via WiFi signal; 0 otherwise. $i \in P \cup S, j \in S$
α_{ij}^{3G}	Coverage matrix with respect to 3G. α_{ij}^{3G} = 1 if 2 devices are installed in the positions PS_j and PS_i and can communicate with each other (given the devices maximum transmission range of 3G signal), or if 1 device is installed in the position PS_j and can communicate with a TS_i via 3G signal; 0 otherwise. $i \in P \cup S, j \in S$
φ_{ij}	Traffic capacity matrix of the wireless links. $i \in P \cup S, j \in S$
θ_i	Traffic generated by TS_i , $\forall i \in P$ θ_i should verify the following constraint: $\theta_i < \sum_{j \in S} \varphi_{ij}$
P_{max}	The maximum transmission power of MS

Table 1. (*Continued*)

Power_WiFi	The discrete set of the power values that a WiFi radio can use
p_{ij}^{3G}	Received power (3G signal) from TS_i or PS_i to PS_j
P_i^{3G}	Transmission power (3G signal) of TS_i or PS_i
P_i^{WiFi}	Transmission power (WiFi signal) of TS_i or PS_i
TR_i	Transmission range of node i
TR_i^{3G}	Transmission range of node i using 3G interface
TR_i^{Wifi}	Transmission range of node i using WiFi interface
f_{ij}^k	Traffic flow routed from TS_i or RS_i to BS_j or RS_j using k
f_{ij}^c	Traffic flow routed from TS_i or RS_i to BS_j or RS_j using code c
x_{ij}	A binary variable that takes 1 if TS_i is assigned to PS_j; 0 otherwise
r_j	A binary variable that takes 1 if a RS is installed in PS_j; 0 otherwise
b_j	A binary variable that takes 1 if a BS is installed in PS_j; 0 otherwise
t_i	A binary variable that takes 1 if TS_i can transmit its traffic to RS/BS; i.e. if TS_i is covered by one or more PSs
y_{ij}^k	A binary variable that takes 1 if there exists a wireless link between PS_i and PS_j or between TS_i and PS_j which uses channel k; 0 otherwise
z_{ij}^c	A binary variable that takes 1 if there exists a 3G connection between 2 PS i and j or between TS_i and PS_j which uses code c; 0 otherwise
u_{ij}	A binary variable that takes 1 if node j is the farthest node from i such that $y_{ij} = 1$; 0 otherwise

2.3 Mathematical Formulation

The objective functions aim at minimizing the cost of the infrastructure deployment and maximizing the throughput of network. We define three objective functions.

Objective Functions

$$\max \sum_{k\le N}\sum_{i\in S\cup P}\sum_{j\in S}\frac{f_{ij}^{k}}{\varphi_{ij}} \tag{a}$$

$$\min \sum_{i\in S\cup P}\sum_{j\in S}\frac{p_{ij}^{3G}}{P_{\max}} \tag{b}$$

$$\min \sum_{i\in S} C_i^{b} b_i + C_i^{r} r_i \tag{c}$$

where (a) maximizes the flow capacity ratio of WiFi links to maximize WiFi throughput; (b) minimizes the power received by the nodes (RSs and BSs) to limit interferences and cell overlapping to maximize 3G throughput; and (c) minimizes the deployment cost.

In this paper, we convert this multi-objectives model to an aggregated form using a single objective model by using a weighted sum of the three objectives. The goal is to resolve the model using common single objective resolution methods. In future work, we plan to resolve the proposed model using multi-objective resolution methods [14].

$$\min(\alpha_1 * \sum_{i\in S}(C_i^{b} b_i + C_i^{r} r_i) + \alpha_2 * M * \sum_{i\in S\cup P}\sum_{j\in S}\frac{p_{ij}}{P_{\max}} - \alpha_3 * M' * \sum_{i\in P\cup S}\sum_{j\in P}\sum_{k\le N}\frac{f_{ij}^{k}}{\varphi_{ij}})$$

where $\alpha_1+\alpha_2+\alpha_3=1$, and $M\in\Re^{+}$ is a penalty coefficient.

M' must respect this constraint: $M' \ge 10^6 * \sum_{i\in S}(C_i^{b}+C_i^{r})$.

In our model, $M = M' = 10^7$.

Constraints

$$\sum_{j\in S}\alpha_{ij} * (r_j + b_j) \ge t_i \qquad \forall i\in P \tag{1}$$

$$x_{ij} \le \alpha_{ij} * (r_j + b_j) \qquad \forall i\in P\cup S, \forall j\in S \tag{2}$$

$$r_j + b_j \le 1 \qquad \forall j\in S \tag{3}$$

$$\sum_{j\in S} x_{ij} \ge t_i \qquad \forall j\in P \tag{4}$$

$$f_{ij}^{k} \le \varphi_{ij} * y_{ij}^{k} \qquad \forall i\in S\cup P, \forall j\in S \quad \forall k\le N \tag{5}$$

$$\sum_{k\le N}\sum_{\substack{j\in S\\ j\notin P}} f_{ij}^{k} + \sum_{c\le L}\sum_{\substack{j\in S\\ j\notin P}} f_{ij}^{c} + f_i = \sum_{k\le N}\sum_{l\in S\cup P} f_{li}^{k} + \sum_{c\le L}\sum_{l\in S\cup P} f_{li}^{c} \qquad \forall i\in S\ (i\notin P) \tag{6}$$

$$\sum_{k\le N}\sum_{\substack{j\in S\\ j\notin P}} f_{ij}^{k} + f_{ij}^{c} \le M_1 * r_i \qquad \forall i\in S\ (i\notin P) \tag{7}$$

$$\sum_{k\le N}\sum_{l\in S\cup P} f_{li}^{k} + f_{li}^{c} \le M_1 * (r_i + b_i) \qquad \forall i\in S \ (i\notin P) \qquad (8)$$

$$\sum_{k\le N}\sum_{j\in S} f_{ij}^{k} + \sum_{c\le L}\sum_{j\in S} f_{ij}^{c} = \theta_i * t_i \qquad \forall i\in S \ (i\notin P) \qquad (9)$$

$$\theta_i + \sum_{k\le N}\sum_{j\in S} f_{ij}^{k} + \sum_{c\le L}\sum_{j\in S} f_{ij}^{c} = \sum_{k\le N}\sum_{l\in S\cup P} f_{li}^{k} + \sum_{c\le L}\sum_{l\in S\cup P} f_{li}^{c} \qquad \forall i\in P\cap S \qquad (10)$$

$$\sum_{j\in S\cup P}\sum_{k\le N} f_{ji}^{k} + \sum_{j\in S\cup P}\sum_{c\le L} f_{ji}^{c} - f_i \le M_2 * (1-b_i) \qquad \forall i\in S \qquad (11)$$

$$f_i \le M_2 * b_i \qquad \forall i\in S \qquad (12)$$

$$\frac{p_{ij}^{3G}}{g_{ij}} \le P_{\max} \qquad \forall i\in S\cup P, \forall j\in S \qquad (13)$$

$$P_i^{3G} \le M * \sum_{c\le L}\sum_{j\in S} z_{ij}^{c} \qquad \forall i\in P\cup S \qquad (14)$$

$$p_{ij}^{3G} \le M_4 * f_{ij}^{c} \qquad \forall i\in S\cup P, \forall j\in S \qquad (15)$$

$$P_i^{3G} \le M * \sum_{c\le L}\sum_{j\in S} z_{ij}^{c} \qquad \forall i\in P\cup S \qquad (16)$$

$$\frac{\dfrac{P_i^{WiFi}}{d(i,n)^{\lambda}}}{\mu + \dfrac{P_m^{WiFi}}{d(m,n)^{\lambda}} * y_{mn}^{k}} * y_{in}^{k} \ge SIR_{\min} \qquad \forall i,m\in P\cup S, \forall n\in S, \forall k\le N \qquad (17)$$

$$P_i^{WiFi} \in Power\_Wifi \qquad \forall i\in P\cup S \qquad (18)$$

$$P_i^{WiFi} \le M * \sum_{k\le N}\sum_{j\in S} y_{ij}^{k} \qquad \forall i\in P\cup S \qquad (19)$$

$$r_i \le \sum_{x}\sum_{y}\sum_{j\in S} \alpha_{i_x j} b_j \alpha_{i_x i_y} \qquad \forall i_x, i_y \in S;\ x,y\in \aleph \qquad (20)$$

$$f_{ij}^{k} \le M_4 * y_{ij}^{k} \qquad (21)$$

$$y_{ij}^{k} \le f_{ij}^{k} \qquad (22)$$

$$f_{ij}^{c} \le M_5 * z_{ij}^{c} \qquad (23)$$

$$z_{ij}^{c} \le f_{ij}^{c} \qquad (24)$$

$$t_i \le \sum_{j\in S} \alpha_{ij} \qquad \forall i\in S\cup P \qquad (25)$$

$$\sum_{j\in S} \alpha_{ij} \le M_6 * t_i \qquad \forall i\in S\cup P \qquad (26)$$

$$y_{ij}^{k} \le \alpha_{ij}^{Wifi} \qquad \forall i\in P\cup S \qquad (27)$$

$$z_{ij}^{c} \le \alpha_{ij}^{3G} \qquad \forall i\in P\cup S \qquad (28)$$

$$f_{ij}^{k}, f_{ij}^{c}, f_i, M_1, M_2, M_3, M_4, M_5 \in \Re^{+} \quad (29)$$

$$x_{ij}, r_i, b_i, y_{ij}^{k}, y_{ij}, z_{ij}^{c}, z_{ij}, u_{ij}, e_{ij} \in \{0,1\} \quad (30)$$

Constraint (1) ensures that each TSi is covered by at least one RS or BS. Constraint (2) ensures that TSi or PSi can be assigned to PSj only if one RS or one BS is installed in PSj and this node covers TSi or PSi. Constraint (3) prevents to have RS and BS installed in the same PSj. Constraint (4) stipulates that each TSi should be assigned to at least one PSj. Constraint (5) prevents a link flow from exceeding the capacity of the link; it also states that a link between 2 nodes i and j using a channel k can exist only if there are 2 nodes installed there, connected and assigned the same channel. Constraints (6) - (10) define the flow balance for each node. Since we consider only the uplink, the flow received by a BS doesn't go out. This traffic is absorbed locally using a new variable f_i which is defined by constraints (11) and (12). Constraint (13) ensures that the emission power of a node cannot exceed the maximum emission power (Pmax). Constraint (14) states that if there is no 3G connection between two nodes i and j, P_i^{3G} must be equal to zero. Constraint (15) forces p_{ij}^{3G} to be equal to zero when there is no flow between two nodes i and j. Constraint (16) forces p_i^{3G} to be equal to zero when there is no flow between node i and other nodes. Constraints (17) and (18) limit the interference on WiFi link. We use a physical model of interference [9] since it is more realistic than a logical model. For example, if $y_{mn}^{k} = 1$ (there exists a wireless link between PSm and PSn or between TSm and PSn which uses channel k) and constraint (17) is satisfied, then a successful transmission is feasible between PSi and PSn or TSi and PSn. For each node i, the smallest value in Power_WiFi that satisfies (17) is selected (constraint (18)). Constraint (19) forces P_i^{WiFi} to be equal to zero when there is no flow between node i and any other node. Constraint (20) stipulates that each node must reach at least one BS via one or multiple hops. Note that this condition is already covered by the flow conservation (6)-(10) and the sink constraints (11)-(12). The traffic must start from MS to reach one BS. It is generated by MS if this MS is covered by one PS (constraints (10) and (11)); in this case, the traffic traverses several nodes (constraint (6)) until it reaches one BS (constraints (7)-(9)). So, these constraints guarantee that every MS is linked to a BS. Constraints (21)-(24) state that when there is no WiFi link (respectively 3G link) between two nodes i and j ($y_{ij}^{k} = 0$, $z_{ij}^{c} = 0$), the flow must be equal to zero ($f_{ij}^{k} = 0$, $f_{ij}^{c} = 0$); and inversely, when the flow between two nodes i and j is equal to zero, that means there is no WiFi link between these two nodes. Constraints (25) and (26) stipulate that for each node i, if there is no node that covers it ($\sum_{j\in S} \alpha_{ij} = 0$), then the traffic cannot be sent ($t_i = 0$); and inversely, if we can't send the traffic of a node i to another node, that means there is no node that covers i. Constraint (27) (resp. constraint (28)) is logical constraint to relate the WiFi (resp. 3G) coverage to the existence of wireless link between two nodes i and j.

Metrics

p_{ij}^{3G} computes the received power from TS_i or PS_i to PS_j [11]

$$p_{ij}^{3G} = \frac{1}{1+\frac{W}{(\frac{E_b}{N_0})*R*v_i}} * I_{total} \quad \text{if } \exists c / z_{ij}^c \neq 0; \qquad p_{ij}^{3G} = 0 \quad \text{otherwise} \tag{31}$$

where W is the WCDMA chip rate which is equal to 3.84 Mcps, R is the bit rate of user i, v_i is the activity factor of user i at physical layer (the recommended values of this variable are 0.67 for speech and 1.0 for data), $\frac{E_b}{N_0}$ is the energy per user bit divided by the noise spectral density, and I_{total} is the total received wideband power including thermal noise power in the base station. I_{total} is formulated as follows:

$$I_{total} = \sum_{i \in F} P_i^{3G} + \eta \tag{32}$$

where F represents the set of nodes i that transmit to the same BS or RS.

g_{ij} used in constraint (15) is the propagation factor of the radio link between TS_i and PS_j or PS_i and PS_j:

$$g_{ij} = \frac{1}{10^{\frac{A(d(i,j))}{10}}} \tag{33}$$

where A(d(i.j)) represents the channel attenuation (in dB) [17]. In this formula d(i,j) must be expressed in Km.

$$A(d(i.j)) = \begin{cases} 143.83 + 38.35 * \log(d(i,j)) \text{ for urban areas} \\ 116.31 + 38.35 * \log(d(i,j)) \text{ for rural areas} \end{cases} \tag{34}$$

The received power of each node i is computed as follows:

$$P_i^{3G} = \max_{j \in S} p_{ij}^{3G} \qquad \forall i \in P \cup S \tag{35}$$

Equations (36) and (37) compute transmission range of WiFi and 3G interfaces respectively. The transmission range of node i consists of the distance between this node and the farthest node j with which it communicates ($\exists k, f_{ij}^k \neq 0$ or $\exists c, f_{ij}^c \neq 0$).

$$TR_i^{Wifi} = \max_{j \in P}(d_{ij} * y_{ij}^k) \qquad \forall i \in P \cup S, k \leq N \tag{36}$$

$$TR_i^{3G} = \max_{j \in P}(d_{ij} * z_{ij}^c) \qquad \forall i \in P \cup S, c \leq L \tag{37}$$

Model linearization

Constraints and metrics (17), (31), (35), (36) and (37) are not linear. In the following, we propose an approach to make them linear. The goal is to produce a linear model for MCN design.

We replace constraint (17) by the following constraints:

$$p\_y_{ij}^{k} \leq M * y_{ij}^{k} \qquad \forall i \in P \cup S, j \in S, k \leq N \qquad (17\text{-}1)$$

$$p\_y_{ij}^{c} - P_i^{WiFi} \leq 1 - y_{ij}^{k} \qquad \forall i \in P \cup S, j \in S, k \leq N \qquad (17\text{-}2)$$

$$p\_y_{ij}^{k} - P_i^{WiFi} \geq M * (y_{ij}^{k} - 1) \qquad \forall i \in P \cup S, j \in S, k \leq N \qquad (17\text{-}3)$$

$$\frac{\dfrac{p\_y_{in}^{k}}{d(i,n)^{\lambda}}}{\mu + \dfrac{p\_y_{mn}^{k}}{d(m,n)^{\lambda}}} \geq SIR_{\min} \qquad \forall i,m \in P \cup S, n \in S, k \leq N \qquad (17\text{-}4)$$

where $p\_y_{ij}^{k} = \begin{cases} P_i^{WiFi} & if \;\; y_{ij}^{k} = 1 \\ 0 & if \;\; y_{ij}^{k} = 0 \end{cases}$

We replace constraint (31) by the following constraints:

$$p\_z_{ij}^{c} \leq M * z_{ij}^{c} \qquad \forall i \in P \cup S, j \in S, c \leq L \qquad (31\text{-}1)$$

$$p\_z_{ij}^{c} - P_i^{3G} \leq 1 - z_{ij}^{c} \qquad \forall i \in P \cup S, j \in S, c \leq L \qquad (31\text{-}2)$$

$$p\_z_{ij}^{c} - P_i^{3G} \geq M * (z_{ij}^{c} - 1) \qquad \forall i \in P \cup S, j \in S, c \leq L \qquad (31\text{-}3)$$

$$p_{ij}^{3G} = \frac{1}{1 + \dfrac{W}{(\dfrac{E_b}{N_0}) * R * v_i}} * (\sum_{i \in F} p\_z_{ij}^{c} + \eta * z_{ij}^{c}) \qquad \forall i \in P \cup S, j \in S, c \leq L \qquad (31\text{-}4)$$

where $p\_z_{ij}^{c} = \begin{cases} P_i^{3G} & if \;\; z_{ij}^{c} = 1 \\ 0 & if \;\; z_{ij}^{c} = 0 \end{cases}$

We replace constraint (35) by the following constraints:

$$P_i^{3G} \geq p_{ij}^{3G} \qquad \forall j \in S, \forall i \in P \cup S \qquad (35\text{-}1)$$

$$P_i^{3G} = p_{ij}^{3G} * e_{ij} \qquad \forall j \in S, \forall i \in P \cup S \qquad (35\text{-}2)$$

$$\sum_{j \notin S} e_{ij} \leq 1 \qquad \forall j \in S \qquad (35\text{-}3)$$

where $e_{ij} = \begin{cases} 1 & if \ p_{ij}^{3G} \geq p_{il}^{3G} \quad \forall l \in S \\ 0 & else \end{cases}$

We replace constraint (36) by the following constraints:

$$TR_i^{Wifi} \geq d_{ij} * y_{ij}^k \qquad \forall i \in P \cup S,\ j \in S, k \leq N \qquad (36\text{-}1)$$

$$TR_i^{Wifi} = \sum_{j \in S} d_{ij} * u_{ij} \qquad \forall i \in P \cup S \qquad (36\text{-}2)$$

$$\sum_{j \in S} u_{ij} \leq 1 \qquad \forall i \in P \cup S \qquad (36\text{-}3)$$

where $u_{ij} = \begin{cases} 1 & if \quad d_{ij} * y_{ij}^k \geq d_{il} * y_{il}^m \quad \forall l \in S, k, m \leq K \\ 0 & else \end{cases}$

We replace constraint (37) by the following constraints:

$$TR_i^{3G} \geq d_{ij} * z_{ij}^c \qquad \forall i \in P \cup S,\ j \in S, c \leq L \qquad (37\text{-}1)$$

$$TR_i^{3G} = \sum_{j \in S} d_{ij} * v_{ij} \qquad \forall i \in P \cup S \qquad (37\text{-}2)$$

$$\sum_{j \in S} v_{ij} \leq 1 \qquad \forall i \in P \cup S \qquad (37\text{-}3)$$

where $v_{ij} = \begin{cases} 1 & if \quad d_{ij} * z_{ij}^c \geq d_{il} * z_{il}^n \quad \forall l \in S, c, n \leq L \\ 0 & else \end{cases}$

3 Problem Resolution

In the beginning, we solved the problem using the linear solver CPLEX. However, since the problem is NP-hard, CPLEX can only solve it for small sized networks. An exact resolution is not feasible for realistic size networks. Thus, we have developed a constructive greedy heuristic to solve this problem and to obtain a good solution (not optimal) within a reasonable amount of computing time.

The heuristic inputs consist of the positions of potential sites (PSs) and traffic spots (TSs), and the expected traffic per TSs. The proposed heuristic aims at placing RSs and BSs in the most appropriate position sites. After running the proposed heuristic, we obtain, as outputs, RSs and BSs positions and their characteristics (coverage range, channel and code assignment, power applied to each node's interface, traffic received and sent by each node and the paths traffic traverse from TSs to BSs).

The operation of the proposed heuristic consists of 2 steps: (1) RSs are first placed since they are cheaper and they are biased towards multi-hop communications; and (2) BSs are installed. Step (1) takes into account the objective functions (a) and (b). It satisfies the constraints of coverage and interference. This step includes two sub-steps: (a) RS_j is placed in PS_j if there exist TS_i covered by only one PS_j, and if there is enough resources to satisfy TS_i demand; and (b) for each PS_j, a set E_j is created of TSs which are not yet assigned to any RS, and covered by this PS_j. Then, these sets E_j

are sorted in decreasing order of their cardinalities. Next, the first set E_j is considered and then a RS is installed in PS_j. All TSs$\in E_j$ are assigned to this RS while satisfying resource and interference conditions. After that, the TSs belonging to more than one set E_j and have been already assigned, are removed from the others sets E_k. The process (1-b) is repeated until all TSs are assigned.

The main objective of Step (2) is to ensure that each node can reach at least one BS via one or multiple hops. This step includes two sub-steps: (a) a RS is replaced by a BS if it is not covered by any other RS or PS to ensure that each MS can reach one BS in one hop or more; note that the maximal number of hops n to reach a BS from a TS must be limited since the verification of the solution feasibility does not scale with the number of hops (in our simulations, n assumes 2); and (b) next, for each RS_j, a set F_j, that consists of covered RSs, is created. The sets F_j are sorted in decreasing order of their cardinalities. The first set F_j is considered and then RS_j is replaced by a BS_j. All RSs$\in F_j$ are assigned to this BS while satisfying interference and capacity constraints. The RSs belonging to more than one set F_j and have been already assigned, are removed from the other sets F_k. The process (2-b) is repeated until all F_k are processed. If there are RSs not assigned to BSs or to other RSs, we add BSs in the appropriate position sites (we mean by that, we repeat step (2) by creating sets of RSs not yet assigned and in the range of PSs instead of other RSs, then install BS in PS when necessary). The pseudo-code of the proposed greedy algorithm (GA) is shown in Algorithm 1.

Algorithm 1. Greedy – Main

Input : *positions of PSs and TSs; TSs' traffic*
Output : *positions of RSs and BSs, nodes' transmission range, channel assignment*

Step 1

a- **for** each TS_i covered by only one PSj **do**
  if available resources satisfy TS_i demand **then**
   place RS_j in PS_j
   assign the appropriate channel
 end if
 end for

b- **for** each PS_j **do**
   create a set E_j
  end for
  sort E_j in decreasing order of their cardinalities
  consider the first set E_j , install RS_i in PS_j
  for each TS$\in E_j$ **do**
   if available resources satisfy TS' demand **then**
    assign TS to RS_i
    assign the appropriate channel
    for each E_k such that $k \neq i$ **do**
     if TS$\in E_k$ such that $k \neq i$ **then**
      remove TS from E_k
     end if
    end for

```
            end if
          end for
   c-  repeat the step (1-b) until all TSs are assigned

Step 2
   a-  for each RS_i do
            if RS_i not covered by any RS_k such that k ≠ i then
               replace RS_i by BS_i
            end if
          end for
   b-  for each RS_j do
             create a set F_j
         end for
       sort F_j in decreasing order of their cardinalities
         consider the first set F_j , replace RS_j by BS_j
       for each RS∈ F_j do
             if available resources satisfy RS' demand then
               assign RS to BS_j
               assign the appropriate channel
               for each F_k such that k ≠ i do
                  if RS∈ F_k then
                     remove RS from F_k
                  end if
               end for
            end if
     end for
   c-  repeat the step (2-b) until all RSs are assigned

Step 3
      for each node i do
           compute the transmission range of  i
      end for
```

In this study, channels are divided in two groups: channels to handle traffic between TSs and RSs or TSs and BSs, and channels used to relay traffic from RSs to other RSs or to BSs. In the case of 3G, we divide the codes similarly in two groups: codes serving to handle traffic between TSs and RSs or TSs and BSs, and codes used to relay traffic from RSs to other RSs or to BSs.

4 Experimentations and Results Analysis

4.1 Experimentation

In this paper, we suppose that the codes are orthogonal and there is no interference in 3G links. In our experiment, the positions of PSs and TSs, and the traffic of each TS are randomly generated. The installation cost of a BS is 5 times bigger than the cost of a RS installation due to the functionalities offered by the BSs compared to RSs and their capacities. We use two radio interfaces (WiFi and 3G) with capacities of 54 Mbs and 2 Mbs respectively (that means each value in matrix φ_{ij} is equal to 54). The number of channels is 12 and the number of codes is 256. The algorithm is coded in C++ programming language.

4.2 Results and Analysis

In this section, we study the performance of the proposed greedy algorithm by comparing its results with those given by CPLEX.

In this study, we increase gradually the number of candidate locations while all other parameters are maintained fixed. We solve the problem for medium size instances ($\leq 1.5km * 1.5km$) using (a) CPLEX and (b) the greedy algorithm, then we compare the objective values returned by each solver. For each problem size, we generate randomly 10 different instances. Fig. 1 shows the average of the objective values obtained by CPLEX and the proposed greedy algorithm with the number of TS.

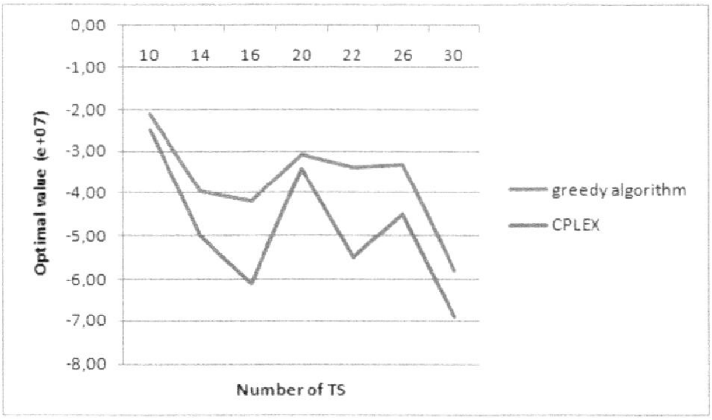

Fig. 1. Comparison between results given by both solvers

"Optimal value" means the best value of the objective function obtained by resolving the problem. In our case, since we try to minimize, more the optimal value is small, more the result is satisfying.

Comparing the results yielded by CPLEX with those obtained by our greedy algorithm, we observe that CPLEX produces smaller objective values (thus better solutions); however, the greedy algorithm produces results that are not too far from the optimal results (returned by CPLEX). In conclusion, the proposed algorithm returns acceptable solutions; this being said, we developed the greedy algorithm just to show the feasibility of the model; we plan to develop more sophisticated resolution methods that produce near optimal solutions.

For 30 TSs, CPLEX takes approximately 6 hours to resolve the problem. However, the proposed algorithm gives a solution in few seconds even for instances of 50, 80 and 100 TSs. In these situations, CPLEX does not return solutions (i.e., response time tends to "infinity").

5 Conclusion

In this paper, we considered the MCN design problem. We proposed a multi-objective MCN design model which is unique. The goal is to minimize the total cost deployment and maximize the flow capacity ratio of the network while satisfying the coverage, the interference constraints and the user requirements. To show the validity of the model, we developed a simple heuristic to resolve the model. We are currently developing a Tabu search algorithm to produce near optimal solutions.

References

[1] Lin, Y.D., Hsu, Y.C.: Multihop cellular: a new architecture for wireless communications. In: Proceedings of IEEE INFOCOM (March 2000)
[2] Vidal, J., Marina, N., Hest-Madsen, A.: Dimensioning Cellular Networks with Fixed Relay Stations. In: ICTEL, pp. 1–5 (2008)
[3] Reininger, P., Iksal, S., Caminada, A., Korczack, J.J.: Multi-Stage Optimisation for Mobile Radio Network Planning. In: IEEE Vehicular Technology Conference, pp. 2034–2038 (1999)
[4] Molina, A., Nix, A.R., Athanasiadou, G.E.: Cellular Network Capacity Planning Using The Combination Algorithm for Total Optimisation. In: VETECS, pp. 2512–2516 (2000)
[5] Amaldi, E., Capone, A.: Planning UMTS Base Station Location: Optimization Models With Power Control and Algorithms. In: TWC, pp. 939–952 (2003)
[6] So, A., Liang, B.: Optimal Placement of Relay Infrastructure in Heterogeneous Wireless Mesh Networks by Bender's Decomposition. In: QShine, pp. 118–121 (2006)
[7] Cho, J., Haas, Z.J.: Throughput Enhancement by Multi-hop Relaying in Cellular Radio Networks with Non-uniform Traffic Distribution. In: VTC, vol. 5, pp. 3065–3069 (2003)
[8] Cho, J., Haas, Z.J.: Throughput Enhancement by Multi-hop Relaying in Cellular Radio Networks with Non-uniform Traffic Distribution. In: VTC, vol. 5, pp. 3065–3069 (2003)
[9] Jacobson, K.R., Krzymie, W.A.: Cell Dimensioning and Network Throughput in Cellular Multi-Hop Relay Networks. In: VTC, pp. 1–5 (2006)
[10] Gupta, P., Kumar, P.R.: The capacity of wireless networks. IEEE Transactions on Information Theory 46, 388–404 (2000)
[11] Holma, H., Toskala, A.: WCDMA for UMTS: Radio Access for Third Generation Mobile Communications, 2nd edn. (2002)
[12] Manoja, B.S., Ananthapadmanabhab, R., Murthya, C.S.R.: Multi-hop cellular networks: Architecture and protocols for best-effort and real-time communication. Journal of Parallel and Distributed Computing 65(6), 767 (2005)
[13] Dixit, S., Yanmaz, E., Tonguz, O.K.: On the Design of Self-Organized Cellular Wireless Networks. IEEE Communication Magazine, 86–93 (2005)
[14] Benyamina, D., Hafid, A., Gendreau, M.: A Multi-objective Optimization Model For Planning Robust and Least Interfered Wireless Mesh Networks. In: IEEE GLOBECOM 2008, pp. 1–6 (2008)

Optimal Oblivious Routing in Hole-Free Networks

Costas Busch[1] and Malik Magdon-Ismail[2]

[1] Louisiana State University, Baton Rouge, LA 70803, USA
busch@csc.lsu.edu
[2] Rensselaer Polytechnic Institute, Troy, NY 12180, USA
magdon@cs.rpi.edu

Abstract. We study oblivious routing algorithms in which the packet paths are constructed independently of each other. Oblivious algorithms are inherently distributed and they can be designed to efficiently balance the network utilization. We give an oblivious routing algorithm for the class of hole-free networks, in which the nodes are topologically embedded in simple areas of the plane. Such networks appear frequently in wireless and sensor network topologies. The algorithm achieves optimal congestion and stretch. The stretch of the resulting paths is constant. The congestion is $O(C^* \log n)$, where C^* is the optimal non-oblivious congestion and n is the number of nodes. This congestion bound is asymptotically worst-case optimal for oblivious routing algorithms.

Keywords: oblivious routing, congestion, path stretch, wireless networks, sensor networks.

1 Introduction

Routing algorithms specify the paths to be followed by packets in a network. A routing algorithm is *oblivious* if the path of every packet is selected independently of the paths of the other packets and without considering the history of the previously routed packets. Oblivious algorithms are by their nature distributed and capable of solving online routing problems, where packets continuously arrive in the network. The objective of this work is to present oblivious algorithms with low congestion and small path stretch. For congestion we consider the bottleneck metric C which is equal to the maximum number of selected paths that use any edge in the network.

Oblivious routing is applicable to wireless and sensor networks. It is particularly suitable to energy and power constraint networks (e.g. battery operated nodes), since it can help to extend the time until some node runs out of power. Lowering the network congestion (lowering edge bottlenecks), results to improved load balancing and thus prolonged lifetime and better utilization of the network. In addition, paths of small stretch (ratio of path length to shortest path) result to low overall energy utilization. Oblivious algorithms are also easy to implement in wireless and sensor networks, on account of their simplicity.

X. Zhang and D. Qiao (Eds.): QShine 2010, LNICST 74, pp. 421–437, 2011.

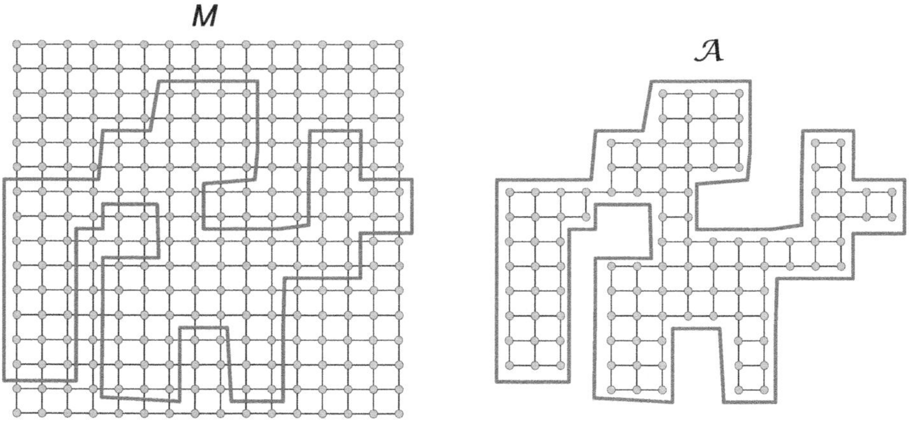

Fig. 1. Hole-free grid-like network G

We give an oblivious routing algorithm designed for *hole-free* networks which are suitable to model wireless network communication environments. A hole-free network is embedded in the 2-dimensional Euclidian plane inside a *simple area* $\mathcal{A}$ (see Figure 1). We are considering *grid-like* networks, which are induced when we apply a simple area $\mathcal{A}$ on top of a $m \times m$ grid M. Hole-free grid-like graphs are interesting because they can model wireless and sensor network topologies where each node is connected with a link with its neighbors at distance 1, within the boundaries of area $\mathcal{A}$.

We give an oblivious routing algorithm on any hole-free grid-like graph G that gives paths with optimal stretch and congestion. In particular, given a set of packets Π with respective sources and destinations, our algorithm returns a set of paths P, one path for each packet in Π, such that $\mathsf{stretch}(P) = \Theta(1)$. In other words, the length of every resulting path is within a constant factor from the length of the respective shortest path. Further, the congestion of paths P is $O(C^* \log n)$, where n is the number of nodes in G, and C^* denotes the optimal congestion that can be achieved for the packets Π. The upper bound is optimal since it is known that any oblivious algorithm has congestion $\Omega(C^* \log n)$ in the worst case for grids [13].

The algorithm we give is oblivious and randomized, which means that each path is computed with some randomized choices in a specific manner that does not depend on other path requests in the network. In particular, a path from a source s to a destination t is computed by using a shortest path q based on which we estimate a sequence of adjacent squares along q. The requested path from s to t is formed by concatenating randomized one-bend paths or two-bend paths formed in each square in the sequence. The squares in the sequence are selected from a hierarchical partition of the grid M into canonical squares of various sizes placed in fixed positions in M. A crucial aspect of the algorithm is that the sum of the side lengths of the squares in the sequence is within a constant factor to the length of the shortest path from s to t. This helps to control the stretch. The congestion is controlled by adjusting the square sizes to the cut sizes (between s and t) which involve the nodes in the squares.

1.1 Related Work

Valiant and Brebner [21,20] are the first to propose the oblivious routing technique of using a single random intermediate node in order to minimize congestion in networks. They give an appropriate general approach for oblivious routing based on solutions to flow problems in the network. Applications are permutation routing problems on the hypercube and butterfly networks.

The motivation for minimizing congestion and stretch simultaneously is because there exist packet scheduling algorithms [12] which deliver the packets along the given paths in time very close to the optimal $O(C + D)$, where D is the maximum path length of the routing algorithm. A trivial lower bound for the total time to transfer all the packets along the selected paths is $\Omega(C+D)$. Hence $C + D$ is a natural metric by which to measure the quality of the paths that are produced by a routing algorithm. Our oblivious routing algorithm provides paths with small $C + D$ because of the low congestion and small stretch.

Maggs *et al.* [13] give an oblivious algorithm for the d-dimensional *mesh* (grid) with congestion $O(d \cdot C^* \cdot \log n)$. However, the stretch factor in that algorithm is unbounded. To control stretch within a constant factor, we generalize in [8] the hierarchical decomposition for the mesh denoted by an *access tree* [13] to a more general *access graph*. Oblivious congestion minimizing algorithms which control the stretch have been also considered by Scheideler [18] for the 2-dimensional mesh, which uses a different approach by routing within a square containing the source and destination and building an access tree specific to this square.

Following the work in [13], there have been extensions to general networks [4,5,10,15,16], where progressively better oblivious algorithms with near optimal congestion. However, in all these algorithms the stretch is unbounded. Further, most of these algorithms are based on a hierarchical decomposition of the network into clusters, which requires a logarithmic number of intermediate nodes. In [7] we present an oblivious routing algorithm for geometric networks, which are special types of networks embedded in the 2-dimensional grid, and it is a very restricted subclass of hole-free networks. That algorithm uses *a single* random intermediate node and doesn't depend on any hierarchical clustering.

Lower bounds on the competitive ratio of oblivious routing has been studied for various types of networks. Maggs et al. [13] give the $\Omega(\frac{C^*}{d} \cdot \log n)$ lower bound on the competitive ratio of an oblivious algorithm on the mesh. Valiant and Brebner [21] perform a worst case theoretical analysis on oblivious routing on specific network topologies such as the hypercube. Borodin and Hopcroft [6] and Kaklamanis et al. [11] showed that deterministic oblivious routing algorithms can not approximate the minimal load on most non-trivial networks, which justifies the necessity for randomization.

Non-oblivious approaches to optimizing $C + D$ have received considerable attention, and near optimal algorithms are discussed in [1,3,17,19]. As already mentioned, such offline algorithms require knowledge of the traffic distribution *a priori* and generally do not scale well with the number of packets. Tradeoffs between stretch and congestion have been studied in wireless networks [9,2].

As an alternative routing scheme for sensor networks, *curve routing* relies on geographic information, which can be obtained through GPS devices [22]. Packets are sent along specified trajectories from sources to destinations. The trajectories are defined in space, and then they are projected to actual paths in the network. Packets contain information about the path trajectory, and each time they are forwarded to the next best node that is closer to the trajectory and the destination. This has the benefit that the actual network path does not need to be precisely defined, but it can be determined on the fly in a similar way as in geographic routing. Thus, such routing methods are suitable for sensor network and wireless ad hoc networks, where the actual graph connectivity may not be known precisely but there is some information about the geographic area that contains the network. Recently, *curveball routing* has been proposed as a routing method to send packets along curves with the benefit of load-balancing the node utilization [14]. All the nodes are projected (uniformly) in the surface of a sphere. Then a routing curve is obtained from the shortest path in the sphere that connects the source to the destination. For the virtual path we only need to know the virtual coordinates of the source and destination. Note that the shortest path in the sphere may not be realized in the actual network, since we may not be able to find nodes closer to the destination in virtual coordinates. In such situations regular geographic routing may be used. Since the surface of a sphere is symmetric, the expected load on the nodes is balanced.

Paper Outline. We begin with some preliminary definitions in Section 2. We continue with describing how to construct sequences of squares for paths in Section 3. We give the oblivious routing algorithm in Section 4. We finish with its congestion and stretch analysis in Section 5.

2 Preliminaries

An area $\mathcal{A}$ on the plane is *simple* (or *hole-free*) if every closed curve in $\mathcal{A}$ can be continuously deformed into a point. Intuitively, the area bounded by any closed curve is completely contained in the area $\mathcal{A}$. For example a disc is hole free, but an annulus is not.

We now consider a 2-dimensional $m \times m$ grid (mesh) graph M consisting of nodes at positions (i, j) where $i, j \in [0, 1, \ldots, m-1]$, and $(0, 0)$ is a the bottom left, such that each node is connected with an edge to any node at distance 1 (there are at most four such nodes). On top of this grid, we draw any simple closed area $\mathcal{A}$. The *induced subgraph* of $\mathcal{A}$ is the subgraph $G_{\mathcal{A}}$ consisting of all nodes and edges which lie in the area $\mathcal{A}$ with the removal of all edges which cross the boundary of $\mathcal{A}$. If the induced subgraph $G_{\mathcal{A}}$ is connected, the we say that $G_{\mathcal{A}}$ is a *simple-area grid-like graph* (or *hole-free grid-like graph*). This construction extends the notion of hole-free from areas on the plane to graphs with natural embeddings on the plane. From now on, G will refer to a simple-area grid-like network.

2.1 Canonical Squares

Let M be an $m \times m$ grid which contains G as a subgraph, where m is a power of 2. We can divide M into $1 + \lg m$ levels of *canonical square subgraphs* as follows. For $0 \leq \ell \leq \lg m$, the canonical square subgraphs at level ℓ partition M into $2^{2(\lg m - \ell)}$ square subgraphs. Each canonical square subgraphs at level ℓ is a $2^\ell \times 2^\ell$ grid subgraph of M, whose bottom left corner node has coordinates $(i \cdot 2^\ell, j \cdot 2^\ell)$, where $0 \leq i, j \leq 2^{\lg m - \ell} - 1$.

For simplicity, we will refer to the canonical square subgraph as canonical squares. The α-partition of M consists of the $\alpha \times \alpha$ canonical squares of M. Note that two canonical squares are either node disjoint or one contains the other. Further, every node of M is contained in exactly $1 + \lg m$ canonical squares. Two canonical squares M_1 and M_2 (not necessarily at the same level) are *adjacent* if they are disjoint and there is an edge from M_1 to M_2, i.e. an edge $(u, v) \in M$, such that $u \in M_1$ and $v \in M_2$.

Given a simple-area grid-like graph G defined in M, a canonical square B is *internal* if B consists only of nodes in G; otherwise, we say that B is *external*, in which case B may consist of both nodes in G and nodes not in G but in M.

2.2 Path Definitions

Consider a simple-area grid-like graph G. The input for a path selection problem is a set of N sources and destinations (i.e. packets), $\Pi = \{s_i, t_i\}_{i=1}^N$ in G. The output is a set of paths in G, $P = \{p_i\}_{i=1}^N$, where each path $p_i \in P$ is from node s_i to node t_i. The length of path p, denoted $|p|$, is the number of edges it uses. We denote the distance from s to t (the length of the shortest path from s to t) by $\mathsf{dist}(s, t)$. The *stretch* of a path p_i, denoted $\mathsf{stretch}(p_i)$, is the ratio of the path length to the shortest path length between its source and destination, $\mathsf{stretch}(p_i) = |p_i|/\mathsf{dist}(s_i, t_i)$. The *stretch factor* for the collection of paths P, denoted $\mathsf{stretch}(P)$, is the maximum stretch of any path in P, $\mathsf{stretch}(P) = \max_i \mathsf{stretch}(p_i)$. We will denote by C the network congestion, which is the maximum number of paths in P that use any edge in the network. We will denote by C^* the optimal congestion incurred by the optimal set of paths the can route the packets in Π.

2.3 Cut Number

Consider a source s and destination t, both in G, and any other node v in G. We now define the *cut number* of node v with respect to s, t, denoted $cut_{s,t}(v)$. Let Q be any connected set of nodes containing v and not both of s, t (Q could contain one or none of s, t). The set Q is an $s - t$ node-cut if every path from s to t uses at least one node in Q. We will say that Q is an $s - t$ node cut for node v. The *cut number* $cut_{s,t}(v)$ is the size (number of nodes) of the smallest $s - t$ node cut for node v.

A shortest path from s to v not containing t or from v to t not containing s give trivial node cuts. Note that at least one (but not necessarily both) of the trivial node cuts above must exist for every node v. Thus we have the following simple lemma,

Lemma 1. $1 \leq cut_{s,t}(v) \leq 1 + \min\{\mathsf{dist}_G(s,v), \mathsf{dist}_G(t,v)\}$.

Intuitively, nodes for which $cut_{s,t}(v) < 1 + \min\{\mathsf{dist}_G(s,i), \mathsf{dist}_G(t,i)\}$ represent bottlenecks in the network (with respect to sending packets from s to t). Consider two nodes u, v. Any node set containing v can be converted to one containing u by including the nodes in a shortest path from u to v. Thus, we have:

Lemma 2. $|cut_{s,t}(u) - cut_{s,t}(v)| \leq \mathsf{dist}_G(u,v)$.

The next lemma will be useful for our later results. It basically states that if a shortest path has to wind a lot, then this path must be passing through a small cut.

Lemma 3. *Suppose that p is a shortest path from s to t which crosses a vertical (or horizontal) line of nodes three times at the vertical (or horizontal) positions $x_1 > x_2 > x_3$ corresponding to the nodes u, v, w. Then $cut_{s,t}(v) < |x_1 - x_3|$.*

3 Square Sequences of Paths

Consider a simple-area grid-like graph G defined in the $m \times m$ grid M. Let p be a shortest path in G from a node s to node t. We define a *square sequence of shortest path* p, to be a sequence $R(p) = M_1, M_2, \ldots, M_k$ ($k \leq |p|$) with the following properties:

- *Canonical-Property:* each M_i is a canonical square (n_i and ℓ_i are the number of nodes and edges, respectively, in a side of M_i), and any two consecutive squares M_i and M_{i+1} are adjacent;
- *Coverage-Property:* p can be written as a concatenation of subpaths $p = p_1 p_2 \cdots p_k$, such that p_i is completely contained in M_i (subpath p_i is from node $v_i \in M_i$ to $w_i \in M_i$);
- *Cut-Property:* $\beta_1 \cdot \ell_i \leq cut_{s,t}(v_i) \leq \beta_2 \cdot n_i$, where $\beta_1, \beta_2 > 0$ are appropriately chosen constants.

The cut-property controls the cut numbers of the nodes in the square sequence of a shortest path. An effect of the cut-property is that adjacent squares do not differ significantly in size. Note that the canonical squares used in $R(p)$ may be internal of external. The same canonical external square may be used multiple times in a square sequence, or with two non-consecutive squares one may include the other, due to the way that the graph is formed. In section 3.3, we describe how to convert a square sequence to consist only of internal canonical squares, but at the expense of not covering all the nodes in the shortest path.

3.1 Square Sequence Construction

Given a shortest path p in G from a node s to node t we describe how to construct square sequence $R(p)$ recursively. We use the following notations: prefix subsequence $R_i(p) = M_1, \ldots, M_i$, prefix subpath $q_i = p_1 \cdots p_i$, and v_i and w_i denote the first and last nodes of p_i, respectively.

At the basis of the recursion we have $R_1(p) = M_1$, which is the canonical 1×1 square that contains s. Suppose that we have constructed the sequence $R_i(p)$. Let v_{i+1} be the first node after w_i in p. Select M_{i+1} to be the *largest* canonical square such that: M_{i+1} contains v_{i+1}, M_{i+1} is adjacent to M_i, and $\beta_1 \cdot \ell_{i+1} \leq cut_{s,t}(v_{i+1})$. The newly selected square M_{i+1} defines the subpath p_{i+1}. The process repeats until the destination node t is included. The following figures illustrate this process.

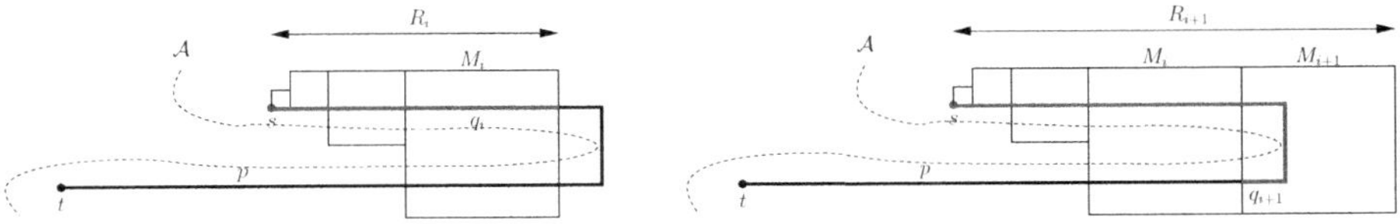

Notice in the above example, M_{i+2} may equal or contain M_i. Thus, the same submesh may be repeated in the canonical decomposition.

3.2 Side Size Bound

In what follows we will focus on providing a bound on the sum of the total side sizes $\sum_{i=1}^{n} n_i$ of the canonical squares in $R(p)$, with respect to the length $|p|$ of the shortest path p. This is useful because it help to bound the stretch of the resulting paths in the oblivious algorithm.

The connection of the square side sizes with the length of the path p comes from the cut-number relations of the squares. In the construction of the canonical squares, the side length ℓ_i is related to the cut number of v_i. We need to find a relation between the $|p_i|$ (the path segment in M_i) and the side length ℓ_i. We will establish this relationship through a sequence of lemmas.

The length of path segment p_i may exceed $2\ell_i$ in case that M_i is external. The first lemma shows that if a path segment is long enough then it has to cross some horizontal or vertical line three times.

Lemma 4. *If $|p_i| \geq 4\ell_i + 2$, then p_i must cross some horizontal or vertical line within M_i at least three times.*

The following result follows from Lemmas 3 and 4 and the cut-property of the square sequence:

Lemma 5. $|p_i| \leq 5\ell_i$.

Next we show that the sequence of canonical squares does not grow or decrease in size too rapidly.

Lemma 6. *$\beta_3 \cdot n_i \leq n_{i+1} \leq \beta_4 \cdot n_i$, for $1 \leq i \leq k-1$, and constants $0 < \beta_3 < 1/2 < \beta_4$.*

We continue with another result with respect to the length of the subpaths p_i inside some M_i.

Lemma 7. *Any segment q' of path p_i starting at v_i of length $|q'| = \alpha \leq \frac{\beta_1 \ell_i}{2}$ cannot cross a horizontal or vertical line 3 times.*

Proof. Suppose that q' intersects a line 3 times. By Lemma 3, there is node $u \in q'$ with cut number $cut_{s,t}(u) < \alpha$, and so $cut_{s,t}(v_i) < 2\alpha$ by Lemma 2. By construction of M_i, $\beta_1 \cdot \ell_i \leq cut_{s,t}(v_i) < 2\alpha \leq \beta_1 \cdot \ell_i$, which is a contradiction.

The next simple observation will be useful.

Lemma 8. *A path of length less than α can use at most 4 different $\alpha \times \alpha$ squares in the α-partition of M. Further all the different squares used lie within a $2\alpha \times 2\alpha$ square.*

Every time the path p moves from one of its canonical squares M_i to the next square M_{i+1}, we will say that *the path makes a canonical step.* Our next lemma basically states that if a path makes many (more than 4) canonical steps, then the path makes significant progress (relative to the length of the canonical squares).

Lemma 9. *If a segment q' of the shortest path p starting at v_i in M_i makes 5 canonical steps, then the length of that segment is at least $|q'| \geq \beta_5 n_i$, for a constant $\beta_5 > 0$.*

Proof. Let $\beta_5 = \beta_3^5$. Suppose that $|q'| < \beta_3^5 n_i$. Note that $|q'| \geq 4$, since q' makes five canonical steps. Since $\beta_3 < 1/2$, it has to be that $n_i > 2^7$. By Lemma 6, each canonical step can decrease the side length by at most a factor β_3. Let n_{min} be the minimum side size (number of nodes) of the canonical squares used over the next 5 canonical steps, then $n_{min} \geq \beta_3^5 n_i > |q'|$. Now consider the n_{min}-partitioning of M. Since $|q'| < n_{min}$, by Lemma 8, q' may only use at most 4 different $n_{min} \times n_{min}$ squares, all enclosed within a single $2n_{min} \times 2n_{min}$ square S. Each canonical transition must cross either the middle horizontal or middle vertical line of S. Since q makes 5 canonical transitions, either the middle horizontal or middle vertical line is crossed at least 3 times. To conclude, since $\beta_3 < 1/2$ and taking $\beta_1 > 2$, we get $|q'| < \beta_3^5 n_i < n_i - 1 = \ell_i < \frac{\beta_1 \ell_i}{2}$, which contradicts Lemma 7.

The next simple lemma follows directly from Lemma 1 and the fact that $(n_i - 1)\beta_1 = \ell_i \beta_1 \leq cut_{s,t}(v_i) \leq 1 + \mathsf{dist}(s,t)$.

Lemma 10. *$n_i \leq \beta_6 \cdot \mathsf{dist}(s,t)$, for some constant $\beta_6 > 0$.*

We are now ready to bound the sum of the sides of the squares in the canonical square decomposition of a shortest $s - t$ path p.

Theorem 1. *For the square sequence of shortest path p, $R(p) = M_1, \ldots, M_k$, the sum of the side sizes of the canonical squares is bounded as $\sum_{i=1}^{k} n_i \leq \beta_7 \cdot \mathsf{dist}(s,t)$, for some constant $\beta_7 > 0$.*

Proof. From Lemma 9, $|p_i| + |p_{i+1}| + |p_{i+2}| + |p_{i+3}| + |p_{i+4}| + |p_{i+5}| \geq \beta_5 n_i$, for $1 \leq i \leq k-5$. Thus,

$$\begin{aligned}\beta_5 \sum_{i=1}^{k-5} n_i &\leq \sum_{i=1}^{k-5} (|q_i| + |q_{i+1}| + |q_{i+2}| + |q_{i+3}| + |q_{i+4}| + |q_{i+5}|), \\ &\leq 6 \cdot \mathsf{dist}(s,t).\end{aligned}$$

By Lemma 10, $\sum_{i=k-4}^{k} n_i \leq 5\beta_6 \cdot \mathsf{dist}(s,t)$, and so by combining these two inequalities, we conclude that $\sum_{i=1}^{k} n_i \leq (6/\beta_5 + 5\beta_6) \cdot \mathsf{dist}(s,t)$.

3.3 Internal Square Sequence

The problem with the canonical decomposition $R(p)$ discussed in the previous section is that it constructs a sequence of squares which may not be completely enclosed in the network G. Here, we construct a sequence with only internal squares.

We sumarize the properties of the canonical square decomposition which will be important for the oblivious path selection algorithm.

1. The cut number of the nodes in the canonical square are proportional to the side of the canonical square.
2. The canonical squares in the sequence do not grow or shrink in size too quickly.
3. The sum of the side sizes of the canonical squares is proportional to the shortest path length $\mathsf{dist}_G(s,t)$.

We briefly sketch why these properties are important. The main idea is that we will construct the final path from random path segments which move from one canonical square to the next. The congestion caused by the paths within any one canonical square is related to the size of the square (how much the packets can spread). The cut number of these nodes gives a lower bound on the congestion for sending packets from s to t. Since the cut number and the side lengths are proportional (property 1 above), this allows us to show that the congestion inside the canonical squares is near optimal. Further, we do not want to create a bottleneck in going from one canonical square to the next, so there should be significant overlap between the intersecting sides of consecutive squares in the decomposition. This is ensured by property 2. Finally, property 3 bounds the stretch, since the path length is proportional to the sum of the side sizes of the canonical squares in the decomposition.

We now show how to convert the canonical square decomposition which may contain some external squares into an internal canonical square decomposition that still satisfies these three required properties. The basic idea is to replace an external square with a sequence of one or more internal squares which are adjacent to it. This may result to a new sequence of squares which may not contain the shortest path p. However, this does not cause a problem since the goal of the oblivious algorithm is to construct new oblivious paths which are formed near the original path p, and not necessarily exactly on top of it.

Let $R(p) = M_1, M_2, \ldots, M_k$ be the canonical square decomposition for p, a shortest s-t path. We will construct an alternative path p' from s to t and a respective square sequence $\overline{R}(s,t) = M'_1, M'_2, \ldots, M'_{k'}$, such that all canonical squares in $\overline{R}(s,t)$ are internal. Similar to $R(p)$, let $\overline{R}_i(s,t)$ be a prefix of the square sequence $\overline{R}(s,t)$, q'_i a prefix of p', and p'_i the subpath of p' in M'_i. The path p' will be very close to p, so that $\mathsf{stretch}(p')$ is constant.

We know that M_1 is internal, by construction of $R(p)$. Let M_j, $j > 1$, be the first external canonical square in $R(p)$. We take the two square sequences to be the same up to M_{j-1}, namely, $\overline{R}_{j-1}(s,t) = R_{j-1}(p)$. The idea is that we will divert the path p from w_{j-1} (the last node in the canonical square before M_j) to v_{j+1} (the first node in the canonical square after M_j) so that the new respective prefix of path p' uses only internal canonical squares. Let $n_{min} = \min(n_{j-1}, n_j, n_{j+1})$ and consider the α-partition of M into squares of side $\alpha = n_{min}/2$. Assume for now that $n_{min} > 1$, and that M_{j-1}, M_j, M_{j+1} are disjoint. Note that M_j is also partitioned into squares of side α, as is M_{j-1} and M_{j+1}.

We now consider the (possibly partial) ring X of $\alpha \times \alpha$ squares in the α-partition of M which are adjacent to M_j but are not contained in M_{j+1}. The main claim, is that there is a path from w_{j-1} to v_{j+1} which uses a sequence of internal squares in the partial ring X, and then enters M_{j+1}. In particular, we consider the paths p_c and p_a from w_{k-1} to M_{j+1} which go around M_j, staying as close as possible M_j. The path p_c goes around M_j clockwise and the path p_a goes around M_j counter-clockwise. The following lemma is crucial in the construction:

Lemma 11. *Either p_a or p_c use exclusively internal squares in X.*

Proof. Suppose that both p_a and p_c use each at least one external square in X. Then, we will show that $cut_{s.t}(v_j) \leq 8n_j$. Let S_a be the first canonical square used by p_a which is external and let v_a be an external node in S_a. Similarly let S_c be the first canonical square used by p_c which is external and let v_c be an external node in S_c. Consider a shortest path from v_j to v_a and let Y_a be the segment of this path up to but excluding the first external node met on the path. Similarly let Y_c be the segment of a shortest path from v_j to v_c up to the first external node on the path. Let $Y = Y_a \cup Y_c$. Since the shortest path from v_j to v_a is entirely in a square of side $n_j + 2\alpha \leq 2n_j$ (since $2\alpha \leq n_j$), $|Y_a| \leq 4n_j$. Similarly, $|Y_c| \leq 4n_j$ and so $|Y| \leq 8n_j$. Clearly, Y is an s-t cut for v_j, since M_{j-1}, M_j, M_{J+1} are taken to be disjoint. Therefore, $cut_{s,t}(v_j) \leq 8n_j$. Since by construction of $R(p)$, $\beta_1 \ell_j \leq cut_{s,t}(v_j)$, and also $n_j = \ell_j + 1$ and we assumed that $n_j \geq 2$, by choosing the constant β_1 appropriately large, we obtain the $cut_{s,t}(v_i) > 8n_j$. A contradiction.

Lemma 11 implies that we can replace p_j with either p_a or p_c (whichever uses only internal $\alpha \times \alpha$ squares) in p'. The corresponding sequence $\overline{R}(s,t)$ is augmented after $\overline{R}_{j-1}(s,t)$ with the $\alpha \times \alpha$ internal canonical squares in ring X that correspond to the chosen p_a or p_c. This way, M_j has been eliminated and replaced by internal squares whose side size is within constant factor from n_j. The process repeats with the next external square in $R(p)$, until all external squares have been eliminated. The process can also be appropriately modified for the cases where $n_{\min} = 1$ and M_{j-1}, M_j, M_{j+1} are not pairwise disjoint. Using Theorem 1 and Lemma 6, we can obtain the following result:

Theorem 2. *Given source s and destination t we can construct a square sequence $\overline{R}(s,t) = M_1, M_2, \ldots, M_k$ such that:*

(i) *each M_i is an internal canonical square (with side size n_i and length ℓ_i),*
(ii) *for any $v \in M_i$, $\gamma_1 \cdot \ell_i \leq cut_{s,t}(v) \leq \gamma_2 \cdot n_i$, for constants $\gamma_1, \gamma_2 > 0$,*
(iii) *M_i and M_{i+1} are adjacent with $\gamma_3 \cdot n_i \leq n_{i+1} \leq \gamma_4 \cdot n_i$, for $1 \leq i \leq k-1$, and constants $\gamma_3, \gamma_4 > 0$, and*
(iv) $\sum_{i=1}^{k} n_i = \Theta(\mathsf{dist}(s,t))$.

4 Oblivious Path Selection Algorithm

Consider a hole-free grid-like network G with n nodes. The input for a path selection problem is a set of N sources and destinations (i.e. packets), $\Pi = \{s_i, t_i\}_{i=1}^{N}$ and the output is a set of paths, $P = \{p_i\}_{i=1}^{N}$, where each path $p_i \in P$ is from node s_i to node t_i. It suffices to describe the algorithm for an arbitrary single s-t source destination pair, and then each packet can use the same algorithm.

We will assume that in the network we have pre-computed the internal square sequence $\overline{R}(s,t) = M_1, M_2, \ldots, M_k$, for every pair s, t, as specified by Theorem 2. A *one-bend path* uses two straight lines in different dimensions, and a *two-bend path* uses three straight lines with alternate dimensions. The path from s to t is computed by using either a one-bend path or a two-bend in each square M_i. The decision for which type of path to use depends on the way that consecutive squares M_{i-1}, M_i, M_{i+1} are aligned. We have the following algorithm.

Consider three adjacent squares M_{i-1}, M_i, M_{i+1}. Let $n_{\min} = \min\{n_i, n_{i+1}\}$. Suppose we have constructed the path from s to t up to square M_{i-1}, and let u_{i-1} be the last node of the path on the side of M_{i-1} adjacent to M_i.
if (M_{i-1} and M_i are adjacent in the right and left sides, respectively, and M_i and M_{i+1} are adjacent in the top and bottom sides, respectively) **then**
    Select a random node u_i among the $n_{\min}$ nodes on the top edge of M_i adjacent M_{i+1}. Construct a one-bend path from u_{i-1} to u_i.
else
    if (M_{i-1} and M_i are adjacent in the right and left sides, respectively, and M_i and M_{i+1} are adjacent in the right and left sides, respectively) **then**

6: Select a random node $x \in M_i$ among n_i nodes on the horizontal line specified by u_{i-1}. Select a random node u_i among the $n_{\min}$ on the right edge of M_i adjacent to M_{i+1}. Construct a two-bend path from u_{i-1} to u_i through x.

7: Every other arrangement of the sides of M_{i-1}, M_i, M_{i+1} can be handled similar to one of the cases above. For the first square we have $u_1 = s$ and for the last square $u_k = t$.

5 Congestion and Stretch Analysis

From the algorithm description in Section 4, the path from s_i to t_i is formed by concatenating one-bend paths or two-bend paths formed in the respective squares in the seuquence $\overline{R}(s_i, t_i)$. By construction, we observe that each subpath in a square M_i has length at most $3n_i$. This observation with combination of property (iv) in Theorem 2 implies that:

Theorem 3 (Stretch). *The stretch of the paths P returned by the oblivious algorithm has* $\mathsf{stretch}(P) = \Theta(1)$.

Thus, we only need to focus on the congestion. We start with a lower bound analysis for the optimal congestion and then we give an upper bound which is within a log factor from the lower bound. In what follows we give the bounds in terms of the *node congestion*, which is the maximum number of paths that use any node. The resulting bounds immediately translate to edge congestion within a factor 4 since each node has at most four adjacent edges. Thus, the asymptotic bounds stay the same for edge congestion.

5.1 Lower Bound on Optimal Congestion

Consider an arbitrary $\alpha \times \alpha$ internal canonical square B. Let $Q \subseteq P$ be the set of paths selected by the algorithm that use B because it is in their square sequence. We establish a lower bound on the optimal congestion C^* due to the paths Q in B.

With respect to any path $p \in Q$ from source s to destination t, Theorem 2 implies that every node $v \in B$ has $cut_{s,t}(v) \leq \gamma_2 \alpha$. Let H be a $(2\gamma_2 + 1)\alpha \times (2\gamma_2 + 1)\alpha$ grid subgraph of nodes in M such that B is in the middle of H. We truncate H wherever it exceeds the network M, resulting to a rectangular grid of maximum side length $(2\gamma_2 + 1)\alpha$. Let H_G denote the subgraph of G in H. Note that H_G may be disconnected. We define the *perimeter* nodes of H_G, denoted $T(H_G)$, as the set of nodes of $u \in H_G$ which have incident edges $(u, v) \in G$, such that $v \notin H_G$.

Lemma 12. *The following properties hold for $T(H_G)$: (i) $|T(H_G)| \leq (8\gamma_2 + 4)\alpha$, and (ii) for any $w_1 \in B$ and $w_2 \in T(H_G)$,* $\mathsf{dist}(w_1, w_2) \geq \gamma_2 \alpha$.

Proof. We define the perimeter nodes of H, and we denote them as $T(H)$, to be those nodes of H which have incident edges $(u, v) \in M$, where $u \in M$ and $v \notin M$.

Since H is a rectangular grid of maximum side size 3α, $|T(H)| \leq 4 \cdot (2\gamma_2 + 1)\alpha = (8\gamma_2 + 4)\alpha$. We will show that $T(H_G) \subseteq T(H)$.

Let $u \in T(H_G)$, and suppose that $u \notin T(H)$. Then there is an edge $(u, v) \in G$, with $v \notin H_G$. Since $u \in H$ and $u \notin T(H)$, we obtain $v \in H$. Thus, v is a node in H and also a node in G, and therefore, by construction of H_G, $v \in H_G$, a contradiction.

Therefore, $T(H_G) \subseteq T(H)$. Consequently, $|T(H_G)| \leq |T(H)| \leq (8\gamma_2 + 4)\alpha$, which proves property (i). Further, since B is chosen to be in the middle of H (before the truncation of H), the smallest distance of any node $w_1 \in B$ to the closest perimeter node of H is at least $\gamma_2\alpha$. Since $T(H_G) \subseteq T(H)$, w_1 is at distance at least $\gamma_2\alpha$ from any node $w_2 \in T(H_G)$, proving property (ii).

We can write $Q = Q_1 \cup Q_2$, where Q_1 are the paths of Q whose both source and destination are outside H_G, and Q_2 are the paths of Q whose either source or destination (or both) are inside H. Note that Q_1 and Q_2 are disjoint. We first relate Q_1 with the lower bound C^*.

Lemma 13. *Given a path $p \in Q_1$ with source s and destination t, every path from s to t uses some node in H_G.*

Proof. Suppose that there is a path $q \in G$ from s to t such that for each node $u \in q$ with $u \notin H_G$. Let $v \in B$. Consider now the smallest (s, t)-cut z, such that $v \in z$. Clearly, $cut_{s,t}(v) = |z|$, where $|z|$ denotes the number of nodes in z. Clearly, z contains a node $u \in q$. Since, $v \in B$ and $u \notin H_G$, z has to use a node in the perimeter $w \in T(H_G)$. Thus, z has two edge-disjoint sets z_1 and z_2, from v to w, and from w to u, respectively.

From Lemma 12, $|z_1| \geq \gamma_2\alpha + 1$. Further, $|z_2| \geq 1$, since $u \neq w$ ($w \in T(H_G)$ and $u \notin T(H_G)$). Consequently, $|z| \geq |z_1| + |z_2| - 1 \geq \gamma_2\alpha + 1$. Therefore, $cut_{s,t}(v) \geq \gamma_2\alpha + 1$. A contradiction, since by Theorem 2, $cut_{s,t}(v) \leq \gamma_2\alpha$.

Lemma 14. *$C^* \geq \gamma_5|Q_1|/\alpha$, for some constant $\gamma_5 > 0$.*

Proof. Consider a path $p \in Q_1$ from s to t. From Lemma 13, each path q from s to t has to use a node in H_G. Since $s \notin H_G$ and $t \notin H_G$, q enters H_G through one of the perimeter nodes in $T(H_G)$.

Thus, any path selection algorithm (including the optimal oblivious or non-oblivious) for the $|Q_1|$ source-destination pairs, has to construct paths that each uses at least one node in $T(H_G)$. Let $U_{total} = |Q_1|$ denote the total node utilization of the nodes in $T(H_G)$ due to the $|Q_1|$ paths that have to be constructed. The average node utilization of the nodes in $T(H_G)$ is $U_{avg} = U_{total}/|T(H_G)|$. The optimal congestion C^* has to be at least as much as the average node utilization U_{avg}, since some node in $T(H_G)$ has to be used by at least U_{avg} paths, by any path selection algorithm. Thus, Lemma 12 gives for a constant γ_5:

$$C^* \geq U_{avg} \geq \frac{U_{total}}{|T(H_G)|} \geq \frac{|Q_1|}{(8\gamma_2 + 4)\alpha} = \frac{\gamma_5|Q_1|}{\alpha}.$$

We continue now to relate the optimal congestion C^* to the paths in Q_2.

Lemma 15. *Let a path $p \in Q_2$ from s to t. Then,* $\mathsf{dist}(s,t) \geq \gamma_6(\alpha - 2)$*, for some constant $\gamma_6 > 0$.*

Proof. Let q be a shortest-path from s to t which is used to construct the sequence of squares $R(q)$ from s to t containing possibly external or internal squares. Path p uses B because it appears in the sequence of internal squares $\overline{R}(s,t)$ that we obtain from $R(q)$. We examine two cases:

- $B \in R(q)$: Then, shortest path q uses some node $v \in B$. From Theorem 2, $\gamma_1(\alpha - 1) \leq cut_{s,t}(v)$. From Lemma 1, $|q| \geq cut_{s,t}(v) - 1 \geq \gamma_1(\alpha - 1) - 1 \geq \gamma_1(\alpha - 2)$.
- $B \notin R(q)$: By the construction of $\overline{R}(s,t)$, B must be adjacent to some $\alpha' \times \alpha'$ canonical square $B' \in R$, where $\alpha' \geq 2\alpha$, such that q goes through B'. Further, if v is the first node of q that uses B', then there is a node $u \in B$ such that $\mathsf{dist}(u,v) \leq \kappa\alpha$, for some appropriate constant κ. From Theorem 2, $cut_{s,t}(u) \geq \gamma_1(a-1)$. From Lemma 2, $cut_{s,t}(v) \geq cut_{s,t}(u) - \mathsf{dist}(u,w) \geq \gamma_1(\alpha - 1) - \kappa\alpha$ From Lemma 1, $|q| \geq cut_{s,t}(v) - 1 \geq (\gamma_1 - \kappa)(\alpha - 2)$.

Considering now both of the cases, $|q| \geq \min\{(\gamma_1 - \kappa)\alpha, \gamma_1(\alpha - 2)\} = \gamma_6(\alpha - 2)$, for some appropriate constant γ_6.

We define the k-neighborhood of a node in graph G as $N_k(v) = \{u \in G : \mathsf{dist}(u,v) \leq k\}$ (note that this includes also v since by default $dist(v,v) = 0$). The k-neighborhood of a set of nodes A in G is $N_k(A) = \bigcup_{v \in A} N_k(v)$.

Lemma 16. *$C^* \geq \gamma_7 |Q_2|/\alpha$, for some constant $\gamma_7 > 0$.*

Proof. Consider the case where $\alpha \geq 4$. Let $A = N_{\gamma_6\alpha}(H_G)$, that is, A is the $\gamma_6\alpha$-neighborhood of H_G in G (note that A includes all the nodes in H_G). Let H' be a square grid of side length $2^{\lceil \log(2\gamma_6\alpha + (2\gamma_2+1)\alpha) \rceil} = \xi\alpha$, for some constant ξ, such that B is in the middle. (Note that the non truncated square grid H with side length $(2\gamma_2 + 1)\alpha$ is also in the middle of H'.) In particular, consider the truncated version (due to the boundaries of M) of H' which is a rectangular grid of maximum side length $\xi'\alpha$ that contains H and B. We have that all the nodes in A are inside H'. Therefore, $|A| \leq (\xi\alpha)^2$.

Consider now a path $p \in Q_2$ from s and t. Suppose, without loss of generality, that $s \in H_G$ (the other case is $t \in H_G$ which is symmetric). By Lemma 15, any path p' from s to t has $|p'| \geq \gamma_6(\alpha - 2)$. Since the source of p' is in H_G, path p' has a prefix p'' which is completely inside A and has $|p''| = \gamma_6(\alpha - 2)$. Thus, p' uses at least $\gamma_6(\alpha - 2)$ nodes of A. Similarly, any path from s to t has to use at least $\gamma_6(\alpha - 2)$ nodes of A.

Thus, any path selection algorithm for the $|Q_2|$ source-destination pairs, has to construct paths that each uses at least $\gamma_6(\alpha - 2)$ nodes in $T(H_G)$. Let $U_{total} = \gamma_6(\alpha - 2)|Q_2|$ denote the total node utilization of the nodes in A due to the $|Q_2|$ paths that have to be constructed. The average node utilization of the nodes in A is $U_{avg} = U_{total}/|A|$. The optimal congestion C^* has to be at least as much as the average node utilization U_{avg}, since some node in A has to be used by at least U_{avg} paths, by any path selection algorithm. Since we took $\alpha \geq 4$, we obtain:

$$C^* \geq U_{avg} \geq \frac{U_{total}}{|A|} \geq \frac{\gamma_6(\alpha-2)|Q_2|}{(\xi\alpha)^2} \geq \frac{\gamma_7|Q_2|}{\alpha},$$

for some constant γ_7. For the case $\alpha < 4$, we can use the trivial bound $C^* > 1$, and adjust appropriately γ_7.

Lemma 17. *$C^* \geq \gamma_8|Q|/\alpha$, for some constant $\gamma_8 > 0$.*

Proof. By combining Lemma 14 and Lemma 16, we obtain $C^* \geq \max\{\gamma_5|Q_1|/\alpha, \gamma_7|Q_2|/\alpha\}$. Let $X = \max\{|Q_1|, |Q_2|\}$. Since Q_1 and Q_2 are disjoint, $X \geq |Q|/2$. Let $\gamma_{\min} = \min\{\gamma_5, \gamma_7\}$, and $\gamma_8 = \gamma_{\min}/2$. We have,

$$C^* \geq \max\left\{\frac{\gamma_{\min}|Q_1|}{\alpha}, \frac{\gamma_{\min}|Q_2|}{\alpha}\right\} \geq \frac{\gamma_{\min}X}{\alpha} \geq \frac{\gamma_{\min}|Q|}{2\alpha} = \frac{\gamma_8|Q|}{\alpha}.$$

5.2 Upper Bound on Algorithm Congestion

We continue with providing an upper bound on the congestion. Let $Q \subseteq P$ be the set of chosen paths by the algorithm that use internal canonical $\alpha \times \alpha$ square B because it is in their internal square sequence.

Lemma 18. *For any $v \in B$, the number of paths in Q which are expected to use v is at most $\varphi|Q|/a$, for some constant φ.*

Proof. We can write $Q = Q_a \cup Q_b$, where Q_a are the chosen paths that follow one-bend paths in B, while Q_b are the chosen paths that follow two-bend paths in B. Note that Q_a and Q_b are disjoint.

Consider a path $p \in Q_a$. Let p' be the one-bend subpath of p in B. We can write p' as the concatenation of two paths $p' = p_1p_2$, where p_1 corresponds to the first part of p' before the bend (the bend node is also included in p_1), and p_2 corresponds to the second part of p' after the bend. Path p may use v either in one of the subpaths p_1 or p_2. Suppose, without loss of generality, that p_1 is horizontal. Let u be the first node in p_1. If u is in the same row with v then p_1 uses v. According to the path selection algorithm, node u is chosen (along a column of nodes) with probability at most $r = \varphi/\alpha$, for some constant $\varphi \geq 1$. Therefore, subpath p_1 uses v with probability at most r. With a similar analysis, subpath p_2 uses v with probability at most r (by adjusting appropriately φ). Therefore, path p uses v with probability at most r.

Consider now the case where $p \in Q_b$. Let p' be the two-bend subpath of p in B. We can write p' as the concatenation of three paths $p' = p_1p_2p_3$, which correspond to the part of p' before the first bend, between the first and second bends, and after the second bend, respectively. Path p may use v in only one of the subpaths p_1 or p_2 or p_3. The probability that either p_1 or p_3 use v is bounded by r, as proven in the one-bend case. Suppose, without loss of generality, that p_2 is horizontal. Let u be the first node of p_2. If u is in the same row with v then

p_2 uses v. According to the path selection algorithm, node u is chosen (along a column of nodes) with probability $r' = 1/\alpha$. Therefore, subpath p_2 uses v with probability at most r'. Therefore, path p uses v with probability at most $\max(r, r') = r$.

Therefore, from the paths in Q_a it is expected that at most $|Q_a|r$ will use v. Similarly, the expected number of paths from Q_b that will use v is at most $|Q_b|r$. Therefore, the expected number of paths from Q that will use v is at most $|Q_b|r + |Q_b|r = |Q|r = \varphi|Q|/a$.

From Lemma 17 and Lemma 18, we obtain the following corollary.

Corollary 1. *For any $v \in B$, the number of paths in P which are expected to use v is $O(C^*)$.*

Theorem 4 (Congestion). *The expected congestion on any node $v \in G$ is $O(C^* \log n)$, where n is the number of nodes in G. (The same result holds also with high probability by applying a Chernoff bound.)*

Proof. Node $v \in G$ may participate to $O(\log m)$ partition levels of internal canonical squares. Since we can choose M to be such that m is at most the diameter D of G, then we have that $m = \Theta(D) = \Theta(n)$, which implies $O(\log n)$ partition levels. From Corollary 1, the internal square at any particular partition level causes $O(C^*)$ expected congestion to v. Thus, the total expected congestion to v is $O(C^* \log n)$.

References

1. Aspens, J., Azar, Y., Fiat, A., Plotkin, S., Waarts, O.: Online load balancing with applications to machine scheduling and virtual circuit routing. In: Proceedings of the 25th ACM Symposium on Theory of Computing, pp. 623–631 (1993)
2. Meyer auf der Heide, F., Schindelhauer, C., Volbert, K., Grünewald, M.: Congestion, dilation, and energy in radio networks. Theory of Computing Systems 37(3), 343–370 (2004)
3. Awerbuch, B., Azar, Y.: Local optimization of global objectives: competitive distributed deadlock resolution and resource allocation. In: Proceedings of 35th Annual Symposium on Foundations of Computer Science, Santa Fe, New Mexico, pp. 240–249 (1994)
4. Azar, Y., Cohen, E., Fiat, A., Kaplan, H., Racke, H.: Optimal oblivious routing in polynomial time. In: Proceedings of the 35th Annual ACM Symposium on Theory of Computing (STOC), San Diego, CA, pp. 383–388. ACM Press, New York (2003)
5. Bienkowski, M., Korzeniowski, M., Räcke, H.: A practical algrorithm for constructing oblivious routing schemes. In: Proceedings of the 15th Annual ACM Symposium on Parallelism in Algorithms and Architectures, pp. 24–33 (June 2003)
6. Borodin, A., Hopcroft, J.E.: Routing, merging, and sorting on parallel models of computation. Journal of Computer and System Science 30, 130–145 (1985)
7. Busch, C., Magdon-Ismail, M., Xi, J.: Oblivious routing on geometric networks. In: Proceedings of the 17th ACM Symposium on Parallelism in Algorithms and Architectures (SPAA), Las Vegas, Nevada, pp. 316–324 (July 2005)

8. Busch, C., Magdon-Ismail, M., Xi, J.: Optimal oblivious path selection on the mesh. IEEE Transactions on Computers 57(5), 660–671 (2008)
9. Gao, J., Zhang, L.: Tradeoffs between stretch factor and load balancing ratio in routing on growth restricted graphs. In: PODC 2004: Proceedings of the Twenty-Third Annual ACM Symposium on Principles of Distributed Computing, New York, USA, pp. 189–196 (2004)
10. Harrelson, C., Hildrum, K., Rao, S.: A polynomial-time tree decomposition to minize congestion. In: Proceedings of the 15th Annual ACM Symposium on Parallelism in Algorithms and Architectures, pp. 34–43 (June 2003)
11. Kaklamanis, C., Krizanc, D., Tsantilas, T.: Tight bounds for oblivious routing in the hypercube. In: Proceedings of 2nd IEEE Symposium on Parallel and Distributed Processing (2nd SPAA 1990), Crete, Greece, pp. 31–36 (July 1990)
12. Leighton, F.T., Maggs, B.M., Rao, S.B.: Packet routing and job-scheduling in $O(congestion + dilation)$ steps. Combinatorica 14, 167–186 (1994)
13. Maggs, B.M.: Meyer auf der Heide, F., Vöcking, B., Westerman, M.: Exploiting locality in data management in systems of limited bandwidth. In: Proceedings of the 38th Annual Symposium on the Foundations of Computer Science, pp. 284–293 (1997)
14. Popa, L., Rostamizadeh, A., Karp, R., Papadimitriou, C., Stoica, I.: Balancing traffic load in wireless networks with curveball routing. In: MobiHoc (2007)
15. Räcke, H.: Minimizing congestion in general networks. In: Proceedings of the 43rd Annual Symposium on the Foundations of Computer Science, pp. 43–52 (November 2002)
16. Räcke, H.: Optimal hierarchical decompositions for congestion minimization in networks. In: Proceedings of the 40th STOC, pp. 255–264 (2008), Co-Winner of Best Paper Award
17. Raghavan, P., Thompson, C.D.: Randomized rounding: A technique for provably good algorithms and algorithmic proofs. Combinatorica 7, 365–374 (1987)
18. Scheideler, C.: Course notes, `http://www14.in.tum.de/lehre/2005WS/na/index.html.en`
19. Srinivasan, A., Teo, C.-P.: A constant factor approximation algorithm for packet routing, and balancing local vs. global criteria. In: Proceedings of the ACM Symposium on the Theory of Computing (STOC), pp. 636–643 (1997)
20. Valiant, L.G.: A scheme for fast parallel communication. SIAM Journal on Computing 11, 350–361 (1982)
21. Valiant, L.G., Brebner, G.J.: Universal schemes for parallel communication. In: Proceedings of the 13th Annual ACM Symposium on Theory of Computing, pp. 263–277 (May 1981)
22. Zhao, F., Guibas, L.J.: Wireless Sensor Networks: An Information Processing Approach. Morgan Kaufmann, San Francisco (2004)

A Novel Buffer Management Architecture for Epidemic Routing in Delay Tolerant Networks (DTNs)

Ahmed Elwhishi[1], Pin-Han Ho[1], K. Naik[1], and Basem Shihada[2]

[1] Dept. of Electrical and Computer Eng., University of Waterloo, Canada
aelwhish@engmail.uwateloo.ca, {pinhan,snaik}@uwaterloo.ca
[2] Department of Computer Science, KAUST University, Thuwal, Saudi Arabia
basem.shihada@kaust.edu.sa

Abstract. Delay tolerant networks (DTNs) are wireless networks in which an end-to-end path for a given node pair can never exist for an extended period. It has been reported as a viable approach in launching multiple message replicas in order to increase message delivery ratio and reduce message delivery delay. This advantage, nonetheless, is at the expense of taking more buffer space at each node. The combination of custody and replication entails high buffer and bandwidth overhead. This paper investigates a new buffer management architecture for epidemic routing in DTNs, which helps each node to make a decision on which message should be forwarded or dropped. The proposed buffer management architecture is characterized by a suite of novel functional modules, including Summary Vector Exchange Module (SVEM), Networks State Estimation Module (NSEM), and Utility Calculation Module (UCM). Extensive simulation results show that the proposed buffer management architecture can achieve superb performance against its counterparts in terms of delivery ratio and delivery delay.

Keywords: Routing, Buffer management.

1 Introduction

One of most important characteristics of a DTN is the lack of an end-to-end path for a given node pair for extended periods [1]. To cope with frequent and long-lived disconnections due to node mobility, a node in a DTN is allowed to buffer a message and wait until it finds an available link to the next hop. The next hop node buffers and forwards the received message accordingly if it is not the destination of the message. This process continues until the message reaches its destination. This model of routing constitutes a significant difference from conventional ad hoc routing, and is usually referred to as encounter-based routing, store-carry-forward routing, or mobility-assisted routing. The names come from the fact that the routing of a message in DTNs has taken the nodal mobility as a critical factor in the decision on whether to forward the message.

X. Zhang and D. Qiao (Eds.): QShine 2010, LNICST 74, pp. 438–453, 2011.

To achieve better robustness, shorter delivery delay, and higher delivery ratio, extensive research efforts have been reported in design of efficient multi-copy routing algorithms [6,3,7]. However, many DTN routing protocols have assumed negligible storage overhead [2,4]. They have not considered the fact that each node could be a hand-held and battery-powered device with stringent power consumption and buffer size limitation. The buffer limitation may cause message drop/loss due to buffer overflow, which leads to a big challenge in the implementation of most previously reported schemes such as those belonging to the class of epidemic (flooding) routing. With Epidemic routing, two nodes simply exchange all messages that are not in common when they encounter. Without an appropriate countermeasure, the message copies could be spread throughout the network like an epidemic and overwhelm the network resources in terms of buffer spaces and bandwidth.

The paper studies a novel buffer management architecture for DTNs under epidemic routing, aiming to enable an effective decision process on which messages should be dropped in the case of buffer overflow. In specific, the proposed buffer management architecture is based on a fluid flow limit model of Markov chain that can simply approximate the solution via ordinary differential equations (ODEs). Note that an extremely high computation complexity is required in directly solving a Markov chain model under epidemic routing even in presence of a small number of nodes [9,8]. The use of ODEs, although serves as an approximation of the Markov chain result, can nonetheless solidly improve the computation efficiency and provide a closed-form expression. On the other hand, the formulation with the proposed fluid flow limit model is very scalable to the network size, where the complexity does not increase when the number of network nodes increases.

The ODE solution gives per-message utility values, which are calculated based on the estimation of two global parameters: the number of message copies, and the number of nodes who have "seen" this message, i.e., the nodes that have either carried the message or rejected the acceptance of this message. The per-message utility values at each node are then used for the decision on whether a buffered message should be dropped in any contact. We will demonstrate a closed-form solution to the proposed ODE approach, such that each the per-message utility can be calculated efficiently. Simulation results confirm the efficiency and effectiveness of the proposed buffer management scheme under the epidemic routing.

The rest of this paper is organized as follows. Section 2 describes the related work in terms of buffer management and scheduling in DTNs. Section 3 provides the background and system description, including a brief of fluid flow model and network model adopted in this study. Section 4 introduces the proposed buffer management scheme under epidemic routing, including a number of key functional modules: Summary Vector Exchange Module (SVEM), Prediction Module (PM), and Utility Calculation Module (UCM). Section 5 provides experiment results which verify the proposed buffer management architecture. Section 6 concludes the paper.

2 Related Work

Although routing issues in DTNs have been extensively researched in resent years, only a few studies have examined the impact of buffer management and scheduling policies on the performance of the routing techniques. Zhang et al. in [9] addressed this issue in the case of epidemic routing by evaluating simple drop policies such as drop-front and drop-tail, and presented an analysis of the situation that occurs when the buffer at a node has a capacity limit. The paper concluded that the drop-front policy outperforms the drop-tail when higher priority is given to the source node. Lindgren et al. in [10] evaluated a set of heuristic buffer management policies and applied them to a number of routing protocols. Since these policies are defined according to locally available network status, the performance could be improved by considering network-wide node status along with a scheduling scheme for the buffered messages at each node. Khrifa et al. in [11] proposed an interesting approach for solving the problem of buffer management by way of a drop policy and a scheduling scheme. This was the first study that explicitly took global knowledge of node mobility as a constraint in the task of buffer management. Specifically, their method estimates the number of copies of message i based on the number of buffered messages that were created before message i. Although interesting, the method may become inaccurate when the number of network nodes is getting larger, especially for newly generated messages. Meanwhile, the effect due to the change of the number of message copies during the remaining lifetime of a message is not considered in the utility function calculation, which means the utility function is only affected by the current message copies and its remaining lifetime. Obviously, the above mentioned studies leave a large room to improve, where a solution for DTN buffer management that can well estimate and manipulate the global status is absent.

3 Background and System Description

The section presents the background of our mathematical model as well as the network model for encounter-based epidemic routing.

3.1 Background on Fluid Flow Model

In a nutshell, the paper has the buffer management task in DTN epidemic routing to be formulated as a fluid-flow Markov-chain process, which is solved by a novel ODE based approach [13,9]. The fluid flow model can then be used to formulate the rate of message propagation among nodes, calculating the expected time until a given node (destination) is infected, and then calculating the delivery ratio (delivery probability). The following notation are used throughout the paper.

- $n_i(t)$ denotes the number of nodes with message i in their buffers (also referred to as "infected" at time t), where t is counted from the creation time of message i. The following relation is used to calculate$n_i(t)$:

$$n_i^{'}(t) = \beta n_i(t)(N - n_i(t))) \tag{1}$$

where N is the number of nodes in the network, and β is the meeting time rate between nodes. Solving equation (1) with the initial condition $n_i(0)$ yields

$$n_i(t) = \frac{Nn_i(0)}{n_i(0) + (N - n_i(0))e^{-\beta Nt}} \tag{2}$$

- $P_i(t) = P_i(T_d < t)$ denotes the cumulative probability (CDF) of message$i$ being delivered at time t, where T_d denotes a random variable for the time instant that the message i is successfully delivered. $P_i(t)$ can be expressed in a differential equation form[9]:

$$P_i^{'}(t) = \beta n_i(t)(1 - P_i(t)) \tag{3}$$

solving the equation (3) with the initial condition $P_i(0) = 0$ yields

$$P_i(T_d < t) = 1 - \frac{N}{N - n_i(0) + n_i(0).e^{\beta Nt}} \tag{4}$$

(1) and (4) are valid only for unlimited buffer space. To extend the above relations to the scenario with limited buffer space, an additional factor should be considered (denoted as P_{fi}), which represents the probability that the nodal buffer space is available and the message can be transferred to an encountered node. Note that P_{fi} can be obtained by historical data of nodal encounters. Thus accordingly, (1) and (4) are reformulated as follows:

$$n_i(t) = \frac{N}{n_i(0) + (N - n_i(0))e^{-P_{f_i}\beta Nt}} \tag{5}$$

$$P_i(T_d < t) = 1 - \left(\frac{N}{N - n_i(0) + n_i(0)e^{\beta P_{fi} Nt}}\right)^{\frac{n_i(0)}{P_{fi}}} \tag{6}$$

3.2 Network Model

In this paper, a homogeneous DTN is modeled as a set of N nodes, all moving according to a specific mobility model in a finite area, where inter-encounter time between each pair of nodes follows *iid*. Let the number of total messages in the network be denoted as $K(t)$, and the buffer capacity of each node be denoted as B. messages. The messages are generated arbitrarily between source and destination nodes. Each message is destined to one of the nodes in the network with a time-to-live (denoted as Tx). A message is dropped if its Tx expires.

For any given node, a, it is assumed that $J_a(t)$ messages are stored in its buffer at time t. Each message $i(t)$,$i \in [1, J_a(t)]$ is denoted by a tuple $< Sr(t)$, $Dst(t)$, T_i, R_i, $n_i(t)$, $m_i(t)$, $P_{f_i} >$, which represents the source, destination, elapsed time since the creation of the message, remaining lifetime of the message ($R_i = Tx_i - T_i$), number of copies of the message, and the number of nodes who have "seen" message i, respectively. Obviously we have $m_i(t) = n_i(t) + s_i(t)$, where $n_i(t)$ is the number of nodes who are carrying a copy of message i, and $s_i(t)$ is the number of nodes who have "seen" but never accepted the message. Thus, P_{f_i} can be calculated as follows: $P_{fi} = \frac{n_i(t)}{m_i(t)}$.

Let the expected inter-encounter time of any two nodes a and b be denoted as EMT_{ab}, which is defined as the time period taken by the two nodes to enter into their transmission again. The encounter (or mixing) rate between a and b, denoted as β_{ab}, is the inverse of the expected inter-encounter time for the two nodes: $\beta_{ab} = \frac{1}{EMT_{ab}}$. We assume that EMT_{ab}, a, $b \in [1, N]$ follows an exponential distribution (or referred to as with an exponential tail [14]). It has been shown that a number of popular mobility models have such exponential tails (e.g., Random Walk, Random Waypoint, Random Direction, Community-based Mobility [5,16]). In practice, recent studies based on traces collected from real-life mobility examples [17] argued that the inter-encounter period and the encounter durations of these traces demonstrate exponential tails after a specific cutoff point. Based on the *iid* of the mobility model of the nodes, the distribution of the inter-meeting time can be predicted. The historical inter-encounter information between nodes a and b can be calculated by averaging cumulatively all inter-encouter times until current time t. Parameter β is calculated as follows:

$$\beta = \frac{1}{E(MT_{a,b})} \approx \frac{1}{\frac{1}{n}\sum MT_{a,b}} \tag{7}$$

The historical information becomes more accurate and the adaptation of the mobility characteristics becomes precise with a greater elapse of time.

4 Proposed Buffer Management Architecture

Fig. 1 provides a whole picture on the proposed DTN buffer management architecture, which illustrates the functional modules and their relations. The summary vector exchange module (SVEM) is implemented at a node during a contact; then the prediction module (PM) is applied to estimate the values of $m_i(T_i)$ and $n_i(T_i)$ according to the most updated network information. The two parameters are further taken as inputs in the calculation of the proposed per-message utility function in the utility calculation module (UCM). The decision of forwarding or dropping the buffered messages is made based on the buffer occupancy status and the utility value of the messages. The rest of the section introduces the details of each module.

4.1 Summary Vector Exchange Module (SVEM)

During each contact, the network information summarized as a "summary vector" is exchanged between the two nodes, which includes the following data: (1) statistics of inter-encounter time of every node pair maintained by the nodes, (2) statistics regarding the buffered messages, including their IDs, remaining time to live (R), destinations, the stored $n_i(T_i)$ and $m_i(T_i)$ values for each message

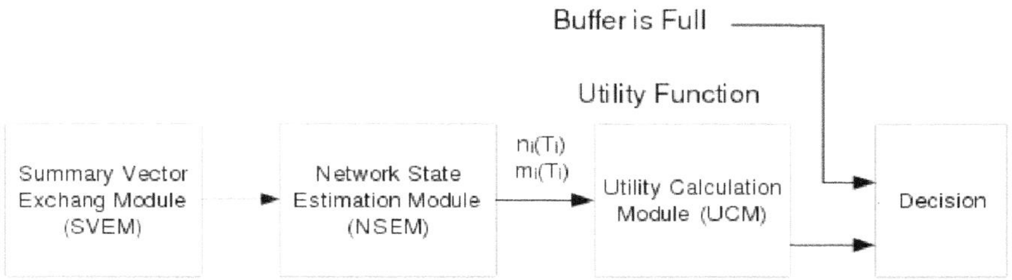

Fig. 1. The buffer management architecture

that were estimated in the previous contact. The SVEM ensures the above information exchange process, and activates PM for the parameter estimation based on the newly obtained network statistics right after each contact.

4.2 Prediction of Message Dissemination

The PM is used to obtain the estimated $m_i(T_i)$ and $n_i(T_i)$ such that the UCM can make decision on the buffer management. For this purpose, we propose a novel estimation approach called Global History-Based Prediction (GHP), which estimates the parameters by considering their statistics since the corresponding message was created.

Let $M_i(T_i)$ and $N_i(T_i)$ denote two random variables that fully describe the two parameters $m_i(T_i)$ and $n_i(T_i)$ at elapsed time T_i, respectively. We have: $E[M_i(T_i)] = \frac{\sum_{i=1}^{j} m_i(T_i)}{j}$ and $E[N_i] = \frac{\sum_{i=1}^{j} n_i(T_i)}{j}$, where j is the total number of messages currently in the buffer of node a and b which are more senior than message i. In the same manner, the average elapsed times for all messages that were generated before message i is calculated as $\widehat{T} = \frac{\sum_{i=1}^{j} T_i}{j}$. Thus, we can have the following estimations for message i: $\widehat{m_i(T_i)}$ and $\widehat{n_i(T_i)}$. These values are then incorporated into the per-message utility metrics, which are calculated as $\widehat{m_i(T_i)} = \frac{T_i E[M_i]}{\widehat{T}}$ and $\widehat{n_i(T_i)} = \frac{T_i E[N_i]}{\widehat{T}}$.

4.3 Utility Calculation Module (UCM)

Based on the problem settings and estimated parameters, the UCM answers the following question at a node during each nodal contact: Given $n_i(T_i)$ and $m_i(T_i)$ and insufficient buffer space for supporting epidemic routing [2], what is an appropriate decision on whether the node should drop any message in its buffer or reject any incoming message from the other node during the contact, such that the average delivery ratio or delivery delay can be optimized? We will describe how this can be achieved in the rest of this section.

Maximization of Delivery Ratio. Let us assume that the buffer is full at node b and there is a message i with elapsed time T_i in a network that has K messages at the moment at which the decision should be made by a node with respect to dropping a message from all messages in its buffer. Further, let $m_i(T_i)$ and $n_i(T_i)$ denote the number of nodes that have "seen" message i since its creation (excluding the source) and those who have a copy of it at this moment, respectively. It is clear that $n_i(T_i) \leq m_i(T_i) + 1$.

Theorem 1. *The best way to maximize the average delivery rate is to drop message i_{min} that satisfies the following:*

$$i_{min} = argmin_i \left[\left[e^{\beta N R_i P_{fi}} \left(\beta R_i n_i(T_i) + \frac{m_i(T_i)}{N} \right) - \frac{m_i(T_i)}{N} \right] \cdot \left(1 - \frac{m_i(T_i)}{N-1} \right)^2 \left(\frac{N}{N - n_i(T_i) + n_i(T_i).e^{\beta N R_i P_f}} \right)^{m_i(T_i)+1} \right] \tag{8}$$

Proof. The probability that a copy of message i will not be delivered by a node is given by the probability that the next meeting time with the destination is greater than its remaining lifetime R_i , assuming that the message i has not yet been delivered. The probability that message i will not be delivered (i.e., none of its copies will be delivered) can be expressed as

$Pr\{message\, i\, not\, delivered \,|\, not\, delivered\, yet\} =$

$$Pr(T_d > T_i + R_i \,|_{T_d > T_i}) = \left(1 - \frac{m_i(T_i)}{N-1} \right) \cdot \left(\frac{N}{N - n_i(T_i) + n_i(T_i) e^{\beta P_{fi} N Ri}} \right)^{\frac{n_i(T_i)}{P_{fi}}} \tag{9}$$

The proof of (9) is provided in the Appendix.

By assuming network homogeneity, there is an equal likelihood that the message is "seen" by each node. Thus, the probability that message i has been already delivered to the destination is equal to

$$Pr\{\, message\, i\, already\, delivered \,\} = \frac{m_i(T_i)}{(N{-}1)} \tag{10}$$

By combining (9) and (10), the probability that message i is successfully delivered before its Tx expires can be calculated as follows:

$Pr_i = 1 - P\{message\, i\, not\, yet\, delivered\}$.

$$P\{message\, i\, will\, not\, be\, delivered\, within\, R_i\}$$

$$Pr_i = 1 - \left(1 - \frac{m_i(T_i)}{N-1} \right)^2 \left(\frac{N}{N - n(T_i) + n(T_i).e^{\beta P_f N Ri}} \right)^{\frac{n_i(T_i)}{P_f}} \tag{11}$$

When a node is operating at its maximum buffer capacity, it should drop one or multiple messages so as to achieve the best gain in the increase of the global delivery ratio $Pr = \frac{1}{K(t)} \sum_{i=1}^{K(t)} Pr_i$. To make the optimal decision locally at the node, Pr_i is differentiated with respect to $n_i(T_i)$, and $\partial n_i(T_i)$ is then discretized and replaced by $\triangle n_i(T_i)$.

The best drop policy is one that maximizes $\triangle Pr_i$:

$\triangle Pr_i = \frac{1}{K(t)} \sum_{i=1}^{K(t)} \frac{\partial Pr_i}{\partial n_i(T_i)} * \triangle n_i(T_i)$
$= \frac{1}{K(t)} \sum_{i=1}^{K(t)} \left[\left[e^{\beta N R_i P_f} \left(\beta R_i n_i(T_i) + \frac{m_i(T_i)}{N}\right) - \frac{m(T_i)}{N}\right] . \left(1 - \frac{m(T_i)}{N-1}\right)^2 \left(\frac{N}{N-n_i(T_i)+n_i(T_i)e^{\beta N R_i P_{fi}}}\right)^{m_i(T_i)+1}\right] \triangle n_i(T_i)$

Thus, the maximum delivery ratio can be achieved if the message that causes the least decrease in $\triangle Pr$ is discarded. On the other hand, when message i is discarded, the number of copies of message i in the network decreases by 1, which results in $\triangle n_i(T_i) = -1$. Thus the optimal buffer dropping policy that can maximize the delivery ratio based on the locally available information at the node is to discard the message with the smallest value of $\frac{\partial Pr_i}{\partial n_i(T_i)}$, which is equivalently to choose a message with a value for i_{min} that satisfies (8). This derivation is an attempt to handle changes in the number of copies of a message that may be increased in the future during new encounters. This goal can be achieved by predicting P_f, the probability of forwarding a copy of message i to any node encountered, which is incorporated into the estimation of the delivery ratio. It is clear that the accuracy of P_f is based mainly on the precision in estimating the values of $m_i(T_i)$ and $n_i(T_i)$.

Minimization of Average Delivery Delay. To minimize the average delivery delay, node b should discard the message such that the expected delivery delay of all messages can be reduced the most.

Theorem 2. *To achieve the minimum average delivery delay, node a should drop the message that satisfies the following:*

$$i_{min} = \left(\frac{N}{N-1+e^{\beta P_{fi} NTi}}\right)^{\frac{1}{P_{fi}}} * \left[\frac{1}{(n_i(T_i))^2 \beta}\left(\frac{n_i(Tx)}{N-1}\right)\right] \quad (12)$$

Proof. The expected delay in delivering a message that still has copies existing in the network can be expressed

$D_i = P\{message\, i\, not\, deliverd\, yet\} * \frac{1}{P_{fi}} E[T_d \mid T_d > T_i]$

$$D_i = \left(\frac{N}{N-1+e^{\beta P_{fi} NTi}}\right)^{\frac{1}{P_{fi}}} * \frac{1}{P_{fi}} E[T_d \mid T_d > T_i] \quad (13)$$

Since we have a homogeneous network, the expected delay of a message can be calculated as

$E[T_d \mid T_d > T_i] = \left[T_i + \frac{1}{P_{fi}} \int_0^{Tx} t f(t) dt\right]$
$= \left[T_i + \frac{1}{P_{fi}} \int_0^{Tx} t \beta e^{-\beta t} dt\right] = T_i + \frac{1}{P_f \beta}\left(1 - e^{-\beta Tx}\right) - \frac{Tx}{P_f}.e^{-\beta Tx}$

If there are $In(T_i)$ messages in the network, $E[T_d \mid T_d > T_i]$ then can be expressed as

$$E[T_d \mid T_d \geq T_i] = T_i + \frac{1}{n_i(T_i)P_{fi}\beta}\left(1 - e^{-\beta n_i(T_i)Tx}\right)$$

$$- \frac{Tx}{P_f}.e^{-\beta n_i(T_i)Tx} \tag{14}$$

The above equation does not take into consideration that new copies of message i might be created during its remaining message life time (R_i). To take this fact into consideration, the total asymptotic number of the copies of message i is calculated during Tx of message i, which is calculated as below:

$$n_i(Tx) = \frac{n_i(T_i)N}{n_i(T_i) + (N - n_i(T_i))e^{-\beta P_{fi} N R_i}} \tag{15}$$

The proof of (15) is given in the Appendix. The second term of (14) can be replaced by $\frac{n_i(Tx)}{N-1}$ which represents the cumulative density function (CDF) of message delivery ratio within Tx. The number of the copies of message i in the third term is substituted by $n_i(Tx)$ as well.

The final expression is written as

$$D_i = \left(\frac{N}{N - 1 + e^{\beta P_{fi} N T_i}}\right)^{\frac{1}{P_{fi}}} * \left[T_i + \frac{1}{n_i(T_i)P_{fi}\beta}\left(\frac{n_i(Tx)}{N-1}\right)\right.$$

$$\left. - \frac{Tx}{P_{fi}}.e^{-\beta n_i(Tx)Tx}\right] \tag{16}$$

The proof of (16) is provided in the Appendix.

When a node buffer is full, the node should make a drop decision that leads to the largest decrease in the global delivery delay of message i, D_i . To find the local optimal decision, D_i is differentiated with respect to $n_i(T_i)$, and ∂D_i is then discritized and replaced by $\triangle D_i$:

$\triangle D_i = \frac{\partial D_i}{\partial n_i(T_i)} * \triangle n_i(T_i)$

$$\triangle D_i = \left(\frac{N}{N - 1 + e^{\beta P_{fi} N Ti}}\right)^{\frac{1}{P_{fi}}} * \left[\frac{-1}{(n_i(T_i))^2 P_f \beta}\right.$$

$$\left.\left(\frac{n_i(Tx)}{N-1}\right)\right] \triangle n_i(T_i)$$

To reduce the delivery delays of all the messages existing in the network, the best decision is to discard the message that maximizes the total average of the delivery delay, $D = \frac{1}{K(t)} \sum_{i=1}^{K(t)} D_i$, among all the messages. Therefore, The optimal buffer-dropping policy that maximizes the delivery delay is thus to discard the message that has the min value of $\mid \frac{\partial D_i}{\partial n_i(T_i)} \mid$, which is equivalently to choose a message with a value for i_{min} that satisfies (12).

Forwarding and Dropping Policy. With the per-message utility, the node firstly sort the buffer messages accordingly from the highest to the lowest. The messages with lower utility values have higher priorities to be dropped when the node's buffer is full, while the messages with higher utility values have higher priorities to be forwarded to the encountered node. Fig. 2 illustrates the forwarding and dropping actions: if the utility u_1 of message j buffered in a is higher than u'_B of message i at node b, then message i is dropped and replaced by a copy of message j if the buffer of b is full during the contact of the two nodes.

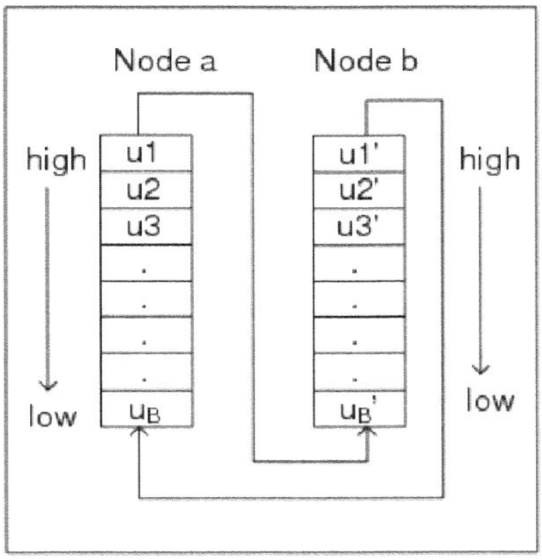

Fig. 2. The forwarding and dropping at a node

5 Simulation

5.1 Experimental Setup

To examine the efficiency of the proposed buffer management architecture, experiments are conducted and presented in this section. To better understand the performance of the proposed estimation strategy–GHP, we also implement two other estimation strategies for the values of $m_i(T_i)$ and $n_i(T_i)$, namely Global Knowledge-based Management (GKM) and Encounter History-Based Prediction (EHP).

The GKM assumes knowing the exact values of $m_i(T_i)$ and $n_i(T_i)$, and is supposed to achieve the best performance. Since such an assumption is not practical[12], the result of GKM is taken as a benchmark for the proposed GHP scheme. With EHP, The two encountered nodes update each other with respect to the messages they have in common, and the values of $m_i(T_i)$ and $n_i(T_i)$ are updated accordingly. This policy of update provides a sub-optimal solution and has been employed in [15] and [11]. In addition to the prediction strategies, we compared the proposed buffer management architecture with a number of counterpart policies listed as follows:

- Drop oldest (DO) drops the message with the shortest remaining time to live.
- Drop front (DF) drops the message that entered the queue the earliest when the buffer is full. This policy obtains the best performance of all the policies used by Lindgren et al. in [10].
- History-based drop (HBD) [11] is based on the history of all messages (on average) in the network after an elapsed time. The variables of the message utility are estimated by averaging the variables of all messages in the network after during the elapsed time.

We assume that a node will never discard a message sourced at the node in favor of a relayed message. It means that the messages issued at a node have the highest priority at the node. If all buffered messages are sourced ones, and the newly arrived message is also a source message at the node, then the oldest one is dropped. This idea was examined in [9] and has been proved with improved delivery ratio. Without loss of generality, random waypoint mobility model is employed in the simulation, where a number of 100 nodes are launched moving e independently on a 500 $\times$ 500 grid with reflective barriers [16]. Each node has a radiation distance as D $\geq$ 0 meters, and each message transmission takes one time unit. Euclidean distance is used to measure the proximity between two nodes (or their positions). A slotted collision avoidance MAC protocol with Clear-to-Send (CTS) and Request-to-Send (RTS) features has been implemented in order to arbitrate between nodes that contend for a shared channel. The message inter-arrival time is uniformly distributed in such a way that the traffic can be varied from low (10 messages generated per node) to high (60 messages generated per node). The bandwidth of the network is assumed to be unlimited. Message delivery ratio and the delivery delay are taken as two performance measures of the simulation. Each data is the average of the results from 30 runs.

5.2 Proposed Policy for Maximizing Delivery Ratio

This section examines the proposed policy for maximizing the average delivery ratio. Two scenarios are performed for each routing scheme: varying the traffic load and fixing the buffer capacity, and fixing the traffic and varying the buffer capacity.

Scenario (1): The Effect of Traffic Load. In this scenario, the traffic load varies from 10 to 70 messages generated per node, and the buffer size is set to a low capacity (10 messages). The plots of the delivery rate obtained for epidemic is shown in Fig. 3.

It can be seen that the GKM gives the best performance for all traffic loads, which meets our expectation. The GHP policy provides the next best result and is competitive with the GKM in the case of low traffic. As the traffic increases, the performance of all policies degrades, while the GHP still outperforms all the other policies except GKM. It can achieve a delivery rate 2.15 times higher than that achieved by DO, 1.7 times higher than DF, 1.22 times higher than HBD, 1.32 times higher than EHP, and only 0.15 times worse than GKM.

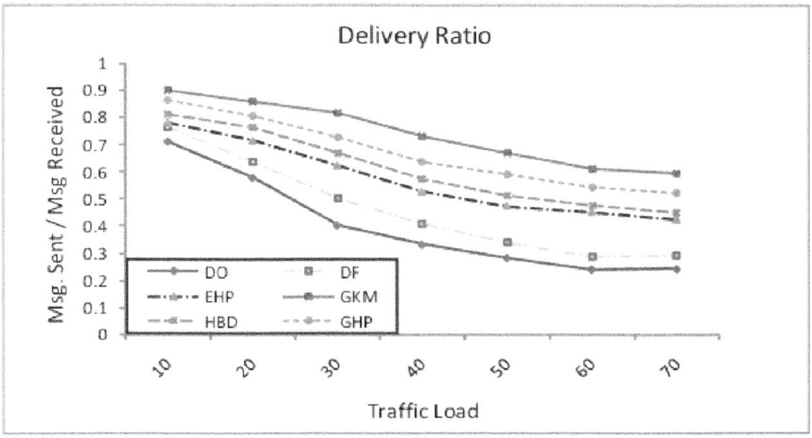

Fig. 3. The effect of traffic load on the delivery ratio

Scenario (2): The Effect of Buffer Size. The results for the second scenario are shown in Fig. 4. The buffer capacity varies from 5 (low capacity) to 600 (high capacity) and the traffic load is fixed at high traffic (70 messages per node). It can be observed that the GKM gives the best performance for all values of buffer capacity, while GHP outperforms all its counterparts except GKM when the buffer capacity is relatively small. As the buffer capacity becomes larger, the performances of all policies improve and become closer to one another. GHP still yields the highest performance in this scenario. For a low buffer capacity, the delivery rate of epidemic routing with its GHP can be 3 times higher than that with DO, 2 times higher than DF, 1.2 times higher than HBD, 1.4 times higher than EHP, and only 0.38 times worse than GKM.

5.3 Proposed Policy for Minimizing Delivery Delay

This section evaluates the effect of the policy of each routing scheme on message delivery delay using the same scenarios in previous section.

Scenario (1): The Effect of the Traffic Load. Figure. 5 shows the results. AS expected, the GKM gives the best performance under all traffic loads for both routing techniques, while the GHP is the second best and is competitive with the GKM in the case of low traffic. As the traffic increases, the demand on the wireless channel and buffers increases, causing a long queuing delays and substantial message loss that negatively affect the performance of all the examined policies. We have observed that the GHP outperforms all other policies, which is better than DO by 1.5 times, DF by 1.40 times, HBD by 1.1 times, EHP by 1.20 times, and a longer delay of only 0.08 of that achieved by GKM.

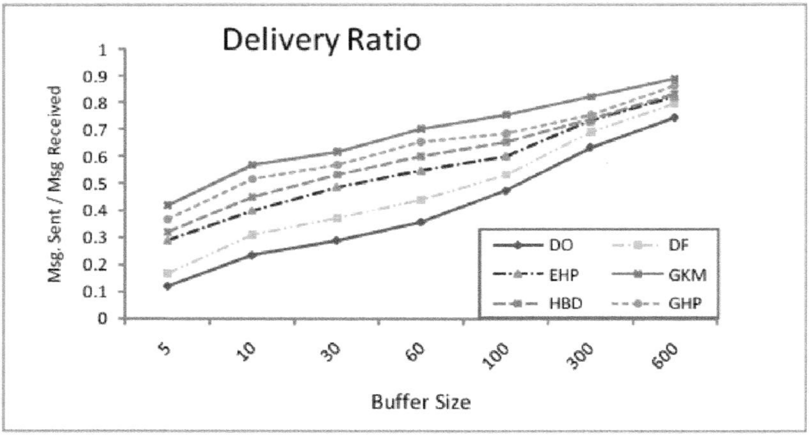

Fig. 4. The effect of buffer capacity on the delivery ratio

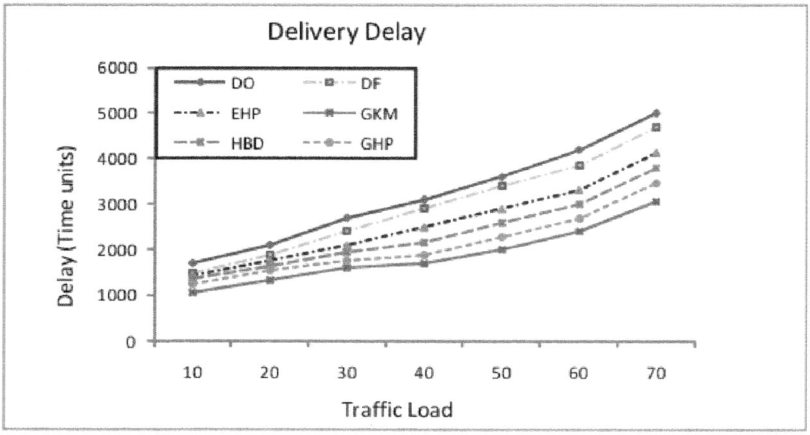

Fig. 5. The effect of traffic load on delivery delay

Scenario (2): The Effect of Buffer Size. Fig. 6 shows the delivery delay obtained for epidemic forwarding with the policies under consideration in the case of varying buffer size and keeping traffic load constant.

As expected again, the GKM of each routing scheme gives the best performance for all values of buffer capacity, while the GHP outperforms all other counterpart schemes except for the GKM when the traffic demand is larger than the buffer capacity. As the buffer capacity became larger, the performances of all policies improve and become closer to one another. GHP still outperforms all other policies. When the buffer capacity is low (5-10 messages), epidemic routing under a GHP policy can achieve a delivery delay shorter than that achieved by

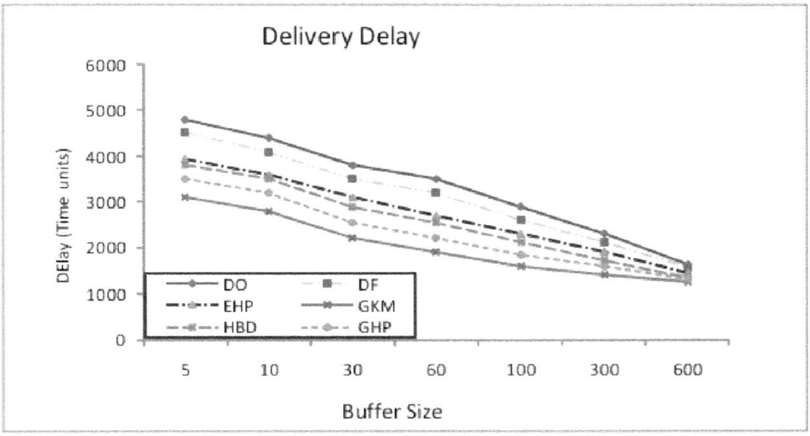

Fig. 6. The effect of buffer capacity on delivery delay

DO by 1.4 times, DF by1.35 times, HBD by 1.20 times, EHP by 1.29 times, and longer delay of only 0.13 of that achieved by GKM. It should be noted that, GHP outperforms all other policies that based on local knowledge about the network.

6 Conclusions and Future Work

This paper has investigated a novel buffer management architecture for epidemic routing in delay tolerant networks, aiming to optimize the message delivery ratio and delivery delay. The proposed architecture incorporates a suite of novel mechanisms for network state estimation and utility derivation, such that a node can obtain the priority for dropping each message in case of buffer full. The simulation results showed that the proposed buffer management architecture can significantly improve the routing performance in terms of the performance metrics of interest under limited network information.

proof of (9): Given $m_i(T_i)$, $n_i(T_i)$,and $P(T_i) = \frac{m(T_i)}{N-1}$, as initial values at T_i , the delivery probability in the interval $t : T_i < t < T_i + R_i$, $P(T_d < T_i + R_i \,|_{T_d > T_i})$, can be constructed using (3) as follows:

$P_i'(t) = \frac{dP}{dt} = \beta n_i(t)(1 - P_i(t))$

$\frac{dP}{1-p} = \beta n_i(t)dt$

$\frac{dP}{1-p} = \beta \frac{N n_i(0)}{n_i(0)+(N-n_i(0))e^{-P_{fi}\beta N t}} dt$

Integrate both sides for the interval R_i, we get

$$P(T_d < T_i + R_i \,|_{T_d > T_i}) = 1 - \left[\left(1 - \frac{m_i(T_i)}{N-1}\right) \left(\frac{N}{N - n_i(T_i) + n_i(T_i)e^{\beta P_{fi} N(Ri)}} \right)^{\frac{n_i(T_i)}{P_f}} \right]$$

Proof of (15): Given $n_i(T_i)$ as initial value, the expected number of message copies within the interval R_i can be constructed using
(1), as follows:

$n_i'(t) = P_{fi}\beta n_i(t)(N - n_i(t))$.

$\beta P_{fi} dt = \frac{dn_i}{n_i(N-n_i)}$, by integrating both sides for the interval R_i we get

$$\beta P_{fi} \int_{0}^{R_i} dt = \int_{n_i(T_i)}^{n_i(T_i+Ri)} \frac{1}{(Nn_i - n_i^2)} dn_i,$$

which lead to:

$$n_i(Tx) = \frac{n_i(T_i)N}{n_i(T_i) + (N - n_i(T_i))e^{-\beta P_{fi} N R_i}}$$

References

1. Delay tolerant networking research group, http://www.dtnrg.org
2. Vahdat, A., Becker, D.: Epidemic routing for partially connected ad hoc networks. Technical Report CS-200006, Duke University (2000)
3. Lindgren, A., Doria, A., Schelen, O.: Probabilistic routing in intermittently connected networks. SIGMOBILE Mobile Computing and Communication Review 7(3) (2003)
4. Spyropoulos, T., Psounis, K., Raghavendra, C.S.: Efficient routing in intermittently connected mobile networks: The multiple-copy case. ACM/IEEE Transactions on Networking (February 2008)
5. Groenevelt, R., Koole, G., Nain, P.: Message delay in manet (extended abstract). In: Proc. ACM Sigmetrics (2005)
6. Elwhishi, A., Ho, P.-H.: SARP - A Novel Multi-Copy Routing Protocol for Intermittently Connected Mobile Networks. In: GLOBECOM 2009, pp. 1–7 (2009)
7. Erramilli, V., Crovella, M., Chaintreau, A., Diot, C.: Delegation forwarding. In: Proceedings of the 9th ACM International Symposium on Mobile Ad Hoc Networking and Computing, Hong Kong, Hong Kong, China (May 26-30, 2008)
8. Ip, Y.-K., Lau, W.-C., Yue, O.-C.: Performance modeling of epidemic routing with heterogeneous node types. In: Proc. of IEEE ICC (2008)
9. Zhang, X., Neglia, G., Kurose, J., Towsley, D.: Performance modeling of epidemic routing. In: Proceedings of IFIP Networking (2006)
10. Lindgren, A., Phanse, K.: Evaluation of Queueing Policies and Forwarding Strategies for Routing in Intermittently Connected Networks. In: First International Conference on Communication System Software and Middleware (2006)
11. Krifa, A., Barakat, C., Spyropoulos, T.: Optimal buffer management policies for delay tolerant networks. In: Proc. of IEEE SECON (2008)
12. Nelson, S.C., Bakht, M., Kravets, R., Harris, A.F.: Encounter based routing in DTNs. SIGMOBILE Mob. Comput. Commun. Rev. 13(1), 56–59 (2009), http://doi.acm.org/10.1145/1558590.1558602
13. Spyropoulos, T., Turletti, T., Obraczka, K.: Routing in Delay-Tolerant Networks Comprising Heterogeneous Node Populations. IEEE Transactions on Mobile Computing 8(8), 1132–1147 (2009), doi:10.1109/TMC.2008.172

14. Aldous, D., Fill, J.: Reversible markov chains and random walks on graphs (monograph in preparation), http://statwww.berkeley.edu/users/aldous/RWG/book.html
15. Balasubramanian, A., Levine, B., Venkataramani, A.: DTN routing as a resource allocation problem. In: Proceedings of the 2007 Conference on Applications, Technologies, Architectures, and Protocols for Computer Communications, Kyoto, Japan (August 27-31, 2007)
16. Spyropoulos, T., Psounis, K., Raghavendra, C.S.: Performance analysis of mobility-assisted routing. In: Proceedings of ACM MOBIHOC (2006)
17. Thomas Karagiannis, M.V., Le Boudec, J.-Y.: Power law and exponential decay of inter contact times between mobile devices. In: Proc. of ACM/IEEE MobiCom (2007)

Self-optimizing Mechanism for Prediction-Based Decentralized Routing

Abutaleb Abdelmohdi Turky, Florian Liers, and Andreas Mitschele-Thiel

Integrated Communication Systems Group
Ilmenau University of Technology
98693 Ilmenau, Germany
{abutaleb-abdelmohdi.turky,florian.liers,mitsch}@tu-ilmenau.de

Abstract. In this paper, we introduce an adaptive traffic prediction approach for self-optimizing the performance of a Prediction-based Decentralized Routing (PDR) algorithm. The PDR algorithm is based on the Ant Colony Optimization (ACO) meta-heuristics in order to compute the routes. In this approach, an ant uses a combination of the link state information and the predicted available bandwidth instead of the ant's trip time to determine the amount of deposited pheromone. A Feed Forward Neural Network (FFNN) is used to build adaptive traffic predictors which capture the actual traffic behavior. Our contribution is a new self-optimizing mechanism which is able to locally adapt the prediction validity period depending on the prediction accuracy in order to efficiently predict the link traffic. We study three performance parameters: the rejection ratio, the percentage of accepted bandwidth and the effect of prediction use. In general, our new algorithm reduces the rejection ratio of requests, achieves higher throughput when compared to the AntNet and Trail Blazer algorithms.

Keywords: Traffic engineering, self-organization, ant-based routing, quality of service, artificial neural network.

1 Introduction

The rapid growth of the Internet forces the Internet Service Providers (ISPs) to search for a new technology which has the capability to maximize the network utilization. They hope to increase their revenues by deploying the concept of service differentiation and offering higher quality services. To support such capabilities, the conventional IP technologies should use the methodology of Traffic Engineering (TE).

TE is defined as that aspect of Internet network engineering dealing with the issue of performance evaluation and performance optimization of operational IP networks [1]. TE aims to cover different optimization issues that are related to the network performance such as providing the requested Quality of Service (QoS), minimizing the total delay and maximizing the network throughput, improving the network resources utilization by optimally distributing the traffic over the network topology and quick recovery in from failures.

X. Zhang and D. Qiao (Eds.): QShine 2010, LNICST 74, pp. 454–468, 2011.

There are different sets of routing classifications that depend on different point of views. Routing algorithms can be classified as static or dynamic. With static routing algorithms such as the Minimum Hop Algorithm (MHA), the administrator computes off-line all possible routes using statistic information and updates the routing table accordingly. Dynamic routing algorithms use information about the current state to compute the requested route on demand. All routing decisions in the dynamic routing approach are performed online to reflect changes of network states. This paper focuses on dynamic routing approaches. Most routing protocols are dynamic such as the Open Shortest Path First (OSPF) protocol [2], the Routing Information Protocol (RIP) and Multi-Protocol Label Switching (MPLS) [3].

The efficiency of TE schemes mainly depends on the routing optimization. Most of the dynamic routing algorithms use the available Bandwidth (BW) to choose the paths between the source and destination pairs. The provided QoS depends on accurate measurements of the available BW. Due to the varying nature of the available BW, updating the link state with the current measured BW is not an efficient approach to represent the link utilization. Therefore, new approaches perform an estimation of the link utilization in the future using the actual traffic profile. The proposed routing mechanism should optimize the network utilization, improve the network survivability and reduce future interference between requests.

Routing algorithms can be classified into centralized or decentralized also. In the centralized routing approach; the source router has all necessary information to compute the routes. In the decentralized routing approach, the routing decisions are taken based on the local state information only and by each network node individually. Most decentralized routing approaches use ant-based mechanisms [4]. These algorithms are based on Ant Colony Optimization (ACO) meta-heuristics. Ant systems represent a self-organizing approach which applies the principle of indirect communication between agents by handling the changes to their environment [5]. Ant routing algorithms are inspired from real ants' behaviors which have the ability of discovering the shortest path to a food source and their nest without any knowledge of geometry but with a keen sense of smell. By applying reinforcement learning techniques, ant routing algorithms can find the optimal or a close-to-optimal path between the source and destination through a positive feedback mechanism.

In this paper, we introduce an adaptive traffic prediction mechanism for self-optimizing the performance of Prediction-based Decentralized Routing (PDR) algorithm which is based on the ACO meta-heuristics to compute the routes. The idea of the PDR algorithm is to use the combination of the link state information and the predicted available BW instead of the ant's trip time to determine the amount of pheromone to deposit. We build our traffic predictor using the FFNN which has proved its accuracy to capture the actual traffic behavior. The proposed predictor uses a new adaptive mechanism to be able to locally adapt the prediction validity period depending on the prediction accuracy in order to efficiently predict the link traffics. Depending on the predicted load value, the algorithm computes the available BW and combines it with the current available BW in the link weight formula that is used to select the optimal path. The remainder of this paper is organized as follows: Section 2 gives an overview of the related work. Section 3 introduces the design details of our approach. Section 4 demonstrates the comparative results and discusses the performance evaluation. Future work and conclusions are presented in section 5.

2 Related Work

The traditionally used algorithm within the MPLS domain is the Shortest Path First (SPF) algorithm proposed by E. Dijkstra [6] to solve the shortest path problem. The basic idea of the SPF algorithm is used in many routing protocols such as the OSPF protocol. Dijkstra's algorithm depends on the fact that states on subsections of shortest paths are also shortest paths. It does not just compute the shortest path to a specific destination, but computes the shortest paths to all possible destinations in the network. Roch A. Guerin [7] has introduced a modification to the shortest path algorithm, called Widest Shortest Path (WSP), which is based on the computation of the shortest paths in the first stage with the extension that if there is more than one shortest path, it chooses the one with the maximum BW. This work is introduced to provide extensions to the OSPF protocol in order to support QoS routing in IP-based networks.

Since all previous algorithms aim to select the best paths without considering future path requests, the related performance does not achieve the best results in maximizing the network utilization or the acceptance rate of requests. The Minimum Interference Routing Algorithm (MIRA) is an example of an advanced routing algorithm [8]. The idea of MIRA is avoiding the routing over links that may interfere with another path request in the future. The definition of interference in MIRA depends on computing the maximum flow (max-flow) value between a given ingress and egress pair. The minimum interference path is the path that maximizes the minimum max-flow between all other ingress-egress pairs.

A.B. Bagula [9] introduced a Least Interference Optimization Algorithm (LIOA) which reduces the interference among competing flows by balancing the number and quantity of flows that are carried by a link to achieve efficient routing of BW guaranteed requests. In general, simulation studies demonstrate that LIOA outperforms many routing algorithm such as MHA, and MIRA algorithms. The comparative study of theses algorithms depends on different performance metrics including the rejection ratio of requests and the successful re-routing of requests upon single link failures.

E. Einhorn and A. Mitschele-Thiel introduced the Reinforcement Learning for Traffic-Engineering (RLTE) algorithm [10]. This work presents a novel distributed and self-organized QoS routing approach that is based on reinforcement learning. We have introduced a primary version of the Predicting of Future Load-based Routing (PFLR) algorithm in [11]. The PFLR algorithm uses the predictions of the future load in order to solve the routing problem. The performance of the PFLR algorithm has been compared to earlier routing approaches like the WSP and CSPF algorithms. The primary version of the PFLR algorithm has reduced the rejection ratio of requests and achieves a higher throughput.

We have proposed an enhanced version of PFLR in [12]. PFLR v.2 combines the predicted future load and current residual BW of each link in a formula to represent the Reciprocal of available BW (*RBW*) and then updates the link weight formula with *RBW*. The PFLRv.2 performance has been compared with current routing approaches like DORA and LIOA algorithms and demonstrated the efficiency of PFLRv.2 by testing three performance criteria: the rejection ratio of requests, the percentage of accepted BW and the rejection ratio of rerouted requests within the link failure scenario.

AntNet [13] is an ACO algorithm for distributed and adaptive best-effort routing in IP networks. AntNet is considered the first algorithm that is inspired by ant colony behavior to solve the routing problem. The behavior of AntNet depends on the mobile agents or the ants' framework. During the forward phase, each ant constructs a path by taking a sequence of decisions based on a stochastic policy parameterized by local pheromones and heuristic information. Once it arrives at the destination, the backward phase starts. The backward ants retrace the route, followed by their forward ant, and update the local routing information with an amount of pheromone in all the intermediate nodes.

Yun and Zincir [14] introduced an adaptive routing algorithm based on the AntNet algorithm. This approach uses a new structure of the routing table to overcome the problem of unrealistic requirements for global information of the original AntNet algorithm. The Trail Blazer (TB) routing algorithm minimizes the network congestion through local decisions, based on latency measurements collected by scout packets [15]. TB is meant to be an extension of existing link-state protocols such as OSPF, which provides shortest-path information to initialize the probability table. Therefore, TB does not require a learning period to discover the network topology. TB is also simpler than the AntNet algorithm.

We have proposed a first version of our TE algorithm named Prediction-based Decentralized Routing (PDR) algorithm [16] that can efficiently enhance the routing performance. This algorithm is a member of a class of traffic-aware routing algorithms based on the behavior of ants. We have compared the performance of the PDR algorithm with WSP and SPF algorithms under two different network load scenarios and have shown that the PDR algorithm performs considerably better.

3 Prediction-Based Decentralized Routing

This section provides a detailed description of our improved version for the Prediction-based Decentralized Routing algorithm (PDRv.2). Figure 1 outlines the operation of PDRv.2. In the algorithm, ants are distributed through the network to discover the best paths. The ants use a combination of the link state information and the predicted link load instead of the ant's trip time to determine the amount of pheromone to deposit. This is simpler and requires less control parameters. After selecting the best path, the routing algorithm forwards the packets through the network and updates the reserved BW of each link that belongs to the best path between the source and the destination.

The idea behind the design of PDR is to consider the future link load to enhance the performance of Ant-based routing algorithms. Therefore, we propose a traffic predictor that able to accurately predict the traffic behavior. ANN offers prediction capability with different types of network traffic and has the ability to learn and adapt dynamically. Experimental results show that, ANN can accurately estimate a complicated network traffic pattern efficiently [17].

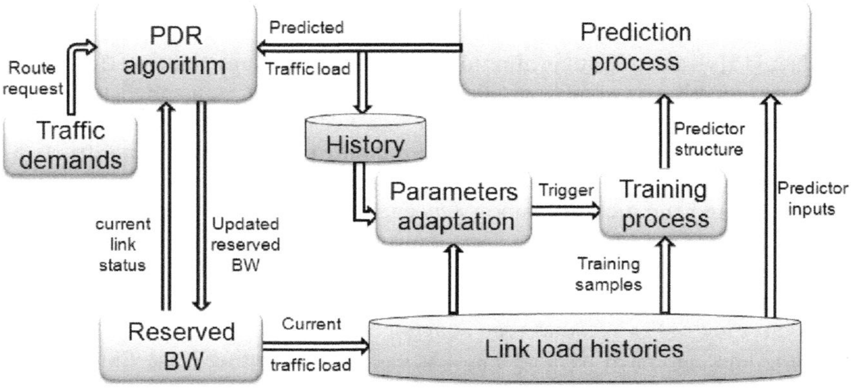

Fig. 1. Prediction-based Decentralized Routing (PDRv.2) algorithm

The proposed predictor has two different processes: the training, and the prediction process. In the training process, the internal structure of FFNN is constructed by a training based on traffic samples of link histories. During the prediction processes, the future link load on every link is estimated after a specified period of time or a specified prediction interval, which is named Window Size (*WS*). This means that, each link has a predictor which is placed on one of the directly connected nodes. Each predictor works on its link history and has its own parameter values. In other words, the predictions are made decentralized to achieve a fast prediction and to conquer the complexity of prediction.

The new proposed feature of the PDRv.2 algorithm is the parameter adaptation process. The Prediction Validity Period (*PVP*) parameter is adapted and self-optimized depending on the prediction accuracy. The *PVP* parameter represents the duration of a period for which the prediction is valid with a high degree of confidence. With the help of this feature, the training of each predictor is triggered independently of each other.

3.1 Training Process

The structure of the used FFNN is shown in Figure 2. It consists of three layers: The input layer contains three neurons; the hidden layer contains fifteen neurons and only one neuron in the output. The Levenberg-Marquardt [18] training algorithm is used because it is the fastest and most accurate one in our case. We have tested different FFNN design and different values of training period size to achieve an efficient predictor. In contrast to the training process in the previous version of PDR algorithm that is event-based, the training process in PDRv.2 algorithm is time-based.

In the event-based approach, if a new path is requested in the network, a new event is generated. During the previous version of PDR, a history of the last thousand events (plus *WS*) of link traffic values is used for training purpose. However in PDRv.2 algorithm, a history of the last hundred time units of link traffic values is used for training purpose. One training pattern contains the minimum, maximum and

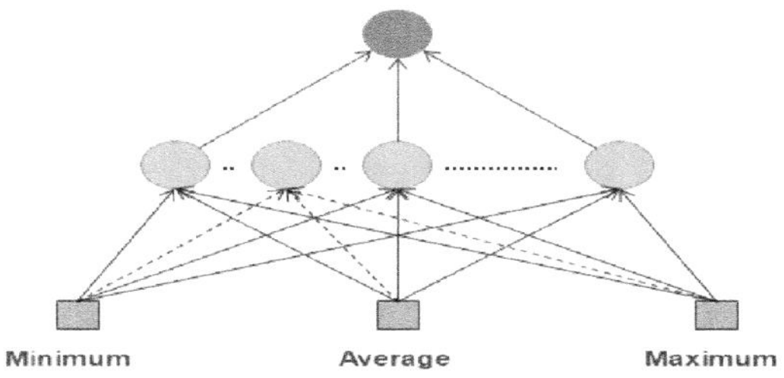

Fig. 2. Feed forward neural network architecture

average of traffic during a time unit. This pattern is formed in a row as input values and one expected output value. The expected output value is a history value *WS* time after the input values. By shifting, one hundred of training patterns are generated. In the previous version of PDR, the training process is triggered every one hundred traffic samples. In the new version, the training process is triggered every *PVP* period which is adapted depending on the prediction accuracy.

3.2 Prediction Process

In the prediction process, the minimum, maximum and average of the traffic during the last time unit are used as input for the FFNN which predicts a value for the link load after a *WS* period. The prediction process is triggered every *WS* period. An analysis study is done to select the best value of *WS*. In other words, the prediction happens every *WS* period and the predictor structure is not changed until the *PVP* period has elapsed.

3.3 Parameter Adaptation Process

In the PDRv.2 algorithm, we propose to use a new adaptive feature called parameter adaptation process. The main objective for this process is to give the predictor the ability to optimize the *PVP* parameter. A *PVP* parameter contains multiple *WS* periods to represent how many times the prediction is done.

$$PVP = WS \times PN, \tag{1}$$

Since *PN* is the Prediction Numbers.

The parameter adaptation process depends on the predictions accuracy that is calculated by comparing the actual and predicted traffic loads. Therefore, archiving processes are required to archive the actual and predicted traffic loads. The prediction accuracy can be represented by the prediction error and there are different error representation methods. In this paper, we use the Root Mean Square Error (*RMSE*) to represent the prediction accuracy. If *AL* is the Actual traffic Load and *PL* is the Predicted traffic Load, then the *RMSE* value is:

$$RMSE = \sqrt{\frac{\sum_{i=1}^{N}(AL_i - PL_i)^2}{N}} \tag{2}$$

During this process, the *PN* parameter is adjusted depending on the comparison between the *RMSE* and Error Threshold (*ETh*) parameters. Then, the *PVP* parameter is updated.

3.4 PDRv.2 Algorithm

The PDR algorithm is built on the principles of the TB routing framework. In the TB design, each router has two tables: a link probability table *Pt* and an average transmission delay table *avg*. *Pt* contains *m* rows, one for each destination node. Each row has *K* entries, one for each outgoing link of the router. The entry *pt[d,i]* is the probability of sending a packet to destination *d* on the outgoing link *i*. The table *avg* has *m* entries, one for each destination node. The entry *avg(d)* is the average transmission delay from the current node to the destination *d*, which is computed from the last *M* scout packets that arrived from *d*. The scout packet is sent from the source to the destination to explore the network. At every intermediate node, the scout packet selects the outgoing link randomly. When scout packets find their destination, they return to their source on the same path they have arrived on and update their accumulated latency *td* in every intermediate node by $td = td + t(i)$,where $t(i)$ is the current latency of the outgoing link *i*. Then, the scout packets use the accumulated latency *td* to update the *pt* table as follows:

$$f(td) = \max(\min(avg(d)/td, 10), 0.1) \tag{3}$$

$$\Delta p = \delta \times f(td) \tag{4}$$

$$pt[d,i] = (pt[d,i] + \Delta p) \;/\; (1 + \Delta p) \tag{5}$$

$$pt[d,j]_{j \neq i} = (pt[d,j]) \;/\; (1 + \Delta p) \tag{6}$$

The average latency *avg(td)* is used to scale the positive reinforcement value of the scout packet. A larger value of *f(td)* indicates a better (shorter) path. *f(td)* is limited to the range [0.1,10] to prevent wide fluctuations in Δp, which is the reinforcement value of *pt[d,j]*. δ defines the learning rate of the algorithm. All entries in *Pt* table of the same destination *d* are scaled by $1+\Delta p$ to ensure that their sum remains 1.

In our approach, an ant uses a combination of the link state information and the predicted link load instead of the ant's trip time to determine the amount of pheromone to deposit, so that it has a simpler process and less control parameters. The current latency *t(i)* of an outgoing link *i* in the TB algorithm is replaced by the Link Weight formula *LW(i)*. *LW(i)* represents a combination of PFLR and LIOA to reduce the interference among competing flows by balancing the number and required BW of flows carried by a link to achieve efficient routing.

The LIOA algorithm represents a cost metric which balances the number and the intensity of the flows offered to the routes. In the LIOA design, $LW(i) = I^{lc} / (Available\ BW)^{(1-lc)}$,whereas *I* is the number of flows carried on the link and *lc* is the least interference control parameter which represents a trade-off between the number and the magnitude of the flows traversing a link. On the other hand, the PFLR algorithm proposes to incorporate the Predicted Available BW (*PABW*) in the link weight formula to optimize the performance of routing. Therefore, we propose to use the *LW(i)* formula as follows:

$$LW(i) = I^{lc}\left(\frac{(1-\alpha)}{(Available\ BW)^{(1-lc)}} + \frac{\alpha}{(PABW)^{(1-lc)}}\right) \tag{7}$$

The *LW(i)* formula is controlled by a parameter called α ,which represents the prediction weight. A low α reduces the influence of the predicted value on the BW. A high value of α increases the influence and suppresses the current value of the available BW.

PDRv.2 Algorithm

---

1) Repeat the following step until the time of the *PVP* has elapsed.
   a) At regular intervals of *WS*, predict the available BW in all links in the network after a specified *WS*.
   b) At regular intervals of *N*, each node generates and sends an ant to a destination.
   c) When a node receives an ant:
      i. It will forward the ant and selects the next link for the ant´s route randomly.
      ii. The ant never selects an outgoing link that leads to a node that has been visited earlier in its path (a loop). If there is no such outgoing link, the ant will die.
   d) When the current node is the destination, then, the ant will return to the source on the same path on which it has arrived.
   e) At each intermediate node :
      i. Compute *LW(i)* of the outgoing link i on every link in the backward path using Equation (7).
      ii. Compute *td*, *td=td+ LW(i)*.
      iii. Update the *pt* and *avg* tables using Equations (3), (4), (5) and (6).
2) Call the parameters adaptation procedure to adapt the *PVP* parameter.
3) Train the predictor on the link load histories.
4) Go to step 1.
5) On the other hand, when a node receives a data packet, which needs to be forwarded, data packets will be routed according to the probabilities in the pt table.

The parameter adaptation procedure consists of three steps. The first step is the computation of *RMSE* using Equation (2). In the second step, the *PN* parameter is adjusted depending on the comparison between the *RMSE* and Error Threshold (*ETh*) parameters. For example, if the *RMSE* value is equal or less than the *ETh* value, this means that the prediction accuracy is very good and the number of predictions should be increased by two. The last procedure step is calculating the new value of the *PVP* parameter using Equation (1).

Parameters adaptation procedure:

---

1) Compute the *RMSE* of prediction using Equation (2).
2) Update the *PN* respect to the following comparisons:
   a) If *RMSE* • *ETh, PN=PN +2.*
   b) If *RMSE > ETh & RMSE* • *ETh*1.5, PN=PN +1.*
   c) If *RMSE > ETh*1.5 & RMSE* • *ETh*2, PN =PN -1.*
   d) If *RMSE > ETh*2, PN=PN -2.*
3) Compute the new *PVP* value, *PVP=WS* PN.*

---

4 Performance Evaluation

In this section, we evaluate the performance of PDRv.2 based on some test scenarios and discuss the results. All test scenarios are implemented using Visual Basic and the ANN toolbox in MATLAB [18]. We modify both, the AntNet and TB algorithm, by replacing the transmission delay with the available BW information to be able to compare the PDRv.2 algorithm with them. Three performances parameters are studied:

1. The rejection ratio of path requests,
2. The percentage of accepted BW and
3. The effect of prediction use.

Our experiment is done on two network topologies. The first one is the MIRA network [8] that is shown in Figure 3, where the thicker links have a capacity of 4800 capacity units while the thinner links have a capacity of 1200 capacity units. The second one is a real network topology that is shown in Figure 4. It is a reference topology suited for an advanced hybrid optical and packet network in the U.S. named Internet2 [19]. In contrast to the performance study of PDRv.1, we consider the requests from all possible combinations of source and destination pairs.

In the MIRA scenario, we examine the performance of the routing algorithms for two generated traffic demands. The first load scenario is a Moderate Load (ML): The arrival of requests follows a Poisson distribution and the holding time of the requests is based on an exponential distribution. The second is a heavy load (HL). In the

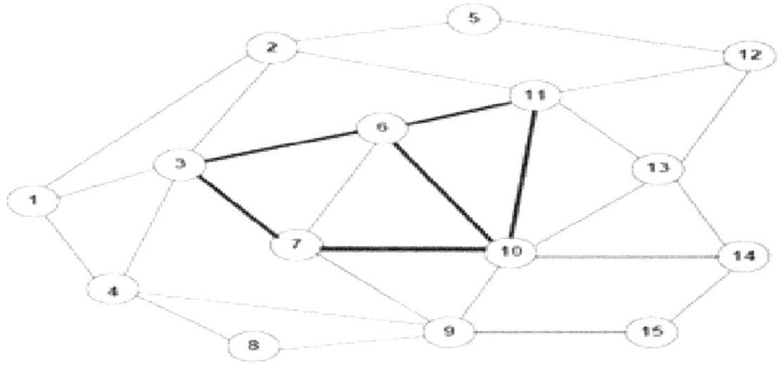

Fig. 3. MIRA network topology

Fig. 4. Internet2 network topology

Internet2 scenario, we examine the performance of the PDRv.2 algorithm for a real traffic demand. The real traffic demands are collected from the trace flies of the NetFlow tool for the first day of 2009 year [19]. Table 1 describes the PDRv.2 parameters and shows the range and used value in our simulation.

Table 1. The PDRv.2 algorithm parameters

Variable	Value
lc (least interference control parameter)	0.1
M (keep the average of the last *M* of *td*)	{15, 20}
δ (learning rate)	{0.01, 0.02}
α (prediction weight)	0.9
WS (window size)	1

4.1 Generated Traffic Scenario

In the next scenario, we consider the MIRA topology and generate two different traffic demands using different values for the Poisson and exponential distributions.

4.1.1 Moderate Load Scenario

Figure 5 shows the rejection ratio of requests for the moderate load scenario. The results show that, the PDRv.2 algorithm rejects approximately 18.36% less requests than the TB algorithm and 49.80% less requests than the AntNet algorithm.

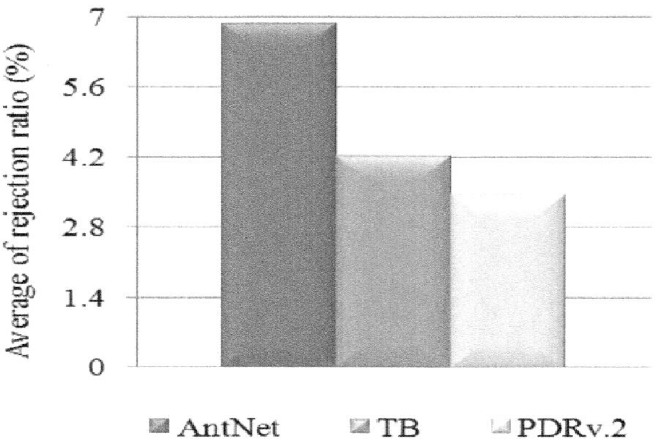

Fig. 5. The rejection ratio of requests for the moderate load scenario

Figure 6 shows the percentage of accepted BW for the moderate load scenario. The PDRv.2 algorithm accepts approximately 1.0% more bandwidth than the TB algorithm and 4.31% more bandwidth than the AntNet algorithm.

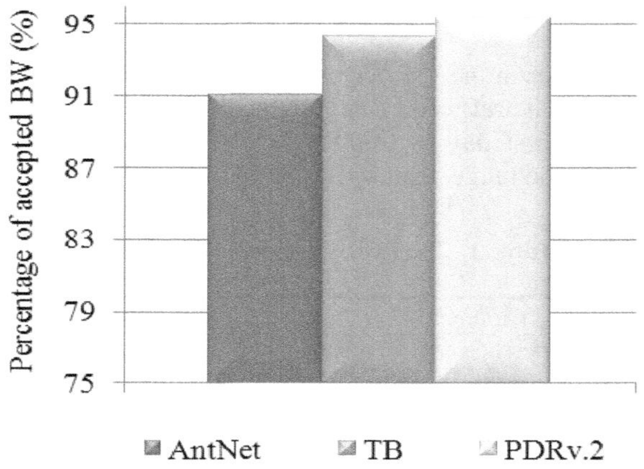

Fig. 6. The percentage of accepted BW for the moderate load scenario

4.1.2 Heavy Load Scenario

Figure 7 shows the rejection ratio of requests for the heavy load scenario. The results show that, the PDRv.2 algorithm rejects approximately 7.55% less requests than the TB algorithm and 44.23% less requests than the AntNet algorithm.

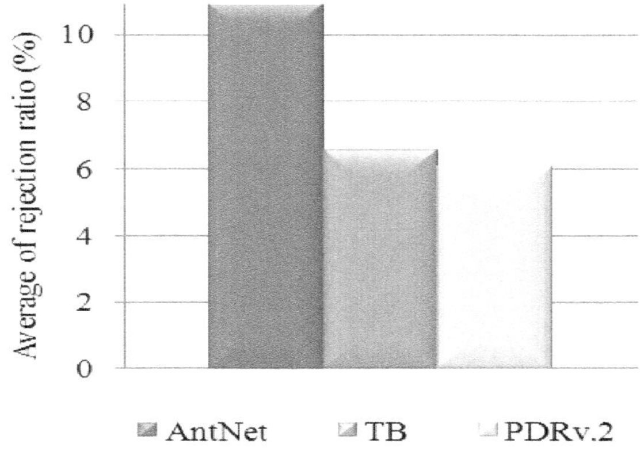

Fig. 7. The rejection ratio of requests for the heavy load scenario

Figure 8 shows the percentage of accepted BW for the heavy load scenario. The results show that, the PDRv.2 algorithm accepts approximately 0.52% more bandwidth than the TB algorithm and 5.93% more bandwidth than the AntNet algorithm.

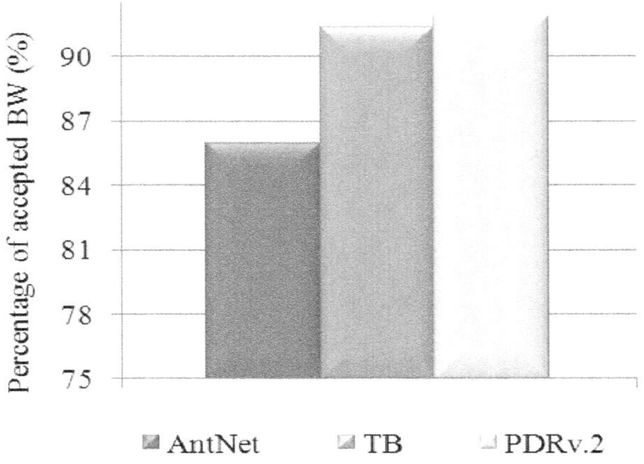

Fig. 8. The percentage of accepted BW for the heavy load scenario

4.2 Real Traffic Scenario

Figure 9 shows the rejection ratio of requests for the real traffic scenario. The results show that, the PDRv.2 algorithm rejects approximately 26.68% less requests than the TB algorithm and 46.14% less requests than the AntNet algorithm.

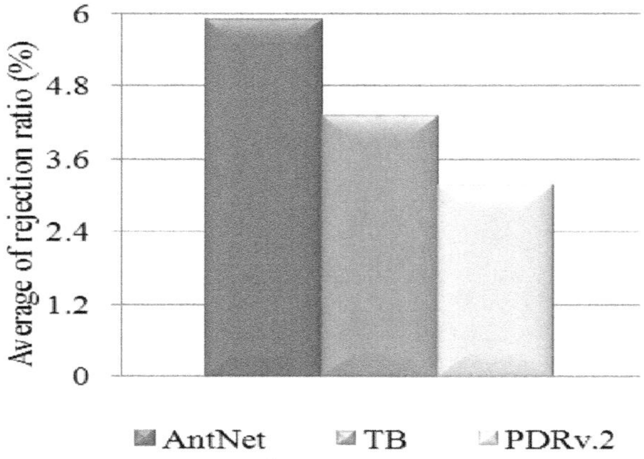

Fig. 9. The rejection ratio of requests for the real traffic scenario

Figure 10 shows the percentage of accepted BW for the real traffic scenario. The results show that, the PDRv.2 algorithm accepts approximately 0.43% more bandwidth than the TB and 2.16% more bandwidth than the AntNet algorithm.

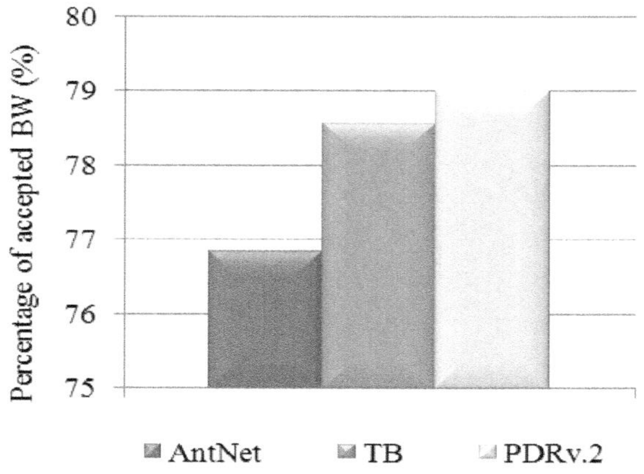

Fig. 10. The percentage of accepted BW for the real traffic scenario

4.3 The Effect of Prediction Use

Table 2 shows the rejection ratio of requests for different prediction weights. In this section, we aim to study the effect of prediction use. Therefore, we run the PDRv.2 algorithm one time with prediction weigh $(\alpha) = 0.9$ and another time with prediction weigh $(\alpha) = 0$. In general, the PDRv.2 algorithm with prediction weigh $(\alpha) = 0.9$ algorithm rejects the least requests in all scenarios of traffic types. In other words, the use of prediction has a positive impact on the routing performance.

Table 2. The rejection ratio of requests (%) for different prediction weighs

Traffic type	Medium load	Heavy load	Real traffic
PDRv.2(α=0.9)	3.45	6.08	3.16
PDRv.2(α=0)	3.79	6.46	3.26

5 Conclusion and Future Work

We have introduced a new self-optimizing mechanism to enhance the performance of the PDR algorithm. The PDRv.2 algorithm is a member of a class of traffic-aware routing algorithms based on the behavior of ants. The main idea of PDR is to let the ants use a combination of the link state information and the predicted available bandwidth instead of the ant's trip time to determine the amount of pheromone to deposit. The new mechanism has the ability to locally adapt the prediction validity period depending on the prediction accuracy in order to efficiently predict the link traffics. We have compared the performance of our proposed PDRv.2 algorithm with the TB and AntNet algorithms in two different networks and with different traffic types. In general, our algorithm performs considerably better than the comparative algorithms with respect to different performance comparison criteria.

In the future, we plan to test the performance of the PDR algorithm with more complex network topologies. We plan to test the performance of PDR with respect to other performance criteria too. In addition, a comparison of the PDR algorithm with other ant algorithms is planned.

References

1. Awduche, D., Chiu, A., Elwalid, A., Widjaja, I., Xiao, X.: Overview and Principles of Internet Traffic Engineering. RFC3272 (2002)
2. Moy, J.: OSPF Version 2. RFC 2328 (1998)
3. Rosen, E., Viswanathan, A., Callon, R.: Multiprotocol Label Switching Architecture. RFC 3031 (2001)
4. Sim, K.M., Sun, W.H.: Ant Colony Optimization for Routing and Load-Balancing: Survey and New Directions. IEEE Trans. on Sys., Man and Cyber. 33(5), 560–572 (2003)
5. Guerin, R., Orda, A., Williams, D.: QoS routing mechanisms and OSPF extensions. J. IEEE Global Telecommunication 3, 1903–1908 (1997)
6. Kunkle, D.R.: Self-organizing Computation and Information Systems: Ant Systems and Algorithms. Technical report, Rochester Inst. of Technology (2001)

7. Dijkstra, E.W.: A note on two problems in connexion with graphs. J. Numerische Mathematik 1(1), 269–271 (1959)
8. Kar, K., Kodialam, M., Lakshman, T.V.: Minimum Interference Routing of Bandwidth Guaranteed Tunnels with MPLS Traffic Engineering Applications. IEEE J. Selected Areas in Comm. 18(2), 2566–2579 (2000)
9. Bagula, A.B., Botha, M., Krzesinski, A.E.: Online Traffic Engineering: The Least Interference Optimization Algorithm. In: ICC 2004, pp. 1232–1236 (2004)
10. Einhorn, E., Mitschele-Thiel, A.: RLTE: Reinforcement Learning for Traffic-Engineering. In: 2nd Inter. Conf. on Autonomous Infrastructure, Man. and Sec., pp. 120–133 (2008)
11. Turky, A.A., Mitschele-Thiel, A.: MPLS Online Routing Optimization Using Prediction. In: Altman, E., Chaintreau, A. (eds.) NET-COOP 2008. LNCS, vol. 5425, pp. 45–52. Springer, Heidelberg (2009)
12. Turky, A.A., Mitschele-Thiel, A.: Use of Load Prediction Mechanism for Dynamic Routing Optimization. In: IEEE Symposium on Comp. and Communications, pp. 782–786 (2009)
13. Caro, G.D., Dorigo, M.: AntNet: Distributed stigmergetic control for communications networks. J. Artificial Intelligence Research 9, 317–365 (1998)
14. Yun, H., Heywood, A.: Intelligent Ants for Adaptive Network Routing. In: CNSR 2004, pp. 255–261 (2004)
15. Gabber, E., Smith, M.A.: Trail Blazer: A Routing Algorithm Inspired by Ants. In: ICNP 2004, pp. 36–47 (2004)
16. Turky, A.A., Mitschele-Thiel, A.: Prediction-based Decentralized Routing Algorithm. In: Self-organizing, Adaptive, Context-Sensitive Distributed Systems, EASST, vol. 17 (2009)
17. Eswaradass, A., Sun, X.H., Wu, M.: Network Bandwidth Predictor (NBP): A System for Online Network performance Forecasting. In: IEEE International Symposium on Cluster Computing and the Grid, pp. 265–268 (2006)
18. Hagan, M.T., Demuth, H.B., Beale, M.H.: Neural Network Design. PWS Publishing, Boston (1996)
19. Neural Network Toolbox, MATLAP version (R2009a), `http://www.mathworks.com/products/neuralnet`
20. Internet2 Observatory Data Collections, `http://www.internet2.edu/observatory/archive/`

HDMRP: An Efficient Fault-Tolerant Multipath Routing Protocol for Heterogeneous Wireless Sensor Networks

Abdelkrim Hadjidj, Abdelmadjid Bouabdallah, and Yacine Challal

Heudiasyc UMR CNRS 6599, University of Technology of Compiegne, France
{ahadjidj,bouabdal,ychallal}@utc.fr

Abstract. In wireless sensor networks, fault tolerance represents a key issue and a design goal of primary concern. Indeed, nodes in wireless sensor networks are prone to failures due to energy depletion or hostile environments. Multipath routing protocols are a category of solutions which enables the network to operate properly despite of faults. In this paper, we present a new multipath routing protocol which provides strong fault tolerance by increasing the number of constructed paths up to three times in some scenarios. Our protocol relies on a new multipath construction paradigm that we have defined specifically for heterogeneous WSN. We call this paradigm: energy-node-disjointness. Our approach leverages a reasonable increase in the cost of the network to a longer network lifetime and a higher resilience and fault tolerance.

Keywords: Fault Tolerance, Multipath Routing, Heterogeneous Wireless Sensor Networks.

1 Introduction

Recent advances in wireless communications and Micro-Electro-Mechanical systems have enabled the development of wireless sensor networks (WSN) which consist of many sensor nodes with sensing, computing and communication capabilities [1]. WSN are deployed over an area to periodically sense and monitor physical or environmental conditions and transmit the sensed data back to a base station. WSN have broad spectrum applications such as environment monitoring, target tracking, military surveillance and healthcare applications.

Heterogeneous WSNs (HWSN) are networks in which nodes have moderated capabilities and some powerful nodes, called masters, contribute to increase the network reliability, lifetime or the delivery ratio [24][11]. WSN Heterogeneity can have several forms: In *Energy Heterogeneity* case, the network is composed of battery-powered sensor nodes and some line-powered master nodes with no energy constraints. Master nodes can also be equipped with ambient energy harvesting technology for incessantly power supplying [21]. In *Link Heterogeneity* case, master nodes have different radio-transceivers in term of distance and bandwidth. This heterogeneity is commonplace in tiered WSN [6][2] where

X. Zhang and D. Qiao (Eds.): QShine 2010, LNICST 74, pp. 469–482, 2011.

cluster-heads have different radio-transceivers for inter-cluster communications and intra-cluster communications. In *Hardware Heterogeneity* case, nodes have different set of sensors and can sense different parameters. This has the benefits of keeping mote's cost relatively low since a node will not need a full sensor set[21]. Finally, in *Security Heterogeneity* case, master nodes are equipped with tamper resistant devices or physical protection.

Due to battery depletion or hostile environments (e.g. wind, rain or high temperature) in which WSN may be deployed, sensor nodes are prone to failure. A part of the network can be disconnected and critical data maybe lost because of faults. Consequently, fault tolerance is a major concern in wireless sensor networks in general and even more in critical applications such as healthcare, forest firefighting or nuclear radiation detection where it is not acceptable to lose sensitive data. Fault tolerance is the capacity to keep a network working correctly despite of failures. Multipath routing protocols are a category of fault tolerance techniques which provides tolerance of faults and increase the network resilience since the probability that all constructed paths fall at the same time because of the same fault is mitigated. If we construct k disjoint paths, we can guarantee that a node remains connected to the Sink even after the failure of up to k-1 paths. Consequently, increasing the number of discovered disjoint paths will improve the fault tolerance of the network. *Node-disjoint multipath routing protocols* [5] construct paths with no common nodes/links. This leads to strong fault tolerance since a node failure impacts only one path. However, node-disjoint routing protocols induce large control message overhead and a lack of scalability. *Braided multipath routing protocols* [5] construct paths with possibly common nodes or links to avoid control messages overhead but provide slightest fault tolerance. Indeed, a failure of a node belonging to several paths will cause the failure of all those paths and may disconnect a large part of the network.

In this paper, we propose a new fault tolerant multipath routing protocol for heterogeneous wireless sensor networks. Simulation results show that our protocol is able to discover up to three times more node disjoint paths compared to existing protocols and improves the fault tolerance by 30% when node failure rate is less than 30%. While most of multipath routing protocols deal with homogenous wireless sensor networks, our protocol HDMRP (Heterogeneous Disjoint Multipath Routing Protocol) is designed especially for HWSN and implements mechanisms to exploit and benefit from robustness and abundant resources in master nodes to significantly increase the network resilience and lifetime.

The rest of the paper is organized as follows. Section 2 gives an overview of existing multipath routing protocols. Section 3 describes the proposed routing protocol. Section 4 discusses the performance evaluation of the proposed routing protocol. Finally, section 5 concludes the paper.

2 Related Works

There has been a host of research works on multipath routing protocols for wireless sensor networks in the last few years. Multipath routing protocols are used for reliability [25][4],load balancing [15][10], QoS provisioning [4][12][22][3][9] and secure communications [19]. In this work, we are mainly interested in fault tolerant multipath routing construction. However, exploitation of these constructed multipath routes is out of the scope of this work.

In [13], Li et al. propose a Node-Disjoint Parallel Multipath Routing (DPMR) which uses one-hop response after a delay time at each node to construct multiple paths simultaneously. To ensure node-disjointness, only nodes that have not been used by other paths forward route requests to their neighbors. In [7], Hou and Shi present LAND, a Localized Algorithm for finding node disjoint paths which constructs a set of minimum cost node-disjoint paths from every node to the Sink. In [15], Lu et al. propose Energy-Efficient Multipath Routing Protocol (EEMRP), a node-disjoint multipath protocol which considers energy and hop count while constructing the multiple paths. EEMRP achieves high energy efficiency without considering network reliability. In [14], Lou and Kwon propose Branch routing protocol (BRP) to improve WSN reliability and security. BRP constructs several trees routed at sink's neighbors which represent branches on the network graph. Each node belongs to one and only one branch but can send data back to the sink on every branch it is aware about. The main drawback of this method is the limited number of discovered paths and therefore the limited fault tolerance.

Node-disjoint multipath routing protocols construct paths with no common nodes/links and provide high resilience and fault tolerance since a node failure impacts only one path. However, they usually suffer from control message overhead and a lack of scalability. Some researchers aim to reduce node-disjoint protocols overhead by relaxing the disjointness requirement. In [5], Ganesan et al. study disjoint and braided paths by comparing their performances and show that braided path protocols overhead is only half the overhead induced by node disjoint protocols. However, braided paths are weaker since a single node failure may cause the failure of multiple routes to the sink. In [23], Yang et al. present NC-RMR, a routing protocol for network reliability which constructs disjoint and braided multipath and uses network coding mechanism to reduce packet redundancy when using multipath delivery. In [16], Nasser et al. propose SEEM protocol (Secure and Energy-Efficient Multipath routing protocol) which finds both braided and disjoint paths and adopt a Client/Server scheme where the Sink (server) executes the paths discovery, paths selection and paths maintenance in a centralized way. As in link-state routing protocols [8], each node in SEEM sends its neighbors list to the sink which consumes so much energy and induces significant overhead.

In [18], Sohrabi et al. propose SAR algorithm (Sequential Assignment Routing) which considers the fact that nodes near the Sink relay more packets and actively participate in communications. As a result, they expend more energy and are more prone to failures due to quick battery depletion. Therefore, SAR requires disjointness only where it has the highest impact (one hop sink neighborhood). In [17], Ouadjaout et al. propose the SMRP protocol which introduces the assumption that Sink's neighbors are powerful master nodes. Therefore, SMRP requires disjointness after two hops sink's neighbors rather than one hop sink's neighbors. As a result, the number of discovered disjoint paths is increased but still limited.

Existing multipath routing protocols balance the tradeoff between fault-tolerance and communication overhead. Indeed, increasing the number of paths for a better fault-tolerance requires more messages exchange and communication overhead. In this paper, we present a new multipath routing protocol able to construct up to three times more multiple paths comparatively to existing protocols by using only one message per node. Furthermore, all above described protocols deal with homogeneous wireless sensor networks and do not consider the power of master nodes. Our protocol is designed especially for heterogeneous WSN and uses abundant resource in powerful master nodes to improve the network fault tolerance.

3 Heterogeneous Disjoint Multipath Routing Protocol (HDMRP)

In this section, we present our new efficient solution which provides fault tolerance in wireless sensor networks. First, we will introduce a new paradigm for heterogeneous wireless sensor networks: energy-node-disjoint paths. Next, we will describe our Heterogeneous Disjoint Multipath Routing Protocol which constructs multiple paths between the sink and each node in the network. Wireless sensor networks (WSN) are typically employed for monitoring and require data collection at a specific node called Sink. We consider a many-to-one traffic pattern where source sensors send measurement data to the Sink. As in several literature works and real-world wireless sensor networks implementations [21][6][20][2], we assume the existence of few robust powerful master nodes in the network.

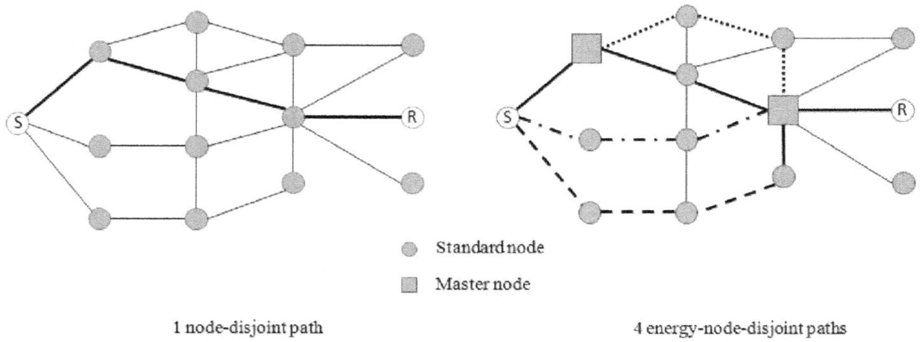

Fig. 1. Node-Disjoint paths and Energy-Node-Disjoint paths

3.1 Energy-Node-Disjointness

In heterogeneous wireless sensor networks, Master nodes are powerful nodes with mitigated energy constraints and able to ensure relatively high connectivity degree and hence relay data on several paths. For this reason, we have introduced a new energy-node-disjointness paradigm as follows:

Definition 1. *n paths are said energy-node-disjoint iff they have no common nodes or all their common nodes are master nodes*

So far, only node-disjointness has been used to build multipath routing topologies:

Definition 2. *n paths are said node-disjoint iff they have no common nodes*

Master nodes are powerful nodes and cannot suffer from battery depletion which is the most frequent node failure cause in WSN. In the case where a master node is common to several paths, the probability that it fails and impacts all paths to which it belongs is mitigated. Consequently, we have introduced a controlled intersection between paths at master nodes which will increase the number of discovered paths. Our protocol builds energy-node-disjoint paths in order to increases the number of alternative paths and therefore the network fault tolerance. For instance, our algorithm discovers four energy-node-disjoint paths between nodes S and D, instead of one node-disjoint path in the network of Fig. 1.

3.2 Protocol Description

In HDMRP, sink neighbors' are called root nodes and root neighbors are called sub-roots. The protocol uses Route REQuest (RREQ) message propagation through nodes to construct multiple energy-node-disjoint paths between each node and the sink. Each non root node maintains a routing table containing an entry for each discovered path. A RREQ corresponds to a path and has the following format: $\{R, S, P_{id}, len, Nmas\}$ where:

- R is the current round number
- S is the sending node id
- P_{id} is the path id
- len is the path length
- $Nmas$ is the number of master nodes in the path

During paths construction, a node may receive several RREQ messages corresponding to one or several paths. To guaranty node disjointness, each node forwards only one RREQ message to its neighbors and acts as a reducing element. However, Master nodes are powerful nodes able to ensure relatively high connectivity degree and can relay data on several paths. Instead of using all nodes as reducing elements belonging to only one path, HDMRP introduces a controlled intersection at master nodes by allowing them to forward several RREQ messages to their neighbors.

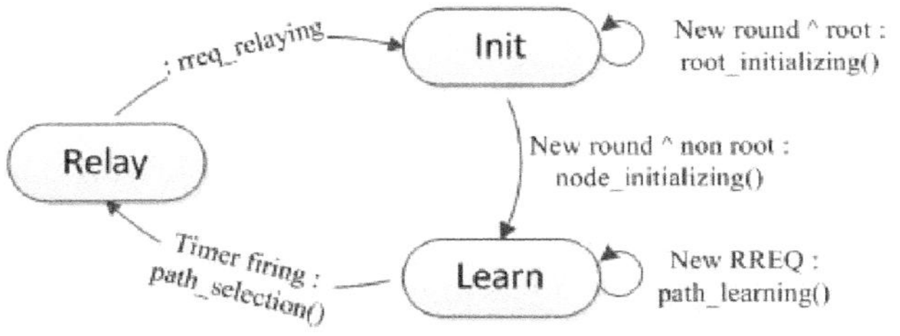

Fig. 2. HDMRP state chart

The Sink starts a new paths construction by broadcasting a route request message RREQ indicating a new round *RREQ.R*. Fig. 2 describes the evolution of a node state in the routing construction process. An arc between two states S_i and S_j has a label "E : P". This label means that when the node is in the state S_i and the event E occurs, the node executes the procedure P and passes to the state S_j. The description of the different procedures that a node must execute in each case is as follows:

root_initializing(): Upon receiving a RREQ indicating a new round *RREQ.R*, a root node relays it to its neighbors and waits for the next round. This terminates the current round construction phase for the root node. The node deduces that it is a root if the received RREQ has a S field equal to the sink ID (cf. Listing 1).

node_initializing(): Upon receiving a RREQ indicating a new round *RREQ.R*, a non root node in the *Init* state removes all previously discovered paths, adds the new received path to its routing table, sets a timer with a length enough sufficient for receiving several RREQs from the network and passes to new routes learning state (*learn state* in Fig. 2). In addition, if the node is a sub-root, it generates a new tagged RREQ by putting its ID in the P_{id} field and broadcasts it to its neighborhood. A node deduces that it is a sub-root if it receives a RREQ message with an empty P_{id} field. See Listing 1, new round initialization part.

path_learning(): Upon receiving a tagged RREQ, a node adds it to its routing table if there's no entry with the same P_{id}. Otherwise, it chooses between the new received RREQ and the one in its routing table based on a cost function. See Listing 1, path learning part. For achieving high energy efficiency, we developed a cost function which ensures having the maximum number of master nodes in each path:

$$cost(RREQ) = \frac{RREQ.len}{RREQ.Nmas} \tag{1}$$

path_selection(): As illustrated in the timer handler of Listing 1, when the timer fires the node has to select one or several RREQs to relay and passes to the *Relay* state. In the case where the node is a master node, it selects all the RREQs stored in its routing table in P2. This increases the connectivity degree of master nodes and hence the number of energy-node-disjoint paths. In the other case, ordinary nodes select one RREQ among those stored in the routing table in P2. This ensures the node-disjointness of paths and hence increases resiliency.

rreq_relaying(): In the *Relay* state, each node broadcasts to its neighbors the selected RREQs and passes to the init state. This terminates the current round construction phase for the node.

Listing 1. HDMRP Algorithm

```
state = init;
RREQ_Processing()
{
newRREQ = receive();
if(currentRound < newRREQ.R)
{
    /* new round initialization */
    currentRound = newRREQ.R
    if(newRREQ.S == 0)
    {
        /* root node */
        newRREQ.S = getNodeID();
        newRREQ.len ++;
        send(newRREQ);
    }
    else
    {
        flush(received_RREQs);
        received_RREQs.add(newRREQ);
        start_timer();
        if(newRREQ.pid == null)
        {
            /* subroot node */
            newRREQ.pid = getNodeID();
            newRREQ.S = getNodeID();
            newRREQ.len ++;
            send(newRREQ);
        }
        state = learn;
    }
}
else if(state == learn && currentRound == newRREQ.rid)
    /* path learning */
    if(cost(newRREQ)<currentCost(newRREQ.pid))
        received_RREQs.add(newRREQ);
}
Timer_Handler()
{
    /* path selection */
    if(nodeIsMaster())
        selected_RREQs = received_RREQs;
    else
    {
        i = random(received_RREQs.size())
        selected_RREQs = received_RREQs[i];
    }
    for (i=0; i<selected_RREQs.size(); i++)
        send(selected_RREQs[i]);
    state = init;
}
```

4 Numerical Results

We have evaluated the performance of our protocol HDMRP by conducting intensive simulations with different protocols and different scenarios. We report the performance results in what follows.

4.1 Simulation Model

We have implemented HDMRP protocol using *networkx*, a graph library for python. To estimate the impact of introducing powerful master nodes to the network, we have considered three variants with different master nodes densities : 5%, 10% and 20%. In addition to our protocol, we have implemented BRP and SMRP protocols which we previously introduced. BRP and SMRP are the only node disjoint protocols which, like HDMRP, use only one message per node in the multipath construction.

The number of constructed paths between a node and the Sink is an important metric to estimate the fault tolerance of our solution. The higher the number of constructed paths, the better the fault tolerance is. For each protocol and each scenario, we measured the average number of discovered paths per node (Y axis) depending on the number of nodes in the network (X axis). For each network size, we have generated 100 random network topologies and calculated the average number of discovered paths over this 100 iterations. The network topology generation parameters are as follows: (i) the network area size is 50 X 50 units, (ii) the Sink node is located in the center of the network, (iii) each node has a communication range of 2 units, (iv) nodes and master nodes are randomly distributed over the network area.

4.2 Number of Paths

Fig. 3 illustrates the average number of discovered paths between a node and the Sink depending on the network size. We notice that in the three variants, our protocol computes more paths than BRP and SMRP. HDMRP constructs up to 110% more paths than BRP and SMRP when 20% of nodes are masters. Furthermore, this performance increase can achieve 35% when only 5% of nodes are master nodes. These results match our expectations: our protocol discovers more paths than BRP and SMRP when master nodes are present. Therefore, it achieves a higher fault tolerance. The difference between HDMRP and the other protocols becomes visible only when the network size is greater than 400 nodes. This can be explained by the limited node density in the case of small networks. Indeed, the probability that a master node receives and relays multiple RREQs is small when the network density is low.

4.3 Master Nodes Deployment Strategy

Master nodes placement strategy is an important factor in HDMRP since the protocol performance is significantly affected by the master nodes position.

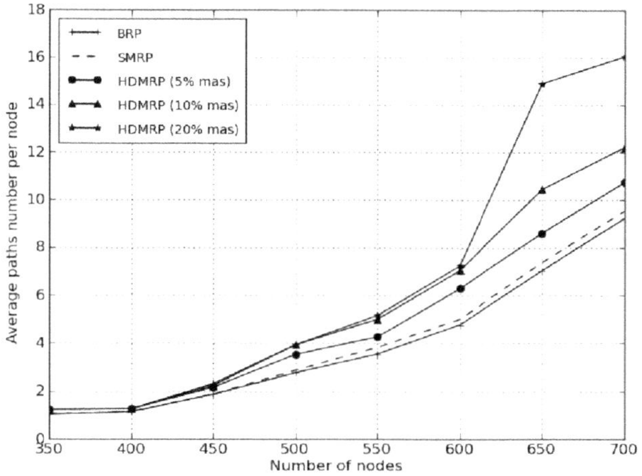

Fig. 3. Average paths number when master nodes are randomly distributed over the network area

To demonstrate this property, we have conducted simulations to compare between three master nodes deployment strategies : Randomized master nodes distribution, alternated level distribution and concentrated area coverage. In randomized distribution, master nodes are deployed uniformly through the whole network area. In alternated level distribution, we place master nodes in even levels and standard nodes in odd levels. This ensures that a node near the sink will be between at least two master nodes. In concentrated area coverage, all master nodes are randomly deployed around the Sink.

Fig. 4 and Fig. 6 illustrates the average number of discovered paths between a node and the Sink depending on network size and the master nodes deployment strategy. Fig. 4 compares between the number of constructed paths in the case of alternated level distribution and the case of randomized distribution. We notice that the number of constructed paths in the alternated level distribution is the highest in every scenario. The results match our expectations and confirm that master nodes deployment strategy has an impact on HDMRP performance. Fig. 5 illustrates the improvement rate of the alternated level deployment strategy comparatively to randomized distribution. We remark an improvement between 5% and 20% in the number of constructed paths when using alternated level deployment strategy. This high improvement rate can be explained by the continuous intensive propagation of tags due to the presence of master nodes every two hops.

To go further, we have compared between the number of constructed paths in the case of concentrated area coverage and the case of randomized distribution. The improvement of the concentrated deployment can be clearly seen in Fig. 6 where the number of discovered paths is always the highest in concentrated deployment case. An interesting observation is that the number of constructed paths when only 5% of nodes are master nodes in the concentrated deployment is almost the same when 20% of nodes are master nodes in randomized distribution. Consequently, with a proper master nodes distribution, HDMRP constructs up to 110% more paths than BRP and SMRP when only 5% of nodes are master nodes. Fig. 7 illustrates the improvement rates of the concentrated area coverage strategy comparatively to randomized distribution. We remark an improvement up to 34% in the number of constructed path when using concentrated area coverage deployment strategy. This can be explained by the fact that masters far from the Sink and situated in the network border have no impact on the number of discovered paths. Indeed, concentrating these master nodes near the sink enable our protocol to achieve very high performance.

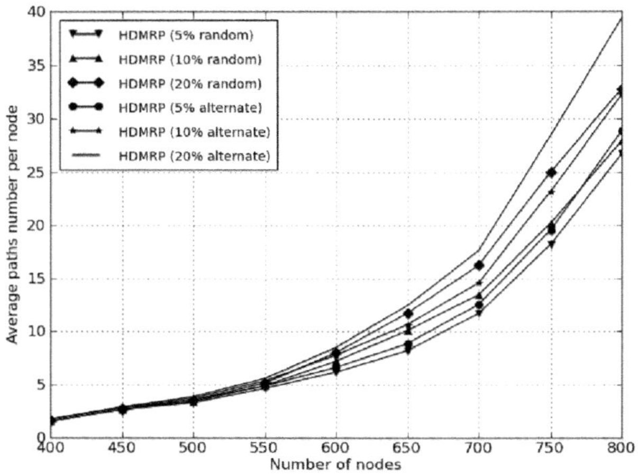

Fig. 4. Average paths number when master nodes are in alternate levels

4.4 Fault Tolerance

To evaluate the fault tolerance of HDMRP, we have considered the impact of node failure on the network connectivity. We have first run BRP, SMRP and HDMRP protocols to construct multiple paths on different network topologies. Then we have varied the node failure rate from 5% to 30% and computed the number of nodes that still have a functional path to the sink (connected nodes). For each protocol, we have executed 100 simulations to estimate the average connected nodes rate (Y axis) depending on node failure rate (X axis). Simulation parameters are as follows: (i) failures are randomly distributed on network nodes (ii) master nodes have the same failure probability as nodes (iii) HDMRP is used with 10% of randomly deployed master nodes.

Fig. 8 illustrates the average number of connected nodes depending on nodes failure rate. We notice that compared to BRP and SMRP, HDMRP is more fault tolerant. For instance, when 10% of nodes fail, 88% of nodes stay connected in HDMRP and only 70% of nodes stay connected in BRP and SMRP. This

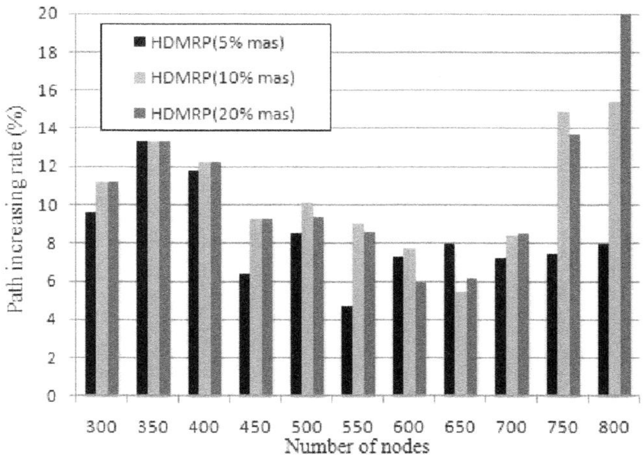

Fig. 5. Increase rate in the number of discovered path in alternate levels deployment strategy

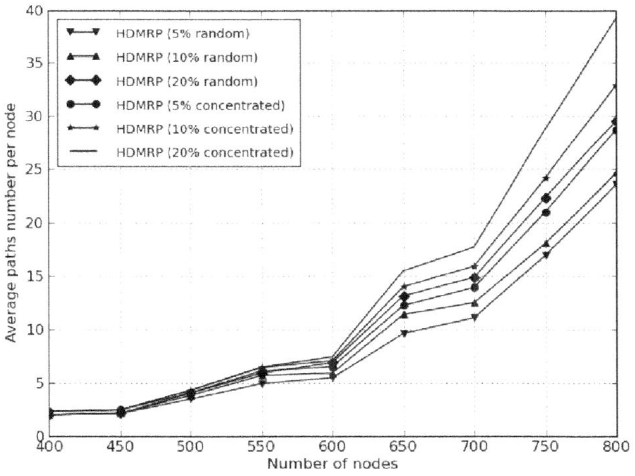

Fig. 6. Average paths number when master nodes are deployed around the sink

is because HDMRP constructs more energy-node-disjoint paths between each node and the sink. Consequently, each node has several alternative paths to the sink in presence of failures. Fig.8 shows that connected nodes ratio in HDMRP is above 80% for node failure rate of up to 17%. Furthermore, the number of connected nodes is almost the same as the number of working nodes for node failure rate of up to 20%. This demonstrates that node failures have a mitigated impact on working nodes and the network. Consequently, HDMRP shows very high robustness and fault tolerance even when the quarter of nodes fail.

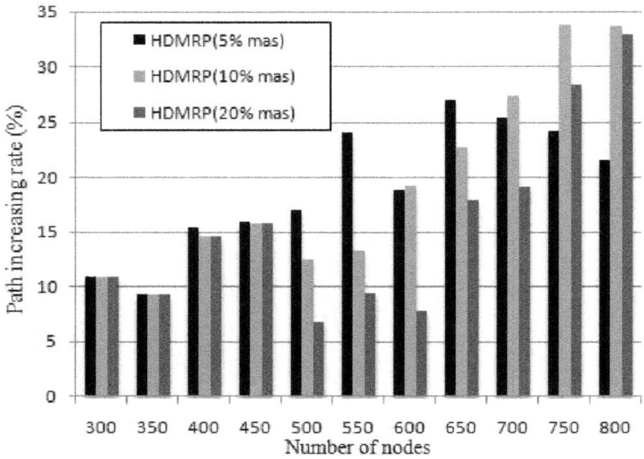

Fig. 7. Increase rate in the number of discovered path in concentrated deployment strategy

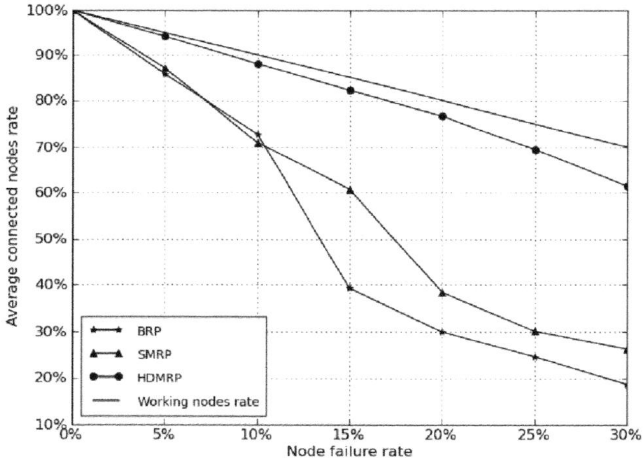

Fig. 8. Average connected nodes rate when network size is 600

5 Conclusion

In this paper, we deal with the challenging issue of providing fault tolerance in wireless sensor networks. We first define a new multipath paradigm for heterogeneous wireless sensor networks. Then, we propose a new fault tolerant multipath routing protocol which discovers an important number of energy node disjoint paths with the slightest overhead of one message per node. We conduct intensive simulations to evaluate our protocol with different scenarios, master nodes densities and deployment strategies. Simulation results are very encouraging and demonstrate the high performance and fault tolerance of our proposed protocol.

References

1. Akyildiz, I.F., Su, W., Sankarasubramaniam, Y., Cayirci, E.: Wireless sensor networks: a survey. Computer Networks 38, 393–422 (2002)
2. Arora, A., Ramnath, R., Ertin, E., Sinha, P., Bapat, S., Naik, V., Kulathumani, V., Zhang, H., Cao, H., Sridharan, M., Kumar, S., Seddon, N., Anderson, C., Herman, T., Trivedi, N., Zhang, C., Nesterenko, M., Shah, R., Kulkarni, S., Aramugam, M., Wang, L., Gouda, M., ri Choi, Y., Culler, D., Dutta, P., Sharp, C., Tolle, G., Grimmer, M., Ferriera, B., Parker, K.: ExScal: Elements of an Extreme Scale Wireless Sensor Network. In: 11th IEEE International Conference on Embedded and Real-Time Computing Systems and Applications, pp. 102–108. IEEE, Los Alamitos (2005)
3. Chen, Y., Nasser, N.: Enabling QoS Multipath Routing Protocol for Wireless Sensor Networks. In: The IEEE International Conference on Communications, pp. 2421–2425. IEEE, Beijing (2008)
4. Felemban, E., Lee, C.-G., Ekici, E.: MMSPEED: Multipath Multi-SPEED Protocol for QoS Guarantee of Reliability and Timeliness in Wireless Sensor Networks. IEEE Transaction on Mobile Computing 5, 738–754 (2006)
5. Ganesan, D., Govindan, R., Shenker, S., Estrin, D.: Highly-resilient, energy-efficient multipath routing in wireless sensor networks. Mobile Computing and Communications Review 5, 11–25 (2001)
6. Gnawali, O., Jang, K.-Y., Paek, J., Vieira, M., Govindan, R., Greenstein, B., Joki, A., Estrin, D., Kohler, E.: The Tenet architecture for tiered sensor networks. In: 4th International Conference on Embedded Networked Sensor Systems, pp. 153–166. ACM, Colorado (2006)
7. Hou, R., Shi, H.: A localized algorithm for finding disjoint paths in wireless sensor networks. IEEE Communications Letters 10, 807–809 (2006)
8. Jacquet, P., Muhlethaler, P., Clausen, T., Laouiti, A., Qayyum, A., Viennot, L.: Optimized link state routing protocol for ad hoc networks. In: The IEEE International Multi Topic Conference, pp. 62–68. IEEE, Los Alamitos (2001)
9. Rok Jung, S., Hoon Lee, J., Hee Roh, B.: An Optimized Node-Disjoint Multipath Routing Protocol for Multimedia Data Transmission over Wireless Sensor Networks. In: The IEEE International Symposium on Parallel and Distributed Processing with Applications, pp. 958–963. IEEE, Sydney (2008)
10. Kim, M., Euihoon, J., Bang, Y.-C., Hwang, S., Shin, C., Jin, G.-J., Kim, B.: An Energy-Aware Multipath Routing Algorithm in Wireless Sensor Networks. IEICE Transaction on Information and Systems 91, 2419–2427 (2008)
11. Lee, J.-J., Krishnamachari, B., Kuo, C.-C.J.: Impact of heterogeneous deployment on lifetime sensing coverage in sensor networks. In: The Annual IEEE Communications Society Conference on Sensor and Ad Hoc Communications and Networks, pp. 367–376. IEEE, Los Alamitos (2004)
12. Li, S., Neelisetti, R.K., Liu, C., Lim, A.: Efficient Multi-Path Protocol for Wireless Sensor Networks. International Journal of Wireless and Mobile Networks 2, 110–130 (2010)
13. Li, S., Wu, Z.: Node-Disjoint Parallel Multi-Path Routing in Wireless Sensor Networks. In: The Second International Conference on Embedded Software and Systems, pp. 432–437. IEEE, Los Alamitos (2005)

14. Lou, W., Kwon, Y.: H-SPREAD: A Hybrid Multipath Scheme for Secure and Reliable Data Collection in Wireless Sensor Networks. IEEE Transaction on Vehicular Technology 55, 1320–1330 (2006)
15. Lu, Y.M., Wong, V.W.S.: An energy-efficient multipath routing protocol for wireless sensor networks. International Journal of Communication Systems 20, 747–766 (2007)
16. Nasser, N., Chen, Y.: SEEM: Secure and energy-efficient multipath routing protocol for wireless sensor networks. Computer Communications 30, 2401–2412 (2007)
17. Ouadjaout, A., Challal, Y., Lasla, N., Bagaa, M.: SEIF: Secure and Efficient Intrusion-Fault Tolerant Routing Protocol for Wireless Sensor Networks. In: The International Conference on Availability, Reliability and Security, pp. 503–508. IEEE, Barcelona (2008)
18. Sohrabi, K., Gao, J., Ailawadhi, V., Pottie, G.J.: Protocols for self-organization of a wireless sensor network. IEEE Personal Communications 7, 16–27 (2000)
19. Stavrou, E., Pitsillides, A.: A survey on secure multipath routing protocols in WSNs. Computer Networks 54, 2215–2238 (2010)
20. Szewczyk, R., Mainwaring, A., Polastre, J., Anderson, J., Culler, D.: An analysis of a large scale habitat monitoring application. In: The 2nd International Conference on Embedded Networked Sensor Systems, pp. 214–226. ACM, Baltimore (2004)
21. Welsh, E., Fish, W., Frantz, J.P.: GNOMES: a testbed for low power heterogeneous wireless sensor networks. In: The International Symposium on Circuits and Systems, pp. 836–839. IEEE, Los Alamitos (2003)
22. Yahya, B., Ben-Othman, J.: An energy efficient and QoS aware multipath routing protocol for wireless sensor networks. In: The IEEE Conference on Local Computer Networks, pp. 93–100. IEEE, Zurich (2009)
23. Yang, Y., Zhong, C., Sun, Y., Yang, J.: Network coding based reliable disjoint and braided multipath routing for sensor networks. Journal of Network and Computer Applications 33, 422–432 (2010)
24. Yarvis, M., Kushafnagar, N., Singh, H., Rangarajan, A., Lin, Y., Singh, S.: Exploiting heterogeneity in sensor networks. In: IEEE INFOCOM, pp. 878–890. IEEE, Los Alamitos (2005)
25. Ye, F., Zhong, G., Lu, S., Zhang, L.: GRAdient broadcast: a robust data delivery protocol for large scale sensor networks. Wireless Networks 11, 285–298 (2005)

Downlink Power Allocation for Stored Variable-Bit-Rate Videos

Yingsong Huang, Shiwen Mao, and Yihan Li

Dept. Electrical & Computer Engineering, Auburn University, Auburn, AL
{yzh0002,yli}@auburn.edu, smao@ieee.org

Abstract. In this paper, we study the problem of power allocation for streaming multiple variable-bit-rate (VBR) videos in the downlink of a cellular network. We consider a deterministic model for VBR video traffic and finite playout buffer at the mobile users. The objective is to derive the optimal downlink power allocation for the VBR video sessions, such that the video data can be delivered in a timely fashion without causing playout buffer overflow and underflow. The formulated problem is a nonlinear nonconvex optimization problem. We analyze the convexity conditions for the formulated problem and propose a two-step greedy approach to solve the problem. We also develop a distributed algorithm based on the dual decomposition technique. The performance of the proposed algorithms are validated with simulations using VBR video traces under realistic scenarios.

Keywords: convex optimization, distributed algorithm, downlink power control, video streaming, variable bit rate video.

1 Introduction

According to a recent study by Cisco, data traffic over wireless networks is expected to increase by a factor of 66 times by 2013. Much of the increase in future wireless data traffic will be video related, as driven by the compelling need for ubiquitous access to multimedia content for mobile users. Such drastic increase in video traffic will significantly stress the capacity of existing and future wireless networks. While new wireless network architectures and technologies are being developed to meet this "grand challenge" [1,2,3,4], it is also important to revisit existing wireless networks, to maximize their potential in carrying real-time video data.

In this paper, we consider the problem of streaming multiple videos in the downlink of a cellular network. The system is interference limited: the capacity of a specific mobile user depends on the Signal to Interference-plus-Noise Ratio (SINR) at the user, which is a function of the power allocation for all the mobile users. Therefore, effective downlink power control is necessary for such a wireless video system to minimize the intra-cell interference for concurrent video sessions.

We consider the challenging problem of streaming concurrent variable-bit-rate (VBR) videos in the cellular network. This is motivated by the fact that

X. Zhang and D. Qiao (Eds.): QShine 2010, LNICST 74, pp. 483–498, 2011.

VBR video offers stable and superior quality over constant bit rate (CBR) videos. Furthermore, many stored video content are VBR. It is important to support such stored VBR videos over existing wireless networks without the need for transcoding. The main challenge in supporting VBR video stems from its high rate variability and complex autocorrelation structure, making it hard for network control and may cause frequent playout buffer underflow or overflow. In this paper, we adopt a deterministic traffic model for stored VBR video, which jointly considers frame size, frame rate, and playout buffer size [5, 6, 7]. Unlike prior work, we exploit effective downlink power control to adjust the downlink capacities based on prior knowledge of frame sizes and palyout buffer occupancies.

Specifically, we present a downlink power control framework for streaming multiple VBR videos in a cellular network. With the deterministic VBR video traffic model, we formulate an optimization problem that jointly considers donwlink power control, intra-cell interference, VBR video traffic characteristics, playout buffer underflow and overflow constraints, and base station (BS) peak power constraint. The objective is to maximize the total throughput, which can achieve high playout buffer utilization. As a result, playout buffer underflow or overflow events can be minimized. We analyze the convex/concave regions of the formulated problem and develop a two-step downlink power allocation algorithm for solving the problem. We also develop a distributed algorithm based on the dual decomposition technique from convex optimization, in order to reduce the control and computation overhead at the BS. We evaluate the performance of the proposed distributed algorithm with simulations using VBR video traces. Our simulation results verify the accuracy of the analysis and demonstrate the efficacy of the proposed algorithms.

The remainder of this paper is organized as follows. The system model is presented in Section 2. We develop a two-step algorithm to solve the power allocation problem in Section 3, and a distributed algorithm based on dual decomposition in Section 4. Simulation results are presented in Section 5 and related work are discussed in Section 6. Section 7 concludes this paper.

2 Network and Video System Model

We consider the downlink of a cellular network. In the cell, a BS streams multiple VBR videos simultaneously to mobile users in the cell, which share the downlink bandwidth. We assume the last-hop wireless link is the bottleneck, while the wired segment of a session path is reliable with sufficient bandwidth. Thus the corresponding video data is always available at the BS before the scheduled transmission time.

VBR video traffic exhibits both strong asymptotic self-similarity and short-range correlation. A stochastic model capturing the complex auto-correlation structure often requires a large number of parameters, and is thus hard to be incorporated for scheduling real-time video data. To this end, we adopt a *deterministic model* that considers frame sizes and playout buffers [6]. Let $D_n(t)$

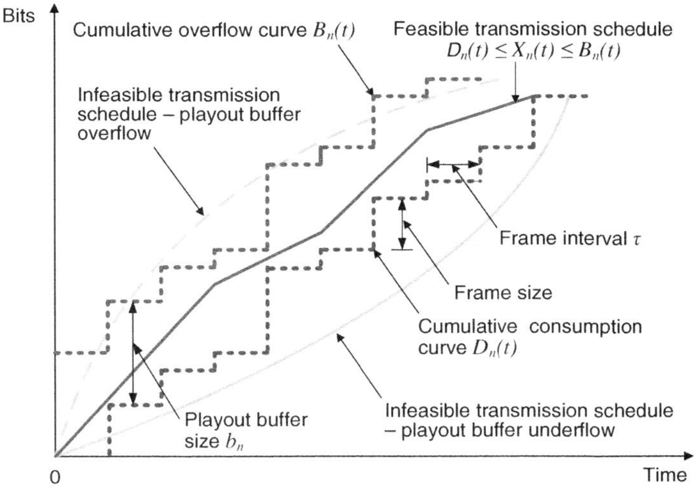

Fig. 1. Transmission schedules for VBR video session n

denote the *cumulative consumption curve* of the n-th mobile user, representing the total amount of bits consumed by the decoder at time t. The cumulative consumption curve is determined by video characteristics such as frame sizes and frame rates. Assume mobile user n has a playout buffer of size b_n bits and its video has T_n frames. We can derive a *cumulative overflow curve* for the user as

$$B_n(t) = \min\{D_n(t-1) + b_n, D_n(T_n)\},\ 0 \le t \le T_n. \tag{1}$$

$B_n(t)$ is the the maximum number of bits that can be received at time t without overflowing user n's playout buffer. Finally we define *cumulative transmission curve* $X_n(t)$ as the cumulative amount of transmitted bits to user n at time t. To simplify notation, we assume the video sessions have identical frame rate and the frame intervals are synchronized. A time slot t is equal to the t-th frame interval, denoted as τ.

The three curves for user n are illustrated in Fig. 1. A feasible transmission schedule will produce a cumulative transmission curve $X_n(t)$ that lies within $D_n(t)$ and $B_n(t)$, i.e., causing neither underflow nor overflow at the playout buffer. In practice, $D_n(t)$'s are known for stored videos and can be delivered to the BS during session setup phase, and $B_n(t)$'s can be derived as in (1).

We consider N subscribers in the cell and let $\mathcal{U}$ denote the set of users. In each time slot t, the BS transmits to each user n with power $P_n(t)$ and the *power allocation* is $\boldsymbol{P}(t) = [P_1(t), \cdots, P_n(t)]^T$. We also consider a *maximum transmit power* constraint $\bar{P}$, i.e., $\sum_{n \in \mathcal{U}} P_n(t) \le \bar{P}$, for all t. When the power allocation $\boldsymbol{P}(t)$ is determined, the SINR at user n can be written as [8, 9]

$$\gamma_n(\boldsymbol{P}(t)) = \frac{L_n G_n P_n(t)}{\beta \sum_{k \neq n} G_n P_k(t) + \eta_n}, \tag{2}$$

where P_n is the power allocated to user n, G_n is the path gain between the BS and user n, η_n is the noise power at user n, L_n is a constant for user n (e.g., processing gain), and β denotes the orthogonality factor, with $0 \le \beta \le 1$. In this paper, we consider the case $\beta = 1$, where the SINR of a user not only depends on its own power allocation but also the power allocations of other users.

We assume slow-fading channels such that the path gains do not change within each time slot [8]. The downlink capacity $C_n(t)$ depends on the SINR at user n, the channel bandwidth B_w, and the transceiver design, such as modulation and channel coding. Without loss of generality, we use the upper bound as predicted by Shannon's Theorem:

$$C_n(\boldsymbol{P}(t)) = B_w \log\left(1 + \gamma_n(\boldsymbol{P}(t))\right). \tag{3}$$

In time slot t, $C_n(t)\tau$ bits of video data will be delivered to user n. The cumulative transmission curve $X_n(t)$ is

$$X_n(0) = 0;\quad X_n(t) = X_n(t-1) + C_n(t)\tau. \tag{4}$$

For a feasible power allocation, the cumulative transmission curves should satisfy

$$D_n(t) \le X_n(t) \le B_n(t), \text{ for all } n, t, \tag{5}$$

i.e., without causing playout buffer underflow or overflow.

From (3)$\sim$(5), the lower and upper limit on the feasible SINR at user n can be derived as

$$\begin{cases} \gamma_n^{min}(t) = \max\left\{\exp\left\{\frac{\max\{0, D_n(t) - X_n(t-1)\}}{B_w\tau}\right\}, \gamma_n^{th}\right\} \\ \gamma_n^{max}(t) = \exp\left\{\frac{B_n(t) - X_n(t-1)}{B_w\tau}\right\}, \end{cases} \tag{6}$$

where γ_n^{th} is the minimum SINR requirement imposed by the transceiver design. $\gamma_n^{min}(t)$ is the SINR that the just empties the buffer at the end of time slot t, without causing underflow; $\gamma_n^{max}(t)$ is the SINR that just fills up the buffer at the end of time slot t, without causing overflow.

Generally, feasible power allocation $\boldsymbol{P}(t)$ is not unique for a given set of VBR video sessions. Among the set of feasible solutions, a schedule that transmits more data is more desirable since it provides more flexibility for optimizing future power allocations. We formulate the problem of optimal downlink power control for VBR videos, termed Problem **A**, as

$$\textbf{(A) maximize } \textstyle\sum_{n\in\mathcal{U}} \log(1 + \gamma_n(t)) \tag{7}$$

subject to:

$$\gamma_n(t) = \frac{L_n G_n P_n(t)}{\sum_{k\neq n} G_n P_k(t) + \eta_n}, \text{ for all } n \tag{8}$$

$$\gamma_n^{min}(t) \le \gamma_n(t) \le \gamma_n^{max}(t), \text{ for all } n \tag{9}$$

$$\textstyle\sum_{n\in\mathcal{U}} P_n \le \bar{P}. \tag{10}$$

In Problem **A**, the objective is to achieve the maximum buffer uitilization at the users, under playout buffer underflow and overflow constraints and BS maximum transmit power constraints. This is a nonlinear nonconvex problem, to which traditional convex optimization techniques cannot directly apply. Due to the large variability of VBR traffic, the SINRs may assume values ranging from very low to very high, to avoid playout buffer underflow and overflow. Thus the existing high SINR approximation [10] and low SINR approximation [11] techniques cannot be directly applied.

3 Two-Step Downlink Power Allocation

In Problem **A**, we consider an interference-limited system, where the capacity of downlink n depends on the power allocations for all the users. In the following, we first derive conditions for the optimal solution, and then present a two-step power allocation algorithm for solving Problem **A**.

Lemma 1. *If there exists a feasible power allocation $\boldsymbol{P}(t)$ that achieves $\gamma_n^{max}(t)$ for all n, the solution is optimal.*

Proof. If the feasible power allocation $\boldsymbol{P}(t)$ achieves $\gamma_n^{max}(t)$ for all n, then all the user buffers are full at the end of the time slot, according to (6). The objective value (7) cannot be further improved without causing buffer overflow. Thus the solution is optimal.

Lemma 2. *If the upper limit $\gamma_n^{max}(t)$ cannot be achieved for every user n, then the optimal power allocation $\boldsymbol{P}(t)$ satisfies $\sum_{n\in\mathcal{U}} P_n(t) = \bar{P}$.*

Proof. Consider a feasible power allocation $\boldsymbol{P}'(t) = [P_1'(t), P_2'(t), \cdots, P_N'(t)]^T$ and $\sum_{n\in\mathcal{U}} P_n'(t) < \bar{P}$. We can construct another feasible power allocation $\boldsymbol{P}''(t) = [P_1''(t), P_2''(t), \cdots, P_N''(t)]^T$, such that $P_n''(t) = \kappa \cdot P_n'(t)$, for all n, and $\kappa \cdot \sum_{n\in\mathcal{U}} P_n'(t) = \sum_{n\in\mathcal{U}} P_n''(t) \leq \bar{P}$, where $\kappa > 1$. For the SINR at user n, we have

$$
\begin{aligned}
\gamma_n(\boldsymbol{P}''(t)) &= \frac{L_n G_n P_n''(t)}{\sum_{k\neq n} G_n P_k''(t) + \eta_n} \\
&= \frac{\kappa L_n G_n P_n'(t)}{\sum_{k\neq n} \kappa G_n P_k'(t) + \eta_n} \\
&> \frac{\kappa L_n G_n P_n'(t)}{\sum_{k\neq n} \kappa G_n P_k'(t) + \kappa\eta_n} \\
&= \gamma_n(\boldsymbol{P}'(t)).
\end{aligned}
$$

It follows that $\sum_{n\in\mathcal{U}} \log(1+\gamma_n(\boldsymbol{P}''(t))) > \sum_{n\in\mathcal{U}} \log(1+\gamma_n(\boldsymbol{P}'(t)))$, since $\log(1+x)$ is an increasing function of x.

Choosing $\kappa = \bar{P}/\sum_{n\in\mathcal{U}} P_n'(t)$, we can construct a feasible solution $\boldsymbol{P}'''(t) = \kappa \cdot \boldsymbol{P}'(t)$, such that $\sum_{n\in\mathcal{U}} P_n'''(t) = \bar{P}$. Then we have $\gamma_n(\boldsymbol{P}'''(t)) > \gamma_n(\boldsymbol{P}'(t))$

and $\sum_{n\in\mathcal{U}} \log(1+\gamma_n(\boldsymbol{P}'''(t))) > \sum_{n\in\mathcal{U}} \log(1+\gamma_n(\boldsymbol{P}'(t)))$. That is, any feasible solution with $\sum_{n\in\mathcal{U}} P'_n(t) < \bar{P}$ will be dominated by feasible solutions with $\sum_{n\in\mathcal{U}} P'''_n(t) = \bar{P}$. We conclude that the optimal solution $\boldsymbol{P}(t)$ must satisfy $\sum_{n\in\mathcal{U}} P_n(t) = \bar{P}$.

We have the following result for the optimal solution of Problem **A**, which directly follows Lemmas 1 and 2.

Theorem 1. *A solution to Problem* **A** *is optimal if (i) it achieves the maximum SINR* $\gamma_n^{max}(t)$ *for all* n*; or (ii) its total transmit power is* $\bar{P}$.

Theorem 1 implies that we can examine the SINR (or buffer) constraints and the peak power constraint separately. In the rest of this section, we present a two-step power allocation algorithm for solving Problem **A**. We first examine Problem **A** under condition (i) in Theorem 1, to obtain Problem **B** as

$$\textbf{(B)}\ \ \gamma_n^{max}(t) = \frac{L_n G_n P_n(t)}{\sum_{k\neq n} G_n P_k(t) + \eta_n}, \text{ for all } n, \tag{11}$$

subject to:

$$\textstyle\sum_{n\in\mathcal{U}} P_n \leq \bar{P}. \tag{12}$$

In Problem **B**, (11) is a system of linear equations of power allocation $\boldsymbol{P}(t)$. Rearranging the terms, we can rewrite (11) in the matrix form as:

$$(\mathbf{I} - \mathbf{F})\,\boldsymbol{P}(t) = \boldsymbol{u}, \ \text{ for } \boldsymbol{P}(t) \succ \mathbf{0}, \tag{13}$$

where $\mathbf{I}$ is the *identity matrix*, $\mathbf{F}$ is a $N \times N$ matrix with

$$F_{nm} = \begin{cases} 0, & \text{if } n = m \\ \gamma_n^{max}/L_n, & \text{otherwise,} \end{cases} \tag{14}$$

and $\boldsymbol{u} = [\eta_1\gamma_1^{max}/L_nG_1, \eta_2\gamma_2^{max}/L_nG_2, \cdots, \eta_N\gamma_N^{max}/L_nG_N]^T$.

Since all the variables are nonnegative, $\mathbf{F}$ is a non-negative matrix. According to the Perron-Frobenius Theorem, we have the following equivalent statements [12]:

Fact 1. *The following statements are equivalent: (i) there exits a feasible power allocation satisfying (13); (ii) the spectrum radius of* $\mathbf{F}$ *is less than 1; (iii) the reciprocal matrix* $(\mathbf{I}-\mathbf{F})^{-1} = \sum_{k=0}^{\infty} (\mathbf{F})^k$ *exists and is component-wise positive.*

Based on Theorem 1 and Fact 1, we derive the *first step* of the two-step power allocation algorithm, as given in Table 1. If Problem **B** is solvable, the Step I algorithm in Table 1 produces the optimal solution for Problem **A** according to Theorem 1. Otherwise, we derive Problem **C** by applying Lemma 2, as

$$\textbf{(C) maximize } \textstyle\sum_{n\in\mathcal{U}} \log(1+\gamma_n(t)) \tag{15}$$

subject to:

$$\gamma_n(t) = \frac{L_n P_n(t)}{\bar{P} - P_n(t) + A_n}, \text{ for all } n \tag{16}$$

$$P_n^{min}(t) \leq P_n(t) \leq P_n^{max}(t), \text{for all } n \tag{17}$$

$$\textstyle\sum_{n\in\mathcal{U}} P_n(t) = \bar{P}, \tag{18}$$

Table 1. Two-Step Power Allocation Algorithm: Step I

1 BS obtains b_n, D_n, and B_n, and computes γ_n^{max} for all user n;
2 BS tests the existence of feasible solutions using (13);
3 IF (13) is solvable, compute its solution $\boldsymbol{P}(t)$;
ELSE, go to Step II of the algorithm, as given in Table 2;
4 IF $\sum_{n \in \mathcal{U}} P_n(t) \le \bar{P}$, stop with the optimal solution $\boldsymbol{P}(t)$;
ELSE go to Step II of the algorithm, as given in Table 2;

where $A_n = \eta_n / G_n$ is the ratio of noise power and channel gain, representing the quality of the user n downlink channel. $P_n^{min}(t)$ and $P_n^{max}(t)$ are solved from (9) and (16), as

$$\begin{cases} P_n^{min}(t) = \gamma_n^{min}(\bar{P} + A_n)/(L_n + \gamma_n^{min}) \\ P_n^{max}(t) = \gamma_n^{max}(\bar{P} + A_n)/(L_n + \gamma_n^{max}). \end{cases} \tag{19}$$

Since the total transmit power is $\bar{P}$, the objective value in (15) and the SINR in (16) for each user only depends on its own power. Note that all the constraints are now linear. To solve Problem **C**, we examine the objective function to see if it is convex. We omit time index t in the following for brevity.

Lemma 3. *The capacity of each user n, C_n, has one* inflection point *P_n^*: when $P_n < P_n^*$, C_n is in concave; when $P_n > P_n^*$, C_n is convex.*

Proof. Taking the first and second derivatives of the objective function (15) with respect to P_n, we have

$$\frac{\partial C_n(P_n)}{\partial P_n} = \frac{L_n(\bar{P} + A_n)}{(\bar{P} - P_n + A_n)[\bar{P} + (L_n - 1)P_n + A_n]} \tag{20}$$

$$\frac{\partial^2 C_n(P_n)}{\partial {P_n}^2} = \frac{-L_n[(L_n - 2)(\bar{P} + A_n) + 2(1 - L_n)P_n](\bar{P} + A_n)}{[(\bar{P} - P_n + A_n)^2 + L_n P_n(\bar{P} - P_n + A_n)]^2}. \tag{21}$$

Since $P_n \le \bar{P}$ and $A_n > 0$, both the first and second derivatives exist. Letting $\frac{\partial^2 C_n(P_n)}{\partial {P_n}^2} = 0$, we derive the unique inflection point

$$P_n^* = \frac{L_n - 2}{2(L_n - 1)}(\bar{P} + A_n). \tag{22}$$

When $P_n < P_n^*$, it can be shown that $\frac{\partial^2 C_n(P_n)}{\partial {P_n}^2} < 0$; when $P_n > P_n^*$, it can be shown that $\frac{\partial^2 C_n(P_n)}{\partial {P_n}^2} > 0$.

The normalized capacities for a two-user system is plotted in Fig. 2, with the inflection points marked. It can been observed that the curves are concave on the left hand side of the inflection points and convex on the right hand side of the inflection points. The processing gain is usually large for practical systems (e.g., $L_n = 128$ in IS-95 CDMA). We assume $L_n \gg 1$ in the following analysis.

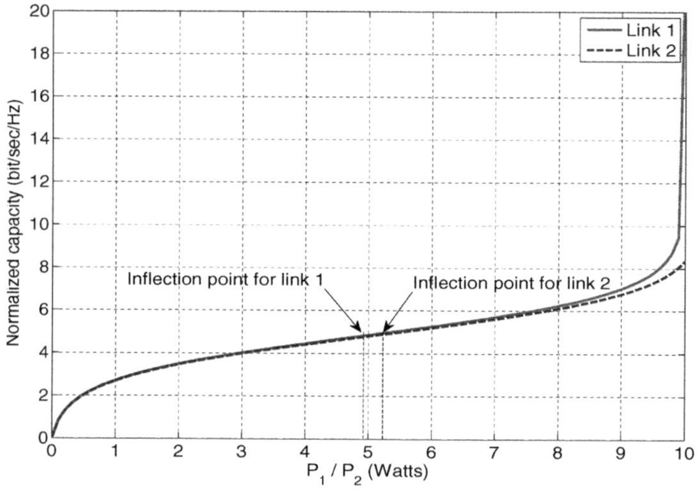

Fig. 2. Normalized capacity curves and inflection points for a two-user system, where link 1 has better quality than link 2, i.e. $A_1 < A_2$

Theorem 2. *For Problem* **C***, there can be at most two links operating in the convex region if* $L_n \geq (4\bar{P} + 6A_n)/(\bar{P} + 3A_n)$.

Proof. The reflection point is $P_n^* = \frac{L_n - 2}{2(L_n - 1)}(\bar{P} + A_n)$. As $L_n \to \infty$, we have $P_n^* = 0.5(\bar{P} + A_n)$. Only one link can operate in the convex region due to constraint (18). Since $\frac{\partial P_n^*}{\partial L_n} > 0$, P_n^* is an increasing function of L_n. When $1 \ll L_n < \infty$, we have $P_n^* < 0.5 \cdot (\bar{P} + A_n)$. Letting $3P_n^* = \bar{P}$, we have $L_n = (4\bar{P} + 6A_n)/(\bar{P} + 3A_n)$.

For a clean channel where $A_n \approx 0$, $L_n \geq 4$ will guarantee at most two links operating in the convex region. The following results are on the impact of channel quality $A_n = \eta_n / G_n$.

Theorem 3. *For a given* L_n*, the inflection point* P_n^* *is an increasing function of* A_n*. For two links* i *and* j *with the same transmit power* P*, if* $A_i < A_j$*, we have* $C_i(P, A_i) > C_j(P, A_j)$ *and* $\frac{\partial C_i(P_i, A_i)}{\partial P_i}|_{P_i = P} > \frac{\partial C_j(P_j, A_j)}{\partial P_j}|_{P_j = P} > 0$.

Proof. The first part can be easily shown by the first derivative of P_n^* with respect to A_n, which is $\frac{\partial P_n^*}{\partial A_n} = \frac{L_n - 2}{2(L_n - 1)} > 0$, for $L_n > 2$. The second part can be easily shown by evaluating (15), (16), and (20).

Theorem 3 shows that, for two links in the convex region with the same initial power P, allocating more power to the link with better quality can achieve larger objective value than alternative ways of splitting the power between the two links (i.e., achieving the multi-user diversity gain). Based on the above analysis, we develop the *second step* of the power allocation algorithm for solving Problem

Table 2. Two-Step Power Allocation Algorithm: Step II

Initialization:
1 BS obtains b_n, D_n, and B_n for all user n;
2 BS computes γ_n^{max}, γ_n^{min}, and P_n^*, for all n;
3 BS computes the minimum required sum power $\bar{P}_{min} = \sum_{n \in \mathcal{U}} P_n^{min}$ and gap $\Delta_P = \bar{P} - \bar{P}_{min}$;
4 IF $\bar{P}_{min} > \bar{P}$, remove links from $\mathcal{U}$, according to descending order of A_n, until $\bar{P}_{min} \leq \bar{P}$;
5 Compute $R_n = \frac{C_n(\min\{P_n^{max}, P_n^{min}+\Delta_P\}) - C_n(P_n^{min})}{\min\{P_n^{max}, P_n^{min}+\Delta_P\} - P_n^{min}}$, for all $P_n^{max} > P_n^*$;
Phase 1:
6 Select all the users satisfying $P_n^{min} < P_n^*$ as a set $\mathcal{U}' \subseteq \mathcal{U}$;
7 Solve Problem **C** under constraints $P_n^{min} \leq P_n \leq \min(P_n^{max}, P_n^*)$ and $\sum_{n \in \mathcal{U}'} P_n \leq \bar{P}' = \bar{P} - \sum_{n \in \bar{\mathcal{U}}'} P_n^{min}$, where $\bar{\mathcal{U}}'$ is the complementary set of $\mathcal{U}'$, and obtain solution $\boldsymbol{P}_1$;
8 Calculate R_n by updating P_n^{min} to the solution in Line 7 and assign the remaining power to the nodes in set $\mathcal{U}$, in descending order of R_n;
9 Obtain the Phase 1 solution, $\boldsymbol{P}_{p_1}$, and objective value f_{p_1};
Phase 2:
10 Select the link with the maximum R_n, and assign all the available power $\bar{P} - \bar{P}_{min}$ to the link, until either all the power is assigned or the link attains power P_n^{max};
11 IF there is still power to allocate, THEN select all the nodes in set $\mathcal{U} \backslash n$ and repeat Lines $5 \sim 8$;
12 Obtain the Phase 2 solution, $\boldsymbol{P}_{p_2}$, and objective value f_{p_2};
Phase 3:
13 Select the first 2 links with the largest R_n's, and assign all the availble power $\bar{P} - \bar{P}_{min}$ to the links, until all the power is assigned or the links attains power P_n^{max}, and repeat Line 11;
14 Obtain the Phase 3 solution, $\boldsymbol{P}_{p_3}$, and objective value f_{p_3};
Decision:
15 Choose the largest objective value among f_{p_1}, f_{p_2} and f_{p_3}, and stop with the corresponding power assignment;

C, as given in Table 2. In Table 2, Lines $3 \sim 4$ tests the feasibility of the power allocation. If the sum of the total minimum required power is larger than the BS peak power, there is no feasible power allocation and there will be buffer underflow. In this case, we select users with "good" channels for transmission and suspend the users with "bad" channels.

The Step II algorithm checks the three possible solution scenarios for Problem **C** depending on the network status and video parameters:

- All links operate in the convex region;
- One link operates in the convex region and the remaining links operate in the concave region
- Two links operate in the convex region and the remaining links operate in the concave region.

Each of the three phases in Table 2 considers the optimality condition for one of the three scenarios. In particular, Phase 1 first optimizes the power allocation in the concave region and then allocates the remaining power to the links that could be moved to the convex region. Phase 2 allocates as much power as possible to the link with the best quality, which could work in the convex region. Phase 3 attempts to move the second best link to the convex region if the total power constraint is not violated. Usually when L_n and n are large, Phase 3 will rarely occur due to the peak power constraint.

In Table 2, Line 7 presents a convex optimization component, for which several effective solution techniques can be applied. In the following section, we describe a distributed algorithm for Line 7 based on dual decomposition.

4 Distributed Algorithm

As discussed in Section 3, the core of the Step II algorithm is to solve Problem **C** in the concave region (see Fig. 2). In this section, we present a distributed algorithm for this purpose, where the users are involved in power allocation to reduce the control and computation overhead on the BS. In the concave region, we have Problem **D** as

$$\textbf{(D) maximize } \sum_{n \in \mathcal{U}} \log(1 + \gamma_n(t)) \tag{23}$$

subject to:

$$\gamma_n(t) = \frac{L_n P_n(t)}{\bar{P} - P_n(t) + A_n}, \text{ for all } n \tag{24}$$

$$P_n^{min}(t) \leq P_n(t) \leq \min\{P_n^{max}, P_n^*\}, \text{for all } n \tag{25}$$

$$\sum_{n \in \mathcal{U}} P_n(t) \leq P_{tot}, \tag{26}$$

where $P_{tot} \leq \bar{P}$ is the total power budget for the links in the concave region. For brevity, we define $P_n^{th} = \min\{P_n^{max}, P_n^*\}$ and drop the time slot index t in the following analysis.

Introducing non-negative Lagrange multipliers λ_n, μ_n, and ν for constraints (25) and (26), respectively, we obtain the Lagrange function as

$$\begin{aligned} &\mathcal{L}(\boldsymbol{P}, \boldsymbol{\lambda}, \boldsymbol{\mu}, \nu) \\ &= \sum_{n \in \mathcal{U}} \left[\log\left(1 + \frac{L_n P_n}{\bar{P} - P_n + A_n}\right) + \lambda_n (P_n - P_n^{min})\right] + \\ &\quad \sum_{n \in \mathcal{U}} \left[\mu_n (P_n^{th} - P_n)\right] + \nu \left(P_{tot} - \sum_{n \in \mathcal{U}} P_n\right) \\ &= \sum_{n \in \mathcal{U}} \left[\mathcal{L}_n(P_n, \lambda_n, \mu_n, \nu) + (\mu_n P_n^{th} - \lambda_n P_n^{min})\right] + \nu P_{tot}, \end{aligned} \tag{27}$$

where

$$\mathcal{L}_n(P_n, \lambda_n, \mu_n, \nu) = \log\left(1 + \frac{L_n P_n}{\bar{P} - P_n + A_n}\right) + (\lambda_n - \mu_n - \nu) P_n. \tag{28}$$

Since $\mathcal{L}_n$ only depends on user n's own parameters, we have the dual decomposition for each user n. For given Lagrange multipliers (or, prices) $\hat{\lambda_n}$, $\hat{\mu_n}$, and $\hat{\nu}$, we have the following subproblem for each user n.

$$\hat{P}_n(\hat{\lambda_n}, \hat{\mu_n}, \hat{\nu}) = \underset{P_n^{min} \leq P_n \leq P_n^{th}}{\arg\max} \ \mathcal{L}_n(P_n, \hat{\lambda_n}, \hat{\mu_n}, \hat{\nu}), \text{ for all } n. \tag{29}$$

Subproblem (29) has a unique optimal solution due to the strict concavity of $\mathcal{L}_n$. We use the gradient method [13] to solve (29), where user n iteratively updates its power P_n as:

$$\begin{aligned} &P_n(l+1) \\ &= [P_n(l) + \theta(l)\nabla_n \mathcal{L}_n(P_n)]^* \\ &= \left[P_n(l) + \theta(l) \frac{L_n(\bar{P} + A_n)}{(\bar{P} - P_n + A_n)(\bar{P} + (L_n - 1)P_n + A_n)} + \theta(l)(\lambda_n - \mu_n - \nu)\right]^*, \end{aligned} \tag{30}$$

where $[\cdot]^*$ denotes the projection onto the range of $[P_n^{min}, P_n^{th}]$. The update stepsize $\theta(l)$ varies in each step l and is determine by the Armijo Rule [13]. Due to the strict concavity of $\mathcal{L}_n$, the series $\{P_n(1), P_n(2), \cdots\}$ will converge to the optimal solution $\hat{P}_n$ as $l \to \infty$.

For a given optimal solution for problem (29), $\hat{\boldsymbol{P}} = [\hat{P}_1, \cdots, \hat{P}_N]^T$, the master dual problem is as follows:

$$\textbf{minimize } \ \mathcal{L}(\hat{\boldsymbol{P}}, \boldsymbol{\lambda}, \boldsymbol{\mu}, \nu) \tag{31}$$

$$\textbf{subject to: } \ \lambda_n, \mu_n, \nu \geq 0, \text{ for all } n. \tag{32}$$

Since the objective function (31) is differentiable, we also apply the gradient method to solve the master dual problem [13], where the Lagrange multipliers are iteratively updated as

$$\begin{cases} \lambda_n(l+1) = [\lambda_n(l) - \alpha_\lambda(l) \cdot \frac{\partial \mathcal{L}(\boldsymbol{\lambda}, \boldsymbol{\mu}, \nu)}{\partial \lambda_n}]^+, \text{ for all } n \\ \mu_n(l+1) = [\mu_n(l) - \alpha_\mu(l) \cdot \frac{\partial \mathcal{L}(\boldsymbol{\lambda}, \boldsymbol{\mu}, \nu)}{\partial \mu_n}]^+, \text{ for all } n \\ \nu(l+1) = [\nu(l) - \alpha_\nu(l) \cdot \frac{\partial \mathcal{L}(\boldsymbol{\lambda}, \boldsymbol{\mu}, \nu)}{\partial \nu}]^+, \end{cases} \tag{33}$$

where $[\cdot]^+$ denotes the projection onto the nonnegative axis. The update stepsizes are also determined by the Armijo Rule [13]. As the dual variables $\boldsymbol{\lambda}(l), \boldsymbol{\mu}(l), \nu(l)$ converge to their stable values as $l \to \infty$, the primal variables $\hat{\boldsymbol{P}}$ will also converge to the optimal solution [14].

The distributed algorithm is given in Table 3, where the above procedures are repeated iteratively. The BS first broadcasts Lagrange multipliers to the users. Each user updates its requested power as in (30), using local information P_n^{min}, P_n^{max}, P_n^*, A_n, L_n, and BS peak power $\bar{P}$. Each user then sends its requested power back to the BS, and the BS will updates the Lagrange multipliers as in (33). And so forth, until the optimal solution is obtained.

Table 3. Distributed Power Control Algorithm

1 BS sets $l = 0$ and prices $\lambda_n(l), \mu_n(l), \nu(l)$ equal to some nonnegative initial values for all n;
2 BS broadcasts the prices to the selected users;
3 Each user locally solves problem (29) as in (30) to obtain its requested power;
4 Each user sends its requested power to the BS;
5 BS updates prices $\lambda_n(l), \mu_n(l), \nu(l)$ as in (33) and broadcasts new prices $\lambda_n(l+1), \mu_n(l+1), \nu(l+1)$ for all n;
6 Set $l = l + 1$ and go to Step 3, until the solution converges;

5 Simulation Results

We evaluate the proposed algorithms with simulations, using a cellular network with 20 users. The downlink bandwidth is 1 MHz. The path gain averages are $G_n = d_n^{-4}$, where d_n is the physical distance from the BS to user n. The downlink channel is modeled as log-normal fading with zero mean and variance 8 dB [8]. The processing gains are set to $L_n = 128$ for all n. The distance d_n is uniformly distributed in [100m, 1000m]. The device temperature is $T_0 = 290$ Kelvin and the equivalent noise bandwidth is $B_w = 1MHz$. The BS peak power constraints is set to $\bar{P} = 10$ Watts. We use three VBR movies traces, *Star Wars*, *NBC News*, and *Tokyo Olympics*, from the Video Trace Lib [15]. Each playout buffer is set to 1.5 times of the largest frame size in the requested VBR video.

In the simulations, the proposed power allocation algorithm is executed at the beginning of each time slot. In Fig. 3, we plot the cumulative consumption, overflow and transmission curves for *NBC News* transmitted to user 2. The top sub-figure is the overview of 10,000 frames. We also plot the curves from frame 2,620 to 2,640 in the bottom sub-figure. We observe that the cumulative transmission curve $X(t)$ is very close to the cumulative overflow curve $B(t)$, indicating that the algorithm always aim to maximize the transmission rate as allowed by the buffer and power constraints. The playout buffers are almost fully utilized most of the time. There is no playout buffer overflow and underflow for the entire range of 10,000 frames. Among the *NBC News* frames, frame 2,625 is the largest frame. We let seven out of the 20 links playout this largest frame simultaneously at time slot 2,625 in the simulation. There is no buffer underflow under such heavy load.

In Fig. 4, we plot the power allocation and price updates for all the 20 links in one of the 10,000 time slots. The power and prices converges in around 70 steps. The converged power vector is $\hat{\boldsymbol{P}} = [0.0022, 1.396, 0.0356, 0.0024, 1.396, 0.0351, 0.0016, 1.396, 0.0356, 0.0026, 1.396, 0.0356, 0.0023, 1.396, 0.0356, 0.0018, 1.396, 0.0356, 0.0034, 1.394]$ Watts.

Finally, we compare the proposed algorithm with a diversity-aware power allocation scheme, where the BS allocates power according to channel quality. With this scheme, the best channel n will be assigned power to achieve its maximum required power $P_n^{max}(t)$. Then the second best channel will be allocated power

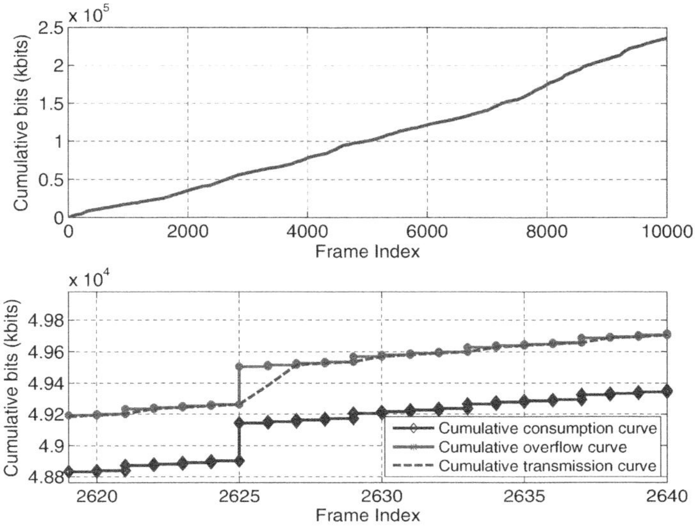

Fig. 3. Transmission schedule for video *NBC News* to user 2

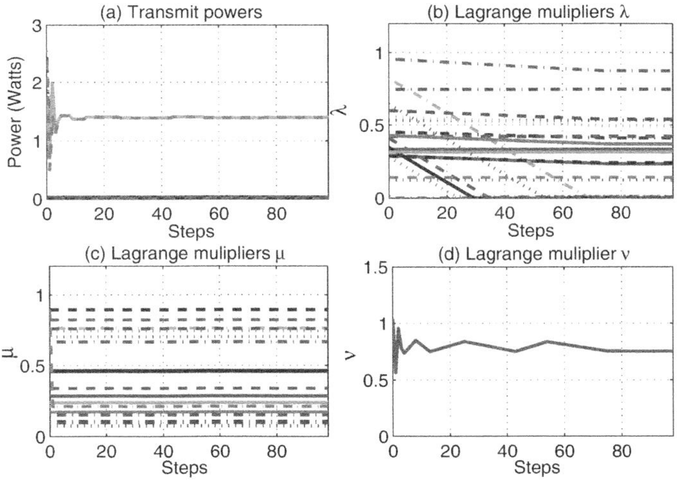

Fig. 4. Convergence of power allocation and Lagrange multipliers

until its maximum required power is achieved, and so forth until all of $\bar{P}$ is allocated.

We simulate 50 users with the same network and video settings. We compare the algorithms by their average playout buffer utilization. In Fig. 5, we plot the average buffer utilization from frame 2,000 to 2,999. It can be seen that the

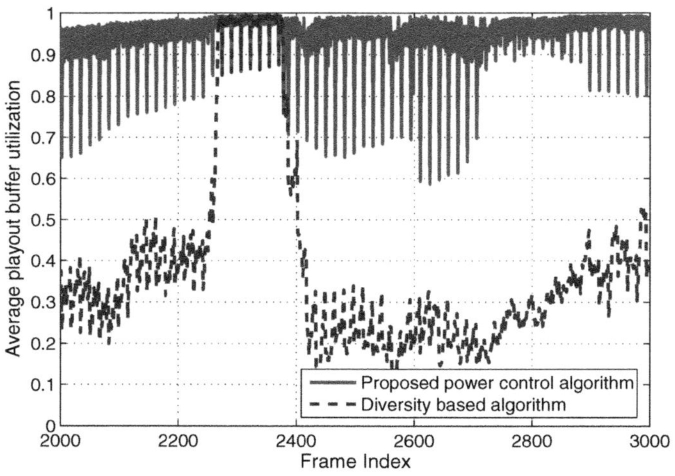

Fig. 5. Average playout buffer utilization

proposed algorithm consistently achieves high buffer utilization, ranging from 60% to 100%. The diversity scheme achieves buffer utilization around 30% except for frames from 2,250 to 2,400. Such considerably higher buffer utilization translates to better video quality: there is no buffer overflow or underflow for proposed algorithm, while there is buffer underflow in 17% of the playout frames for the diversity scheme.

6 Related Work

There have been several papers on VBR video over wired network. Due to long-range-dependent (LRD) VBR video traffic, the piecewise-constant-rate transmission and transport (PCRTT) method was used to optimize certain objectives while preserving continuous video playout. In [5], Liew and Chan developed bandwidth allocation schemes for multiple VBR videos to share a CBR channel. In [6], Salehi et al. presented an optimal algorithm for smoothing VBR video over a CBR link. Feng and Liu [7] introduced a critical bandwidth allocation algorithm to reduce the number of bandwidth variations and to maximize receiver buffer utilization. Due to the fundamental difference between wireless and wired links, these techniques cannot be directly applied to the problem of VBR video over wireless networks.

The downlink power allocation problem was studied in [8, 9], aiming to obtain the power allocation that maximizes a properly defined system utility. A distributed algorithm based on dynamic pricing and partial cooperation was proposed. Deng, Webera, and Ahrens [16] studied the achievable maximum sum rate of multi-user interference channels. These papers provide the theoretical foundation and effective algorithms for utility maximization of downlink traffic, but the techniques used cannot be directly applied for VBR video over wireless networks with buffer and delay constraints.

In [17, 18], the authors studied the problem of one VBR stream over a given time-varying wireless channel. In [17], it was shown that the separation between a delay jitter buffer and a decoder buffer is in general suboptimal, and several critical system parameters were derived. In [18], the authors studied the frequency of jitters under both network and video system constraint and provided a framework for quantifying the trade-offs among several system parameters. In this paper, we jointly consider power control in wireless networks, playout buffers, and video frame information, and address the more challenging problem of streaming multiple VBR videos, and present a cross-layer optimization approach that does not depend on any specific channel or video traffic models.

7 Conclusion

We developed a downlink power allocation model for streaming multiple VBR videos in a cellular network. The model considers interactions among downlink power control, channel interference, playout buffers, and VBR video traffic characteristics. The formulated problem aims at maximizing the total transmission rate under both peak power and playout buffer overflow/underflow constraints. We presented a two-step approach for solving the problem and a distributed algorithm based on the dual decomposition technique. Our simulation studies validated the efficacy of the proposed algorithms.

Acknowledgment. This work is supported in part by the US National Science Foundation (NSF) under Grants CNS-0953513, ECCS-0802113, IIP-1032002, and IIP-1127952, and through the NSF Wireless Internet Center for Advanced Technology at Auburn University. Any opinions, findings, and conclusions or recommendations expressed in this material are those of the author(s) and do not necessarily reflect the views of the Foundation.

References

1. Zhao, Y., Mao, S., Neel, J., Reed, J.H.: Performance evaluation of cognitive radios: metrics, utility functions, and methodologies. Proc. IEEE 97(4), 642–659 (2009)
2. Su, H., Zhang, X.: Cross-layer based opportunistic MAC protocols for QoS provisionings over cognitive radio mobile wireless networks. IEEE J. Sel. Areas Commun. 26(1), 118–129 (2008)
3. Alay, O., Liu, P., Guo, Z., Wang, L., Wang, Y., Erkip, E., Panwar, S.: Cooperative layered video multicast using randomized distributed space time codes. In: Proc. IEEE INFOCOM Workshops 2009, Rio de Jeneiro, Brazil, pp. 1–6 (2009)
4. Mao, S., Lin, S., Wang, Y., Panwar, S., Li, Y.: Multipath video transport over wireless ad hoc networks. IEEE Wireless Commun. 12(4), 42–49 (2005)
5. Liew, S., Chan, H.: Lossless aggregation: a scheme for transmitting multiple stored VBR video streams over a shared communications channel without loss of image quality. IEEE J. Sel. Areas Commun. 15(6), 1181–1189 (1997)

6. Salehi, J., Zhang, Z.-L., Kurose, J., Towsley, D.: Supporting stored video: reducing rate variability and end-to-end resource requirements through optimal smoothing. IEEE/ACM Trans. Networking 6(4), 397–410 (1998)
7. Feng, W.-C., Liu, M.: Critical bandwidth allocation techniques for stored video delivery across best-effort networks. In: Proc. IEEE ICDCS 2000, Taipei, Taiwan, pp. 56–63 (2000)
8. Lee, J., Mazumdar, R., Shroff, N.: Downlink power allocation for multi-class wireless systems. IEEE/ACM Trans. Networking 13(4), 854–867 (2005)
9. Lee, J., Kwon, J.: Utility-based power allocation for multiclass wireless systems. IEEE Trans. Vehic Tech. 58(7), 3813–3819 (2009)
10. Chiang, M.: Balancing transport and physical layers in wireless multihop networks: jointly optimal congestion control and power control. IEEE J. Sel. Areas Commun. 23(1), 104–116 (2005)
11. Gjendemsj, A., Gesbert, D., Oien, G., Kiani, S.: Binary power control for sum rate maximization over multiple interfering links. IEEE Trans. Wireless Commun. 7(8), 3164–3173 (2008)
12. Mitra, D.: An asynchronous distributed algorithm for power control in cellular radio system. In: Proc. WINLAB Workshop on 3G Wireless Information Networks, New Brunswick, NJ, pp. 249–257 (1993)
13. Bertsekas, D.: Nonlinear Programming. Athena Scientific, Belmont (1995)
14. Palomar, D.P., Chiang, M.: A tutorial on decomposition methods for network utility maximization. IEEE J. Sel. Areas Commun. 24(8), 1439–1451 (2006)
15. Reisslein, M.: Video trace library. Arizona State University, `http://trace.eas.asu.edu/`
16. Deng, S., Webera, T., Ahrens, A.: Capacity optimizing power allocation in interference channels. AEU Int. J. Electronics Commun. 63(2), 139–147 (2009)
17. Stockhammer, T., Jenkac, H., Kuhn, G.: Streaming video over variable bit-rate wireless channels. IEEE Trans. Multimedia 6(2), 268–277 (2004)
18. Liang, G., Liang, B.: Balancing interruption frequency and buffering penalties in VBR video streaming. In: Proc. IEEE INFOCOM 2007, Anchorage, AK, pp. 1406–1414 (2007)

Cross-Layer Multipath Transmission of Elastic and Streaming Traffic over Heterogeneous Wireless Networks and Its Performance Analysis*

Wei Song

Faculty of Computer Science
University of New Brunswick
Fredericton, NB, Canada
wsong@unb.ca

Abstract. Next-generation wireless networks are expected to be heterogeneous by integrating multiple broadband access technologies. Popular wireless devices become equipped with various network interfaces. Multihoming support can be enabled to allow for multiple simultaneous associations with heterogeneous networks. Taking advantage of multihoming capability, we investigate multipath transmission of elastic and streaming traffic over heterogeneous wireless networks. A new flow splitting and multipath transmission scheme is proposed by exploiting cross-layer information of bulk data and video streams. The heavy-tailedness of elastic flow size is mitigated by balancing the traffic load, while the large bandwidth requirement of streaming flows is satisfied by aggregating fractional bandwidth available in multiple networks. Based on the leaky bucket algorithm and a nearly decomposable Markov process, the flow-level performance is evaluated in terms of flow blocking probabilities and data loss probability. Also, we analyze the fine-granular packet-level performance with respect to transfer delay using a batch arrival queueing model. Numerical results demonstrate the performance gain of the multipath transmission scheme.

Keywords: Heterogeneous wireless networking, multipath streaming, multihoming support, elastic and streaming services.

1 Introduction

Nowadays, a variety of broadband access options are offered by the proliferating wireless networks such as the third-generation (3G) cellular networks, IEEE 802.11 wireless local area networks (WLAN), and IEEE 802.15 wireless personal area networks (WPAN). Aiming at different application environments, these

* This research was supported by a Discover Grant from Natural Sciences and Engineering Research Council (NSERC) of Canada and a Start-up Grant from New Brunswick Innovation Foundation (NBIF).

X. Zhang and D. Qiao (Eds.): QShine 2010, LNICST 74, pp. 499–514, 2011.

wireless networks will coexist and provide an integrated heterogeneous access to mobile users. To exploit the complementary strengths of heterogeneous networks, multi-radio and smart wireless devices will be the mainstream for future wireless networks. To coordinate the heterogeneous wireless access and multi-radio devices, network selection is one of the major issues that are researched intensively in the literature. Depending on a decision algorithm, an incoming traffic flow is automatically assigned to a best available network. An ongoing traffic flow can also be dynamically migrated between different networks via vertical handoff by monitoring available bandwidth, channel status, and topology change. Many centralized and distributed network selection algorithms are proposed to provide the *always-the-best* connectivity and enhance quality-of-service (QoS) [14,9,16]. Basically, access selection aims to sharing the heterogeneous network resources along the time scale at the flow level.

Taking one step further, we can exploit the multihoming and multi-streaming support and apply flow splitting and data stripping for multipath transmission over heterogeneous wireless links. Multihoming enables a wireless device to maintain multiple simultaneous associations with more than one attachment point. Multi-streaming allows data to be partitioned into multiple streams and delivered independently to the application at the receiver. Multi-streaming can prevent head-of-line blocking problem that occurs in the regular transport control protocol (TCP). The stream control transmission protocol (SCTP) is one of the well-known transport-layer specifications that offer multihoming and multi-streaming capabilities. The original SCTP is designed to improve throughput and reliability by exploiting multiple paths. It is further extended to support host mobility and even interworking functionality [7]. If multihoming and multi-streaming capabilities are enabled for multi-radio devices, a traffic flow can be split into multiple streams and delivered simultaneously over multiple network interfaces. As such, the access selection problem is addressed from a different perspective.

In this paper, we study bandwidth sharing for integrated heterogeneous wireless networks by means of multipath transmission and flow splitting. Taking advantage of multihoming and multi-streaming capabilities of multi-radio devices, a novel cross-layer multipath transmission scheme is proposed to make use of any fractional bandwidth available in the integrated networks. Traffic flows are broadly classified into two primary categories: elastic and streaming [2]. Here, we consider two representative services, i.e., bulk data transfer and video streaming. The proposed scheme exploits the application-layer knowledge such as data file size and coding and compression structure to enable simultaneous data stripping across multiple networks. Video frames arriving in a burst are dispatched to heterogeneous links of different capacities based on frame types and frame grouping.

Also, we evaluate the flow-level performance of the multipath transmission scheme in terms of flow blocking probabilities and transfer delay. The analysis is based on a straightforward two-dimensional Markov chain, which is nearly decomposable in the quasi-stationary regime. At the packet level, a batch arrival queueing process [6] is employed to analyze the statistics of packet transfer delay. As demonstrated in the numerical results, multipath transmission significantly outperforms a randomized access selection with respect to bandwidth aggregation.

The remainder of this paper is organized as follows. In Section 2, we introduce the network model and traffic model for this study. A novel multipath transmission is proposed and analyzed in Section 3 for elastic and streaming traffic, exploiting application-layer information and multihoming capability of multi-radio devices. Numerical results are presented in Section 4, followed by conclusions in Section 5.

2 System Model

In this study, we consider a heterogeneous wireless infrastructure integrating multiple access options, as shown in Fig. 1. Multi-radio devices are considered to be multihoming capable with an extended SCTP protocol. We focus on a scenario that three network interfaces can be activated simultaneously for power efficiency and interference mitigation. The study can be extended to relax this limitation. Moreover, we assume that a middleware deployed at both the application server and user device deals with the splitting and merging of traffic flows across available networks. The middleware employs both the application-layer and network-layer information in load splitting. Specifically, the available bandwidth over each associated network can be estimated by exploiting the multiple interfaces of user devices.

Nowadays, the wireless network capacity is boosted significantly with advanced techniques such as adaptive modulation and coding, power control, and multiple input and multiple output (MIMO). A variety of new services become proliferating, such as video streaming, Web browsing, file transfer, and conversational video. These applications are broadly classified into two categories, i.e.,

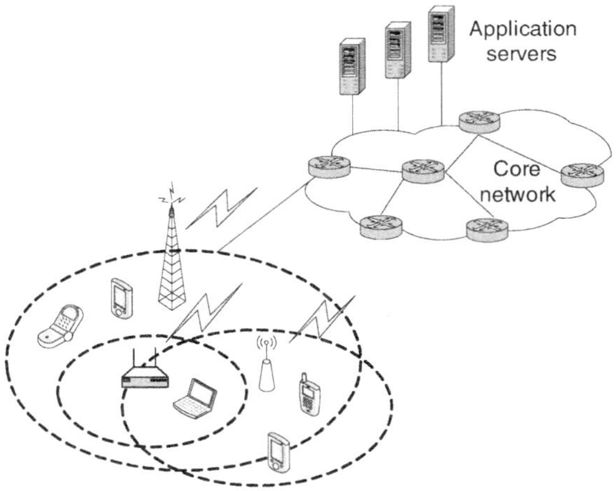

Fig. 1. System model of integrated heterogeneous wireless networks

elastic and streaming services. Elastic services such as file transfer and Web browsing can accept varying rates depending on available bandwidth. The main performance metric is the transfer delay or equivalently the throughput of traffic flows. On the other hand, streaming services require to preserve an intrinsic rate. For instance, video streaming plays back the video content at the receiver during the delivery. A stringent rate requirement needs to be satisfied to prevent data overflow and depletion at the playout buffer. In this study, we focus on flow splitting and multipath transmission for two representative elastic and streaming services: bulk data transfer and video streaming.

2.1 Elastic Traffic

An elastic flow is characterized by the size of bulk data to be transferred. It is widely observed that the size of Web documents and data files is heavy-tailed and presents high variability. Here, we model the elastic flow size L_e with a Weibull distribution as in [10], whose probability density function (PDF) is given by

$$f_e(x) = \frac{\alpha_e}{\beta_e}\left(\frac{x}{\beta_e}\right)^{\alpha_e - 1} e^{-(x/\beta_e)^{\alpha_e}}$$
$$0 < \alpha_e \leq 1, \qquad \beta_e > 0, \qquad x > 0 \tag{1}$$

where α_e is the shape parameter and β_e is the scale parameter. The exponential distribution is a special case of the Weibull distribution with $\alpha_e = 1$, while the Weibull distribution is heavy-tailed when $0 < \alpha_e < 1$. The smaller the value of α_e, the heavier the tail of a Weibull distribution.

2.2 Streaming Traffic

The essential traffic characteristics of streaming flows are the flow duration and variable rate. In this study, we take video streaming flows as an example. It is known that video traffic is inherently long-range dependent and highly correlated due to compression coding. In the third-generation (3G) cellular networks, H.264 Advanced Video Coding (AVC) is recommended for high-quality video [1]. To remove temporal redundancy, intracoded (I) frames are interleaved with predicted (P) frames and bidirectionally coded (B) frames. I frames are compressed versions of raw frames independent of other frames, whereas P frames only refer preceding I/P frames and B frames can refer both preceding and succeeding frames. A sequence of video frames from a given I frame up to the next I frame comprise a group of pictures (GoP). Because P and B frames are encoded with reference to preceding and/or succeeding I/P frames, the transmission traffic follows the batch arrivals shown in Fig. 2. Here, the GoP follows a structure of size 16 such as "$I_0P_4B_1B_2B_3P_8B_5B_6B_7P_{12}B_9B_{10}B_{11}I_{16}B_{13}B_{14}B_{15}\ldots$" In contrast, video frames are decoded and displayed at the receiver in a reorganized order.

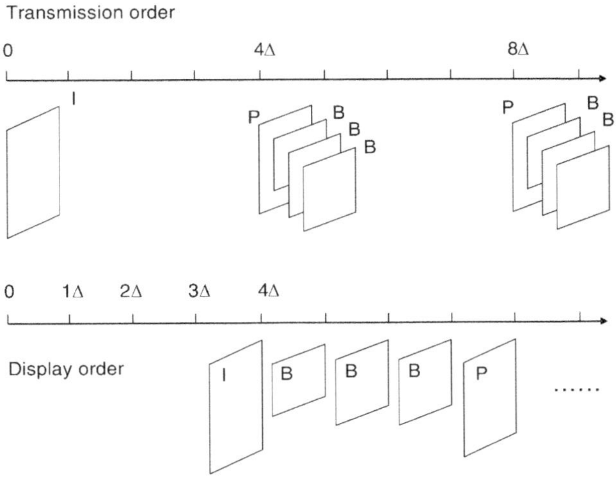

Fig. 2. Transmission and display orders of video frames

In the literature, there has been extensive work modeling the varying rate and frame size of video traffic [11,5]. To capture both frame size variation and autocorrelation, we extend the Markov-modulated Gamma-based model (MMG) proposed in [11] for performance analysis. For a video stream consisting of a GOP sequence, video clips are grouped into a small number of shot classes depending on the GOP size. The size of I, P, and B frames in a class is modeled by an axis-shifted Gamma distribution. In the original MMG model, the GOP size boundaries for classification are geometrically separated. As observed in [8], the size of video frames based on H.264 exhibits heavy-tailed property. That is, extremely large frames exist with a non-negligible probability. To discern differences for large-size video clips in classification, we propose to use the following sigmoid function to determine the class boundaries:

$$x_i = \frac{1}{1 + e^{-\alpha_s \cdot (i - \beta_s)}}, \qquad i = 1, 2, ..., K + 1 \tag{2}$$

where K is the number of video classes. As this sigmoid function takes values within $(0, 1)$, we map the frames size in the range of $\left[X_{min}, X_{max}\right]$ such that

$$x_1 = \frac{X_{min}}{\theta_s \cdot X_{max}}, \qquad x_{K+1} = \frac{1}{\theta_s} \tag{3}$$

where θ_s $(0 < \theta_s < 1)$ is a scale factor.

Take a video trace coded with single layer H.264/AVC as an example [13]. We choose the video sequence of *Tokyo Olympics* with a common intermediate format (CIF) resolution (352×288), a fixed frame rate at 30 frames/s, a GoP size of 16 with 3 B frames between I/P key pictures, and a quantization step size indexed at 24. Fig. 3 shows the size boundaries to classify GOPs according to a

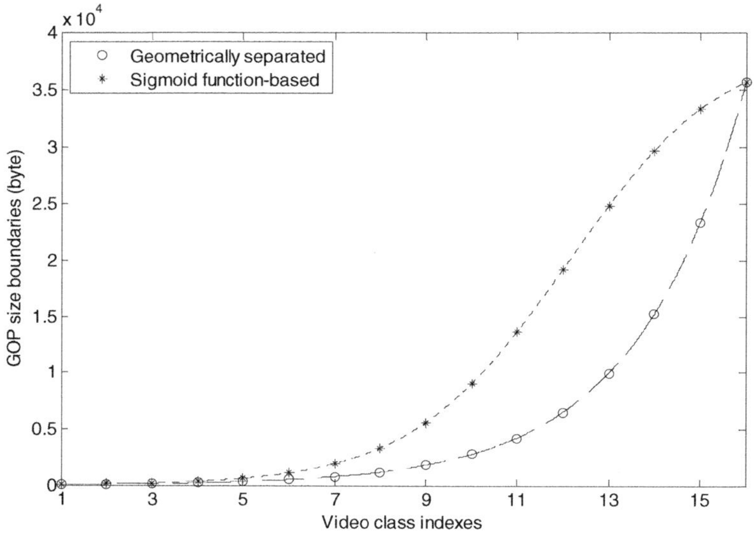

Fig. 3. Size boundaries to classify video clips

geometric function or a sigmoid function, respectively. As seen, the S-shaped size boundaries can also differentiate differences when the GOP size is very large.

According to the size boundaries, video clips are classified into K shot classes and the transition probability p_{ij} from class i to class j can be estimated from normalized relative frequency of transitions:

$$p_{ij} = f_{ij}/f_i \tag{4}$$

where f_{ij} is the total number of transitions from state i to j and f_i is the total number of transitions out of state i. The resulting matrix of transition probabilities, denoted by P, can be translated into a corresponding infinitesimal generating matrix in a continuous-time domain, denoted by M, as follows

$$M = g(P - I) \tag{5}$$

where g is the rate of GOPs and I is the identity matrix.

In the original MMG model [11], the size of each type of frames in a class is modeled with an axis-shifted Gamma distribution. To render tractable analysis, we decouple the flow-level and packet-level traffic models. For each video class (state), the video traffic is considered as a fluid flow of rate γ_i $(i = 1, 2, ..., K)$. At the finer packet-level, video frames generated in a burst are fragmented into packets for transmission. We use a batch arrival process to model the packet-level traffic.

3 Multipath Transmission Exploiting Multihoming Capability

As shown in Fig. 1, a multihomed wireless device is assumed to capable of accessing three different networks simultaneously. In this section, we will discuss how to split an elastic or streaming flow across multiple access networks. A novel cross-layer multipath transmission scheme is proposed to aggregate available bandwidth and enhance QoS provisioning.

3.1 Flow Splitting for Elastic Traffic

For an elastic flow defined in (1), due to the heavy-tailedness and high variability, it is difficult to determine how much bandwidth is required exactly to satisfy a transfer delay bound. To mitigate the variability and bandwidth requirement, we can split an elastic flow across three available access depending on a probability q_i. The purpose is to decompose a heavy-tailed flow into the superposition of a number of light-tailed substreams. A recursive algorithm is proposed in [3] to fit heavy-tailed distributions with a finite mixture of exponentials, i.e., a hyper-exponential distribution. Here, we approximate the heavy-tailed Weibull distribution in (1) with a three-stage hyper-exponential distribution, in which each exponential component corresponds to the size of a substream flow over an available access. That is,

$$\hat{f}_e(x) = \sum_{i=1}^{3} q_i e^{-\varphi_i x}$$
$$\sum_{i=1}^{3} q_i = 1, \quad 0 < \varphi_i < 1, \quad x > 0. \tag{6}$$

The parameters q_i and φ_i can be derived by fitting the first three moments. For the Weibull distribution in (1), the kth raw moment is given by

$$m_k = \mathrm{E}\left[L_e^k\right] = \alpha_e^k \Gamma\left(1 + \frac{k}{\beta_e}\right). \tag{7}$$

The approximation is sufficient for engineering applications.

3.2 Data Stripping for Video Streaming

According to video codec, a number of B frames are generated between two key I/P frames depending on the target encoding bit rate. As shown in Fig. 2, each video traffic burst consists of an I or P frame and a number of B frames. For example, a video trace with a GOP size of 16 can have 0 to 15 B frames in a traffic burst. Suppose three networks are available with data rates at R_1, R_2, and R_3, respectively, where $R_1 \geq R_2 \geq R_3$. Since I/P frames are usually of a large size and important reference to decode other related frames, we consider assigning the single I/P frame in each burst toward the network with the highest data rate R_1. Next, we sort the B frames of a burst in an ascending order and denote the ith B frame size by B_i, where $i = 1, ..., S$ and S is the number B frames in a burst. Then, the first k of these B frames are distributed to the second network such that

$$\frac{\sum_{i=1}^{k} B_i}{\sum_{i=1}^{S} B_i} \approx \frac{R_2}{R_2 + R_3}. \tag{8}$$

The remaining B frames go to the third network.

Exploiting cross-layer information such as frame type and size and channel rate, we split a video streaming flow into three substreams. Each substream requires a smaller bandwidth from the associated network. In other words, the fractional resources available in each network are aggregated to support a bandwidth-demanding streaming flow. As discussed in Section 2.2, a Markov-modulated Gamma-based model can be used to characterize the video substream. Here, we use the leaky bucket algorithm in [12] to derive the effective bandwidth of video streams to bound data loss probability. The leaky bucket algorithm emulates data transmission over a channel via introducing a virtual token pool which has a finite size B_T and generates tokens at a rate r. A data buffer of size B_D can also be involved to mitigate data loss due to traffic variations. A token is required from the token pool to transmit a data unit and data loss occurs if the data buffer is full and the token pool is empty. Since data loss probability $P_{L,s}$ depends on token generation rate r, the effective bandwidth of a given video substream can be derived to bound $P_{L,s} \leq \varepsilon$.

3.3 Flow-Level Performance

For each available network, we consider a restricted access mechanism as in [16] to share the bandwidth between elastic and streaming substreams. For presentation brevity, we take one network as an example and omit subscripts for the network in the following. According to the restricted access mechanism, for a network with a capacity C, streaming flows are offered a preemptive priority over elastic flows and only occupies up to a minimum amount of bandwidth R ($R < C$). Based on the leaky bucket analysis, an effective bandwidth of b is derived for streaming substreams to bound data loss probability. Thus, the network can admit N_s streaming flows at most, where $N_s = \lfloor \frac{R}{b} \rfloor$. As elastic flows can adapt to varying bandwidth, all the bandwidth unused by streaming traffic is shared equally by active elastic flows. Totally N_e elastic flows are admissible to bound flow transfer delay.

For the flow-level performance analysis, it is assumed that streaming and elastic substreams arrive as Poisson processes with a mean rate λ_s and λ_e, respectively. The duration of streaming flows is considered to be exponentially distributed with a mean $1/\mu_s$. A heavy-tailed elastic flow is split into substreams of an exponentially distributed size, as discussed in Section 3.1. Hence, the flow-level performance can be evaluated with a two-dimensional Markov chain, which is nearly decomposable under a quasi-stationary assumption [2]. That is, the number of elastic flows is assumed to evolve rapidly with respect to the streaming traffic and attain a stationary regime. This is due to the fact that elastic and streaming flows evolve at different time scales. The time required to transfer a data file such as a Web page should be bounded within seconds, whereas the mean duration of video streaming flows is usually in the order of minutes. In the quasi-stationary regime, the two-dimensional Markov chain is decomposed into an $M/M/K/K$ queue for streaming flows and an $M/M/1$ processor sharing (PS) queue for elastic flows. Thus, we have the streaming and elastic flow blocking probabilities ($P_{B,s}$ and $P_{B,e}$, respectively) and flow transfer delay (T_e) as follows

$$P_{B,s} = \psi_0 \, \frac{\rho_s^{N_s}}{N_s!} \tag{9}$$

$$P_{B,e} = \sum_{i=0}^{N_s} \psi_i \frac{\left[1-\rho_e(i)\right]\rho_e(i)^{N_e}}{1-\rho_e(i)^{N_e+1}} \tag{10}$$

$$T_e = \sum_{i=0}^{N_s} \psi_i \frac{\rho_e(i)^{N_e+1}\left[N_e\rho_e(i) - N_e - 1\right] + \rho_e(i)}{\lambda_e\left[1-\rho_e(i)^{N_e}\right]\left[1-\rho_e(i)\right]} \tag{11}$$

$$\rho_s = \frac{\lambda_s}{\mu_s}, \quad \rho_e(i) = \frac{q\lambda_e}{\varphi(C - i\cdot b)}, \quad i = 0, 1, ..., N_s \tag{12}$$

where ψ_i denotes the steady-state probability of i ongoing streaming flows, given by

$$\psi_i = \frac{\rho_s^i}{i!}\left[\sum_{i=0}^{N_s}\frac{\rho_s^i}{i!}\right]^{-1}. \tag{13}$$

The traffic parameters q, φ, and b are derived as in Section 2.1 and 2.2. Considering flow splitting across multiple available networks, we have the overall flow blocking probabilities and flow transfer delay

$$P_{B,s} = 1 - \prod_{i=1}^{3}\left[1 - P_{B,s}^{(i)}\right] \tag{14}$$

$$P_{B,e} = 1 - \prod_{i=1}^{3}\left[1 - P_{B,e}^{(i)}\right] \tag{15}$$

$$T_e = \max\left[T_e^{(i)}\right]. \tag{16}$$

3.4 Packet-Level Performance

In Section 3.3, we analyze the flow-level performance of the proposed scheme and the discreteness of data units are neglected. For elastic services such as file transfer, user QoS experience is more concerned with flow-level dynamics. In contrast, streaming services are also sensitive to finer packet-level performance due to real-time playback during delivery. In this section, we further evaluate the packet-level performance of video streaming traffic.

As seen in Fig. 2, video frames are generated in burst according to coding and compression algorithms. For each traffic burst, a random number of video frames are disseminated toward each network channel. We assume that these application data are segmented into fixed-size packets for transmission. The transmission time of the packets, denoted by τ, is used to discretize the time scale for analysis purpose. Hence, the packet transmission process can be modeled by a $D^{[A]}/D/1$ queueing system, which has a constant batch interarrival time, general batch-size distribution, and deterministic service rate. As the size of video frames can be modelled with a Gamma distribution [8,11], we characterize the batch size with a negative binomial distribution, which is a discrete analog of Gamma distribution. The probability mass function (PMF) of the batch size (A) is then given by

$$f_s(k) = \mathrm{P}[A = k] = \binom{k+r-1}{r-1}(1-\eta)^r \eta^k \tag{17}$$
$$r > 0, \quad 0 < \eta < 1, \quad k = 0, 1, ...$$

where the binomial coefficient

$$\binom{k+r-1}{r-1} = \frac{(k+r-1)(k+r-2)...(r)}{k!}. \tag{18}$$

The parameters r and η can be obtained by fitting the mean and variance of the batch size:

$$\mathrm{E}[A] = r\frac{\eta}{1-\eta}, \qquad \mathrm{Var}[A] = r\frac{\eta}{(1-\eta)^2}. \tag{19}$$

Following the probability generating function (PGF) technique in [6], we can evaluate the statistics of packet transfer delay through a $D^{[A]}/D/1$ queueing system. Let Q_i denote the number of backlog packets in the transmission buffer at the end of time slot i. The evolution of buffer occupancy is thus given by

$$Q_i = \left[Q_{i-1} + A_i - 1\right]^+. \tag{20}$$

When the system converges to an equilibrium ($i \to \infty$), the queue occupancy Q reaches a steady state defined by a PMF function $q_s(k)$. The following generating function [6] can be obtained by multiplying (20) by z^k and summing over k

$$Q(z) = \sum_{k=0}^{\infty} q_s(k) z^k = \frac{(1-\zeta)(z-1)}{z - A(z)} \tag{21}$$

where ζ is the batch (traffic burst) arrival rate per time unit and $A(z)$ is the PGF of the batch arrival size A, given by

$$A(z) = \sum_{k=0}^{\infty} f_s(k) z^k = \left[\frac{\eta}{1-(1-\eta)z}\right]^r. \tag{22}$$

Hence, the kth factorial moment of queue occupancy (Q) can be obtained from (21) as follows

$$\Omega_k = \mathrm{E}\Big[Q(Q-1)...(Q-k+1)\Big] = \lim_{z \to 1^-} \frac{d^k Q(z)}{dz^k}. \tag{23}$$

According to the generalized Little's formula [4], we have the kth raw moment of the queueing delay

$$W_k = \frac{\Omega_k}{\hat{\zeta}^k}. \tag{24}$$

In particular, the mean and variance of packet transfer delay (in seconds) is obtained as

$$T_s = \tau \left(\frac{Q^{'}(1^-)}{\hat{\zeta}} + 1 \right) \tag{25}$$

$$\sigma_{T_s}^2 = \tau^2 \left[\frac{Q^{''}(1^-) + Q^{'}(1^-) - \left(Q^{'}(1^-)\right)^2}{\hat{\zeta}^2} \right]. \tag{26}$$

Table 1. System parameters for numerical analysis

Symbol	**Value**	**Definition**
$\mathrm{E}[L_e]$	1280	Mean elastic flow size (KB)
α_e	0.7	Shape parameter of elastic flow size L_e
β_e	1011.2	Scale parameter of elastic flow size L_e
α_s	0.5856	Sigmoid function parameter for segmentation
β_s	12.0683	Sigmoid function parameter for segmentation
θ_s	1.1	Scale factor to segment video streaming flow
S	7	Number of B frames between two I/P frames
K	6	Number of video classes for traffic modelling
ε	0.01	Upper bound for data loss probability
λ_s	$0.001 \sim 0.02$	Mean arrival rate of streaming flows (/s)
μ_s^{-1}	30	Mean duration of streaming flows (min)
λ_e	$0.8 \sim 0.1$	Mean arrival rate of elastic flows (/s)
C_1	40	Available bandwidth of network 1 (Mbit/s)
C_2	20	Available bandwidth of network 2 (Mbit/s)
C_3	12	Available bandwidth of network 3 (Mbit/s)

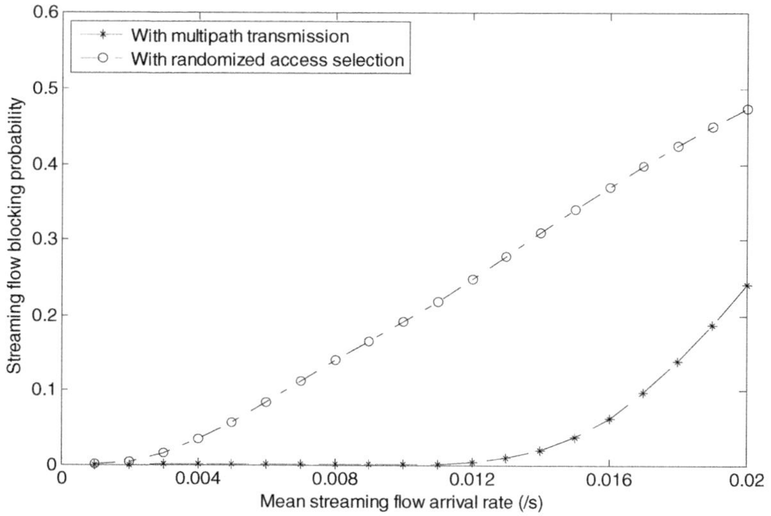

Fig. 4. Flow blocking probability of streaming traffic ($P_{B,s}$)

4 Numerical Results

In this section, we present numerical results to evaluate the performance of our proposed multipath transmission scheme. Table 1 gives the system parameters for numerical analysis. Assuming the average elastic flow size is $\mathrm{E}[L_e] = 1280$ KB, we take the shape and scale parameter in (1) as $\alpha_e = 0.7$ and $\beta_e = 1011.2$. Fitting the first three moments, we can obtain the parameters of the hyper-exponential distribution for the superposition of substreams: $q_1 = 0.1186$, $q_2 = 0.4314$, $q_3 = 0.4500$, $\varphi_1 = 2.7799 \times 10^{-4}$, $\varphi_2 = 0.0010$, and $\varphi_3 = 0.0011$ (kbit^{-1}).

For video streaming flows, we consider H.264/AVC video sequences of *Tokyo Olympics* from the video trace library of Arizona State University [13]. These video sequences have a CIF resolution, a fixed frame rate at 30 frames/s, a GoP size of 16, and 7 B frames between two I/P key pictures. The quantization level varies with the step size and a higher quantization index (between 0 and 51) results in a lower encoding bit rate. The proposed data stripping algorithm in Section 3.2 exploits the coding and compression structure to enable multipath transmission. To bound data loss probability at the flow level, the effective bandwidth can be derived with the leaky bucket algorithm [12]. For instance, at the quantization level 42, the effective bandwidth requirements of the three video substreams to upper bound data loss probability by 0.01 are $b_1 = 658.9$ kbit/s, $b_2 = 413.5$ kbit/s, and $b_3 = 215.3$ kbit/s, respectively.

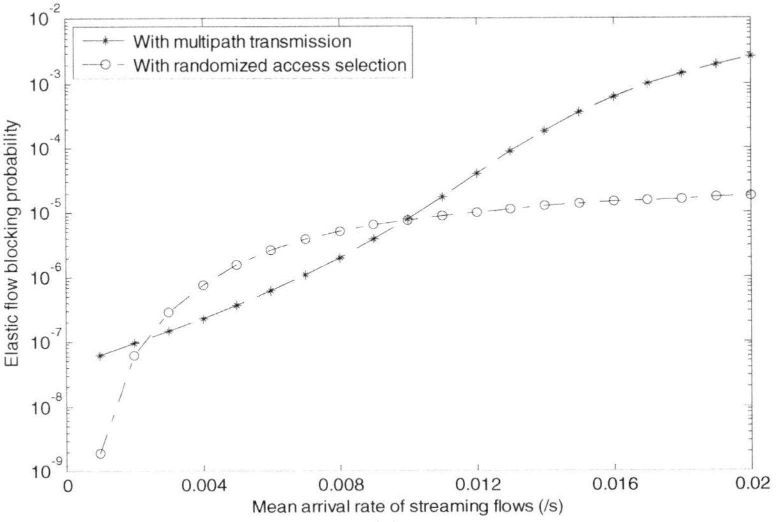

(a)

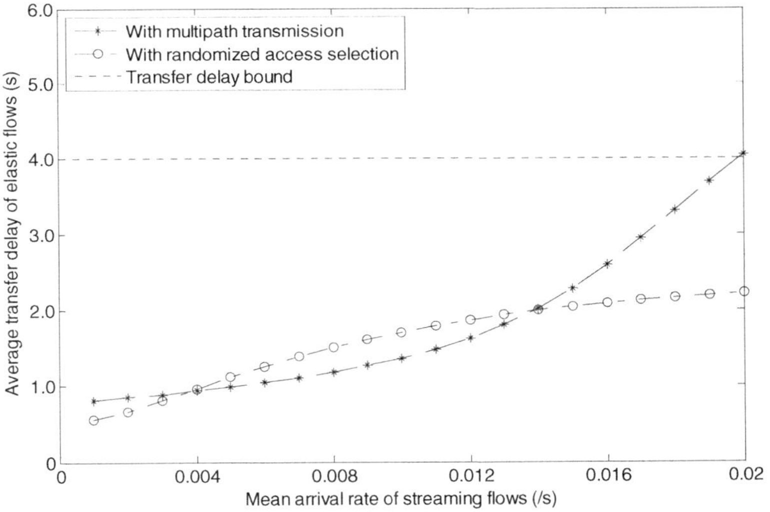

(b)

Fig. 5. Flow-level performance of elastic traffic. (a) Flow blocking probability ($P_{B,e}$). (b) Flow transfer delay (T_e).

4.1 Flow-Level Performance

Fig. 4 and Fig. 5 illustrate the flow-level performance of the proposed multipath transmission scheme evaluated by (9 - 14) and that of a randomized access selection algorithm [15]. For the randomized selection algorithm, an incoming

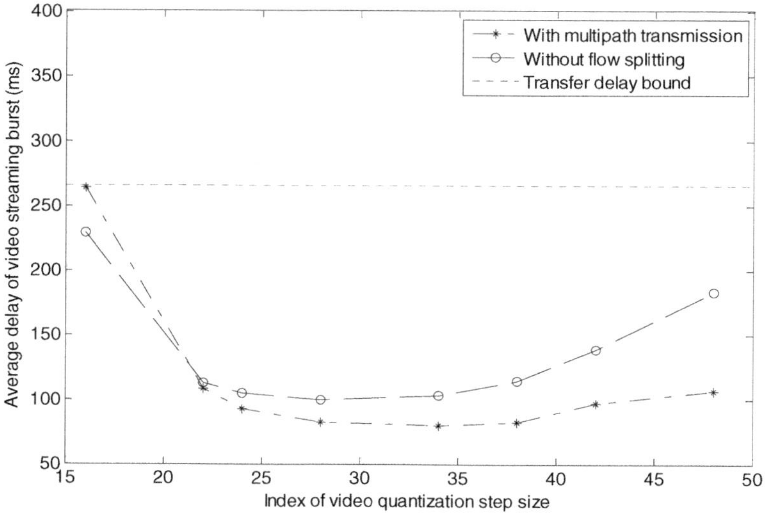

Fig. 6. Average transfer delay of video streaming bursts at different quantization levels

flow requests admission to an available network with a probability ϑ_i. These selection probabilities ϑ_i can be determined to maximize the acceptable traffic load to the overall integrated system. As seen in Fig. 4, multipath transmission significantly reduces the blocking probability for streaming flows. Although an ideal access selection is expected to choose a most desirable network for an incoming traffic flow, it is possible that none individual network satisfies the overall bandwidth requirement. Multipath transmission can effectively aggregate available bandwidth across multiple networks to serve split substreams requiring smaller bandwidth and achieves a much lower blocking probability.

As shown in Fig. 5, the randomized selection provides a better performance for elastic flows in some cases. With the randomized selection, streaming flows are more likely to be declined from the system. Consequently, more bandwidth unused by streaming traffic can be shared by elastic flows. As a result, a lower blocking probability and transfer delay is achievable for elastic flows at the expense of a significantly higher blocking probability for streaming traffic. Nevertheless, the performance of elastic flows with multipath transmission is well acceptable.

4.2 Statistics of Packet Transfer Delay

As shown in Fig. 2, video frames are generated in burst and segmented into packets for transmission. To ensure smooth playback at the receiver, the burst of application data need to be delivered at a speed higher than the burst arrival rate. Considering the video traces at 30 frames/s, we have the delay bound at

266.7 ms, given 7 B frames between two key I/P frames. Fig. 6 shows the average transfer delay of video streaming bursts. The statistics of packet transfer delay can be evaluated with the the batch arrival queueing process and PGF technique in Section 3.4. It can be seen in Fig. 6 that multipath transmission effectively reduces the average transfer delay of video frames. A primary reason for the performance gain is that the traffic burstiness is balanced and mitigated with flow splitting.

5 Conclusions and Future Work

In this paper, we propose a cross-layer multipath transmission scheme for elastic and streaming flows. The proposed scheme exploits application-layer and network-layer information to enable flow splitting and data stripping across multiple heterogeneous networks. A heavy-tailed elastic flow is thereby decomposed into multiple substreams of an exponentially distributed size. A much smaller effective bandwidth is required for multiple associated networks to support video substreams. Moreover, we evaluate the flow-level and packet-level performance in terms of flow blocking probabilities and transfer delays. Our analytical approach takes into account the heavy-tailed size of elastic flows and burstiness of video streaming flows. The $D^{[A]}/D/1$ queuing process with batch arrivals is considered to analyze packet transfer delay of video streaming traffic, which is sensitive to this finer packet-level performance due to real-time playback.

Numerical results are presented to demonstrate the performance of the proposed multipath transmission scheme. A significant performance gain is particularly observed for bandwidth-demanding video streaming traffic. The QoS enhancement is attributed to the effective aggregation of available bandwidth across multiple networks. In our future work, we will investigate how to adapt multipath transmission with network variations. Especially, in high mobility conditions, link capacities may fluctuate rapidly due to severe fading or switch of network attachment points. Hence, the flow splitting needs to be dynamically adapted and cope with network variations.

References

1. 3GPP. Transparent end-to-end packet-switched streaming service (PSS); protocols and codecs. 3GPP TS 26.234 V9.3.0 (June 2010)
2. Delcoigne, F., Proutière, A., Régnié, G.: Modeling integration of streaming and data traffic. Perform. Eval. 55(3-4), 185–209 (2004)
3. Feldmann, A., Whitt, W.: Fitting mixtures of exponentials to long-tail distributions to analyze network performance models. Perform. Eval. 31(3-4), 245–279 (1998)
4. Gross, D., Harris, C.M.: Fundamentals of Queueing Theory. Wiley, Chichester (1974)
5. Heyman, D.P.: The GBAR source model for VBR videoconferences. IEEE/ACM Trans. Networking 5(4), 554–560 (1997)
6. Lin, A.Y.-M., Silvester, J.A.: On the performance of an ATM switch with multichannel transmission groups. IEEE Trans. Commun. 41(5), 760–770 (1993)

7. Ma, L., Yu, F., Leung, V.C.M., Randhawa, T.: A new method to support UMTS/WLAN vertical handover using SCTP. IEEE Wireless Commun. Mag. 11(4), 44–51 (2004)
8. Masi, D.M.B., Fischer, M.J., Garbin, D.A.: Video frame size distribution analysis. The Telecommunications Review 19, 74–86 (2008)
9. Niyato, D., Hossain, E.: Dynamics of network selection in heterogeneous wireless networks: An evolutionary game approach. IEEE Trans. Veh. Technol. 58(4), 2008–2017 (2009)
10. Rezaul, K.M., Pakšstas, A.: Web traffic analysis based on EDF statistics. In: Proc. 7th Annual PostGraduate Symposium on the Convergence of Telecommunications, Networking and Broadcasting (PGNet) (June 2006)
11. Sarkar, U.K., Ramakrishnan, S., Sarkar, D.: Markov-modulated Gamma-based framework. IEEE/ACM Trans. Networking 11(4), 638–649 (2003)
12. Schwartz, M.: BroadBand Integrated Networks. Prentice-Hall, Englewood Cliffs (1996)
13. Seeling, P., Reisslein, M., Kulapala, B.: Network performance evaluation with frame size and quality traces of single-layer and two-layer video: a tutorial. IEEE Communications Surveys & Tutorials 6(3), 58–78, third quarter (2004)
14. Si, P., Ji, H., Yu, F.R.: Optimal network selection in heterogeneous wireless multimedia networks. ACM/Springer Wireless Networks 16(5), 1277–1288 (2009)
15. Song, W., Cheng, Y., Zhuang, W.: Improving voice and data services in cellular/WLAN integrated network by admission control. IEEE Trans. Wireless Commun. 6(11), 4025–4037 (2007)
16. Song, W., Jiang, H., Zhuang, W.: Performance analysis of the WLAN-first scheme in cellular/WLAN interworking. IEEE Trans. Wireless Commun. 6(5), 1932–1952 (2007)

Battery-Aware Wireless Video Delivery

Jianxin Sun, Dalei Wu, and Song Ci

Department of Computer and Electronics Engineering,
University of Nebraska-Lincoln, NE 68182, USA
{jianxin.sun,dwu}@huskers.unl.edu, sci@engr.unl.edu

Abstract. Feasibility and popularity of mobile multimedia have made video communications between mobile devices a rising trend with a wide range of applications. However, two main problems have emerged. First: in most of mobile devices, the power-hungry multimedia processor relies on the battery as the only form of power resource and the characteristics of a battery keep changing as the discharging operation goes. How to wisely utilize the energy stored in batteries on mobile device becomes a critical issue in designing a wireless video communication system. Second: in order to achieve QoS on mobile devices, the requested video chip has to be displayed under a given standard of quality. Therefore, it is also necessary for the received video to satisfy the constraint of an acceptable level of distortion. To analyze and optimize the communication quality and energy consumption behavior of battery-driven wireless video communication systems, we propose an optimization framework which takes into account the characteristics of battery driven devices by considering the relation between energy consumption and capacity discharging behavior of battery. In our framework, the video coding and transmission parameters are jointly optimized to minimize the battery capacity consumption under a predefined level of expected received video distortion. Experimental results indicate the efficiency and effectiveness of the proposed optimization framework.

Keywords: multimedia, video, battery, wireless communication system, QoS, distortion.

1 Introduction

Technologies in video compression and transmission over wireless communication networks have enabled mobile multimedia on portable wireless devices, such as cellular phones, laptop computers connected to WLANs, and cameras in surveillance and environmental tracking systems. Video coding and streaming are also envisioned in an increasing number of applications in the areas of battlefield intelligence, reconnaissance, and telemedicine. Present 3G and emerging 4G wireless systems, and IEEE 802.11 WLAN standards have dramatically increased the transmission bandwidth, and therefore, resulted in a great amount of personal communication users on video streaming applications. Although wireless video communications is highly desirable, a primary limitation in wireless systems is

X. Zhang and D. Qiao (Eds.): QShine 2010, LNICST 74, pp. 515–529, 2011.

the basic design architecture that most mobile devices are typically powered by batteries with limited energy capacity. This limitation is of fundamental importance due to the high energy consumption rate in the duality of encoding and transmitting video bit streams during multimedia communications. From the perspective of battery-aware design and power management, how to wisely decide the energy allocation is a critical issue in order to efficiently use battery energy to guarantee a required lifetime and avoid task failure or malfunction due to the exhaustion of battery capacity before the whole video delivery task is finished.

The lifetime, or time-to-failure, of one battery is the time when it becomes fully discharged. Once the battery is exhausted, the mobile system shuts down. According to different acceptable distortion levels on wireless video communication and characteristics of one specific type of battery, we need to jointly consider a series of configurations in related modules and operating procedures during source coding and transmission to minimize the battery capacity under a certain constraint on distortion.

So far, there has been no dedicated analytical framework investigating an entire wireless video communication system where mobile devices use regular battery as the only form of power source in literature. And no experimental analysis on battery performance or optimization under a specific wireless multimedia platform has been performed. An analytical framework was presented to address the Power-Rate-Distortion relationship of a generic video encoder in [1]. However, video transmission was not considered in the evaluation of distortion and power consumption. Although some work [2,3,4] analyzed the energy efficiency of both video coding and transmission, the issue of power consumption was addressed without specifying the underlying characteristics of battery driven devices and no solution to minimize capacity consumption under the constraint of video quality is presented.

In this paper, we develop an optimization framework for wireless video delivery under the constraint of the video distortion required in a wireless video communication system. We first discuss the experimental methods and models to analyze the energy consumption in video encoding and transmission. Based on the analytical results, the problem of battery-aware wireless video coding and delivery is formulated to jointly select the video coding and transmission parameters to minimize the battery capacity consumption under the constraint of expected end-to-end received video distortion imposed by the desired video quality requirement. Our framework aims at the joint optimization of video coding and wireless transmission from the perspective of battery capacity condition.

This paper is organized as follows. Section 2 presents the formulation of the problem to solve. In section 3, measurements of energy consumption for video encoder and models for video stream transmission are introduced. Section 4 presents the method to calculate the expected distortion of a wireless video communication system. Analysis of working status of battery driven equipment is discussed in Section 5. Optimization method and framework are proposed in

Section 6. Section 7 presents the experimental results. Some concluding remarks are given in Section 8.

2 Problem Statement

2.1 Energy Consumption on Video Coding

To analyze the power consumption on video coding in a portable device, first, we need to determine the computational complexity of video coding at the encoder. Here, the computational complexity is measured by the running time of processor when video coding processes are under operation. Then, based on the power management technology of the underlying microprocessor in the mobile device, such as DVS CMOS circuits design technology [5], we can measure the energy used in those processes.

As shown in Figure 1, the major modules in a typical video encoding system include motion estimation (ME) and compensation, DCT, quantization, entropy encoding of the quantized DCT coefficients, inverse quantization, inverse DCT, picture reconstruction, and interpolation. In the literature, plenty of research results have been reported to evaluate and reduce the computational complexity, and thereby the power consumption of these modules [1,6,7]. It has been shown that, for each module in Figure 1, one or more control parameters together with the specific characteristics of a video chip can be extracted or selected to control the computational complexity of the module. For example, according to [1], the ME module could use the number of sum of absolute difference (SAD) as the complexity control parameter, while the modules of DCT, quantization, inverse quantization, inverse DCT, picture reconstruction may use a same complexity control parameter – the number of macroblocks (MB) which has nonzero DCT coefficients after quantization in a video frame. Let $\Lambda = [\lambda_1, \lambda_2, \cdots, \lambda_I]$ be the set of control parameters to control the computational complexity of these modules.

Therefore, the overall encoder complexity (or processor workload) ξ is a function of video processing parameters Λ, denoted by $\xi(\Lambda)$. Hence, the energy consumption of the underlying microprocessor to compress and encode one video clip, denoted by E_e, is a function of processor workload x_i, therefore, is also a function of Λ, denoted by

$$E_e = \Phi(x_i) \cdot t = E_e(\Lambda) = E_e(\lambda_1, \lambda_2, \cdots, \lambda_I), \tag{1}$$

where $\Phi(\cdot)$ is the power consumption model of the microprocessor [8], which can be obtained by measurement. For example, the power consumption model of the Intel PXA255 XScale processor is well approximated by $\Phi(x_i) = \beta \times x_i^\gamma$, where $\gamma = 2.5$, and β is a constant [9].

2.2 Energy Consumption on Video Transmission

To analyze the energy consumption on video transmission, we need to consider both the transmission scheme and the power control technology adopted by

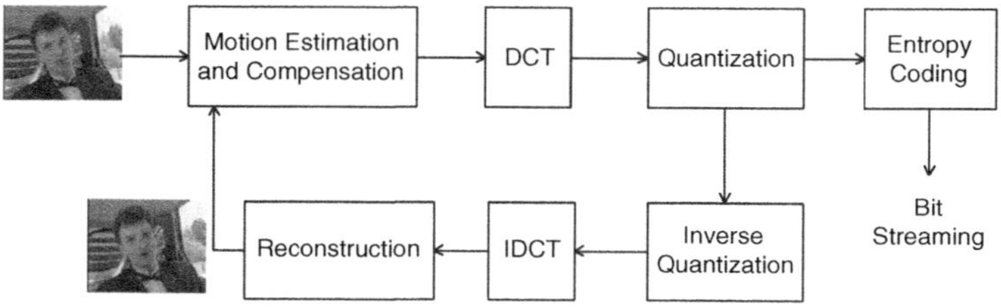

Fig. 1. Block diagram of a typical video encoder. For INTRA MB or frames, motion estimation and compensation are not needed.

the transmitter on mobile device. General energy E_t used in transmitting one video stream after video coding not only depends on wireless channel conditions, such as, instantaneous channel fading factor and channel noise power density, but also on transmission parameters, such as, frequency bandwidth, desirable packet error rate (PER), and modulation and coding schemes. Without loss of generality, let $\Theta = [\theta_1, \theta_2, \cdots, \theta_J]$ be the set of parameters affecting or controlling the transmission energy level. Therefore, the energy used in transmission can be represented by

$$E_t = E_t(\Theta) = E_t(\theta_1, \theta_2, \cdots, \theta_J). \tag{2}$$

2.3 Problem Formulation

Based on equations (1) and (2), the total energy consumption required from mobile device battery to deliver a video clip is

$$E_{tot} = E_e + E_t, \tag{3}$$

Because our framework aims at minimizing the capacity consumed from battery, it is necessary to convert energy into electric quantity. Once the hardware platform is set, we can derive the relation between the energy used for video delivery and the electric quantity consumed from the battery. The consumed battery capacity measured as the electric quantity can be express as

$$C_{tot} = f(E_{tot}), \tag{4}$$

Let D be the expected video distortion, and it depends on the video processing parameters Λ and the transmission parameters Θ. Different power supply results in different battery lifetime since the capacity of a given battery at certain status is fixed. Therefore, based on (3) and (4), the objective of the proposed framework is, for a specific video application, to determine the optimal values of the parameters $\{\Lambda, \Theta\}$ for a series of current video frames to minimize the battery capacity under the constraint on distortion D_{max} required by the specific application, which can be formulated as

$$\min_{\{\Lambda,\ \Theta\}} C_{tot}(\Lambda, \Theta)$$
$$s.t.: D(\Lambda,\ \Theta) \leq D_{\max}. \tag{5}$$

3 Energy Consumption Models and Measurement

3.1 Encoder Energy Consumption Model

Video coding or compression is a basic technology that enables the storage and transmission of a large amount of digital video data. Many standard video encoder systems, including MPEG-1/-2/-4, H.26x, and H.264/AVC employ a hybrid coding architecture based on DCT and Motion Estimation Compensation (ME/MC) scheme. Existing processes during wireless video communication include modules of ME (motion estimation), PROCODING (including DCT, inverse DCT, quantization, inverse quantization, and reconstruction) and ENC (entropy encoding). During video coding process, with the help of electric quantity measure equipment, the energy used in battery-powered multimedia processor can be derived from the operating time of CPU which enables the hardware resources to accomplish series of steps for video coding. On a specific video coding platform, the operating time of CPU depends on the characteristics of the specific video clip and complexity parameters selected in the steps of video coding. In the process of PROCODING, Quantization Parameter (QP), a parameter that controls the quality and bit rate of video compression, is a key factor to affect the number of nonzero MBs (NZMB) in one video frame which are needed to be coded. While the computation for quantization is independent of the bit rate, with a smaller quantization step size, more computation for variable length coding (VLC) is needed due to the increased number of nonzero coefficients. Many experiments have shown that all the steps together in PROCODING employ a large proportion of CPU occupancy and eat up more than 50% of the total energy consumption on encoder. Compared with other modules in source coding, PROCODING energy consumption is almost twice of the energy consumed in ME and six times of the energy consumed in ENC. Quantization step is the key complexity control parameter in the processes of PROCODING, therefore, it is reasonable for us to set quantization step q as the main optimal complexity control parameter in the video coding processes to calculate the energy used to encode a specific video clip on a certain hardware platform. The total time used in coding the whole video clip depends on the CPU running time spent on encoding every frame of this video. Denote the total number of frames in a video clip as n and the time used to encode kth frame as t_k. Then the total operating time of CPU T_e^{tot} to encode the a video clip can be expressed as

$$T_e^{tot} = \sum_{k=1}^{n} t_e^k, \tag{6}$$

On the other hand, total energy used in coding the whole video clip also depends on the energy E_e^k, which is used to encode the kth frame of this video. So the total operating energy of CPU E_e^{tot} used for coding can be written as

$$E_e^{tot} = \sum_{k=1}^{n} E_e^k, \tag{7}$$

It is possible for us to directly find the amount of electric quality used to encode every frame of a video by referring the corresponding experiment test result from equipment of energy measurement. In this way, once the hardware platform is decided, the energy in encoder for one frame is a function of the CPU running time t_k spent on coding this frame and the key complexity control parameter q_k chosen to compress the same frame. The function of total energy used in encoder can be given by

$$E_e^{tot} = \sum_{k=1}^{n} f_e(t_k, q_k). \tag{8}$$

where $f_e(\cdot)$ is the uniform way to calculate the energy consumption of one frame during coding processes.

3.2 Transmission Energy Consumption Model

The total transmission energy can be calculated by adding together the energy used on transmitting every frame of a video. This is given by

$$E_t^{tot} = \sum_{k=1}^{n} E_t^k, \tag{9}$$

where E_t^k is the energy consumption used for transmitting the kth frame. The energy used to transmit a frame depends on the compressed bits of this frame and channel transmission rate. Compressed bits of one frame is the size of a stream which is generated after coding processes of a video frame and mostly decided by the characteristics of the input video and the quantization step q applied on this frame. Channel transmission rate depends on the transmission bandwidth and adaptive modulation and coding (AMC) scheme. Different choice of AMC scheme applied in the transmitter will result in different transmission rates and spectral efficiency. Let W be the underlying channel bandwidth, and K_i be the transmission rate of AMC scheme i. Then, the resulting transmission rate when data is transmitted by using the ith AMC scheme is

$$R_i = K_i \cdot W, \tag{10}$$

If we use F_i to represent the compressed bits size of the ith frame after coding processes, then the energy used for kth frame can be denoted as

$$E_t^k = P \cdot \frac{F_i}{R_i} = P \cdot \frac{F_i}{K_i \cdot W}, \tag{11}$$

Where P is the transmission power. F_i can also be determined by referring the corresponding experiment test result. In general, the energy in transmission for one frame is a function of time spent on coding processes t_k, AMC scheme i and the complexity control parameter q_k chosen to compress the current frame.

Therefore the total energy consumption in transmitting a video clip can be written as

$$E_t^{tot} = \sum_{k=1}^{n} f_t(t_k, q_k, i). \tag{12}$$

where $f_t(\cdot)$ is the function to calculate the energy consumption used to transmit one frame.

4 Expected End-to-End Distortion

During the wireless video communication process, the total expected end-to-end distortion is caused during source coding and transmission. In order to acquire an accurate result of the distortion through the whole wireless video transmission system, in this work, we consider and calculate the overall end-to-end distortion instead of just simply adding the coding introduced-distortion and the transmission introduced-distortion together. Because a robust error concealment technique is necessary to avoid significant visible error in the reconstructed frames at the decoder, we consider a simple but efficient temporal concealment scheme used in our previous research [10]: a lost macroblock is concealed using the median motion vector candidate of its received neighboring mocroblocks (the topleft, top, and top-right) in the preceding row of macroblacks. The candidate motion vector of a macroblock is defined as the median motion vectior of all 4×4 blocks in the macroblock. If the preceding row of macroblocks is also lost, then the estimated motion vector is set to zero and the macroblock in the same spatial location is the previously reconstructed frame is used to conceal the current loss. Although some straight-forward error concealment strategies do not cause packet dependencies, as a generic framework, the more complicated scenario is considered here as a superset for the simpler cases. Due to the difficulty in computing the actual video quality perceived by the end users, in this work the received video quality is evaluated as the expected end-to-end distortion by using the ROPE method. The expected distortion is accurately calculated in real-time at the source node by taking all related parameters into account, such as source codec parameters (e.g., quantization, packetization, and error concealment) and network parameters (e.g., packet loss rate and throughput). Therefore, given the dependencies introduced by the above error concealment scheme, the expected distortion of slice/packet π_i can be calculated at the encoder as

$$E[D_i] = (1 - p_i)E[D_i^R] + p_i(1 - p_{i-1})E[D_i^{LR}] + p_i Pi - 1E[D_i^{LL}], \tag{13}$$

where p_i is the loss probability of packet π_i, $E[D_i^R]$ is the expected distortion of packet π_i if received, and $E[D_i^{LR}]$ and $E[D_i^{LL}]$ are respectively the expected distortion of the lost packet π_i after concealment when packet π_{i-1} is received or lost. The expected distortion of the whole video frame which contents m packets, denoted by $E[D]$, can be written as

$$E[D] = \sum_{i=1}^{m} E[D_i], \tag{14}$$

Generally multiple modulation and coding schemes are available to wireless stations in a wireless data network to achieve a good tradeoff between the transmission rate and transmission reliability. Modulation schemes that allow a larger number of bits per symbol, have symbols closer to each other in the constellation diagram, which may result in more error in decoding. Varying code rates can be employed with each modulation scheme to adapt to changing channel conditions by allowing more redundancy bits for channel coding (lower code rate k/n) as channel conditions deteriorate. As the code rate decreases, the effective data rate is reduced, and hence the achievable throughput decreases. We set the term scheme i to refer to a specific choice of AMC scheme. The probability of error in a packet of L bytes, for a given AMC scheme i, as a function of the bit error probability $p_{b,i}$, can be express as $p_{e,i}(L) = 1 - (1 - p_{b,i})^{8L}$. Moreover, $p_{e,i}$ can also be approximated with sigmoid functions [11,12] in the form of

$$p_{e,i}(L) = \frac{1}{1 + e^{\lambda(x-\delta)}}, \tag{15}$$

where x is the Signal-to-Interference-Noise-Ratio (SINR). Table 1 shows the sigmoid parameters (λ, δ) for the 8 AMC schemes in modeling packet transmissions over an 802.11a WLAN network. From this table and (15), it is easy to see that $p_{e,i}$ depends on the specific AMC scheme i and so is the overall distortion since the end-to-end distortion is the function of $p_{e,i}$. Once the packet error probability is calculated, the expected end-to-end distortion can be derived based on equations (13) and (14).

We have noted that, except for the characteristics of the input video, the quantization parameters (QP) applied in source coding procedure play another critical role in contributing the total distortion since the larger the quantization step size is, the more small DCT coefficients will be lost. Thus, from (14), different levels of distortion will be achieved under different levels of QP. In other words, the value of quantization step q needs to be considered as another parameter to control the total distortion. Therefore, for a specific platform, the total expected distortion associated with AMC scheme i and QP q can be denoted as

$$E[D]_{tot} = D(q, i). \tag{16}$$

Table 1. Approximation of packet error probability for different AMC schemes

Mod Scheme	δ (dB)	λ (dB^{-1})	$CodeRate$ $(bits/symbol)$	$AMCScheme$ (i)
BPSK	2.3	0.640	0.5	1
BPSK	6.1	0.417	0.75	2
QPSK	5.3	0.461	1	3
QPSK	9.3	0.444	1.5	4
16-QAM	10.9	0.375	2	5
16-QAM	15.1	0.352	3	6
64-QAM	18.2	0.625	4	7
64-QAM	21.2	0.419	4.5	8

5 Battery Capacity Measurement

Mobile devices are mostly driven by battery. Once the battery becomes fully discharged, the battery-powered electronic system goes off-line. Available battery capacity has a nonlinear relationship with its discharging current due to the battery current effect. That means a battery tends to provide more energy at a lower discharge current. Some battery lifetime and capacity prediction models are available, like the analytical model in [13]. But considering that analysis and optimization of our research is mainly base on the profiles of distortion and energy consumption, precise data measurement is necessary to make the work more practical and applicable. So we choose to execute a series of real experiments to build the experimental profile, which includes all the real time data information of the parameters we need for optimization and analysis. Once such profile is set up, electronic parameters, like voltage, current and electric quantity used for every steps of the operation, can be derived from this profile. By describing the expression of (4), the capacity consumed from the battery can be calculated from the energy consumption. The total capacity needed from battery to achieve wireless video delivery can be derived from the follows

$$C_{tot} = C_e + C_c = C_e + \frac{E_t}{V}, \tag{17}$$

where C_e is the electric quantity consumed to execute the video coding, and this value can be derived from the profile result of experiment. C_c is the electric quantity consumed for video transmission, and this can be calculated from the transmission energy consumption model introduced in section 3 if the transmitter operating voltage V is known. Combine with (11) and (12), we get

$$\begin{aligned} C_{tot} &= C_e + \frac{\sum_{k=1}^{n} P \cdot \frac{F_i}{K_i \cdot W}}{V} \\ &= C_e + \frac{\sum_{k=1}^{n} f_t(t_k, q_k, i)}{V}. \end{aligned} \tag{18}$$

6 Optimized Battery Capacity Framework under Distortion Constraint

In order to solve the the formulated optimization problem in Section 2, two profiles need to be established in advance. The first profile is about how the pattern of the expected received frame distortion changes according to different set of choices on quantization parameters q and AMC scheme i. The second is how the battery capacity used for the delivery of one frame is decided by the same set of choice on QP and AMC scheme. Therefore, in our framework, we choose quantization parameters q and AMC scheme i to form a two dimensional independent vectorial variable. Let Q be the total options of QP, I the total optional AMC schemes. So every video frame has $Q \cdot I$ options of this vector. Because both parameters are the key control variables to determine the working pattern in both coding and transmission, we name this vector as the control vector, and denote it as (q, i) according to the definitions of previous sections.

Video is fundamentally different from other multimedia resources, for it basically comprises a group of separated video frames. When we deal with the optimization of battery capacity under the constraint of video distortion, it is not reasonable to figure out only one set of optimal control vector to process all the video frames uniformly, because the dynamics in video content and channel conditions make it necessary to adjust the QP and AMC parameters frame by frame. Therefore, we apply the optimization to figure out the best control vector toward each frame to minimize the battery capacity used to deliver each frame under the constraint of expected received frame distortion. From (16) and (18), we can see that the total expected distortion and total energy consumption are based on the choice of control vector used to code and transmit each video frames. For a specific video, after the execution of experiment and model applications, the expected received video distortion and battery capacity used for delivering the whole video clip can be calculated by referring to the received frame distortion and battery capacity used for delivering each frame.

7 Experimental Result

We conducted experiments to show the performance of the proposed framework. Four video sequences with varied contents (Carphone, Foreman, Coastguard, Mobile) in QCIF format are considered in our work. An Imote2 wireless sensor node which applies PXA271 XScale processor is used in the experiment. Arbin measurement system is in charge of monitoring and recording all the desired

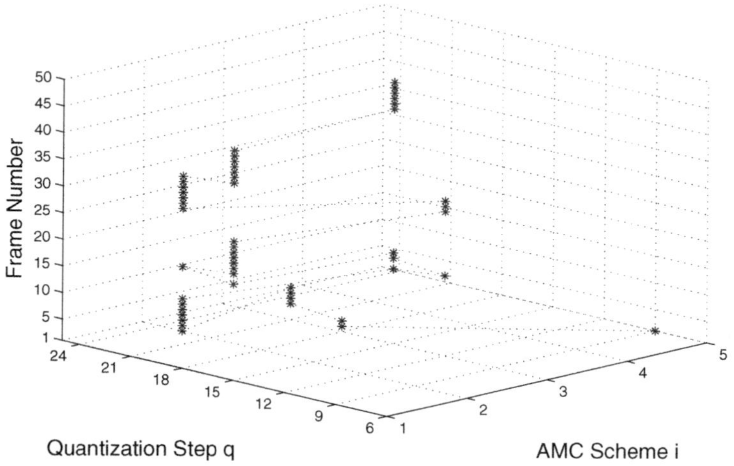

Fig. 2. The optimal solution achieving the minimized distortion without the battery capacity constraint

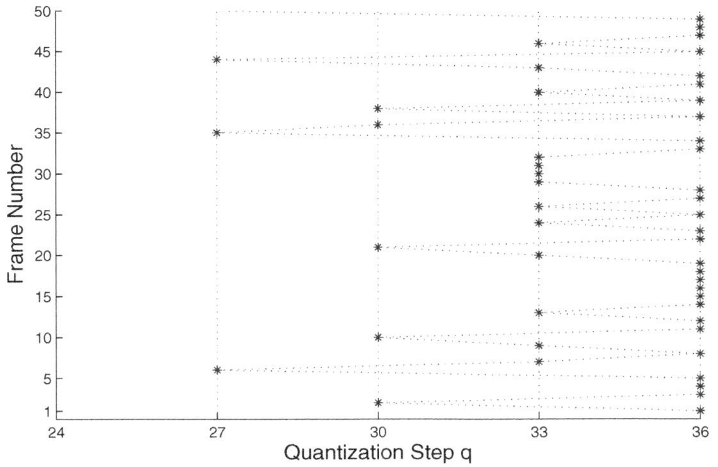

Fig. 3. The optimal solution achieving the minimized battery capacity consumption without the distortion constraint. (AMC Scheme i=8).

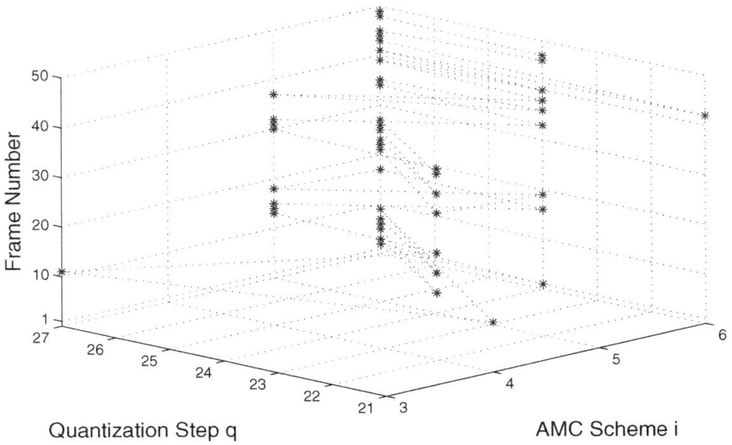

Fig. 4. The optimal solution under the constraint of 36 dB average frame distortion

electronic data. The Y-component of the first 50 frames of each video sequence is encoded with H.264 codec (JVT reference software, JM 16.2 [14]). We choose the quantization step size (QP) and AMC schemes listed in table 1 as the tunable source coding and transmission parameters. The permissible QP values are [9, 12, 15 ..., 36]. According to table 1, the permissible AMC scheme i values are [1, 2,

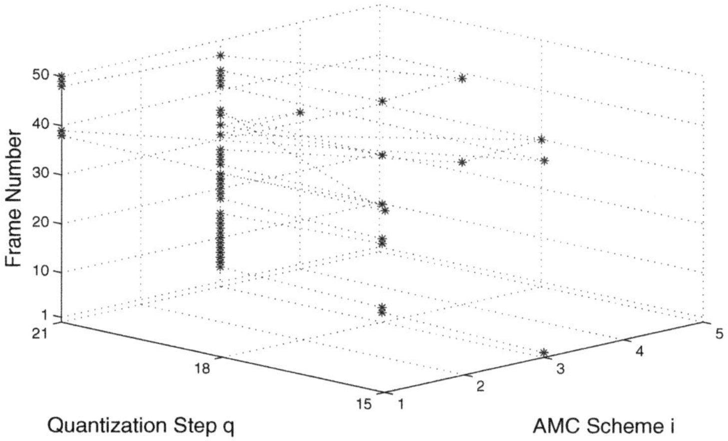

Fig. 5. The optimal solution under the constraint of 42 dB average frame distortion

3, . . ., 8]. Because different QP results in different bit rate. To maintain a smooth date rate and thereby a relatively constant power consumption on transmission to extend the battery lifetime, the difference of the selected QP for neighboring slices is limited within a threshold of 3. All frames except the first one are coded as inter frames. To reduce error propagation due to packet loss, 10 random I Macroblocks were inserted into each frame. The frames are packetized such that each packet/slice contains one row of MBs, which enables a good balance between error robustness and compression efficiency.

During an experiment by using the video clip Foreman, figure 2 shows the condition that no capacity constraint for video delivery was applied, and the optimization process chose the optimal solution which has the minimized distortion in each frame. In this case, the solution has a total PSNR of 2247.1 dB and average PSNR of 44.9 dB for each frame, battery capacity consumption is 0.0119 Ah. From the figure we can see that the control vector of each frame concentrates in the lower range of AMC scheme i and quantization step q. Figure 3 shows the scenario where no distortion constraint is applied for video delivery, and the optimization process chose the optimal solution which has the minimized battery capacity consumption in each frame. In this case, the solution has a total PSNR of 1462.4 dB and average PSNR of 29.2 dB for each frame, the battery capacity consumption is 0.0101 Ah. We can also see that the control vector of each frame concentrates in the higher range of the quantization step q, and all the frames are transmitted under the AMC scheme of number 8.

The optimization framework proposed in this paper was tested by three experiments under different values of distortion constraint. In the first experiment we applied the optimization toward the first 50 frames of the Foreman video clip, and set the average frame distortion constraint as 36 dB. After executing

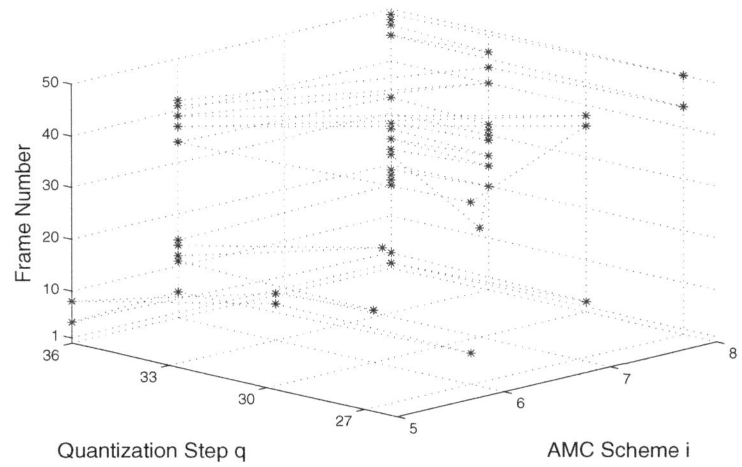

Fig. 6. The optimal solution under the constraint of 29 dB average frame distortion

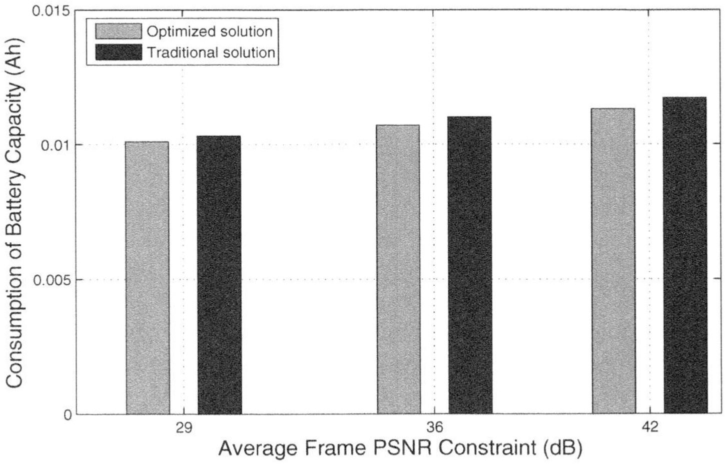

Fig. 7. Comparison of battery capacity consumption under three average frame distortion constraints

our framework, the control vector for each frame can be decided to minimize the battery capacity consumption and results in a received frame which has a distortion under 36 dB. Figure 4 shows the 50 control vectors corresponding to the first 50 frames of the tested video clip. Every point in the space represents an optimized control vector of one frame to satisfy the constraint. All the optimized

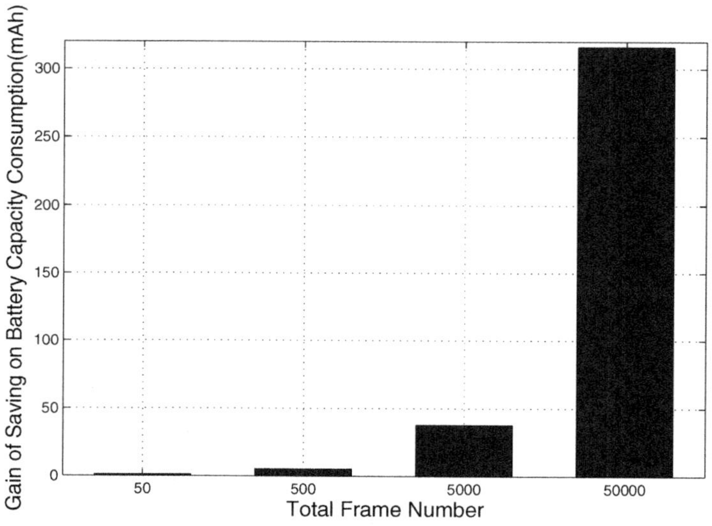

Fig. 8. Comparison of battery capacity saving in delivering 50, 500, 5000 and 50000 frames

control vectors of these 50 frames has formed an optimized solution for this video clip. By applying our optimization, this solution can still be figured out as the number of the video frames grows. Figure 5 and figure 6 show the other two testing results of establishing the optimized solutions under the average frame distortion constraints of 42 dB and 29 dB.

Figure 7 shows battery capacity consumption comparison between the optimal solution and traditional solution but satisfies the received framed distortion constraint. We can see that the solution selected by applying the framework has the most minimized battery capacity consumption under a certain video quality constraint. Figure 8 represents how much battery capacity can be saved according to the total number of video frames with proposed framework. In the figure, the gain of saving on battery capacity increases in an exponential fashion when the total number of video frames needs to be delivered increases. As a result, the proposed optimization can save considerable amount of battery capacity if it is applied to a relatively long video delivering case.

8 Conclusion

We developed an analytical framework for mobile wireless video communication systems driven by battery. A method to optimize the battery capacity consumption under a constraint of expected received video distortion is proposed. Based on analytical results, the video coding and transmission are jointly considered to minimize the battery capacity used for one video frame under the constraint of expected received frame distortion. Experimental results verified the efficiency and effectiveness of the proposed optimization framework.

Acknowledgements. This material is based upon work partially supported by the National Science Foundation under Grant No. 0801736.

References

1. He, Z., Liang, Y., Ahmad, I., Wu, D.: Power-Rate-Distortion Analysis for Wireless Video Communication Under Energy Constraints. IEEE Trans. Circuits Syst. Video Technol. 5, 645–658 (2005)
2. He, Z., Wu, D.: Resource Allocation and Performance Analysis of Wireless Video Sensors. IEEE Trans. Circuits Syst. Video Technol. 5, 590–599 (2006)
3. Katsaggelos, A.K., Zhai, F., Eisenberg, Y., Berry, R.: Energy-Efficient Wireless Video Coding and Delivery. IEEE Journal Wireless Communications 12, 24–30 (2005)
4. Lu, X., Wang, Y., Erkip, E.: Power Efficient H.263 Video Transmission Over Wireless Channels. In: Proc. International Conference Image Processing, pp. I-533–I-536 (2002)
5. Min, R., Furrer, T., Chandrakasan, A.: Dynamic Voltage Scaling Techniques for Distributed Microsensor Networks. In: Proc. IEEE Computer Society Workshop VLSI, pp. 43–46 (2000)
6. Kwon, D.N., Driessen, P.F., Basso, A., Agathoklis, P.: Performance and Computational Complexity Optimization in Configurable Hybrid Video Coding System. IEEE Trans. Circuts Syst. Video Technol. 16, 31–42 (2006)
7. Lian, C.-J., Chien, S.-Y., Lin, C.-P., Tseng, P.-C., Chen, L.-G.: Power-Aware Multimedia: Concepts and Design Perspectives. IEEE Circuits and Systems Magazine 7, 26–34 (2007)
8. Burd, T., Broderson, R.: Processor Design for Portable System. J. VLSI Signal Process. 13, 203–222 (1996)
9. He, Z., Cheng, W., Chen, X.: Energy minimization of Portable Video Communication Devices Based on Power-Rate-Distortion optimization. IEEE Trans. Circuits Syst. Video Technol. 18, 596–608 (2008)
10. Wu, D., Ci, S., Luo, H., Wang, H., Katsaggelos, A.K.: Application-Centric Routing for Video Streaming Over Multi-hop Wireless Networks. In: Proc. IEEE Communications Society Conference Sensor. Mesh and Ad Hoc Communications and Networks, pp. 1–9 (2009)
11. Krishnaswamy, D.: Game Theoretic Formulations for Network-Assisted Resource Management in Wireless Networks. In: Proc. IEEE Vehicular Technology Conference, vol. 3, pp. 1312–1316 (2002)
12. Krishnaswamy, D.: Network-Assisted Link Adaptation with Power Control and Channel Reassignment in Wireless Networks. In Proc. 3G Wireless Conference, pp. 165–170 (2002)
13. Rakhamtov, D.N., Vrudhula, S.B.K.: An Analytical High-Level Battery Model for Use in Energy Management of Portable Electronic Systems. In Proc. IEEE/ACM International Conference on Computer-Aided Design, pp. 488–493 (2001)
14. H.264 JM 16.2 reference software, `http://iphome.hhi.de/suehring/tml/download`

MERIT: P2P Media Streaming with High Content Diversity and Low Delay

Shiyun Zhang, Abhishek Bhattacharya, Deng Pan, and Zhenyu Yang

School of Computing and Information Sciences
Florida International University, Miami, FL 33199
{szhan001,abhat002,pand,yangz}@cis.fiu.edu

Abstract. P2P is successful in various multimedia applications such as On-demand/live streaming due to the efficient upload bandwidth usage among participating peers which offloads server request thereby saving bandwidth as system size scales up. Many designs were proposed for P2P multimedia streaming systems, including the most promising tree/mesh overlays. In this paper, we propose *MERIT* as an integrated framework for scalable mesh-based P2P multi-streaming whose design objective is to preserve content diversity as well as optimizing start-up delay while satisfying the in-/out- bound bandwidth constraints. We formulate our design goals as an optimization problem and start with a centralized heuristic exploiting the global knowledge of peers. We then present a decentralized version of our algorithm which is scalable and follows similar design principles as the centralized one. Simulation results indicate that our heuristics outperform state-of-the-art approaches by improving streaming quality and start-up delay with efficient utilization of bandwidth resources at each peer.

Keywords: peer-to-peer systems, multi-streaming, delay management, content dissemination, bandwidth.

1 Introduction

Peer-to-peer (P2P) overlay networks provide a promising approach for live/on-demand streaming of multimedia content in comparison to traditional approaches due to the absence of IP multicast support and the limited scalability of client-server based model. Recently, P2P-based multi-streaming systems are gradually emerging with their potentials illustrated in the literature, e.g., [6,3,13]. Our work is formulated on a P2P multi-streaming i.e. MDC model consisting of a single source that transmits video content organized in independent multiple streams to a large set of receivers with the goal of delivering high quality media in a scalable fashion by effectively utilizing the contributed resources of participating peers. The content dissemination architecture is constructed using mesh-based overlay which is proved to have better performance due to its adaptability for network dynamics and bandwidth heterogeneity [7].

Some of the important objectives for designing an efficient high-bandwidth P2P multi-streaming system are listed as follows: (1) accommodate bandwidth

X. Zhang and D. Qiao (Eds.): QShine 2010, LNICST 74, pp. 530–543, 2011.

heterogeneity, e.g., peers with diverse up-/down- link bandwidth constraints, (2) maximize multi-streaming quality, e.g., the *content coverage* as the number of disjoint streams, (3) minimize start-up delay (STPD), i.e., the delay to each respective parent along the data delivery path. Our main contribution is to incorporate the importance of *content diversity* by preserving rare streams in MDC-layered streaming which helps to improve quality in the system as a whole. The concept of content diversification using rarest-first strategy is not new and already very popular in file-distribution applications invoked initially by BitTorrent [4]. We utilize this strategy to be an important component in MERIT framework which represents the first effort of providing a careful integration for all the above mentioned objectives to improve QoS of real-time P2P multi-streaming applications.

Figure 1 shows an instance of a mesh-based streaming session where the streaming server, S, is involved in disseminating the multi-stream content (divided into streams a, b, c, d) to the participating peers labeled as A, B, C, D, and E. For simplicity, we assume each stream takes similar bandwidth and the residual outgoing bandwidth is represented as the number of supportable streams. Meanwhile, all the overlay links are marked with delays. The figure also illustrates the streams received at each peer. Given that, the start-up delay (STPD) is calculated as the maximum delay from its respective parent over associated multiple delivery paths. For example, the start-up delays of peers in the figure are as follows: $\text{STPD}_A = 8$, $\text{STPD}_B = 11$, $\text{STPD}_C = 7$, $\text{STPD}_D = 5$, and $\text{STPD}_E = 13$.

As an illustration of the problem, consider the MDC-based multi-stream planning for peer E as we want to maximize its content coverage (number of disjoint streams received) by using rarest-first stream ordering while reducing the start-up delay. We assume that the incoming bandwidth for E is 2 (i.e. can support

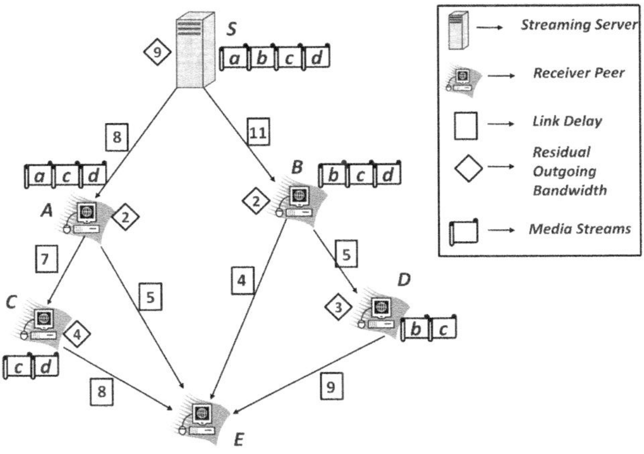

Fig. 1. A simple mesh overlay based P2P streaming model

a maximum of 2 streams). There are several choices depending on various optimization objectives. In [5], a greedy algorithm is proposed which always utilizes peers with maximal outgoing residual bandwidth. Thus, peer C will be selected as parent with the delivery path S→A→C→E achieving a content coverage of 2 (streams c, d from C) with a STPD of 8. In [9], a multiple-tree overlay based approach is proposed with the purpose of maintaining high streaming quality when nodes leave and join frequently. Since a parent is randomly chosen which has enough residual outgoing bandwidth to deliver a particular stream, the delivery paths could be S→A→C→E and S→B→D→E. The content coverage for this plan is 2 (c from D and d from C) with a startup delay of 9. We observe that currently there is no integrated solution which combines delay management, bandwidth constraints and rarity-ordered content coverage under one framework.

As motivated, we propose *MERIT* (media streaming with integrated framework) in this paper which aims at considering content coverage and start-up delay in an integrated fashion while satisfying the in-/out- bound bandwidth constraints. In the aforementioned example, the solution derived by *MERIT* will select A and B as parents with the delivery paths as S→A→E and S→B→E with a content coverage of 2 (a from A and b from B) and STPD of 5. *MERIT* consists of the delivery mesh construction and the content planning methods in a more tightly coupled manner to obtain better results than previously proposed schemes. Basically, we exploit rarest-first strategy and a *network-power* metric for decision making in the mesh construction and content planning stages to achieve optimal solutions. Simulation results further confirm the effectiveness of our solution in comparison with other state-of-the-art approaches.

The remaining paper is organized as follows. Related work is discussed in Section 2. The formal description of our problem is introduced in Section 3. Section 4 describes our proposed heuristic solution. Experimental evaluations are discussed in Section 5 and we conclude in Section 6.

2 Related Work

The concept of P2P streaming and its utility in video distribution was first proposed by [14]. Initial approaches involved the construction of a single multicasting tree overlay [12] and the content is distributed to all the nodes by organizing them into clusters/layers. Single tree approach faces some serious issues regarding vulnerability as the failure of an upper-level node cuts off the entire subtree and repairing the tree is difficult and costly with extreme conditions of dynamic peer participation. Multiple-tree approach was proposed in [2,9] to overcome this difficulty. [9] proposed a hybrid client/server architecture where the peers relay data to each other to protect the server thereby achieving scalability and robustness. [2] employs multiple description coding to break down the original stream into a number of sub-streams and then pushing each sub-stream through a specific stripe tree. The multiple trees are organized in an internal-node disjoint manner which achieves resiliency since the failure of a peer will only affect a single stripe tree thereby having lesser effect on streaming quality.

Mesh-based overlays were initially introduced as application layer multicast solutions but later it was incorporated in P2P video streaming domain and its effectiveness was proved over multiple tree approaches in [7]. The robustness of mesh overlays is mainly due to its randomized construction with each peer having multiple parent/children which helps for better adaptations during peer churn. According to related literature, research on mesh-based P2P streaming system can be divided into two important phases: (a) mesh construction [10,11], and (b) content scheduling [1,5]. [10] focused on the mesh construction problem and proposed a minimum delay mesh with the objective of minimizing end-to-end delay whereas [11] builds the mesh in a way for efficient utilization of peer bandwidth. On the other hand, [1] provided various selection schemes for push-based live streaming application and found a particular strategy to be optimal for both delay and rate whereas [5] proposed a content scheduling scheme for layered video streams for maximizing streaming quality. [15] is one of the commercially successful internet p2p live streaming systems which employs a mesh-based approach for connecting the peers and then selecting proper parents for receiving specific sub-streams. [7] also proposed a random mesh overlay by optimizing bandwidth-per-flow which will maximize the utilization of both incoming/outgoing bandwidth of all peers followed by a pull-based content dissemination mechanism.

Our main distinction from the previous work is that the organization of delivery mesh and content planning is done in a more integrated fashion with the desired objective of overarching rarity-ordered content coverage, start-up delay and bandwidth constraint under one common framework. We believe an integrated solution is critical for real-time multi-streaming applications.

3 Problem Formulation

We present a formal description of the problem which considers the delivery mesh construction and content planning phases across a set of receiving peers for streaming content from the server with asymmetric incoming and outgoing bandwidths. We introduce the following terminologies which will be helpful in formulating the problem:

- ***Mesh graph***, G=⟨V, E⟩ is modeled as a directed overlay network where V is the set of vertices representing the server/peers and E is the set of overlay edges. We denote v_0 as the streaming server and $P = \{v_1, v_2, .., v_{n-1}\}$ the set of receiving peers. Each peer v_i possesses an incoming bandwidth I_i and an outgoing bandwidth O_i quantified as the number of supportable streams. For v_0, we only consider its outgoing bandwidth.
- ***Link Delay***, d_{ij} is associated with every edge $e_{i \to j} \in E$ and $v_i, v_j \in V$ which represents the underlying unicast path delay from peer v_i to peer v_j in the physical network.
- ***Candidate parents***, H_i is the set of all possible parents derived from the mesh overlay for each receiving peer v_i where $H_i \subseteq V$ and let $|H_i| = l_i$.

– ***Selected parents***, H'_i is defined as the set of chosen parents from H_i for receiving peer v_i such that $H'_i \subseteq H_i$ according to some content planning schemes while respecting the incoming/outgoing bandwidth constraints. Let $|H'_i| = m_i$.
– ***Content set***, $C = \{C_1, C_2, .., C_c\}$ is the set of total content streams generated at server v_0 and the number of total content streams is $|C| = c$. We assume the entire video is divided into substreams using Multiple Description Coding (MDC) such that each substream can be rendered independently with the overall video quality improves as the total number of disjoint substreams increases. However, with modification our scheme can also be used with layered coding such as SVC where the higher layers are dependent on the lower layers. As a general idea of integrating delay management, content coverage and bandwidth constraint, our proposed framework is not restricted to any specific coding technique.
– ***Content coverage***, CC_i is the set of content received by peer v_i and is defined as follows:

$$CC_i = \bigcup\nolimits_{\forall v_k \in H'_i} C_i^k \tag{1}$$

where C_i^k denotes the content set received by v_i from parent v_k and each set C_i^k is disjoint in CC_i (i.e., $C_i^a \bigcap C_i^b = \emptyset$ for $\forall v_a, v_b \in H'_i$). The ideal objective will be to achieve $CC_i = C$ for each v_i.
– ***Rarity Index***, RI_i for each peer v_i is defined as: Suppose each content stream C_i collected in CC_i has a *rarity factor* r_i^j ($j \leftarrow 1$ to $|CC_i|$) which is defined by the total number of parents in H'_i that currently possess stream C_i. The Rarity Index is defined as follows:

$$RI_i = \sum\nolimits_{\forall j \in CC_i} r_i^j \tag{2}$$

– ***Startup delay***, Startup delay D_i for each peer v_i is defined as the maximum link delay with its neighbor among the possible delivery paths and is formulated as follows:

$$D_i = max_{\forall v_j \in H'_i} d_{ji} \tag{3}$$

where v_j is a selected parent of v_i (define $D_0 = 0$).

Our problem is to construct a content delivery mesh (i.e., deriving H'_i for each peer v_i) and a content planning scheme (i.e., C_i^k) such that the following objectives are satisfied:

$$\textbf{minimize:} \quad \sum_{i=1}^{n-1} RI_i \tag{4}$$

$$\textbf{minimize:} \quad max_{\forall v_i \in P}(D_i) \tag{5}$$

but subjected to the following constraints:

$$\forall v_i \in P : |CC_i| \leq I_i \tag{6}$$

$$\forall v_k \in V : \sum_{v_i \in P} C_i^k \leq O_k \tag{7}$$

4 refers to the minimization of the rarity-ordered content coverage which basically ensures to preserve rare streams thus helping to promote content diversity, and 5 indicates the minimization of the average startup delay among all the peers. 6 and 7 refers to the inbound/outbound bandwidth constraints respectively.

4 Proposed Solution

Given a mesh graph G, our objective is to create a delivery mesh so that requirements of content, delay and bandwidth are fulfilled. We employ a pull-based or receiver driven content dissemination mechanism where the initial process consists of selecting m_i parents in H_i' from l_i candidate parents in H_i. The naive solution for the problem of choosing m_i from l_i is to try all possible combinations till we exhaust and then finding the optimal one, which is computationally intensive. To avoid such strained process, we devised a simple mechanism for our heuristic solution which is divided into two phases: (1) *Stream Selection Policy*, followed by (2) *Parent Selection Policy*. Next, we describe the details of the two phases with an illustrative example as shown in Figure 2.

The Stream Selection Policy enforces the rarest-first strategy for preserving content diversity. The policy is simple and described as follows: Compute the *rarity factor*, r_i^j of each stream by summing up the number of parents in H_i' currently possessing the particular stream. Now, sort them in ascending order so that the rarest stream is at the top of the list. The sorted stream list is passed to the Parent Selection Policy for picking suitable parents with desired streams. The Parent Selection Policy scans the sorted list from the top and schedules a

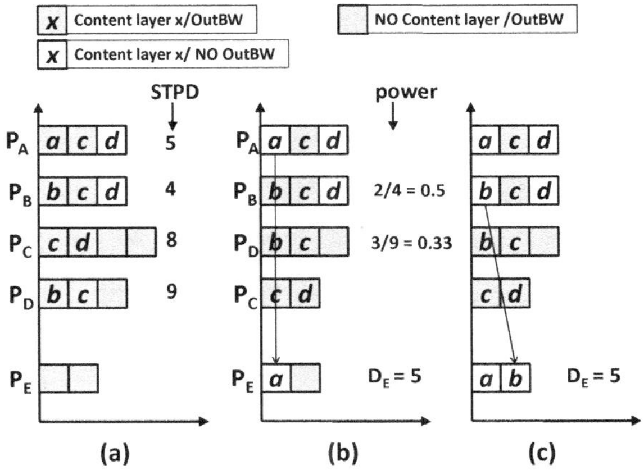

Fig. 2. Content planning in MERIT based on rarity and power ordering

single parent for each stream so that the content can be pulled. If there are more than one parent entry for a particular stream, then we need to make a decision for selecting one parent. In such situations, we utilize the concept of *network-power* as an optimization metric for ranking the parents in descending order and selecting the one with the highest power. Traditionally in networking, *power* is defined as the bandwidth divided by delay and is formulated as follows [10]:

$$pow_i^j = \frac{R_j \Uparrow}{d_{ji}} \tag{8}$$

where pow_i^j is the *power* of parent v_j with respect to child v_i, $R_j^{\Uparrow}$ is the residual outgoing bandwidth of parent v_j. The *power* metric essentially tends to prefer parents with higher available outgoing bandwidth and lower start-up delay.

Going back to the example in Figure 1, we show how this scheme translates the ordering to reach the solution computed by *MERIT* as illustrated in Figure 2. First, we compute the *rarity factor* of each stream with respect to peer E, namely, r_E^a, r_E^b, r_E^c, and r_E^d. For example, $r_E^a = 1$ since stream a is present with only parent A, and likewise, $r_E^b = 2$, $r_E^c = 4$, and $r_E^d = 3$. Next, we sort the stream list based on ascending order of *rarity factor* values, followed by the parent selection process. We can immediately assign stream a from parent A since there is only one provider. Since the in-coming bandwidth restriction for E is 2, we have chance to select one more stream for improving content coverage. *MERIT* will choose b from the sorted list since stream b is the second rarest content among the parents. Now, stream b is present with parents B and D, so we need to calculate the power (pow_E^B, pow_E^D) and select the parent with the highest value. For example, $pow_E^B = 0.5$ and $pow_E^D = 0.33$. Thus, B will be selected as the new parent and the final delivery paths computed are: S→A→E and S→B→E, as derived earlier in Section 1. The justification for this content planning mechanism is derived from the fact that it increases the content coverage by selecting the streams in the rarest-first order which will ensure content diversity in the whole system and thus, will lead to an overall improvement in streaming quality of all the peers. The *network-power* based parent selection mechanism tends to prefer parents with higher bandwidth and lower delay which will improve overall QoS by generating a high quality content delivery mesh.

4.1 Centralized Solution

In this section, we develop a centralized heuristic based on the complete knowledge of the peers and the entire mesh topology including bandwidth/delay for all peers/links. Our algorithm is shown in Table 1. The centralized algorithm is a greedy heuristic where it always chooses the parent hosting the stream with the highest *rarity factor* r_i^s as the first choice. The algorithm starts from the source and in each iteration it pushes the children (v_i) with the least startup delay that have not received any stream till now to expand the delivery mesh (Line 11). This is a favorable choice as it tries to greedily reduce the average startup delay among all the peers. In the content planning phase, peer v_i chooses

Table 1. Centralized MERIT algorithm

1. Initialize a set, $Suppliers \leftarrow \emptyset$
2. $Suppliers \leftarrow Suppliers \cup \{v_0\}$ // v_0 is source
3. Initialize a priority queue, $Neighbors$ ordered by increasing start-up delay
4. $Neighbors \leftarrow Neighbors \cup$ children of v_0
5. for each $i \leftarrow 1$ to $n-1$
6. $R_i^{\Uparrow} \leftarrow O_i$ // set residual outgoing BW to full
7. $R_i^{\Downarrow} \leftarrow I_i$ // set residual incoming BW to full
8. $change \leftarrow true$
9. while $(change == true)$
10. $change \leftarrow false$
11. $v_i \leftarrow Neighbors.dequeue$
12. $H_i \leftarrow$ up-link neighbors of $v_i \cap Suppliers$
13. Intialize a candidate stream list Z_i
14. for each $v_j \in H_i$
15. for each stream $s \in CC_j$
16. if$(s \notin Z_i)$
17. $Z_i \leftarrow Z_i \cup \{s\}$
18. Compute r_i^s for each $s \in Z_i$
19. Sort Z_i in ascending order based on r_i^s values
20. while$(|CC_i| < R_i^{\Downarrow})$
21. Extract next $s \in Z_i$ in sorted order
22. if$(r_i^s == 1)$//assume peer v_j have stream s
23. $parent \leftarrow v_j$
24. else if$(r_i^s > 1)$ //assume peer set X_i having //stream s
25. for each $x \in X_i$ Compute pow_i^x
26. Select $parent \leftarrow x$ with highest pow_i^x
27. if$(s \notin CC_i)$ and $R_{parent}^{\Uparrow} \geq 1$
28. $CC_i \leftarrow CC_i \cup \{s\}$
29. $change \leftarrow true$
30. if $(D_i > d_{ji})$ // update startup delay
31. $D_i \leftarrow d_{ji}$
32. $R_i^{\Downarrow} \leftarrow R_i^{\Downarrow} - 1$
33. $R_j^{\Uparrow} \leftarrow R_j^{\Uparrow} - 1$
34. if $(CC_i \neq \emptyset)$
35. $Suppliers \leftarrow Suppliers \cup v_i$
36. $Neighbors \leftarrow Neighbors \cup$ children of v_i

streams with increasing rarity (Line 21). Initially, the rarest stream with least r_i^s value is chosen and the number of parent's currently hosting the stream is found. If a single parent only host the stream, then the parent is selected for stream s provided bandwidth constraints are met (Lines 22, 23). If there are multiple parent's hosting stream s, then pow_i^j is computed for every candidate parent j, and the parent with the highest *network-power* value is selected for stream s

(Lines 24 to 26). After that, the startup delays, the in-/out- bound bandwidths are updated (Lines 30 to 33). Peer v_i continues to select parents until no more streams can be allocated thereby maximizing content coverage/streaming quality. The algorithm keeps pushing peers and adding paths into the delivery mesh according to the order of *rarity* and *network-power* metrics. It terminates when all the peers have been tested for receiving and forwarding streams to all other neighboring peers.

4.2 Decentralized Solution

A scalable and practical solution cannot be centralized since we cannot afford to store and maintain the global information at the server which will be overburdened with queries from increasing number of peers in the system. Moreover, it also creates a single point of failure and bottleneck related problems as the size of the system increases. To accommodate such situations, we implement a decentralized solution which is realistic and scalable but follows the same guiding principles as the centralized protocol.

To start with, we assume the existence of a rendezvous point for facilitating the new peer to join the system. The rendezvous point keeps a small random subset of live peers in the system. A joining peer contacts the rendezvous point upon entry and is provided with a candidate list of live peers. The joining peer then probes all the candidate parents in the list and in return receives the information regarding content streams, residual outgoing bandwidth and average start-up delay from them. To avoid waiting too long for the unresponsive parents due to slow processing or network congestion/packet loss, the joining peer waits for a fixed interval of time and then it removes the parent from the candidate list. The joining peer calculates the *rarity factor* scores of each stream from the newly formed neighborhood of parent set with the information received. The joining peer sorts the streams based on decreasing *rarity factor* value and then follows a greedy selection method. The joining peer chooses the parent with the highest *network-power* score from Equation 8 and streams from it based on the bandwidth/content availability. If this parent cannot serve the joining peer, then the next parent is contacted and the process is repeated till either all the streams are received by the joining peer or there are no more new parents to scan. Once the parents are selected, the next part of content planning and delivery path generation is the same as the centralized protocol.

Peer leave can be of two types: (a) graceful leave, when the leaving peer informs its parents and children beforehand so that they can have sufficient time to readjust the delivery mesh as quickly as possible, (b) peer failure, when the peer voluntarily/involuntarily leaves the system without informing others. Failure is hard to overcome as it can happen at any time in any volume without prior information. To tackle this problem, each peer probes its parents and children at periodic intervals, and if any peer doesn't respond back within the specified period then it drops the respective entry from its list and frees the allocated resources. If a peer looses many parents and its streaming quality is sufficiently degraded, then it can query its neighbor's neighbor or the rendezvous point in search of potential parents.

We can observe that the initial peer selection for joining the mesh and stream selection with the highest rarity value in the whole system, are optimal choices by centralized protocol considering the global information availability. For a decentralized protocol, the selection space is pruned down to only the neighborhood parent set, which are sub-optimal choices in comparison to the centralized version. Still, a decentralized solution is desirable due to its higher system efficiency in requirement to maintain only a small set of neighborhood information. Moreover, the shortfall in QoS performance metrics of decentralized in comparison to centralized is within acceptable limits (as explored in the following section) making it a more suitable and practical choice.

5 Experimental Evaluation

In this section we discuss about the various simulation experiments that are performed to test the different algorithms under various scenarios. We use BRITE [8] to generate the topologies for our experiments. Each topology was generated from top-down hierarchical models with autonomous systems and routers. Peers are generated randomly and attached to the router nodes with the total number of peers varying from 100 to 1000. The access link bandwidths are set in the range of 100 kbps and 500 kbps with a mixture of exponential, uniform and heavy-tailed distributions. We intentionally limit the maximum bandwidth within 500 kbps and take a pessimistic approach to test the *MERIT* algorithms in resource scarce conditions. We keep the streaming rate of each stream at a constant of 100 kbps unless otherwise noted. The number of total streams is kept at a constant value of 6(unless otherwise specified) for all experiments. We plot all the results by taking an average of 10 runs for each point in the graph with a confidence interval of 95% and different BRITE topologies for each run. To evaluate the effectiveness of rarity in MDC-based layered streaming, we experimented with two different versions of decentralized MERIT: (1) MERIT_R i.e.,considering the rarity function as described in the pseudocode, and (2) MERIT_N i.e., replacing the rarity based stream selection with a randomized one to improve streaming quality and a power-based parent selection process. We compare our results with CoopNet (Multiple-tree approach as proposed in [9]). We build distribution trees as described in CoopNet and employ a random parent selection scheme respecting bandwidth constraints for each tree which allocates corresponding stream.

We first compare the content coverage (i.e. the total number of distinct streams received) of our approach with CoopNet as shown in Figure 3. Clearly, the content coverage of MERIT is better than CoopNet. The important observation in Figure 3 is that even MERIT_R performed better than MERIT_N which states the effectiveness of preserving rarity in MDC-based layered streaming solutions. Each point in the graph represents the average content coverage over the total number of peers. The average content coverage over all the different network sizes for MERIT_R comes to 4.499 (and 3.427 for MERIT_N) whereas

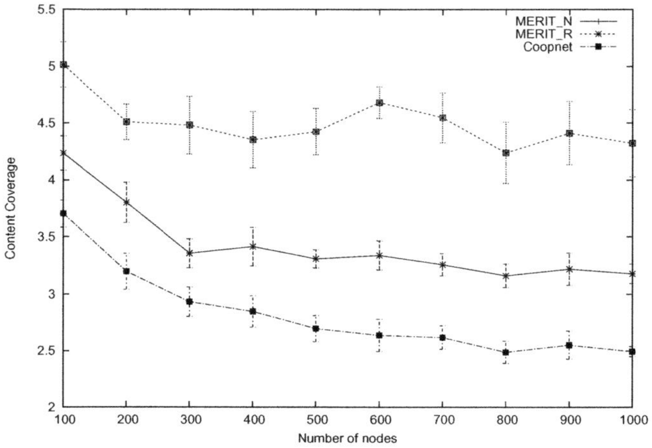

Fig. 3. Plot representing average content coverage over different network sizes

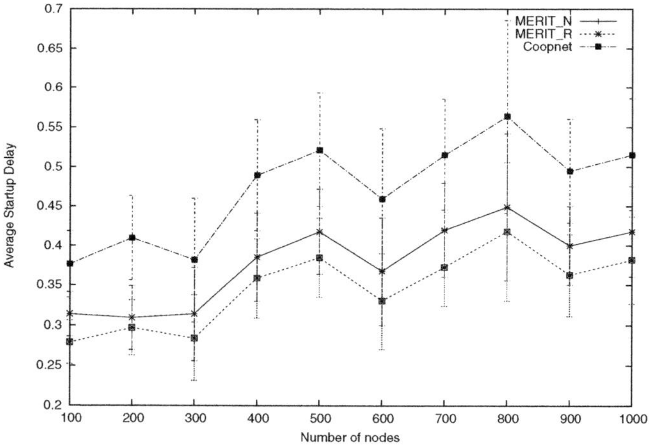

Fig. 4. Comparsion Plot of average start-up delay over different network sizes

CoopNet accounts for 2.816. The rarity-based MERIT_R achieves a 59.8% performance gain over CoopNet which proves that the concept of *content diveristy* and *network-power* helps to improve the streaming quality to a large extent. We also observe that the performance curve for MERIT_R does not drop and remain within acceptable limits with the increase in network size which indicates the scalability of the protocol.

Minimizing startup delay for each peer is one of MERIT's optimization objectives whereby the peers with low startup delay are given preference in joining the content delivery mesh and parent selection policy as evident in the expression of

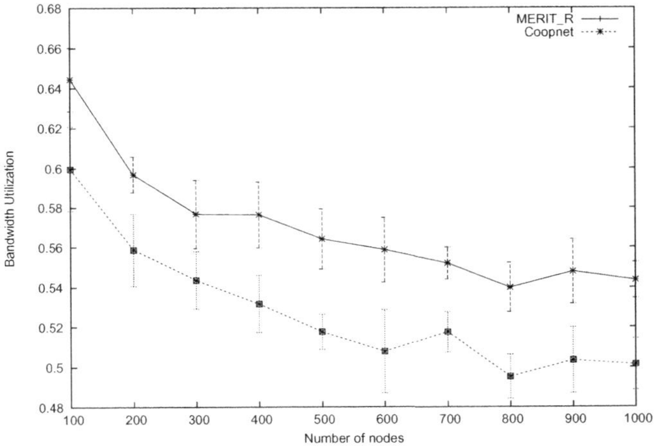

Fig. 5. Percentage bandwidth utilization over different network sizes

power. We plot the average startup delay achieved by each algorithm in Figure 4 where each point in the graph represents the value collected from all peers. From Figure 4, it is clear that the start-up delay of MERIT_R and MERIT_N is better than CoopNet which is quite natural since CoopNet does not make any attempt to reduce start-up delay explicitly. The *power* based parent selection helps to improve the start-up delay to a considerable extent which is an important objective for live, interactive streaming applications. MERIT_R achieves about 24.8% startup delay reduction over CoopNet. The average startup delay across all the network sizes for MERIT_R is 0.379 and for MERIT_N it comes to 0.346 which is comparatively better than that of CoopNet at 0.473. MERIT_R and MERIT_N are within comparable limits since rarity does not help in improving average start-up delay.

We plot the percentage of bandwidth utilization of MERIT_R and CoopNet over different network sizes as shown in Figure 5. It is observed that outgoing bandwidth is a scarce resource and needs to be efficiently utilized by the system. On the other hand, the incoming bandwidth is more abundant since a peer generally has its incoming bandwidth to be multiple times of its outgoing bandwidth. In our experimental setting we have assigned incoming/outgoing bandwidths to be between 100 kbps and 500 kbps and so it will be more relevant to compute the aggregated outgoing/incoming bandwidth utilization. The average bandwidth utilization factor for MERIT_R is around 58% which is generally acceptable and does not seem to vary much by increasing network size. The average bandwidth utilization for CoopNet comes to 52.7% indicating an average of 5.3% performance enhancement of MERIT over CoopNet. We are currently investigating further to improve the bandwidth utilization and more even distribution among the peers which will help to uniformly share the streaming quality.

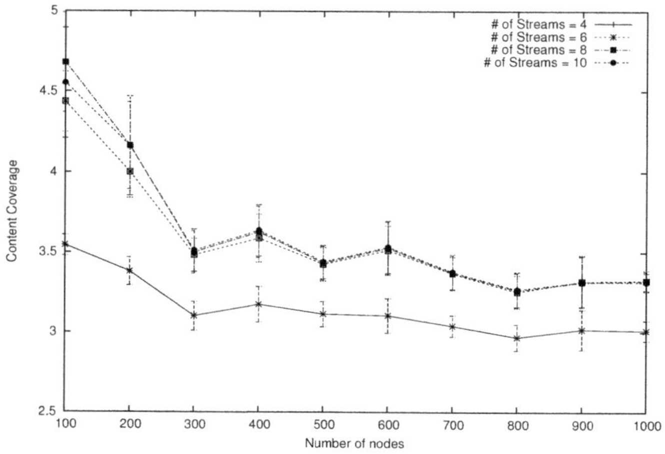

Fig. 6. Plot of streaming quality for number of layers=4, 6, 8, 10

Next, we investigate the effect of number of streams on content coverage for centralized MERIT with results plotted in Figure 6. Increasing the number of streams will provide diversification of content availability but places more demand on the system as each peer will strive to receive the maximum number of streams. Analyzing Figure 6, we observe that the percentage of content coverage decreases consistently with the increase in the number of streams. We discover that the percentage of content coverage with 4 streams is 78.6% which decreases to 59.5% for 6 streams and further decreases to 45.3% with 8 streams and 36.1% with 10 streams for constant bandwidth restrictions of 100-500 kbps. So, increasing the number of streams generates more system load in MERIT which reacts by pulling down the streaming quality of the peers.

6 Conclusion

In this paper, we presented MERIT, which is an integrated mesh-based P2P media dissemination solution under the scenario of MDC-based multi-streaming and heterogeneous resource distribution of peers. We argued the importance of content rarity preserving property in layered media streaming application which helps to improve diversity, thus achieving an overall improvement in quality over all peers as verified through simulation studies. Moreover, we also incorporated the concept of network-power to improve the start-up delay which is a crucial parameter for live interactive applications. We proposed the MERIT framework composed of Stream Selection Policy and Parent Selection Policy to generate a high-quality content delivery mesh in layered media streaming solutions. We have described a formal framework of the problem and the various desired objectives with constraints that are required to be met. Based on our approach, we proposed a centralized heuristic with complete knowledge as a baseline for design principle, and then followed it by presenting a decentralized solution using similar conventions. Simulation results have shown that MERIT achieved a high content coverage with low start-up delay compared to the state-of-art approach. We envision the usefulness of MERIT as a p2p multi-streaming solution for various interactive multimedia applications.

References

1. Bonald, T., Massoulie, L., Mathieu, F., Perino, D., Twigg, A.: Epidemic live streaming: Optimal performance trade-offs. In: Proc. of SIGMETRICS, pp. 325–336 (2008)
2. Castro, M., Druschel, P., Kermarrec, A.M., Nandi, A., Rowstron, A., Singh, A.: Splitstream: High-bandwidth multicast in co-operative environments. In: Proc. of SOSP, pp. 298–313 (2003)
3. Cheng, W., Liu, D., Ooi, W.T.: Peer-assisted view-dependent progressive mesh streaming. In: Proc. of ACM Multimedia, pp. 441–450 (2009)
4. Cohen, B.: Incentives build robustness in bittorrent. In: Proc. Workshop on Economics of Peer-to-Peer Systems (2003)
5. Cui, Y., Nahrstedt, K.: Layered peer-to-peer streaming. In: Proc of ACM NOSSDAV, pp. 162–171 (2003)
6. Hu, S.Y., Huang, T.H., Chang, S.C., Sung, W.L., Chen, B.Y.: Flod: A framework for peer-to-peer 3d streaming. In: Proc. of IEEE INFOCOM, pp. 1373–1381 (2008)
7. Magharei, N., Rejaie, R.: Prime: Peer-to-peer receiver driven mesh-based streaming. In: Proc. of IEEE INFOCOM, pp. 1415–1423 (2007)
8. Medina, A., Lakhina, A., Matta, I., Byers, J.: Brite: An approach to universal topology generation. In: Proc. of IEEE MASCOTS (2001)
9. Padmanabhan, V.N., Wang, H.J., Chou, P.A., Sripanidkulchai, K.: Distributing streaming media content using co-operative networking. In: Proc. of ACM NOSSDAV, pp. 177–186 (2002)
10. Ren, D., Li, Y.T.H., Chan, S.H.G.: On reducing mesh delay for peer-to-peer live streaming. In: Proc. of IEEE INFOCOM, pp. 1058–1066 (2008)
11. Small, T., Li, B., Liang, B.: Outreach: Peer-to-peer topology construction towards minimized server bandwidth costs. IEEE Journal on Selected Areas in Communication, 35–45 (2007)
12. Tran, D.A., Hua, K.A., Do, T.T.: Zigzag: An efficient peer-to-peer scheme for media streaming. In: Proc of IEEE INFOCOM, pp. 2102–2111 (2003)
13. Xiao, X., Shi, Y., Gao, Y., Zhang, Q.: Layerp2p: A new data scheduling approach for layered streaming in heterogeneous networks. In: Proc. of IEEE INFOCOM, pp. 603–611 (2009)
14. Xu, D., Hefeeda, M., Hambrusch, S.E., Bhargava, B.K.: On peer-to-peer media streaming. In: Proc of ICDCS, pp. 363–371 (2002)
15. Zhang, X., Liu, J., Li, B., Yum, T.K.P.: Coolstreaming/donet: A data-driven overlay network for peer-to-peer live media streaming. In: Proc. of IEEE INFOCOM, pp. 2102–2111 (2005)

An Overview of the DSRC/WAVE Technology

Yunxin (Jeff) Li

NICTA, 13 Garden Street
Eveleigh, NSW 2015, Australia
Jeff.Li@nicta.com.au

Abstract. Wireless vehicular networks operating on the dedicated short-range communications (DSRC) frequency bands are the key enabling technologies for the emerging market of intelligent transport system (ITS). The wireless access in vehicular environments (WAVE) is significantly different from the Wi-Fi and cellular wireless networking environments. The specifications defined by IEEE802.11P and IEEE1609 represent the most mature set of standards for DSRC/WAVE networks. This paper provides an overview of the current state of the art, and analyses the potential differences between application requirements and what can be offered by the current WAVE solutions. It is shown that the current solutions may be inadequate for large-scale deployment. The primary challenge is to develop scalable, robust, low-latency and high-throughput technologies for safety applications that will significantly reduce collisions and save lives and property loss. Further research ideas are proposed to address this challenge.

Keywords: WAVE, DSRC, IEEE802.11P, IEEE1609, VANET, OFDM, QoS, scalability, latency, throughput.

1 Introduction

There has been tremendous investment from government, academia and industry under the big umbrella of intelligent transport systems (ITS), leading to the development of safety and traffic management technologies in vehicles and road infrastructure. Wireless vehicular communications and networking is a key enabling technology for future ITS services. The International Organization for Standardization (ISO) TC204 WG16 is developing a family of international standards and architecture on communications access for land mobiles (CALM). It is expected that the future CALM system will make use of a wide range of technologies including satellite, cellular (GSM, 3G and 4G/WiMAX), Wi-Fi wireless local area network (WLAN) and its wireless access in vehicular environments (WAVE) evolutions (IEEE802.11P and IEEE P1609), Bluetooth wireless personal area network (WPAN), mm-Wave, infrared and radio frequency identification (RFID). In addition, many sensing technologies such as radar, imaging and video processing will be integrated into the CALM architecture. Most of the communication technologies in the CALM family are borrowed from other mature applications, with the exception of the recently proposed WAVE standards on the dedicated short range communications (DSRC) frequency band. DSRC/WAVE is the only wireless technology that can potentially

X. Zhang and D. Qiao (Eds.): QShine 2010, LNICST 74, pp. 544–558, 2011.

meet the extremely short latency requirement for road safety messaging and control. The unique feature of low latency secures the role of DSRC, as an essential communication technology, in future CALM networks that will make use of multi-radios on multi-bands. However, the current DSRC solutions are not fully field proven. There are significant DSRC-related social and technical challenges that have to be dealt with before large-scale deployment.

There are two classes of devices in a WAVE system[1-2]: on-board unit (OBU) and roadside unit (RSU). They are equivalent to the mobile station (MS) and base station (BS) in the cellular systems respectively. There are two classes of communications enabled by the OBUs and RSUs: vehicle to vehicle (V2V) and vehicle to infrastructure (V2I). While a MS in the cellular environment normally communicates with another MS via the BS, the OBU in a vehicle normally directly communicates with other OBUs within the radio coverage area. This direct V2V communication reduces the message latency and low latency is an essential requirement for safety applications such as collision avoidance[3]. Another difference is that an OBU is more likely to be embedded and connected with other electronic systems of the vehicle via in-vehicle networking such as controller area network (CAN) and FlexRay, while a MS is normally detached from the CAN. In addition to improving safety, WAVE networks can play major roles in travel plan, traffic management, navigation, fleet and asset management, environment monitoring, logistics, congestion and emission reduction, toll collection, smart parking, emergency services and a wide range of other location-based services.

WAVE networks have a set of technical challenges not encountered in other wireless networks. One challenge is to use WAVE technology in collision avoidance between fast moving vehicles. For example, it can be used to warn the drivers at the crossing between roads and railways if there is a danger of collision. In V2V communication, the relative velocity between two vehicles moving in the opposite direction is the sum of their individual speeds. In addition, such V2V communication system has to be robust in extremely abnormal situations, as accidents and collisions are less likely to happen in normal situations. One example of such an abnormal situation is when two cars are traveling on a narrow two-way street towards each other at fast speed. Therefore V2V communication has deal with much faster fading and much more Doppler frequency spread than any other wireless systems. On the other hand, most other wireless communication systems such as the Wi-Fi and cellular systems are designed to work in well anticipated and even controlled environments. Fundamentally WAVE networks have to be extremely robust as their failure may cause the loss of life and property. Some messages transmitted on a WAVE network have a tight latency requirement, and a decision based on delayed information could be quite harmful. The WAVE networks may operate in a wide range of hash environments. The density can vary from a few vehicles to perhaps tens of thousands of vehicles in the radio coverage area. To meet these challenging requirements, a WAVE solution must be scalable, robust, low-latency, high-throughput and cognitive[4].

The primary objective of this paper is to provide an overview of the current DSRC/WAVE technologies, identify their strength and weakness, and stimulate further research. The rest of the paper is organized as follows. In Section 2, we provide a short review of the spectrum allocation and its associated regulatory requirements. Section 3 is dedicated to the introduction of the IEEE802.11P and IEEE

1609 standards. We then analyze the latency, capacity and other requirements for large-scale DSRC deployment in Section 4. The requirement analysis is compared with the current DSRC solutions to identify their gap. Further research ideas are proposed to address this gap. Finally summary and conclusion are drawn in Section 5.

2 Regulatory Requirements

In US, 75 MHz of spectrum in the 5.9 GHz frequency band has been allocated for DSRC applications. Out of the 75 MHz spectrum, 5 MHz is reserved as the guard band and seven 10-MHz channels are defined as in shown in Fig. 1. The available spectrum is configured into 1 control channel (CCH) and 6 service channels (SCHs). The CCH is reserved for carrying high-priority short messages or management data, while other data are transmitted on the SCHs. The pair of channels (channel 174 and 176, and channel 180 and 182) can be combined to form a single 20-MHz channel, channel 175 and 181 respectively. The channel number (*CN*) is derived by counting the number of 5-MHz spectrum in the frequency band from 5000 MHz to the center frequency $f(CN)$ of the channel *CN*, i.e.,

$$f(CN)=5000 + 5CN \quad \text{(MHz)}.$$

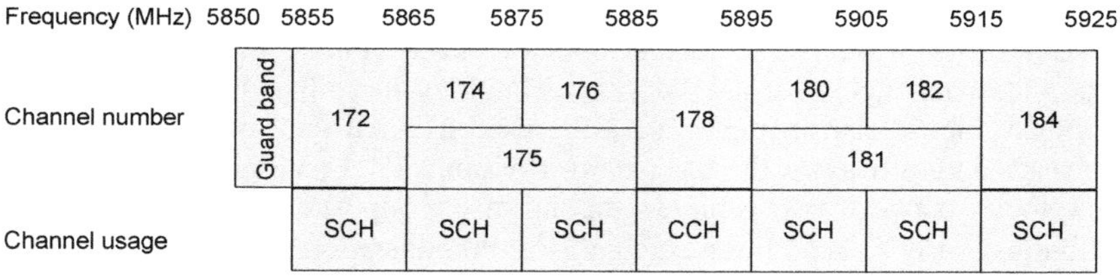

Fig. 1. The DSRC Frequency Allocation in US

In terms of transmitter (TX) power, four classes of devices have been defined whose maximum TX power ranges from 0 dBm to 28.8 dBm. The associated coverage distance by a single radio link depends on the channel environment, the TX power and the modulation and coding schemes (MCS) used. This distance may range from 10m to 1km. The details of OBU and RSU TX limits of equivalent isotropically radiated power (EIRP) also depend on the operating CN and applications. It is worth noting that the current FCC code of federation regulations (CFR) heavily refers to the American Society for Testing and Materials (ASTM) standard E2213-03, while the industry is adopting the IEEE802.11P and IEEE 1609 standard. The IEEE standard on the other hand refers to the FCC CFR for regulatory requirements. This means that implementers should address the channel and power limit defined in the ASTM standard. For example, FCC CFR specifies that the channel 172 and 184 shall be used for "public safety applications involving safety of life and property"[5] and this requirement is not fully compatible with the current IEEE1609.4 multi-channel operation where it is more natural to use channel 178 (i.e., the CCH) for such applications. Other than some minor differences in power level and spectrum mask requirements, ASTM standard E2213-03 and IEEE802.11P are both based on IEEE802.11A and they are effectively compatible.

Other frequency bands have also been used for DSRC applications even before the 5.9 GHz band allocation. They were typically used for highway or city central business district (CBD) toll collection. Of particular interest are the frequency bands defined in Table 1. It is worth noting that the DSRC regulatory requirements in many parts of the world are in the process of being finalized. There is a chance that similar spectrum allocation and requirements will be adopted world wide for DSRC applications. Spectrum harmonization is desirable for global inter-operability and low-cost DSRC services.

Table 1. Spectrum Allocation for WAVE/DSRC Applications

Country/Region	Frequency Bands (MHz)	Reference Documents
ITU-R (ISM band)	5725-5875	Article 5 of Radio Regulations
Europe	5795-5815, 5855/5875-5905/5925	ETS 202-663, ETSI EN 302-571, ETSI EN 301-893
North America	902-928, 5850-5925	FCC 47 CFR
Japan	715-725, 5770-5850	MIC EO Article 49

3 DSRC/WAVE Standards

Collectively the IEEE 1609 family, IEEE802.11p and the Society of Automotive Engineers (SAE) J2735[6] form the key parts of the currently proposed WAVE protocol stack. The WAVE protocol architecture with its major components is shown in Fig. 2, and they are summarized as follows.

- IEEE P1609.0 "Draft Standard for Wireless Access in Vehicular Environments (WAVE) – Architecture."
- IEEE 1609.1 "Trial Use Standard for Wireless Access in Vehicular Environments (WAVE) - Resource Manager."
- IEEE 1609.2 "Trial Use Standard for Wireless Access in Vehicular Environments (WAVE) - Security Services for Applications and Management Messages."
- IEEE 1609.3 "Trial Use Standard for Wireless Access in Vehicular Environments (WAVE) - Networking Services."
- IEEE 1609.4 "Trial Use Standard for Wireless Access in Vehicular Environments (WAVE) - Multi-Channel Operations."
- IEEE P1609.11 "Over-the-Air Data Exchange Protocol for Intelligent Transportation Systems (ITS)."
- IEEE802.11p Part 11: "Wireless LAN Medium Access Control (MAC) and Physical Layer (PHY) specifications -- Amendment: Wireless Access in Vehicular Environments."

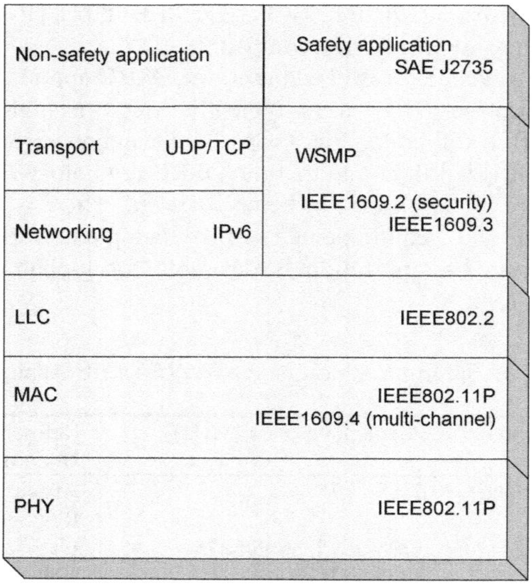

Fig. 2. The WAVE Protocol Stack and Its Associated Standards

3.1 The WAVE PHY

The current FCC CFR still refers the ASTM E2213 as the PHY standard. ASTM E2213 was published in 2003, and was based on the IEEE802.11A OFDM PHY. Since 2007 IEEE has consolidated all older versions of the PHY and the MAC into the IEEE 802.11–2007 edition. IEEE802.11P is an amendment to IEEE802.11-2007 for WAVE applications. Compared to IEEE802.11-2007, minimum change has been proposed in IEEE802.11P. In particular, IEEE802.11P only adopts the OFDM PHY on 10-MHz channels in the 5.9 GHz frequency band. On the other hand, the Wi-Fi industry normally implements the OFDM PHY on the 20-MHz channels, even though 5/10/20 MHz channels have been specified in IEEE802.11-2007. Compared to the 20-MHz Wi-Fi OFDM PHY, the subcarrier spacing and the supported data rate of IEE802.11P are halved while its symbol interval including cyclic prefix (CP) is doubled. Other parameter comparisons are shown in Table 2. In addition, IEEE802.11P requires the signal spectrum to decay faster to further reduce the adjacent channel interference. Different TX filtering may impact other TX performances such as error vector magnitude (EVM) to which a designer should pay attention.

The receiver design is typically out of the scope of the standard specification. However, due to significantly different channel environments, it is expected that a WAVE receiver may attract special design considerations. Compared to Wi-Fi OFDM receivers, the WAVE receivers have to deal with much higher Doppler spread which causes inter-carrier interference (ICI). The fact that the subcarrier spacing has been halved means that the WAVE OFDM receiver is more sensitive to carrier frequency offset and Doppler shift. Due to the higher Doppler spread and higher multi-path

Table 2. Comparison of WAVE and Wi-Fi OFDM Parameters

Parameters	WAVE	Wi-Fi
Frequency Band	5.9 GHz	5/2.4 GHz
Channel Bandwidth	10 MHz	20 MHz
Supported Data Rate (Mbps)	3, 4.5, 6, 9, 12, 18, 24, and 27	6, 9, 12, 18, 24, 36, 48 and 54
Modulation	Same as Wi-Fi	BPSK, QPSK, 16QAM and 64QAM
Channel Coding	Same as Wi-Fi	Convolutional coding rate: 1/2, 2/3 and 3/4
No. of Data Subcarriers	Same as Wi-Fi	48
No. of Pilot Subcarriers	Same as Wi-Fi	4
No. of Virtual Subcarriers	Same as Wi-Fi	12
FFT/IFFT Size	Same as Wi-Fi	64
FFT/IFFT Interval	6.4 μS	3.2 μS
Subcarrier Spacing	0.15625 MHz	0.3125 MHz
CP Interval	1.6 μS	0.8 μS
OFDM Symbol Interval	8 μS	4 μS

delay spread, the channel coherence bandwidth and channel coherence time become smaller, or in other words, the channel becomes more frequency-selective and faster fading. The following questions have to be addressed in designing a high-performance WAVE receiver.

- Is the CP sufficient to remove the inter-symbol interference (ISI) in the most hash WAVE environment with large multi-path delay spread?
- Can the fundamental requirement that the channel remains constant (or time invariant) during one OFDM symbol interval be met?
- Is the channel coherence bandwidth large enough so that the channel can be estimated on the pilot subcarriers and the estimates can be effectively interpolated to data subcarriers?
- Is the channel coherence time large enough to enable effective channel tracking? The fact that the OFDM symbol interval has been doubled and the channel is faster fading work against this condition.

Further field measurement, research and even change of the existing standardized parameters might be needed to achieve the satisfactory outcome.

3.2 WAVE MAC

In the architecture of IEEE802.11 networks, three kinds of service set (SS) are defined: basic service set (BSS), independent BSS (IBSS) and extended service set (ESS). The IBSS is formed by stations (STAs) without infrastructure, generally called an ad-hoc network. A BSS includes an access point (AP) that behaves as the controller/master STA. The ESS is the union of two or more BSSs connected by a distribution system (DS). A STA in the IBSS acting as the controller or the access point (AP) in the BSS periodically broadcasts a beacon that contains the service set ID (SSID) and other information. Other STAs in the SS receives the beacon and synchronize their time and frequency with those contained in the beacon. STAs can communicate with each other only if they are the members of the same SS. The same architecture of SS can be used for WAVE applications. However, forming a SS takes several steps including time and frequency synchronization, authentication and association. These steps take a time interval that is not affordable in some safety applications. In a vehicle traffic flow, two vehicles may be within the reach of the wireless link for less than a second. To minimize the message latency, a mode called "outside the context of BSS (OCB)" is introduced. The OCB mode applies to any two or more devices within the coverage area of a single radio link. A STA in OCB mode can send and receive data and control frames any time without forming or being a member of any SS. While enjoying the benefit of low-latency, the OCB mode does not receive the authentication, association or data confidentiality services on the MAC layer. The equivalent services have partially been moved to the higher layer as defined in IEEE1609.2.

From the receiver's point of view, care must be taken on frequency synchronization. Currently IEEE802.11P does not specify a different frequency accuracy for the WAVE transceiver oscillators. In IEEE802.11-2007, the frequency accuracy is specified to be ±20 ppm, i.e., the maximum frequency difference between the TX and RX oscillator frequencies can be up to 40 ppm. For example, the center frequency difference of transceivers operating on channel 184 could be as high as $5920\times40\times10^{-6}$ = 0.2368 MHz, greater than the subcarrier spacing of 0.15625 MHz. Without good frequency correction, the receiver error rate cannot be guaranteed even if the signal to noise ratio (SNR) is high. The preamble can be used to estimate the frequency offset. It is likely that WAVE radios may have access to other more accurate frequency sources such as the frequency derived form the GPS signal. The WAVE standard also support a timing advertisement frame, which can replace some of the features lost due to the lack of periodically sent SS beacon.

To transmit a frame either as a member of a SS or in OCB mode, the STA shall compete for the channel using the IEEE802.11 carrier sense multiple access/collision avoidance (CSMA/CA) mechanism. Another major MAC layer extension for WAVE applications is the multi-channel operation defined by IEEE1609.4. This extension makes use of the concept of frequency/time division multiple access (FDMA/TDMA), where the 7 FDMA channel frequencies are as shown in Fig. 1. The TDMA channel is shown in Fig. 3. The time is divided into repetitive periods of 100 ms. During each 100ms, 50 ms is allocated to CCH and another 50 ms is allocated for SCH, including 4 ms guarding interval (GI) for switching between CCH and SCH. The motivation is to accommodate single-channel radios in the WAVE system to access both CCH and SCH services. A single-channel radio is defined as a radio that can either transmit or receive on a single 10-MHz channel but not simultaneously. On the CCH frequency (i.e., channel 178) and during CCH time (i.e., the 46 ms), only two kinds of messages can be sent.

- Short messages, primarily for safety applications, as defined by the WAVE short message protocol (WSMP).
- WAVE service advertisement (WSA) messages used to announce the services available on other SCH frequency channels.

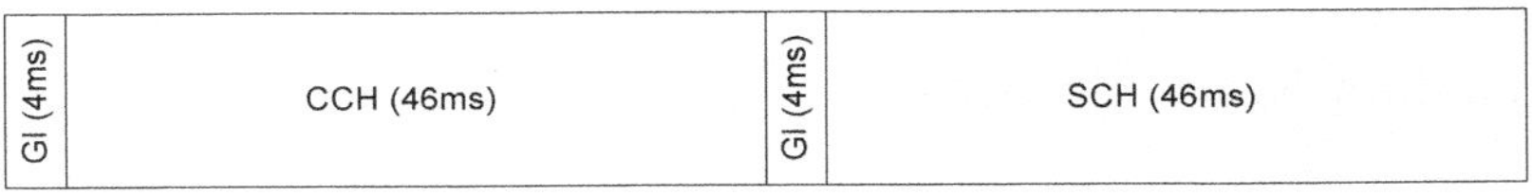

Fig. 3. The TDMA Extension of WAVE MAC

A single-channel radio can switch to other SCH frequencies for services during the SCH time and switch back to the CCH frequency during the CCH time for sending or receiving safety and other critical messages. The TDMA extension does not mean that the channels are idle for about 50% of the time. All SCHs can be active all the time to exchange service data and the CCH channel can be active all the tine to exchange control, management and other short messages. Practically it does mean a loss of time for safety messages sent on the CCH frequency. It cannot be guaranteed that all other STAs shall be listening on the CCH frequency during the SCH time, and as such safety messages sent on the CCH frequency during SCH time can be ineffective. Therefore it would be desirable to concentrate all safety messages on the CCH frequency and during CCH time, i.e., only 460 ms out of a second. This represents a capacity reduction for safety messages.

3.3 The WAVE WSMP

In addition to the standard IPv6 networking protocols operating over the SCHs, a WAVE-specific protocol called WSMP has been developed to carry messages on both the CCH and the SCHs. Unlike the standard IP protocol, the WSMP allows the applications to directly control the lower-layer parameters such as transmit power, data rate, channel number and receiver MAC addresses. To further shorten the latency, WSMP over the CCH can skip the steps for forming a WAVE BSS (WBSS) that delivers IP and WAVE short message (WSM) traffic on the SCHs.

The primary motivation for developing the WSMP is to reduce the overload. A WSMP packet is shown in Fig. 4. The overhead is 11 bytes, compared to a minimum of 52 bytes of a UDP/IPv6 packet. If a device receives a WSMP packet that has a WSM Version number not supported by the device, the received WSMP packet shall be discarded. The Security Type identifies if the packet is Unsecured, Signed or Encrypted. The Channel Number, Data Rate and TX Power allow the WSMP to directly control the radio parameters. The purpose of the provider service ID (PSID) field serves the similar role as the port number of the UDP/TCP packet, i.e., to identify the application that will process the WSM Data. The Length field indicates the number of bytes in the WSM Date, which might have been security-protected as specified in IEEE1609.2.

WSM Version (1 Octet)	Security Type (1 Octet)	Channel Number (1 Octet)	Data Rate (1 Octet)	TX Power (1 Octet)	PSID (4 Octets)	Length (2 Octets)	WSM Data (Variable)

Fig. 4. The Format of a WSMP Packet

4 Further Research Ideas

In this section we look at how the DSCR/WAVE technologies can be used to solve practical problems and specially safety applications. We identify the areas that need further research and verification.

4.1 Latency and Capacity Requirement

There have been a lot of safety applications proposed and studied in many projects. However a full scientific analysis of the latency and capacity requirements imposed to DSRC systems is still lacking. Currently the IEEE 1609.4 specifies the reoccurrence of the CCH at the rate of every 100 ms. If the OBU on each vehicle can capture the CCH during each CCH time, it can send its beacon and update its status to its neighbors at the rate of 10 Hz. The question is if the capacity of the DSRC system is large enough to accommodate all vehicles with a 10 Hz beacon. Additionally, have the latency requirements of all safety applications been met if the DSRC system can guarantee that all vehicles can update their beacons at 10 Hz? We believe that these questions have not been properly answered.

Fig. 5 shows the current WAVE over-the-air frame format. Let's assume that we can pack the 46 ms of CCH time with frames one after another with 58μs distributed inter-frame space (DIFS) between them. The combination of DIFS, Preamble and Signal fields takes 98μs, which can be considered as the minimum overhead to send a frame over the air. The Payload fields of Service, PSDU, Tail and Pad in Fig. 5 take a variable time depending on the size of the protocol service data unit (PSDU) and the MCS. Let's define the minimum air-time overhead ratio (OR) as the minimum overhead (98μs) divided by the time used for sending the Payload.

Preamble (32 μs)	Signal (8 μs)	Service (16 bits)	PSDU (variable)	Tail (6 bits)	Pad (variable)

Fig. 5. WAVE Over-the-Air Frame Format

Table 3 shows the capacity in terms of the maximum number of frames that can be sent over the air during the 46 ms of the CCH time that is allocated for short safety messages and service advertisement. To minimize the air time overhead, we prefer longer frames, and as a result less number of frames can be sent. This simple analysis shows that between 67 and 234 frames can be packed into this 46-ms time period. This analysis did not consider the time that will be lost when STAs compete for the wireless medium under the CSMA/CA MAC scheme, and therefore it represents the most optimistic capacity limit.

Table 3. The Capacity Analysis of the CCH

Payload Duration	Min Overhead Ratio	Max No. of Frames
98 μs	100%	234
196 μs	50%	156
294 μs	33.3%	117
392 μs	25%	93
490 μs	20%	78
588 μs	16.7%	67

Table 4 shows the number of vehicles that can exist in the coverage area of a single-link radio system. By comparing Table 3 and 4, we can easy identify many scenarios where the capacity is insufficient to allow each vehicle a chance to send its status every 100 ms. Is 100 ms latency good enough for all safety applications? Let's imagine a scenario of emerging collision of two cars speeding at 180 km/h on a two-way street towards each other in two adjacent lanes and one of them suddenly loses control. Within 100 ms, their distance can be shortened by 10m and if their starting clearance were less than 10m they would have collided within less than 100ms. In a practical system, the total latency is the sum of the time taken for sensing the event, communicating the event to the drivers and taking reactive control by the drivers. The total latency shall definitely be more than 100 ms of the network latency. We therefore conclude that the current DSRC solutions may be inadequate for large-scale deployment. Future research should focus on scalable solutions in terms of latency and capacity.

Table 4. The Number of Vehicles in a Radio Coverage Area

Coverage Radius	Area per Vehicle	No. of Vehicles
100 m	300 m^2	105
200 m	250 m^2	503
400 m	200 m^2	2513
600 m	150 m^2	7540
800 m	100 m^2	20106
1000 m	50 m^2	62832

4.2 The Social Science of DSRC

DSRC together with other technologies can enable smart applications we as a society never experienced before. Starting with simple applications like automatic toll collection, smart infrastructure can be built out that will significantly reduce traffic congestion, travel time, on-road accident, emission, driver stress and other unhealthy social behaviors. Unlike a human driver who can become bored, distracted and tired, a digital driving assistant (DDA) can stay alert and remain concentrated all the time. It can also have much better sensing and therefore information of the immediate and even far-reaching environment. For example, the DDA can see and understand the all-around 3-D environment with multiple video cameras and advanced signal processing algorithms. An aging/faulty component can alert the DDA that a break-down is about to happen unless corrective actions are taken. With the advancement of microelectronics and digital signal processing (DSP), a DDA can capture, store, process information and react to unexpected events faster than human drivers. Gradually the DDA will alleviate the driver more and more from the chores of driving until the vehicle becomes totally autonomous. Another scenario is remote driving. A driver could drive the vehicle or perhaps multiple vehicles from the comfortable office. DSRC networks enable the transparent control of the vehicles from anywhere in the world.

It is usually true that technologies that bring huge benefits if used appropriately can cause huge damage if abused. Therefore how to prevent the abuse of DSRC-enabled systems is a critical factor for the success of this new technology. PCs and mobiles phones are the two most popular devices for mass communication, and they are subject to abuse and malicious attacks. However, the damage caused by such attacks on PCs and phones cannot be compared with the fact that a terror organization can remotely drive a fleet of heavy trucks through DSRC networks. Physical and information security, user privacy[7] and other DSRC-related social sciences are the key issues to be addressed before large-scale deployment can be accepted by the government and the public. The legal impact of DSRC-enabled systems is another area needing further investigation. For example, we have to address the insurance policy, the legal procedures after the happening of on-road accidents and the legal liability of autonomous vehicle owners before such vehicles can be put on road. The psychology of driver behavior in interacting with DDA is another research area that will ultimately create the true requirements for DDA products and DSRC system design. We do believe that the social and technology problems related to DSRC systems can be solved and as a result future drivers will experience a fundamentally different driving experience.

4.3 Cross-Layer Optimization of DSRC Protocol Stack

The different layers of the DSRC/WAVE protocol stack and its regulatory requirements are developed by several working groups at different times. There is not enough evidence that the protocol stack is optimized in terms of scalable capacity and latency[8-9].

The current PHY was effectively a re-use of the 10-MHz channel Wi-Fi OFDM PHY without any change of parameters. It is advantageous to have compatibility between WAVE and Wi-Fi systems, and such re-use obviously has shortened the time to define the standard. However, the WAVE and Wi-Fi operating environments are significantly different. Ideally the PHY should be designed based on the measurement and modeling of the wireless channel[10-12]. The WAVE channel environment is very much diversified in terms of vehicle density, vehicle velocity, TX power, path-loss, multipath components, coverage range and the environmental electromagnetic interference (EMI). There are conflicting requirements, for example, longer CP is good for removing the ISI due to strong multipath, but bad for spectrum efficiency.

One improvement is to adapt the OFDM parameters to the real-time operating environment. DSRC radio is usually integrated with other sensing technologies such as GPS and therefore has sufficiently reliable information to optimize the PHY parameters real-time. Currently the application layer has control of the channel number, data rate and TX power. It can be expanded to include other parameters such as antenna gain, directivity and polarization. From example, the electronic braking signal should be sent to the vehicles behind rather than in front. By physically focusing the beam to the desired directions, we can save energy, and the same spectrum can be simultaneously reused in other directions. Orthogonal frequency division multiplexing (OFDM) has a disadvantage of large peak to average power ratio (PAPR) and almost all transmitters are peak power limited. This peak power limit, as an important parameter of the TX, is readily available to the application layer. When the application layer determines the average power used for the sending the message, it has to make a pre-defined back off from the peak power. In other words, the average power limit is less than the peak TX power limit by, e.g., 10 dB. The current application layer will never try to exceed the average power limit. Such an approach is sub-optimum, and a better approach might be for the application layer to set the coverage range and latency requirement and let the lower layers figure out the optimum set of parameters to achieve the coverage and latency requirement.

We use the example of peak power reduction of OFDM signals to demonstrate the advantages of having the lower layers to control the power. The PAPR of an OFDM signal is a random variable whose values are generally greater than one, but depends on the data that created the modulated waveform. On many occasions the peak power limit of the TX is not reached, either because the required average power to reach the coverage area is low or because the particular set of data happen to create a low-PAPR OFDM signal. Occasionally the peak power exceeds the limit, and if not properly dealt with, will cause in-band distortion that will reduce the error performance of the concerned channel and cause signal spillage into adjacent channels that will reduce the error performance of other channels. There are many PAPR reduction schemes that can be applied to reduce the peak power and as a result to boost the average power if required. One simple approach is to use the tone reservation (TR) scheme that trades the combination of data tones and tones reserved for PAPR reduction. It is better to let the PHY make the trade-off decision, and by doing so a more scalable dynamic range in the average power (and consequently the coverage area) can be achieved. In Fig. 6 we show an example of peak reduction scheme. The average power has been normalized to 1 mw (or 0dBm). The X-axis shows a threshold for the peak power. The y-axis shows the probability that the peak power of an OFDM modulated signal is higher than the threshold. A variable number of iterations (0, 5, 10, 15 and 20) using the TR scheme[13] has been illustrated with 0 standing for the original OFDM signal. It is shown that more than 3 dB of peak power reduction can easily be achieved. In the PAPR reduction experiment we have used only two data tones and the unused virtual tones for peak power reduction, causing minimum sacrifice in net data rate loss. It has been shown that the peak power can be reduced by more than 6 dB by sacrificing about 20% of the data tones for peak power reduction[13]. There is an apparent trade-off between effective data rate and peak power reduction.

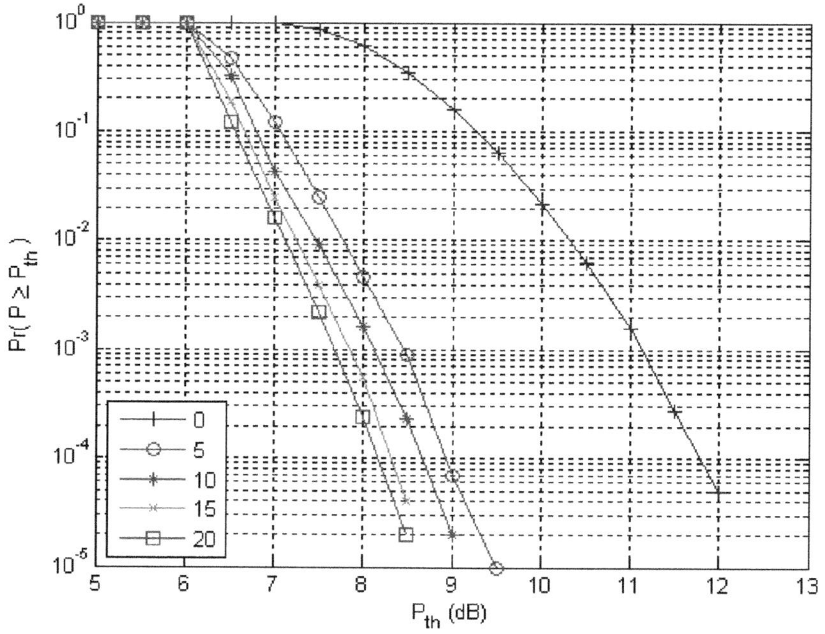

Fig. 6. PAPR Reduction of OFDM Signals by TR method

The objective of cross-layer optimization is to develop a scalable system that can capture the full diversity gain of frequency, time, space and antenna polarization. In general, from a capacity point of view, it would be more advantageous to adopt a broad-band approach. It is worthwhile to research OFDMA approach, treating the 70-MHz spectrum as a single channel, which might produce more improvement than treating it as seven separate channels[14-16]. Such an approach may increase the DSP complexity by requiring faster data converters and processors. However, DSP and application-specific integrated circuit (ASIC) technologies are now advanced enough to handle such broad-band channels. In fact, much wider channel bandwidth at about 2.16 GHz has been proposed in other gigabit wireless systems such as IEEE802.11AD.

It is hard to verify the impact of cross-layer optimization without a physical network. As a result, computer modeling and simulation will continue to be the primary tools for verifying the research results[17-18]. This is especially true if we have to verify the costs and benefits of a national ITS deployment to convince the government and the general public to invest in the infrastructure. Therefore large-scale system modeling and simulation including applications, communication protocol stacks and macro and micro controls will continue to be an outstanding research challenge.

5 Summary and Conclusions

We have provided a high-level overview of the current DSRC/WAVE regulatory requirements, protocol stacks and related standards of IEEE802.11P and IEEE1609. The key differences between WAVE and Wi-Fi or cellular wireless networking environments have been high-lighted. WAVE requires much more secure operation due to its role in safe-guarding human life and public property from road accidents. The strongest measure has to be taken to ensure that a DSRC-enabled ITS system cannot be abused or attacked. WAVE devices have to cope with fast frequency-selective fading due to faster mobility, larger Doppler spread and multipath delay spread. We have analysed the differences between application requirements and what can be offered by the current WAVE solutions. It is shown that the current solutions may be inadequate for large-scale deployment. The primary challenge is to develop scalable, robust, low-latency and high-throughput technologies for safety applications that will significantly reduce collisions and save lives and property loss. To address this challenge, further future research will include application requirement analysis, ITS-related social science study, cross-layer protocol stack optimization and large-scale system modelling and simulation.

Acknowledgments. The author would like to thank Casey Alford at EST Embedded Systems Technology for reading the manuscript and providing constructive feedback.

References

1. Morgan, Y.L.: Notes on DSRC & WAVE Standards Suite: Its Architecture, Design and Characteristics. Accepted by IEEE Comm. Surveys & Tutorials
2. Hartenstein, H., Laberteaux, K. (eds.): VANET: Vehicular Applications and Inter-Networking Technologies. John Wiley and Sons, Chichester (2010)
3. Shladover, S.E., Tan, S.K.: Analysis of Vehicle Positioning Accuracy Requirements for Communication–Based Cooperative Collision Warning. Journal of ITS, 131–140 (2006)
4. Jamalipour, A.: Cognitive Heterogeneous Mobile Networks. IEEE Trans. on Wireless Comm., 2–3 (June 2008)
5. FCC Code of Federation Regulations 47, Part 95 – Personal Radio Services (2009)
6. DSRC Implementation Guide – A guide to users of SAE J2735 message sets over DSRC. SAE International (2010)
7. Hubaux, J., Apkun, S., Luo, J.: The Security and Privacy of Smart Vehicles. IEEE Security & Privacy, 49–55 (May/June 2004)
8. Yang, K., Wang, X.: Cross-Layer Network Planning for Multi-radio Multi-channel Cognitive Wireless Networks. IEEE Trans. on Comm., 1705–1714 (October 2008)
9. Bai, F., Krishnamachari, B.: Exploring the Wisdom of the Crowd: Localized, Distributed Information-centric VANETs. IEEE Comm. Mag., 138–146 (May 2010)
10. Matolak, D.: Channel Modeling for Vehicle-to-Vehicle Communications. IEEE Comm. Mag., 76–83 (May 2008)
11. Sen, I., Matolak, D.: Vehicle–Vehicle Channel Models for the 5-GHz Band. IEEE Trans. on ITS, 235–244 (June 2008)
12. Fernández, Ó.: Flexible Rays. IEEE Vehic. Tech. Mag., 18–27 (March 2008)

13. Tellado, J., Cioffi, J.M.: Revisiting DMT's Peak-to-Average Ratio. VDSL ETSI/ANSI TD08 (1998)
14. Chang, C., et al.: Maximum Freedom Last Scheduling Algorithm for Downlinks of DSRC Networks. IEEE Trans. on ITS, 223–232 (June 2007)
15. Su, H., Zhang, X.: Clustering-Based Multichannel MAC Protocols for QoS Provisionings Over Vehicular Ad Hoc Networks. IEEE Trans. on Vehic. Tech., 3309–3323 (November 2007)
16. Bilogrevic, I.: Revocation Games in Ephemeral Networks. Master Thesis (2009)
17. Martinez, F.J., et al.: A Survey and Comparative Study of Simulators for Vehicular Ad Hoc Networks (VANETs). Wireless Comm. and Mobile Comp., Special Issue (2009)
18. Bononi, L., et al.: MoVES: A Framework for Parallel and Distributed Simulation of Wireless Vehicular Ad Hoc Networks. ACM Computer Networks (January 2008)

A Comparison of 802.11a and 802.11p for V-to-I Communication: A Measurement Study

Wei-Yen Lin[1], Mei-Wen Li[1], Kun-Chan Lan[1], and Chung-Hsien Hsu[2]

[1] Computer Science and Information Engineering
No.1, University Road, Tainan City 701, Taiwan (R.O.C.)
{q5697107,p78981304}@mail.ncku.edu.tw, klan@csie.ncku.edu.tw
[2] Information and Communications Research Laboratories
Bldg. 14, 195, Sec. 4, Chung Hsing Rd., Chutung, Hsinchu, Taiwan. 31040
stanleyhsu@itri.org.tw

Abstract. 802.11p, also known as WAVE, is a standard protocol intended for future traffic systems in order to support safety and commercial non-safety applications for vehicular communication. 802.11p is modified from 802.11a, and both are based on OFDM. The main difference between 802.11a and 802.11p is that the latter is proposed to use 10 MHz frequency bandwidth (half of bandwidth of 802.11a) in order to make the signal more robust against fading and increase the tolerance for multipath propagation effects of signals in a vehicular environment. In this paper, we investigate the performance difference between 802.11a and 802.11p for Vehicle-to-Infrastructure communication through real-world experiments. We measure contact duration and losses of 802.11p and 802.11a in both LOS and NLOS environments. In addition, we investigate their throughput with different modulations over various distances between OBU and RSU to evaluate the feasibility of using rate adaptation for non-safety V-to-I applications.

Keywords: 802.11p, guard interval, multipath, modulation.

1 Introduction

To decrease the number of traffic accidents, the U.S. Federal Communication Commission (FCC) has allocated a 75MHz spectrum at 5.9GHz for vehicle-to-vehicle and vehicle-to-infrastructure communications. The proposed vehicular communication technology, known as the Dedicated Short Range Communication (DSRC), is currently being standardized by the IEEE [1, 2]. Many major car manufacturers have responded positively, and are actively working together in bringing this promising technology into reality [3, 4]. An IEEE 802.11 standard, called IEEE 802.11p [5], is designed as the basic model for DSRC, and can be used to provide safety and service applications for Intelligent Transportation Systems (ITS) in the vehicular environment.

ITS include telematics and all types of communications in vehicles, between vehicles (V-to-V), and between vehicles and fixed locations (V-to-I). In general, the various types of ITS rely on radio services for communication and use specialized technologies [18]. DSRC is currently one of the most promising technologies related to automotive ITS.

X. Zhang and D. Qiao (Eds.): QShine 2010, LNICST 74, pp. 559–570, 2011.

There are two basic types of nodes in a DSRC (also called WAVE) networks: an On Board Unit (OBU) is located on vehicles and acts as an IEEE 802.11 station, while a Road Side Unit (RSU) is deployed on the road side and serves as an IEEE 802.11 access point. WAVE network can be operated in either infrastructure or ad-hoc modes. In the infrastructure mode, OBU accesses the network via WBSS (Wave Basic Service Set), which is consists of by OBUs and RSUs. The stations that create the WBSS to provide the service are called "providers," while those joining the service are called "users". WAVE is comprised of two protocol stacks: the standard Internet Protocol (IPv6) and the WAVE short message protocol (WSMP), as shown in Figure 1. WSMP allows applications to directly control physical layer characteristics used in transmitting the messages, e.g., channel number and transmitter power. WSMs (WAVE short messages) are delivered to the corresponding application at a destination based on the Provider Service Identifier (PSID) [6]. In this paper, we perform our experiments using WAVE short messages to send and receive packets.

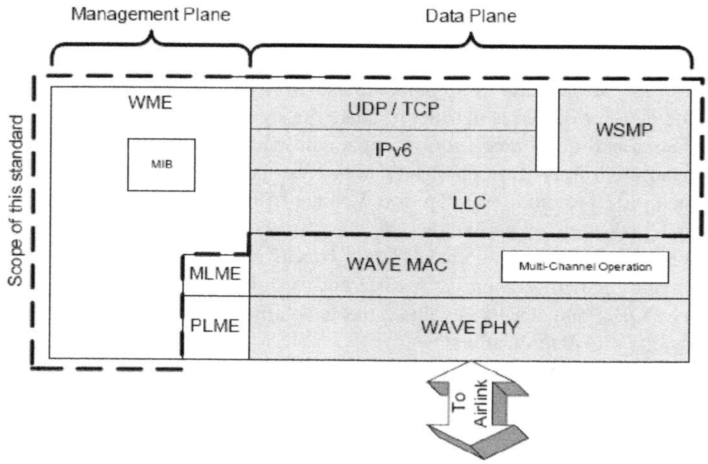

Fig. 1. WAVE protocol stack. WAVE accommodates two protocol stacks: the standard Internet Protocol (IPv6) and the unique WAVE short message protocol (WSMP).

The performance of 802.11a for vehicular communication has been studied extensively in simulations as well as on various testbed. However, as far as we know, there very few studies have been performed to understand the performance of 802.11p in a real-life scenario. Theoretically, the PHY layer of IEEE 802.11p PHY is pretty similar to that of IEEE 802.11a. The important PHY parameters for both the protocols are listed in Table 1 [8, 9]. In this paper, we set out to answer the following question: How much better is 802.11p for V-to-I communication compared to 802.11a in a real-life situation? And does the performance of 802.11p match the expectations that people have for it?

Table 1. Comparison of the physical layer implementations used in IEEE 802.11a and IEEE 802.11p

Parameters	IEEE 802.11a	IEEE 802.11p half clocked mode	Changes
Bit rate (Mbit/s)	6, 9, 12, 18, 24, 36, 48, 54	3, 4.5, 6, 9, 12, 18, 24, 27	Half
Modulation mode	BPSK, QPSK, 16QAM, 64QAM	BPSK, QPSK, 16QAM, 64QAM	No change
Code rate	1/2, 2/3, 3/4	1/2, 2/3, 3/4	No change
Number of subcarriers	52	52	No change
Symbol duration	4 μs	8 μs	Double
Guard time	0.8 μs	1.6 μs	Double
FFT period	3.2 μs	6.4 μs	Double
Preamble duration	16 μs	32 μs	Double
Subcarrier spacing	0.3125 MHz	0.15625 MHz	Half

2 Related Work

WAVE (Wireless Access in Vehicular Environments) or, 802.11p, is an IEEE standard which provides enhancements to the physical (PHY) and medium access control (MAC) layers of 802.11a for vehicular communication. Many studies have proposed different wireless technologies for vehicle communications based on existing standardized wireless technologies, such as Infrared, GSM, DSRC, Wi-Fi, Wi-Max, Bluetooth, RFID, for communication under one umbrella for ITS [10].

The V-to-I architecture allows vehicles to communicate with some roadway infrastructure and can enable many promising ITS applications. For example, the speed and location of a vehicle to be transmitted to a central server directly or indirectly connected to the road side unit. This server will track the speed and location of all vehicles and will aggregate this data for ITS applications, such as determining the fastest path from a vehicle's current location to its destination or identifying the location of an incident, among other applications [22]. There are also many applications that can be supported by a V-to-I communication network, such as web surfing, multimedia streaming, and real time car navigation. For example, some prior studies [23] have focused on the use of V-to-I for web applications.

There have been a lot of studies focusing on DSRC, but most of them have used simulations for their evaluations. [13] proposed a vehicle-to-infrastructure (V-to-I) communication solution by extending IEEE 802.11p. They introduced a collision-free MAC phase with an enhanced prioritization mechanism based on vehicle positions and the overall road traffic density, and evaluated their protocol's performance in Matlab. In [9, 12], the researchers used an NS-2 simulator to undertake a detailed simulation study of the performance of both DSRC and 802.11 for vehicular networks, and proposed a practical approach for IEEE 802.11 rate adaptation. In [14], an OMNeT++ simulator was used to evaluate the collision probability, throughput and delay of 802.11p. In [15], they used a QualNet simulator to compare two systems (WiMAX and 802.11p) for V-to-I communication under different vehicle speeds, traffic data rates, and network topologies.

To the best of the authors' knowledge, few DSRC testbeds have been implemented. In [16], a GNURadio platform was used for the data transmission in a vehicular network. [10] implemented XBee/Zigbee OBE (on-board equipment) and RSE (road-side equipment) for vehicle communication (V-to-V and V-to-I), while in [17], they used an Aeroflex 3416 to collect measurement data. All of the above studies required an extra connection to a computer/laptop equipped GPS for their experiments. CAMP (Crash Avoidance Metrics Partnership) is a working group comprised of seven automotive companies for providing vehicle-based safety systems. [24] So far, their task is considering the crash imminent braking to develop and validate performance requirements and objective tests for imminent crash automatic braking systems. The test methods are evaluated with vehicle-to-vehicle crashes and vehicle-to-object crashes. Cohda Wireless [25] has completed more than 700 DSRC trials, for 15 distinct DSRC use-case scenarios, in the USA, Italy and Australia. Their V-to-I scenario was executed with the RSU mounted in closed intersections and open intersections while a car traveled at 60 km/h, through and beyond an intersection with buildings on all corners. They compared the performance of the Cohda Radio to radios using WiFi chipsets configured for DSRC operation from several different manufacturers. They showed that the connection provided by the radio using a WiFi chipset is limited in range and capacity for data to be transmitted, making it unattractive for multiple user access. In contrast, we performed our measurement to measure the contact duration with varying speed in fixed range to identify how fast 802.11p can connect to roadside station. In this work, we use an IWCU (ITRI WAVE/DSRC Communications Unit) [11] made by Industrial Technology Research Institute (ITRI), which is loaded with a Linux kernel 2.6.30 built-in processor and has 16MB Flash and 64MB SDRAM for system memory. It implements IEEE 802.11p standard and operates at 5.85-5.925 GHz with built-in GPS capability.

Finally, rate adaptation is one of the key mechanisms at the link layer that can be used to improve network performance. Several rate adaptation algorithms have been proposed in the literature. However, all the existing work in rate adaptation is all based on WiFi. In [12], a rate adaptation algorithm based on 802.11a was evaluated in an NS2 simulation for a single hop topology. [19] performed a series of 802.11a-based outdoor experiments to compare the existing rate adaptation algorithms. In [20], the authors collected measurements from their 802.11-based mesh network to understand the correlation between SNR and distance, given the same modulation. In this work, we set out to examine the performance of 802.11p when different modulations are used. Specifically, we measure the throughput of an 802.11p link when different modulations are used at different distances.

3 Trace Collection

In this paper, we use the IWCU from ITRI. The IWCU serial products are meant to provide V-to-V and V-to-I communication enabling ITS applications ranging from safety to infotainment, as shown in Table 2 [11]. We look at three performance metrics of 802.11p in our experiments: contact duration, loss distribution and throughput when different modulations are used. The scenarios and parameter settings are explained in the following three subsections.

Table 2. IWCU Specifications

Component	RSU	OBU
Processor	IXP422 266MHz	
Processing power	266MIPS	
System Memory	16MB Flash, 64MB SDRAM	
DSRC Radio	IEEE 802.11p 5.85-5.925 GHz, 12dBm TX, -90 dBm RX Sensitivity	
Channel Width	10MHz/20MHz	
Antenna	5.9GHz, 2×5dBi	
Ethernet	10/100 Mbps (RJ-45) port × 1 with Auto Uplink™, Full-duplex	
GPS	Receiver Type: 50 Channels(GPS L1 frequency, C/A Code, GALILEO Open Service L1 frequency) A-GPS support、 Active antenna ×1 Sensitivity: -160 dBm (Tracking & Navigation)	
Dimension	300 × 250 × 80 mm	193 × 150 × 47 mm
Pre-Standards Compliance	IEEE 802.11p D6.0, IEEE 1609.3, IEEE 1609.4	
OS	Linux, kernel 2.6.28	
System Services	FTP, SSH, Telnet, HTTP	

3.1 Contact Duration

In this experiment, we measure how long a car can maintain a connection with the road-side unit (RSU) when it passes the unit. We define the contact duration as the longest interval between when the first packet was sent by the car and the time when the last packet was received by the road-side unit. We start the car at a position where it is out of the radio range of RSU, gradually increasing the car's speed and achieving the desired speed at the 'start' point. The same speed is maintained from the 'start' to 'end' points. The distance between 'start' and 'end' is 200m, and the scenario is shown as in Figure 2.

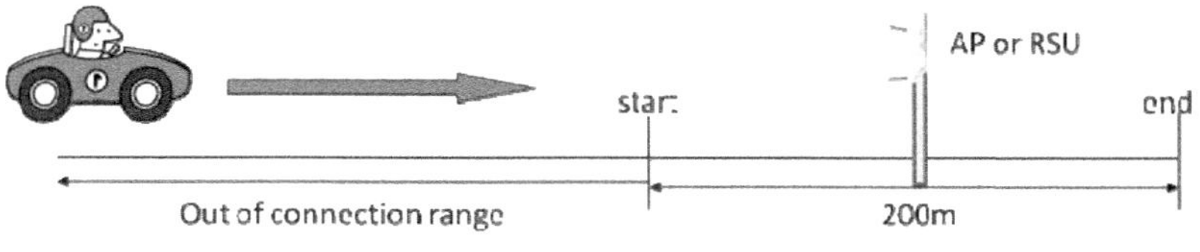

Fig. 2. Measurement of contact time

3.2 Loss Comparison in LOS and NLOS Environments

In DSRC the vehicles should send safety messages every 100ms and QPSK has been proposed for use as the desired data rate [2, 21]. Based on these suggestions, we set up our experiments, as shown in Table 3.

Multipath propagation is one of the most important characteristics in vehicular communication. 802.11p employs a channel bandwidth of 10 MHZ (which is different from 802.11a) and can affect its ability to cope with multipath systems [7]. We perform two sets of experiment, line-of-sight (LOS) and non-line-of-sight (NLOS), to validate whether 802.11p is better than 802.11a in an NLOS environment.

To make a fair comparison, in our experiments we use the same power level, data rate, packet size, and sending rate for both 802.11a and 802.11p. We developed some C programs to transmit WSMP packets between OBU and RSU to collect the packet loss. For the 802.11a experiments, we used a laptop equipped with a 802.11a wireless card as the sender to generate the traffic and an 802.11a AP as the receiver.

Table 3. Comparison between 802.11a and 802.11p experiment parameter settings

Parameter	Setting
Modulation	QPSK (1/2)
Transmit power	20 dbm (including antenna gain)
Sampling rate	10 packets/s
Sending time	3 minutes
Packet size	100 bytes
Packet type	WSMP for 802.11p UDP for 802.11a
Tool	Implemented programs
Traffic	CBR, Round-trip
Number of times	10

3.2.1 Location and Methodology

The experiments were performed on our campus. For the LOS experiment, we collected our data at a campus field which is approximate a 100 x 200 m^2 area, as shown in Figure 3. For the NLOS experiment, we set up our testbed around a pond with some rocks and trees at its center, and the transmitter and receiver were placed at opposite sides of the pond, as shown in Figure 4. The devices were placed at a height of 3m on top of a pole. The distance between the transmitter and the receiver was 70m in both experiments.

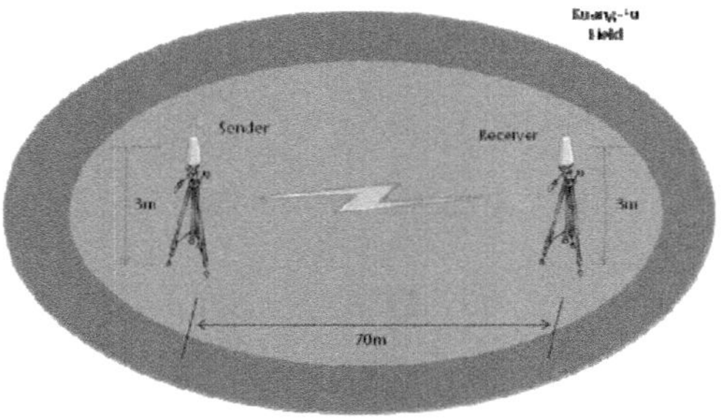

Fig. 3. For the LOS experiment, we measured on a campus field with no obstacles

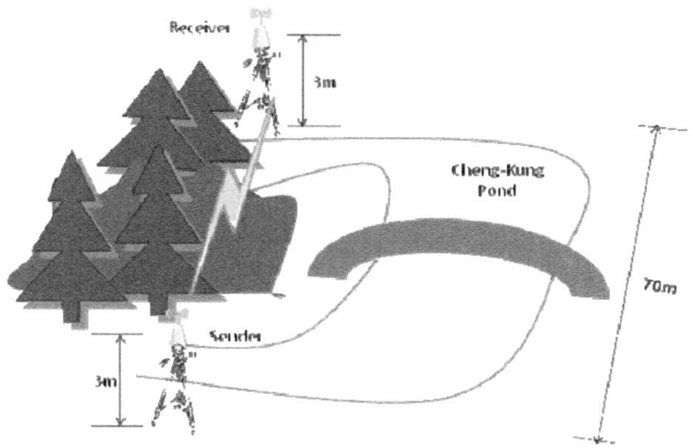

Fig. 4. For the NLOS experiment, we measured at campus pond with a large rocky area and trees as an obstacle to introduce a multipath system

3.3 Measurements for Different Modulations of 802.11p

Four different modulations are supported by 802.11p, namely BPSK (1/2), QPSK (1/2), 16QAM (1/2), and 64QAM (3/4). Theoretically, their corresponding data rates are 3Mbps, 6Mbps, 12Mbps, and 27Mbps, respectively. We measured the throughput of 802.11p based on different modulation schemes for various distances. We used 16 dbm (with 5 dbi omnidirectional antenna gain) as the transmission power level due to hardware limitations. The parameter settings are shown in Table 4.

Table 4. Parameter settings

Parameter	Setting
Modulation	BPSK, QPSK, 16QAM, 64QAM
Transmit power	16 dbm
*1 Sending rate	27 Mbps
Sending time	3 minutes
*2 Packet size	1400 bytes
Packet type	WSMP
Tool	Implemented programs
Traffic	CBR, one-way
Number of times	6

*1 maximum data rate. *2 maximum packet size of WSM explained on [6]

To avoid the disturbance of passers-by and moving cars, we collected our measurements at midnight. We placed the 802.11p devices at a height of 3m on top of a pole.

4 Results

4.1 Contact Duration

As there is no authentication/association process in 802.11p, it can set up a connection with the AP much faster than 802.11a can, as shown in Table 5. This means that a 802.11p-equipped car can send much more data to the road-side unit than when 802.11a is used. At the speed of 60Km/hr, the contact duration can be as long as 14 seconds, which is able to support some TCP-based applications like Email or instant messaging (e.g. MSN).

Table 5. Contact time

802.11a	Time (sec)	802.11p	Time (sec)
20 Km/h	4.5	20 Km/h	38.5
40 Km/h	0	40 Km/h	19
60 Km/h	0	60 Km/h	14

4.2 Loss Distributions in LOS and NLOS Environments

OFDM symbol provides a cyclic prefix called Guard Interval (GI). If the duration of GI is longer than the duration of all multipath signals following the first signal, the symbols can be restored and thus prevent the symbols suffering from inter-symbol interference (ISI). We repeated our experiments ten times to collect the data loss figures, as shown in Figure 5. In the LOS environment, the losses of 802.11p are close to zero. The highest loss rate among the ten experiments for 802.11p is 2.68%, even in the NLOS environment. Our results show that 802.11p is more robust against the multipath effect as compared to 802.11a. This is because 802.11p has doubled its GI (1.6 us), and, as a result, the multipath effect can be effectively mitigated, and hence it has less loss compared with 802.11a.

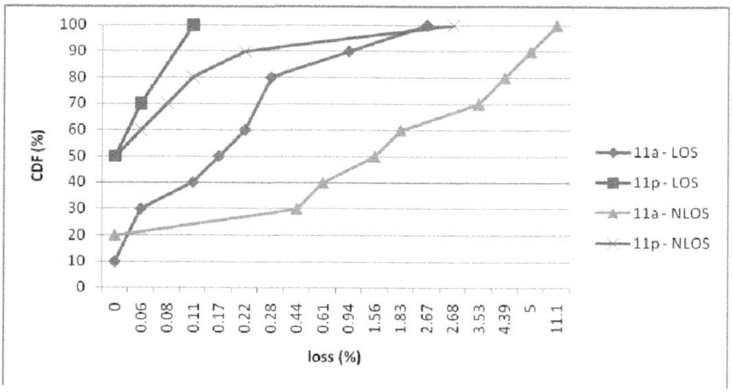

Fig. 5. The distribution of consecutive packet losses

In addition, by counting the sequence number of transmitted packets, we can investigate the distribution of consecutive packet losses, which indicate how bursty the loss is. As shown in Figure 6, we can find that the packet losses in 802.11a are a bit burstier than that in 802.11p.

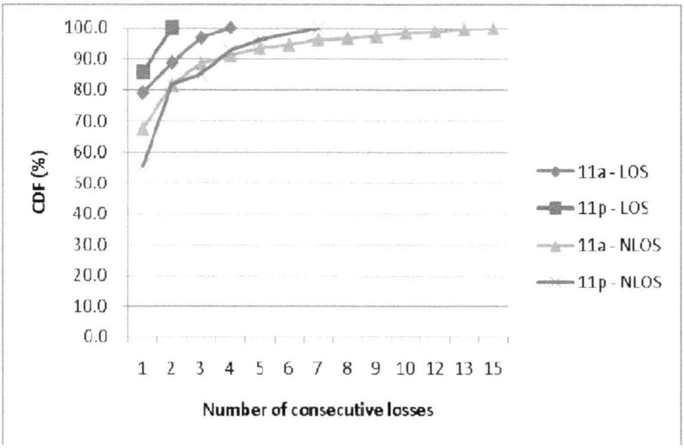

Fig. 6. The cumulative distribution function for consecutive losses and the number of occurrences

4.3 Measurements for Different Modulations of 802.11p

Finally, to understand the performance of 802.11p under different bit rates, we measure the throughput of 802.11p over various distances. Theoretically, different modulations have different sensitivity levels.

The sensitivity level is defined as

$$\text{Sensitivity} = \text{Rx noise floor} + SNR_{min} \quad (1)$$

where the SNR_{min} is the minimum SNR needed to obtain the wanted BER. A higher bit-rate requires a higher SNR; hence, the transmitted signal experiences degradation as the distance between the sender and receiver increases, as a receiver with a high bit-rate is not available to decode bits and this may introduce high bit error rate. As shown in Figure 7 and Figure 8, the achievable throughput of 802.11p is significantly lower than its theoretical counterpart (here we use the log-normal path loss model to model the radio propagation and calculate the theoretical throughput), although their curves are quite similar. The highest achievable data rate using 64QAM is about 18M when the car is very close to the road side unit (< 25m). The longest distance we can achieve with a throughput greater than 2M is 150m, when BPSK is employed. Such a distance and data rate might be sufficient to use the road-side unit as a gateway to the Internet and provide some Internet-based applications for the car. In addition, we observe the variation of throughput becomes larger when we increase the data rate from BPSK to QAM, which is particularly obvious when the car and the road side unit are at a closer distance, as shown in Figure 8.

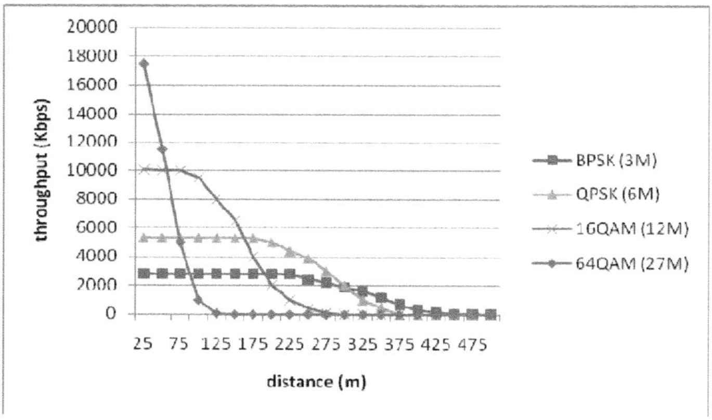

Fig. 7. Theoretical throughput using different modulation over various distances

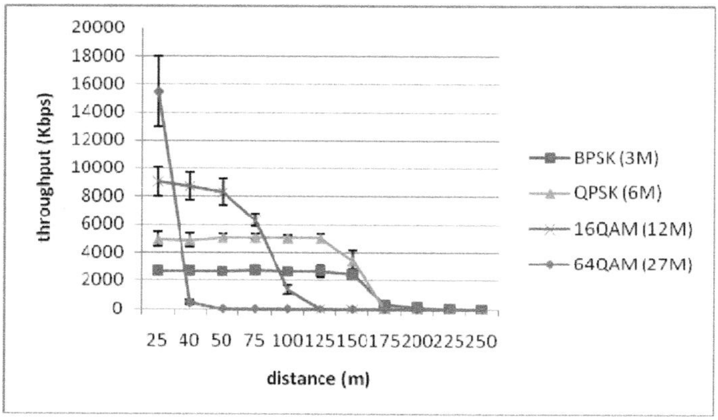

Fig. 8. The 802.11p throughput using different modulations from our testbed

For the non-safety communication of WAVE, the adaptation of the data rate should be carefully considered to achieve acceptable performance. In this work, we measured the throughput of different modulations over different distances. Our results suggest that, when using DSRC, if the car is far from the road side unit (> 150m), sending the data using a low data rate could achieve better network performance. When the car is close to the road side unit (< 25m), the system might then want to switch to a higher data rate so that it can send more data.

5 Conclusions and Future Work

In this paper, we perform extensive experiments to understand the performance of 802.11p in a real-world scenario, as compared to traditional WiFi, which has been used by most researchers working on V-to-I communication. We find that, as compared to 802.11a, the contact duration between car and the road side until is much longer when 802.11p is used, because the latter does not require any authentication process before setting up a connection. We show that the losses of 802.11p are also significantly lower than those of 802.11a in both LOS and NLOS environments. Finally, we find that the throughput of 802.11p is only around half of its highest theoretical rate. When the highest rate is selected, the throughput could quickly drop to zero when a car moves away from RSU (e.g. > 40m). On the other hand, when the BPSK is used, a stable throughput of at least 2MB can be maintained between the car and the road side until for as far as 150m. In future work, we will look at the effects of other parameters, such as transmission power, user mobility, and interference, on the performance of 802.11p for V-to-I communication.

Acknowledgment. This work of Dr. Lan described in this paper was supported by NSC under Grant No. NSC99-2220-E-006 -023.

References

1. Family of Standards for Wireless Access in Vehicular Environments (WAVE). IEEE 1609
2. Standard specification for telecommunications and information exchange between roadside and vehicle systems -5.9 GHz Band Dedicated Short Range Communications (DSRC) Medium Access Control (MAC) and Physical Layer (PHY) Specifications. In: ASTM, (E2213-03) (2003)
3. Bai, F., El Batt, T., Holland, G., Krishnan, H., Sadekar, V.: Towards characterizing and classifying communication-based automotive applications from a wireless networking perspective. In: First IEEE Workshop on Automotive Networking and Applications (AutoNet 2006), in conjunction with Globecom (2006)
4. Jiang, D., Taliwal, V., Meier, A., Holfelder, W., Herrtwich, R.: Design of 5.9 GHz DSRC-based vehicular safety communication. IEEE Wireless Communications, 36–43 (2006)
5. Draft P802.11p/D3.0: the IEEE 802.11 Working Group of the IEEE 802 Committee (2007)
6. IEEE 1609.3 Trial-Use Standard for Wireless Accesses in Vehicular Environments (WAVE) - Networking Services. IEEE Vehicular Technology Society (2006)
7. Hartenstein, H., Laberteaux, K.: VANET Vehicular Applications and Inter-Networking Technologies. John Wiley & Sons Inc., US (2010)
8. Müller, M.: WLAN 802.11p measurements for vehicle to vehicle (V2V) DSRC Application Note. In: Rohde & Schwarz (2009)
9. Khan, A., Sadhu, S., Yeleswarapu, M.: A comparative analysis of DSRC and 802.11 over Vehicular Ad hoc Networks. Dept. of Computer Science, University of California (2008)
10. Keeratiwintakorn, P., Thepnorarat, E., Russameesawang, A.: Ubiquitous Communication for V2V and V2I for Thailand Intelligent Transportation System. In: NTC International Conference, Thailand (2009)
11. ITRI WAVE/DSRC Communication Unit (IWCU) User's Guide, Version 1.03
12. Lacage, M., Manshaei, M.H., Turletti, T.: IEEE 802.11 Rate Adaptation: A Practical Approach. In: ACM MSWiM (2004)
13. Böhm, A., Jonsson, M.: Position-Based Data Traffic Prioritization in Safety-Critical, Real-Time Vehicle-to-Infrastructure Communication. In: Proc. IEEE Vehicular Networking and Applications Workshop (VehiMobil 2009) in conjunction with the IEEE International conference on Communication (ICC), Dresden, Germany (2009)

14. Eichler, S.: Performance Evaluation of the IEEE 802.11p WAVE Communication Standard. In: IEEE 66th Vehicular Technology Conference, VTC 2007, pp. 2199–2203 (2007)
15. Msadaa, I.C., Cataldi, P., Filali, F.: A Comparative Study between 802.11p and Mobile WiMAX-based V2I Communication Networks. In: 2010 Fourth International Conference on Next Generation Mobile Applications, Services and Technologies (NGMAST), pp. 186–191 (2010)
16. Fuxjäger, P., et al.: IEEE 802.11p Transmission Using GNURadio. In: Proceedings of the IEEE 6th Karlsruhe Workshop on Software Radios (WSR), pp. 83–86 (2010)
17. Tan, I.N.L., Tang, W., Laberteaux, K., Bahai, A.: Measurement and analysis of wireless channel impairments in DSRC vehicular communications. Elect. Eng. Comput. Sci. Dept., Univ. California, Berkeley, Berkeley, CA, Tech. Rep. UCB/EECS-2008-33 (2008)
18. Intelligent Transport Systems, `http://www.etsi.org/website/Technologies/IntelligentTransportSystems.aspx`
19. Shankar, P., Nadeem, T., Rosca, J., Iftode, L.: CARS: Context Aware Rate Selection for Vehicular Networks. In: ICNP (2008)
20. Lee, K.C., Navarro, J.M., Chong, T.Y., Uichin, L., Gerla, M.: Trace-based Evaluation of Rate Adaptation Schemes in Vehicular Environments. In: Vehicular Technology Conference (VTC 2010), pp. 16–19 (Spring 2010)
21. Wang, Z., Hassan, M.: How much of DSRC is available for non-safety use. In: ACM VANET 2008, pp. 23–29 (2008)
22. Miller, J., Horowitz, E.: FreeSim – A Free Real-Time Freeway Traffic Simulator. In: IEEE 10th Intelligent Transportation Systems Conference (2007)
23. Lan, K.C., Huang, C.M., Tsai, C.Z.: On the locality of vehicle movement for vehicle-infrastructure communication. In: Eight International Conference on ITS Telecommunications, pp. 116–120 (October 2008)
24. `http://www.its.dot.gov/cicas/cicas_current_act.htm`
25. `http://www.cohdawireless.com/`

Analysis of Information Fusion for Low Cost, Precise and Reliable Vehicular Cooperative Positioning with DSRC

Mahmoud Efatmaneshnik[1], Allison Kealy[2], Samsung Lim[1], and Andrew G. Dempster[1]

[1] School of Surveying and Spatial Information Systems, The University of New South Wales, Australia
[2] Department of Geomatics, The University of Melbourne, Australia
{mahmoud.e,s.lim}@unsw.edu.au,
allison.kealy@umelb.edu.au, a.dempster@unde.edu.au

Abstract. Information from digital maps and measurements from a low cost digital compass and the vehicle ABS odometer can be fused with those available from an onboard GPS receiver to bridge the outages frequently experienced by GPS when navigating in urban environments. This fusion strategy can also be used to improve the positioning accuracy of GPS. The emergence of infrastructure such as dedicated short range communications (DSRC) allows us to establish ad hoc vehicular networks in which range measurements between the moving vehicles can be measured and included as part of the integrated positioning solution. All these information sources together form a platform for robust position estimation that can meet the strict performance requirements of a range of road safety systems and services. In this paper, a Cramer Rao Lower Bound analysis is performed to describe the performance of such a comprehensive cooperative positioning system.

Keywords: DSRC, Cooperative Positioning, Cramer Rao Lower Bound, GPS outage, Information Fusion, Positioning error, Network Localization.

1 Introduction

The delivery of robust, technology based systems and services that can improve road safety, traffic congestion and optimize road use and management fundamentally rely on a robust infrastructure that accommodates for information communication between vehicles, vehicles and pedestrians and with road infrastructure. 5.9GHz band of Dedicated Short Range Communication (DSRC) technology is being established globally as a means of delivering this enhanced operational capability. The DSRC architecture comprises three integrated domains:

1. Application domain: the interface with drivers/pedestrians, e.g. the type of applications, the form of warning and recommendation/assistance messages, etc.
2. Communication domain: DSRC boards, related electronic devices, and hardware/software are considered in this domain, using three layers of the ISO model: the application, physical, and MAC (or communication protocol) layers.

X. Zhang and D. Qiao (Eds.): QShine 2010, LNICST 74, pp. 571–583, 2011.

3. Positioning domain: Accurate and Reliable positioning technologies are the first and foremost requirement of all DSRC Location-based services such as collision avoidance systems.

The focus of this paper is on meeting the positioning performance requirements for tasks that are within the DSRC applications domain. In addition, the approach for developing a robust positioning capability has to be one that can be practically delivered across the established communications domain. In particular, we focus on the positioning requirements for safety applications as these are considered to be the most stringent. With regards to reliable positioning for safety purposes three issues must be addressed:

1. Position availability
2. Position accuracy
3. Position frequency

All the safety applications require continual position estimation and for the most important applications, such as collision avoidance, position must be estimated at a frequency of 10Hz with an accuracy of no less than 50cm [1]. Yet the issue that is often neglected is that single frequency GPS receivers that are assumed to be used for vehicular navigation are typically a position estimation engine with 10m accuracy at 1Hz frequency. This means that the positioning requirements of safety applications are not met with conventional GNSS (Global Navigation Satelite Systems) positioning systems in urban areas.

In addition to the accuracy issue, GNSS are not reliable in the deep urban canyons typical of most cities due to lack of satellite visibility. The likelihood of position solution outage is high near buildings and multipathing effects (reflection of the satellite signal by obstacles such as buildings) can also deteriorate position accuracy. Cooperative Network Positioning or Cooperative Positioning (CP) is a cost-effective solution to these problems as it requires no major additional equipment other than the DSRC boards. CP is cited in the literature as a competent provider of safe and reliable position information for road safety applications [2-6].

In this paper we analyze the reliability of CP. The next section gives an overview of GPS performance, and then the traditional low-cost information fusion approaches to GPS performance improvement are discussed. The rest of the paper is dedicated to CP analysis via Cramer Rao Lower Bound (CRLB).

1.1 GPS Performance

GPS-alone positioning with a conventional low cost L1single frequency antenna is often used for vehicular positioning and has a Circular Error Probable (CEP) of 10m and a directional error of 7m. In addition, within urban and other difficult environments, it can also include a variable bias due to multipath. For example a set of GPS data collected on 25/11/2009 in the Sydney central business district (CBD) with a typical GPS navigation receiver Ublox^{TM} showed a maximum error of 90m from the centre line of the road (Fig. 1). In other areas where high rise buildings were absent (i.e. insignificant multipath) the maximum error was 5.5m from the centre line of the travelling road (these parts of the data are excluded from Fig. 2). The accuracy assessment used GIS software. The positional errors are assumed to be the nearest distances from the GPS coordinates to the road, i.e. the easting and northing of the GPS coordinates were compared with the existing road network from the NSW Land and Property Management Authority (LPMA) and each point's nearest distance to the corresponding road line was computed. The minimum and maximum errors were 0.001m and 90.699m, respectively, and the mean and the standard deviation were 7.783m and 9.881m, respectively.

In addition to the multipath, the high rise buildings create interruptions in the GPS measurements, because they can totally mask the satellite. As is evident in Fig. 1 there are some outages i.e. gaps in the GPS trajectory. Fig. 2 shows the frequency of the masked area lengths in meters. Most of the masked areas are segments of the road of less than 25m length and only a few more that 35m. The total distance of the travelled path shown on the map is 2.7km. Most of the outages cover very short lengths of the streets, and the longer the length of the masked area the lower its likelihood is. The total number of outages that have lengths greater than 15m are 25 which means there are roughly 10 outages of those lengths in every kilometre of the CBD area. The outages may also be studied in terms of the duration of position unavailability, which, given the slow movement of vehicles due to traffic in the Sydney CBD, are of the order of 5 to 35 seconds. The frequency of the outage durations peaks at about 10 seconds. There are very few outages at durations more than 20s.

Based on these observations, we consider a simple model for outages in the next sections and in the simulation. We assume 10 uniformly distributed outages randomly allocated over every kilometre of the road. The uniform distribution of the outages has a lower bound of 15m and an upper bound of 35m. This outage model represents a kind of worst case scenario.

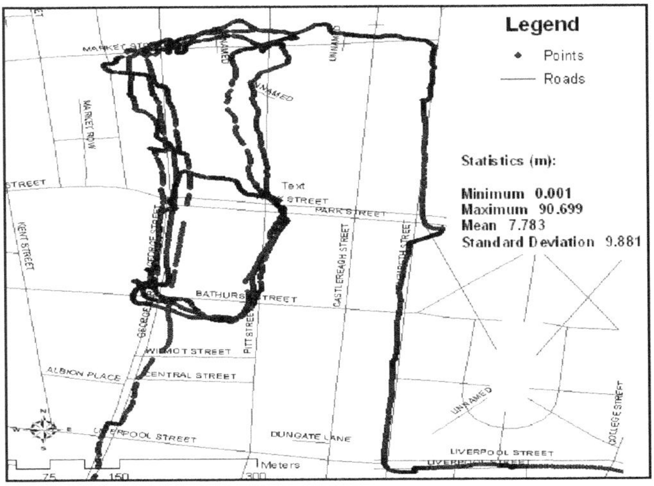

Fig. 1. The trajectory of the collected GPS data in Sydney CBD area

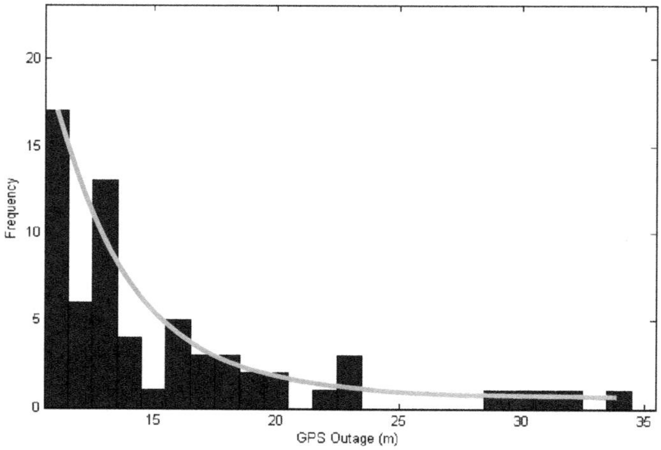

Fig. 2. The masked areas and their frequencies

1.2 Information Fusion for Vehicular Positioning

Data fusion from multiple sensors has always been a popular challenge for vehicular positioning. Emerging techniques for vehicular positioning include the use of digital road maps, Automatic Braking Systems (ABS) odometeric velocity solutions, and gyro or compass heading solutions. The purpose of these techniques is to increase the accuracy of the GPS, in particular in places where the multipath has diminished the accuracy or there is an outage.

Map information is extracted from the Geographic Information System (GIS) which has coordinates with typical error of 2m. Digital maps are very useful in removing the multipath-induced bias. Odometry builds an incremental model of the motion using measurements of the elementary wheel rotations. Odometers are already available in the front and rear wheels of vehicles equipped with an Anti-lock Breaking Systems (ABS). The accuracy of the ABS velocity measurements is about 0.1m/s but require digital conversion. Because the velocity derived from ABS does not indicate heading a heading solution is required that can be readily extracted from GPS or other means such as digital compass, low cost gyros or a combinations of these. A digital compass can provide a heading solution of accuracy 0.5° at relatively low cost. The combined heading and velocity information can effectively fill the gaps of GPS outages.

A method to provide an estimated location of an outdoor vehicle relative to a digital road map using Belief Theory and Kalman Filtering is described in [7]. They described an Extended Kalman Filter that combined GPS and odometer measurements from ABS sensors to produce an approximation of the vehicle position. By using Belief Theory and based on distance, direction and velocity measurements, a most likely segment from a road network database was chosen. Then a new observation was built from the selected segment, and the approximate position was adjusted in a second Kalman filter.

A similar work [8] introduced the odometric technique using the four ABS sensors available on four wheels and GPS. Due to the redundancy of the measurements, better precision was achieved relative to when only one of the rear wheels' odometers was deployed. The results showed a maximum of 4m directional error. Also when there were small masks of 10s in the GPS measurements, the results showed position errors of no more than 5m.

In [9], an integrated multi-sensor vehicle navigation system with the use of a low cost rate gyro and differential odometry to supplement GNSS availability under signal masking is described. The purpose of the system was to provide GPS accuracy of 10m nearly 100% of the time even when there are large outages of 100m to 200m. Two information fusion set-ups were used: ABS only and ABS derived speed (ABSS) with gyro heading (no ABS heading). The results are shown in Table 1.

Table 1. Position error average statistics of two tests reported in [9]

Outage (s)	ABS Only		ABSS/Gyro	
	Max(m)	RMS(m)	Max(m)	RMS(m)
100	124	20	10	7
200	143	27	40	10

Autonomous vehicle navigation by integrating the measurements of IMU, GPS, and digital compass was reported in [10]. They used a sophisticated dynamics model which considered Earth rotation, for IMU bias measurement and system noise estimation. Then a sigma Kalman filter was used for system state estimation, which has a higher accuracy than the extended Kalman Filter, was devised for the integration part.

2 Cooperative Positioning

CP is an idea originating from sensor networks research where each node (or even a central unit) uses range information between nodes to localize the network of nodes as a whole. One advantage of CP is that the localization accuracy is expected to increase with the node density (number of neighbour vehicles used in collective localization). In VANETs, Cooperative Network Localization utilizes distance measurements between nodes with partially known or unknown positions. This is of course the case when some or all of the nodes (vehicles) are equipped with GPS. Currently our research is examining ranging methods using DSRC [2, 11]. The network collective positioning is done once the ranges and the position information are exchanged between the nodes. The reliability of the positioning with this technique to some great extents is dependent on accurate ranging between vehicles.

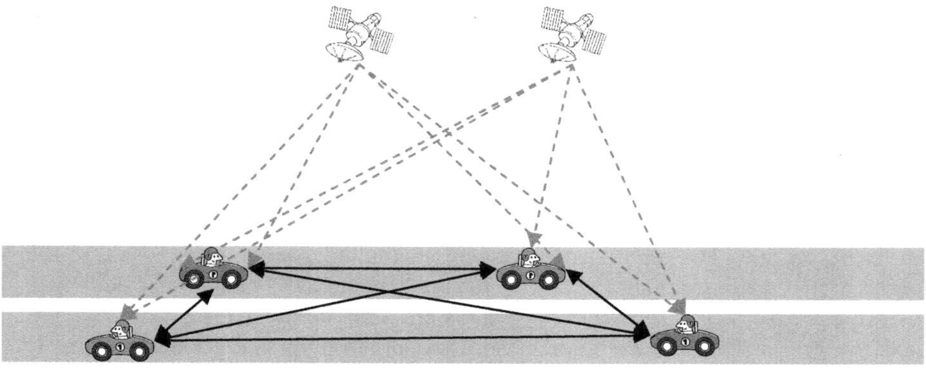

Fig. 3. Cooperative Positioning concept. Arrows indicate measured pseudoranges

2.1 Cramer Rao Lower Bound for Cooperative Positioning

The CRLB sets the minimum variance of any unbiased estimation of a random variable, and is the inverse of the Fisher Information matrix. Thus, CRLB is a means of estimating the performance of the cooperative positioning technique as well as benchmarking the performance of any localization algorithm. The CRLB has been derived for the general case of Adhoc Networks in [12]. Consider a vehicle with n-1 other vehicles in its neighborhood. The position of the vehicles can be unknown or partially known with known accuracy. Assume the ground truth positions of vehicles are $x = \{[X_i \quad Y_i]\}_{i=1}^{n}$ for localization in a two dimensional plane. Let $\hat{\delta}_{i,j} = \delta_{i,j} + e_{i,j}$ be the inter-node range measurements between pairs of n nodes with a standard noise $e_{i,j} \sim N(0, \sigma_{i,j}^2)$, all stacked in a column matrix δ which has n×(n-1) elements. Under the normality assumption, $f_\delta(\delta \mid x)$ is a normal distribution with $\sigma_{i,j}^2$ on its diagonal covariance matrix Σ which underlines the assumption of inter-nodes measurements independence then

$$f_\delta(\delta \mid x) = N(\delta, \Sigma) \tag{1}$$

The Fisher Information matrix $F(X \mid \delta)$ for the estimation of position vector X is:

$$F(x \mid \delta) = E\left[(\nabla_X f_\delta(\delta \mid x)) \times (\nabla_X f_\delta(\delta \mid x))^T\right] \tag{2}$$

Where ∇_X denotes the gradient with respect to elements of vector X. The inverse of $F(X \mid \delta)$ in its current form gives the CRLB for the case when only the range measurements are used without the a priori information about the node locations. Assuming that Σ_P is the covariance matrix of the a priori position information $\boldsymbol{X}$ **(e.g.** provided by GPS), a hybrid CRLB for cooperative positioning can be introduced as:

$$P = \left(\Sigma_P^{-1} + F(x \mid \delta) \right)^{-1} \tag{3}$$

For mobile network localization, the motion information of the nodes and their likely routes can be used to refine the position estimates. A computational algorithm that facilitates tracking and fusion of spatial information across time is referred to as a filter. By the filtering mechanism the position estimation at each instant *k* in time is refined by the estimation at instant *k-1* based on a mobility model. Assume the velocity based mobility:

$$x_{k+1} = x_k + T_k u_k + T_k \xi_k \tag{4}$$

Where,

$$x_k = [X_1^k, X_2^k, \cdots, X_n^k, Y_1^k, Y_2^k, \cdots, Y_n^k] \tag{5}$$

$$u_k = [U_1^k, U_2^k, \cdots, U_n^k, V_1^k, V_2^k, \cdots, V_n^k] \tag{6}$$

U is velocity in *X* direction and *V* in *Y*. T_k is the time interval between measurements. ξ_k is a standard error with zero mean and variance of σ_v^2. Thus the hybrid CRLB for mobile networks becomes:

$$P_{k+1} = \left((P_k + T_k \Sigma_v)^{-1} + \Sigma_P^{-1} + F(x \mid \delta) \right)^{-1} \tag{7}$$

Where $\boldsymbol{P}_{k+1}$ and $\boldsymbol{P}_k$ are the hybrid CRLB matrices at instants $k+1$ and k, T_k is the scalar elapsed time, Σ_v is the diagonal covariance matrix of the measured velocity of vehicles. From equation 8 it is clear that noisier velocity measurements, thus larger σ_v^2, lead to a smaller contribution from the previous instant's position information and if it is too large filtering has no effect on the accuracy. An indicator of cooperative positioning error for each node over a period of time (*K*) can be readily obtained based on Circular Error Probability (CEP):

$$C = \left(\sum_{k=1}^{K} \frac{\sigma_{x_i}^{k\,2} + \sigma_{y_i}^{k\,2}}{K} \right)^{\frac{1}{2}} \tag{8}$$

Where $\sigma_{x_i}^{k\,2}$ and $\sigma_{y_i}^{k\,2}$ respectively are the estimated error variance of node *i* at time *k* in directions *x* and *y*.

2.2 Simulation and Performance Analysis

A simulation study was set up to demonstrate the performance of CP based on the error indicator in equation (8) averaged over time. A street traffic network was simulated by Netlogo™ that is an agent-based and Java-based programming platform. Fig. 4 is a snapshot of the traffic network. The length of the horizontal road was set at

2km. From this simulation, only the ground truth values of the vehicles' positions were obtained. The rest of the calculations were performed in the Matlab™ environment. Four traffic conditions were considered that were controlled by the total number of vehicles in the traffic simulation street network. The considered traffic rates were 300vehicles/h in the main streets for sparse/fast traffic, 600vehicles/hour, 900 vehicles/hour and 1200vehicles/hour for heavy/slow traffic. In all four cases a target vehicle was set to travel a particular path of about 2km. The simulation time in each case thus depends on the traffic condition, for example for sparse traffic it takes a shorter time than for heavy traffic to travel the same distance. The DSRC range was set at 250 m. The underlying assumption for this test and all the following tests are that the inter-node ranging error has a mean of zero and variance of σ_R^2, which was changed to show the sensitivity of the CRLB to σ_R^2. The positioning directional error was set at 7m.

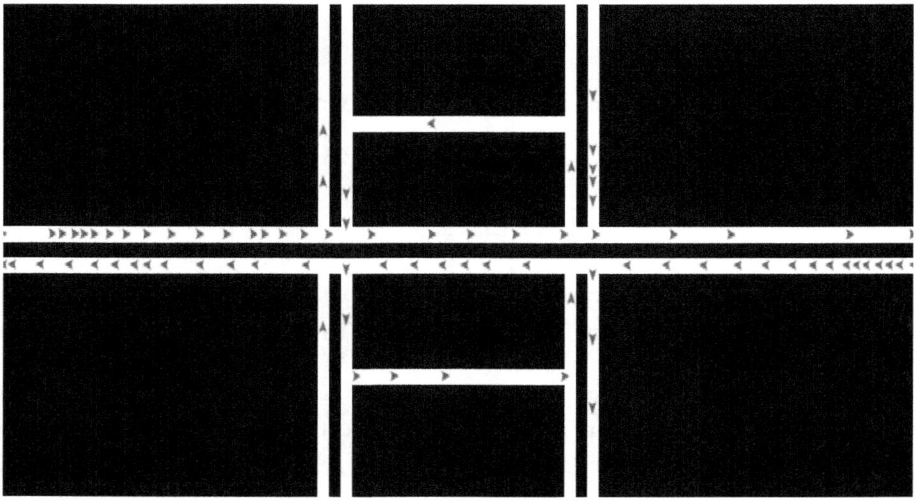

Fig. 4. Netlogo agent based software was used for the traffic simulation

Fig. 5(a) shows the expected error for CP based on CRLB analysis for heavy traffic conditions. The errors are averaged over the simulation run (200s). For this test σ_R varied between 1m and 10m; σ_v, the standard deviation of velocity, varied between 0.1m/s and 2.1m/s. From this figure, it can be seen that CP error has high sensitivity to the velocity error. Below 1 meter accuracies are typical for when the velocity measurement has error standard deviation of about 0.1m/s. Remember that this velocity error margin can be met with ABS-based odometers. Fig. 5(b) shows the typical CP error in all types of traffic conditions. This figure proves that the velocity measurement accuracy contributes more to the CP accuracy than heavier traffic conditions.

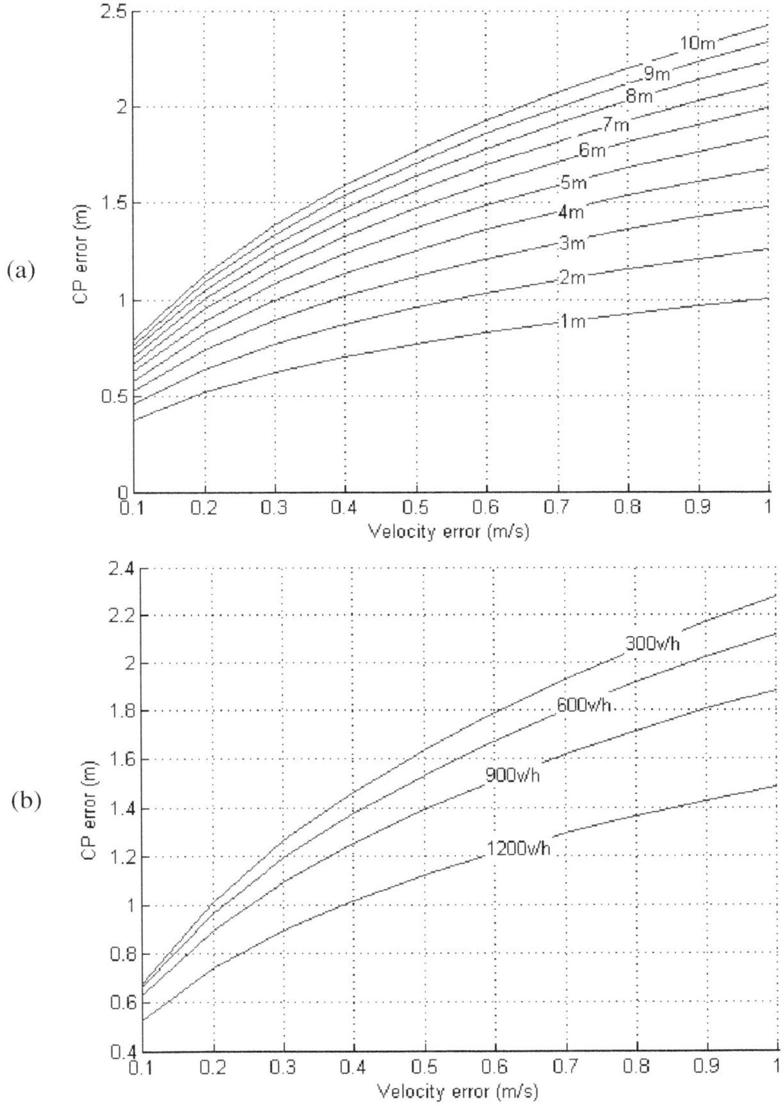

Fig. 5. Error bounds for heavy traffic conditions with several range measurements error standard deviation (from 1 to 10) and velocity error standard deviation **(a)**. The range errors are indicated on the error curves. Average CRLB based error over the simulation run for $\sigma_R=3m$. The traffic conditions in terms of rate of vehicles/hour (v/h) are indicated on the error curves **(b)**.

In the next simulation experiment 20 outages of 15m to 35m long were randomly allocated throughout the simulated street network. If a vehicle entered the outage zones, its position error deviation was set at infinity (or a very large number) which literally means the absence of position information in those zones. Fig.6 shows the

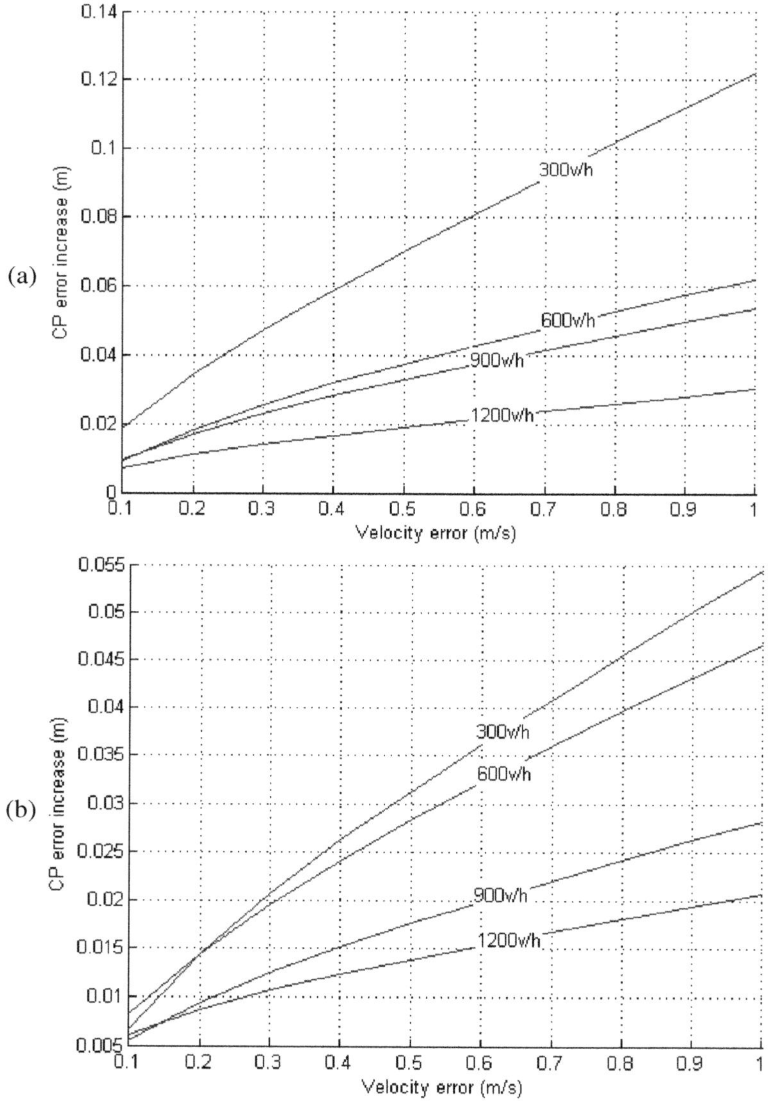

Fig. 6. The increase in CP error for σ_R=*10m* due to 10 outages of 15m to 35m in every kilometre of the road **(a)**. The same increase for σR=*3m* **(b)**. The traffic rates are printed on the error increase curves.

results in terms of the increase in average CP error relative to the no-outage case (the previous test). From this figure it is evident that even for sparse traffic, the GPS outages do not have significant effect on the CP error, as long as ranges can be measured.

In Fig. 7 the error bounds for two cases as well as the GPS drop out times are shown. Case 1 is the worst case scenario for CP performance where traffic conditions are very heavy. Also the ranging and velocity errors are set at their highest, $\sigma_R=10m$ and $\sigma_v=1m/s$. Case 2 has the same traffic conditions and ranging error but better velocity measurements $\sigma_v=0.1m/s$. This figure shows that in case 1, CP error has some sensitivity in the form of sharp short spikes to the outages. However, case 2 has a very smooth error even during the outages. The smoothness particularly is due to better velocity measurements. This figure illustrates the important positive impact of the velocity accuracies on both smoothness of the CP solution as well as its accuracy.

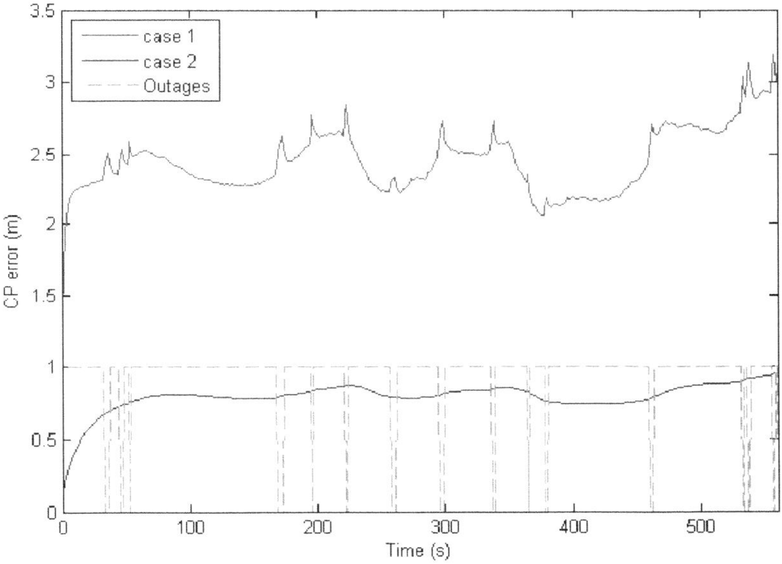

Fig. 7. CP error for two cases. Both cases are for heavy traffic condition and $\sigma_R=10m$. The velocity error for case 1 (red) is $\sigma_v=1m/s$ and for case 2 (blue) is $\sigma_v=0.1m/s$. The green dashed lines show the outages. A value of *1* indicates GPS availability and *0* the GPS unavailability.

The last case considered here is the hypothetical case where other information fusion techniques using GPS, odometers, compass and digital map are available to all vehicles in the network. In this condition we assume that the position information has an error of *4m* and that during outages has an error of *5m*; the errors are chosen according to the results reported in [8] that had similar outage conditions. In this case, CP always delivers the desired accuracy of less than *50cm* for all types of traffic and any ranging accuracy between *1m* to *10m*. The simultaneous implementation of CP with the results of [9] in Table 1 more or less would similar to the CP results without any outages.

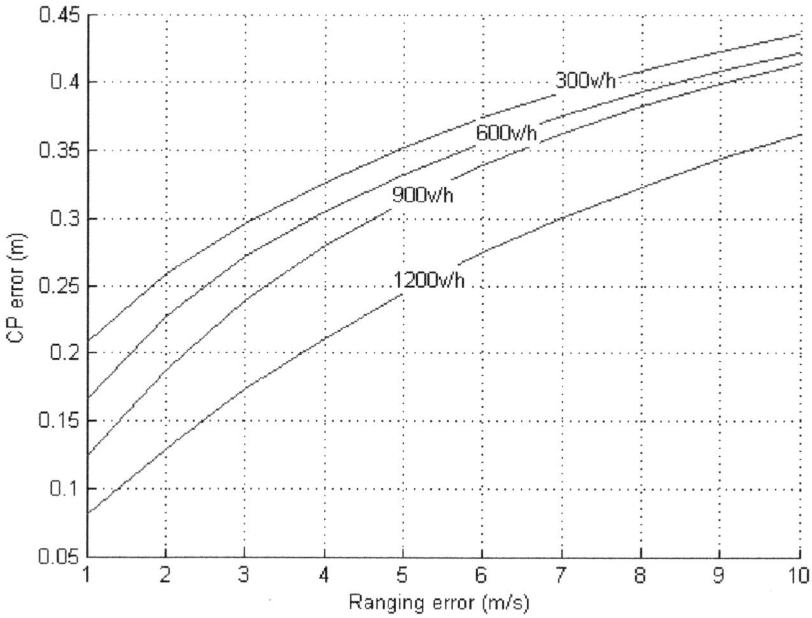

Fig. 8. CP and other information fusion approaches together. The hypothetical CP performance if ranging error $\sigma_R=3m$, .the position information error is $\sigma_p=4m$, and the same during the outages is $\sigma_p=5m$.

3 Conclusion

In this paper the CP performance was simulated by means of CRLB analysis. The results are promising and indicate the capacity of CP to provide reliable and robust vehicular positioning. CP as a positioning solution has no major cost implications to DSRC system developers. The only concerns are the particular protocols for ranging between the vehicles. These are most likely to be based on a Time of Arrival technique. Future VANET architectures will also have Road Side Units that can be used as fixed position beacons that will improve the CP accuracy even further. In a The effect of multipath driven position bias on the CP performance can also evaluated by more or less similar technique of this paper. However, as noted before the digital maps are very helpful in reducing the bias so the multipath effect can be eliminated especially if accurate velocity measurements are available. The odometer is again important for this case because although the GPS bias can be corrected by digital map, the GPS velocity accuracy that severely deteriorates by the multipath cannot.

References

1. Vander, J., Shladover, W.S., Miller, M.A.: Conceptual Development and Performance Assessment for the Deployment Staging of Advanced Vehicle Control and Safety Systems. Institute of Transportation Studies California Partners for Advanced Transit and Highways (PATH), University of California, Berkeley (2004)

2. Alam, N., Balaei, A.T., Dempster, A.: Ranging with DSRC: Facts and Challenges. In: International Global Navigation Satellite Systems Society IGNSS Symposium 2009Holiday Inn Surfers Paradise, Qld, Australia (2009)
3. Benslimane, A.: Localization in Vehicular Ad-hoc networks. ICW/ICHSN/ICMCS/SENET, pp. 19–25 (2005)
4. Efatmaneshnik, M., Balaei, A.T., Alam, N., Dempster, A.G.: A modified multidimensional scaling with embedded particle filter algorithm for cooperative positioning of vehicular networks. In: 2009 IEEE International Conference on Vehicular Electronics and Safety (ICVES), pp. 7–12 (2009)
5. Efatmaneshnik, M., Balaei, A.T., Dempster, A.G.: A Channel Capacity Perspective on Cooperative Positioning Algorithms for VANET ION GNSS, Savannah, Georgia (2009)
6. Parker, R., Valaee, S.: Vehicular Node Localization Using Received-Signal-Strength Indicator. IEEE Transactions on Vehicular Technology 56, 3371–3380 (2007)
7. El Najjar, M.E., Bonnifait, P.: A Road-Matching Method for Precise Vehicle Localization Using Belief Theory and Kalman Filtering. Autonomous Robots 19, 173–191 (2005)
8. Bonnifait, P., Bouron, P., Crubille, P., Meizel, D.: Data fusion of four ABS sensors and GPS for an enhanced localization of car-like vehicles. In: Proceedings of 2001 ICRA IEEE International Conference on Robotics and Automation, vol. 2, 1592, pp. 1597–1602 (2001)
9. Stephen, J.E.: Development of a GNSS-based Multi-Sensor Vehicle Navigation System. Geomatics Engineering. University of Calgary (2000)
10. Zhang, P., Gu, J., Milios, E.E., Huynh, P.: Navigation with IMU/GPS/digital compass with unscented Kalman filter. In: 2005 IEEE International Conference on Mechatronics and Automation, vol. 3, 1493, pp. 1497–1502 (2005)
11. Alam, N., Balaie, A.T., Dempster, A.G.: Dynamic Path Loss Exponent Estimation in a Vehicular Network using Doppler Effect and Received Signal Strength. In: VTC Vehicular Technology Conference (2010)
12. Savvides, A., Garber, W., Adlakha, S., Moses, R., Srivastava, M.: On the error characteristics of multihop node localization in ad-hoc sensor networks. In: Zhao, F., Guibas, L.J. (eds.) IPSN 2003. LNCS, vol. 2634, pp. 317–332. Springer, Heidelberg (2003)

A Distributed Challenge Detection System for Resilient Networks

Yue Yu

School of Information Technologies
University of Sydney
NSW 2006, Australia
tinayu@it.usyd.edu.au

Abstract. The network has become essential to our daily life. With the increase in dependence, challenges to the normal operation of the network bear ever more severe consequences. Challenges include malicious attacks, misconfigurations, faults, and operational overloads. Understanding challenges is needed to build resilience mechanism. A crucial part of resilience strategy involves real-time detection of challenges, followed by identification to initiate appropriate remediation. We observe that the state-of-art to challenge detection is insufficient. Our goal is to advocate a new autonomic, distributed challenge detection approach. In this paper, we present a resilient distributed system to identify the challenges that have severe impact on the wired and wireless mesh network (WMN). Our design shows how a challenge (malicious attack) is handled initially by lightweight network monitoring, then progressively applying more heavyweight analysis in order to identify the challenge. Non-malicious challenges could also be simulated by our network failure module. Furthermore, WMNs are an interesting domain to consider network resilience. Automatic detection and mitigation is a desirable property of a resilient WMN. We present guidelines to address the challenge of channel interferences in the WMN. The feasibility of our framework is demonstrated through experiment. We conclude that our proof-of-concept case study has provided valuable insight into resilient networks, which will be useful for further research.

1 Introduction

With the growth of networks and the integration of services, increasingly severe consequences come from the disruption of networked services. Quality of life, the economic viability of businesses, and the security of nations are directly linked to the resilience, survivability, and dependability of the global network. However, the network becomes vulnerable with the increased dependence and sophistication of services. The scale of growth and deregulation bringing numerous service providers has resulted in a network that is difficult to manage. There is a pressing need for better resilience, manageability, and security for the future network [1]. Therefore, our research in distributed challenge detection is aimed to make networks more resilient to various challenges. Resilience means the ability of the network to provide an acceptable level of service in the face of challenges to normal operation. This

X. Zhang and D. Qiao (Eds.): QShine 2010, LNICST 74, pp. 584–597, 2011.

service includes the ability for users to access information, the maintenance of end-to-end communication, the operation of distributed processing and networked storage. The challenges that may impact normal operation include unintentional misconfigurations; malicious attacks; environmental challenges; unusual but legitimate traffic; provider failure. Therefore, the definition of resilience is a superset of commonly used definitions for survivability, dependability, and fault tolerance [1].

The main objective of the EU ResumeNet (Resilience and Survivability for Future Networking: Framework, Mechanisms and Experimental Evaluation) is to propose a multilevel, systemic, and systematic approach to network resilience. To achieve this, our approach addresses the challenge diagnosis problem as followings: it first monitors traffic for anomalies in real-time, which is online, when traffic traverses the network, rather than processing trace files offline. It is too costly to undertake detection operate all the time to perform the root cause analysis of ongoing challenges. They should only be enabled after the detection of the basic symptoms that may lead to the anomalies arise. Therefore, for example, once a link monitor detects a possible anomaly, the alert will be generated in the core router of the infected sub-network, which effectively shrinks the network range for detection. The detection is followed by instantaneously locating the victim. In addition, the network failure scenario is considered. Such a challenge based on the node and link failures can impact single or multiple network components [2], area-based challenges could affect multiple network elements.

In wireless mesh networks (WMNs), challenges are different from those in other networks, and can also have more severe impact than wired network. This is due to the inherently less reliable (compared to Ethernet) wireless technology and high reliance between the mesh elements. Furthermore, it is more vulnerable than wired networks due to the broadcast nature. WMN is an appealing technology for flexibly interconnecting computers. In contrast to wired network, it is a cost effective and simpler solution for rural areas. The WMN could build a resilient infrastructure via a combination of wireless network and ad-hoc routing protocols. With such deployment, problems on the physical and link layer are more likely to occur and have bigger impact. In this paper, we focus on wireless interference, as the big impact challenge needs to be met since it could affect the QoS of individual applications and bring severe trouble to the network [3]. Our work investigates a more systematic approach to meeting challenges to the network than has occurred before.

The paper is organized as follows. In Section 2, we review the shortcomings with the current state-of-art to challenge detection, and highlight aspects requiring additional work. The framework of the distributed challenge detection system is introduced in Section 3. In Section 4, we compare different experimental platforms, and explain why OMNeT is the most suitable network simulator. We demonstrate the simulation of our distributed challenge detection system in Section 5, and explain how we are populating our resilience strategy with the new network monitor, network failure and anomaly detection mechanisms. In Section 6, our approach is validated on the OMNeT and the interference challenge in the WMN will be discussed. The current work is summarized and future directions are concluded in section 7.

2 Related Work

The vulnerabilities of the current network and the need for resilience are widely acknowledged. There has been considerable research into network monitoring, anomaly detection, fault tolerance, attacks, anomaly modeling engine separately. We reviewed the latest technologies in these areas. With the current network monitoring techniques, threshold based random walks for fast portscan detection is unscalable [4]. The proposed (threshold crossing alerts) TCAs requires the cooperation of the manufacturers to run on the network devices, which will be difficult. Or they can run on separate hardware, which will be complex [5]. Jackson et al. [6] cope with the distributed monitor problem in internetworks, but the capability to monitor every link cannot be assumed. Today most detection use signature-based IDS that detect known attacks only. In contrast, anomaly detection is effective in foiling known and unknown attacks. Real time volume based anomaly detection is resource challenge [7]. The causes for DoS attacks and mechanisms for defending is surveyed [8], however it is not yet practical to identify attack paths and we require global cooperation to combat (Distributed denial of service)DDoS attacks. To evaluate the impact of faults, fault injection is considered as the first stage. It offers a cost and time effective way to test system. The method to inject faults into the real network is proposed in [9]. Other approaches are presented to study network survivability. Random events affect node and link availability, so cause the failures. There is a complete survey of fault localization [10]. Open research problems still remain with multi-layer fault localization, temporal correlation, and distributed diagnosis.

In the wireless domain, the completely decentralized networks WMN depend on every node to provide packet forwarding services for normal operation. The need for collaboration is highlighted in such network due to the lack of a central entity to supervise the activity of the nodes. The challenges that are inherent in the wireless domain include weakly connected channels, mobility of nodes in an ad-hoc network, and unpredictably long delays, radio interference and error-prone links [2]. Furthermore, WMNs are particularly susceptible to node selfishness. The key challenge discussed in this paper is interference and two general approaches could be applied to simulate interference and radio propagation, by using a complex and computational expensive model [11] or a simple one with the risk of gaining misleading conclusions [12]. Therefore, we need a method to recreate interference traffic from real measurements and seamlessly build into the network simulator. To conclude, so far yet little has been done to systematically embed resilience into the future network or develop novel, distributed mechanisms for monitoring resilience to detect challenges as they occur. Therefore, our research contributes towards the development of new experimental systems to perform challenge detection.

3 Distributed Challenge Detection Framework

In the ResumeNet project, we are investigating a framework for resilient networking. Resilience is required to be a key property of future networks because of our unrelenting demand for network services, the challenging environments, and the continued existence of intelligent adversaries. Two corresponding approaches are adopted: network and service resilience. Our work focuses on the network resilience, which is adding resilience to the services the system provides [13]. From the project's outset, we should understand the complexity of systems, the various challenges and its root cause. Assessing which challenge affect the system in which way is essential to deciding the corresponding mitigation strategy. For example, a web server could be overloaded in the short term with many requests, greater than what is provisioned for, and this could be a DDoS attack or a 'flash crowd' event, which is the unusual but legitimate demand for service. We need to distinguish between challenges that have similar symptoms, since they require different remediation. To do this, the first task is to construct a network that could tolerate foreseen adversarial events. The optimal topology needs to be designed. It utilizes the best possible way to interconnect the access nodes with the corresponding transmission technology. The fine tuned routing mechanism allow the good conditioned connection be built so that they could react to the failure quickly. Then we need a distributed monitoring and assessment platform that could detect network anomalies. This is because no perfect protection is provided by the defensive measures [13]. The unforeseen challenge will degrade the service.

Our proposed resilient network system could be applied in the heterogeneous network, which covers the wired and wireless network. In rural areas, WMNs are often deployed as the affordable and simple way to access internet, challenges could lead to a service such as internet connectivity being impaired or unacceptable depends on the severity. The in-depth understanding with the nature of the system and its challenges is significant to find the matching identification method and improve the resilient network. In WMN, various challenges could arise to the normal operation. They are particular vulnerable to infrastructure-based attack as a result of the relative simplicity of physical access to the mesh. Misconfiguration of devices could be a major issue when non-expert users manage the network. In addition, WMNs are particular vulnerable to the elements. Figure 1 listed the potential challenges that could affect WMN [14]. The hardware resources that can be used to detect the challenges may have wildly varying computational capabilities [3]. Nevertheless, it is not easy to prevent these challenges from leading to significant network outages.

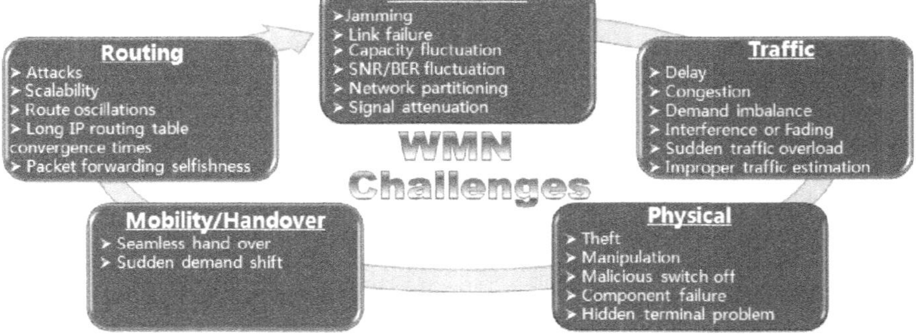

Fig. 1. Wireless Mesh Network Challenges

4 Comparison of Experimental Platforms

Building a distributed challenge detection system in the real world is a challenge. Firstly, a large topology network is required to get meaningful results. Secondly, such large topology will increase the hardware cost and administration effort. Thirdly, it is not convenient to experiment with attacks on the current available real world platform, e.g. AARNET, PlanetLab, since it's hard to ensure the experiment will not disrupt the normal operational network, which will cause more severe effect. Fourthly, we need to have full control with all the system nodes so that they could easily be configured to suit our case [15]. However, this is also difficult to realize.

Another possibility is to use off-line IDS datasets such as DARPA [16], KDD Cup, which could help to relieve from the real world difficulties but with the realistic dataset. DARPA intrusion detection is collected between 1998 and 2000 from Lincoln Lab. The 1998 version covers 38 attack types. The data slightly improved in 1999, so that 201 instances of 56 attack types distributed. Whilst the 2000 scenario specific datasets include two attack situations. KDD Cup dataset gathered in 1999 with 41 features. After investigation we decided not to use them. Because first, the datasets are not up to date, so that the most recent and the unforeseen attacks couldn't be measured. Secondly, the DARPA dataset was fundamentally broken due to numerous irregularities [17]. Thirdly, the performance measure applied in DARPA'98 evaluation, ROC curves, has been widely criticized [18]. Lastly, it was still useful to evaluate the true positive however any false positive results were meaningless [19].

Network simulators could overcome all the mentioned hurdles and meet the needs by integrating real world applications. However, it still requires us to compare different simulators to recognize the most appropriate environment. So we surveyed the widely applied simulators, NS-2, NS-3, OMNeT, SSFNet, JiST/SWANS and J-Sim. NS-2 and OMNeT are continuously supported today. NS-3 is the latest updated platform developed from 2008. While SSFNet, JiST/SWANS and J-Sim are nearly inactive since 2004, 2005, 2006 respectively. Our project needs the updated simulator, so NS-2, NS-3, OMNeT could better suit. Moreover, our work needs large topology, but NS-2 has the scalability issue with memory usage and simulation run-time [20]. However, NS-3 and OMNeT are scalable. Considering the run time, JiST/SWANS are the fastest, whilst J-Sim is the slowest. OMNeT is slower than NS-3 but faster than NS-2 [20]. NS-3 has lowest computational and less memory demands whereas JiST/SWANS exhaust memory [21]. OMNeT consume more memory than NS-3 but less than NS-2. In terms of GUI, NS-3 and NS-2 are relying on source code but OMNeT has a rich GUI with online visualization.

NS-2, NS-3 and OMNeT are widely used in wireless network simulation. Dedicated Short Range Communications (DSRC) is a good medium for inter-vehicle communications. The features of the simulators applied in DSRC are listed in [22]. Many researchers did DSRC related simulation based on NS-2 [23], NS-3[24] or OMNeT [25]. Among them, OMNeT and NS-2 are the most mature ones. We also studied the survey on the performance of wireless network simulators [26]. OMNeT reduce the complexity, and become an excellent tool for wireless network simulation as a result of its scalability, efficiency and the simplicity of modifying the network properties. OMNeT simulation API is more powerful than NS-2's. NS-2 is only lightly maintained now. NS-3 will eventually replace NS-2, but it is not backward compatible. It is trying to avoid some problems with NS-2. The NS-3 goals include some features (e.g. real-life protocol, parallel simulation) that have already proven to be useful with OMNeT. Moreover, the new animators, configuration tools and etc. are still in work. In addition, OMNeT offer basic modules, which is extensible using C++, whilst NS-2 is not easily modifiable. OMNeT provides well online documentation and active discussion forum. Overall, our preferred platform should be updated and reliable, so the flexible and generic simulator OMNeT best suits us.

5 Distributed Challenge Detection Simulation

To simulate various challenges, complex simulation scripts are needed to model the network protocols, topology, and the challenges. The challenges are separated into malicious and nonmalicious challenges. The malicious challenge (e.g. DDoS attack) could be monitored by network monitoring models and detected by anomaly detection models. The challenges like operational mistakes, unintentional misconfiguration, accidental fiber cuts, and node failures could be grouped as nonmalicious challenges. This type of challenge represents most adverse events observed in practice and could be simulated as random node and link failures [27], which cause the network failure.

5.1 Network Monitoring

Distributed network monitoring is needed to detect coordinated attacks. The module we developed to perform the traffic monitor on the link is called linkmonitor module. The monitor could continuously collect traffic information so that values display on the link in real time. Our development based on the OMNeT cDatarateChannel. After programming and debugging with the ThruputMeteringChannel, we further extend its function to monitor threshold. To gain a comprehensive overview of the link we monitor, the display could be customized with different attributes. Properties such as link color, propagation delay, data rate, can be assigned to connections. The characters could be monitored include packets number, current packet/sec, average packet/sec, current bandwidth, average bandwidth, channel utilization, average utilization, traffic volume, threshold. Channel utilization is the ratio of current traffic to the maximum traffic, which assists to understand the network performance and troubleshoot failures. This module implemented as the channel so that offers the flexibility to collect information from any link within any network to gain the clear detailed view of its typical behavior. The threshold monitor could notify that a certain parameter has exceeded a certain threshold and direct attention to those areas, so we could be immediately alert. It could effectively evaluate the network traffic to pinpoint the sub-network where victim located, and meanwhile prevent superfluous and incorrect alerting. The traffic values not only display on the link and shown in the module output in real time, but also recorded into the output vector file in OMNeT, which could be traced back to analyze why and when the anomalies behavior happens. The output vector captures traffic over time. The collected historic data demonstrate the network behavior in terms of performance and reliability. In addition, real-time statistics are important for detailed in-depth analysis. To ensure no false alarm created by the flash crowd also the alert could be raised immediately after the attack, we embed the timer function into the link monitor. The alert won't be generated unless the traffic above threshold for the continuous reasonable interval time. This method effectively avoids the events caused by the flash crowd.

5.2 Network Failure

Part of the network resilience strategy is enhancing the resilience to the network failures also modern networks should design to be fault-tolerant. The fault of the systems component could be another source of event pose the threat to the normal operation. It could lead to the network failure. There is a potential large set of faults. We simulate the faults that could be triggered by the nonmalicious challenges. So the flexible network could offer quick and efficient fault management techniques to provide network survivability. The network failure will result in the packet loss. With a broken link, the packets will be discarded until a new connection is rebuilt. Then the fault restoration will be used, once the network failure is identified, the backup path will be built immediately with the dynamically allocated spare capacity. The overall process of failure recovery shouldn't cause long delay so that could ensure the network robustness. With the above understanding, we simulate the network failure situation. The connection failure could appear in any place within the network structure. We designed connection failure channel based on the cDatarateChannel, which has the flexibility to be placed as a channel between any network objects. We could schedule connection failure event happen at certain simulation time, after it recovers, we could schedule another failure at another time as occasional failure could occur more than once on the same connection in the real network. We could include multiple concurrent connection failure channels in the network. When the connection failure happens, no packet could send through. The monitoring system should quickly raise the alarm once the broken link discovered. It means the failure detection time should be short. Then a real-time solution will be triggered.

5.3 Anomaly Detection

The anomaly detection module has been developed for the evaluation of attack detection and traffic analysis. As the linkmonitor offers the chance to get close to the victim by raising the alarm on the sub network where victim positioned, hence could perform efficient filtering. So the anomaly detection module use the simple algorithm to identify the victim, this effectively reduce the computation complexity and cost. The anomaly detection module implemented as the cSimpleModule and built into the INET compound module Router. In OMNeT, modules communicate by messages, which contain usual attributes as timestamp and arbitrary data. The cPacket class extends cMessage with fields to represent network packets (frames, datagrams, transport packets etc.) [28]. Simple module sends message through output gate. The output gate and input gate linked by a channel. Therefore, the message travels through the channel and arrives at the input gate of another simple module. The Compound module consists of several simple modules and transparently relaying messages between their inside and the outside world. The Router includes the modules NetworkLayer, Routing Table etc. The anomaly detection module interconnected with NetworkLayer, TCP, and UDP by incoming and outgoing gates through channel.

Every packet transfered into the router will pass the anomaly detection module for processing. The anomaly detection module will transparently process different network packets in a unified way. In addition, we use a hashing table to store the objects into the IDS table, and the table could iterate through. The IDS table could be monitored in real time as the module output, and the event be created immediately after identifying the destination IP address of the victim.

6 Experimental Results and Analysis

In this section, the system will be validated. The attack is injected across time to evaluate the performance. The interference challenge in the WMN will be discussed. The results demonstrate the accuracy, flexibility, scalability and efficiency of our method.

6.1 OMNET++/INET

OMNeT is a public source C++ based object oriented discrete event simulator for modeling communication networks, multiprocessors and other distributed or parallel systems [28]. It applies in diverse domains and written in two languages, NED designed for the network topology and C++ programmed for the modules. The compound module assembles from reusable simple modules. OMNeT utilize Tkenv as the GUI and it's easily debugging and trace. It could animate the flow of messages and present the node state changes in the network charts. Build on OMNeT, INET extends it by package of network protocols and offers objects, which combined with the channels to complete the network. Testing our system contains two steps, create various attacks and detect the anomalies. When consider the background traffic generation, IDS testing is classified into four categories. Compared to no background traffic, real or sanitized background traffic, testing by generating background traffic approach has benefits such as data freely distributed, no unknown attack and repeatable simulated traffic [29]. So ReaSE is chosen as our realistic background traffic and DDoS attack generator. It extends INET by server and client entities.

6.2 Network Topology and Attack Implementation

To build our network, firstly the realistic AS level topologies is generated to connect several separate administrative domains. Each AS is categorized as stub AS or transit AS. One transit AS is built to provide connections through itself to other networks. The stub AS is connected to only one other AS. This ensures each AS is accessible by crossing transit AS only. Two stub ASes and one transit AS are configured, named SAS1, SAS2 and TAS0. SAS1 connect to SAS2 through TAS0. Secondly, the router level topology within each AS is specified. Each AS has core, edge and gateway routers placed. The distinction between different routers is realized by allocating different bandwidth. Within the AS, it has total min 8 routers and max 15 routers. A few meshed core routers with low node degree that forward aggregated traffic of a high number of gateway routers with high node degree [30]. Each edge router connects between 2 and 13 hosts to the network complete the hierarchical topology. Therefore, each AS has different topology sizes and fills with nodes independently.

Thirdly, the network built with different traffic profiles to ensure the reasonable mixture of various protocols. The traffic profiles covers web, Interactive, mail, misc and ping traffic, which are based on transport protocols TCP, TCP, TCP, UDP and ICMP respectively. The router level topology's host systems are classified into clients and servers. Clients correspond to the ReaSE module InetUserHost, whilst servers represents by Web, Mail, and Interactive server. Fourthly, the bandwidth between different types of nodes are assigned from ReaSEGUI, also we configure the server fraction value, which specify the percentage of all router modules of each router-level topology are replaced by special server nodes. Overall, 136 hosts and servers are placed cross the AS in our network. At last, since ReaSE integrate the real attack tool tribe flood network to conduct the DDoS attack, so it is utilized to perform a random distribution by replacing randomly selected clients InetUserHost with DDoS zombies. The compound module DDoSZombie contains simple module TribeFloodNetwork with other INET modules that are essential to achieve the functionality of an attacking system [30]. In our experiment, total 30 DDoS zombies are located across the AS, at simulation time 120s, the zombies conduct the attack based on TCP SYN packets, and 90% of the zombies collectively launch the attack by sending a fix rate TCP SYN packets to the victim Webserver27 which is in SAS1. Figure 2 shows the linkmonitor result within SAS1 right after attack. Two linkmonitor position on the ingress link core router0 to the gateway1 and gateway2 to core router0, as highlighted in red. When the threshold value turns to 1, it generates the alarm, and informs that the attack is detected on the ingress link core router0 to gateway1. The gateway1 connect to edge router7, that is also the router victim connect to.

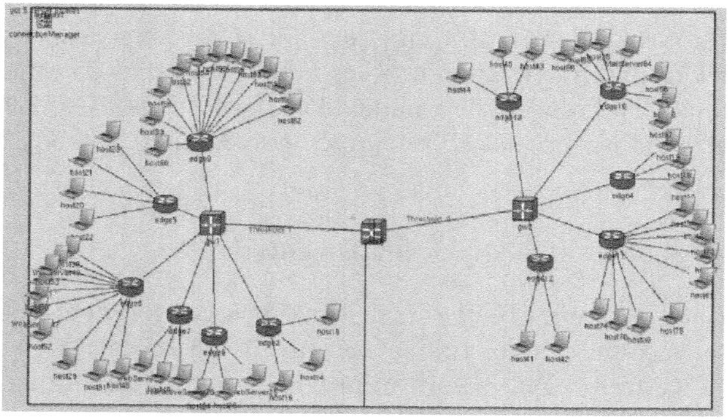

Fig. 2. Link Monitor on SAS1

In service level monitoring, except false alarm, another way to assess the quality of threshold detection is measure the delay between the time a crossing is reported and its actual occurrence [31]. Figure 3 demonstrates the result of the threshold monitor on ingress link core0 to gateway1 router. The alarm is arisen by linkmonitor at 127s.

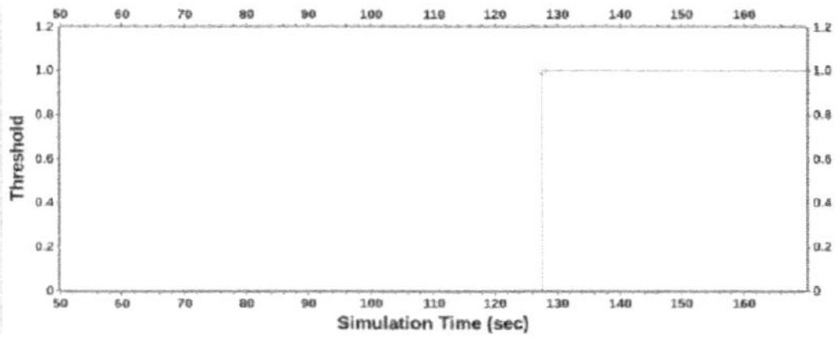

Fig. 3. Threshold Monitor

Our detection technique could identify the victim in a short time. When use linkmonitor to monitor the traffic on the ingress link as shown in Fig 4, the DDoS ramp-up behavior could clearly be observed between 120s and 140s, as each zombie is configured to delay its start for a uniformly distributed time from 0 to 20s. After the threshold raises alarm at 127s, the attack identified the victim IP 0.2.0.28 at 132s.

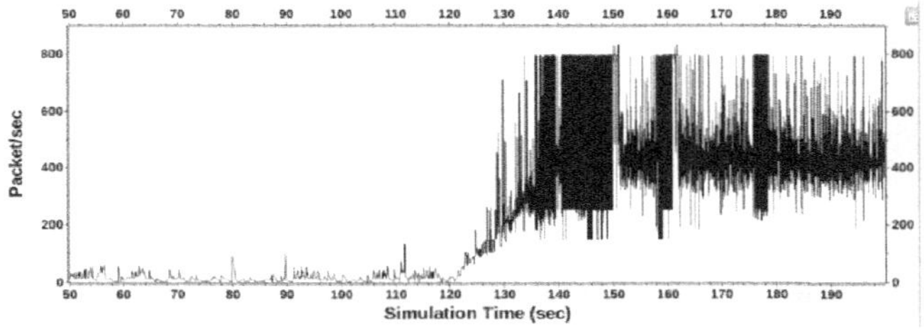

Fig. 4. SAS1 Core Router Traffic Monitor

6.3 Challenges in the Wireless Mesh Networks

There are numerous challenges need to be met in WMNs. It is impossible to look into all the challenges at once, so here we focus on one key challenge, that is interference, and demonstrate the approach to effectively address it in WMN. Because of interference, a challenge causes high traffic between two nodes could affect the available bandwidth between two other nodes. One or more root causes could result in interference which is not easy to identify from the initial symptom. Because of the mesh structure, interference could not only cause by the network itself but also other APs in proximity. Therefore, the detection will request the neighboring nodes to support each other for identifying the challenge then optimize the whole system state as the mitigation strategy. The interferer could easily and correctly detected by investigating how it appears for different nodes and the remediation may request the changing of wireless channel in the whole system [3]. Usually the interference could occur at either the receiver or sender. Interference at receiver caused by the appearance of another transmission that results in a sent frame could not receive directly. Whereas the interference at the sender is the result of the continuing transmissions from other nodes preventing the sender from sending a frame. The interference could not easily be reduced to the physical interference between multiple frequencies, and complex interactions between MAC protocols also involved especially since some of them are very particular to certain manufacturers. There are diverse set of symptoms that indicate the interference, such as SSID mismatch, element unknown, high on-chip retries, bad CRC, channel mismatch [32].

OMNeT relies on external extensions to implement the wireless ad hoc networks. The two major ones are the INET Framework (IF) and the Mobility Framework (MF). The latter is an extension explicitly designed for mobile ad hoc networks [26]. To build the WMN in OMNeT, the simulation is produced by combining the approaches of several existing frameworks into one: the protocol library is obtained from the MAC simulator and the MF; the mobility support, connection management and general structure is taken from the MF module [33]. There are several current available modules could be integrated into the WMN, which include nodes, radio propagation models for multiple signal dimensions, physical layer, receivers and an extensive library of MAC and network protocols. A key element used for the radio channel communication is the channel controller module, which handle the radio propagation, record on-going transmission and offer information for radio devices to use the reception and interference model. To simulate the interference challenge, we need to find a method to integrate the interferences generated from real measurements in a transparent way into the OMNeT INET framework. This could base on the work in [34] where the interference scenarios are represented in two dimensions. On the spatial dimension, the traffic injected is received with a calculated reception power by the wireless nodes. On the temporal dimension, two interference scenarios are produced: 1st, the simulated system respond to the interfering traffic, but has no interaction with the interfering sources; 2nd, there is mutual interactions between the simulated system and the interfering traffic. The typical WMN contains several wireless access points (APs) that route packets from clients to their destination, usually a set of egress points to the internet. Figure 5 shows the WMN simulation, it has four types of nodes, AP1 and AP2 are the clients' access points, and each connects five clients. AP4 is a mesh node. AP3 is the Internet gateway. The simple interference model is depicted in Figure 6. Host1 connects to AP1, Host2 communicates to AP2 separately. AP1 and Host1 combined as an interferer.

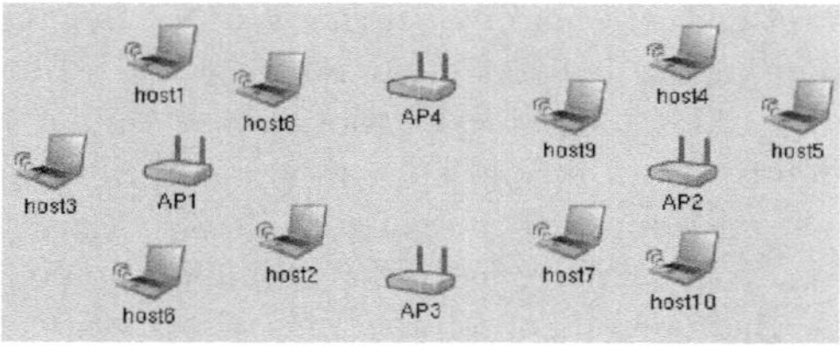

Fig. 5. WMN Simulation

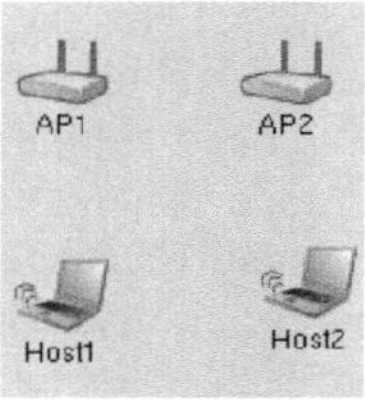

Fig. 6. Interference Model

Considering the WMN resource constraints and to minimize message overhead, we separate detection and remediation into stages. The complex investigation which involves other hosts is triggered to fully recognize the challenge only after local detection of the symptom. The initial detection could use the lightweight machine-learning classification algorithm with low false alarm rate to identify as much interference as possible. We need to find the best solution to minimize the interference by comparing different solutions and accessing its influence. After this stage, further analysis will perform to gain more understanding with the challenges. All the available networks will be detected by network scan locally. In addition, with the distributed cooperation, we could gain accurate estimation of the node utilization in the detected networks. This analysis could effectively avoid the collision caused by different nodes switched to the same channel, so only one neighboring node is responsible for finding a new channel allocation. The two connected nodes are required to have different channels. To select the best channel, we need to evaluate the cost associate with interference and channel changes. We also need to consider the communication on one channel could affect the transmission on the adjacent channel. When minimizing the interference, the channels should be changed in all the nodes simultaneously. Otherwise one channel change could cause the interference at another node, which result in once more channel change and interference, produce a circle. However, changing the channel on an AP need clients to follow the channel that might causes the wireless connection unusable for a short time. Therefore, it is suboptimal to simply switching to the current best channel at every occurrence [33].

7 Conclusions and Future Work

In this paper, we present a new distributed challenge detection system for network resilience. Currently, three activities are carried out by our system: real time network monitoring, detection of the challenge symptoms, and challenge identification. We have surveyed the state-of-art and highlighted their shortcomings. We propose that a systematic approach to resilience is required to consider the complete socio-technical system and the challenges it may face [13].

Our experimentation defined with the aim of addressing various resilience issues in the context of different types of networks and service provision settings. We present a simulation framework to simulate realistic challenges, specifically intelligent attacks and non-malicious challenges that go well beyond the network failure. Currently we validate our system on OMNeT with the wired network and propose an approach to address the significant challenge interference in the WMN. Future work will involve further developing the scenarios we want to evaluate our work through, e.g. we may examine how various types of challenges can influence the wireless networks at local and global level. As WMNs have to cope with a much wider range of challenges than the wired network, the ongoing work for WMN will require further implementation of the current architecture so that corresponding remediation could be carried out. In addition, besides the appropriate identification of a resource starvation attack on an ISP's infrastructure caused by high volumes of traffic from a DDoS attack, other types of attacks will also be considered. In future, fault management is one of the major components of the network management suite. We need to introduce innovative concepts for fault detection, root cause analysis and self-healing architectures. We hope the system could implement root-cause analysis to detect faults once they occur, and also to identify the source for performing automatic fault recovery. Different types of network service faults will be considered: they range from node misbehavior at different network layers, to software misconfigurations.

This paper elaborates an initial proof-of-concept implementation with the understanding of how to ensure resilience for a future network. Our work will persist in the context of strong experimental scenarios that we believe will feature in a future network. Through these scenarios, we will evaluate the validity of our strategies for resilience. We hope this project could have a broader socio-economic impact by contributing to the development for the future internet.

Acknowledgements. This work has been supported by NICTA (National ICT Australia). This research is carried out under the European Union Research Framework Programme 7 via the ResumeNet project with contract no. FP7–224619. The author is grateful for the guidance of Michael Fry. Also the author would like to acknowledge Aruna Seneviratne for his support and thank all the partners of the ResumeNet project for their contribution to this work. In addition, I thank Thomas Gamer and J.C. Maureira for their useful discussions on the simulation.

References

[1] ResumeNet, `http://www.resumenet.eu/`
[2] Doerr, C., Omic, J., et al.: Resilience and Survivability for future networking: framework, mechanisms, and experimental evaluation, ResumeNet Deliverable D2.1b (2010)
[3] Smith, P., Fry, M., et al.: Resilience and Survivability for future networking: framework, mechanisms, and experimental evaluation, ResumeNet Deliverable D2.2a (2010)
[4] Jung, J., Paxson, V., Berger, A., Balakrishnan, H.: Fast portscan detection using sequential hypothesis testing, pp. 211–225. IEEE, Los Alamitos (2004)
[5] Wuhib, F., Stadler, R.: Decentralised Service-Level Monitoring Using Network Threshold Alerts. IEEE Communications Magazine, 44 (2006)
[6] Jackson, A.W., Milliken, W., Santivanez, C.a., Condell, M., Strayer, W.T.: A Topological Analysis of Monitor Placement, pp. 169–178. IEEE, Los Alamitos (2007)
[7] Fry, M., Fischer, M., Karaliopoulos, M., Smith, P., Hutchison, D.: Challenge identification for network resilience. IEEE, Los Alamitos (2010)
[8] Peng, T., Leckie, C., Ramamohanarao, K.: Survey of Network-Based Defense Mechanisms Countering the DoS and DDoS Problems. ACM Computing Surveys 1, 39 (2007)
[9] Labovitz, C., Ahuja, A., Bose, A., Jahanian, F.: Delayed internet routing convergence. IEEE/ACM Transactions Networking 9, 293–306 (2001)
[10] Steinder, M., Sethi, A.S.: A survey of fault localization techniques in computer networks. Science of Computer Programming 53, 165–194 (2004)
[11] Qiu, L., Zhang, Y., Wang, F., Han, M.K., Mahajan, R.: A general model of wireless interference, pp. 171–182. ACM, NY (2007)
[12] Kotz, D., Newport, C., Gray, R. S., Liu, J., Yuan, Y., Elliott, C.: Experimental evaluation of wireless simulation assumptions, Technical Report, Dartmouth College (2004)
[13] Fessi A., Plattner, B., et al.: Resilience and Survivability for future networking: framework, mechanisms, and experimental evaluation, ResumeNet Deliverable D1.5 (2009)
[14] Doerr, C., Smith, P., et al.: Resilience and Survivability for future networking: framework, mechanisms, and experimental evaluation, ResumeNet Deliverable D2.3a (2010)

[15] Mayer, C.P., Gamer, T.: Integrating real world applications into OMNeT, Institute of Telematics, University of Karlsruhe, Karlsruhe, Germany (2008)
[16] Lippmann, R., et al.: The 1999 DARPA Off-Line Intrusion Detection Evaluation. Computer Networks 34(4), 579–595 (2000)
[17] Mahoney, M.V., Chan, P.K.: An analysis of the 1999 DARPA/Lincoln Laboratory Evaluation Data for network anomaly detection. In: Vigna, G., Krügel, C., Jonsson, E. (eds.) RAID 2003. LNCS, vol. 2820, pp. 220–237. Springer, Heidelberg (2003)
[18] Tavallaee, M., Bagheri, E., Lu, W., Ghorbani, A.: A Detailed Analysis of the KDD CUP 99 Data Set. IEEE, Los Alamitos (2009)
[19] Brugger, T.: KDD Cup 1999 dataset considered harmful, White Paper, Department of Computer Science, University of California Davis (2007)
[20] Weingartner, E., vom Lehn, H., Wehrle, K.: A performance comparison of recent network simulators, pp. 1–5. IEEE, Germany (2009)
[21] Kargl, F., Schoch, E.: Simulation of MANETs: A qualitative comparison between JiST/SWANS and NS-2. In: International Workshop on MobiEval (2007)
[22] Young, C.P., Chang, B.R., Chen, S.Y., Wang, L.C.: A Highway Traffic Simulator with Dedicated Short Range Communications Based Cooperative Collision Prediction and Warning Mechanism. IEEE, Los Alamitos (2008)
[23] Schmidt-Eisenlohr, F., et al.: Cumulative Noise and 5.9GHz DSRC Extensions for ns-2.28, University of Karlsruhe, Tech. Rep. (2006)
[24] Johansson B., et al.: Highway Mobility And Vehicular Ad-Hoc Networks In NS-3, CiteSeerX (2010)
[25] Eichler, S.: Strategies for pseudonym changes in vehicular ad hoc networks depending on node mobility. In: Proceedings of the 2007 IEEE Intelligent Vehicles Symposium (2007)
[26] Orfanus, D., Lessmann, J., Janacik, P., Lachev, L.: In Performance of wireless network simulators: a case study, pp. 59–66. ACM, New York (2008)
[27] Cetinkaya, E.K., Jabbar, A., Mahmood, R., Sterbenz, J.P.G.: Modelling Network Attacks and Challenges: A Simulation-based Approach. In: EDCC, Valencia, Spain (2010)
[28] Varga, A.: OMNeT++ User Manual, `http://www.omnetpp.org/doc/manual/usman.html`
[29] Mell, P., Hu, V., Lipmann, R., et al.: An Overview of Issues in Testing Intrusion Detection Systems, Technical Report, National Institute of Standard and Technology (2003)
[30] Gamer, T., Scharf, M.: Realistic Simulation Environments for IP-based Networks. In: ICTS (2008)
[31] Wuhib, F., Stadler, R.: Decentralised Service-Level Monitoring Using Network Threshold Alerts. IEEE Communications Magazine, 44 (2006)
[32] Smith, P., Fry, M., et al.: Resilience and Survivability for future networking: framework, mechanisms, and experimental evaluation, ResumeNet Deliverable D2.2b (2010)
[33] Rasheed, T.: Wireless Mesh Network Simulation Framework for OMNeT++, Create-Net Technical Report (2007)
[34] Maureira, J.C., Dalle, O., Dujovne, D.: Generation of Realistic 802.11 Interferences in the Omnet++ INET Framework Based on Real Traffic Measurements. In: ICST (2009)

Maximizing Mix Zone Effectiveness for the Mitigation of De-anonymization Threats in the Traffic Probe Message Service

Jeremy J. Blum and Peter O. Okosun

777 W. Harrisburg Pike
Middletown, PA 17057 USA
{jjb24,poo102}@psu.edu

Abstract. The Traffic Probe Message Service uses vehicle-to-roadside wireless communication to collect kinematic and other state data from participating vehicles. The draft standard requires vehicles to use pseudonymous identifiers in order to hide their identity. Whenever vehicles transmit state data to base stations called roadside equipment, the vehicles change their identifier and halt the collection of state data for a random period. These changes are designed to prevent a de-anonymization attack from reconstructing a vehicle's path through the road network. Thus, the roadside equipment creates mix zones, which given enough vehicles within a zone and sufficient changes in vehicle mobility patterns, can reduce the success of de-anonymization attacks. In highway scenarios, optimal mixing is likely in the regions near highway interchanges. This paper hypothesizes that given the rules snapshot generation, the optimal place for pseudonym changes is upstream of the middle of an interchange. Simulations of various traffic conditions in a large highway scenario support this hypothesis, and suggest that roadside equipment be placed such that they create pseudonym changes at these locations in order to maximize the ability of mix zones to mitigate de-anonymization threats.

1 Introduction

The deployment of Digital Short-Range Communications equipment in vehicles and on the roadside will quickly be able to support a range of applications, known as Day-One applications. One of the applications, the Traffic Probe Message Service, will enable traffic managers to gather roadway state data via wireless communication with vehicles [1]. Whenever equipped vehicles pass by base stations called roadside equipment, the vehicles will transmit a series of recent snapshots containing their locations and kinematic information.

The willingness of consumers to accept limited privacy protection in other mobile service contexts has been a function of the direct benefits realized from the service [2]. While the Traffic Probe Message Service provides system-wide benefits, the benefits to participating drivers are indirect. Given these, privacy concerns have been acknowledged as a potential hurdle for consumer acceptance of this service.

Consequently, the Probe Message Service includes mechanisms to protect the privacy of participating vehicles. The most important of these mechanisms is the use

X. Zhang and D. Qiao (Eds.): QShine 2010, LNICST 74, pp. 598–611, 2011.

of short-lived pseudonymous identifiers, which change at specified intervals during a vehicle's journey, including whenever the vehicles communicate with a newly encountered roadside unit. These short-lived identifiers are designed to mitigate the threat of de-anonymization attacks, which would seek to reconstruct a vehicle's trip through a road network. Given these short lived pseudonyms, such an attack would need to link the different pseudonyms used by a vehicle during one trip.

This timing and location of pseudonymous identifier affects on the ability of these short-lived pseudonyms to provide privacy protection. Intuitively, the pseudonym changes should be coordinated in the mix zones that occur in roadway scenarios, for example, near highway interchanges. The lane changes, changes in speeds, and entering and exiting of vehicles from a roadway increase privacy protection that can be provided through coordinated changes of identifiers in these areas. Using a model of a highway interchange, this paper hypothesizes that in general the pseudonym changes should occur upstream of an interchange, rather than downstream of the interchange.

Simulations and analysis of vehicle movement on a section of highway of I-880 in California, USA, support this hypothesis. Against a simple, multi-target de-anonymization attack, areas just upstream of interchanges provide better privacy protection than areas downstream from interchanges. In these areas, the de-anonymization algorithm experienced a misclassification rate that was 15.2% higher on average. These results support guidelines that suggest roadside equipment should be placed such that it creates pseudonym changes upstream, rather than downstream, of interchanges.

2 Background and Related Work

After an overview of the Traffic Probe Message Service, this section describes common mechanisms used to provide privacy protection in mobile services, including short-lived pseudonymous identifiers and obfuscation. Then, the section describes the privacy preserving mechanisms for the probe message service that have been proposed in the draft standard and previous research.

2.1 The Traffic Probe Message Service

Vehicles participating in the Traffic Probe Message Service collect snapshots due to three distinct types of triggers. When a vehicle is within range of a roadside unit, it transmits messages to the roadside with these collected snapshots in groups of one to four snapshots per message.

Event-Triggered Snapshots are generated sporadically by one-time events, such as roadway hazards. For example, the activation of the traction control system, indicative of poor roadway conditions, could trigger this type of snapshot [1]. These snapshots are sent in messages completely anonymously without any pseudonymous ID.

Start/Stop Snapshot Messages are triggered initially whenever a vehicle stops for more than a set period of time and then again when the vehicle exceeds a threshold velocity. By default, the stop snapshot is triggered whenever a vehicle stops for more than five seconds, and a start snapshot is triggered when a vehicle exceeds the speed of 4.5 m/s. Unlike the event-triggered snapshots, these snapshots are sent with a pseudonymous ID.

While participating vehicles are moving, they collect Periodic Snapshot Messages. These snapshots are generated at intervals specified by the roadside. By default, a vehicle travelling 20 miles per hour would generate a snapshot every 4 seconds, while a vehicle travelling 60 miles per hour would generate a snapshot every 20 seconds.

The Periodic Snapshots include data on the location and kinematics of the vehicle including latitude, longitude, heading, velocity, acceleration, and yaw rate. In addition, snapshots can contain a range of other data elements, including control systems state, e.g. brake applied pressure, steering wheel angle, and traction control state; vehicle type, tire pressure, wiper rate, rain sensors, sun sensors, ambient air pressure, and temperature.

2.2 Privacy for Mobile Users

A common approach to provide privacy protection for mobile users relies on pseudonymous identifiers and obfuscation of mobile users' paths and data. The effectiveness of these approaches can be measured by their ability to prevent de-anonymization attacks that can link an individual to the path of a mobile user.

Pseudonyms that last for an entire trip or longer provide limited protection for mobile users. For example, vehicle traces with an update period of one record per minute have been used in de-anonymization attacks in which researchers were able to identify 85% of the homes of the mobile users [3]. In another de-anonymization attack with a shorter update period of 6 seconds, researchers were able to determine the location of users' homes within 61 meters [2].

From a traffic manager's point of view, these long-term pseudonyms would provide valuable data, for example, for the reconstruction of origin-destination matrices. However, due to the limited protection provided by these pseudonyms, the Traffic Probe Message Service uses short-lived pseudonyms, in which a vehicle changes pseudonyms during a single trip.

However, inference attacks can reconstruct a user's path even if the user changes pseudonyms during a trip. In order to link together messages produced under different pseudonyms, these attacks can use trajectory-based linking, relying on the tendency of users to continue moving in the same direction; map-based linking, relying on constraints imposed by a road network to aid in the linking; and empirical linking, relying on previous mobility patterns to aid in linking [4].

In order to limit the effectiveness of these linking attacks, often a pseudonym change will be accompanied by a random period of silence [5]. If users are transmitting messages at precise time intervals, the timing of the first message with a new pseudonym may be sufficient to link to the old pseudonym. A random period of silence can also increase the distance between transmissions with different pseudonyms increasing the difficulty in linking together the transmissions.

In addition, changes in pseudonyms can be synchronized by having users change identifiers in the same geographical area, called a mix zone. For mobile services offered by base stations, these mix zones fall naturally in areas outside of the range of application providers [6]. For other services, intersections can serve as attractive mix zones because of the difficulty in linking that arises due to the changes in paths that occur there [7]. In addition, pseudonym changes can be coordinated in an ad hoc fashion via direct user-to-user communication, e.g. with direct vehicle-to-vehicle communication in vehicular networks [8].

2.3 Privacy Protection in the Traffic Probe Message Service

The Traffic Probe Message Service includes two primary mechanisms to preserve participant privacy. The probe message service is designed to obfuscate the beginning and end of every trip. The first snapshot is generated only after a vehicle has travelled 500 meters in the beginning of a trip. In addition, the vehicle deletes the snapshots collected at the end of a trip between the last roadside equipment and the final destination.

Given typical travel patterns for many users, a vehicle's path in the middle portion of the trip may allow for an inference attack. In order to prevent this type of attack, the probe service also requires vehicles to periodically change their pseudonymous identifiers. After changing their identifier, the vehicles wait a random period of silence time before creating the next snapshot. Two random numbers are generated, one for a distance between 50 and 250 meters, and another for a time between 3 and 13 seconds. The next snapshot is generated when after the random distance is travelled or the random time elapses, whichever comes first.

After a vehicle transmits a message to new roadside equipment, it must change its pseudonym. Therefore, the roadside equipment creates a mix zone, in which the changing of pseudonymous ids is synchronized geographically.

In addition, vehicles are required to change their pseudonyms every 120 seconds or 1 km, whichever comes last. Therefore, a vehicle may use multiple pseudonyms in messages to a single RSE. The snapshots generated under different pseudonyms cannot appear in the same message. Otherwise, linking an old pseudonym and new pseudonym would be straightforward if both appeared in a single message.

Previous research has focused on changes in the Traffic Probe Message Service that can improve privacy protection for participating vehicles [9]. These key changes include promoting the geographic coordination of pseudonym changes. Pseudonym changes at roadside equipment are coordinated geographically. However, other changes occur every 120 seconds or 1 km, whichever comes first. The authors suggest that vehicles be forced to change their pseudonyms at fixed distances, so that all pseudonym changes occur in fixed geographic locations. In addition, the authors suggest that snapshots include limited additional data, for example, vehicle type, that could aid in de-anonymization attacks. This paper extends this work by investigating locations that maximize the effectiveness of mix zones occurring at roadside equipment.

3 A De-anonymization Attack on Traffic Probe Message Service in Highway Scenarios

A de-anonymization attack on vehicles travelling in highway scenarios must link together pseudonyms used by a particular vehicle. This section describes a linking attack that attempts to link the last snapshot transmitted by a vehicle with an old pseudonym with the first snapshot generated by a vehicle under a new pseudonym. In order to increase the complexity of a de-anonymization attack, the snapshot messages are assumed to contain a minimum of information, only a timestamp, position information and velocity.

The attack uses a multi-target tracking approach to link together an old pseudonym with a new pseudonym. It first determines the feasibility of a match between the last snapshot calculated under an old pseudonym and the first snapshot calculated under a new pseudonym. In order for a match to be feasible, three constraints must be satisfied.

3.1 Snapshot Time Difference Constraint

The *Snapshot Time Difference Constraint* ensures that the timing of snapshot *j* is such that it could have generated by the same vehicle that generated snapshot *i*. At the low end of this time difference, a vehicle could have generated its last snapshot immediately before changing its pseudonym. After changing its pseudonym, the vehicle could then have chosen the smallest delay (3 seconds) before generating its next snapshot. At the high end, a vehicle could have been just about to generate a snapshot when changing its pseudonym, i.e. it generated its last snapshot just less than 20 seconds prior to the pseudonym change. This vehicle could then choose the longest delay (13 seconds) before generating its next snapshot. Thus, the *Snapshot Time Difference Constraint* is defined as follows:

$$3 \leq \Delta t < 33 \tag{1}$$

Where:

Δt is the difference between the timestamp of snapshot *j*, generated under a new pseudonym, and the timestamp of snapshot *i*, the last snapshot generated under an old pseudonym (in seconds)

3.2 Maximum Distance Travelled Constraint

The maximum distance that could be travelled between two snapshots is determined by two different cases. In the first case, the vehicle, starting at a velocity of v_i at the time of snapshot *i* accelerates as fast and as long as possible and at the last possible moment decelerates at the fastest possible rate to achieve a velocity of v_j at the time of snapshot *j*. The maximum velocity during this case never exceeds a maximum possible velocity for this section of roadway.

In cases where the maximum velocity would be exceeded, the maximum distance is calculated as follows. It is assumed that the vehicle accelerates as fast as possible in order to raise its velocity from v_i at time of snapshot *i* until it reaches the maximum possible velocity. The vehicle than travels at this maximum velocity until the last possible moment at which point it brakes as hard as possible to lower its velocity to v_j at time of snapshot *j*. Thus, the *Maximum Distance Travelled Constraint* is specified as follows.

Upper Bound on $\|l_j - l_i\|$	Condition
$\frac{(v_i + a_{max}^{+} t_a)}{2} t_a + \frac{(v_i + a_{max}^{+} t_a + v_j)}{2} t_b,$ $t_a = \frac{v_j - v_i - a_{max}^{-} \Delta t}{a_{max}^{+} - a_{max}^{-}}$ $t_b = \frac{v_j - v_i - a_{max}^{+} \Delta t}{a_{max}^{-} - a_{max}^{+}}$	if $(v_i + a_{max}^{+} t_a) \leq v_{max}$
$\frac{(v_i + v_{max})}{2} t_a + v_{max} t_c + \frac{(v_{max} + v_j)}{2} t_b,$ $t_a = \frac{v_{max} - v_i}{a_{max}^{+}}$ $t_b = \frac{v_j - v_{max}}{a_{max}^{-}}$ $t_c = \Delta t - t_a - t_b$	Otherwise

(2)

Where:

$\|l_j - l_i\|$ is the distance travelled along the roadway between snapshot i and snapshot j (in meters)

v_i is the velocity reported in snapshot i (in meters/second)

v_j is the velocity reported in snapshot j (in meters/second)

v_{max} is the maximum possible velocity on this section of roadway (in meters/second)

a_{max}^{+} is the maximum possible positive acceleration possible on this section or roadway (in meters/second^2)

a_{max}^{-} is the maximum possible negative acceleration (deceleration) possible on this section or roadway (in meters/second^2)

Δt is the difference between the timestamp of snapshot j and the timestamp of snapshot i (in seconds).

t_a is a possible time spent accelerating in the interval between snapshots (in seconds).

t_b is a possible time spent braking in the interval between snapshots (in seconds).

t_c is a possible time spent travelling at a constant speed in the interval between snapshots (in seconds)

3.3 Minimum Distance Travelled Constraint

Like the maximum distance constraint, the *Minimum Distance Travelled Constraint* is determined by two different cases. In the first case, the vehicle, starting at a velocity of v_i at time of snapshot i brakes as fast and as long as possible and at the last possible moment accelerated at the fastest possible rate to achieve a velocity of v_j at the time of snapshot j. The minimum velocity during this case never goes below a minimum possible for the given section of roadway.

In cases where the minimum velocity would be reached, the distance is calculated as follows. It is assumed that the vehicle decelerates as fast as possible in order to lower its velocity from v_i at time of snapshot i until it reaches the minimum possible. The vehicle than travels at the minimum velocity (or stops) until the last possible moment at which point it accelerates as fast as possible to raise its velocity to v_j at time of snapshot j. Thus, the *Minimum Distance Travelled Constraint* is specified as follows.

Lower Bound on $\|l_j - l_i\|$	**Condition**
$\frac{(v_i + a_{max}^{-} t_b)}{2} t_b + \frac{(v_i + a_{max}^{-} t_b + v_j)}{2} t_b,$ $t_b = \frac{v_j - v_i - a_{max}^{+} \Delta t}{a_{max}^{-} - a_{max}^{+}}$ $t_b = \frac{v_j - v_i - a_{max}^{-} \Delta t}{a_{max}^{+} - a_{max}^{-}}$	if $(v_i - a_{max}^{-} t_b) \geq v_{min}$
$\frac{(v_i + v_{min})}{2} t_b + v_{min} t_c + \frac{(v_{min} + v_j)}{2} t_a,$ $t_b = \frac{v_{min} - v_i}{a_{max}^{-}}$ $t_b = \frac{v_j - v_{min}}{a_{max}^{+}}$ $t_c = \Delta t - t_a - t_b$	Otherwise

(4)

Where:

$\|l_j - l_i\|$ is the distance travelled along the roadway between snapshot i and snapshot j (in meters)

v_i is the velocity reported in snapshot i (in meters/second)

v_j is the velocity reported in snapshot j (in meters/second)

v_{min} is the minimum possible velocity on this section of roadway (in meters/second)

a_{max}^{+} is the maximum possible positive acceleration possible on this section or roadway (in meters/second^2)

a_{max}^{-} is the maximum possible negative acceleration (deceleration) possible on this section or roadway (in meters/second^2)

Δt is the difference between the timestamp of snapshot j and the timestamp of snapshot i (in seconds)

t_a is a possible time spent accelerating in the interval between snapshots (in seconds)
t_b is a possible time spent braking in the interval between snapshots (in seconds).
t_c is a possible time spent travelling at a constant speed in the interval between snapshots (in seconds)

3.4 Snapshot Pairing

Snapshot pairs that satisfy the three constraints are then ranked in ascending order according to the following score:

$$s_{i,j} = \left| \left\| l_j - l_i \right\| - \frac{v_i + v_j}{2} \Delta t \right| \tag{5}$$

Where:
$s_{i,j}$ is the score assigned to the snapshot pair (snapshot *i*, snapshot *j*)
$\left\| l_j - l_i \right\|$ is the distance that would be travelled along the roadway between the location reported in snapshot *i* and the location reported in snapshot *j* (in meters)
v_i is the velocity reported in snapshot *i* (in meters/second)
v_j is the velocity reported in snapshot *j* (in meters/second)
Δt is the difference between the timestamp of snapshot *j* and the timestamp of snapshot *i* (in seconds)

The de-anonymization attack then iterates through the list of sorted snapshot pairs. If neither the snapshot *i* nor the snapshot *j* in the next pair has been linked already then a link is established between these pairs. The algorithm then repeats this process until all snapshots have been linked.

By eliminating snapshots that have already been paired, the de-anonymization takes a multi-target tracking approach, which has been more effective than single-target tracking attacks in similar inference attacks on mobile users [4]. The approach could be further improved by expanding the information used. For example, the current attack uses only the first and last snapshot generated by a vehicle under a given pseudonym. A stronger attack could use all of the snapshots generated under a pseudonym to create a profile of a driver, and then use these profiles to aid in the linking. In addition, if the snapshots include attributes other than pseudonym, timestamp, position, and velocity, these attributes could also be used in the linking attack.

4 Optimizing Mix Zone Location in the Traffic Probe Message Service

Roadside equipment, because of the pseudonym change that occurs after an exchange of messages with a vehicle, should be placed where they can create the most effective mix zones. Based on the mixing that occurs near highway interchanges and the rules of snapshot generation, the paper hypothesizes that causing pseudonym changes upstream of interchanges will be more effective than downstream changes.

Simulation of the Traffic Probe Message Service in a large highway scenario is then used to test this hypothesis.

4.1 Mix Zones in Highway Scenarios

Highway scenarios offer ideal locations for mix zone locations promoting probe participation privacy. Intersections on arterial roadways also provide for opportunities to "create confusion at the crossroads," due to the possibility that a vehicle could turn [7]. However, in the probe message service, the periodic snapshots and start/stop snapshot messages can aid in the tracking of vehicles through an intersection.

In highway scenarios, one would intuitively expect mix zones to occur near highway entrances and exits. Upstream of off-ramps, there is mixing of traffic as exiting vehicles move from their chosen travel lane to the exit lane. These lane changes create and remove gaps between vehicles causing other vehicles to change their velocity. Highway on-ramps similarly create mixing downstream of the ramp as new vehicles enter the highway.

Figure 1 shows two potential locations for pseudonym changes near a simple highway interchange. As shown upper part of the figure, these changes could occur upstream of the interchange or the changes could occur after the interchange. In both

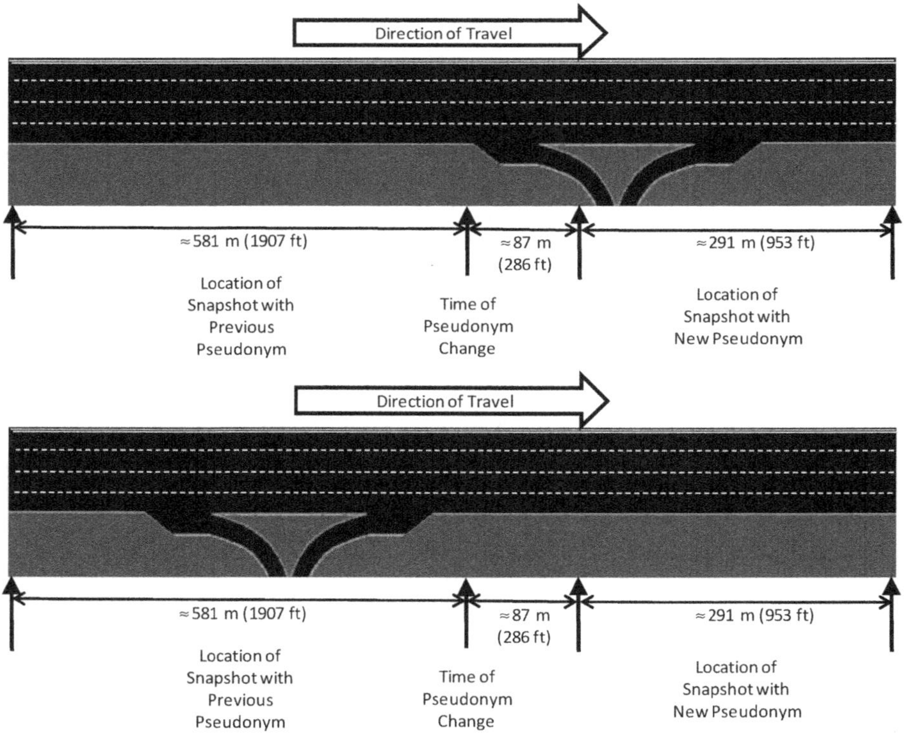

Fig. 1. Scenarios for Pseudonym Changes Near Interchanges

scenarios, the change can take advantage of mixing that occurs upstream of off ramps and downstream of on ramps. Because snapshots are generated at highway speeds every 20 s, the last snapshot under the old pseudonym will be generated between 0 and 20 s prior to the pseudonym change. At highway speeds, this location will be between 0 and approximately 580 m. At these speeds, the vehicle then generates the first snapshot under the new pseudonym between 3 and 13 seconds after the pseudonym change, i.e. after travelling approximately 90 to 380 m.

There is, however, a significant difference between the mixing that occurs upstream of off ramps and downstream of on-ramps. Because of the uncertainty involved in when gaps in a target lane will occur, drivers, who need to exit a highway, will on average perform their lane changes well in advance of the off ramp. Drivers entering a highway, on the other hand, will tend to make their lane changes as soon as a suitable gap in the desired lane appears.

Therefore, the mixing upstream of an off-ramp will tend to occur at father distances from an interchange than the mixing that occurs after an on-ramp. The scenario at the top of Figure 1, therefore, should provide better privacy protection because it has a better chance of covering both the mixing that occurs at longer distances upstream of the off-ramp and the mixing that occurs at closer distances downstream from the on-ramp.

Interchanges in highway scenarios can be significantly more complex that the ones considered here. In a more complex scenario, though, there is likely to be more mixing throughout area covered by an interchange. Therefore, given this more significant mixing, the precise location of the pseudonym change is likely to be less important in this type of scenario from the point of view of privacy protection.

4.2 Simulation of De-anonymization Attacks on the Traffic Probe Message Service

The simulation of de-anonymization attacks on the probe message service has three distinct pieces. The first component is the highway environment, which is a simulated highway based on I-880 in Hayward, California, USA, with traffic demand for rush hour and also for off-peak times. The second component is the simulation of the traffic probe message service. The third component is the de-anonymization algorithm that attempts to link the snapshot with the old pseudonym and the snapshot with the new pseudonym. The attack is the one described in Section 3.

The simulation of vehicular mobility was done with the microscopic vehicle traffic simulator CORSIM, a validated and widely used simulation program [10]. This program tracks each individual vehicle as it travels through the road network. The vehicle's mobility patterns are a function of driver behavior, vehicle performance characteristics, and constraints imposed by the roadway geometry and surrounding vehicles.

CORSIM was used to model the roadway geometry of a 9.2 mile section of highway similar to I-880 in Hayward, California [11]. This section of highway contains ten off-ramps and ten on-ramps. The scenarios are based on traffic counts obtained by loop detectors and reported in the Freeway Service Patrol Evaluation Project, University of California. The highway contains between 8 and 10 lanes, and the scenarios model average traffic, without HOV lanes, during peak hours and off peak hours. In peak traffic scenario, the average density of vehicles was approximately 170 vehicles per mile. In the off-peak scenario, the average density was approximately 70 vehicles per mile.

The mobility of vehicles provided by the CORSIM simulation was used in the simulation of the traffic probe message service. The percentage of the vehicles participating in the service was varied from 5% and 15%. These vehicles begin participating in the service after passing roadside equipment placed either upstream from an off-ramp or downstream from an on-ramp. The vehicles create their first snapshot after a random interval chosen from between 3 and 13 seconds, the defaults defined in the standard. Thereafter, vehicles create new snapshots every 20 seconds until the vehicles pass another piece of roadside equipment. At this point, the vehicle changes its pseudonym, waits another randomly chosen interval between 3 and 13 seconds, and then begins to create snapshots with a new pseudonym.

4.3 Simulation Results

The simulation results show that the location of the most effective mix zones in highway environments is a function of the roadway geometry. Indeed, the results support the hypothesis that in simple interchanges, better privacy protection arises from having vehicles change their pseudonyms upstream of an interchange rather than downstream.

The error in the de-anonymization attack can be measured as the misclassification rate, i.e. the number of correctly linked snapshots generated under old and new pseudonyms divided by the total number of old snapshots. The pool of old snapshots is drawn only from participating vehicles currently on the highway when passing roadside equipment. The pool of new snapshots, though, includes all participating vehicles. This larger pool is necessary because vehicles may not be generating snapshots prior to entering the highway when they may be close to the start of their trips.

Figure 2 shows the misclassification rate of the de-anonymization attack as a function of the traffic scenario, level of participation in the service, and location of pseudonym change. As shown in this figure, as the participation rates increase, the misclassification rate increases. Higher participation rates result in higher density of participating vehicles, which increase the likelihood of an incorrect pairing by the inference attack. Similarly, in these scenarios, both of which contain free flowing traffic, the higher traffic densities in the peak hour scenario resulted in a higher misclassification rate than in the off peak scenario.

More importantly, the results shown in Figure 2 suggest that the optimal mix zones are a function of roadway geometry. In all but the scenario with the lowest density of participating vehicles, the de-anonymization attack performs worse when pseudonym changes occur upstream of an interchange. The misclassification rate in these scenarios is 15.2% higher on average. Therefore, pseudonym changes in these locations provide better privacy protection.

Figure 3 shows the misclassification rate when pseudonyms are changed at different locations upstream from an off-ramp. As shown in this figure, the privacy

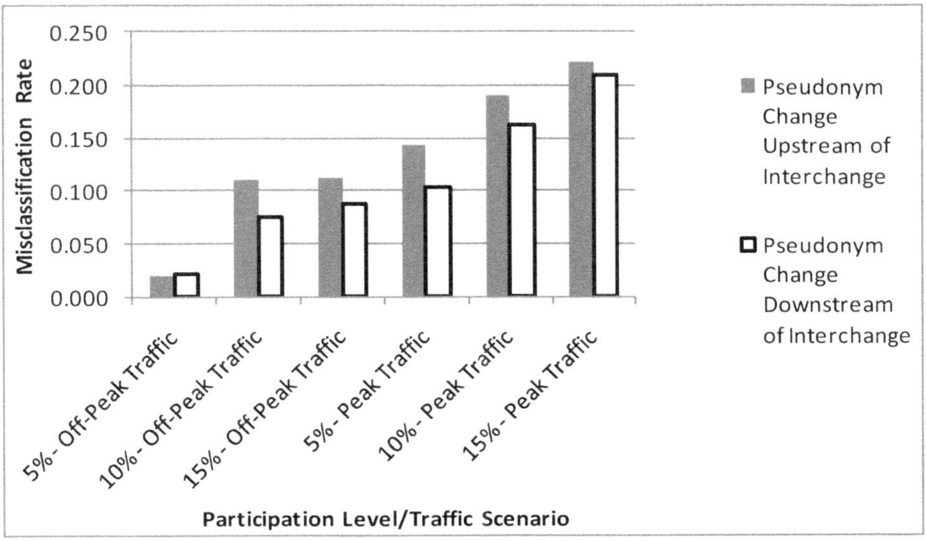

Fig. 2. De-anonymization Attack Misclassification Rate by Traffic Scenario, Participation Level (%), and Location of Pseudonym Change

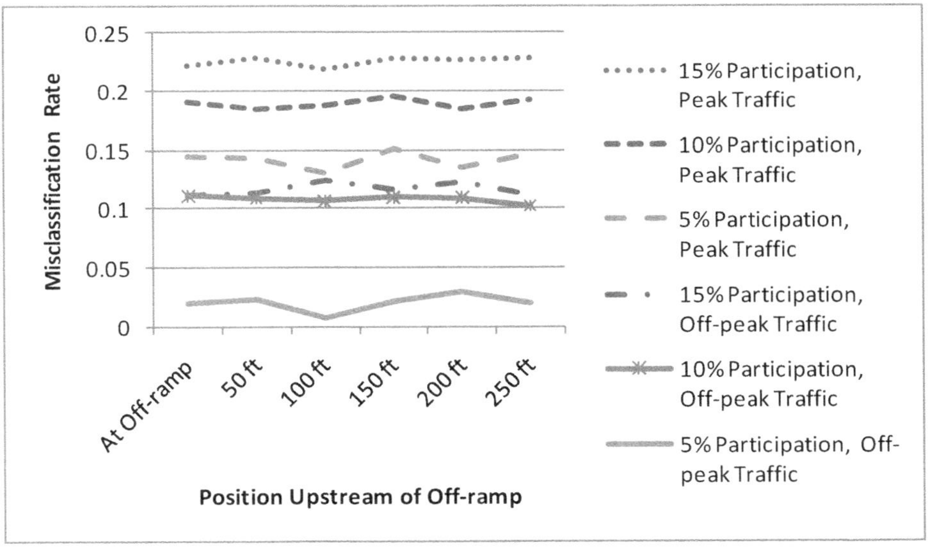

Fig. 3. Effect of Location of Pseudonym Change on the De-anonymization Attack Misclassification Rate at Various Locations Upstream of Off-ramps

protection provided by upstream changes in pseudonyms is not significantly affected by the precise location of the change. Particularly as vehicle density increases, the misclassification rates are fairly constant across all of these locations between 0 and 250 ft upstream of an off-ramp.

These results provide guidance for the placement of roadside equipment to increase privacy protection for vehicles participating in the Traffic Probe Message Service. In order to obtain the most effective mix zones, this equipment should be placed such that vehicles change their pseudonyms upstream, rather than downstream, of interchanges. These locations appear to provide better mixing regardless in most traffic conditions. Moreover, as participation in the probe message service grows, the advantages provided by these locations will increase.

5 Conclusions and Future Work

The Traffic Probe Message Service promises a wealth of additional information to help traffic operators efficiently manage road networks. However, widespread participation in the service is needed in order to realize the full promise of the system. Privacy concerns have been identified as a potential obstacle to achieving this widespread participation.

This paper examined the effect of roadside equipment placement on the privacy protection for users of the service. The placement of this equipment affects privacy protection because vehicles change their pseudonymous identifiers every time they pass this equipment. Therefore, for privacy protection, the equipment should be placed in areas where maximum mixing can occur.

In highway environments, simulation of a large highway suggests that in order to capture the mixing that occurs both upstream and downstream from an interchange, the pseudonym changes should happen upstream rather than downstream of the interchange.

This current study does have two significant limitations. First, the vehicle mobility patterns were generated from a microscopic vehicle simulator. Although this simulation program has been validated, it likely does not produce the same range of variable driver behavior as in the real world. Moreover, this study examined a limited set of scenarios, focusing on only one highway with a limited variety of interchanges.

The authors currently plan future work to extend this work and address these limitations. The authors intend to confirm the results of simulations with results from field observations of vehicle mobility patterns. In addition, the authors intend to study additional roadway scenarios.

References

1. DSRC Committee. DRAFT SAE J2735 Dedicated Short Range Communications (DSRC) Message Set Dictionary: Annex B: Traffic Probe Message Use and Operation. Society of Automotive Engineers, Warrendale, PA (2007)
2. Krumm, J.: A survey of computational location privacy. Personal and Ubiquitous Computing 13(6), 391–399 (2009)
3. Hoh, B., Gruteser, M., Xiong, H., Alrabady, A.: Enhancing security and privacy in traffic-monitoring systems. IEEE Pervasive Computing Magazine, 38–46 (2006)
4. Gruteser, M., Hoh, B.: On the anonymity of periodic location samples. In: Hutter, D., Ullmann, M. (eds.) SPC 2005. LNCS, vol. 3450, pp. 179–192. Springer, Heidelberg (2005)

5. Huang, L., Matsuura, K., Yamane, H., Sezaki, K.: Towards modeling wireless location privacy. In: Danezis, G., Martin, D. (eds.) PET 2005. LNCS, vol. 3856, pp. 59–77. Springer, Heidelberg (2006)
6. Beresford, A.R., Stajano, F.: Location privacy in pervasive computing. IEEE Pervasive Computing Magazine, 46–55 (2003)
7. Meyerowitz, T.J., Choudhury, R.R.: Realtime location privacy via mobility prediction: creating confusion at crossroads. In: 10th Workshop on Mobile Computing Systems and Applications, pp. 1–6 (2009)
8. Li, M., Sampigethaya, K., Huang, L., Poovendran, R.: Swing & swap: user-centric approaches towards maximizing location privacy. In: ACM Workshop on Privacy in Electronic Society, pp. 19–28 (2006)
9. Blum, J.J., Okusun, P.O.: Privacy Implications of the Traffic Probe Message Service. In: IEEE Intelligent Transportation Systems Conference, pp. 342–347 (2010)
10. Owen, L.E., Zhang, Y., Rao, L., McHale, G.: Traffic flow simulation using CORSIM. In: Winter Simulation Conference, vol. 2, pp. 1143–1147 (2000)
11. Petty, K.: FSP 1.1: The Analysis Software for the FSP Project. University of California, Berkeley, Berkeley, CA (1994)

Layered and Service-Dependent Security in CSMA/CA and Slotted Vanets

Luca Pilosu*, Hector Agustin Cozzetti, and Riccardo Scopigno

Istituto Superiore Mario Boella
Via P.C. Boggio 61, 10138 Torino, Italy
{pilosu,cozzetti,scopigno}@ismb.it

Abstract. The implications and challenges of security in vehicular ad-hoc networks are huge for several reasons and, basically, for human safety effects and due to the complex and highly dynamic setting. Additionally security, being a cross-layer topic, can be managed at different layers of the network stack (e.g. at MAC-level - with encryption and authentication, at network-layer - as with IPsec, at transport-layer TLS, SSL). A rich scientific literature has addressed the issue of vanet security, however, all the proposed solutions (i) focus on a specific layer and (ii) offer either robust but not scalable solutions (such as the PKI infrastructure, hardly managed under mobility for all the nodes and services) or weak ones (at least weak if applied to human safety). For this reason, in the present paper, security for vanet is faced with a layered approach which lets envisage several solutions, properly and hierarchically differentiated for distinct services. Additionally, what introduces an even stronger novelty, the dissertation covers both CSMA/CA and slotted MAC protocols, having the latter recently encountered a certain scientific favour.

1 Security in Vehicular Ad-Hoc Networks

In vehicular ad-hoc networks (vanets), each vehicle contributes to create the network structure by establishing dynamic links to different nodes in its neighborhood. Therefore, possible vulnerabilities affecting any node in the network can dramatically lower the security level of the whole network, which does not have a pre-defined architecture (and hierarchy).

Moreover, the typical characteristics of the wireless medium make it difficult to implement strong and strict controls on the communications. The main problem is that any node within the transmission range can easily eavesdrop and participate in communications. All in all, it is early clear that security constitutes one of the main criticalities in vanets and, as such, requires a deep analysis.

Vanets are meant to support a wide range of services and each demands specific guarantees on security. For instance, safety messages include critical

* This research work has been carried out within *IoT - Piattaforma Tecnologica Innovativa per l'IoT* Project, supported by regional fundings of Regione Piemonte - Italy.

X. Zhang and D. Qiao (Eds.): QShine 2010, LNICST 74, pp. 612–622, 2011.

alerts aimed at delivering real-time life-services (including accident prevention and reactive cooperative driving): consequently they require a very high level of security - in terms of integrity, reliability and overall authentication of the network architecture. Other applications (e.g., infotainment and location-based services) can tolerate more relaxed requisites however they can benefit too from a controlled and traced access - for example for user profiling.

Finally, independently on the delivered services, also the MAC protocol itself can gain benefits from integrity and consistency checks, to infer possible misbehaving nodes – this also can be considered security-related.

As a result security can be managed at different layers of the network stack (*e.g.* MAC-level WPA2 encryption and authentication, network-layer IPsec, transport-layer TLS, SSL) - this reflects the more general attitude of security, being a cross-layer topic [4]. The most suitable solution depends on the characteristics of the application being protected, but also on the kind of the attack which has to be faced.

Concerning this point, literature shows that the effects on the network can be different if caused from internal or external, persistent or random, accidental or malicious attacks. Some of the most important attacks that can be carried on in vehicular networks are [9]:

- **jamming:** the attacker generates voluntarily interfering signals, in order to block all the communications in his transmission range;
- **poison traffic attack:** the attacker injects fake information into the network;
- **alteration traffic attack:** in this attack, the information sent regularly from other vehicles are altered and forwarded;
- **stolen identity:** the attacker personifies a different identity (e.g., a police vehicle), with the goal of propagating fake information without being caught.

So far security has been mentioned as a cross-layer and service-dependent issue. But also the underlying protocols may play a meaningful role. The emerging international standard is based on CSMA/CA (in the WAVE stack, the 802.11p standard [1] defines enhancements to 802.11 required to support Intelligent Transportation Systems (ITS) applications in the licensed band of 5.9 GHz (5.85-5.925 GHz)). However recent scientific literature has shown the potentialities of slotted approaches (among them MS-Aloha [5] seems to be the most extensively studied one).

The present paper analyzes what solutions can restrain the security-threats affecting vanets. These solutions are investigated together with their dependency on the underlying MAC protocol: as depicted in Fig. 2, the solutions are manifolds and not all of them are suitable for all MAC protocols.

1.1 Slotted Vanets – MS-Aloha

MS-Aloha is a slotted protocol which specifically addresses the issue of determinism in vehicular communications exploiting distributed contention methods. The wireless medium is managed in a slotted frame structure, with a fixed number of slots and fixed slot length. A global synchronization is required and is supposed

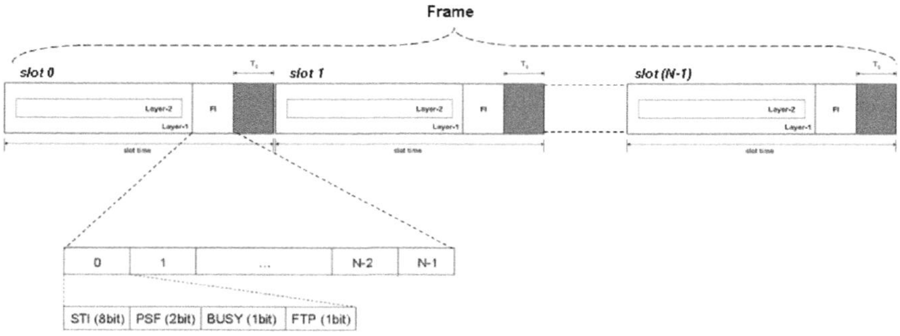

Fig. 1. Two-layer description of a MS-Aloha slot (PHY and MAC), including *FI* format

to be provided, for example, by a GPS/Galileo receiver. Concerning wireless physical layer, any one could be adopted and, to facilitate possible integration with IEEE 1609 stack, that of IEEE 802.11p is supposed to be adopted.

In MS-Aloha, before transmitting, each station has to sense the medium in order to identify unassigned slots and randomly select one of them. Collisions may still occur in this "contention" phase, especially in case of hidden stations. While WiFi foresees the well-known RTS/CTS handshake to overcome this issue, MS-Aloha uses message broadcasting: each "active" node continuously propagates its view of frame allocation in a fixed trailer called FI (Fig. 1). If the frame includes N slots, each FI contains $N \times FI_i$ fields, each 12-bit wide and specifying if the i-th slot is perceived free or busy, and to which node it is allocated (in its view)- a short node identifier is used for this purpose [5]. Each sub-field FI_i (Fig. 1) is made up by the following bits:

- **STI (*source temporary identifier*) - 8 bit:** the short label identifying the node heard on slot i.
- **PSF (*priority status field*) - 2 bit:** priority bits used for pre-emption mechanisms.
- **BUSY/CLS - 2 bit:** describing the slot i as free, busy, collision, or third hop (to limit the propagation of the FI information at third hop).

In practice, if a terminal M needs to transmit, it carries out a *contention phase*: it tries to gain access to a slot (chosen among those perceived "free"), after sensing the channel for a whole frame period. The node sends then a broadcast frame on the chosen slot (say slot j). All the nodes which have already been allocated a slot broadcast back their view of the frame, specifying if they have "heard" node M in slot j. This mechanism acknowledges node M of its transmission: if it receives just an FI containing contradicting information on slot j, it infers a collision and selects a new slot.

Each node refreshes the information on slot i when the frame has reached again the position i (the information on slot allocation expires after a frame-time). In order to improve slot re-use and make the protocol more deterministic [5], all the information on channel state which has not been directly detected by a node – and comes only from the analysis of the received FIs – is not forwarded over more than one hop. This way the same slot is *not* announced "busy" too far from the node which is using it, improving slot reuse.

MS-Aloha has been defined and simulatively demonstrated to work under mobility with encouraging results.

2 Security Solutions and Vanet Protocols

Some solutions can be designed in a protocol independent way (as the layer-2 authentication solution proposed in sect. 2.1). Other well-known solutions have been designed for a specific protocol but can be easily extended to other ones: it is typical of the upper layers in the protocol stack. For instance IEEE 1609.2 (*Security Services for Applications and Management Messages*) [3] covers methods for securing WAVE management messages and application messages: despite what has been defined for IEEE 1609 [2] stack, they may also apply to slotted approaches, as in case of the Public-Key Infrastructure presented in 2.3.

	MS-Aloha	CSMA/CA	
Application Layer *"Full" Security*	✓	✓	Identity (non repudiation) Message authentication (integrity) Privacy (pseudonym)
MAC Layer *MAC_DoS Prevention*	✓	✗	Prevention of DoS exploiting MAC
Link Layer *Baseline Privacy*	✓	✓	Broadcast Domain Segmentation

Fig. 2. Cross-layer protocol-dependent and protocol independent mechanisms for vanet e-security

Finally there are mechanisms which apply only to a specific protocol class, as the one presented in sect.2.2 for slotted approaches: since such protocols include rich set of acknowledgements sent by each node, this gives the opportunity to enforce security (in terms of DoS prevention) by consistency checks.

Additionally, not only different services require proper security mechanisms, but also, reversely, each mechanism may suite particularly well a specific need. This perspective indeed is the one adopted in this paper, and the different available mechanisms are investigated in a bottom-up sequence.

The analysis leads to conclusions which can be summarized as follows: the only way to guarantee message and node authentication, as required by vanet safety mechanisms, is to exploit the PKI infrastructure (as widely published in literature); however also link layer encryption can play its role, mainly in segmenting broadcast domains without involving heavy protocol overheads; moreover some Denial-of-Service attacks can be counteracted only at MAC layer.

All in all, application layer security is the essential security tool but it is not the only required and not always affordable.

2.1 Link-Layer: Lightweight Security

The WiFi traditional authentication and encryption methods (such as WEP, WPA and WPA2) are often disregarded for vanets for several, sensible reasons. Among them the following ones: *(i)* they rely on a concept of centralized authentication by the infrastructure (AP) – which can hardly fit the ad-hoc paradigm – and the provided security suites more unicast than multicast and broadcast traffic; *(ii)* the overall authentication scheme relies on the sharing of a common key among a large number of nodes or, alternatively, it requires an EAP-like protocol which can hardly scale for all the nodes and face their mobility; *(iii)* the identification of the users is not guaranteed especially for broadcast transmissions; *(iv)* given the large number of cars participating in a vanet and sharing the secret access keys and the large number of road-infrastructures, the keys should have geographical and temporal terms.

In particular the third point has recently been appointed as the main weakness of the WiFi Protected Access [6] in the so called *Hole 196* vulnerability. Central to this vulnerability is the Group Temporal Key (GTK), shared among all authorized clients in a WPA2 network. Typically only an AP is supposed to transmit group-addressed data traffic (encrypted using the GTK), however, nothing in the standard stops a malicious authorized client from injecting spoofed GTK-encrypted packets. It means that an insider (authorized user) can sniff and decrypt data from other authorized users and inter-user data privacy among authorized users is inherently absent. For sake of clarity, the Advanced Encryption Standard (AES), on which WPA2 is based, has not been cracked by *Hole 196.*

So far the following scheme is proposed for link-layer baseline security:

- the encryption mechanism of WPA2 (derived from AES block cipher) is preserved;
- since vanet traffic is mainly broadcasted, encryption uses only GTK; consequently GTK must change often over time; for simplicity GTK *may* may be derived from a Group Master Key (GMK) as in 802.11 std, for instance enforced by *rotation* mechanisms;
- all the communications must be encrypted and the key is shared within a given area; as a result encryption is useful to insulate traffic among separate regions: for instance a highway and a contiguous urban street can define logically separate vanets and, correspondingly, separate broadcast domains are defined;
- the key(s) may obtained by authentication with central authority over a PKI infrastructure; in this case the process falls under the application-layer security umbrella and: *(i)* the infrastructure gains a more scalable control on the paths (without overhearing all the wireless traffic) and a more precise control with respect to the ingress-egress gates; *(ii)* the nodes collect transactions receipts which also certify their paths. Such certification could also be used for advanced traffic management schemes (for instance to award credits as prizes to *green routes*);
- the nodes are supposed to be provided with more keys (they could also be contemporaneously used for decryption), in order to manage temporal

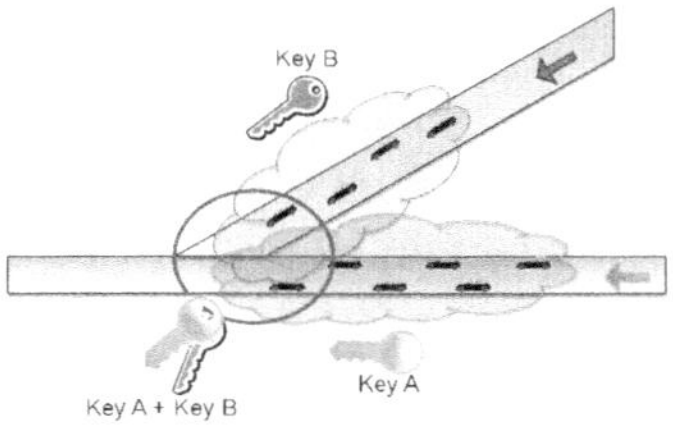

Fig. 3. Key management scenario: an intersection between areas where different keys are used. Vehicles in the red circle must know both Key A and Key B.

validity and spatial hand-off; the proposed scheme has several infrastructure checkpoints that manage key distribution in the different areas. This segmentation helps to split the broadcast domains in cases (i.e. highway near a city, or overpasses) where different logical areas are physically close to each other. However, a vehicle could happen to be in the transmission range of a confining area it is approaching to, in which a different key is used. Moreover, vanets are typically affected by fading [16] [17], which could make it difficult to define strict region boundaries. In order to avoid decoding problems and improve security, every mobile node is supposed to know and preserve also the keys used in the neighbor areas, and to be able to use all of them to decrypt the received messages (see Fig.3). The spatial validity of a key is supposed to be associated with a geographic area (e.g. city or region) that can be hypothesized to be wide and well defined;
- aspects such as message and node authentication and encryption of unicast traffic are left to upper layers; to say it more formally, in order to evaluate the messages received a node can build its own trust model about the sender, and the certification of the message has a predominant role in this;
- considering the limited scope of the encryption, the only issue concerns what happens if a malicious user is forwarded the GTK, skipping the authentication; notably its transmissions could not be prevented but it would be discouraged by the existing upper-layer security mechanisms.

In this way, this first level of security allows to segment a single vanet domain into smoothly separate areas; additionally the infrastructure achieves a first (not fully certified, but simplified) knowledge of the identity of the vehicles in the networked area. Finally, the proposed link-layer security does not constitute a true security solution but something worth when upper security layers are provided.

2.2 MAC Layer: DoS Prevention

Also Medium Access Control protocols can introduce weaknesses which can be exploited for Denial-of-Service (DoS) attacks - pure jamming approaches will not be considered because they destroy every kind of radio communication in the attacker's frequency band. This is, evidently, a MAC-protocol dependent issue: for this reason it will be analyzed separately for CSMA/CA and for MS-Aloha.

CSMA/CA implies that all the nodes are fair in computing collision-avoidance waiting time. If any station did not follow the CSMA/CA rules, it could *(i)* gain a higher bandwidth than it should (*unfairness*); *(ii)* prevent transmission by the other nodes (*pure DoS*). Unfortunately such behaviors could be hardly discovered in CSMA/CA, because hidden station cannot be excluded and each node has a thorough knowledge only of its own view of the channel: as a result it cannot infer possible violations to the CSMA/CA rules.

In MS-Aloha instead, the violations are more evident (unless pure jamming is considered) and, reversely, the attack can be twofold: a node can either attempt to access more slots than it can (*unfairness*) or cause logical collisions in order to block any transmissions (*pure DoS*).

Fortunately in MS-Aloha there is plenty of redundant information. For each slot, indeed, every node can receive up to N (number of slots per frame) state indications. In the typical configuration, every frame has 224 slots, resulting in a large number of information about the channel state. So, conceptually, it is possible to leverage on such redundant information to identify and manage nodes trying to disturb/disrupt the communication with fake information about the slot occupation.

More in practice, it is easy to understand how FIs with wrong slot indications can cause a complete blocking of the available resources. Also malfunctioning nodes can cause similar effects.

The initial hypothesis is that the channel state perceived by close nodes is almost identical. In an area of about 80 meters around the node (where the PDR -Packet Delivery Rate- is still about 90%), the involved nodes should see the same channel status. From this starting point, it is possible to infer possible inconsistent slot allocations. Depending on the way FIs are modified, two kinds of attack can be driven: 1 - the FIs sent are completely random, inconsistent with the other ones; 2 - the FIs are generated starting from a genuine one, changing only some fields. The more an FI is not coherent with the other ones, the easier the identification of suspicious behaviors is. Intuitively, if an attacker transmitted FI information only slightly counterfeited, it would be harder to identify it.

An example is shown in Fig.4: node D wants to unfairly gain control on the full channel. For this purpose it can mimic collisions by itself and by other nodes: thus it announces an FI containing the sequence D,J,M,E,F and then spooves such node identifiers (independently of their real existence): this way the full channel would be first freed and then maliciously engaged by D. However spatial and temporal correlation (of the received FI) can be exploited to prevent this by evaluating the dependability of the FI information received.

The simplest case is a node at one-hop distance which collects a certain number of coherent FIs before signaling a collision (thresholds should vary with the density of vehicles in the network). If a single node signals a collision in slot i, in the following frame the collision is validated only if other nodes confirm a direct sensing of the same collision, otherwise the FI is discarded as a fake. Additionally, correlation among FIs can take place also in the time domain.

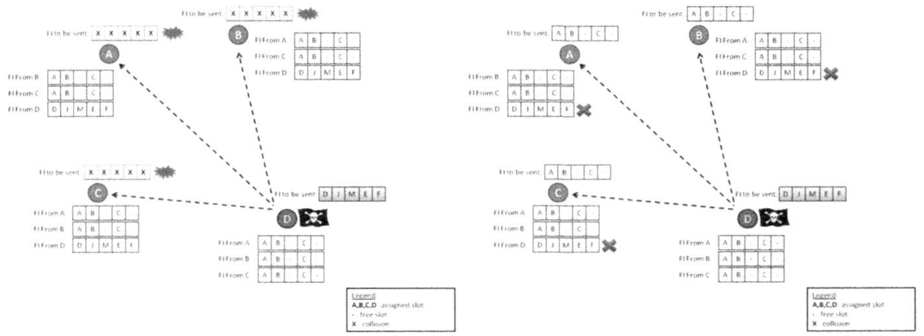

Fig. 4. Exploitation of correlation among FIs for the identification of misbehaving nodes: the case of Malicious Station with node D causing fake collisions

The logic can be further refined and summarized in the following way:

- each FI subfield can be evaluated separately;
- any FI information is "accepted" only if it is confirmed by a given number of nodes; this also prevents problems due to the sudden appearance of a node, as for a car exiting a garage: such node is supposed to accept the FI status of the other nodes (outside the garage) and not to disrupt the overall current slot assignment. Such approach would additionally increase MS-Aloha stability and robustness in general, it does not just prevent security threats;
- the mechanism of correlation is enriched with a weight for information, which takes into account what has been announced in the past from the same node and slot: for instance one approaching node can be inferred by FI and slot collisions validated 'straightforward' as soon as they take place (this recalls some concept derived from literature on reputation [10], [11]);
- each slot state transition (*e.g. free-busy, busy-collision, ...*) can be assigned a specific evaluation criterion. Here the novelty and the potential is high and is enabled once more by the redundant information available in MS-Aloha: the topic requires an analysis which can hardly fit the limit of this paper. However, just to make a practical example, the trasition busy-collision is mentioned again. If node C causes a collision on slot J, it is anyway expected to have been previously sensed by a set of nodes (at two-hop distance). Such nodes can enforce it, increasing the number of the nodes notifying the same transition. Similarly each transition could be differently enforced by the distributed intelligence of the vanet.

It is worth mentioning that, in principle, the FI classification should distinguish and differently manage the cases of malicious and malfunctioning nodes (which should be flagged but not banned).

Finally, for sake of completeness, at MAC layer, a recent advance should be mentioned: a side aspect of security is privacy and in [15] it has been demonstrated to be enforced by leveraging on the use of pseudonyms.

2.3 Application Layer: Vehicular Security by PKI

With the previously described methods it is not possible to authenticate the information from a given vehicle, but only to deal with DoS attacks or malfunctioning vehicles. Especially for critical information (e.g. ambulance, police) the security becomes mandatory and strict, therefore it is necessary to add more robust security levels, providing authentication, integrity and non-repudiation.

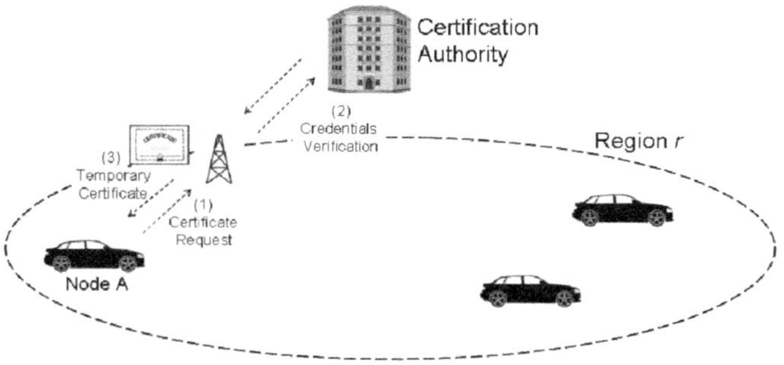

Fig. 5. PKI architecture applied to Vanet

The ideal architecture for such services is a public key infrastructure (PKI) with digital certificates. This approach presents some problems as it is introduced on mobile networks with high density [12]. Management of cryptographic keys and the CA hierarchy, control and revocation of certificates [13] are some of the open issues to be solved. Altogether they also set the issue of scalability: if all the messages were authenticated by the PKI infrastructure, each node should continuously - for each message - get, decrypt and check certificates, and check the appended hash. This can be hardly managed at wirespeed for all transmissions and cannot be either deferred (due to safety implications) or pre-computed (because the nodes are too many).

For this reason it is here proposed that this process is carried out only for critical information and entities (e.g. ambulance, police). This brings also some benefits in terms of protocol overheads, which are particularly relevant for slotted protocols, due to the fixed lengths and the already existing overheads. Considering the well known PKI [14], trailers are appended in the following way:

$$\mathrm{A} \longrightarrow \mathrm{B} : M \parallel \mathrm{E}_{KR_a}\left[\mathrm{H}(M)\right] \parallel \mathrm{E}_{KR_{as}}\left[T \parallel ID_A \parallel KU_a\right]; \qquad (1)$$

A is sender, B the receiver, M the message, $\mathrm{H}(M)$ its hash and E_{KR_a} and $\mathrm{E}_{KR_{as}}$ respectively the private key of A and of the CA.

On one hand keys need to be short in order not to make the overhead too heavy; on the other hand, for the sake of driver's security, the shorter a key is, the shorter its life-cycles must be too. However a short life-cycle has its drawbacks: the PKI consists of several parts, and includes logical components (protocols) to manage the keys (distribution and revocation) – such protocols are exchanged more frequently with shorter life-cycle and may impact on the delays and even affect available bandwidth.

Fortunately some quite recent approaches, such as Elliptic Curve Cryptography (ECC) [14], offer an excellent level of security with shorter session keys. For instance 160-bit ECC are comparable – in terms of robustness – to 1024-bit RSA/DSA, which offer a widely accepted level of security.

The hypothesis of ECC seems particularly reasonable for vanets at least for two reasons: longer packets are more likely to be discarded (due to higher number of errored bits which cannot be recovered by CRC); messages are required to be short for medium access efficiency in CSMA/CA. Concerning the latter aspect, it has however also beneficial effects on slotted approaches, such as MS-Aloha. In fact a 160-bit key would generate a node's certificate (the last part in chain (1)) which could be housed in the FI (which is supposed to be 224x12 bits). Reversely the message and its hash would be sent in the ordinary way (inside the payload). Consequently if it is accepted that the emergency messages substitute the FI with the node's certificate, a slotted approach may gain even more, in terms of efficiency, than it has been demonstrated so far [18].

2.4 Concluding Remarks

In summary, the paper has analyzed one of the most important aspects about vanets: the security of communications. For the IEEE 802.11p standard, a scheme based on CSMA/CA has been chosen, but some promising approaches, based on TDMA schemes, are emerging for MAC layer in vehicular networks. This paper has addressed the main topics, open issues and possible solutions for securing both asynchronous and slotted vanets.

Notably the analysis reveals that slotted approaches (MS-Aloha in particular) can exploit the redundant information carried by the protocol itself. This feature allows to implement security mechanisms on top of the original MAC protocol, without requiring substantial modifications. In particular, the proposed mechanism is based on a lightweight authentication of messages by means of correlation of the information broadcasted from every node. Conversely, such benefits do not hold for CSMA/CA.

Additionally MS-Aloha and CSMA/CA can both support the same security mechanisms at lower and upper layers, respectively with link-layer encryption and a proper adaptation of the PKI architecture.

In conclusion vanets are well setup in terms of security tools and slotted solution may offer some additional perspectives.

References

1. IEEE 802.11p TG. IEEE 802.11p/D9.0, Draft Amendment for Wireless Access in Vehicular Environments (WAVE) (September 2009)
2. IEEE 1609 WAVE Standards, `vii.path.berkeley.edu/1609_wave/`
3. IEEE 1609.2 Trial Use Standard for Wireless Access in Vehicular Environments (WAVE) - Security Services for Applications and Management Messages, `vii.path.berkeley.edu/1609_wave/`

4. Shakkottai, S., Rappaport, T.S., Karlsson, P.C.: Cross-layer design for wireless networks. IEEE Commun. Mag. 41(10) (October 2003)
5. Scopigno, R., Cozzetti, H.A.: Mobile Slotted Aloha for Vanets. In: IEEE VTC Fall, pp. 1–5 (September 2009)
6. Ahmad, S.: Hole 196, Black Hat (2010), http://www.blackhat.com/html/bh-us-10/bh-us-10-home.html
7. Papadimitratos, P., Buttyan, L., Holczer, T., Schoch, E., Freudiger, J., Raya, M., Ma, Z., Kargl, F., Kung, A., Hubaux, J.-P.: Secure Vehicular Communication Systems: Design and Architecture. IEEE Communications Magazine, 100–109 (November 2008)
8. Aslam, B., Zou, C.: Distributed Certificate and Application Architecture for Vanets. In: MILCOM, pp. 1–7 (October 2009)
9. Yang, H., Luo, H., Ye, F., Lu, S., Zhang, L.: Security in Mobile Ad Hoc Networks: Challenges and Solutions. IEEE Wireless Communications, 38–47 (February 2004)
10. Despotovic, Z., Aberer, K.: P2P reputation management: Probabilistic estimation vs. social networks. The International Journal of Computer and Telecommunications Networking, 485–500 (March 2006)
11. Mui, L., Mohtashemi, M., Halberstadt, A.: A Computational Model of Trust and Reputation. In: Proceedings of the 35th Hawaii International Conference on System Science, HICSS (2002)
12. Raya, M., Hubaux, J.-P.: The Security of Vehicular Ad Hoc Networks. In: Proceedings of the 3rd ACM Workshop on Security of Ad Hoc and Sensor Networks, pp. 11–21 (2005)
13. Raya, M., Papadimitratos, P., Hubaux, J.-P.: Securing Vehicular Communications. IEEE Wireless Communications Magazine (October 2006)
14. Stallings, W.: Cryptography and Network Security, 3rd edn. Prentice Hall, Englewood Cliffs (2003)
15. Schaub, F., Kargl, F., Ma, Z., Weber, M.: V-tokens for Conditional Pseudonymity in VANETs. In: Scopigno, R., Cozzetti, H.A. (eds.) Proceedings of IEEE Wireless Communications and Networking Conference, WCNC 2010 (April 2010)
16. Nakagami, M.: The m-Distribution, a general formula of intensity of rapid fading. In: Statistical Methods in Radio Wave Propagation: Proc. Symposium at the University of California. Permagon Press, Oxford (1960)
17. Sklar, B.: Rayleigh Fading Channels in Mobile Digital Communication Systems Part I: Characterization. IEEE Communication Magazine, 90–100 (July 1997)
18. Scopigno, R., Cozzetti, H.A.: Evaluation of Time-Space Efficiency in CSMA/CA and Slotted Vanets. In: IEEE 71th Vehicular Technology Conference, VTC Fall (October 2010)

Smart Elliptic Curve Cryptography for Smart Dust

Johann Großschädl[1], Matthias Hudler[2], Manuel Koschuch[2], Michael Krüger[2], and Alexander Szekely[3]

[1] University of Luxembourg,
Laboratory of Algorithmics, Cryptology and Security,
6, rue Richard Coudenhove-Kalergi, L–1359 Luxembourg, Luxembourg
johann.groszschaedl@uni.lu

[2] FH Campus Wien – University of Applied Sciences,
Competence Centre for IT-Security,
Favoritenstraße 226, A–1100 Vienna, Austria
firstname.lastname@fh-campuswien.ac.at

[3] Graz University of Technology,
Institute for Applied Information Processing and Communications,
Inffeldgasse 16a, A–8010 Graz, Austria
alexander.szekely@iaik.tugraz.at

Abstract. Wireless ad-hoc and sensor networks play a vital role in an ever-growing number of applications ranging from environmental monitoring over vehicular communication to home automation. Security and privacy issues pose a big challenge for the widespread adoption of these networks, especially in the automotive domain. The two most essential security services needed to maintain the proper functioning of a wireless network are authentication and key establishment; both can be realized with Elliptic Curve Cryptography (ECC). In this paper, we introduce an efficient ECC implementation for resource-restricted devices such as sensor nodes. Our implementation uses a 160-bit Optimal Prime Field (OPF) over which a Gallant-Lambert-Vanstone (GLV) curve with good cryptographic properties can be defined. The combination of optimized field arithmetic with fast group arithmetic (thanks to an efficiently computable endomorphism) allows us to perform a scalar multiplication in about $5.5 \cdot 10^6$ clock cycles on an 8-bit ATmega128 processor, which is significantly faster than all previously-reported ECC implementations based on a 160-bit prime field.

Keywords: Ad-hoc network, elliptic curve cryptography, performance evaluation, arithmetic in finite fields, endomorphism.

1 Introduction

The term "smart dust" refers to miniature computing devices with sensing and wireless networking capabilities [26]. Current-generation micro-sensors, such as the Crossbow MICA2DOT mote [4], have a volume of a few cubic-centimeters

X. Zhang and D. Qiao (Eds.): QShine 2010, LNICST 74, pp. 623–634, 2011.

including battery. However, it can be expected that progress in miniaturization will reduce the size of motes significantly over the next couple of years. These tiny devices can form a Wireless Sensor Network (WSN) and undertake certain tasks such as battlefield surveillance or the monitoring of weather and/or road conditions for a traffic-control system [1]. Kristofer Pister, who first coined the term "smart dust," forecasted in the mid-90s that "in 2010 MEMS sensors will be everywhere and sensing virtually everything" [18]. Today, much of Pister's vision from some 15 years ago has become reality; we may think of miniature sensor nodes used for environmental surveillance, home automation, or medical monitoring [1]. Widespread deployment of sensors for intelligent transportation systems is expected in the near future, e.g. WSNs to monitor traffic conditions and report these conditions to smart vehicles.

In October 1999, the Federal Communications Commission (FCC) allocated a 75 MHz band in the 5.9 GHz frequency range for exclusive use by automotive applications and intelligent transportation systems [5]. Dedicated Short Range Communications (DSRC) is a suite of protocols and standards for wireless networking within this 75 MHz band [2]. The main purpose of DSRC is to enable short-range communication among vehicles and between vehicles and roadside infrastructure (e.g. traffic condition sensors) with the goal of increasing safety on roads and improving traffic flow [13]. Higher-level protocols operating above DSRC (or IEEE 802.11p) are specified in the IEEE standard 1609 [24]. Typical fields of application of DSRC (resp. WAVE) include safety measures (e.g. road condition warning, collision avoidance), traffic management (e.g. variable speed limits, intelligent traffic lights), driver assistance (e.g. parking aids, cruise control), and electronic payment (e.g. toll collection, parking fees). Several of these applications pose significant challenges to security and privacy, as was pointed out in [12,15,19]. To address these issues, the IEEE 1609.2 standard contains a number of measures to ensure the confidentiality, integrity, and authenticity of messages exchanged over DSRC. The public-key cryptosystems specified in IEEE 1609.2 are based on Elliptic Curve Cryptography (ECC) [10] to achieve a balance between efficiency and security. More precisely, IEEE 1609.2 defines ECIES for asymmetric encryption and ECDSA as signature primitive [15].

In this paper, we introduce an optimized ECC implementation for wireless ad-hoc and sensor networks. Our implementation uses an Optimal Prime Field (OPF) as underlying algebraic structure to facilitate fast modular reduction on different platforms [7]. Furthermore, we take advantage of a Gallant-Lambert-Vanstone (GLV) elliptic curve with an efficiently computable endomorphism to accelerate the scalar multiplication [6]. The focus of our implementation lay on high performance, low memory footprint, and low register usage. We evaluated the performance of our ECC software on a Crossbow Micaz mote featuring an 8-bit ATmega128 processor clocked at 7.37 MHz. The combination of fast field arithmetic with fast curve arithmetic allowed us to perform a full scalar multiplication over a 160-bit OPF in roughly $5.5 \cdot 10^6$ clock cycles, which represents a new speed record for 160-bit ECC on 8-bit processors. Our ECC software can be easily ported to other platforms and achieves excellent performance also on processors with few general-purpose registers as we do not rely on the hybrid multiplication technique [9]. The results we present in this paper demonstrate that strong public-key cryptography is feasible for resource-constrained devices such as sensor nodes.

2 Elliptic Curve Cryptography

In this section (which is largely based on our previous work [16]), we introduce the basic concepts of ECC with a focus on implementation aspects. In short, an elliptic curve E over a prime field $\mathbb{F}_p$ can be formally defined as the set of all tuples $(x, y) \in \mathbb{F}_p \times \mathbb{F}_p$ satisfying a Weierstraß equation of the form

$$y^2 = x^3 + ax + b \quad \text{with} \quad a, b \in \mathbb{F}_p \tag{1}$$

These tuples are called *points* with x and y referred to as coordinates. The set of points together with a special point $\mathcal{O}$ (the so-called *point at infinity*) allows one to form a commutative group with $\mathcal{O}$ being the identity element. The group operation is the addition of points, which can be performed through arithmetic operations (addition, subtraction, multiplication, squaring, and inversion) in the underlying field $\mathbb{F}_p$ according to well-defined formulae (see e.g. [10]). Adding a point $P = (x, y)$ to itself is referred to as point doubling and can also be done through a well-defined sequence of operations in $\mathbb{F}_p$. In general, point doubling requires fewer field operations than the addition of two points.

The *order* of an elliptic curve group $E(\mathbb{F}_p)$ is the number of $\mathbb{F}_p$-rational points on the curve E, plus one for the point at infinity. It is well known from Hasse's theorem that $\#E(\mathbb{F}_p)$ has the following bounds:

$$p + 1 - 2\sqrt{p} \leq \#E(\mathbb{F}_p) \leq p + 1 + 2\sqrt{p} \tag{2}$$

For cryptographic applications, $\#E(\mathbb{F}_p)$ should have a large prime factor; in the ideal case it is a prime. Before ECIES encryption (or any other elliptic curve scheme) can be carried out, the involved parties have to agree upon a common set of so-called *domain parameters*, which specify the finite field $\mathbb{F}_p$, the elliptic curve E (i.e. the coefficients $a, b \in \mathbb{F}_p$ defining E according to Equation (1)), a base point $P \in E(\mathbb{F}_p)$ generating a cyclic subgroup of large order, the order n of this subgroup, and the co-factor $h = \#E(\mathbb{F}_p)/n$. Consequently, elliptic curve domain parameters over $\mathbb{F}_p$ are simply a sextuple $D = (p, a, b, P, n, h)$ [10]. In elliptic curve cryptography, a private key is an integer k chosen randomly from the interval $[1, n-1]$. The corresponding public key is the point $Q = k \cdot P$ on the curve. Given k and P, the point $Q = k \cdot P$ can be obtained by means of an operation called *scalar multiplication* [10]. Numerous algorithms for scalar multiplication have been proposed; the simplest way to compute $k \cdot P$ is to perform a sequence of point additions and doublings, similar to the square-and-multiply algorithm for modular exponentiation.

While a scalar multiplication of the form $Q = k \cdot P$ can be calculated quite efficiently, the inverse operation, i.e. finding k when P and Q are given, is a hard mathematical problem known as the *Elliptic Curve Discrete Logarithm Problem (ECDLP)*. To date, the best algorithm known for solving the ECDLP requires fully exponential time if the domain parameters were chosen with care [10]. In contrast, the best algorithm for solving the Discrete Logarithm Problem (DLP) in $\mathbb{Z}_p^*$ or the Integer Factorization Problem (IFP) has a sub-exponential running time. As a consequence, elliptic curve cryptosystems can use much shorter keys compared to their "classical" counterparts based on the DLP or IFP. A common rule of thumb states that a properly designed 160-bit ECC scheme is about as secure as 1024-bit RSA.

2.1 Scalar Multiplication

The computationally expensive part of virtually all elliptic curve cryptosystems is scalar multiplication, an operation of the form $k \cdot P$ where k is an integer and P is a point on the curve. A scalar multiplication can be performed by means of repeated point additions and point doublings, both of which, in turn, involve a sequence of arithmetic operations (i.e. addition, multiplication, squaring, and inversion) in the underlying finite field. Inversion is by far the most expensive operation in prime fields [10]. However, it is possible to add points on an elliptic curve without the need to perform costly inversions, e.g. by representing the points in *projective coordinates* [10]. In Section 2 we described the conventional (i.e. affine) coordinate system in which a point P is associated with an x and a y coordinate, i.e. a tuple $(x, y) \in \mathbb{F}_p \times \mathbb{F}_p$. By contrast, in projective coordinate systems, a point is represented by a triplet (X, Y, Z), which corresponds to the affine coordinates $(X/Z^u, Y/Z^v)$ when $Z \neq 0$ (u and v depend on the specific coordinate system chosen). For example, the projective point $P = (X, Y, Z)$ in Jacobian coordinates corresponds to the affine point $P = (X/Z^2, Y/Z^3)$. It is also possible to add two points when one is given in projective coordinates and the other in affine coordinates [10]. In fact, such mixed coordinates often lead to very efficient point addition formulae. For example, adding a point in Jacobian coordinates to an affine point requires eight multiplications and three squarings in $\mathbb{F}_p$ (but no inversion). Doubling a point given in Jacobian coordinates takes four multiplications and four squarings.

The *double-and-add* algorithm performs a scalar multiplication via repeated point additions and doublings, analogous to the multiply-and-square algorithm for modular exponentiation. It uses the binary expansion of the integer k and computes $k \cdot P$ as follows: For each bit k_i of k, the current intermediate result is doubled, and the base point P is added if bit $k_i = 1$ (no addition is performed when $k_i = 0$). Given an l-bit scalar k, the double-and-add algorithm executes exactly l point doublings, whereas the number of point additions depends on the Hamming weight of k. In the average case $l/2$ additions are carried out; the worst-case number of additions is l. The conventional double-and-add method can be easily improved by using a signed-digit representation of k. One option is the *non-adjacent form (NAF)*, which reduces the number of additions (of either P or $-P$) to $l/3$ in the average case and $l/2$ in the worst case [10]. However, the number of point doublings remains the same.

2.2 Arithmetic in Prime Fields

As mentioned before, the execution time of a scalar multiplication depends on the efficiency of the arithmetic in the underlying finite field. The elements of a prime field $\mathbb{F}_p$ are the integers $0, 1, \ldots, p-1$, and the arithmetic operations are addition and multiplication modulo p. In ECC, primes of a length of between 160 and 512 bits are typically used. Consequently, the field elements can not be directly processed, but must be represented by arrays of single-precision words (e.g. arrays of unsigned n-bit integers when working on an n-bit processor). In our case, i.e. 160-bit ECC on an 8-bit processor, each field element is stored in an array of 20 bytes. Arithmetic in $\mathbb{F}_p$ is similar to that in $\mathbb{Z}_p^*$ as needed for the implementation of RSA and other "classical" public-key schemes. Therefore, all algorithms for modular arithmetic in $\mathbb{Z}_p^*$ are directly applicable to $\mathbb{F}_p$ as well (e.g. Montgomery reduction [14]). However, it is possible and common practice to use special primes in ECC for which optimized modular reduction methods exist; a typical example are generalized-Mersenne (GM) primes [10].

There are two basic algorithms for multi-precision multiplication: one is the *operand-scanning method* (also called row-wise multiplication) and the other is the *product-scanning method* (column-wise multiplication) [9,10]. Both require the same number of single-precision multiplications (i.e. `mul` instructions on an ATmega128), namely 400 in our case of 160-bit operands, but they differ in the number of memory accesses and single-precision additions. We first describe the original operand and product scanning methods, which operate on 8-bit words (i.e. bytes) when implemented on an ATmega processor. Then, we sketch the hybrid multiplication method of Gura et al [9], which combines the advantages of operand scanning and product scanning to reduce the total number of load instructions. The operand-scanning method has a nested-loop structure with a relatively simple inner loop. Each iteration executes an operation of the form $a \cdot b + c + d$ with a, b, c, d denoting 8-bit words (i.e. bytes). On an ATmega this operation requires one `mul` instruction to produce the partial product $a \cdot b$, and a total of four `add` (resp. `adc`) instructions to add the two bytes c and d to the 16-bit quantity $a \cdot b$. Furthermore, two load (`ld`) instructions and a store (`st`) are executed in each iteration. On the other hand, the product-scanning method performs a multiply-accumulate operation in its inner loop, i.e. two bytes are multiplied and the 16-bit partial product is added to a cumulative sum held in three registers. The product-scanning method also executes two `ld` instructions per iteration, but no store.

The execution time of the conventional product-scanning technique can be vastly improved when the processor features a large number of general-purpose registers, which is the case with the ATmega128 [3]. The *hybrid* multiplication method, introduced by Gura et al in [9], works similar as the product-scanning technique, but processes $d \geq 2$ bytes of the operands at a time, which reduces the number of required loop iterations by a factor of d. In each iteration of the inner loop, $d \geq 2$ bytes of the operands are loaded from RAM, then multiplied together using d^2 `mul` instructions, and added to a running sum. However, the hybrid method can not be applied on processors with few registers.

3 Our Implementation

In this section, we describe our ECC implementation in detail and analyze its performance on an 8-bit ATmega128 processor [3]. Our implementation differs from previous work (e.g. [9,16,17,22,23,25]) in two main aspects. First, we combine efficient finite-field arithmetic with fast group arithmetic; the former is achieved by using an Optimal Prime Field (OPF) [7] as "underlying" algebraic structure, while the scalar multiplication profits from an efficiently computable endomorphism provided by a so-called Gallant-Lambert-Vanstone (GLV) curve [6]. The basic concepts of OPFs (including a formal definition of OPFs, how to do arithmetic in OPFs, and how to construct a GLV curve over an OPF) can be found in [7]; here in this paper we show the applicability of OPFs to secure ad-hoc and sensor networks. A second difference to related work is the concrete implementation of the multi-precision multiplication. We do not use the hybrid multiplication method from [9], but apply a special loop unrolling technique to achieve high performance on a wide variety of platforms, including processors with a small number of general-purpose registers. The hybrid multiplication, on the other hand, can only be implemented when a large number of registers is available, which is not always the case. Even though the ATmega128 features a large register file with 32 registers (which means that hybrid multiplication is possible), we decided to use it as experimental platform for our implementation in order to allow for direct comparison with previous work. The ATmega128 is a simple 8-bit RISC processor based on the AVR instruction set, i.e. the usual arithmetic and logical instructions are supported, including an integer-multiply instruction with a 2-cycle latency [3].

3.1 Optimal Prime Field (OPF)

An Optimal Prime Field (OPF) [7] is a finite field defined by a prime p of the form $u \cdot 2^k + v$, where u and v are constants that fit into a single register of the target processor (or, more precisely, $0 < u, v < 2^w$ where w denotes the word size of the target processor). In our case, both u and v are 8-bit constants since our implementation is optimized for an 8-bit processor. The specific prime we chose is $p = 232 \cdot 2^{152} + 99$, which happens to be a 160-bit prime that looks as follows when written in hex notation:

$$p = \text{0xE800000000000000000000000000000000000063} \tag{3}$$

OPFs are characterized by a low Hamming weight [7]. In particular, when p is stored in an array of w-bit words, only the most and least significant words are non-zero; all words between them are zero. The low weight of these primes allows for efficient software implementation of the modular reduction operation because only the two non-zero words of p need to be processed [7]. Well-known modular reduction methods, including Montgomery and Barrett reduction, can be well optimized for low-weight primes such that the reduction operation has linear complexity, similar to generalized-Mersenne (GM) and pseudo-Mersenne

(PM) primes [10,20]. A particular advantage of OPFs over GM prime fields is their flexibility; there exist a large number of OPFs for a given bitlength, while the number of GM primes with "good" arithmetic properties is rather limited [20]. The large number of OPFs for a given bitlength facilitates the construction of a GLV curve with an efficiently computable endomorphism suitable for the implementation of ECC, which is not possible with the GM primes specified by the NIST (see [7] for further details).

We implemented multiplication and squaring in OPFs using Montgomery's modular reduction method. More precisely, we optimized the Finely Integrated Operand Scanning (FIPS) method [14] for Montgomery multiplication with respect to the low weight of our OPF so that only the non-zero bytes $p_{19} = 232$ and $p_0 = 99$ of the prime $p = 232 \cdot 2^{152} + 99$ are processed. An implementation of the FIPS technique for "arbitrary" primes executes $2s^2 + s$ single-precision multiplications (i.e. `mul` instructions) when the operands consist of s words. In our case of processing 160-bit operands on an 8-bit processor (i.e. $s = 20$), this amounts to 820 `mul` instructions for a single Montgomery multiplication. However, after optimizing the FIPS technique taking into account that $s - 2$ bytes of p are 0, only only $s^2 + 3s$ single-precision multiplications have to be carried out, which results in 460 `mul` instructions for 160-bit operands. A conventional multiplication of two s-word operands (without reduction operation) requires s^2 `mul` instructions [10,14]; consequently, the overhead of modular reduction in our OPF is $3s$ `mul` instructions, i.e. scales linearly with the operand length.

The optimized FIPS method for Montgomery multiplication in an OPF has a nested-loop structure with a simple inner loop in which a multiply-accumulate operation is performed, i.e. two bytes are multiplied and the 16-bit product is added to a running sum held in three registers. More precisely, each iteration of the inner loop executes two `ld`, one `mul`, one `add`, and two `adc` instructions on an Atmega128, which takes nine clock cycles altogether. Furthermore, three clock cycles of loop overhead (i.e. increment or decrement of a loop counter and branch instruction) contribute to the execution time. Since this overhead constitutes 25% of the overall execution time of the inner loop, it makes sense to apply techniques for reducing loop overhead, such as loop unrolling or hybrid multiplication. The basic idea of hybrid multiplication, as described in [9], is to process $d > 1$ bytes of the operands in each iteration of the inner loop (instead of just a single byte), which reduces the overall number of loop iterations by a factor of d. However, hybrid multiplication is only advantageous on processors featuring a large number of registers; for example, when $d = 4$ (i.e. four bytes of each operand are processed per loop iteration), 14 registers are necessary to store the bytes of the operands and running sum [9].

As already mentioned, we did not apply the hybrid multiplication technique since we aimed for a flexible implementation that can be easily ported to various different platforms, including processors with a small number of registers. Instead, we reduced the loop overhead of our optimized FIPS method for OPFs by unrolling the inner loop. In general, loop unrolling is done by replicating the loop body a certain number of times and adjusting (i.e. reducing) the iteration

Table 1. Execution times (in clock cycles) of arithmetic operations in OPFs of length 128, 144, 160, 176, and 192 bits on an ATmega128 processor

Operation	128 bit	144 bit	160 bit	176 bit	192 bit
Addition	506	560	614	668	722
Subtraction	514	562	610	672	731
Multiplication	3,598	4,358	5,239	6,143	7,070
Squaring	2,967	3,488	4,086	4,642	5,277
Inversion	110,414	133,698	157,840	185,361	214,687

count accordingly. Loop unrolling improves the performance since the overhead for updating and testing the loop counter and branching back to the beginning of the loop is executed less frequently. However, this gain in performance comes at the expense of an increase in code size. To achieve a proper balance between these two metrics of interest, we decided to unroll only the inner loop. Unfortunately, the number of inner-loop iterations of both the standard FIPS method and our optimized OPF-variant is not constant, but varies between 1 and s. An efficient unrolling technique for such loops is Duff's device [11], as was shown in [8] for the standard FIPS method. We applied the Duff device from [8] to the optimized FIPS method for our 160-bit OPF, which means that we replicated the body of the inner loop 20 times. This approach for loop unrolling achieves almost the same performance as "full" loop unrolling (i.e. unrolling of both the inner and the outer loop), but does so at a faction of the code size.

Table 1 summarizes the execution time of addition, subtraction, multiplication, squaring, and inversion in OPFs of different size, ranging from 128 to 192 bits. The addition in an OPF is performed by first calculating the sum of two field elements, followed by a conditional subtraction of the prime p. Hence, the addition time (and also the subtraction time) is not constant but depends on whether or not a subtraction (resp. addition) of p is carried out. Also both the multiplication and squaring may require a final subtraction of p, which impacts the execution time. Our OPF multiplication with unrolled inner loop is a little slower than the hybrid multiplication from [9], but occupies only five registers for operands and the running sum, while the hybrid technique with $d = 4$ needs 14 registers.

3.2 Gallant-Lambert-Vanstone (GLV) Curve

In order to speed up the scalar multiplication $k \cdot P$, we use a so-called Gallant-Lambert-Vanstone (GLV) curve as introduced in [6]. This family of curves can be described by a Weierstraß equation of the form $y^2 = x^3 + b$ (i.e. $a = 0$ and $b \neq 0$) or $y^2 = x^3 + ax$ (i.e. $a \neq 0$ and $b = 0$) over a prime field $\mathbb{F}_p$ and features an efficiently computable endomorphism ϕ whose characteristic polynomial has small coefficients. Such an endomorphism provides the possibility to calculate a scalar multiplication as $k \cdot P = k_1 \cdot P + k_2 \cdot \phi(P)$, which is more efficient than a straightforward calculation of $k \cdot P$ since k_1 and k_2 have typically only half the bitlength of k and the two multiplications $k_1 \cdot P$ and $k_2 \cdot \phi(P)$ can be performed simultaneously (i.e. in an interleaved fashion) using Shamir's trick as explained in [10, Section 3.3.3]. However, as already mentioned, a GLV curve with good cryptographic properties can not be constructed over any prime field; in particular, it is not possible to find a GLV curve of prime order over the GM prime fields standardized by the NIST (see [7] for further details).

Since there exist a large number of OPFs, it is not hard to find an OPF of a given bitlength such that a GLV curve with good cryptographic properties can be constructed over it. Our implementation uses the GLV curve

$$E : y^2 = x^3 + 3 \tag{4}$$

(i.e. $a = 0$ and $b = 3$) over the prime field $\mathbb{F}_p$ defined by $p = 232 \cdot 2^{152} + 99$. The group of points on this curve has prime order, namely

$$\#E(\mathbb{F}_p) = n = 1324485858831130769622088630463083182986367428713 \tag{5}$$

and satisfies all properties listed in [21, Section 3.1.1.1], which means that this curve can be securely used for the implementation of ECC. Consequently, this specific pair of OPF and GLV curve offers good arithmetic and cryptographic properties. In the following, we summarize some basic facts about our special curve, analogously to [10, page 125]. Since $a = 0$ and $p \equiv 1 \bmod 3$, our curve is of the type described in Example 4 in [6]. The underlying field $\mathbb{F}_p$ contains an element β of order 3 (since $p \equiv 1 \bmod 3$); our implementation uses

$$\beta = 1039364585860691323337591166412095487330325497064 \tag{6}$$

According to [6, Example 4], the map $\phi : E \to E$ defined by

$$\phi : (x, y) \mapsto (\beta x, y) \quad \text{and} \quad \phi : \mathcal{O} \mapsto \mathcal{O} \tag{7}$$

is an endomorphism of E defined over $\mathbb{F}_p$. The characteristic polynomial of ϕ is $\lambda^2 + \lambda + 1$. In order to exploit this endomorphism for scalar multiplication, we need a root modulo n of the characteristic polynomial, i.e. we need a solution to the equation $\lambda^2 + \lambda + 1 \equiv 0 \bmod n$; for our implementation we use

$$\lambda = 893685873620479505526293352198704065242719655609 \tag{8}$$

The solution λ has the property that $\phi(P) = \lambda P$ for all $P \in E(\mathbb{F}_p)$ [6,10]. Note that computing $\phi(P)$ for a point $P = (x, y)$ requires only one multiplication in $\mathbb{F}_p$, namely $\beta \cdot x$.

When using a GLV curve [6], the common strategy for computing $k \cdot P$ is to decompose the scalar k into two "half-length" integers k_1 and k_2 (referred to as balanced length-two representation of k) such that $k = k_1 + k_2\lambda \bmod n$. This decomposition of k into k_1 and k_2 is described in detail in [6] and [10]. Because $k \cdot P = k_1 \cdot P + k_2 \cdot \lambda \cdot P = k_1 \cdot P + k_2 \cdot \phi(P)$, the result of $k \cdot P$ can be obtained by first computing $\phi(P)$ (which takes just a single field multiplication) and then using simultaneous multiple point multiplication ("Shamir's trick") to perform

Table 2. Execution times (in million clock cycles) of 160-bit scalar multiplication

Implementation	Field type	Fixed P.	Rand. P.	Notes
Wang and Li [25]	PM prime	9.14	9.95	Sliding window
Szczechowiak et al. [22]	GM prime	9.38	9.38	Comb method
Ugus et al. [23]	PM prime	5.09	7.59	MOV, 1 prec. point
Liu and Ning [17]	PM prime	15.05	15.05	Sliding window
Gura et al. [9]	PM prime	6.48	6.48	NAF
Our implementation	OPF	5.48	5.48	GLV curve

these two half-length scalar multiplications in an interleaved fashion. A conventional computation of $k \cdot P$ using the double-and-add approach requires a total of l point doublings and on average $l/2$ point additions, whereby l refers to the bitlength of k. The number of additions can be reduced to $l/3$ by representing the scalar k in Non-Adjacent Form (NAF). On the other hand, our GLV curve allows us to obtain $k \cdot P$ with only $l/2$ doublings and $l/4$ additions when the two half-length scalars k_1 and k_2 are represented in Joint Sparse Form (JSF) as described in [10]. Consequently, the endomorphism of our GLV curve halves the number of point doublings and reduces the number of point additions by some 8.3% on average.

Our implementation performs the point (i.e. curve) arithmetic using mixed Jacobian-affine coordinates as detailed in [10, Section 3.2.2]. More precisely, we implemented the point addition and doubling according to Algorithm 3.2.2 and Algorithm 3.2.1, respectively, whereby we optimized the latter with respect to $a = 0$ such that a point doubling can be carried out with three multiplications in $\mathbb{F}_p$ instead of four [7]. The addition of points over our 160-bit OPF requires roughly 57,760 clock cycles on an ATmega128 processor, while a point doubling is executed in about 35,450 cycles. Our implementation of the point arithmetic is written in ANSI C and directly calls the Assembly-language functions for the OPF arithmetic. A scalar multiplication executes in $5.48 \cdot 10^6$ clock cycles on an ATmega128 processor when exploiting Shamir's trick and representing the two half-length scalars in JSF. The comparison with related work in Table 2 shows that our implementation sets a new speed record for ECC on an ATmega128.

4 Conclusions

In this paper, we introduced an efficient ECC implementation for DSRC-based ad-hoc and sensor networks that realize security services (e.g. authentication) according to the IEEE standard 1609.2. Our implementation is able to perform a 160-bit scalar multiplication on an ATmega128 processor in slightly less than $5.5 \cdot 10^6$ clock cycles, which establishes a new performance record for ECC on 8-bit platforms. Compared to previous work, our implementation advances the state-of-the-art in two main aspects. First, we combine efficient field arithmetic (thanks to the use of an OPF) with fast scalar multiplication on a GLV curve by exploiting an efficiently computable endomorphism. Second, we do not use Gura's hybrid multiplication technique, but unroll the inner loops of our field multiplication and squaring operations following "Duff's device" to reduce the execution time. Therefore, our ECC software can be easily ported to various other platforms, even to processors with very few general-purpose registers on which hybrid multiplication would not work.

Acknowledgements. The authors are grateful to Stefan Mendel for his contributions to the Assembly-language implementation of the OPF arithmetic.

Manuel Koschuch, Matthias Hudler and Michael Krüger have been supported by the MA27 – EU-Strategie und Wirtschaftsentwicklung – in the course of the funding programme "Stiftungsprofessuren und Kompetenzteams für die Wiener Fachhochschul-Ausbildungen."

References

1. Akyildiz, I.F., Su, W., Sankarasubramaniam, Y., Çayirci, E.: Wireless sensor networks: A survey. Computer Networks 38(4), 393–422 (2002)
2. ASTM International: ASTM E2213-03 Standard Specification for Telecommunications and Information Exchange Between Roadside and Vehicle Systems — 5 GHz Band Dedicated Short Range Communications (DSRC) Medium Access Control (MAC) and Physical Layer (PHY) Specifications (2003)
3. Atmel Corporation: 8-bit ARV® Microcontroller with 128K Bytes In-System Programmable Flash: ATmega128, ATmega128L. Datasheet (June 2008), `http://www.atmel.com/dyn/resources/prod_documents/doc2467.pdf`
4. Crossbow Technology, Inc.: MICA2DOT Wireless Microsensor Mote. Data sheet (January 2006), `http://www.xbow.com/Products/Product_pdf_files/Wireless_pdf/MICA2DOT_Datasheet.pdf`
5. Federal Communications Commission (FCC): FCC Allocates Spectrum in 5.9 GHz Range for Intelligent Transportation Systems Uses. News release (October 1999), `http://www.fcc.gov/Bureaus/Engineering_Technology/News_Releases/1999/nret9006.html`
6. Gallant, R.P., Lambert, R.J., Vanstone, S.A.: Faster point multiplication on elliptic curves with efficient endomorphisms. In: Kilian, J. (ed.) CRYPTO 2001. LNCS, vol. 2139, pp. 190–200. Springer, Heidelberg (2001)
7. Großschädl, J., et al.: Optimal prime fields for use in elliptic curve cryptography (2010) (preprint, submitted for publication)
8. Großschädl, J., Tillich, S., Szekely, A.: Performance evaluation of instruction set extensions for long integer modular arithmetic on a SPARC V8 processor. In: Proceedings of the 10th Euromicro Conference on Digital System Design (DSD 2007), pp. 680–689. IEEE Computer Society Press, Los Alamitos (2007)
9. Gura, N., Patel, A., Wander, A.S., Eberle, H., Chang Shantz, S.: Comparing elliptic curve cryptography and RSA on 8-bit CPUs. In: Joye, M., Quisquater, J.J. (eds.) CHES 2004. LNCS, vol. 3156, pp. 119–132. Springer, Heidelberg (2004)
10. Hankerson, D.R., Menezes, A.J., Vanstone, S.A.: Guide to Elliptic Curve Cryptography. Springer, Heidelberg (2004)

11. Holly, R.: A reusable Duff device. Dr. Dobb's Journal 30(8), 73–74 (2005)
12. Hubaux, J.P., Ĉapkun, S., Luo, J.: The security and privacy of smart vehicles. IEEE Security & Privacy 2(3), 49–55 (2004)
13. Jiang, D., Taliwal, V., Meier, A., Holfelder, W., Herrtwich, R.G.: Design of 5.9 GHz DSRC-based vehicular safety communication. IEEE Wireless Communications 13(5), 36–43 (2006)
14. Koç, Ç.K., Acar, T., Kaliski, B.S.: Analyzing and comparing Montgomery multiplication algorithms. IEEE Micro 16(3), 26–33 (1996)
15. Laurendeau, C., Barbeau, M.: Threats to security in DSRC/WAVE. In: Kunz, T., Ravi, S.S. (eds.) ADHOC-NOW 2006. LNCS, vol. 4104, pp. 266–279. Springer, Heidelberg (2006)
16. Lederer, C., Mader, R., Koschuch, M., Großschädl, J., Szekely, A., Tillich, S.: Energy-efficient implementation of ECDH key exchange for wireless sensor networks. In: Markowitch, O., Bilas, A., Hoepman, J.H., Mitchell, C.J., Quisquater, J.J. (eds.) WISTP 2009. LNCS, vol. 5746, pp. 112–127. Springer, Heidelberg (2009)
17. Liu, A., Ning, P.: TinyECC: A configurable library for elliptic curve cryptography in wireless sensor networks. In: Proceedings of the 7th International Conference on Information Processing in Sensor Networks (IPSN 2008), pp. 245–256. IEEE Computer Society Press, Los Alamitos (2008)
18. Pister, K.S.: My view of sensor networks in 2010 (2010), `http://robotics.eecs.berkeley.edu/~pister/SmartDust/in2010`
19. Raya, M., Hubaux, J.P.: The security of vehicular ad hoc networks. In: Atluri, V., Ning, P., Du, W. (eds.) Proceedings of the 3rd ACM Workshop on Security of Ad Hoc and Sensor Networks (SASN 2005), pp. 11–21. ACM Press, New York (2005)
20. Solinas, J.A.: Generalized Mersenne numbers. Tech. Rep. CORR-99-39, Centre for Applied Cryptographic Research (CACR), University of Waterloo, Waterloo, Canada (1999)
21. Standards for Efficient Cryptography Group (SECG): SEC 1: Elliptic Curve Cryptography. Working draft, version 1.7 (November 2006), `http://www.secg.org/download/aid-631/sec1_1point7.pdf`
22. Szczechowiak, P., Oliveira, L.B., Scott, M., Collier, M., Dahab, R.: NanoECC: Testing the limits of elliptic curve cryptography in sensor networks. In: Verdone, R. (ed.) EWSN 2008. LNCS, vol. 4913, pp. 305–320. Springer, Heidelberg (2008)
23. Ugus, O., Westhoff, D., Laue, R., Shoufan, A., Huss, S.A.: Optimized implementation of elliptic curve based additive homomorphic encryption for wireless sensor networks. In: Wolf, T., Parameswaran, S. (eds.) Proceedings of the 2nd Workshop on Embedded Systems Security (WESS 2007), pp. 11–16 (2007), `http://arxiv.org/abs/0903.3900`
24. U.S. Department of Transportation: IEEE 1609 – Family of standards for wireless access in vehicular environments (WAVE). ITS standards fact sheet (September 2009), `http://www.standards.its.dot.gov/fact_sheet.asp?f=80`
25. Wang, H., Li, Q.: Efficient implementation of public key cryptosystems on mote sensors. In: Ning, P., Qing, S., Li, N. (eds.) ICICS 2006. LNCS, vol. 4307, pp. 519–528. Springer, Heidelberg (2006)
26. Warneke, B., Last, M., Liebowitz, B., Pister, K.S.: Smart dust: Communicating with a cubic-millimeter computer. Computer 34(1), 44–51 (2001)

Author Index

GPSR Compliance

The European Union's (EU) General Product Safety Regulation (GPSR) is a set of rules that requires consumer products to be safe and our obligations to ensure this.

If you have any concerns about our products, you can contact us on ProductSafety@springernature.com

In case Publisher is established outside the EU, the EU authorized representative is:

Springer Nature Customer Service Center GmbH
Europaplatz 3
69115 Heidelberg, Germany

Batch number: 10844611

Printed by Printforce, the Netherlands